KRANTZ

Several Complex Variables is a central area of mathematics with strong interactions with partial differential equations, algebraic geometry, number theory, and differential geometry. The 1995–96 MSRI program on several complex variables emphasized these interactions and concentrated on developments and problems of current interest that capitalize on this interplay of ideas and techniques.

This collection provides a remarkably clear and complete picture of the status of research in these overlapping areas and will provide a basis for significant continued contributions from researchers. Several of the articles are expository or have extensive expository sections, making this an excellent introduction for students to the use of techniques from these other areas in several complex variables.

Thanks to its distinguished list of contributors, this volume provides a representative sample of the best work done this decade in several complex variables.

Mathematical Sciences Research Institute
Publications

37

Several Complex Variables

Mathematical Sciences Research Institute
Publications

Volumes 1–4 and 6–27 are available from Springer-Verlag

Several Complex Variables

Edited by

Michael Schneider

Universität Bayreuth

Yum-Tong Siu

Harvard University

Michael Schneider (1942–1997)
Universität Bayreuth
95440 Bayreuth
Germany

Yum-Tong Siu
Department of Mathematics
Harvard University
Cambridge, MA 02138
United States

Series Editor
Silvio Levy

Mathematical Sciences
  Research Institute
1000 Centennial Drive
Berkeley, CA 94720
United States

MSRI Editorial Committee
Joe P. Buhler (chair)
Alexandre Chorin
Silvio Levy
Jill Mesirov
Robert Osserman
Peter Sarnak

The Mathematical Sciences Research Institute wishes to acknowledge
support by the National Science Foundation.

PUBLISHED BY THE PRESS SYNDICATE OF THE UNIVERSITY OF CAMBRIDGE
The Pitt Building, Trumpington Street, Cambridge, United Kingdom

CAMBRIDGE UNIVERSITY PRESS
The Edinburgh Building, Cambridge CB2 1RP, UK   http://www.cup.cam.ac.uk
40 West 20th Street, New York, NY 10011-4211, USA   http://www.cup.org
10 Stamford Road, Oakleigh, Melbourne 3166, Australia
Ruiz de Alarcón 13, 28014 Madrid, Spain

Printed in the United States of America

Library of Congress Cataloguing in Publication data is available.

A catalogue record for this book is available from the British Library.

ISBN 0 521 77086 6 hardback

Several Complex Variables
MSRI Publications
Volume **37**, 1999

Contents

Several Complex Variables
MSRI Publications
Volume **37**, 1999

Preface

This volume consists of sixteen articles written by participants of the 1995–96 Special Year in Several Complex Variables held at the Mathematical Sciences Research Institute in Berkeley, California.

The field of Several Complex Variables is a central area of mathematics with strong interactions with partial differential equations, algebraic geometry and differential geometry. The 1995–96 MSRI program on Several Complex Variables emphasized these interactions and concentrated on developments and problems of current interest that capitalize on this interplay of ideas and techniques.

This collection provides a picture of the status of research in these overlapping areas at the time of the conference, with some updates. It will serve as a basis for continued contributions from researchers and as an introduction for students. Most of the articles are surveys or expositions of results and techniques from these overlapping areas in several complex variables, often summarizing a vast amount of literature from a unified point of view. A few articles are more oriented toward researchers but nonetheless have expository sections.

On August 29, 1997 Michael Schneider, one of the two editors of this volume, died in a rock-climbing accident in the French Alps. This volume is dedicated to his memory. The front matter includes his portrait, a listing of the major events in his mathematical career, and a selection of his mathematical contributions.

Michael Schneider

May 18, 1942 to August 29, 1997

- Studied mathematics and physics with O. Forster and K. Stein at the University of Munich, with one semester in Geneva.
- Diploma, University of Munich, 1966.
- Doctorate, University of Munich, 1969, with a dissertation on complete intersections in Stein manifolds.
- Assistant, University of Regensburg, 1969–1974.
- Habilitation, University of Regensburg, 1974.
- Professor, University of Göttingen, 1975–1980.
- Chaired Professor, University of Bayreuth, 1980–1997.
- Editor, *Journal für die reine und angewandte Mathematik*, 1984–1995.
- Served as Fachgutachter of the Deutsche Forschungsgemeinschaft and on the Wissenshaftlichen Beirat of the Forschungszentrum Oberwolfach.

Selected Mathematical Contributions

- A complex submanifold Y of a Stein manifold X with $\dim Y = \frac{1}{2} \dim X$ is Stein if and only if the normal bundle of Y is trivial and the fundamental class defined by Y vanishes as a homology class in X.
- If the normal bundle of a complex submanifold Y of codimension k in a compact complex manifold X is positive in the sense of Griffiths, then $X - Y$ is k-convex.
- (with Badescu) If the normal bundle of a d-dimensional complex submanifold Y in a projective manifold X is $(d-1)$-ample in the sense of Sommese, then the field of formal meromorphic functions along Y is a finite field extension of the field of meromorphic functions on X.
- A stable vector bundle of rank 2 on $\mathbb{P}_n$ is ample if and only if $c_1 \geq 2c_2 - \frac{c_1^2}{2}$.
- (with Elencwajg and Hirschowitz) If E is a holomorphic vector bundle of rank $r \leq n$ on $\mathbb{P}_n$ which has the same splitting type on every line, then E either is split or is isomorphic to $\Omega^1_{\mathbb{P}_n}(a)$ or $T_{\mathbb{P}_n}(b)$.
- For a stable vector bundle of rank 3 on $\mathbb{P}_n$ with $c_1 = 0$ one has $|c_3| \leq c_2^2 + c_2$.
- For a semistable bundle E of rank 3 on $\mathbb{P}_n$ ($n \geq 3$) and a general hyperplane H in $\mathbb{P}_n$, the restriction $E|H$ is semistable unless $n = 3$ and E is a twist of the tangent or cotangent bundle of $\mathbb{P}_3$.
- (with Catanese) If X is an n-dimensional Cohen–Macaulay projective variety which is nonsingular outside a subvariety of codimension at least 2 and H is a very ample divisor in X and E is a vector bundle on X, then there exists a polynomial function $P_{n,h,E}$ in the first h Chern classes of E and the first two Chern classes of X such that for every nonzero section of E whose scheme of zeroes Z has codimension h, one has $\deg(Z) \leq P_{n,h,E}$.

- (with Beltrametti and Sommese) Complete classification of all threefolds of degree 9, 10, and 11 in $\mathbb{P}_5$.
- (with Braun, Ottaviani and Schreyer) There are only finitely many families of threefolds in $\mathbb{P}_5$ which are not of general type.
- There are only finitely many families of m-dimensional submanifolds in $\mathbb{P}_n$ not of general type if $m \geq \frac{n+2}{2}$.
- An n-dimensional compact complex manifold with ample cotangent bundle cannot be embedded into $\mathbb{P}_{2n-1}$.
- (with Demailly and Peternell) If X is a compact Kähler manifold whose tangent bundle is numerically effective, then the Albanese map of X is a surjective submersion and, after possibly replacing X by a finite étale cover, the fibers of the Albanese map of X are Fano manifolds with numerically effective tangent bundles and the fundamental group of X agrees with that of its Albanese.
- (with Demailly and Peternell) Let X be a compact Kähler manifold with numerically effective anticanonical line bundle. If the anticanonical line bundle of X admits a metric with semipositive curvature, then the universal cover of X is a product whose factors are either Euclidean complex spaces, Calabi–Yau manifolds, symplectic manifolds, or manifolds with the property that no positive tensor powers of the cotangent bundle admit a nonzero holomorphic section. In particular, the fundamental group of X has an abelian subgroup of finite index. If the anticanonical line bundle of X is only numerically effective, then the fundamental group of X has subexponential growth and, in particular, X does not admit a map onto a curve of genus at least 2.
- (with Barlet and Peternell) If X is a $\mathbb{P}_2$-bundle over a compact algebraic surface, then any two nonsingular surfaces with Griffiths-positive normal bundle in X must intersect.
- (with Peternell) Let X be a compact complex threefold and Y be a complex hypersurface in X whose complement is biholomorphic to $\mathbb{C}^3$. Then X is projective if Y is normal or if the algebraic dimension $a(X)$ of X is 2, or if $a(Y) = 2$. Moreover, the cases $a(X) = 2$ or $a(X) = 1$ with a meromorphic function on X whose pole-set is Y cannot occur.
- (with Catanese) Let X be a projective n-dimensional manifold of general type and G be an abelian group of birational automorphisms of X. If the canonical line bundle K_X of X is numerically effective, then the order of G is bounded by $C_n(K_X^n)^{2^n}$, where C_n is a constant depending only on n. If $n = 3$ and if m is a positive integer admitting a G-eigenspace in $H^0(X, mK_X)$ of dimension at least 2, then the order of G is bounded by the maximum of $6P_2(X)$ and $P_{3m+2}(X)$, where $P_q(X)$ is the dimension of $H^0(X, qK_X)$.

Several Complex Variables
MSRI Publications
Volume **37**, 1999

Local Holomorphic Equivalence
of Real Analytic Submanifolds in $\mathbb{C}^N$

M. SALAH BAOUENDI AND LINDA PREISS ROTHSCHILD

ABSTRACT. This paper presents some recent results of the authors jointly with Peter Ebenfelt concerning local biholomorphisms which map one real-analytic or real-algebraic submanifold of $\mathbb{C}^N$ into another. It is shown that under some optimal conditions such mappings are determined by their jets of a predetermined finite order at a given point. Under these conditions, if the manifolds are algebraic, it is also shown that the components of the holomorphic mappings must be algebraic functions. The stability group of self mappings is shown to be a finite dimensional Lie group for most points in the case of real-analytic holomorphically nondegenerate real hypersurfaces in $\mathbb{C}^N$. The notion of Segre sets associated to a point of a real-analytic CR submanifold of $\mathbb{C}^N$ is one of the main ingredients in this work. Properties of these sets and their relationship to minimality of these manifolds are discussed.

Introduction

We consider here some recent results concerning local biholomorphisms which map one real analytic (or real algebraic) subset of $\mathbb{C}^N$ into another such subset of the same dimension. One of the general questions studied is the following. Given $M, M' \subset \mathbb{C}^N$, germs of real analytic subsets at p and p' respectively with $\dim_{\mathbb{R}} M = \dim_{\mathbb{R}} M'$, describe the (possibly empty) set of germs of biholomorphisms $H : (\mathbb{C}^N, p) \to (\mathbb{C}^N, p')$ with $H(M) \subset M'$.

Most of the new results stated here have been recently obtained in joint work with Peter Ebenfelt. We shall give precise definitions and specific references in the text. One of the results (Theorem 2) states that if $M \subset \mathbb{C}^N$ is a connected real analytic holomorphically nondegenerate CR manifold which is minimal at some point, then at most points $p \in M$, a germ H of a biholomorphism at p mapping M into M', another submanifold of $\mathbb{C}^N$, is determined by its jet at p of a finite order, depending only on M. This result is used to prove (Theorem 3)

Both authors were partially supported by National Science Foundation Grant DMS 95-01516.

that the real vector space of infinitesimal CR automorphisms of M is finite dimensional at every point.

Denote by $\text{Aut}(M, p)$ the group of germs of biholomorphisms of $\mathbb{C}^N$ at p, fixing p and mapping M into itself. Theorem 4 and its corollaries show that if M is a holomorphically nondegenerate hypersurface, then for most points $p \in M$, $\text{Aut}(M, p)$, equipped with its natural topology, is a finite dimensional Lie group parametrized by a subgroup of the jet group of $\mathbb{C}^N$ at 0 of a certain finite order. The proof of Theorem 4 gives an algorithm to determine all germs of biholomorphisms at p mapping the hypersurface M into another hypersurface M' and taking p to p'. The set of all such biholomorphisms (possibly empty) is parametrized by a real analytic, totally real submanifold of a finite order jet group of $\mathbb{C}^N$ at 0.

Section 6 deals with the special case where the real submanifolds M and M' are real algebraic, that is, defined by the vanishing of real valued polynomials. In particular, Theorem 8 implies that if M and M' are holomorphically nondegenerate generic algebraic manifolds of the same dimension, and if M is minimal at p, then any germ of a biholomorphism at p mapping M into M' is algebraic. Theorem 9 shows that holomorphic nondegeneracy and minimality are essentially necessary for the algebraicity of all such mappings.

A main ingredient in the proofs of the results stated in this paper is the use of the *Segre sets* associated to every point of a real analytic CR submanifold in $\mathbb{C}^N$. The description of these sets and their main properties is given in Section 1 and Theorem 1. One of these properties is that the complexification of the CR orbit of a point $p \in M$ coincides with the maximal Segre set at p. In particular, a real analytic generic submanifold M is minimal at p if and only if the maximal Segre set is of complex dimension N.

Bibliographical references relevant to the results given in this paper can be found at the end of each section of the text.

We shall give now some basic definitions. Most of the results described here can be reduced to the case where M and M' are real analytic generic submanifolds. Recall that a real analytic submanifold $M \subset \mathbb{C}^N$ is *generic* if near every $p \in M$, we may write

$$(0.1) \qquad M = \{Z \in \mathbb{C}^N : \rho_j(Z, \bar{Z}) = 0 \text{ for } j = 1, \ldots, d\},$$

where $\rho_1, \ldots, \rho_d$ are germs at p of real-valued real analytic functions satisfying $\partial\rho_1(p) \wedge \cdots \wedge \partial\rho_d(p) \neq 0$. Here

$$\partial\rho = \sum_{j=1}^{N} \frac{\partial\rho}{\partial Z_j} dZ_j.$$

More generally, we say that M is *CR* if $\dim_{\mathbb{R}}(T_pM \cap JT_pM)$ is constant for $p \in M$, where T_pM is the real tangent space of M at p, and J the anti-involution of the standard complex structure of $\mathbb{C}^N$. If M is CR, then $\dim_{\mathbb{R}} T_pM \cap JT_pM = 2n$

is even and n is called the *CR dimension* of M. In particular, if M is generic of codimension d, then $n = N - d$.

We say that a real submanifold of $\mathbb{C}^N$ is *holomorphically nondegenerate* if there is no germ of a nontrivial vector field

$$\sum_{j=1}^{N} c_j(Z) \frac{\partial}{\partial Z_j},$$

with $c_j(Z)$ holomorphic, tangent to an open subset of M. Another criterion of holomorphic nondegeneracy, which can be checked by a simple calculation, is the following. Let $L = (L_1, \ldots, L_n)$ be a basis for the CR vector fields of a generic manifold M near p. For any multi-index α put $L^\alpha = L_1^{\alpha_1} \ldots L_n^{\alpha_n}$. Introduce, for $j = 1, \ldots, d$ and any multi-index α, the $\mathbb{C}^N$-valued functions

(0.2) $$V_{j\alpha}(Z, \bar{Z}) = L^\alpha \rho_{jZ}(Z, \bar{Z}),$$

where ρ_{jZ} denotes the gradient of ρ_j with respect to Z, with ρ_j as in (0.1). We say that M is *finitely nondegenerate* at $p \in M$ if there exists a positive integer k such that the span of the vectors $V_{j\alpha}(p, \bar{p})$, for $j = 1, \ldots, d$ and $|\alpha| \leq k$, equals $\mathbb{C}^N$. If k is the smallest such integer we say that M is *k-nondegenerate* at p. These definitions are independent of the coordinate system used, the defining functions of M, and the choice of basis L. One can then check that if a generic manifold M is connected, then M is holomorphically nondegenerate if and only if it is finitely nondegenerate at some point $p \in M$. Another equivalent definition is that M is holomorphically nondegenerate if and only if it is essentially finite at some point $p \in M$.

It can also be shown that a connected, generic manifold M is holomorphically nondegenerate if and only if there exists a positive integer $l(M)$, $1 \leq l(M) \leq N - 1$, such that M is $l(M)$-nondegenerate at every point outside a proper real analytic subset of M. We shall call $l(M)$ the *Levi number* of M. Hence to determine holomorphic nondegeneracy, one need compute (0.2) for only finitely many multi-indices α. In particular, a connected real analytic hypersurface is Levi nondegenerate at some point if and only it its Levi number is 1. For a connected hypersurface in $\mathbb{C}^2$, Levi nondegeneracy at some point is equivalent to holomorphic nondegeneracy. However, in $\mathbb{C}^N$, for $N > 2$, there exist connected, real analytic holomorphically nondegenerate hypersurfaces which are nowhere Levi nondegenerate.

1. Segre Sets of a Germ of a CR Manifold

In this section, we introduce the Segre sets of a generic real analytic submanifold in $\mathbb{C}^N$ and recall some of their properties. We refer the reader to [Baouendi et al. 1996a] for a more detailed account of these sets; see also [Ebenfelt 1998]. Let M denote a generic real analytic submanifold in some neighborhood $U \subset \mathbb{C}^N$

of $p_0 \in M$. Let $\rho = (\rho_1, \ldots, \rho_d)$ be defining functions of M near p_0 as in (0.1), and choose holomorphic coordinates $Z = (Z_1, \ldots, Z_N)$ vanishing at p_0. Embed $\mathbb{C}^N$ in $\mathbb{C}^{2N} = \mathbb{C}_Z^N \times \mathbb{C}_\zeta^N$ as the real plane $\{(Z, \zeta) \in \mathbb{C}^{2N} : \zeta = \bar{Z}\}$. Denote by pr_Z and pr_ζ the projections of $\mathbb{C}^{2N}$ onto $\mathbb{C}_Z^N$ and $\mathbb{C}_\zeta^N$, respectively. The natural anti-holomorphic involution $\sharp$ in $\mathbb{C}^{2N}$ defined by

$$(1.1) \qquad\qquad \sharp(Z, \zeta) = (\bar{\zeta}, \bar{Z})$$

leaves the plane $\{(Z, \zeta) : \zeta = \bar{Z}\}$ invariant. This involution induces the usual anti-holomorphic involution in $\mathbb{C}^N$ by

$$(1.2) \qquad\qquad \mathbb{C}^N \ni Z \mapsto \mathrm{pr}_\zeta(\sharp \mathrm{pr}_Z^{-1}(Z)) = \bar{Z} \in \mathbb{C}^N.$$

Given a set S in $\mathbb{C}_Z^N$ we denote by $^*S$ the set in $\mathbb{C}_\zeta^N$ defined by

$$(1.3) \qquad\qquad ^*S = \mathrm{pr}_\zeta(\sharp \mathrm{pr}_Z^{-1}(S)) = \{\zeta : \bar{\zeta} \in S\}.$$

We use the same notation for the corresponding transformation taking sets in $\mathbb{C}_\zeta^N$ to sets in $\mathbb{C}_Z^N$. Note that if X is a complex analytic set defined near Z^0 in some domain $\Omega \subset \mathbb{C}_Z^N$ by $h_1(Z) = \cdots = h_k(Z) = 0$, then $^*X$ is the complex analytic set in $^*\Omega \subset \mathbb{C}_\zeta^N$ defined near $\zeta^0 = \bar{Z}^0$ by $\bar{h}_1(\zeta) = \cdots = \bar{h}_k(\zeta) = 0$. Here, given a holomorphic function $h(Z)$ we use the notation $\bar{h}(Z) = \overline{h(\bar{Z})}$.

Let $\mathcal{M} \subset \mathbb{C}^{2N}$ be the complexification of M given by

$$(1.4) \qquad\qquad \mathcal{M} = \{(Z, \zeta) \in \mathbb{C}^{2N} : \rho(Z, \zeta) = 0\}.$$

This is a complex submanifold of codimension d in some neighborhood of 0 in $\mathbb{C}^{2N}$. We choose our neighborhood U in $\mathbb{C}^N$ so small that $U \times {}^*U \subset \mathbb{C}^{2N}$ is contained in the neighborhood where $\mathcal{M}$ is a manifold. Note that $\mathcal{M}$ is invariant under the involution $\sharp$ defined in (1.1).

We associate to M at p_0 a sequence of germs of sets $N_0, N_1, \ldots, N_{j_0}$ at p_0 in $\mathbb{C}^N$ — the Segre sets of M at p_0 — defined as follows. Put $N_0 = \{p_0\}$ and define the consecutive sets inductively (the number j_0 will be defined later) by

$$(1.5) \qquad N_{j+1} = \mathrm{pr}_Z\big(\mathcal{M} \cap \mathrm{pr}_\zeta^{-1}({}^*N_j)\big) = \mathrm{pr}_Z\big(\mathcal{M} \cap {}^\sharp \mathrm{pr}_Z^{-1}(N_j)\big).$$

We shall assume that the open set U is fixed sufficiently small and make no further mention of it. These sets are, by definition, invariantly defined and arise naturally in the study of mappings between submanifolds, as will be seen in Section 2.

The sets N_j can also be described in terms of the defining equations $\rho(Z, \bar{Z}) = 0$. For instance,

$$(1.6) \qquad\qquad N_1 = \{Z : \rho(Z, 0) = 0\}$$

and

$$(1.7) \qquad\qquad N_2 = \{Z : \exists \zeta^1 : \rho(Z, \zeta^1) = 0, \, \rho(0, \zeta^1) = 0\}.$$

We have the inclusions

(1.8)
$$N_0 \subset N_1 \subset \cdots \subset N_j \subset \cdots$$

and j_0 is the largest number j such that the generic dimension of N_j is the same as that of N_{j-1}. (The generic dimensions of the Segre sets stabilize for $j \geq j_0$.)

To show that the Segre sets are images of holomorphic mappings, it is useful to make use of appropriate holomorphic coordinates. Recall that we can find holomorphic coordinates $Z = (z, w)$, with $z \in \mathbb{C}^n$ and $w \in \mathbb{C}^d$, vanishing at p_0 and such that M near p_0 is given by

$$w = Q(z, \bar{z}, \bar{w}) \quad \text{or} \quad \bar{w} = \bar{Q}(\bar{z}, z, w),$$

where $Q(z, \chi, \tau)$ is holomorphic in a neighborhood of 0 in $\mathbb{C}^{2n+d}$, valued in $\mathbb{C}^d$ and satisfies $Q(z, 0, \tau) \equiv Q(0, \chi, \tau) \equiv \tau$. The coordinates $Z = (z, w)$ satisfying the above properties are called *normal coordinates* of M at p_0. In normal coordinates (z, w) one may use the definition above to express the Segre sets N_j for $j = 1, \ldots, j_0$ as images of germs at the origin of certain holomorphic mappings

(1.9)
$$\mathbb{C}^n \times \mathbb{C}^{(j-1)n} \ni (z, \Lambda) \mapsto (z, v^j(z, \Lambda)) \in \mathbb{C}^N.$$

We have
$$N_1 = \{(z, 0) : z \in \mathbb{C}^n\},$$
$$N_2 = \{(z, Q(z, \chi, 0)) : z, \chi \in \mathbb{C}^n\},$$

and so forth. Thus, we can define the generic dimension d_j of N_j as the generic rank of the mapping (1.9).

So far we have considered only generic submanifolds. We may reduce to this case, since any real analytic CR manifold M is a generic manifold in a complex holomorphic submanifold $\mathcal{V}$ of $\mathbb{C}^N$, called the *intrinsic complexification* of M. The Segre sets of M at a point $p_0 \in M$ can be defined as subsets of $\mathbb{C}^N$ by the process described above just as for generic submanifolds, or they can be defined as subsets of $\mathcal{V}$ by identifying $\mathcal{V}$ near p_0 with $\mathbb{C}^K$ ($K = \dim \mathcal{V}$) and considering M as a generic submanifold of $\mathbb{C}^K$. It can be shown that these definitions are equivalent.

If M is a real analytic CR submanifold of $\mathbb{C}^N$ and $p_0 \in M$, then by Nagano's theorem [1966] there exists a real analytic CR submanifold of M through p_0 of minimum possible dimension and the same CR dimension as M. Such a manifold is called the *CR orbit* of p_0.

The main properties concerning the Segre sets that we shall use are summarized in the following theorem of the authors jointly with Ebenfelt.

THEOREM 1 [Baouendi et al. 1996a; 1998]. *Let M be a real analytic CR submanifold in $\mathbb{C}^N$, and let $p_0 \in M$. Denote by W the CR orbit of p_0 and by X the intrinsic complexification of W.*

(a) *The maximal Segre set N_{j_0} of M at p_0 is contained in X and contains an open subset of X arbitrarily close to p_0. In particular, $d_{j_0} = \dim_{\mathbb{C}} X$.*

(b) *There are holomorphic mappings defined near the origin* $Z_0(t_0), Z_1(t_1), \ldots,$
$Z_{j_0}(t_{j_0})$ *and* $s_0(t_1), \ldots, s_{j_0-1}(t_{j_0})$ *with*

$$(1.10) \qquad \mathbb{C}^{d_j} \ni t_j \mapsto Z_j(t_j) \in \mathbb{C}^N, \quad \mathbb{C}^{d_j} \ni t_j \mapsto s_{j-1}(t_j) \in \mathbb{C}^{d_{j-1}},$$

such that $Z_j(t_j)$ *is an immersion at the origin,* $Z_j(t_j) \in N_j$, *and such that*

$$(1.11) \qquad \big(Z_j(t_j), \bar{Z}_{j-1}(s_{j-1}(t_j))\big) \in \mathcal{M},$$

for $j = 1, \ldots, j_0$. *In addition* $Z_j(0)$, $j = 1, \ldots, j_0$, *can be chosen arbitrarily close to* p_0.

PROOF. Part (a) is contained in [Baouendi et al. 1996a, Theorem 2.2.1], and the mappings in part (b) are constructed in the paragraph following Assertion 3.3.2 of the same reference. □

REMARK 1.12. For each j with $j = 0, 1, \ldots, j_0$, the holomorphic immersion $Z_j(t_j)$, in part (b) above provides a parametrization of an open piece of N_j. However, this piece of N_j need not contain the point p_0. Indeed, N_j need not even be a manifold at p_0.

Recall that a CR submanifold M is said to be *minimal* at a point $p_0 \in M$ if there is no proper CR submanifold of M through p_0 with the same CR dimension as M. Equivalently, M is minimal at p_0 if the CR orbit of p_0 is all of M. For a real analytic submanifold, this notion coincides with the notion of finite type in the sense of [Bloom and Graham 1977]; that is, M is of finite type at p_0 if the Lie algebra generated by the CR vector fields and their complex conjugates span the complex tangent space to M at p_0. It is easy to determine whether a hypersurface M is of finite type at p_0 by using a defining function for M near p_0. Furthermore, if a connected hypersurface M is holomorphically nondegenerate, it is of finite type at most points. (The converse is not true, however.) One of the main difficulties in higher codimension is that it is cumbersome to describe finite type in local coordinates. Furthermore, unlike in the hypersurface case, in general, holomorphic nondegeneracy does not imply the existence of a point of finite type.

One can check that if M is connected then M is minimal almost everywhere if and only if M is minimal at some point. The following is an immediate consequence of the theorem.

COROLLARY 1.13. *Let* M *be a real analytic generic submanifold in* $\mathbb{C}^N$ *and* $p_0 \in M$. *Then* M *is minimal at* p_0 *if and only if* $d_{j_0} = N$ *or, equivalently, if and only if the maximal Segre set at* p_0 *contains an open subset of* $\mathbb{C}^N$.

We note if M is a hypersurface, then $j_0 = 1$ if M is not minimal at p_0, and $j_0 = 2$ otherwise. We now describe the Segre sets at 0 for two generic manifolds in $\mathbb{C}^3$ of codimension 2, one minimal and one not minimal.

EXAMPLE 1.14. Consider $M \subset \mathbb{C}^3$ defined by

$$\operatorname{Im} w_1 = |z|^2, \quad \operatorname{Im} w_2 = \operatorname{Re} w_2 |z|^4.$$

In this example M is not minimal at 0. We have $j_0 = 2$ and the maximal Segre set of M at 0 is given by

$$N_2 = \{(z, w_1, w_2) : z \neq 0, w_2 = 0\} \cup \{0, 0, 0\}.$$

Here $d_2 = 2$, and N_2 is not a manifold at 0. However, the intersection of (the closure of) N_2 with M equals the CR orbit of 0.

EXAMPLE 1.15. Let $M \subset \mathbb{C}^3$ be the generic submanifold defined by

$$\operatorname{Im} w_1 = |z|^2, \quad \operatorname{Im} w_2 = |z|^4.$$

Then M is of finite type at 0. The Segre set N_2 at 0 is the manifold given by

$$N_2 = \{(z, w_1, w_2) : w_2 = -iw_1^2/2\}.$$

We have here $j_0 = 3$, and N_3 is given by

$$N_3 = \{(z, w_1, w_2) : w_2 = iw_1(w_1/2 - 2z\chi), \chi \in \mathbb{C}\}$$

and hence N_3 contains $\mathbb{C}^3$ minus the planes $\{z = 0\}$ and $\{w_1 = 0\}$.

Before concluding this section we point out that the Segre set N_1, introduced above, coincides with the so-called Segre variety, introduced in [Segre 1931] and used in [Webster 1977a; Diederich and Webster 1980; Diederich and Fornæss 1988; Chern and Ji 1995] and elsewhere. The subsequent Segre sets N_j are all unions of Segre varieties. We believe that the results described above are the first to explore Segre sets for manifolds of higher codimension and to use them characterize minimality. The notion of minimality as described in this section, was first introduced by Tumanov [1988a].

2. Holomorphic Mappings and Segre Sets

In this section we describe how the Segre sets constructed in Section 1 can be used to prove that mappings between CR manifolds are determined by their jets of a fixed order, under appropriate conditions on the manifolds. The main result of this section is the following.

THEOREM 2 [Baouendi et al. 1998]. *Let $M \subset \mathbb{C}^N$ be a connected real analytic, holomorphically nondegenerate CR submanifold with Levi number $l(M)$, and let d be the (real) codimension of M in its intrinsic complexificiation. Suppose that there is a point $p \in M$ at which M is minimal. Then for any $p_0 \in M$ there exists a finite set of points $p_1, \ldots, p_k \in M$, arbitrarily close to p_0, with the following property. If $M' \subset \mathbb{C}^N$ is another real analytic CR submanifold with*

$\dim_{\mathbb{R}} M' = \dim_{\mathbb{R}} M$, *and* F, G *are smooth CR diffeomorphisms of* M *into* M' *satisfying in some local coordinates* x *on* M

(2.1) $\dfrac{\partial^{|\alpha|} F}{\partial x^{\alpha}}(p_l) = \dfrac{\partial^{|\alpha|} G}{\partial x^{\alpha}}(p_l)$ *for* $l = 1, \ldots, k$ *and* $|\alpha| \le (d+1)l(M)$,

then $F \equiv G$ *in a neighborhood of* p_0 *in* M. *If* M *is minimal at* p_0, *then one can take* $k = 1$. *If, in addition,* M *is* $l(M)$-*nondegenerate at* p_0, *then one may take* $p_1 = p_0$.

REMARKS. (i) Condition (2.1) can be expressed by saying that the $(d+1)l(M)$-jets of the mappings F and G coincide at all the points $p_1, \ldots, p_k$.

(ii) The choice of points $p_1, \ldots, p_k$ can be described as follows. Let $U_1, \ldots, U_k$ be the components of the set of minimal points of M in U, an arbitrarily small neighborhood of p_0 in M, which have p_0 in their closure. For each $l = 1, \ldots, k$, we may choose any p_l from the dense open subset of U_l consisting of those points which are $l(M)$-nondegenerate.

We shall give an indication of the proof of Theorem 2 only for the case where M is generic and is $l(M)$-nondegenerate and minimal at p_0. We start with the following proposition.

PROPOSITION 2.2. *Let* $M, M' \subset \mathbb{C}^N$ *be real analytic generic submanifolds, and* $p_0 \in M$. *Assume that* M *is holomorphically nondegenerate and* $l(M)$-*nondegenerate at* p_0. *Let* H *be a germ of a biholomorphism of* $\mathbb{C}^N$ *at* p_0 *such that* $H(M) \subset M'$. *Then there are* $\mathbb{C}^N$ *valued functions* Ψ^{γ}, *holomorphic in all of their arguments, such that*

(2.3) $\dfrac{\partial^{|\gamma|} H}{\partial Z^{\gamma}}(Z) = \Psi^{\gamma}\left(Z, \zeta, \left(\dfrac{\partial^{|\alpha|} \bar{H}}{\partial \zeta^{\alpha}}(\zeta) \right)_{|\alpha| \le l(M)+|\gamma|} \right),$

for all multi-indices γ *and all points* $(Z, \zeta) \in \mathcal{M}$ *near* $(p_0, \bar{p}_0)$. *Moreover, the functions* Ψ^{γ} *depend only on* M, M' *and*

(2.4) $\dfrac{\partial^{|\beta|} H}{\partial Z^{\beta}}(p_0)$,   *with* $|\beta| \le l(M)$.

PROOF. This follows from the definition of $l(M)$-nondegeneracy at p_0 and the use of the implicit function theorem. For details, see the proof of [Baouendi et al. 1996a, Assertion 3.3.1] and also [Baouendi and Rothschild 1995, Lemma 2.3]. $\square$

We shall use Proposition 2.2 to give an outline of the proof of Theorem 2 under the more restrictive assumptions indicated above. Let N_j, for $j = 0, 1, \ldots, j_0$, be the Segre sets of M at p_0, and let $Z_0(t_0), \ldots, Z_{j_0}(t_{j_0})$ be the canonical parametrizations of the N_j's and $s_0(t_1), \ldots, s_{j_0-1}(t_{j_0})$ the associated maps as in Theorem 1. Since M is minimal at p_0, it follows from [Tumanov 1988a] that F and G extend holomorphically to a wedge with edge M near p_0. Hence, by a theorem of the first author jointly with Jacobowitz and Treves [Baouendi et al. 1985], F and G extend holomorphically to a full neighborhood of p_0 in $\mathbb{C}^N$, since finite nondegeneracy at p_0 implies essential finiteness at p_0. We again denote by F and G their holomorphic extensions to a neighborhood. Assumption (2.1) with $p_l = p_0$ then implies that

$$(2.5) \qquad \frac{\partial^{|\alpha|} F}{\partial Z^\alpha}(p_0) = \frac{\partial^{|\alpha|} G}{\partial Z^\alpha}(p_0), \quad \text{for } |\alpha| \le (d+1)l(M).$$

By Proposition 2.2, there are functions Ψ^γ such that both F and G satisfy the identity (2.3) for $(Z, \zeta) \in \mathcal{M}$. Substituting (Z, ζ) in (2.3) by the left hand side of (1.11) and recalling that $Z_0(t_0) \equiv p_0$ (that is, it is the constant map), we deduce that F and G, as well as all their derivatives of all orders less than or equal to $dl(M)$ are identical on the first Segre set N_1. Note that since each N_j is the holomorphic image of a connected set, if two holomorphic functions agree on an open piece, they agree on all of N_j. Inductively we deduce that the restrictions of the mappings F and G to the Segre set N_j, as well as their derivatives of orders at least $((d+1) - j)l(M)$, are identical. The conclusion of Theorem 2 now follows from Theorem 1, since M minimal at p_0 implies that N_{j_0} contains an open piece of $\mathbb{C}^N$.

Theorem 2 is optimal in the sense that holomorphic nondegeneracy is necessary for its conclusion and that the condition that M is minimal almost everywhere is necessary in model cases. We have the following result.

PROPOSITION 2.6 [Baouendi et al. 1998]. *Let $M \subset \mathbb{C}^N$ be a connected real analytic CR submanifold.*

(i) *If M is holomorphically degenerate, then for any $p \in M$ and any integer $K > 0$ there exist local biholomorphisms F and G near p mapping M into itself and fixing p such that*

$$(2.7) \qquad \frac{\partial^{|\alpha|} F}{\partial Z^\alpha}(p) = \frac{\partial^{|\alpha|} G}{\partial Z^\alpha}(p), \quad \text{for } |\alpha| \le K,$$

but $F \not\equiv G$ on M.

(ii) *If M is defined by the vanishing of weighted homogeneous polynomials, and nowhere minimal then for any $p \in M$ and any integer $K > 0$ there exist local biholomorphisms F and G near p mapping M into itself and fixing p such that (2.7) holds for all $|\alpha| \le K$, but $F \not\equiv G$ on M.*

In case M is a Levi-nondegenerate hypersurface (that is, $d = 1$ and $l(M) = 1$), Theorem 2 reduces to the result of Chern and Moser [1974] that a germ of a CR diffeomorphism is uniquely determined by its derivatives of order ≤ 2 at a point. Generalizations of this result for Levi nondegenerate manifolds of higher codimension were found later [Tumanov and Khenkin 1983; Tumanov 1988b]. More precise results for Levi nondegenerate hypersurfaces have been given by Beloshapka [1979] and Loboda [1981]. The notion of holomorphic nondegeneracy for hypersurfaces is due to Stanton [1995], who showed that it is a necessary and sufficient condition for the finite dimensionality of the space of "infinitesimal holomorphisms", as will be mentioned in Section 3. The notion of finite non-degeneracy was first introduced in the case of a hypersurface in our joint work with X. Huang [Baouendi et al. 1996b].

3. Infinitesimal CR Automorphisms

A smooth real vector field X defined in a neighborhood of p in M is an *infinitesimal holomorphism* if the local 1-parameter group of diffeomorphisms $\exp tX$ for t small extends to a local 1-parameter group of biholomorphisms of $\mathbb{C}^N$. More generally, X is called an *infinitesimal CR automorphism* if the mappings $\exp tX$ are CR diffeomorphisms. We denote by $\mathrm{hol}(M, p)$ the Lie algebra generated by the germs at p of the infinitesimal holomorphisms, and by $\mathrm{aut}(M, p)$ the one generated by the germs of infinitesimal CR automorphisms. Since every local biholomorphism preserving M restricts to a CR diffeomorphism of M into itself, it follows that $\mathrm{hol}(M, p) \subset \mathrm{aut}(M, p)$.

The following result gives the finite dimensionality of the larger space $\mathrm{aut}(M, p)$ not only for hypersurfaces, but also for CR manifolds of higher codimension.

THEOREM 3 [Baouendi et al. 1998]. *Let $M \subset \mathbb{C}^N$ be a real analytic, connected CR submanifold. If M is holomorphically nondegenerate, and minimal at some point, then*

$$(3.1) \qquad \dim_{\mathbb{R}} \mathrm{aut}(M, p) < \infty$$

for all $p \in M$.

PROOF. Let $p_0 \in M$ and let $X^1, \ldots, X^m \in \mathrm{aut}(M, p_0)$ be linearly independent over $\mathbb{R}$. Let $x = (x_1, \ldots, x_r)$ be a local coordinate system on M near p_0 and vanishing at p_0. In this coordinate system, we may write

$$(3.2) \qquad X^j = \sum_{l=1}^{r} \tilde{X}_l^j(x) \frac{\partial}{\partial x_l} = \tilde{X}^j(x) \cdot \frac{\partial}{\partial x}.$$

For $y = (y_1, \ldots, y_m) \in \mathbb{R}^m$, we denote by $\Phi(t, x, y)$ the flow of the vector field $y_1 X_1 + \cdots + y_m X_m$, that is, the solution of

$$(3.3) \qquad \frac{\partial \Phi}{\partial t}(t, x, y) = \sum_{i=1}^{m} y_i \tilde{X}^i(\Phi(t, x, y)),$$

$$\Phi(0, x, y) = x.$$

Using elementary ODE arguments, one can show that by choosing $\delta > 0$ sufficiently small, there exists $c > 0$ such that the flows $\Phi(t, x, y)$ are smooth (C^∞) in $\{(t, x, y) \in \mathbb{R}^{1+r+m} : |t| \le 2, |x| \le c, |y| \le \delta\}$. Denote by $F(x, y)$ the corresponding time-one maps, that is,

$$(3.4) \qquad F(x, y) = \Phi(1, x, y).$$

LEMMA 3.5. *There is δ', $0 < \delta' < \delta$, such that for any fixed y^1, y^2 with $|y^1|, |y^2| \leq \delta'$, if $F(x, y^1) \equiv F(x, y^2)$ for all x, $|x| \leq c$, then necessarily $y^1 = y^2$.*

PROOF. It follows from (3.3) and (3.4) that

$$(3.6) \qquad \frac{\partial F}{\partial y_i}(x, 0) = \tilde{X}^i(x).$$

Thus, denoting by $\tilde{X}(x)$ the $r \times m$-matrix with column vectors $\tilde{X}^i(x)$, we have

$$(3.7) \qquad \frac{\partial F}{\partial y}(x, 0) = \tilde{X}(x).$$

By Taylor expansion we obtain

$$(3.8) \qquad \|F(x, y^2) - F(x, y^1)\| \geq \left\| \frac{\partial F}{\partial y}(x, y^1) \cdot (y^2 - y^1) \right\| - C|y^2 - y^1|^2,$$

where $C > 0$ is some uniform constant for $|y^1|, |y^2| \leq \delta$. The linear independence of the vector fields $X^1, \ldots, X^m$ over $\mathbb{R}$ implies that there is a constant C' such that

$$(3.9) \qquad \|\tilde{X}(x) \cdot y\| \geq C'|y|.$$

The lemma follows by using (3.6), (3.9) and a standard compactness argument. $\square$

We proceed with the proof of Theorem 3. Denote by U the open neighborhood of p on M given by $|x| < c$. We make use of Theorem 2 with M replaced by U. Let $p_1, \ldots, p_k$ be the points in U given by the theorem. By choosing the number $\delta' > 0$ in Lemma 3.5 even smaller if necessary, we may assume that the maps $x \mapsto F(x, y)$, for $|y| < \delta'$, are CR diffeomorphisms of M. Consider the smooth mapping from $|y| < \delta'$ into $\mathbb{R}^\mu$ defined by

$$(3.10) \qquad y \mapsto \left(\frac{\partial^{|\alpha|} F(p_l, y)}{\partial x^\alpha} \right)_{\substack{|\alpha| \leq (d+1)l(M) \\ 1 \leq l \leq k}} \in \mathbb{R}^\mu,$$

where μ equals $k \cdot r$ times the number of monomials in r variables of degree less than or equal to $(d+1)l(M)$. This mapping is injective for $|y| < \delta'$ in view of Theorem 2 and Lemma 3.5. Consequently, we have a smooth injective mapping from a neighborhood of the origin in $\mathbb{R}^m$ into $\mathbb{R}^\mu$. This implies that $m \leq \mu$ and hence the desired finite dimensionality of the conclusion of Theorem 3. $\square$

As in the case of Theorem 2, here again the condition of holomorphic nondegeneracy is necessary for the conclusion of Theorem 3 to hold. Also, if M is not minimal at any point, but is defined by weighted homogeneous polynomials, then $\dim_\mathbb{R} \text{hol}(M, p)$ is either 0 or ∞. This can be viewed as an analogue of Proposition 2.6.

We conclude this section by some bibliographical notes. Tanaka [1962] proved that $\text{hol}(M, p)$ is a finite dimensional vector space if M is a real analytic Levi nondegenerate hypersurface. More recently Stanton [1995; 1996] proved that if M is a real analytic hypersurface, $\mathrm{hol}(M, p)$ is finite dimensional for any $p \in M$ if and only if M is holomorphically nondegenerate. Theorem 3 above generalizes Stanton's result. It should be also mentioned that the methods outlined here are quite different from those of [Stanton 1996].

4. Parametrization of Local Biholomorphisms Between Hypersurfaces

In this section and the next, we shall restrict ourselves to the case of hypersurfaces. Let $M \subset \mathbb{C}^N$ be a real analytic hypersurface and $p_0 \in M$. Denote by $\mathrm{hol}_0(M, p_0)$ the elements of $\mathrm{hol}(M, p_0)$ that vanish at p_0. Also denote by $\mathrm{Aut}(M, p_0)$ the set of all germs of biholomorphisms at p_0, fixing p_0 and mapping M into itself. Under the assumption that M is holomorphically nondegenerate the finite dimensionality of $\mathrm{hol}(M, p_0)$, which follows from Theorem 3 (and, as just mentioned, is in fact proved in [Stanton 1996]), implies that there is a unique topology on $\mathrm{Aut}(M, p_0)$, considered as an abstract group, such that the latter is a Lie transformation group with $\mathrm{hol}_0(M, p_0)$ as its Lie algebra (see [Kobayashi 1972, p. 13], for example). On the other hand $\mathrm{Aut}(M, p_0)$ has a natural inductive limit topology corresponding to uniform convergence on compact neighborhoods of p_0 in $\mathbb{C}^N$. One of the main results of this section (Corollary 4.2) implies that for almost all $p_0 \in M$ the two topologies on $\mathrm{Aut}(M, p_0)$ must coincide.

We shall first introduce some notation. Let k be a positive integer and $J_p^k = J^k(\mathbb{C}^N)_p$ the set of k-jets at p of holomorphic mappings from $\mathbb{C}^N$ to $\mathbb{C}^N$ fixing p. J_0^k can be identified with the space of holomorphic polynomial mappings of degree $\leq k$, mapping 0 to 0. Let $G^k = G^k(\mathbb{C}^N)$ be the complex Lie group consisting of those holomorphic mappings in J_0^k with nonvanishing Jacobian determinant at 0. We take the coefficients $\Lambda = (\Lambda_\alpha)$ of the polynomials corresponding to the jets to be global coordinates of G^k. The group multiplication in G^k consists of composing the polynomial mappings and dropping the monomial terms of degree higher than k.

For $p, p' \in \mathbb{C}^N$, denote by $\mathcal{E}_{p,p'}$ the space of germs of holomorphic mappings $H : (\mathbb{C}^N, p) \to (\mathbb{C}^N, p')$, (that is, $H(p) = p'$) with Jacobian determinant of H nonvanishing at p equipped with the natural inductive limit topology corresponding to uniform convergence on compact neighborhoods of p. We define a mapping $\eta_{p,p'} : \mathcal{E}_{p,p'} \to G^k$ as follows. For $H \in \mathcal{E}_{p,p'}$, let $F \in \mathcal{E}_{0,0}$ be defined by $F(Z) = H(Z + p) - p'$. Then $j_k(F)$, the k-jet of F at 0, is an element of G^k. We put $\eta_{p,p'}(H) = j_k(F)$. In local holomorphic coordinates Z near p we have $\eta_{p,p'}(H) = (\partial_Z^\alpha H(p))_{1 \leq |\alpha| \leq k}$. The mapping $\eta_{p,p'}$ is continuous; composition of mappings is related to group multiplication in G^k by the identity

$$(4.1) \qquad \eta_{p,p''}(H_2 \circ H_1) = \eta_{p',p''}(H_2) \cdot \eta_{p,p'}(H_1)$$

for any $H_1 \in \mathcal{E}_{p,p'}$ and $H_2 \in \mathcal{E}_{p',p''}$, where $\cdot$ denotes the group multiplication in G^k. We write η for $\eta_{p,p'}$ when there is no ambiguity.

If M and M' are two real analytic hypersurfaces in $\mathbb{C}^N$ with $p \in M$ and $p' \in M'$, denote by $\mathcal{F} = \mathcal{F}(M, p; M', p')$ the subset of $\mathcal{E}_{p,p'}$ consisting of those germs of mappings which send M into M', and equip $\mathcal{F}$ with the induced topology.

THEOREM 4 [Baouendi et al. 1997]. *Let M and M' be two real analytic hypersurfaces in $\mathbb{C}^N$ which are k_0-nondegenerate at p and p' respectively and let $\mathcal{F} = \mathcal{F}(M, p; M', p')$ as above. Then the restriction of the map $\eta : \mathcal{E}_{p,p'} \to G^{2k_0}$ to $\mathcal{F}$ is one-to-one; in addition, $\eta(\mathcal{F})$ is a totally real, closed, real analytic submanifold of G^{2k_0} (possibly empty) and η is a homeomorphism of $\mathcal{F}$ onto $\eta(\mathcal{F})$. Furthermore, global defining equations for the submanifold $\eta(\mathcal{F})$ can be explicitly constructed from local defining equations for M and M' near p and p'.*

With the notation above, we put $\mathrm{Aut}(M, p) = \mathcal{F}(M, p; M, p)$ and refer to it as the *stability group* of M at p. When $\mathrm{Aut}(M, p)$ is a Lie group with its natural topology, it is easy to show that $\mathrm{hol}_0(M, p)$, as defined above, is its Lie algebra. We have the following corollary of Theorem 4.

COROLLARY 4.2. *If, in addition to the assumptions of Theorem 4, $M = M'$ and $p = p'$, then $\eta(\mathcal{F})$ is a closed, totally real Lie subgroup $G(M, p)$ of G^{2k_0}. Hence the stability group $\mathrm{Aut}(M, p)$ of M at p has a natural Lie group structure. In general, for different (M, p) and (M', p'), $\eta(\mathcal{F})$ is either empty or is a coset of the subgroup $G(M, p)$.*

In the next section we shall give an outline of a proof of Theorem 4 which gives an algorithm to calculate $G(M, p)$ and, in particular, to determine whether two hypersurfaces are locally biholomorphically equivalent.

Since a connected, real analytic, holomorphically nondegenerate hypersurface M is $l(M)$-nondegenerate at every point outside a proper real analytic subset $V \subset M$, the following is also a consequence of Theorem 4.

COROLLARY 4.3. *Let M be a real analytic connected real hypersurface in $\mathbb{C}^N$ which is holomorphically nondegenerate. Let l be the Levi number of M. Then there is a proper real analytic subvariety $V \subset M$ such that for any $p \in M \backslash V$, η is a homeomorphism between $\mathrm{Aut}(M, p)$ and a closed, totally real Lie subgroup of G^{2l}.*

One may also generalize Theorem 4 to the case where p and p' are varying points in M and M' respectively. We first introduce some notation. If X and Y are two complex manifolds and k a positive integer, we denote by $J^k(X, Y)$ the complex manifold of k-jets of germs of holomorphic mappings from X to Y, that is,

$$J^k(X, Y) = \bigcup_{x \in X, y \in Y} J^k(X, Y)_{(x,y)}$$

where $J^k(X, Y)_{(x,y)}$ denotes the k-jets of germs at x of holomorphic mappings from X to Y and taking x to y. (See [Malgrange 1967; Golubitsky and Guillemin

1973], for example.) With this notation, $J^k(X, X)_{(x,x)}$ is the same as $J^k(X)_x$ introduced above with $X = \mathbb{C}^N$.

Denote by $E(X, Y)$ the set of germs of holomorphic mappings from X to Y equipped with its natural topology defined as follows. If $H_x \in E(X, Y)$ is a germ at x of a holomorphic mapping from X to Y which extends to a holomorphic mapping $H : U \to Y$, where $U \subset X$ is an open neighborhood of x, then a basis of open neighborhoods of H_x is given by

$$N_{U', V'} = \{F_p \in E(X, Y) : p \in U', \ F : U' \to V'\},$$

where U' is a relatively compact open neighborhood of x in U and V' is an open neighborhood of $H(x)$ in Y. In particular, a sequence $(H_j)_{x_j}$ converges to H_x if x_j converges to x and there exists a neighborhood U of x in X to which all the (H_j) and H extend, for sufficiently large j, and the H_j converge uniformly to H on compact subsets of U. This topology restricted to $E(X, Y)_{(x,y)}$ (the germs at x mapping x to y) coincides with the natural inductive topology mentioned above.

For every k there is a canonical mapping $\sigma_k : E(X, Y) \to J^k(X, Y)$. Note that $\sigma_k|_{E(X,Y)_{(p,p')}}$ is the same as the mapping $\eta_{p,p'}$ with $X = Y = \mathbb{C}^N$. It is easy to check that σ_k is continuous. If $\dim_{\mathbb{C}} X = \dim_{\mathbb{C}} Y$ then we denote by $G^k(X, Y)$ the open complex submanifold of $J^k(X, Y)$ given by those jets which are locally invertible. Similarly, we denote by $\mathcal{E}(X, Y)$ the open subset of $E(X, Y)$ consisting of the invertible germs. It is clear that the restriction of σ_k maps $\mathcal{E}(X, Y)$ to $G^k(X, Y)$.

If $M \subset X$ and $M' \subset Y$ are real analytic submanifolds, we let $E_{(M,M')}(X, Y)$ be the set of germs $H_p \in E(X, Y)$ with $p \in M$ which map a neighborhood of p in M into M'. Similarly, we denote by $\mathcal{E}_{(M,M')}(X, Y)$ those germs in $E_{(M,M')}(X, Y)$ which are invertible. Note that with $X = Y = \mathbb{C}^N$ we have

$$\mathcal{E}_{(M,M')}(X, Y)_{(p,p')} = \mathcal{F}(M, p; M', p').$$

We may now state a generalization of Theorem 4 with varying points p, p'.

THEOREM 5 [Baouendi et al. 1997]. *Let X and Y be two complex manifolds of the same dimension, $M \subset X$ and $M' \subset Y$ two real analytic hypersurfaces, and k_0 a positive integer. Suppose that M and M' are both at most k_0-nondegenerate at every point. Then the mapping*

$$\sigma_{2k_0} : \mathcal{E}_{(M,M')}(X, Y) \to G^{2k_0}(X, Y)$$

is a homeomorphism onto its image Σ. Furthermore, Σ is a real analytic subset of $G^{2k_0}(X, Y)$, possibly empty, and each fiber $\Sigma \cap G^{2k_0}(X, Y)_{(p,p')}$, with $p \in M$, $p' \in M'$ is a real analytic submanifold.

From Theorem 5, together with some properties of subgroups of Lie groups, one may obtain the following result on the discreteness of $\mathrm{Aut}(M, p)$ in a neighborhood of p_0.

THEOREM 6 [Baouendi et al. 1997]. *Let M be a real analytic hypersurface in $\mathbb{C}^N$ finitely nondegenerate at p_0. If $\mathrm{Aut}(M, p_0)$ is a discrete group, then $\mathrm{Aut}(M, p)$ is also discrete for all p in a neighborhood of p_0 in M. Equivalently, if $\mathrm{hol}_0(M, p_0) = \{0\}$ then $\mathrm{hol}_0(M, p) = \{0\}$ for all p in a neighborhood of p_0 in M.*

EXAMPLE 4.4. Let M be the hypersurface given by

$$\mathrm{Im}\, w = |z|^2 + (\mathrm{Re}\, z^2)|z|^2.$$

Then by using the algorithm described in Section 5 below, one can show that $\mathrm{Aut}(M, 0)$ consists of exactly two elements, namely the identity and the map $(z, w) \mapsto (-z, w)$. In particular, $\mathrm{hol}_0(M, 0) = \{0\}$. Hence, by Theorem 6, $\mathrm{hol}_0(M, p) = \{0\}$ for all $p \in M$ near 0.

We mention here that there is a long history of results on transformation groups of Levi nondegenerate hypersurfaces, beginning with the seminal paper [Chern and Moser 1974]. (See also [Burns and Shnider 1977; Webster 1977b].) In particular, the fact that $\mathrm{Aut}(M, p)$ is a Lie group follows from [Chern and Moser 1974] when M is Levi nondegenerate at p. Further contributions were made by the Russian school (see for example the survey papers [Vitushkin 1985; Kruzhilin 1987], as well as the references therein). Results for higher-codimensional quadratic manifolds were obtained by Tumanov [1988b]. We point out that even for Levi nondegenerate hypersurfaces the approach given here is not based on [Chern and Moser 1974].

5. An Algorithm for Constructing the Set of All Mappings Between Two Real Analytic Hypersurfaces

Even in the case of a hypersurface, the parametrization of the Segre sets given by Theorem 1, is in general not an immersion onto a neighborhood of the base point p_0. Hence in the proof of Theorem 2, one goes to a nearby point to verify the uniqueness of the holomorphic mapping (which is already assumed to exist). By contrast, in the proof of Theorem 4, this method can no longer be used, because one has to know when a particular value of a parameter actually corresponds to a holomorphic mapping between the hypersurfaces in question.

In this section we outline the proof of Theorem 4, which actually gives an algorithm to construct the defining equations of the manifold

$$\Sigma_{p,p'} = \eta(\mathcal{F}(M, p; M', p'))$$

from defining equations of M and M' near p and p'. Moreover, for each $\Lambda \in G_0^{2k_0}$ the algorithm constructs a mapping which is the unique biholomorphic mapping H sending (M, p) into (M', p') with $\eta(H) = \Lambda$ for $\Lambda \in \Sigma_{p,p'}$. We give here the main steps of this algorithm.

STEP 1. We choose normal coordinates (z, w) and (z', w') for M and M' vanishing respectively at p and p'. We may write any $H \in \mathcal{F}(M, p; M', p')$ in the form $H = (f, g)$, such that the map is defined by $z' = f(z, w)$ and $w' = g(z, w)$. Note that it follows from the normality of the coordinates that $g(z, 0) \equiv 0$. For each fixed k we choose coordinates Λ in G^k with $\Lambda = (\lambda_{z^\alpha w^j}, \mu_{z^\alpha w^j})$, where $0 < |\alpha| + j \leq k$, such that if $H = (f, g) \in \mathcal{F}$, then the coordinates of $\eta(H)$ are defined by $\lambda_{z^\alpha w^j} = \partial_{z^\alpha w^j} f(0)$ and $\mu_{z^\alpha w^j} = \partial_{z^\alpha w^j} g(0)$. We identify an element in G^k with its coordinates Λ. We shall denote by G_0^k the submanifold of G^k consisting of those $\Lambda = (\lambda, \mu)$ for which $\mu_{z^\alpha} = 0$ for all $0 < |\alpha| \leq k$. It is easily checked that G_0^k is actually a subgroup of G^k and hence a Lie group.

We apply (2.3) with $Z = (z, 0)$ and $\zeta = 0$. We obtain the following. There exist $\mathbb{C}^N$-valued functions $\Psi_j(z, \Lambda)$, with $j = 0, 1, 2, \ldots$, each holomorphic in a neighborhood of $0 \times G_0^{k_0 + j}$ in $\mathbb{C}^n \times G_0^{k_0 + j}$, such that if $H(z, w) \in \mathcal{F}(M, p; M', p')$ with $(\partial^\alpha H(0))_{|\alpha| \leq k_0 + j} = \Lambda_0 \in G_0^{k_0 + j}$, then

$$(5.1) \qquad \partial_w^j H(z, 0) = \Psi_j(z, \overline{\Lambda_0}), \quad \text{for } j = 0, 1, 2, \ldots.$$

Furthermore, we have $\Psi_{0, N}(z, \Lambda) \equiv 0$, where $\Psi_{0, N}$ is the last component of the mapping Ψ_0. The fact that the Ψ_j do not depend on H follows from a close analysis of (2.3).

STEP 2. By taking $\gamma = 0$, $Z = (z, Q(z, \chi, 0))$ and $\zeta = (\chi, 0)$ in (2.3), we find a $\mathbb{C}^N$-valued function $\Phi(z, \chi, \Lambda)$, holomorphic in a neighborhood of $0 \times 0 \times G_0^{2k_0}$ in $\mathbb{C}^n \times \mathbb{C}^n \times G_0^{2k_0}$, such that for $H \in \mathcal{F}(M, p; M', p')$ with $(\partial^\alpha H(0))_{|\alpha| \leq 2k_0} = \Lambda_0$, we have

$$(5.2) \qquad H(z, Q(z, \chi, 0)) \equiv \Phi(z, \chi, \Lambda_0).$$

Again here the fact that Φ does not depend on H follows from a close analysis of the proof of (2.3).

STEP 3. We begin with the following lemma.

LEMMA 5.3. *Assume $(M, p), (M', p')$ are as above. There exists a $\mathbb{C}^N$-valued function $F(z, t, \Lambda)$ holomorphic in a neighborhood of $0 \times 0 \times G_0^{2k_0}$ in $\mathbb{C}^n \times \mathbb{C} \times G_0^{2k_0}$ and a germ at 0 of a nontrivial holomorphic function $B(z)$, such that for a fixed $\Lambda_0 \in G_0^{2k_0}$ there exists $H \in \mathcal{F}(M, p; M', p')$ with*

$$(5.4) \qquad (\partial^\alpha H(0))_{|\alpha| \leq 2k_0} = \Lambda_0$$

if and only if all three following conditions hold:

(i) *$(z, w) \mapsto F(z, w/B(z), \Lambda_0)$ extends to a function $K_{\Lambda_0}(z, w)$ holomorphic in a full neighborhood of 0 in $\mathbb{C}^N$.*
(ii) *$(\partial^\alpha K_{\Lambda_0}(0))_{|\alpha| \leq 2k_0} = \Lambda_0$.*
(iii) *$K_{\Lambda_0}(M) \subset M'$.*

If (i), (ii), (iii) hold, then the unique mapping in $\mathcal{F}(M, p; M', p')$ satisfying (5.4) is given by $H(Z) = K_{\Lambda_0}(Z)$.

PROOF. From the k_0-nondegeneracy, we have $Q_{\chi_1}(z,0,0) \not\equiv 0$ and we set

(5.5) $$A(z) = Q_{\chi_1}(z,0,0).$$

We write $\chi = (\chi_1, \chi')$; we shall solve the equation

(5.6) $$w = Q(z,(\chi_1,0),0)$$

for χ_1 as a function of (z,w) and analyze the solution as z and w approach 0. We have

(5.7) $$Q(z,(\chi_1,0),0) = \sum_{j=1}^{\infty} A_j(z)\chi_1^j,$$

with $A_1(z) = A(z)$ and $A_j(0) = 0$, for $j = 1, \ldots$. Dividing (5.6) by $A(z)^2$, we obtain

$$\frac{w}{A(z)^2} = \frac{\chi_1}{A(z)} + \sum_{j=2}^{\infty} A_j(z) \frac{\chi_1^j}{A(z)^2}.$$

We set $C_j(z) = A_j(z)A(z)^{j-2}$, with $j \geq 2$, and let

(5.8) $$\psi(z,t) = t + \sum_{j=2}^{\infty} v_j(z)t^j$$

be the solution in u given by the implicit function theorem of the equation $t = u + \sum_{j=2}^{\infty} C_j(z)u^j$, with $\psi(0,0) = 0$. The functions ψ and v_j are then holomorphic at 0 and $v_j(0) = 0$. A solution for χ_1 in (5.6) is then given by

(5.9) $$\chi_1 = \theta(z,w) = A(z)\psi\left(z, \frac{w}{A(z)^2}\right).$$

The function $\theta(z,w)$ is holomorphic in an open set in $\mathbb{C}^{n+1}$ having the origin as a limit point.

Now define F by

(5.10) $$F(z,t,\Lambda) = \Phi(z,(A(z)\psi(z,t),0),\Lambda),$$

where Φ is given by Step 2, and let $B(z) = A(z)^2$, with $A(z)$ given by (5.5). Then (i) follows from Step 2. The rest of the proof of the lemma is now easy and is left to the reader. $\qquad\square$

It follows from Lemma 5.3 and its proof that if $H(z,w)$ is a biholomorphic mapping taking (M,p) into (M',p') with $\Lambda = \eta(H)$ then

(5.11) $$H(z,w) = F\left(z, \frac{w}{A(z)^2}, \Lambda\right),$$

where $A(z)$ is defined by (5.5), and $F(z,t,\Lambda)$ is defined by (5.10). Note again here that F and A are independent of H.

STEP 4. In this last step, the following lemma and its proof give the construction of the real analytic functions defining $\Sigma_{p,p'} = \eta(\mathcal{F})$.

LEMMA 5.12. *Under the hypotheses of Theorem 4 there exist functions b_j, for $j = 1, 2, \ldots$, holomorphic in $G_0^{2k_0} \times G_0^{2k_0}$ such that there is $H \in \mathcal{F}(M, p; M', p')$ satisfying (5.4) if and only if $b_j(\Lambda_0, \overline{\Lambda_0}) = 0$, for $j = 1, 2, \ldots$.*

The proof of Lemma 5.12 will actually give an algorithm for the construction of the functions b_j from the defining equations of M and M'.

PROOF. We first construct a function $K(Z, \Lambda)$ holomorphic in a neighborhood of $0 \times G_0^{2k_0}$ in $\mathbb{C}^N \times G_0^{2k_0}$ such that (i) of Lemma 5.3 holds for a fixed $\Lambda_0 \in G_0^{2k_0}$ if and only if $F(z, w/B(z), \Lambda_0) \equiv K(z, w, \Lambda_0)$. Recall that $F(z, t, \Lambda)$ is holomorphic in a neighborhood of $0 \times 0 \times G_0^{2k_0}$ in $\mathbb{C}^n \times \mathbb{C} \times G_0^{2k_0}$. Hence we can write

$$(5.13) \qquad F(z, t, \Lambda) = \sum_{\alpha, j} F_{\alpha j}(\Lambda) z^\alpha t^j,$$

with $F_{\alpha j}$ holomorphic in $G_0^{2k_0}$. For each compact subset $L \subset G_0^{2k_0}$ there exists $C > 0$ such that the series $(z, t) \mapsto \sum_{\alpha, j} F_{\alpha j}(\Lambda) z^\alpha t^j$ converges uniformly for $|z|, |t| \leq C$ and for each fixed $\Lambda \in L$. For $|z| \leq C$ and $|w/B(z)| \leq C$ we have

$$(5.14) \qquad F\left(z, \frac{w}{B(z)}, \Lambda\right) = \sum_{j=0}^{\infty} \frac{F_j(z, \Lambda)}{B(z)^j} w^j$$

with $F_j(z, \Lambda) = \sum_\alpha F_{\alpha, j}(\Lambda) z^\alpha$. After a linear change of holomorphic coordinates if necessary, and putting $z = (z_1, z')$, we may assume, by using the Weierstrass Preparation Theorem, that

$$B(z)^j = U_j(z) \left(z_1^{K_j} + \sum_{p=0}^{K_j - 1} a_{jp}(z') z_1^p \right),$$

with $U_j(0) \neq 0$ and $a_{jp}(0) = 0$. By the Weierstrass Division Theorem we have the unique decomposition

$$(5.15) \qquad F_j(z, \Lambda) = Q_j(z, \Lambda) B(z)^j + \sum_{p=0}^{K_j - 1} r_{jp}(z', \Lambda) z_1^p,$$

where $Q_j(z, \Lambda)$ and $r_{jp}(z', \Lambda)$ are holomorphic in a neighborhood of $0 \times G_0^{2k_0}$ in $\mathbb{C}^n \times G_0^{2k_0}$. It then suffices to take

$$(5.16) \qquad K(z, w, \Lambda) = \sum_j Q_j(z, \Lambda) w^j.$$

Moreover, (i) of Lemma 5.3 holds if and only if $z' \mapsto r_{jp}(z', \Lambda_0)$ vanishes identically for all j, p. By taking the coefficients of the Taylor expansion of the r_{jp} with respect to z' we conclude that there exist functions c_j, $j = 1, 2, \ldots$, holomorphic in $G_0^{2k_0}$ such that (i) holds if and only if $c_j(\Lambda_0) = 0$, $j = 1, 2, \ldots$.

It follows from the above that we have $K_{\Lambda_0}(Z) \equiv K(Z, \Lambda_0)$, where $K_{\Lambda_0}(Z)$ is given by Lemma 5.3. By taking $d_j(\Lambda)$, for $1 \leq j \leq J$, as the components of

$(\partial_Z^\alpha K(0, \Lambda))_{|\alpha| \leq 2k_0} - \Lambda$, we find that if (i) is satisfied then (ii) holds if and only if $d_j(\Lambda_0) = 0$, for $1 \leq j \leq J$. Similarly, we note that (iii) is equivalent to

$$(5.17) \qquad \rho'\big(K(z, w, \Lambda_0),\, \overline{K}(\chi, \overline{Q}(\chi, z, w), \overline{\Lambda_0})\big) \equiv 0,$$

where ρ' is a defining function for M'. By expanding the left hand side of (5.17) as a series in z, w, χ with coefficients which are holomorphic functions of $\Lambda_0, \overline{\Lambda_0}$, we conclude that there exist functions e_j, for $j = 1, 2, \ldots$, holomorphic in $G_0^{2k_0} \times G_0^{2k_0}$ such that if (i) is satisfied then (iii) holds if and only if $e_j(\Lambda_0, \overline{\Lambda_0}) = 0$ for $j = 1, 2, \ldots$. $\qquad\square$

The main points in the proof of Theorem 4 follow from Steps 1–4 above. The proof that $\Sigma_{p,p'}$ is a manifold is first reduced to the case where $M = M'$, $p = p'$; for that case one uses the fact that a closed subgroup of a Lie group is again a Lie group; see [Varadarajan 1974], for example. We shall omit the rest of the details of the proof.

REMARK 5.18. We have stated Theorems 4–6 only for hypersurfaces. The proofs of these results do not generalize to CR manifolds of higher codimension. In fact, the proofs given here are based on an analysis of the behavior of the Segre set N_2 near the origin; see (5.7)–(5.9). Such a precise analysis for higher codimension seems much more complicated. It would be interesting to have analogues of Theorems 4–6 in higher codimension.

6. Holomorphic Mappings Between Real Algebraic Sets

In this section we shall consider the case where the submanifolds M and M' are algebraic. Recall that a subset $A \subset \mathbb{C}^N$ is a *real algebraic set* if it is defined by the vanishing of real valued polynomials in $2N$ real variables; we shall always assume that A is irreducible. By A_{reg} we mean the regular points of A: see [Hodge and Pedoe 1947–53], for example. Recall that A_{reg} is a real submanifold of $\mathbb{C}^N$, all points of which have the same dimension. We write $\dim A = \dim_{\mathbb{R}} A$ for the dimension of the real submanifold A_{reg}. A germ of a holomorphic function f at a point $p_0 \in \mathbb{C}^N$ is called *algebraic* if it satisfies a polynomial equation of the form $a_K(Z)f^K(Z) + \cdots + a_1(Z)f(Z) + a_0(Z) \equiv 0$, where the $a_j(Z)$ are holomorphic polynomials in N complex variables with $a_K(Z) \not\equiv 0$. In this section we give conditions under which a germ of a holomorphic map in $\mathbb{C}^N$, mapping an irreducible real algebraic set A into another such set of the same dimension, is actually algebraic, that is, all its components are algebraic functions.

The first result deals with biholomorphic mappings between algebraic hypersurfaces. The following theorem gives a necessary and sufficient condition for algebraicity of mappings in this case.

THEOREM 7 [Baouendi and Rothschild 1995]. *Let M, M' be two connected, real algebraic, hypersurfaces in $\mathbb{C}^N$. If M is holomorphically nondegenerate and H is a biholomorphic mapping defined in an open neighborhood in $\mathbb{C}^N$ of a point*

$p_0 \in M$ satisfying $H(M) \subset M'$, then H is algebraic. Conversely, if M is holomorphically degenerate, then for every $p_0 \in M$ there exists a germ H of a nonalgebraic biholomorphism of $\mathbb{C}^N$ at p_0 with $H(M) \subset M$ and $H(p_0) = p_0$.

For generic manifolds of higher codimension, holomorphic nondegeneracy is no longer sufficient for the algebraicity of local biholomorphisms. For example, one can take $M = M'$ to be the generic submanifold of $\mathbb{C}^3$ given by $\operatorname{Im} Z_2 = |Z_1|^2, \operatorname{Im} Z_3 = 0$, then the biholomorphism $Z \mapsto (Z_1, Z_2, e^{Z_3})$ maps M into itself, but is not algebraic. Here M is holomorphically nondegenerate, but nowhere minimal. However:

THEOREM 8 [Baouendi et al. 1996a]. *Let $M, M' \subset \mathbb{C}^N$ be two real algebraic, holomorphically nondegenerate, generic submanifolds of the same dimension. Assume there exists $p \in M$, such that M is minimal at p. If H is a biholomorphic mapping defined in an open neighborhood in $\mathbb{C}^N$ of a point $p_0 \in M$ satisfying $H(M) \subset M'$ then H is algebraic.*

The conditions for Theorem 8 are almost necessary, as is shown by the following converse.

THEOREM 9 [Baouendi et al. 1996a]. *Let $M \subset \mathbb{C}^N$ be a connected real algebraic generic submanifold. If M is holomorphically degenerate then for every $p_0 \in M$ there exists a germ of a nonalgebraic biholomorphism H of $\mathbb{C}^N$ at p_0 mapping M into itself with $H(p_0) = p_0$. When M is defined by weighted homogeneous real-valued polynomials, the existence of such a nonalgebraic mapping also holds if M is not minimal at any point (even if M is holomorphically nondegenerate).*

We do not give here the details of the proofs of Theorems 7–9. The main ingredient in proving the algebraicity of H in Theorems 7 and 8 is the fact that the closure of each of the Segre sets N_j (described in Section 1) of a generic real algebraic manifold is actually a complex algebraic set in $\mathbb{C}^N$. In particular, the following result is a consequence of Theorems 1 and 8 and the algebraicity of the maximal Segre set N_{j_0}.

COROLLARY 6.1 [Baouendi et al. 1996a]. *Let M be a real algebraic CR submanifold of $\mathbb{C}^N$ and $p_0 \in M$. Then the CR orbit of p_0 is a real algebraic submanifold of M and its intrinsic complexification, X, is a complex algebraic submanifold of $\mathbb{C}^N$. For any germ H of a biholomorphism at p_0 of $\mathbb{C}^N$ into itself mapping M into another real algebraic manifold of the same dimension as that of M, the restriction of H to X is algebraic.*

It is perhaps worth mentioning here that the CR orbit of p_0 is the Nagano leaf passing through p_0 ([Nagano 1966]) and hence can be obtained by solving systems of ODE's. In general, the solution of such a system is not algebraic, even when the coefficients of the differential equations are algebraic.

EXAMPLE 6.2. Consider the algebraic holomorphically nondegenerate generic submanifold $M \subset \mathbb{C}^4$ given by

$$(6.3) \qquad \mathrm{Im}\, w_1 = |z|^2 + \mathrm{Re}\, w_2 |z|^2, \quad \mathrm{Im}\, w_2 = \mathrm{Re}\, w_3 |z|^4, \quad \mathrm{Im}\, w_3 = 0.$$

Here M is holomorphically nondegenerate, but nowhere minimal. For all $0 \neq r \in \mathbb{R}$ the orbit of the point $(0,0,0,r)$ is the leaf $M \cap \{(z,w) : w_3 = r\}$, and its intrinsic complexification is $\{(z,w), w_3 = r\}$. By Corollary 6.1, if H is a germ of a biholomorphism at $0 \in \mathbb{C}^4$ mapping M into an algebraic submanifold of $\mathbb{C}^4$ of dimension 5, then $(z, w_1, w_2) \mapsto H(z, w_1, w_2, r)$ is algebraic for all $r \neq 0$ and small. The orbit of the point $0 \in \mathbb{C}^4$ is $M \cap \{z,w) : w_2 = w_3 = 0\}$ and its intrinsic complexification is $\{(z,w) : w_2 = w_3 = 0\}$ and hence again by Corollary 6.1, the mapping $(z, w_1) \mapsto H(z, w_1, 0, 0)$ is algebraic. By further results in [Baouendi et al. 1996a] on propagation of algebraicity, one can also show that the mapping $(z, w_1, w_2) \mapsto H(z, w)$ is algebraic for all fixed $w_3 \in \mathbb{C}$, sufficiently small. This result is optimal. Indeed, the nonalgebraic mapping $H : \mathbb{C}^4 \mapsto \mathbb{C}^4$, defined by

$$H(z, w_1, w_2, w_3) = (z e^{i w_3}, w_1, w_2, w_3),$$

is a biholomorphism near the origin, and maps M into itself.

The following statement extends Theorem 8 to more general real algebraic sets.

THEOREM 10 [Baouendi et al. 1996a]. *Let $A \subset \mathbb{C}^N$ be an irreducible real algebraic set, and p_0 a point in $\overline{A_{\mathrm{reg}}}$, the closure of A_{reg} in $\mathbb{C}^N$. Suppose the following two conditions hold.*

(i) *The submanifold A_{reg} is holomorphically nondegenerate.*

(ii) *If f is a germ, at a point in A, of a holomorphic algebraic function in $\mathbb{C}^N$ such that the restriction of f to A is real valued, then f is constant.*

Then if H is a holomorphic map from an open neighborhood in $\mathbb{C}^N$ of p_0 into $\mathbb{C}^N$, with $\mathrm{Jac}\, H \not\equiv 0$, and mapping A into another real algebraic set A' with $\dim A' = \dim A$, necessarily the map H is algebraic.

For further results on algebraicity and partial algebraicity, see [Baouendi et al. 1996a].

We will end with a brief history of some previous work on the algebraicity of holomorphic mappings between real algebraic sets. Early in this century Poincaré [Poincaré 1907] proved that if a biholomorphism defined in an open set in $\mathbb{C}^2$ maps an open piece of a sphere into another, it is necessarily a rational map. This result was extended by Tanaka [1962] to spheres in higher dimensions. Webster [1977a] proved a general result for algebraic, Levi-nondegenerate real hypersurfaces in $\mathbb{C}^N$; he proved that any biholomorphism mapping such a hypersurface into another is algebraic. Later, Webster's result was extended in some cases to Levi-nondegenerate hypersurfaces in complex spaces of different dimensions (see, for example, [Webster 1979; Forstnerič 1989; Huang 1994] and their references). See also [Bedford and Bell 1985] for other related results. We also refer the reader to [Tumanov and Khenkin 1983; Tumanov 1988b], which contain results on mappings of higher codimensional quadratic manifolds. See also related results of Sharipov and Sukhov [1996] using Levi form criteria; some of these results are special cases of Theorem 8.

Added in Proof. Since this paper was submitted, the authors jointly with Ebenfelt, have published a book [Baouendi et al. 1999]. The reader is referred to this book for background material as well as further results related to the subject of this article.

References

[Baouendi and Rothschild 1995] M. S. Baouendi and L. P. Rothschild, "Mappings of real algebraic hypersurfaces", *J. Amer. Math. Soc.* **8**:4 (1995), 997–1015.

[Baouendi et al. 1985] M. S. Baouendi, H. Jacobowitz, and F. Treves, "On the analyticity of CR mappings", *Ann. of Math.* (2) **122**:2 (1985), 365–400.

[Baouendi et al. 1996a] M. S. Baouendi, P. Ebenfelt, and L. P. Rothschild, "Algebraicity of holomorphic mappings between real algebraic sets in $\mathbb{C}^n$", *Acta Math.* **177**:2 (1996), 225–273.

[Baouendi et al. 1996b] M. S. Baouendi, X. Huang, and L. Preiss Rothschild, "Regularity of CR mappings between algebraic hypersurfaces", *Invent. Math.* **125**:1 (1996), 13–36.

[Baouendi et al. 1997] M. S. Baouendi, P. Ebenfelt, and L. P. Rothschild, "Parametrization of local biholomorphisms of real analytic hypersurfaces", *Asian J. Math.* **1**:1 (1997), 1–16.

[Baouendi et al. 1998] M. S. Baouendi, P. Ebenfelt, and L. P. Rothschild, "CR automorphisms of real analytic manifolds in complex space", *Comm. Anal. Geom.* **6**:2 (1998), 291–315.

[Baouendi et al. 1999] M. S. Baouendi, P. Ebenfelt, and L. P. Rothschild, *Real submanifolds in complex space and their mappings*, Press, Princeton Math. Series **47**, Princeton Univ. Press, Princeton, NJ, 1999.

[Bedford and Bell 1985] E. Bedford and S. Bell, "Extension of proper holomorphic mappings past the boundary", *Manuscripta Math.* **50** (1985), 1–10.

[Belošapka 1979] V. K. Belošapka, "On the dimension of the group of automorphisms of an analytic hypersurface", *Izv. Akad. Nauk SSSR Ser. Mat.* **43**:2 (1979), 243–266, 479. In Russian; translated in *Math USSR Izv.* **14** (1980), 223–245.

[Bloom and Graham 1977] T. Bloom and I. Graham, "On "type" conditions for generic real submanifolds of C^n", *Invent. Math.* **40**:3 (1977), 217–243.

[Burns and Shnider 1977] D. Burns, Jr. and S. Shnider, "Real hypersurfaces in complex manifolds", pp. 141–168 in *Several complex variables,* part 2 (Williamstown, MA, 1975), edited by R. O. Wells, Jr., Proc. Sympos. Pure Math. **30**, Amer. Math. Soc., Providence, 1977.

[Chern and Ji 1995] S. S. Chern and S. Ji, "Projective geometry and Riemann's mapping problem", *Math. Ann.* **302**:3 (1995), 581–600.

[Chern and Moser 1974] S. S. Chern and J. K. Moser, "Real hypersurfaces in complex manifolds", *Acta Math.* **133** (1974), 219–271. Erratum in **150**:3–4 (1983), p. 297.

[Diederich and Fornæss 1988] K. Diederich and J. E. Fornæss, "Proper holomorphic mappings between real-analytic pseudoconvex domains in C^{n}", *Math. Ann.* **282**:4 (1988), 681–700.

[Diederich and Webster 1980] K. Diederich and S. M. Webster, "A reflection principle for degenerate real hypersurfaces", *Duke Math. J.* **47**:4 (1980), 835–843.

[Ebenfelt 1998] P. Ebenfelt, "Holomorphic mappings between real analytic submanifolds in complex space", pp. 35–69 in *Integral geometry, Radon transforms and complex analysis* (Venice, 1996), edited by E. Casadio Tarabusi et al., Lecture Notes in Math. **1684**, Springer, Berlin, 1998.

[Forstnerič 1989] F. Forstnerič, "Extending proper holomorphic mappings of positive codimension", *Invent. Math.* **95**:1 (1989), 31–61.

[Golubitsky and Guillemin 1973] M. Golubitsky and V. Guillemin, *Stable mappings and their singularities*, Graduate Texts in Math. **14**, Springer, New York, 1973.

[Hodge and Pedoe 1947–53] W. H. D. Hodge and D. Pedoe, *Methods of algebraic geometry*, Cambridge University Press, Cambridge, 1947–53.

[Huang 1994] X. Huang, "On the mapping problem for algebraic real hypersurfaces in the complex spaces of different dimensions", *Ann. Inst. Fourier (Grenoble)* **44**:2 (1994), 433–463.

[Kobayashi 1972] S. Kobayashi, *Transformation groups in differential geometry*, Ergebnisse der Mathematik und ihrer Grenzgebiete **70**, Springer, New York, 1972.

[Kruzhilin 1987] N. G. Kruzhilin, "Description of the local automorphism groups of real hypersurfaces", pp. 749–758 in *Proceedings of the International Congress of Mathematicians* (Berkeley, 1986), vol. 1, Amer. Math. Soc., Providence, RI, 1987.

[Loboda 1981] A. V. Loboda, "Local automorphisms of real-analytic hypersurfaces", *Izv. Akad. Nauk SSSR Ser. Mat.* **45**:3 (1981), 620–645. In Russian; translated in *Math. USSR Izv.* **18** (1982), 537–559.

[Malgrange 1967] B. Malgrange, *Ideals of differentiable functions*, Tata Research Studies in Mathematics **3**, Oxford University Press and Tata Institute, Bombay, 1967.

[Nagano 1966] T. Nagano, "Linear differential systems with singularities and an application to transitive Lie algebras", *J. Math. Soc. Japan* **18** (1966), 398–404.

[Poincaré 1907] H. Poincaré, "Les fonctions analytiques de deux variables et la représentation conforme", *Rend. Circ. Mat. Palermo, II Ser.* **23** (1907), 185–220.

[Segre 1931] B. Segre, "Intorno al problem di Poincaré della rappresentazione pseudoconforme", *Rend. Acc. Lincei* **13** (1931), 676–683.

[Sharipov and Sukhov 1996] R. Sharipov and A. Sukhov, "On CR-mappings between algebraic Cauchy-Riemann manifolds and separate algebraicity for holomorphic functions", *Trans. Amer. Math. Soc.* **348**:2 (1996), 767–780.

[Stanton 1995] N. K. Stanton, "Infinitesimal CR automorphisms of rigid hypersurfaces", *Amer. J. Math.* **117**:1 (1995), 141–167.

[Stanton 1996] N. K. Stanton, "Infinitesimal CR automorphisms of real hypersurfaces", *Amer. J. Math.* **118**:1 (1996), 209–233.

[Tanaka 1962] N. Tanaka, "On the pseudo-conformal geometry of hypersurfaces of the space of n complex variables", *J. Math. Soc. Japan* **14** (1962), 397–429.

[Tumanov 1988a] A. E. Tumanov, "Extending CR functions on manifolds of finite type to a wedge", *Mat. Sb. (N.S.)* **136(178)**:1 (1988), 128–139. In Russian; translated in *Math. USSR Sb.* **64**:1 (1989), 129–140.

[Tumanov 1988b] A. E. Tumanov, "Finite-dimensionality of the group of CR-auto-morphisms of a standard CR-manifold, and characteristic holomorphic mappings of Siegel domains", *Izv. Akad. Nauk SSSR Ser. Mat.* **52**:3 (1988), 651–659, 672. In Russian; translated in *Math. USSR Izvestia*, **32** (1989), 655–662.

[Tumanov and Khenkin 1983] A. E. Tumanov and G. M. Khenkin, "Local charac-terization of holomorphic automorphisms of Siegel domains", *Funktsional. Anal. i Prilozhen.* **17**:4 (1983), 49–61. In Russian; translated in *Functional Anal. Appl.* **17** (1983), 285–294.

[Varadarajan 1974] V. S. Varadarajan, *Lie groups, Lie algebras, and their representa-tions*, Prentice-Hall, Englewood Cliffs, N.J., 1974. Reprinted as Graduate Texts in Mathematics **102**, Springer, New York, 1984.

[Vitushkin 1985] A. G. Vitushkin, "Holomorphic mappings and the geometry of hypersurfaces", pp. 159–214 in *Several Complex Variables I*, edited by A. G. Vitushkin, Encycl. Math. Sci. **7**, Springer, Berlin, 1985.

[Webster 1977a] S. M. Webster, "On the mapping problem for algebraic real hyper-surfaces", *Invent. Math.* **43**:1 (1977), 53–68.

[Webster 1977b] S. M. Webster, "On the transformation group of a real hypersurface", *Trans. Amer. Math. Soc.* **231**:1 (1977), 179–190.

[Webster 1979] S. M. Webster, "On mapping an n-ball into an $(n+1)$-ball in complex spaces", *Pacific J. Math.* **81**:1 (1979), 267–272.

M. SALAH BAOUENDI
DEPARTMENT OF MATHEMATICS 0112
UNIVERSITY OF CALIFORNIA, SAN DIEGO
LA JOLLA, CA 92093-0112
UNITED STATES
sbaouendi@ucsd.edu

LINDA PREISS ROTHSCHILD
DEPARTMENT OF MATHEMATICS 0112
UNIVERSITY OF CALIFORNIA, SAN DIEGO
LA JOLLA, CA 92093-0112
UNITED STATES
lrothschild@ucsd.edu

Several Complex Variables
MSRI Publications
Volume **37**, 1999

How to Use the Cycle Space in Complex Geometry

DANIEL BARLET

ABSTRACT. In complex geometry, the use of n-convexity and the use of ampleness of the normal bundle of a d-codimensional submanifold are quite difficult for $n > 0$ and $d > 1$. The aim of this paper is to explain how some constructions on the cycle space (the Chow variety in the quasiprojective setting) allows one to pass from the n-convexity of Z to the 0-convexity of $C_n(Z)$ and from a $(n+1)$-codimensional submanifold of Z having an ample normal bundle to a Cartier divisor of $C_n(Z)$ having the same property. We illustrate the use of these tools with some applications.

1. Basic Definitions

Let Z be a complex manifold; recall that an n-*cycle* in Z is a locally finite sum

$$X = \sum_{j \in J} n_j X_j,$$

where the X_j are distinct nonempty closed irreducible n-dimensional analytic subsets of Z, and where $n_j \in \mathbb{N}^*$ for any $j \in J$. The *support* of the cycle X is the closed analytic set $|X| = \bigcup_{j \in J} X_j$ of pure dimension n. The integer n_j is the *multiplicity* of the irreducible component X_j of $|X|$ in the cycle X. The cycle X is *compact* if and only if each X_j is compact and J is finite. We shall consider mainly compact cycles, but to understand problems which are of local nature on cycles it will be better to drop this assumption from time to time. We shall make it explicit when the cycles are assumed to be compact.

Topology of the cycle space. For simplicity we assume here that cycles are compact. The continuity of a family of cycles $(C_s)_{s \in S}$ consists of two conditions:

– Geometric continuity of the supports: This is the fact that $\{s \in S / |C_s| \subset U\}$ is open in S when U is an open set in Z.

This text is an expanded version of a series of two lectures of the same title given at MSRI at the end of March 1996.

– Continuity of the volume: For any choice of a continuous positive hermitian $(1,1)$ form on Z, the volume function

$$\mathrm{vol}_h(C_s) = \int_{C_s} h^{\wedge n}$$

is continuous on S.

It is not quite obvious that, when the first condition is fulfilled, the second one can be expressed in the following way:

– If Y is a locally closed submanifold of codimension n in Z such that, for $s_0 \in S$, $\partial Y := \bar{Y} - Y$ does not intersect C_{s_0} and such that $C_{s_0} \cap Y$ has exactly k points (counting multiplicities[1]), then for s near enough to s_0, we have again $\sharp (C_s \cap Y) = k$ (counting multiplicities) and the intersection map $s \to (C_s \cap Y)$ with value in the symmetric product $\mathrm{Sym}^k Y$ is continuous near s_0.

For more information on the relationship between volume and intersection multiplicities, see [Barlet 1980c].

A main tool in the topological study of cycles is E. Bishop's compactness theorem (see [Bishop 1964; Barlet 1978a; Lieberman 1978; Fujiki 1978; SGAN 1982]):

THEOREM 1. *Let Z a complex analytic space and $C_n(Z)$ the (topological) space of compact n-cycles of Z. A subset $\mathcal{A}$ of $C_n(Z)$ is relatively compact if and only if*

(1) *there is a compact subset K compact of Z such that $|C| \subset K$ for every $C \in \mathcal{A}$, and*

(2) *there is a positive definite hermitian metric of class C^0 in Z and $\Gamma = \Gamma(h, \mathcal{A})$ such that*

$$\mathrm{vol}_h(C) = \int_C h^{\wedge n} \leq \Gamma \quad \text{for all } C \in \mathcal{A}. \qquad \square$$

REMARK. If Z is a Kähler manifold and if we choose h to be the Kähler metric on Z, the function vol_h is locally constant on $C_n(Z)$ so the condition (2) is satisfied for any connected set $\mathcal{A}$ in $C_n(Z)$. See [Barlet 1978a, Prop. 1].

---

[1]Multiplicities are counted as follows: locally we can assume that $Z \simeq U \times Y$ where U and Y are open polydiscs in $\mathbb{C}^n$ and $\mathbb{C}^p$, such that $|C_{s_0}| \cap \bar{U} \times \partial Y = \varnothing$, because $|C_{s_0}| \cap Y$ is finite (compare to the definition of "écaille adapté" in [Barlet 1975, Chapter 1]). Then C_{s_0} defines a branched coverings of U via the projection $U \times Y \to U$ and we have the following classification theorem for degree k branched coverings in such a situation [Barlet 1975, Chapter 0]: *There exists a natural bijection between degree k branched coverings of U in $U \times Y$ and holomorphic maps $f : U \to \mathrm{Sym}^k Y$.* So if C_{s_0} corresponds to f and Y is $\{t_0\} \times Y$ in Z, the intersection $C_{s_0} \cap Y$ is the k-uple $f(t_0)$.

Analytic families of cycles. Consider a family of compact n-dimensional cycles $(C_s)_{s \in S}$ of the complex manifold Z parametrized by a *reduced* complex space S. Assume that this family is continuous and let Y be a locally closed complex submanifold of Z such that in an open neighbourhood S' of $s_0 \in S$ we have

$$|C_s| \cap \partial Y = \varnothing \quad \text{and} \quad \sharp(Y \cap C_s) = k.$$

Then we require that the intersection map

$$I_Y : S' \to \operatorname{Sym}^k Y$$

be holomorphic, where $\operatorname{Sym}^k Y$, the k-th symmetric product of Y, is endowed with the normal complex-space structure given by the quotient Y^k / σ_k. We say that $(C_s)_{s \in S}$ is analytic near $s_0 \in S$ if, for *any* such choice of Y, the map I_Y is analytic near s_0.

For an analytic family $(C_s)_{s \in S}$ the graph

$$|G| = \{(s, z) \in S \times Z / z \in |C_s|\}$$

is a closed analytic subset of $S \times Z$ which is proper and n-equidimensional over S by the first projection.

Though it is quite hard to prove that a given family $(C_s)_{s \in S}$ is analytic using our definition, for normal S we have the following very simple criterion:

THEOREM 2. *Let Z a complex manifold and S a normal complex space. Let $G \subset S \times Z$ a analytic set which is proper and n-equidimensional over S. Then there is a unique analytic family of compact n-dimensional cycles $(C_s)_{s \in S}$ of Z satisfying these conditions:*

(i) *For s generic in S, we have $C_s = |C_s|$ (so all multiplicities are equal to one).*
(ii) *For all $s \in S$, we have $\{s\} \times |C_s| = G \cap (\{s\} \times Z)$ (as sets).* □

REMARKS. (i) Of course for nongeneric $s \in S$ in the theorem, we could have $C_s \neq |C_s|$, and one point in the proof is to explain what are the multiplicities on the irreducible components on $|C_s|$ we have to choose. The answer comes in fact from the continuity property of the intersection with a codimension n submanifold Y as explained before.
(ii) In fact the notion of analytic family of cycles is invariant by local embeddings of Z, so it is possible to extend our definition to singular Z by using a local embedding in a manifold. Then, the previous theorem extends to any Z.
(iii) To decide if a family of cycles is analytic, when S has wild singularities, could be delicate (see for instance the example in [Barlet 1975, p. 44]).
(iv) A flat family of compact n-dimensional *subspaces* of Z gives rise to an analytic family of n-cycles. More precisely, if $G \subset S \times Z$ is a S-flat and S-proper *subspace* of $S \times Z$ (with S reduced) which is n-equidimensional on S, then the family of cycles of Z associated to $\pi^{-1}(s)$ where $\pi : G \to S$ is the first projection (and $\pi^{-1}(s)$ is a *subspace* of $\{s\} \times Z$) is an analytic family of

cycles [Barlet 1975, Chapter 5]. For cycles of higher codimension one has to take care that for each cycle C there exists a lot of *subspaces* of Z such the associated cycle is C.

To conclude this section, recall that the functor

$$S \to \{\text{analytic families of } n\text{-compact cycles of } Z\}$$

is representable in the category of finite-dimensional reduced complex analytic spaces [Barlet 1975, Chapter 3]. This means that it is possible to endow the (topological) space $C_n(Z)$ with a reduced locally finite-dimensional complex analytic structure in such a way that we get a natural bijective correspondence between holomorphic maps $f : S \to C_n(Z)$ and analytic families of compact n-cycles of Z parametrized by S (any reduced complex space). This correspondence is given by the pull back of the (so called) universal family on $C_n(Z)$ (each compact n-cycle of Z is parametrized by the corresponding point in $C_n(Z)$).

2. Holomorphic Functions on $C_n(Z)$

The idea for building holomorphic functions on $C_n(Z)$ by means of integration of cohomology classes in $H^n(Z, \Omega^n_Z)$ comes from the pioneering work [Andreotti and Norguet 1967]. It was motivated by the following question, which comes up after the famous paper [Andreotti and Grauert 1962]: vanishing (or finiteness) theorems for $H^{n+1}(Z, \mathcal{F})$ for any coherent sheaf $\mathcal{F}$ on Z allow one to produce cohomology classes in $H^n(X, \mathcal{F})$. But what to do with such cohomology classes when $n > 0$?

The answer given in [Andreotti and Norguet 1967] is: produce a lot of holomorphic functions on $C_n(Z)$ in order to prove the holomorphic convexity of the components of $C_n(Z)$.

If we assume Z smooth and allow us to normalize $C_n(Z)$, the following theorem is an easy consequence of Stokes theorem.

THEOREM 3. *There exists a natural linear map*

$$\rho : H^n(Z, \Omega^n_Z) \to H^0\big(C_n(Z), \mathcal{O}\big)$$

given by $\rho(\omega))(C) = \int_C \widetilde{\omega}$, where $\widetilde{\omega}$ is a Dolbeault representative (so a (n,n) C^∞ form on $Z, \bar{\partial}$ closed) of $\omega \in H^n(Z, \Omega^n_Z)$.   □

For nonnormal parameter space, this result is much deeper and is proved in [Barlet 1980b]. For general Z (not necessarily smooth) and general S (reduced) this result was proved later, in [Barlet and Varouchas 1989].

Let me sketch now the main idea in [Andreotti and Norguet 1967] (in a simplified way). Assume that Z is a n-complete manifold. (In this terminology from Andreotti and Norguet, 0-complete is equivalent to Stein, so the n-completeness of Z implies that $H^{n+1}(Z, \mathcal{F}) = 0$ for any coherent sheaf $\mathcal{F}$ on Z.) Let $C_1 \neq C_2$

two compact n-dimensional cycles in Z. Let $X = |C_1| \cup |C_2|$; it is easy to find $\omega \in H^n(X, \Omega_X^n)$ (X is compact n-dimensional) such that

$$\int_{C_1} \omega \neq \int_{C_2} \omega.$$

The long exact sequence of cohomology for

$$0 \to \mathcal{F} \to \Omega_Z^n \to \Omega_X^n \to 0$$

and the vanishing of $H^{n+1}(Z, \mathcal{F})$ give an $\Omega \in H^n(Z, \Omega_Z^n)$ inducing ω on X. Then the global holomorphic function F on $C_n(Z)$ defined by

$$F(C) = \int_C \Omega$$

satisfies $F(C_1) \neq F(C_2)$ and $C_n(Z)$ is holomorphically separable!

Proving the next theorem, which is an improvement of [Andreotti and Norguet 1967] and [Norguet and Siu 1977] obtained in [Barlet 1978a], requires much more work.

THEOREM 4. *Let Z a strongly n-convex analytic space. Assume that the exceptional compact set (that is, the compact set where the exhaustion may fail to be n-convex) has a kählerian neighbourhood. Then $C_n(Z)$ is holomorphically convex.*

If Z is compact, Z is strongly n-convex but the conclusion may be false if Z is not Kähler; see [Barlet 1978a, Example 1].

3. Construction of Plurisubharmonic Functions on $C_n(Z)$

One way to pass directly from the n-convexity of Z to the 0-convexity of $C_n(Z)$ is to build up a strictly plurisubharmonic function on $C_n(Z)$ from the given n-convex exhaustion of Z. One important tool for that purpose is the following:

THEOREM 5 see [Barlet 1978a, Theorem 3]. *Let Z be an analytic space and φ a real differential form on Z of class C^2 and type (n, n). Assume that $i\partial\bar\partial\varphi \geq 0$ on Z and $i\partial\bar\partial\varphi \gg 0$ on the open set U (positivity is here in the sens of Lelong; it means positivity on totally decomposed vectors of $\Lambda^n T_Z$ for smooth Z). Then the function F_φ defined on $C_n(Z)$ by*

$$F_\varphi(C) = \int_C \varphi$$

is continuous and plurisubharmonic on $C_n(Z)$.

Moreover, when each irreducible component of the cycle C_0 meets U, F_φ is strongly plurisubharmonic near C_0 (that is, it stays plurisubharmonic after any small local C^2 perturbation). □

The strong plurisubharmonic conclusion is sharp: such a property is not stable by base change, so the conclusion can only be true in the cycle space itself!

As a consequence, we obtained the following nice, but not very usefull, result:

THEOREM 6 [Barlet 1978a]. *If Z is a n-complete space, $C_n(Z)$ is a 0-complete space (i.e., Stein).*   □

In fact it is possible to give some intermediate statement between Theorem 4 and Theorem 6 in order to obtain the following application:

THEOREM 7 [Barlet 1983]. *Let V be a compact connected Kähler manifold and let $F \to V$ a vector bundle on V such that*

(1) *F is a n-convex space, and*
(2) *through each point in F passes a compact n-dimension analytic subset of F.*

Then the algebraic dimension $a(V)$ of V (that is, the transcendance degree over $\mathbb{C}$ of the field of meromorphic function on V) satisfies $a(V) \geq \dim_{\mathbb{C}} V - n$.   □

For $n = 0$ this reduce to a variant of Kodaira's projectivity theorem.

Note that if V is a compact Kähler manifold admitting a smooth fibration with n-dimensional fibers on a projective manifold X, say $f : V \to X$, we can choose $F = f^*L$ where L is a positive line bundle on X to satisfy the hypothesis in the previous theorem.

To give an idea of how the meromorphic functions on V are built, I merely indicate that, in an holomorphically convex space which is a proper modification of its Remmert reduction, any compact analytic subspace is Moišezon (this is a consequence of Hironaka's flattening theorem [1975]). But Theorem 5 gives a way to show that an irreducible component Γ of $C_n(Z)$ is a proper modification of its Remmert reduction: it is enough to have a plurisubharmonic function on Γ that is strongly plurisubharmonic at one point.

4. Construction of a Kähler Metric on $C_n(Z)$

As an illustration of the idea presented in the previous paragraph, I will explain the following beautifull result of J. Varouchas (see [Varouchas 1984], [Varouchas 1989] + [Barlet and Varouchas 1989]):

THEOREM 8. *If Z is a Kähler space, $C_n(Z)$ is also a Kähler space.*   □

REMARK. Here "Kähler space" is being used in the strong sense: there exists an open covering $(U_\alpha)_{\alpha \in A}$ of Z and $\varphi_\alpha \in C^\infty(U_\alpha)$ such that φ_α is strongly plurisubharmonic and $\varphi_\alpha - \varphi_\beta = \mathrm{Re}(f_{\alpha\beta})$ on $U_\alpha \cap U_\beta$ with $f_{\alpha\beta} \in H^0(U_\alpha \cap U_\beta, \mathcal{O}_Z)$. An important fact, proved by J. Varouchas [1984] using Richberg's Lemma [1968], is that you obtain an equivalent definition by assuming that the φ_α are only continuous and strongly plurisubharmonic.

To give the idea of the construction, assume that Z is smooth and fix a compact n-cycle C_0 in Z. Let ω be the given Kähler form on Z. The first step is to explain that, in an open neighbourhood U of $|C_0|$ one can write $\omega^{\wedge n+1} = i\, \partial\bar{\partial}\alpha$, where α is a real C^∞ (n,n)-form on U. This is acheived by using the following result:

THEOREM 9 [Barlet 1980a]. *Let Z a complex space and let C a n-dimensional compact analytic set in Z. Then C admits a basis of open neighbourhoods that are n-complete.*   □

The next step is to use the Theorem 5 to get the strict plurisubharmonicity of the continuous function $C \to \int_C \alpha$ using the strong Lelong positivity of $\omega^{\wedge n+1}$. The third step is then to prove that the difference of two such local strongly plurisubharmonic continuous functions on $C_n(Z)$ is the real part of an holomorphic function. This is delicate and uses the integration Theorem 3.

A very nice corollary of this result is the following theorem, which explains that Fujiki's class $\mathcal{C}$ (consisting of holomorphic images of compact complex Kähler manifolds: see [Fujiki 1980]) is the class of compact complex spaces which are bimeromorphic to compact Kähler manifolds.

THEOREM 10 [Varouchas 1989]. *Let Z a compact connected Kähler manifold and let $\pi : Z \to X$ a surjective map on a complex space X. Then there exists a compact Kähler manifold W and a surjective modification $\tau : W \to X$.*

PROOF. Sketch of proof Denote by n the dimension of the generic fiber of π and let $\Sigma \subset X$ a nowhere dense closed analytic subset such that

$$Z - \pi^{-1}(\Sigma) \to X - \Sigma$$

is n-equidimensional with $X - \Sigma$ smooth. By Theorem 2 the fibers of π restricted to $\left(Z - \pi^{-1}(\Sigma)\right)$ give an analytic family of n cycles of Z parametrized by $X - \Sigma$. So we have an holomorphic map $f : X - \Sigma \to C_n(Z)$.

In fact, f is meromorphic along Σ [Barlet 1980c]. Let $Y \subset X \times C_n(Z)$ the graph of this meromorphic map. Then $Y \to X$ is a surjective modification (along Σ) and $Y \to C_n(Z)$ is generically injective.

Let $\tilde{Y}$ the image of Y in $C_n(Z)$. This is a compact Kähler space (being a closed subspace of Kähler space) and we have a diagram of modifications:

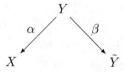

Using [Hironaka 1964] we can find a projective modification $W \xrightarrow{\gamma} \tilde{Y}$ such that W is smooth (and Kähler because γ is projective and $\tilde{Y}$ is Kähler) with a

commutative diagram

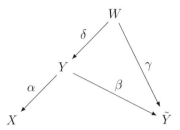

Then $\alpha \circ \delta$ is a modification and the theorem is proved! $\square$

5. Higher Integration

Already in [Andreotti and Norguet 1967] there appears the idea of considering "higher integration" maps

$$\rho^{p,q} : H^{n+q}(Z, \Omega_Z^{n+p}) \to H^q\big(C_n(Z), \Omega^p_{C_n(Z)}\big). \tag{1}$$

For a family of compact n-cycles in a smooth Z parametrized by a smooth S, it is easy to deduce such a map from the usual direct image of currents and the Dolbeault–Grothendieck lemma.

First remark that the case $p = 0$ is a rather standard consequence (in full generality) of Theorem 3.

But it is clear that the case $p \geq 1$, $q = 0$ allows one to hope for a way to build up holomorphic p-forms on $C_n(Z)$. Some relationship between intermediate Jacobian of Z and Picard groups of components of $C_n(Z)$ looks very interesting!

But this is not so simple: M. Kaddar [1995] has shown that such a map $\rho^{p,0}$ does not exist in general for $p \geq 1$. But if one replaces the sheaf $\Omega^p_{C_n(Z)}$ by the sheaf $\omega^p_{C_n(Z)}$ of $\bar{\partial}$ closed $(p,0)$ currents on $C_n(Z)$ (modulo torsion), the existence of

$$\rho^{p,q} : H^{n+q}(Z, \Omega_Z^{n+p}) \to H^q\big(C_n(Z), \omega^p_{C_n(Z)}\big). \tag{2}$$

is proved in the same reference.

For a reduced pure dimensional space X, the sheaf ω^p_X has been introduced in [Barlet 1978b]. It is a coherent sheaf, it satisfies the analytic extension property in codimension 2 and coincides with Grothendieck sheaf in maximal degree (which is the dualizing sheaf for X Gorenstein).

But, of course, something is lost in this higher integration process because we begin with $\Omega_Z^{\cdot}$ and we end with $\omega^{\cdot}_{C_n(Z)}$. Again M. Kaddar has given an example to show that the map (2) does not factorize by $H^{n+q}(Z, \omega_Z^{n+p})$.

The good point of view is to work with L_2 p-holomorphic forms (a meromorphic p-form on X is L_2 if and only if its pull back in a desingularization of X is holomorphic; this is independent on the chosen desingularization).

The sheaf L_2^p is again a coherent sheaf without torsion on any reduced space X and we have natural inclusions of coherent sheaves (for any $p \geq 0$)

$$\Omega^p_{X/\mathrm{torsion}} \hookrightarrow L_2^p \hookrightarrow \omega_X^p,$$

which coincide on the regular part of X.

Kaddar [1996b] also proved the following result:

THEOREM 11. *The higher integration map* $\rho^{p,q}$ *can be factorized through a natural map*

$$R^{p,q} : H^{n+q}(Z, L_2^{n+p}) \to H^q\big(C_n(Z), L_2^p\big). \qquad \square$$

The main difficulty in this "final version" of the higher integration map is to prove that the $L_2^{\cdot}$ holomorphic forms can be restricted to subspaces (in a natural way). Of course the bad case is when the subspace is included in the singular set of the ambient space. To handle this difficulty, the idea is again to use higher integration via the map (2), to define, at generic points first, the desired restriction from a suitable desingularization. Of course one has to show that it satisfies the L_2-condition that this does not depend on choices; then Kaddar shows that this construction has nice functorial properties.

6. An Application

In this section I shall present a famous conjecture of R. Hartshorne [1970] which is a typical problem where the reduction of convexity gives a nice strategy to solve the problem. Unfortunately, in the general case, it is not known how to build up a convenient family of compact cycles in order to reach the contradiction.

This is related to the following difficult problem:

PROBLEM. Let X a projective manifold and $A \subset X$ a compact submanifold of dimension $d \geq 2$ with an ample (or positive) normal bundle. Is it possible to find an irreducible analytic family of $(d-1)$-cycles in X which fills up X and such at least one member of the family is contained in A (as a set)?

Even for $\dim A = 2$ and $\dim X = 4$ I do not know if this is possible in general (though the particular case where $\dim X = 4$ and X is, in a neighbourhood of A, the normal bundle of the surface A follows from [Barlet et al. 1990, Theorem (1.1)].) The easy but interesting case I know is when X is an hypersurface of an homogeneous manifold W: it is enough to use the family $gA \cap X$ where $g \in \mathrm{Aut}_0(W)$.

Let me recall a transcendental variant of Hartshorne's conjecture:

CONJECTURE (H). *Let* X *be a compact connected Kähler manifold. Let* A *and* B *two compact submanifolds of* X *with positive normal bundles. Assume* $\dim_{\mathbb{C}} A + \dim_{\mathbb{C}} B \geq \dim_{\mathbb{C}} X$. *Then* $A \cap B \neq \emptyset$.

Before explaining the general strategy used in [Barlet 1987] (and [Barlet et al. 1990]), let me give a proof in the case where A is a curve and B is a divisor:

By the positivity of the normal bundle of B in X we know (from [Schneider 1973]) that $X - B$ is strongly 0-convex. So $H^0(X-B, \mathcal{O}_X)$ is an infinite-dimensional vector space. Using again [Schneider 1973] the positivity of the normal bundle of A in X we can find arbitrary small open neighbourhoods of A in X which are $(\dim X - 1)$-concave. This implies $\dim_{\mathbb{C}} H^0(\mathcal{U}, \mathcal{O}_X) < +\infty$ by [Andreotti and Grauert 1962] for any such $\mathcal{U}$. Assume now $A \cap B = \varnothing$ and choose $\mathcal{U} \subset X - B$. Now, by analytic continuation, the restriction map: $H^0(X-B, \mathcal{O}_X) \to H^0(\mathcal{U}, \mathcal{O}_X)$ is injective and this gives a contradiction.

The main idea to understand what is going on in this proof is to observe that a point can get out of A and go to reach B to make the analytic continuation.

In the general case, assume $\dim A + \dim B = \dim X$ (to simplify notations) and that we get an irreducible analytic family $(C_s)_{s \in S}$ of compact n-cycles in X such that

(1) $n = \dim_{\mathbb{C}} A - 1$,
(2) there exists $s_0 \in S$ such that $|C_{s_0}| \subset A$, and
(3) there exists $s_\infty \in S$ such that any component of $|C_{s_\infty}|$ meets B.

Then we argue along the same lines:

Assume $A \cap B = \varnothing$ and let $S_\infty = \{s \in S/|C_s| \cap B \neq \varnothing\}$. Then S_∞ is a nowhere dense, closed analytic subset of S.

Using [Schneider 1973] we get: $X - B$ is strongly n-convex; so integration of cohomology classes in $H^n(X-B, \Omega_X^n)$ will produce enough holomorphic functions on $S - S_\infty$ to separate points near infinity in the Remmert reduction of $S - S_\infty$.

There exists again an $\left(\dim X - (n+1)\right)$-concave open set $\mathcal{U} \supset A$ contained in $X - B$. Then by [Andreotti and Grauert 1962] we have $\dim_{\mathbb{C}} H^n(\mathcal{U}, \Omega_X^n) < +\infty$.

Now, because of the irreducibility of S, any holomorphic function on $S - S_\infty$ is uniquely determined by its restriction to the (nonempty) open set $V = \{s \in S - S_\infty/|C_s| \subset \mathcal{U}\}$. Now we have the following commutative diagram:

$$
\begin{array}{ccc}
H^n(X-B, \Omega_X^n) & \xrightarrow{\text{res}} & H^n(\mathcal{U}, \Omega_X^n) \\
\downarrow{\scriptstyle \int} & & \downarrow{\scriptstyle \int} \\
H^0(S - S_\infty, \mathcal{O}) & \xrightarrow{\text{res}} & H^0(V, \mathcal{O}).
\end{array}
$$

This does not yet give the contradiction.

To obtain one, we have to consider the family of cycles parametrized by $\mathrm{Sym}^k(S) \simeq S^k/\sigma_k$ defined by

$$
(s_1 \ldots s_k) \to \sum_{i=1}^{k} C_{s_i}.
$$

When $k \to +\infty$ the dimension of the Remmert'reduction of $\mathrm{Sym}^k(S - S_\infty)$ goes to $+\infty$, but the dimension of $H^n(\mathcal{U}, \Omega_X^n)$ does not change and that gives the contradiction.

THEOREM 12 [Barlet 1987]. *Hartshorne's conjecture* (H) *is true for X a compact connected Kähler smooth hypersurface of an homogeneous complex manifold.* $\square$

We now discuss how to algebrize this strategy in order to reach the initial formulation of R. Harshorne.

CONJECTURE (H). *Let X a smooth projective compact connected variety and let A and B two submanifolds with ample normal bundles such that*

$$\dim_{\mathbb{C}} A + \dim_{\mathbb{C}} B \geq \dim_{\mathbb{C}} X.$$

Then $A \cap B \neq \varnothing$.

Now the ampleness assumption does not imply the positivity (it is not yet known if ampleness implies positivity for rank ≥ 2) and so the convexity and concavity fail in the previous proof. The concavity part will be replaced by the following theorem:

THEOREM 13 [Barlet et al. 1994]. *Let X a complex manifold and $(C_s)_{s \in S}$ an analytic family of n-cycles. Fix $s_0 \in S$ and let $|C_{s_0}| = Y$. Then there exists an increasing sequence $(\alpha_k)_{k \in \mathbb{N}}$ such that $\lim_{k \to \infty} \alpha_k = +\infty$ and if $\omega \in H^n(X, \Omega_X^n)$ is in the kernel of the restriction map*

$$H^n(X, \Omega_X^n) \to H^n(Y, \Omega_X^n / \mathcal{J}_Y^k \Omega_X^n)$$

then $\rho(\omega)_{s_0} \in \mathcal{M}_{S,s_0}^{\alpha_k}$, where $\mathcal{M}_{S,s_0}$ is the maximal ideal of $\mathcal{O}_{S,s_0}$. $\square$

Here $\mathcal{J}_Y$ is the defining ideal sheaf of Y and $\rho(\omega)(s) = \int_{C_s} \omega$.

So this shows that the $(\alpha_k - 1)$-jet at s_0 of the function $s \to \int_{C_s} \omega$ is determined by the restriction of ω in $H^n(Y, \Omega_X^n / \mathcal{J}_Y^k \Omega_X^n)$.

Now if $Y \subset A$, where A is a submanifold with an ample normal bundle, the previous restriction will factorize through $H^n(A, \Omega_X^n / \mathcal{J}_A^k \Omega_X^n)$; and the ampleness of $N_{A/X}$ gives the fact that this vector space stabilizes for $k \gg 1$.

This shows that the image of

$$\tilde{\rho} : H^n(X, \Omega_X^n) \to \mathcal{O}_{S,s_0}$$

is finite-dimensional. Again the irreducibility of S allows to conclude that $\rho(H^n(X, \Omega_X^n)) \subset H^0(S, \mathcal{O}_S)$ is finite-dimensional.

7. Construction of Meromorphic Functions on $C_n(X)$

To replace the convexity part we introduce a new idea:

Let B be a smooth manifold of codimension $n + 1$ in X and consider the algebraic analogue of the exact sequence

$$\cdots \to H^n(X, \Omega_X^n) \to H^n(X - B, \Omega_X^n) \to H_B^{n+1}(X, \Omega_X^n) \to \cdots$$

denoted by

$$\cdots \to H^n(X, \Omega_X^n) \to H_{\mathrm{alg}}^n(X - B, \Omega_X^n) \to H_{[B]}^{n+1}(X, \Omega_X^n) \to \cdots.$$

So $H_{\mathrm{alg}}^n(X - B, \Omega_X^n)$ is the subspace of $H^n(X - B, \Omega_X^n)$ of cohomology classes having a meromorphic singularity along B,

$$\left(H_{[B]}^{n+1}(X, \Omega_X^n) := \varinjlim_k \mathrm{Ext}^{n+1}(\mathcal{O}_X / \mathcal{I}_B^k, \Omega_X^n)\right).$$

Let $(C_s)_{s \in S}$ be a family of compact n-cycles in X and set

$$S_\infty = \{s \in S / |C_s| \cap B \neq \varnothing\}.$$

We want to investigate the behaviour of the function $s \to \int_{C_s} \omega$ when $s \to S_\infty$ assuming that

$$\omega \in H_{\mathrm{alg}}^n(X - B, \Omega_X^n).$$

This question is solved in [Barlet and Magnusson 1998] in a rather general context. Here I give a simpler statement.

THEOREM 14 [Barlet and Magnusson 1998]. *Let X be a complex manifold and B a submanifold of codimension $n + 1$. Let $(C_s)_{s \in S}$ an analytic family of compact n-cycles in X parametrized by the reduced space S. Let*

$$|G| = \{(s, x) \in S \times X / x \in |C_s|\}$$

be the graph of the family and denote by p_1 and p_2 the projection of $|G|$ on S and X respectively. Assume that $p_2^*(B)$ is proper over S by p_1 and denote by $\Sigma = (p_1)_* p_2^*(B)$ as a complex subspace of S. Then there exists a natural integration map

$$\sigma : H_B^{n+1}(X, \Omega_X^n) \to H_\Sigma^1(S, \mathcal{O}_S)$$

which induces a filtered map

$$\sigma : H_{[B]}^{n+1}(X, \Omega_X^n) \to H_{[\Sigma]}^1(S, \mathcal{O}_S)$$

compatible with the usual integration map, so that the following diagram is commutative:

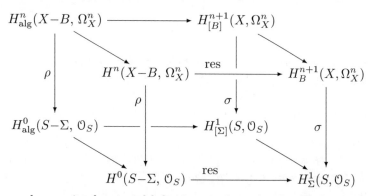

This map σ has a nice functorial behaviour and can be sheafified in a filtered sheaf map

$$(p_1)_* p_2^* \big(\underline{H}^{n+1}_{[B]}(\Omega^n_X) \big) \to \underline{H}^1_{[\Sigma]}(\mathcal{O}_S). \qquad \square$$

REMARKS. (1) This result asserts that a cohomology class ω in $H^n_{\mathrm{alg}}(X-B, \Omega^n_X)$ having a pole of order $\leq q$ along B $\big($i.e., $\mathcal{I}^q_B.\omega$ is locally zero in $\underline{H}^{n+1}_{[B]}(\Omega^n_X)\big)$ will give a meromorphic function on S with an order $\leq q$ pole along Σ $\big($where m has an order $\leq q$ pole along Σ means that $\mathcal{I}^q_\Sigma.m$ is locally zero in $H^1_{[\Sigma]}(\mathcal{O}_S)\big)$. Of course the ideal $\mathcal{I}_\Sigma$ is associated to $(p_1)_* p_2^*(B) = \Sigma$ as a subspace of S (B is reduced).

(2) The compactness of cycles is not important in the previous result, but of course keeping the assumption that $p_2^* B$ is proper over S.

In the noncompact case, we have to add a family of support on X in order to have an integration map $\rho : H^n_\Phi(X-B, \Omega^n_X) \to H^0(S - \Sigma, \mathcal{O}_S)$ where $F \in \Phi$ if F is closed in X and F is proper over S; so $B \in \Phi$ and we have again compatibility between ρ and σ via the commutative diagram

$$
\begin{array}{ccc}
H^n_\Phi(X-B, \Omega^n_X) & \xrightarrow{\ \rho\ } & H^0(S - \Sigma, \mathcal{O}_S) \\
\big\downarrow{\scriptstyle \mathrm{res}} & & \big\downarrow{\scriptstyle \mathrm{res}} \\
H^{n+1}_B(X, \Omega^n_X) & \xrightarrow{\ \sigma\ } & H^1_\Sigma(S, \mathcal{O}_S)
\end{array}
$$

Our next result is to show that, in fact, the closed analytic subset $|\Sigma|$ can be endowed with a natural "Cartier divisor" structure in S (remember that S is any reduced space; or to say it in an other way: we have no control on the singularities of $C_n(X)!$).

THEOREM 15 [Barlet and Magnusson 1998]. *Let Z be a complex manifold and $Y \subset Z$ a closed analytic subspace of Z of codimension $n + 1$ which is locally a complete intersection in Z. Let $(C_s)_{s \in S}$ be an analytic family of n-cycles in Z (not necessarily compact) such that we have the following property:*

Let $|G| \subset X \times Z$ the graph of the family $(C_s)_{s \in S}$ and denote by p_1 and p_2 the projections of $|G|$ on S and Z respectively. Assume that

$$p_1 : p_2^{-1}(|Y|) \to S$$

is finite. Then there exists a natural Cartier divisor structure Σ_Y on the closed analytic set $|\Sigma| = p_1(p_2^{-1}|Y|)$, that is, a locally principal $\mathfrak{I}_{\Sigma_Y}$ ideal of $\mathcal{O}_S$ defining $|\Sigma|$.

This Cartier structure is characterized by the following properties:

(1) Let $s_0 \in |\Sigma|$ and set $|C_{s_0}| \cap |Y| = \{y_1, \ldots, y_l\}$, where $y_i \neq y_j$ when $i \neq j$. Let $U_1 \ldots U_l$ be disjoint open sets in Z such that $y_j \in U_i$ for $j \in [1, l]$. Let $\mathfrak{I}_j$ be the Cartier structure on $|\Sigma_j| = p_1(p_2^{-1}(|Y| \cap U_j))$ near s_0. Then $\mathfrak{I}_{\Sigma_Y} \prod_{j=1}^{l} \mathfrak{I}_j$ near s_0.

(2) Let $s_0 \in |\Sigma|$ and let U be an open set in Z such that $|C_{s_0}| \cap |Y| \subset U$ and such that $\mathfrak{I}_Y = \pi^*(m_{\mathbb{C}^{n+1}}, 0)$ where $\pi : U \to \mathbb{C}^{n+1}$ is an holomorphic flat map (that is, $\pi := (z_0 \ldots z_n)$ where $z_0 \ldots z_n$ is a generator of $\mathfrak{I}_Y$ on U). Then a generator of $\mathfrak{I}_{\Sigma_Y}$ near s_0 is given by the holomorphic function

$$s \to N(z_0)\big(C_s \cap (z_1 = \cdots = z_n = 0)\big),$$

where $C_s \cap (z_1 = \cdots = z_n = 0)$ in $\mathrm{Sym}^k Z$ is defined via [Barlet 1975, Theorem 6 (local)] and where $N(z_0) : \mathrm{Sym}^k U \to \mathbb{C}$ is the norm of the holomorphic function $z_0 : U \to \mathbb{C}$ (so $N(z_0)(x_1 \ldots x_k) = \prod_{j=1}^{k} z_0(x_j)$).

(3) The construction of Σ_Y is compatible with base change: so if $\tau : T \to S$ is holomorphic (with T reduced) the Cartier structure associated to the family $(C_{\tau(t)})_{t \in T}$ and $Y \subset Z$ is the Cartier divisor $\tau^*(\Sigma_Y)$ in T.

Now the filtration by the order of poles in the Theorem 14 is related to the Cartier divisor structure of Theorem 15 by this result:

PROPOSITION. In the situation of the Theorem 15 we have $(p_1)_*\big(p_2^*(Y)\big)$ is a subspace of Σ_Y. So the morphism of Theorem 14 gives a filtered sheaf map

$$(\rho_1)_* \big(\underline{H}^{n+1}_{[p_2^*Y]}(p_2^* \Omega_Z^n)\big) \to \underline{H}^1_{[\Sigma_Y]}(\mathcal{O}_S).$$

This means that the order of poles along $|\Sigma|$ for meromorphic function on S, obtained by integration, is now defined by the "natural" equation of $|\Sigma|$ given by the Theorem 15.

REMARK. Let $Z = \mathbb{P}_N(\mathbb{C})$ and Y be any codimension $n + 1$ cycle in Z (in this special case we don't need a local complete intersection subspace!); take S to be the Grassmann manifold of n-planes of $\mathbb{P}_N(\mathbb{C})$. Then we find here Chow and Van der Waerden construction, the Cartier divisor on S gives, in a Plücker embedding, the Cayley form of the cycle Y.

Now we come to the main result in [Barlet and Magnusson 1999], which asserts that, assuming moreover that Z is n-convex and that Y is a compact submanifold of codimension $n+1$ with ample normal bundle, the line bundle associated to the Cartier divisor Σ_Y of S is ample near Σ when the family $(C_s)_{s \in S}$ is sufficiently nice. So it transfers in this case the ampleness from $N_{Y/Z}$ to $N_{\Sigma/S}$.

THEOREM 16 [Barlet and Magnusson 1999]. *Let Z be a complex manifold which is n-convex. Let $(Y, \mathfrak{I}_Y)$ a compact subspace of Z which is locally a complete intersection of codimension $n + 1$ in Z and such that $N_{Y/Z}$ is ample. Let S a reduced analytic space and $(C_s)_{s \in S}$ an analytic family of cycles and let $|G| \subset S \times Z$ the graph of this family, with projections p_1 and p_2 on S and Z respectively. Assume that*

(1) *$p_1 : p_2^{-1}(|Y|) \to S$ is proper and injective;*
(2) *for all $s \in |\Sigma|$, C_s and Y are smooth and transverse at z_s, where $z_s = C_s \cap Y$; and*
(3) *there exists a closed analytic set $\Theta \subset |\Sigma| \times |\Sigma|$, symmetric, finite on $|\Sigma|$, and such that for all $(s, s') \notin \Theta$, we have either $z_s = z_{s'}$ or $T_{C_s, z_s} \neq T_{C_{s'}, z_s}$, where $T_{C,z}$ is the tangent space to C at z (C has to be smooth at z!), when $z_s = z_{s'}$.*

Then, denoting by Σ_Y the Cartier divisor structure on $|\Sigma| = p_1\big(p_2^{-1}(|Y|)\big)$ given by Theorem 15, the line bundle $[\Sigma_Y]$ is locally ample in S, that is, there exists $\nu \in \mathbb{N}$ such that $E_\nu = H^0(S, [\Sigma_Y]^\nu)$ gives an holomorphic map $S \to \mathbb{P}(E_\nu^*)$, finite in a neighbourhood of Σ. □

To conclude, I will quote Kaddar's application [1996a] of his construction of a relative fundamental class in Deligne cohomology for an analytic family of cycles in a complex manifold Z. Using this class, he can associate to a codimension $n + 1$ cycle Y in Z a line bundle on S (the parameter space) by integration at the level of Deligne cohomology. Moreover he proved that this gives, say in a projective setting, an holomorphic map from cycles of codimension $n + 1$ to the Picard group of the cycle space of n-cycles. This was a first motivation for me to prove Theorem 15 which produces also in a rather wide context a Cartier divisor on S, and so a line bundle.

The idea that a result such as Theorem 16 was possible goes back to F. Campana's work [1980; 1981] on algebraicity of the cycle space. He notices that for a compact analytic subset S of the cycle space, the analytic subset Σ of these cycles which meet a given Moišezon subspace Y in Z is again Moišezon. So this transfer of algebraicity is a rough basis for Theorem16 where we transfer the ampleness of the normal bundle of Y in Z to the ampleness of the normal bundle of Σ in S. This of course gives not only algebraicity on Σ but also information on S: we build up enough meromorphic functions on S to prove that S is Moišezon (when S is compact) in fact with a weaker assumption than stated here (see the weak version of Theorem 16 in [Barlet and Magnusson 1999]) and we also describe the line bundle which gives meromorphic functions on S.

Acknowledgments

I thank MSRI for its hospitality and the organizers of the Special Year in Several Complex Variables for inviting me. I also thank Silvio Levy for his remarkable work in preparing this article for publication.

The commutative diagrams in this article were made using Paul Taylor's diagrams.sty package.

References

[Andreotti and Grauert 1962] A. Andreotti and H. Grauert, "Théorème de finitude pour la cohomologie des espaces complexes", *Bull. Soc. Math. France* **90** (1962), 193–259.

[Andreotti and Norguet 1967] A. Andreotti and F. Norguet, "La convexité holomorphe dans l'espace analytique des cycles d'une variété algébrique", *Ann. Scuola Norm. Sup. Pisa* (3) **21** (1967), 31–82.

[Barlet 1975] D. Barlet, "Espace analytique réduit des cycles analytiques complexes compacts d'un espace analytique complexe de dimension finie", pp. 1–158 in *Fonctions de plusieurs variables complexes II*, edited by F. Norguet, Lecture Notes in Math. **482**, Springer, Berlin, 1975.

[Barlet 1978a] D. Barlet, "Convexité de l'espace des cycles", *Bull. Soc. Math. France* **106**:4 (1978), 373–397.

[Barlet 1978b] D. Barlet, "Le faisceau $\omega_X^{\cdot}$ sur un espace analytique X de dimension pure", pp. 187–204 in *Fonctions de plusieurs variables complexes III*, edited by F. Norguet, Lecture Notes in Math. **670**, Springer, Berlin, 1978.

[Barlet 1980a] D. Barlet, "Convexité au voisinage d'un cycle", pp. 102–121 in *Fonctions de plusieurs variables complexes IV*, edited by F. Norguet, Lecture Notes in Math. **807**, Springer, Berlin, 1980.

[Barlet 1980b] D. Barlet, "Familles analytiques de cycles et classes fondamentales relatives", pp. 1–24 in *Fonctions de plusieurs variables complexes IV*, edited by F. Norguet, Lecture Notes in Math. **807**, Springer, Berlin, 1980.

[Barlet 1980c] D. Barlet, "Majoration du volume des fibres génériques et forme géométrique du théorème d'aplatissement", pp. 1–17 in *Séminaire Pierre Lelong — Henri Skoda (Analyse) 1978/79*, edited by P. Lelong and H. Skoda, Lecture Notes in Math. **822**, Springer, Berlin, 1980.

[Barlet 1983] D. Barlet, "Un théorème d'algébricité intermédiaire", *Invent. Math.* **71**:3 (1983), 655–660.

[Barlet 1987] D. Barlet, "À propos d'une conjecture de R. Hartshorne", *J. Reine Angew. Math.* **374** (1987), 214–220.

[Barlet and Magnusson 1998] D. Barlet and J. Magnusson, "Intégration de classes de cohomologie méromorphes et diviseurs d'incidence", *Ann. Sci. École Norm. Sup.* (4) **31**:6 (1998), 811–842.

[Barlet and Magnusson 1999] D. Barlet and J. Magnusson, "Transfert de l'amplitude du fibré normal au diviseur d'incidence", *J. reine angew. Math.* **513** (1999). to appear.

[Barlet and Varouchas 1989] D. Barlet and J. Varouchas, "Fonctions holomorphes sur l'espace des cycles", *Bull. Soc. Math. France* **117**:3 (1989), 327–341.

[Barlet et al. 1990] D. Barlet, T. Peternell, and M. Schneider, "On two conjectures of Hartshorne's", *Math. Ann.* **286**:1-3 (1990), 13–25.

[Barlet et al. 1994] D. Barlet, L. Doustaing, and J. Magnússon, "La conjecture de R. Hartshorne pour les hypersurfaces lisses de $\mathbf{P}^n$", *J. Reine Angew. Math.* **457** (1994), 189–202.

[Bishop 1964] E. Bishop, "Conditions for the analyticity of certain sets", *Michigan Math. J.* **11** (1964), 289–304.

[Campana 1980] F. Campana, "Algébricité et compacité dans l'espace des cycles d'un espace analytique complexe", *Math. Ann.* **251**:1 (1980), 7–18.

[Campana 1981] F. Campana, "Coréduction algébrique d'un espace analytique faible-ment kählérien compact", *Invent. Math.* **63**:2 (1981), 187–223.

[Fujiki 1978] A. Fujiki, "Closedness of the Douady spaces of compact Kähler spaces", *Publ. Res. Inst. Math. Sci.* **14**:1 (1978), 1–52.

[Fujiki 1980] A. Fujiki, "Some results in the classification theory of compact complex manifolds in $\mathcal{C}$", *Proc. Japan Acad. Ser. A Math. Sci.* **56**:7 (1980), 324–327.

[Hartshorne 1970] R. Hartshorne, *Ample subvarieties of algebraic varieties*, Lecture Notes in Math. **156**, Springer, Berlin, 1970. Notes written in collaboration with C. Musili.

[Hironaka 1964] H. Hironaka, "Resolution of singularities of an algebraic variety over a field of characteristic zero", *Ann. of Math.* (2) **79** (1964), 109–203, 205–326.

[Hironaka 1975] H. Hironaka, "Flattening theorem in complex-analytic geometry", *Amer. J. Math.* **97** (1975), 503–547.

[Kaddar 1995] M. Kaddar, "Intégration partielle sur les cycles", *C. R. Acad. Sci. Paris Sér. I Math.* **320**:12 (1995), 1513–1516.

[Kaddar 1996a] M. Kaddar, "Classe fondamentale relative en cohomologie de Deligne et application", *Math. Ann.* **306**:2 (1996), 285–322.

[Kaddar 1996b] M. Kaddar, "Intégration sur les cycles et formes de type L^2", *C. R. Acad. Sci. Paris Sér. I Math.* **322**:7 (1996), 663–668.

[Lieberman 1978] D. I. Lieberman, "Compactness of the Chow scheme: applications to automorphisms and deformations of Kähler manifolds", pp. 140–186 in *Fonctions de plusieurs variables complexes III*, Lecture Notes in Math. **670**, Springer, Berlin, 1978.

[Norguet and Siu 1977] F. Norguet and Y. T. Siu, "Holomorphic convexity of spaces of analytic cycles", *Bull. Soc. Math. France* **105**:2 (1977), 191–223.

[Richberg 1968] R. Richberg, "Stetige streng pseudokonvexe Funktionen", *Math. Ann.* **175** (1968), 257–286.

[Schneider 1973] M. Schneider, "Über eine Vermutung von Hartshorne", *Math. Ann.* **201** (1973), 221–229.

[SGAN 1982] *Séminaire: Géométrie analytique* (2eme partie), Rev. Inst. Élie Cartan **5**, 1982.

[Varouchas 1984]   J. Varouchas, "Stabilité de la classe des variétés kählériennes par certains morphismes propres", *Invent. Math.* **77**:1 (1984), 117–127.

[Varouchas 1989]   J. Varouchas, "Kähler spaces and proper open morphisms", *Math. Ann.* **283**:1 (1989), 13–52.

DANIEL BARLET
INSTITUT ÉLIE CARTAN (NANCY)
CNRS/INRIA/UHP, UMR 7502
UNIVERSITÉ HENRI POINCARÉ (NANCY I) ET INSTITUT UNIVERSITAIRE DE FRANCE
FACULTÉ DES SCIENCES, B.P. 239
54506 VANDOEUVRE-LES-NANCY CEDEX
FRANCE
   barlet@iecn.u-nancy.fr

Several Complex Variables
MSRI Publications
Volume **37**, 1999

Resolution of Singularities

EDWARD BIERSTONE AND PIERRE D. MILMAN

ABSTRACT. This article is an exposition of our algorithm for canonical resolution of singularities in characteristic zero (*Invent. Math.* **128** (1997), 207–302), with an essentially complete proof of the theorem in the hypersurface case. We define a local invariant for desingularization whose values are finite sequences that can be compared lexicographically. Our invariant takes only finitely many maximum values (at least locally), and we get an algorithm for canonical desingularization by successively blowing up its maximum loci. The invariant can be described by a local construction that provides equations for the centres of blowing up. Our construction is presented here in parallel with a worked example.

1. Introduction

Resolution of singularities has a long history that goes back to Newton in the case of plane curves. For higher-dimensional singular spaces, the problem was formulated toward the end of the last century, and it was solved in general, for algebraic varieties defined over fields of characteristic zero, by Hironaka in his famous paper [1964]. (That paper includes the case of real-analytic spaces; Hironaka's theorem for complex-analytic spaces is proved in [Hironaka 1974; Aroca et al. 1975; 1977].) But Hironaka's result is highly non-constructive. His proof is one of the longest and hardest in mathematics, and it seems fair to say that only a handful of mathematicians have fully understood it. We are not among them! Resolution of singularities is used in many areas of mathematics, but even certain aspects of the theorem (for example, *canonicity:* see 1.11 below) have remained unclear.

This article is an exposition of an elementary constructive proof of canonical resolution of singularities in characteristic zero. Our proof was sketched in the hypersurface case in [Bierstone and Milman 1991] and is presented in detail in [Bierstone and Milman 1997].

When we started thinking about the subject almost twenty years ago, our aim was simply to understand resolution of singularities. But we soon became convinced that it should be possible to give simple direct proofs of at least those aspects of the theorem that are important in analysis. In [Bierstone and Milman 1988], for example, we published a very simple proof that any real-analytic variety is the image by a proper analytic mapping of a manifold of the same dimension. The latter statement is a real version of a local form of resolution of singularities, called *local uniformization*.

It is the idea of [Bierstone and Milman 1988, Section 4] that we have developed (via [Bierstone and Milman 1989]) to define a new local invariant for desingu-larization that is the main subject of this exposition. Our invariant $\text{inv}_X(a)$ is a finite sequence (of nonnegative rational numbers and perhaps ∞, in the case of a hypersurface), defined at each point a of our space X. Such sequences can be compared lexicographically. $\text{inv}_X(\cdot)$ takes only finitely many maximum values (at least locally) and we get an algorithm for canonical resolution of sin-gularities by successively blowing up its maximum loci. Moreover, $\text{inv}_X(\cdot)$ can be described by local computations that provide equations for the centres of blowing up.

We begin with an example to illustrate the meaning of resolution of singular-ities:

EXAMPLE 1.1. Let X denote the quadratic cone $x^2 - y^2 - z^2 = 0$ in affine 3-space — the simplest example of a singular surface.

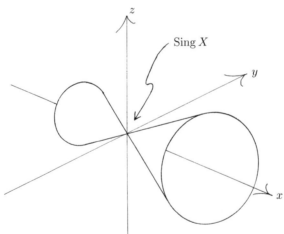

$$X : \ x^2 - y^2 - z^2 = 0$$

X can be desingularized by making a simple quadratic transformation of the ambient space:

$$\sigma : \quad x = u, \ y = uv, \ z = uw.$$

The inverse image of X by this mapping σ is given by substituting the formulas for x, y and z into the equation of X:

$$\sigma^{-1}(X) : \quad u^2(1 - v^2 - w^2) = 0.$$

Thus $\sigma^{-1}(X)$ has two components: The plane $u = 0$ is the set of critical points of the mapping σ; it is called the *exceptional hypersurface*. (Here $E' := \{u = 0\}$ is the inverse image of the singular point of X.) The quotient after completely factoring out the "exceptional divisor" u defines what is called the *strict transform* X' of X by σ. Here X' is the cylinder $v^2 + w^2 = 1$.

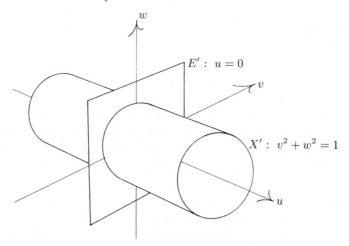

In this example, $\sigma|X'$ is a *resolution of singularities* of X: X' is smooth and $\sigma|X'$ is a proper mapping onto X that is an isomorphism outside the singularity. But the example illustrates a stronger statement, called *embedded resolution of singularities*: X is desingularized by making a simple transformation of the ambient space, after which, in addition, the strict transform X' and the exceptional hypersurface E' have only *normal crossings*; this means that each point admits a coordinate neighbourhood with respect to which both X' and E' are coordinate subspaces.

The quadratic transformation σ in Example 1.1 is also called a *blowing-up* with *centre* the origin. (The centre is the set of critical values of σ.) More accurately, the blowing-up of affine 3-space with centre a point is covered in a natural way by three affine coordinate charts, and σ above is the formula for the blowing-up restricted to one chart.

Sequences of quadratic transformations, or point blowings-up, were first used to resolve the singularities of curves by Max Noether in the 1870's [Brill and Noether 1892–93].

The more general statement of "embedded resolution of singularities" seems to have been formulated precisely first by Hironaka. But it is implicit already in the earliest rigorous proofs of local desingularization of surfaces, as a natural generalization prerequisite to the inductive step of a proof by induction on dimension (compare Sections 2 and 3 below). For example, in one of the earliest proofs of local desingularization or uniformization of surfaces, Jung used embedded desingularization of curves by sequences of quadratic transformations (applied to the branch locus of a suitable projection) to prove uniformization for surfaces [Jung 1908]. Similar ideas were used in the first proofs of global resolution of singularities of algebraic surfaces, by Walker [1935] and Zariski [1939]. (The latter was the first algebraic proof, by sequences of normalizations and point blowings-up.)

From the point of view of subsequent work, however, Zariski's breakthrough came in [1943], where he localized the idea of the centre of blowing-up, thus making possible an extension of the notion of quadratic transformation to blowings-up with centres that are not necessarily 0-dimensional. This led him to a version of embedded resolution of singularities of surfaces, and to a weaker (non-embedded) theorem for 3-dimensional algebraic varieties [Zariski 1944]. It was the path that led to Hironaka's great theorem and to most subsequent work in the area, including our own. Among the references not otherwise cited in this article we mention [Abhyankar 1966; 1982; 1988; Bierstone and Milman 1990; Giraud 1974; Hironaka 1977; Lipman 1975; Moh 1992; Villamayor 1989; 1992; Youssin 1990]. (Added in proof: [Encinas and Villamayor 1998].)

From a general viewpoint, some important features of our work in comparison with previous treatments are: (1) It is canonical (see 1.11). (2) We isolate simple local properties of an invariant (Section 4, Theorem B) from which global desingularization is automatic. (3) Our proof in the case of a hypersurface (a space defined locally by a single equation) does not involve passing to higher codimension, as does the inductive procedure of [Hironaka 1964].

Very significant results on resolution of singularities over fields of nonzero characteristic have recently been obtained by de Jong [1996] and have been announced by Spivakovsky.

1.2. BLOWING UP. We first describe the blowing-up of an open subset W of r-dimensional affine space with centre a point a. (Say $a = 0 \in W$.) The *blowing-up* σ with *centre* 0 is the projection onto W of a space W' that is obtained by replacing the origin by the $(r-1)$-dimensional projective space $\mathbb{P}^{r-1}$ of all lines through 0:

$$W' = \{(x, \lambda) \in W \times \mathbb{P}^{r-1} : x \in \lambda\}$$

and $\sigma \colon W' \to W$ is defined by $\sigma(x, \lambda) = x$. (Outside the origin, a point x belongs to a unique line λ, but $\sigma^{-1}(0) = \mathbb{P}^{r-1}$. Clearly, σ is a proper mapping.) W' has a natural algebraic structure: If we write x in terms of the affine coordinates $x = (x_1, \ldots, x_r)$, and λ in the corresponding homogeneous coordinates $\lambda = [\lambda_1, \ldots, \lambda_r]$, then the relation $x \in \lambda$ translates into the system of equations $x_i \lambda_j = x_j \lambda_i$, for all i, j.

These equations can be used to see that W' has the structure of an algebraic manifold: For each $i = 1, \ldots, r$, let W_i' denote the open subset of W' where $\lambda_i \neq 0$. In W_i' we have $x_j = x_i \lambda_j / \lambda_i$, for each $j \neq i$, so we see that W_i' is smooth: it is the graph of a mapping in terms of coordinates $(y_1, \ldots, y_r)$ for W_i' defined by $y_i = x_i$ and $y_j = \lambda_j / \lambda_i$ if $j \neq i$. In these coordinates, σ is a quadratic transformation given by the formulas

$$x_i = y_i, \qquad x_j = y_i y_j \text{ for all } j \neq i,$$

as in Example 1.1.

Once blowing up with centre a point has been described as above, it is a simple matter to extend the idea to blowing up a manifold, or smooth space, M with centre an arbitrary smooth closed subspace C of M: Each point of C has a product coordinate neighbourhood $V \times W$ in which $C = V \times \{0\}$; over this neighbourhood, the blowing-up with centre C identifies with $\mathrm{id}_V \times \sigma$: $V \times W' \to V \times W$, where id_V is the identity mapping of V and $\sigma: W' \to W$ is the blowing-up of W with centre $\{0\}$. The blowing-up $M' \to M$ with centre C is an isomorphism over $M \setminus C$. The preceding conditions determine $M' \to M$ uniquely, up to an isomorphism of M' commuting with the projections to M.

EXAMPLE 1.3. Let X denote the surface $z^3 - x^2 yz - x^4 = 0$.

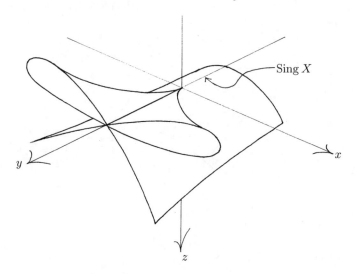

This surface is particularly interesting in the real case because, as a *subset* of $\mathbb{R}^3$, it is singular only along the nonnegative y-axis. But resolution of singularities is an algebraic process: it applies to spaces that include a functional structure (given here by the equation for X). As a *subspace* of $\mathbb{R}^3$, X is singular along the entire y-axis.

In general, for a hypersurface X — defined locally, say, by an equation $f(x) = 0$ — to say that a point a is *singular* means there are no linear terms in the Taylor expansion of f at a; in other words, the order $\mu_a(f) > 1$. (The *order* or *multiplicity* $\mu_a(f)$ of f at a is the degree of the lowest-order homogeneous part of the Taylor expansion of f at a. We will also call $\mu_a(f)$ the *order* $\nu_{X,a}$ of the hypersurface X at a.)

The general philosophy of our approach to desingularization — going back to [Zariski 1944] — is to blow up with smooth centre as large as possible inside the locus of the most singular points. In our example here, X has order 3 at each point of the y-axis. In general, order is not a delicate enough invariant to determine a centre of blowing-up for resolution of singularities, even in the hypersurface case. (We will refine order in our definition of inv_X.) But here let us take the blowing-up σ with centre the y-axis:

$$\sigma: \quad x = u, \ y = v, \ z = uw.$$

(Again, this is the formula for blowing up in one of two coordinate charts required to cover our space. But the strict transform of X in fact lies completely within this chart.) The inverse image of X is

$$\sigma^{-1}(X): \quad u^3(w^3 - vw - u) = 0;$$

$\{u = 0\}$ is the exceptional hypersurface E' (the inverse image of the centre of blowing up) and the strict transform X' is smooth. (It is the graph of a function $u = w^3 - vw$.)

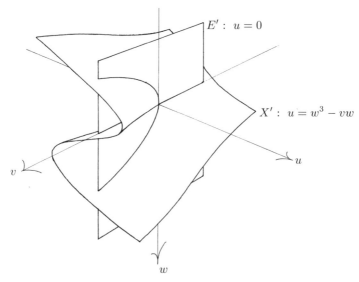

X' is a desingularization of X, but we have not yet achieved an embedded resolution of singularities because X' and E' do not have normal crossings at the origin. Further blowings-up are needed for embedded resolution of singularities.

1.4. EMBEDDED RESOLUTION OF SINGULARITIES. Let X denote a (singular) space. We assume, for simplicity, that X is a closed subspace of a smooth ambient space M. (This is always true locally.) The goal of embedded desingularization, in its simplest version, is to find a proper morphism σ from a smooth space M' onto M, in our category, with the following properties:

(1) σ is an isomorphism outside the singular locus $\operatorname{Sing} X$ of X.

(2) The strict transform X' of X by σ is smooth. (See 1.6 below.) X' can be described geometrically (at least if our field $\mathbb{K}$ is algebraically closed: compare [Bierstone and Milman 1997, Remark 3.15]) as the smallest closed subspace of M' that includes $\sigma^{-1}(X \setminus \operatorname{Sing} X)$.

(3) X' and $E' = \sigma^{-1}(\operatorname{Sing} X)$ simultaneously have only normal crossings. This means that, locally, we can choose coordinates with respect to which X' is a coordinate subspace and E' is a collection of coordinate hyperplanes.

We can achieve this goal with σ the composite of a sequence of blowings-up; a finite sequence when our spaces have a compact topology (for example, in an algebraic category), or a locally-finite sequence for non-compact analytic spaces. (A sequence of blowings-up over M is *locally finite* if all but finitely many of the blowings-up are trivial over any compact subset of M. The composite of a locally-finite sequence of blowings-up is a well-defined morphism σ.)

1.5. THE CATEGORY OF SPACES. Our desingularization theorem applies to the usual spaces of algebraic and analytic geometry over fields $\mathbb{K}$ of characteristic zero — algebraic varieties, schemes of finite type, analytic spaces (over $\mathbb{R}$, $\mathbb{C}$ or any locally compact $\mathbb{K}$) — but in addition to certain categories of spaces intermediate between analytic and C^∞ [Bierstone and Milman 1997]. In any case, we are dealing with a category of local-ringed spaces $X = (|X|, \mathcal{O}_X)$ over $\mathbb{K}$, where $\mathcal{O}_X$ is a coherent sheaf of rings. We are intentionally not specific about the category in this exposition because we want to emphasize the principles involved, and the main requirement for our desingularization algorithm is simply that a smooth space $M = (|M|, \mathcal{O}_M)$ in our category admit a covering by *(regular) coordinate charts* in which we have analogues of the usual operations of calculus of analytic functions; namely:

The coordinates $(x_1, \ldots, x_n)$ of a chart U are *regular functions* on U (i.e., each $x_i \in \mathcal{O}_M(U)$) and all partial derivatives $\partial^{|\alpha|}/\partial x^\alpha = \partial^{\alpha_1 + \cdots + \alpha_n}/\partial x_1^{\alpha_1} \cdots \partial x_n^{\alpha_n}$ make sense as transformations $\mathcal{O}_M(U) \to \mathcal{O}_M(U)$. Moreover, for each $a \in U$, there is an *injective* "Taylor series homomorphism" $T_a \colon \mathcal{O}_{M,a} \to \mathbb{F}_a[[X]] = \mathbb{F}_a[[X_1, \ldots, X_n]]$, where $\mathbb{F}_a$ denotes the residue field $\mathcal{O}_{M,a}/\underline{m}_{M,a}$, such that T_a induces an isomorphism

$$\widehat{\mathcal{O}}_{M,a} \xrightarrow{\cong} \mathbb{F}_a[[X]]$$

and T_a commutes with differentiation: $T_a \circ (\partial^{|\alpha|}/\partial x^\alpha) = (\partial^{|\alpha|}/\partial X^\alpha) \circ T_a$, for all $\alpha \in \mathbb{N}^n$. ($\underline{m}_{M,a}$ denotes the maximal ideal and $\widehat{\mathcal{O}}_{M,a}$ the completion of $\mathcal{O}_{M,a}$. $\mathbb{N}$ denotes the nonnegative integers.)

In the case of real- or complex-analytic spaces, of course, $\mathbb{K} = \mathbb{R}$ or $\mathbb{C}$, $\mathbb{F}_a = \mathbb{K}$ at each point, and "coordinate chart" means the classical notion. Regular coordinate charts for schemes of finite type are introduced in [Bierstone and Milman 1997, Section 3].

Suppose that $M = (|M|, \mathcal{O}_M)$ is a manifold (smooth space) and that $X = (|X|, \mathcal{O}_X)$ is a closed subspace of M. This means there is a coherent sheaf of ideals $\mathfrak{I}_X$ in $\mathcal{O}_M$ such that $|X| = \operatorname{supp} \mathcal{O}_M/\mathfrak{I}_X$ and $\mathcal{O}_X$ is the restriction to $|X|$ of $\mathcal{O}_M/\mathfrak{I}_X$. We say that X is a *hypersurface* in M if $\mathfrak{I}_{X,a}$ is a principal ideal, for each $a \in |X|$. Equivalently, for every $a \in |X|$, there is an open neighbourhood U of a in $|M|$ and a regular function $f \in \mathcal{O}_M(U)$ such that $|X|U| = \{x \in U : f(x) = 0\}$ and $\mathfrak{I}_X|U$ is the principal ideal (f) generated by f; we write $X|U = V(f)$.

1.6. STRICT TRANSFORM. Let X denote a closed subspace of a manifold M, and let $\sigma: M' \to M$ be a blowing-up with smooth centre C. If X is a hypersurface, then the strict transform X' of X by σ is a closed subspace of M' that can be defined as follows: Say that $X = V(f)$ in a neighbourhood of $a \in |X|$. Then, in some neighbourhood of $a' \in \sigma^{-1}(a)$, $X' = V(f')$, where $f' = y_{\text{exc}}^{-d} f \circ \sigma$, y_{exc} denotes a local generator of $\mathfrak{I}_{\sigma^{-1}(C)} \subset \mathcal{O}_{M'}$, and $d = \mu_{C,a}(f)$ denotes the *order* of f *along* C at a: $d = \max\{k : (f) \subset \mathfrak{I}_{C,a}^k\}$; d is the largest power to which y_{exc} factors from $f \circ \sigma$ at a'.

The strict transform X' of a general closed subspace X of M can be defined locally, at each $a' \in \sigma^{-1}(a)$, as the intersection of all hypersurfaces $V(f')$, for all $f \in \mathfrak{I}_{X,a}$. We likewise define the strict transform by a sequence of blowings-up with smooth centres.

Each of the categories listed in 1.5 above is closed under blowing up and strict transform [Bierstone and Milman 1997, Proposition 3.13 ff.]; the latter condition is needed to apply the desingularization algorithm in a given category.

1.7. THE INVARIANT. Let X denote a closed subspace of a manifold M. To describe inv_X, we consider a sequence of transformations

$$(1.8) \qquad \begin{array}{ccccccccc} \longrightarrow & M_{j+1} & \xrightarrow{\sigma_{j+1}} & M_j & \longrightarrow & \cdots & \longrightarrow & M_1 & \xrightarrow{\sigma_1} & M_0 & = & M \\ & X_{j+1} & & X_j & & & & X_1 & & X_0 & = & X \\ & E_{j+1} & & E_j & & & & E_1 & & E_0 & = & \varnothing \end{array}$$

where, for each j, $\sigma_{j+1}: M_{j+1} \to M_j$ denotes a blowing-up with smooth centre $C_j \subset M_j$, X_{j+1} is the strict transform of X_j by σ_{j+1}, and E_{j+1} is the set of exceptional hypersurfaces in M_{j+1}; i.e., $E_{j+1} = E_j' \cup \{\sigma_{j+1}^{-1}(C_j)\}$, where E_j' denotes the set of strict transforms by σ_{j+1} of all hypersurfaces in E_j.

Our invariant $\operatorname{inv}_X(a)$, where $a \in M_j$ and $j = 0, 1, 2, \ldots$, will be defined inductively over the sequence of blowings-up; for each j, the invariant $\operatorname{inv}_X(a)$, for $a \in M_j$, can be defined provided that the centres C_i, $i < j$, are *admissible* (or inv_X-*admissible*) in the sense that

(1) C_i and E_i simultaneously have only normal crossings, and
(2) $\operatorname{inv}_X(\cdot)$ is locally constant on C_i.

The condition (1) guarantees that E_{i+1} is a collection of smooth hypersurfaces having only normal crossings. We can think of the desingularization algorithm in the following way: $X \subset M$ determines $\operatorname{inv}_X(a)$, for $a \in M$, and thus the first admissible centre of blowing up $C = C_0$; then $\mathrm{inv}_X(a)$ can be defined on M_1 and determines C_1, etc.

The notation $\mathrm{inv}_X(a)$, where $a \in M_j$, indicates a dependence not only on X_j, but also on the original space X. In fact $\mathrm{inv}_X(a)$, for $a \in M_j$, is invariant under local isomorphisms of X_j that preserve $E(a) = \{H \in E_j : H \ni a\}$ and certain subcollections $E^r(a)$ (which will be taken to encode the history of the resolution process). To understand why some dependence on the history should be needed, we consider how, in principle, it might be possible to determine a *global* centre of blowing up using a *local* invariant:

EXAMPLE 1.9. It is easy to find an example of a surface X whose singular locus, in a neighbourhood of a point a, consists of two smooth curves with a normal crossing at a, and where X has the property that, if we blow up with centre $\{a\}$, there are points a' in the fibre $\sigma^{-1}(a)$ where the strict transform X' has the *same* local equation (in suitable coordinates) as that of X at a, or an even more complicated equation (as in Example

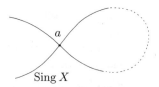

3.1 below). This suggests that to simplify the singularities in a neighbourhood of a by blowing up with smooth centre in Sing X, we should choose as centre one of the two smooth curves. But our surface may have the property that neither curve extends to a global smooth centre, as illustrated. So there is no choice but to blow up with centre $\{a\}$, although it seems to accomplish nothing: The figure shows the singular locus of X'; there are two points $a' \in \sigma^{-1}(a)$ where the singularity is the same as or worse than before. But what has changed at each of these points is the status of one of the curves, which is now *exceptional.* The moral is that, although the singularity of X at a has not been simplified in the strict transform, an invariant which takes into account the history of the resolution process as recorded by the accumulating exceptional hypersurfaces might nevertheless measure some improvement.

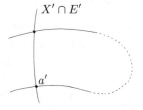

Consider a sequence of blowings-up as before. For simplicity, *we will assume that $X \subset M$ is a hypersurface.* Then $\mathrm{inv}_X(a)$, for $a \in M_j$, is a finite sequence beginning with the order $\nu_1(a) = \nu_{X_j,a}$ of X_j at a:

$$\mathrm{inv}_X(a) = \big(\nu_1(a), s_1(a); \nu_2(a), s_2(a); \ldots, s_t(a); \nu_{t+1}(a)\big).$$

(In the general case, $\nu_1(a)$ is replaced by a more delicate invariant of X_j at a, the *Hilbert–Samuel function* $H_{X_j,a}$ — see [Bierstone and Milman 1997] — but the remaining entries of $\mathrm{inv}_X(a)$ are still rational numbers (or ∞) as we will describe, and the theorems below are unchanged.) The $s_r(a)$ are nonnegative integers counting exceptional hypersurfaces that accumulate in certain blocks

$E^r(a)$ depending on the history of the resolution process. And the $\nu_r(a)$, for $r \geq 2$, represent certain "higher-order multiplicities" of the equation of X_j at a; $\nu_2(a), \ldots, \nu_t(a)$ are quotients of positive integers whose denominators are bounded in terms of the previous entries of $\mathrm{inv}_X(a)$. (More precisely, we have $e_{r-1}!\nu_r(a) \in \mathbb{N}$ for $r = 1, \ldots, t$, where $e_0 = 1$ and $e_r = \max\{e_{r-1}!, e_{r-1}!\nu_r(a)\}$.) The pairs $(\nu_r(a), s_r(a))$ can be defined successively using data that depends on $n - r + 1$ variables (where n is the ambient dimension), so that $t \leq n$ by exhaustion of variables; the final entry $\nu_{t+1}(a)$ is either 0 (the order of a nonvanishing function) or ∞ (the order of the function identically zero).

EXAMPLE 1.10. Let $X \subset \mathbb{K}^n$ be the hypersurface $x_1^{d_1} + x_2^{d_2} + \cdots + x_t^{d_t} = 0$, where $1 < d_1 \leq \cdots \leq d_t$ for $t \leq n$. Then

$$\mathrm{inv}_X(0) = \left(d_1, 0; \frac{d_2}{d_1}, 0; \ldots; \frac{d_t}{d_{t-1}}, 0; \infty\right).$$

This is $\mathrm{inv}_X(0)$ in "year zero" (before the first blowing up), so there are no exceptional hypersurfaces.

THEOREM A (Embedded desingularization.) *There is a finite sequence of blowings-up* (1.8) *with smooth* inv_X-*admissible centres* C_j *(or a locally finite sequence, in the case of noncompact analytic spaces) such that*:

(1) *For each* j, *either* $C_j \subset \mathrm{Sing}\, X_j$ *or* X_j *is smooth and* $C_j \subset X_j \cap E_j$.
(2) *Let* X' *and* E' *denote the final strict transform of* X *and exceptional set, respectively. Then* X' *is smooth and* X', E' *simultaneously have only normal crossings.*

If σ denotes the composite of the sequence of blowings-up σ_j, then E' is the critical locus of σ and $E' = \sigma^{-1}(\mathrm{Sing}\, X)$. In each of our categories of spaces, $\mathrm{Sing}\, X$ is closed in the *Zariski topology* of $|X|$ (the topology whose closed sets are of the form $|Y|$, for any closed subspace Y of X; see [Bierstone and Milman 1997, Proposition 10.1]). Theorem A resolves the singularities of X in a meaningful geometric sense provided that $|X| \setminus \mathrm{Sing}\, X$ is Zariski-dense in $|X|$. (For example, if X is a *reduced* complex-analytic space or a scheme of finite type.) More precise desingularization theorems (for example, for spaces that are not necessarily reduced) are given in [Bierstone and Milman 1997, Chapter IV].

This paper contains an essentially complete proof of Theorem A in the hypersurface case, presented though in a more informal way than in [Bierstone and Milman 1997]. We give a constructive definition of inv_X in Section 3, in parallel with a detailed example. In Section 4, we show that inv_X is indeed an invariant, and we summarize its key properties in Theorem B. (The terms $s_r(a)$ of $\mathrm{inv}_X(a)$ can, in fact, be introduced immediately in an invariant way; see 1.12 below.) It follows from Theorem B(3) that the maximum locus of inv_X has only normal crossings and, moreover, each of its local components extends to a global smooth subspace. (See Remark 3.6.)

The point is that each component is the intersection of the maximum locus of inv_X with those exceptional hypersurfaces containing the component; the exceptional divisors serve as global coordinates. We can obtain Theorem A by successively blowing up with centre given by any component of the maximum locus.

1.11. UNIVERSAL AND CANONICAL DESINGULARIZATION. The exceptional hypersurfaces (the elements of E_j) can be ordered in a natural way (by their "years of birth" in the history of the resolution process). We can use this ordering to extend $\mathrm{inv}_X(a)$ by an additional term $J(a)$ that will have the effect of picking out one component of the maximum locus of $\mathrm{inv}_X(\cdot)$ in a canonical way; see Remark 3.6. We write $\mathrm{inv}_X^{\mathrm{e}}(\cdot)$ for the extended invariant $\big(\mathrm{inv}_X(\cdot); J(\cdot)\big)$. Then our embedded desingularization theorem A can be obtained as follows:

ALGORITHM. *Choose as each successive centre of blowing up C_j the maximum locus of $\mathrm{inv}_X^{\mathrm{e}}$ on X_j.*

The algorithm stops when our space is "resolved" as in the conclusion of Theorem A. In the general (not necessarily hypersurface) case, we choose more precisely as each successive centre C_j the maximum locus of $\mathrm{inv}_X^{\mathrm{e}}$ on the nonresolved locus Z_j of X_j; in general, $\{x : \mathrm{inv}_X(x) = \mathrm{inv}_X(a)\} \subset Z_j$ (as germs at a), so that again each C_j is smooth, by Theorem B(3), and the algorithm stops when $Z_j = \varnothing$.

The algorithm applies to a category of spaces satisfying a compactness assumption (for example, schemes of finite type, restrictions of analytic spaces to relatively compact open subsets), so that $\mathrm{inv}_X(\cdot)$ has global maxima. Since the centres of blowing up are completely determined by an invariant, our desingularization theorem is automatically *universal* in the following sense: To every X, we associate a morphism of resolution of singularities $\sigma_X : X' \to X$ such that any local isomorphism $X|U \to Y|V$ (over open subsets U of $|X|$ and V of $|Y|$) lifts to an isomorphism $X'|\sigma_X^{-1}(U) \to Y'|\sigma_Y^{-1}(V)$ (in fact, lifts to isomorphisms throughout the entire towers of blowings-up).

For analytic spaces that are not necessarily compact, we can use an exhaustion by relatively compact open sets to deduce *canonical* resolution of singularities: Given X, there is a morphism of desingularization $\sigma_X : X' \to X$ such that any local isomorphism $X|U \to X|V$ (over open subsets of $|X|$) lifts to an isomorphism $X'|\sigma_X^{-1}(U) \to X'|\sigma_X^{-1}(V)$. See [Bierstone and Milman 1997, Section 13].

1.12. THE TERMS $s_r(a)$. The entries $s_1(a)$, $\nu_2(a)$, $s_2(a)$, ... of $\mathrm{inv}_X(a) = \big(\nu_1(a), s_1(a); \ldots, s_t(a); \nu_{t+1}(a)\big)$ will themselves be defined recursively. We write inv_r for inv_X truncated after s_r (with the convention that $\mathrm{inv}_r(a) = \mathrm{inv}_X(a)$ if $r > t$). We also write $\mathrm{inv}_{r+1/2} = (\mathrm{inv}_r; \nu_{r+1})$ (with the same convention), so that $\mathrm{inv}_{1/2}(a)$ means $\nu_1(a) = \nu_{X_j,a}$ (in the hypersurface case, or $H_{X_j,a}$ in general). For each r, the entries s_r, ν_{r+1} of inv_X can be defined over a sequence of blowings-up (1.8) whose centres C_i are $(r - \frac{1}{2})$-*admissible* (or $\mathrm{inv}_{r-1/2}$-*admissible*) in the sense that:

(1) C_i and E_i simultaneously have only normal crossings.

(2) $\mathrm{inv}_{r-1/2}(\,\cdot\,)$ is locally constant on C_i.

The terms $s_r(a)$ can be introduced immediately, as follows: Write $\pi_{ij} = \sigma_{i+1} \circ \cdots \circ \sigma_j$, for $i = 0, \ldots, j-1$, and $\pi_{jj} = $ identity. If $a \in M_j$, set $a_i = \pi_{ij}(a)$, for $i = 0, \ldots, j$. First consider a sequence of blowings-up (1.8) with $\frac{1}{2}$-admissible centres. ($\mathrm{inv}_{1/2} = \nu_1$ can only decrease over such a sequence; see, for example, Section 2 following.) Suppose $a \in M_j$. Let i denote the "earliest year" k such that $\nu_1(a) = \nu_1(a_k)$, and set $E^1(a) = \{H \in E(a) : H$ is the strict transform of some hypersurface in $E(a_i)\}$. We define $s_1(a) = \#E^1(a)$.

The block of exceptional hypersurfaces $E^1(a)$ intervenes in our desingularization algorithm in a way that can be thought of intuitively as follows. (The idea will be made precise in Sections 2 and 3.) The exceptional hypersurfaces passing through a but not in $E^1(a)$ have accumulated during the recent part of our history, when the order ν_1 has not changed; we have good control over these hypersurfaces. But those in $E^1(a)$ accumulated long ago; we have forgotten a lot about them in the form of our equations (for example, if we restrict the equations of X to these hypersurfaces, their orders might increase) and we recall them using $s_1(a)$.

In general, consider a sequence of blowings-up (1.8) with $(r + \frac{1}{2})$-admissible centres. ($\mathrm{inv}_{r+1/2}$ can only decrease over such a sequence; see Section 3 and Theorem B.) Suppose that i is the smallest index k such that $\mathrm{inv}_{r+1/2}(a) = \mathrm{inv}_{r+1/2}(a_k)$. Let $E^{r+1}(a) = \{H \in E(a) \setminus \bigcup_{q \leq r} E^q(a) : H$ is transformed from $E(a_i)\}$. We define $s_{r+1}(a) = \#E^{r+1}(a)$.

It is less straightforward to define the multiplicities $\nu_2(a), \nu_3(a), \ldots$ and to show they are invariants. Our definition depends on a construction in local coordinates that we present in Section 3. But we first try to convey the idea by describing the origin of our algorithm.

2. The Origin of our Approach

Consider a hypersurface X, defined locally by an equation $f(x) = 0$. Let $a \in X$ and let $d = d(a)$ denote the order of X (or of f) at a; i.e., $d = \nu_1(a) = \mu_a(f)$. We can choose local coordinates $(x_1, \ldots, x_n)$ in which $a = 0$ and $(\partial^d f / \partial x_n^d)(a) \neq 0$; then we can write

$$f(x) = c_0(\tilde{x}) + c_1(\tilde{x})x_n + \cdots + c_{d-1}(\tilde{x})x_n^{d-1} + c_d(x)x_n^d$$

in a neighbourhood of a, where $c_d(x)$ does not vanish. ($\tilde{x}$ means $(x_1, \ldots, x_{n-1})$.) Assume for simplicity that $c_d(x) \equiv 1$ (for example, by the Weierstrass preparation theorem, but see Remark 2.3 below). We can also assume that $c_{d-1}(\tilde{x}) \equiv 0$, by "completing the d-th power" (i.e., by the coordinate change $x_n' = x_n +$

$c_{d-1}(\tilde{x})/d)$; thus

$$(2.1) \qquad f(x) = c_0(\tilde{x}) + \cdots + c_{d-2}(\tilde{x})x_n^{d-2} + x_n^d.$$

Our aim is to simplify f by blowing up with smooth centre in the *equimultiple locus* of $a = 0$; i.e., in the locus of points of order d,

$$S_{(f,d)} = \{x : \mu_x(f) = d\}.$$

The representation (2.1) makes it clear that the equimultiple locus lies in a smooth subspace of codimension 1; in fact, by elementary calculus,

$$(2.2) \qquad S_{(f,d)} = \{x : x_n = 0 \text{ and } \mu_{\tilde{x}}(c_q) \geq d - q, \ q = 0, \ldots, d-2\}.$$

The idea now is that the given data $\big(f(x), d\big)$ involving n variables should be equivalent, in some sense, to the data $\mathcal{H}_1(a) = \big\{(c_q(\tilde{x}), d - q)\big\}$ in $n-1$ variables, thus making possible an induction on the number of variables. (Here in year zero, before we begin to blow up, $\nu_2(a) = \min_q \mu_a(c_q)/(d-q)$.)

REMARK 2.3. For the global desingularization algorithm, the Weierstrass preparation theorem must be avoided for two important reasons: (1) It may take us outside the given category (for example, in the algebraic case). (2) Even in the complex-analytic case, we need to prove that inv_X is semicontinuous in the sense that any point admits a coordinate neighbourhood V such that, given $a \in V$, $\{x \in V : \mathrm{inv}_X(x) \leq \mathrm{inv}_X(a)\}$ is Zariski-open in V (i.e., is the complement of a closed analytic subset). We therefore need a representation like (2.2) that is valid in a Zariski-open neighbourhood of a in V. This can be achieved in the following simple way that involves neither making $c_d(x) \equiv 1$ nor explicitly completing the d-th power: By a linear coordinate change, we can assume that $(\partial^d f/\partial x_n^d)(a) \neq 0$. Then in the Zariski-open neighbourhood of a where $(\partial^d f/\partial x_n^d)(x) \neq 0$, we let $N_1 = N_1(a)$ denote the submanifold of codimension one (in our category) defined by $z = 0$, where $z = \partial^{d-1}f/\partial x_n^{d-1}$, and we take $\mathcal{H}_1(a) = \big\{\big((\partial^q f/\partial x_n^q)|N_1, \ d-q\big)\big\}$. As before, we have $S_{(f,d)} = \{x : x \in N_1 \text{ and } \mu_x(h) \geq \mu_h, \text{ for all } (h, \mu_h) = \big((\partial^q f/\partial x_n^q)|N_1, \ d-q\big) \in \mathcal{H}_1(a)\}$.

We now consider the effect of a blowing-up σ with smooth centre $C \subset S_{(f,d)}$. By a transformation of the variables $(x_1, \ldots, x_{n-1})$, we can assume that in our local coordinate neighbourhood U of a, C has the form

$$(2.4) \qquad Z_I = \{x : x_n = 0 \text{ and } x_i = 0, \ i \in I\},$$

where $I \subset \{1, \ldots, n-1\}$. According to 1.2 above, $U' = \sigma^{-1}(U)$ is covered by coordinate charts U_i', for $i \in I \cup \{n\}$, where each U_i' has coordinates $y = (y_1, \ldots, y_n)$ in which σ is given by

$$\begin{aligned}
x_i &= y_i, \\
x_j &= y_i y_j, && \text{if } j \in (I \cup \{n\}) \setminus \{i\}, \\
x_j &= y_j, && \text{if } j \notin I \cup \{n\}.
\end{aligned}$$

In each U_i', we can write $f(\sigma(y)) = y_i^d f'(y)$; the strict transform X' of X by σ is defined in U_i' by the equation $f'(y) = 0$. (To be as simple as possible, we continue to assume $c_d(x) \equiv 1$, though we could just as well work with the set-up of Remark 2.3; see [Bierstone and Milman 1997, Proposition 4.12].) By (2.1), if $i \in I$, then

$$(2.5) \qquad f'(y) = c_0'(\tilde{y}) + \cdots + c_{d-2}'(\tilde{y})y_n^{d-2} + y_n^d,$$

where

$$(2.6) \qquad c_q'(\tilde{y}) = y_i^{-(d-q)}c_q(\tilde{\sigma}(\tilde{y})), \qquad \text{for } q = 0, \ldots, d-2.$$

The analogous formula for the strict transform in the chart U_n' shows that f' is invertible at every point of $U_n' \setminus \bigcup_{i \in I} U_i' = \{y \in U_n' : y_i = 0,\ i \in I\}$; in other words, $X' \cap U' \subset \bigcup_{i \in I} U_i'$.

The formula for $f'(y)$ above shows that the representation (2.2) of the equimultiple locus (or that of Remark 2.3) is stable under ν_1-admissible blowing up when the order does not decrease; i.e., at a point $a' \in U_i'$ where $d(a') = d$,

$$S_{(f',d)} = \{y : y_n = 0 \text{ and } \mu_{\tilde{y}}(c_q') \geq d - q \text{ for } q = 0, \ldots, d-2\},$$

where $N_1(a') = \{y_n = 0\}$ is the strict transform of $N_1(a) = \{x_n = 0\}$ and the c_q' are given by the transformation law (2.6). The latter is not strict transform, but something intermediate between strict and total transform $c_q \circ \sigma$. It is essentially for this reason that some form of embedded desingularization will be needed for the coefficients c_q (i.e., in the inductive step) even to prove a weaker form of resolution of singularities for f.

$N_1(a)$ is called a smooth hypersurface of *maximal contact* with X; this means a smooth hypersurface that contains the equimultiple locus of a, stably (i.e., even after admissible blowings-up as above). The existence of $N_1(a)$ depends on characteristic zero. A maximal contact hypersurface is crucial to our construction by increasing codimension. (In 1.12 above, $E^1(a)$ is the block of exceptional hypersurfaces that do not necessarily have normal crossings with respect to a maximal contact hypersurface; the term $s_1(a)$ in $\mathrm{inv}_X(a)$ is needed to deal with these exceptional divisors.)

We will now make a simplifying assumption on the coefficients c_q: we assume that one of these functions is a monomial (times an invertible factor) that divides all the others, but in a way that respects the different "multiplicities" $d - q$ associated with the transformation law (2.6). In other words, we make the monomial assumption on the $c_q^{1/(d-q)}$ (to equalize the "assigned multiplicities" $d - q$) or on the $c_q^{d!/(d-q)}$ (to avoid fractional powers). We assume, then, that

$$(2.7) \qquad c_q(\tilde{x})^{d!/(d-q)} = (\tilde{x}^{\Omega})^{d!}c_q^*(\tilde{x}), \qquad \text{for } q = 0, \ldots, d-2,$$

where $\Omega = (\Omega_1, \ldots, \Omega_{n-1})$ with $d!\Omega_i \in \mathbb{N}$ for each i, $\tilde{x}^{\Omega} = x_1^{\Omega_1} \cdots x_{n-1}^{\Omega_{n-1}}$, and the c_q^* are regular functions on $\{x_n = 0\}$ such that $c_q^*(a) \neq 0$ for some q. We also write $\Omega = \Omega(a)$.

We can regard (2.7) provisionally as an assumption made to see what happens in a simple test case, but in fact we can reduce to this case by a suitable induction on dimension (as we will see below). (Assuming (2.7) in year zero, $\nu_2(a) = |\Omega|$, where $|\Omega| = \Omega_1 + \cdots + \Omega_{n-1}$. But from the viewpoint of our algorithm for canonical desingularization as presented in Section 3, the argument following is analogous to a situation where the variables x_i occurring in $\tilde{x}^\Omega$ are exceptional divisors in $E(a) \backslash E^1(a)$; in this context, $|\Omega|$ is an invariant we call $\mu_2(a)$ (Definition 3.2) and $\nu_2(a) = 0$.)

Now, by (2.2) and (2.7),

$$S_{(f,d)} = \{x : x_n = 0 \text{ and } \mu_{\tilde{x}}(\tilde{x}^\Omega) \geq 1\}.$$

(The order of a monomial with rational exponents has the obvious meaning.) Therefore (using the notation (2.4)), $S_{(f,d)} = \bigcup Z_I$, where I runs over the *minimal* subsets of $\{1, \ldots, n-1\}$ such that $\sum_{j \in I} \Omega_j \geq 1$; i.e., where I runs over the subsets of $\{1, \ldots, n-1\}$ such that

$$(2.8) \qquad 0 \leq \sum_{j \in I} \Omega_j - 1 < \Omega_i, \qquad \text{for all } i \in I.$$

Consider the blowing-up σ with centre $C = Z_I$, for one such I. By (2.7), in the chart U_i' we have

$$(2.9) \qquad c_q'(\tilde{y})^{d!/(d-q)} = \left(y_1^{\Omega_1} \cdots y_i^{\sum_I \Omega_j - 1} \cdots y_{n-1}^{\Omega_{n-1}} \right)^{d!} c_q^*(\tilde{\sigma}(\tilde{y})),$$

$q = 0, \ldots, d-2$. Suppose $a' \in \sigma^{-1}(a) \cap U_i'$. By (2.5), $d(a') \leq d(a)$. Moreover, if $d(a') = d(a)$, then by (2.8) and (2.9), $1 \leq |\Omega(a')| < |\Omega(a)|$. In particular, the order d must decrease after at most $d!|\Omega|$ such blowings-up.

The question then is whether we can reduce to the hypothesis (2.7) by induction on dimension, replacing (f, d) in some sense by the collection $\mathcal{H}_1(a) = \{(c_q, d - q)\}$ on the submanifold $N_1 = \{x_n = 0\}$. To set up the induction, we would have to treat from the start a collection $\mathcal{F}_1 = \{(f, \mu_f)\}$ rather than a single pair (f, d). (A general X is, in any case, defined locally by several equations.) Moreover, since the transformation law (2.6) is not strict transform, we would have to reformulate the original problem to not only desingularize X: $f(x) = 0$, but also make its total transform normal crossings. To this end, suppose that $f(x) = 0$ actually represents the strict transform of our original hypersurface in that year in the history of the blowings-up involved where the order at a first becomes d. (We are following the transforms of the hypersurface at a sequence of points "a" over some original point.) Suppose there are $s = s(a)$ accumulated exceptional hypersurfaces H_p passing through a; as above, we can also assume that H_p is defined near a by an equation

$$x_n + b_p(\tilde{x}) = 0,$$

$1 \leq p \leq s$. (Each $\mu_a(b_p) \geq 1$.) The transformation law for the b_p analogous to (2.6) is

$$b'_p(\tilde{y}) = y_i^{-1} b_p(\tilde{\sigma}(\tilde{y})), \qquad \text{for } p = 1, \ldots, s.$$

Suppose now that in (2.7) we also have

(2.10) $$b_p(\tilde{x})^{d!} = (\tilde{x}^\Omega)^{d!} b_p^*(\tilde{x}), \qquad \text{for } p = 1, \ldots, s$$

(and assume that either some $c_q^*(a) \neq 0$ or some $b_p^*(a) \neq 0$). Then the argument above shows that $(d(a'), s(a')) \leq (d(a), s(a))$ (with respect to the lexicographic ordering of pairs), and that if $(d(a'), s(a')) = (d(a), s(a))$ then $1 \leq |\Omega(a')| < |\Omega(a)|$. ($s(a')$ counts the exceptional hypersurfaces H_p' passing through a'. As long as d does not drop, the new exceptional hypersurfaces accumulate simply as $y_i = 0$ for certain $i = 1, \ldots, n-1$, in suitable coordinates $(y_1, \ldots, y_{n-1})$ for the strict transform $N' = \{y_n = 0\}$ of $N = \{x_n = 0\}$.)

The induction on dimension can be realized in various ways. The simplest — the method of [Bierstone and Milman 1988, Section 4] — is to apply the inductive hypothesis within a coordinate chart to the function of $n-1$ variables given by the product of all nonzero $c_q^{d!/(d-q)}$, all nonzero $b_q^{d!}$, and all their nonzero differences. The result is (2.7) and (2.10) (with $c_q^*(a) \neq 0$ or $b_p^*(a) \neq 0$ for some q or p; see [Bierstone and Milman 1988, Lemma 4.7]). Pullback of the coefficients c_q by a blowing-up of N with smooth centre C, say of the form (2.4) above, corresponds to strict transform of f by the blowing-up with centre $\{x_i = 0 : i \in I\}$. Thus we sacrifice the condition that each centre lie in the equimultiple locus (or even in X!). But we do get a very simple proof of local uniformization. In fact, we get the conclusion (2) of our desingularization theorem A, using a mapping σ: $M' \to M$ which is a composite of mappings that are either blowings-up with smooth centres or surjections of the form $\coprod_j U_j \to \bigcup_j U_j$, where the latter is a locally-finite open covering of a manifold and $\coprod$ means disjoint union.

To prove our canonical desingularization theorem, we repeat the construction above in increasing codimension to obtain $\mathrm{inv}_X(a) = (\nu_1(a), s_1(a); \nu_2(a), \ldots)$ — here $(\nu_1(a), s_1(a))$ is $(d(a), s(a))$ above — together with a corresponding local "presentation". The latter means a local description of the locus of constant values of the invariant in terms of regular functions with assigned multiplicities, that survives certain blowings-up. ($N_1(a), \mathcal{H}_1(a)$ above is a presentation of ν_1 at a.)

3. The Desingularization Algorithm

In this section we give a constructive definition of inv_X together with a corresponding presentation (in the hypersurface case). We illustrate the construction by applying the desingularization algorithm to an example — a surface whose desingularization involves all the features of the general hypersurface case. We will use horizontal lines to separate from the example the general considerations that are needed at each step.

EXAMPLE 3.1. Let $X \subset \mathbb{K}^3$ denote the hypersurface $g(x) = 0$, where $g(x) = x_3^2 - x_1^2 x_2^3$:

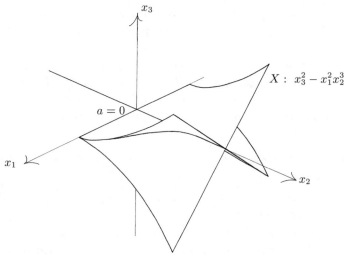

Let $a = 0$. Then $\nu_1(a) = \mu_a(g) = 2$. Of course, $E(a) = \varnothing$, so that $s_1(a) = 0$. (This is "year zero"; there are no exceptional hypersurfaces.) Thus $\mathrm{inv}_1(a) = (\nu_1(a), s_1(a)) = (2, 0)$. Let $\mathcal{G}_1(a) = \{(x_3^2 - x_1^2 x_2^3, 2)\}$. We say that $\mathcal{G}_1(a)$ is a *codimension* 0 *presentation of* $\mathrm{inv}_{1/2} = \nu_1$ at a. (Here where $s_1(a) = 0$, we can also say that $\mathcal{G}_1(a)$ is a codimension 0 presentation of $\mathrm{inv}_1 = (\nu_1, s_1)$ at a.)

———————

In general, consider a hypersurface $X \subset M$. Let $a \in M$ and let $S_{\mathrm{inv}_{1/2}}(a)$ denote the germ at a of $\{x : \mathrm{inv}_{1/2}(x) \geq \mathrm{inv}_{1/2}(a)\}$, which coincides with the germ at a of $\{x : \mathrm{inv}_{1/2}(x) = \mathrm{inv}_{1/2}(a)\}$. If $g \in \mathcal{O}_{M,a}$ generates the local ideal $\mathcal{I}_{X,a}$ of X and $d = \nu_1(a) = \mu_a(g)$, then $\mathcal{G}_1(a) = \{(g, d)\}$ is a codimension 0 presentation of $\mathrm{inv}_{1/2} = \nu_1$ at a. This means $S_{\mathrm{inv}_{1/2}}(a)$ coincides with the germ of the "equimultiple locus"

$$S_{\mathcal{G}_1(a)} = \{x : \mu_x(g) = d\},$$

and that the latter condition survives certain transformations.

More generally, suppose that $\mathcal{G}_1(a)$ is a finite collection of pairs $\{(g, \mu_g)\}$, where each g is a germ at a of a regular function (i.e., $g \in \mathcal{O}_{M,a}$) with an "assigned multiplicity" $\mu_g \in \mathbb{Q}$, and where we assume that $\mu_a(g) \geq \mu_g$ for every g. Set

$$S_{\mathcal{G}_1(a)} = \{x : \mu_x(g) \geq \mu_g, \text{ for all } (g, \mu_g) \in \mathcal{G}_1(a)\};$$

$S_{\mathcal{G}_1(a)}$ is well-defined as a germ at a. To say that $\mathcal{G}_1(a)$ is a *codimension* 0 *presentation of* $\mathrm{inv}_{1/2}$ at a means that

$$S_{\mathrm{inv}_{1/2}}(a) = S_{\mathcal{G}_1(a)}$$

and that this condition survives certain transformations:

To be precise, we will consider triples of the form $\big(N = N(a), \mathcal{H}(a), \mathcal{E}(a)\big)$, where:

N is a germ of a submanifold of codimension p at a (for some $p \geq 0$).

$\mathcal{H}(a) = \{(h, \mu_h)\}$ is a finite collection of pairs (h, μ_h), where $h \in \mathcal{O}_{N,a}$, $\mu_h \in \mathbb{Q}$ and $\mu_a(h) \geq \mu_h$.

$\mathcal{E}(a)$ is a finite set of smooth (exceptional) hyperplanes containing a, such that N and $\mathcal{E}(a)$ simultaneously have normal crossings and $N \not\subset H$, for all $H \in \mathcal{E}(a)$.

A *local blowing-up* $\sigma: M' \to M$ over a neighbourhood W of a, with smooth centre C, means the composite of a blowing-up $M' \to W$ with centre C, and the inclusion $W \hookrightarrow M$.

DEFINITION 3.2. We say that $\big(N(a), \mathcal{H}(a), \mathcal{E}(a)\big)$ is a *codimension p presentation* of $\mathrm{inv}_{1/2}$ at a if:

(1) $S_{\mathrm{inv}_{1/2}}(a) = S_{\mathcal{H}(a)}$, where $S_{\mathcal{H}(a)} = \{x \in N : \mu_x(h) \geq \mu_h,$ for all $(h, \mu_h) \in \mathcal{H}(a)\}$ (as a germ at a).

(2) Suppose that σ is a $\frac{1}{2}$-admissible local blowing-up at a (with smooth centre C). Let $a' \in \sigma^{-1}(a)$. Then $\mathrm{inv}_{1/2}(a') = \mathrm{inv}_{1/2}(a)$ if and only if $a' \in N'$ (where $N' = N(a')$ denotes the strict transform of N) and $\mu_{a'}(h') \geq \mu_{h'}$ for all $(h, \mu_h) \in \mathcal{H}(a)$, where $h' = y_{\mathrm{exc}}^{-\mu_h} h \circ \sigma$ and $\mu_{h'} = \mu_h$. (y_{exc} denotes a local generator of $\mathcal{I}_{\sigma^{-1}(C)}$.) In this case, we will write $\mathcal{H}(a') = \{(h', \mu_{h'}) : (h, \mu_h) \in \mathcal{H}(a)\}$ and $\mathcal{E}(a') = \{H' : H \in \mathcal{E}(a)\} \cup \{\sigma^{-1}(C)\}$.

(3) Conditions (1) and (2) continue to hold for the transforms X' and $\big(N(a'), \mathcal{H}(a'), \mathcal{E}(a')\big)$ of our data by sequences of morphisms of the following three types, at points a' in the fibre of a (to be also specified).

The three types of morphisms allowed are the following. (Types (ii) and (iii) are not used in the actual desingularization algorithm. They are needed to prove invariance of the terms $\nu_2(a), \nu_3(a), \ldots$ of $\mathrm{inv}_X(a)$ by making certain sequences of "test blowings-up", as we will explain in Section 4; they are not explicitly needed in this section.)

(i) $\frac{1}{2}$-*admissible local blowing-up* σ, and $a' \in \sigma^{-1}(a)$ such that $\mathrm{inv}_{1/2}(a') = \mathrm{inv}_{1/2}(a)$.

(ii) *Product with a line.* σ is a projection $M' = W \times \mathbb{K} \to W \hookrightarrow M$, where W is a neighbourhood of a, and $a' = (a, 0)$.

(iii) *Exceptional blowing-up.* σ is a local blowing-up $M' \to W \hookrightarrow M$ over a neighbourhood W of a, with centre $H_0 \cap H_1$, where $H_0, H_1 \in \mathcal{E}(a)$, and a' is the unique point of $\sigma^{-1}(a) \cap H_1'$.

The data is transformed to a' in each case above, as follows:

(i): $X' = $ strict transform of X; $\big(N(a'), \mathcal{H}(a'), \mathcal{E}(a')\big)$ as defined in 3.2(2) above.

(ii) and (iii): $X' = \sigma^{-1}(X)$, $N(a') = \sigma^{-1}(N)$, $\mathcal{H}(a') = \{(h \circ \sigma, \mu_h)\}$.

$$\mathcal{E}(a') = \begin{cases} \{\sigma^{-1}(H) : H \in \mathcal{E}(a)\} \cup \{W \times 0\} & \text{in case (ii);} \\ \{H' : H \in \mathcal{E}(a),\ a' \in H'\} \cup \{\sigma^{-1}(C)\} & \text{in case (iii).} \end{cases}$$

If $\big(N(a), \mathcal{H}(a), \mathcal{E}(a)\big)$ is a presentation of $\mathrm{inv}_{1/2}$ at a, then $N(a)$ is called a subspace of *maximal contact*; compare Section 2.

Suppose now that $\mathcal{G}_1(a)$ is a codimension 0 presentation of $\mathrm{inv}_{1/2}$ at a. (Implicitly, $N(a) = M$ and $\mathcal{E}(a) = \varnothing$.) Assume, moreover, that there exists $(g, \mu_g) = (g_*, \mu_{g_*}) \in \mathcal{G}_1(a)$ such that $\mu_a(g_*) = \mu_{g_*}$ (as in Example 3.1).

We can always assume that each $\mu_g \in \mathbb{N}$, and even that all μ_g coincide: Simply replace each (g, μ_g) by $(g^{e/\mu_g}, e)$, for suitable $e \in \mathbb{N}$.

Then, after a linear coordinate change if necessary, we can assume that $(\partial^d g_* / \partial x_n^d)(a) \neq 0$, where $d = \mu_{g_*}$. Set

$$z = \frac{\partial^{d-1} g_*}{\partial x_n^{d-1}} \in \mathcal{O}_{M,a}, \qquad N_1 = N_1(a) = \{z = 0\},$$

$$\mathcal{H}_1(a) = \left\{ \left(\frac{\partial^q g}{\partial x_n^q} \Big|_{N_1}, \mu_g - q \right) : 0 \leq q < \mu_g,\ (g, \mu_g) \in \mathcal{G}_1(a) \right\}.$$

Then $\big(N_1(a), \mathcal{H}_1(a), \mathcal{E}_1(a) = \varnothing\big)$ is a codimension 1 presentation of $\mathrm{inv}_{1/2}$ at a: This is an assertion about the way our data transform under sequences of morphisms of types (i), (ii) and (iii) above. The effect of a transformation of type (i) is essentially described by the calculation in Section 2. The effect of a transformation of type (ii) is trivial, and that for type (iii) can be understood in a similar way to (i): see [Bierstone and Milman 1997, Propositions 4.12 and 4.19] for details.

DEFINITION 3.3. We define

$$\mu_2(a) = \min_{\mathcal{H}_1(a)} \frac{\mu_a(h)}{\mu_h}.$$

Then $1 \leq \mu_2(a) \leq \infty$. If $\mathcal{E}(a) = \varnothing$ (as in year zero), we set

$$\nu_2(a) = \mu_2(a)$$

and $\mathrm{inv}_{1 1/2}(a) = \big(\mathrm{inv}_1(a); \nu_2(a)\big)$. Then $\nu_2(a) \leq \infty$. Moreover, $\nu_2(a) = \infty$ if and only if $\mathcal{G}_1(a) \sim \{(z, 1)\}$. (This means that the latter is also a presentation of $\mathrm{inv}_{1/2}$ at a.) If $\nu_2(a) = \infty$, then we set $\mathrm{inv}_X(a) = \mathrm{inv}_{1 1/2}(a)$. $\mathrm{inv}_X(a) = (d, 0, \infty)$ if and only if X is defined (near a) by the equation $z^d = 0$; in this case, the desingularization algorithm can do no more, unless we blow-up with centre $|X|$!

In Example 3.1, $\mu_a(g) = 2 = \mu_g$, and by the construction above we get the following codimension 1 presentation of $\mathrm{inv}_{1/2}$ (or inv_1) at a:

$$N_1(a) = \{x_3 = 0\}, \qquad \mathcal{H}_1(a) = \{(x_1^2 x_2^3, 2)\}.$$

Thus $\nu_2(a) = \mu_2(a) = \frac{5}{2}$. As a codimension 1 presentation of $\mathrm{inv}_{1 1/2}$ (or inv_2) at a, we can take

$$N_1(a), \qquad \mathcal{G}_2(a) = \{(x_1^2 x_2^3, 5)\}.$$

In general, "presentation of inv_r" (or "of $\text{inv}_{r+1/2}$") means the analogue of "presentation of $\text{inv}_{1/2}$" above. Suppose that $\big(N_1(a), \mathcal{H}_1(a)\big)$ is a codimension 1 presentation of inv_1 at a $(\mathcal{E}_1(a) = \varnothing)$. Assume that $1 \le \nu_2(a) < \infty$. (In year zero, we always have $\nu_2(a) = \mu_2(a) \ge 1$.) Let

$$\mathcal{G}_2(a) = \big\{ (h, \nu_2(a)\mu_h) : (h, \mu_h) \in \mathcal{H}_1(a) \big\}.$$

Then $\big(N_1(a), \mathcal{G}_2(a)\big)$ is a codimension 1 presentation of $\text{inv}_{11/2}$ at a (or of inv_2 at a, when $s_2(a) = 0$ as here). Clearly, there exists $(g_*, \mu_{g_*}) \in \mathcal{G}_2(a)$ such that $\mu_a(g_*) = \mu_{g_*}$.

This completes a cycle in the recursive definition of inv_X, and we can now repeat the above constructions: Let $d = \mu_{g_*}$. After a linear transformation of the coordinates $(x_1, \ldots, x_{n-1})$ of $N_1(a)$, we can assume that $(\partial^d g_* / \partial x_{n-1}^d)(a) \ne 0$. We get a codimension 2 presentation of inv_2 at a by taking

$$N_2(a) = \left\{ x \in N_1(a) : \frac{\partial^{d-1} g_*}{\partial x_{n-1}^{d-1}}(x) = 0 \right\},$$

$$\mathcal{H}_2(a) = \left\{ \left(\frac{\partial^q g}{\partial x_{n-1}^q} \bigg|_{N_2(a)}, \mu_g - q \right) : 0 \le q < \mu_g, \; (g, \mu_g) \in \mathcal{G}_2(a) \right\}.$$

In our example, the calculation of a codimension 2 presentation can be simplified by the following useful observation: Suppose there is $(g, \mu_g) \in \mathcal{G}_2(a)$ with $\mu_a(g) = \mu_g$ and $g = \prod g_i^{m_i}$. If we replace (g, μ_g) in $\mathcal{G}_2(a)$ by the collection of (g_i, μ_{g_i}), where each $\mu_{g_i} = \mu_a(g_i)$, then we obtain an (equivalent) presentation of inv_2.

In our example, therefore,

$$N_1(a) = \{x_3 = 0\}, \qquad \mathcal{G}_2(a) = \{(x_1, 1), (x_2, 1)\}$$

is a codimension 1 presentation of inv_2 at a. It follows immediately that

$$N_2(a) = \{x_2 = x_3 = 0\}, \qquad \mathcal{H}_2(a) = \{(x_1, 1)\}$$

is a codimension 2 presentation of inv_2 at a. Then $\nu_3(a) = \mu_3(a) = 1$ and, as a codimension 3 presentation of $\text{inv}_{21/2}$ (or of inv_3) at a, we can take

$$N_3(a) = \{x_1 = x_2 = x_3 = 0\}, \qquad \mathcal{H}_3(a) = \varnothing.$$

We put $\nu_4(a) = \mu_4(a) = \infty$. Thus we have

$$\text{inv}_X(a) = (2, 0; \tfrac{5}{2}, 0; 1, 0; \infty)$$

and $S_{\text{inv}_X}(a) = S_{\text{inv}_3}(a) = N_3(a) = \{a\}$. The latter is the centre C_0 of our first blowing-up $\sigma_1 \colon M_1 \to M_0 = \mathbb{K}^3$; M_1 can be covered by three coordinate charts U_i, $i = 1, 2, 3$, where each U_i is the complement in M_1 of the strict transform of the hyperplane $\{x_i = 0\}$. The strict transform $X_1 = X'$ of X lies in $U_1 \cup U_2$. To illustrate the algorithm, we will follow our construction at a sequence of points over a, choosing after each blowing-up a point in the fibre where inv_X has a maximum value in a given coordinate chart.

Year one. U_1 has a coordinate system (y_1, y_2, y_3) in which σ_1 is given by the transformation

$$x_1 = y_1, \quad x_2 = y_1 y_2, \quad x_3 = y_1 y_3.$$

Then $X_1 \cap U_1 = V(g_1)$, where

$$g_1 = y_1^{-2} g \circ \sigma_1 = y_3^2 - y_1^3 y_2^3.$$

Consider $b = 0$. Then $E(b) = \{H_1\}$, where H_1 is the exceptional hypersurface $H_1 = \sigma_1^{-1}(a) = \{y_1 = 0\}$. Now, $\nu_1(b) = 2 = \nu_1(a)$. Therefore $E^1(b) = \varnothing$ and $s_1(b) = 0$. We write $\mathcal{E}_1(b) = E(b) \setminus E^1(b)$, so that $\mathcal{E}_1(b) = E(b)$ here. Let $\mathcal{F}_1(b) = \mathcal{G}_1(b) = \{(g_1, 2)\}$. Then $\big(N_0(b) = M_1, \mathcal{F}_1(b), \mathcal{E}_1(b)\big)$ is a codimension 0 presentation of inv_1 at b. Set

$$N_1(b) = \{y_3 = 0\} = N_1(a)', \qquad \mathcal{H}_1(b) = \{(y_1^3 y_2^3, 2)\};$$

$\big(N_1(b), \mathcal{H}_1(b), \mathcal{E}_1(b)\big)$ is a codimension 1 presentation of inv_1 at b. As before,

$$\mu_2(b) = \min_{\mathcal{H}_1(b)} \frac{\mu_b(h)}{\mu_h} = \frac{6}{2} = 3.$$

But, here, in the presence of nontrivial $\mathcal{E}_1(b)$, $\nu_2(b)$ will involve first factoring from the $h \in \mathcal{H}_1(b)$ the exceptional divisors in $\mathcal{E}_1(b)$ (taking, in a sense, "internal strict transforms" at b of the elements of $\mathcal{H}_1(a)$).

In general, we define

$$\mathcal{F}_1(b) = \mathcal{G}_1(b) \cup \big(E^1(b), 1\big),$$

where $\big(E^1(b), 1\big)$ denotes $\{(y_H, 1) : H \in E^1(b)\}$, and y_H means a local generator of the ideal of H. Then $\big(N_0(b), \mathcal{F}_1(b), \mathcal{E}_1(b)\big)$ is a codimension 0 presentation of $\text{inv}_1 = (\nu_1, s_1)$ at b, and there is a codimension 1 presentation $\big(N_1(b), \mathcal{H}_1(b), \mathcal{E}_1(b)\big)$ as before. The construction of Section 2 above shows that we can choose the coordinates $(y_1, \ldots, y_{n-1})$ of $N_1(b)$ so that each $H \in \mathcal{E}_1(b) = E(b) \setminus E^1(b)$ is $\{y_i = 0\}$, for some $i = 1, \ldots, n-1$; we again write $y_H = y_i$. (In other words, $\mathcal{E}_1(b)$ and $N_1(b)$ simultaneously have normal crossings, and $N_1(b) \not\subset H$, for all $H \in \mathcal{E}_1(b)$.)

DEFINITION 3.4. For each $H \in \mathcal{E}_1(b)$, we set

$$\mu_{2H}(b) = \min_{(h, \mu_h) \in \mathcal{H}_1(b)} \frac{\mu_{H,b}(h)}{\mu_h},$$

where $\mu_{H,b}(h)$ denotes the *order* of h *along* H at b; i.e., the order to which y_H factors from $h \in \mathcal{O}_{N,b}$, $N = N_1(b)$, or $\max\{k : h \in \mathcal{I}_{H,b}^k\}$, where $\mathcal{I}_{H,b}$ is the ideal of $H \cap N$ in $\mathcal{O}_{N,b}$. We define

$$\nu_2(b) = \mu_2(b) - \sum_{H \in \mathcal{E}_1(b)} \mu_{2H}(b).$$

In our example,

$$\nu_2(b) = \mu_2(b) - \mu_{2H_1}(b) = 3 - \tfrac{3}{2} = \tfrac{3}{2}.$$

Write

$$D_2(b) = \prod_{H \in \mathcal{E}_1(b)} y_H^{\mu_{2H}(b)}.$$

Suppose, as before, that all μ_h are equal: say all $\mu_h = d \in \mathbb{N}$. Then $D^d = D_2(b)^d$ is the greatest common divisor of the h that is a monomial in the exceptional coordinates y_H, $H \in \mathcal{E}_1(b)$. For each $h \in \mathcal{H}_1(b)$, write $h = D^d g$ and set $\mu_g = d\nu_2(b)$; then $\mu_b(g) \geq \mu_g$. Clearly, $\nu_2(b) = \min_g \mu_b(g)/d$. Moreover, $0 \leq \nu_2(b) \leq \infty$, and $\nu_2(b) = \infty$ if and only if $\mu_2(b) = \infty$.

If $\nu_2(b) = 0$ or ∞, we put $\text{inv}_X(b) = \text{inv}_{11/2}(b)$. If $\nu_2(b) = \infty$, then $S_{\text{inv}_X}(b) = N_1(b)$. If $\nu_2(b) = 0$ and we set $\mathcal{G}_2(b) = \{(D_2(b), 1)\}$, then $(N_1(b), \mathcal{G}_2(b), \mathcal{E}_1(b))$ is a codimension 1 presentation of inv_X at b; in particular,

$$S_{\text{inv}_X}(b) = \{y \in N_1(b) : \mu_y(D_2(b)) \geq 1\}$$

(compare Section 2).

Consider the case that $0 < \nu_2(b) < \infty$. Let $\mathcal{G}_2(b)$ denote the collection of pairs $(g, \mu_g) = (g, d\nu_2(b))$ for all $(h, \mu_h) = (h, d)$, as above, together with the pair $(D_2(b)^d, (1 - \nu_2(b))d)$ *provided that* $\nu_2(b) < 1$. Then $(N_1(b), \mathcal{G}_2(b), \mathcal{E}_1(b))$ is a codimension 1 presentation of $\text{inv}_{11/2}$ at b.

In the latter case, we introduce $E^2(b) \subset \mathcal{E}_1(b)$ as in 1.12, and we set $s_2(b) = \#E^2(b)$, $\mathcal{E}_2(b) = \mathcal{E}_1(b) \setminus E^2(b)$. Set

$$\mathcal{F}_2(b) = \mathcal{G}_2(b) \cup (E^2(b), 1).$$

Then $(N_1(b), \mathcal{F}_2(b), \mathcal{E}_1(b))$ is a codimension 1 presentation of inv_2 at b, and we can pass to a codimension 2 presentation $(N_2(b), \mathcal{H}_2(b), \mathcal{E}_2(b))$. Here it is important to replace $\mathcal{E}_1(b)$ by the subset $\mathcal{E}_2(b)$, to have the property that $\mathcal{E}_2(b)$, $N_2(b)$ simultaneously have normal crossings and $N_2(b) \not\subset H$, for all $H \in \mathcal{E}_2(b)$. (Again, the main rôle of $\mathcal{E}$ in a presentation is to prove invariance of the $\mu_{2H}(\cdot)$ and in general of the $\mu_{3H}(\cdot), \ldots$, as in Section 4.)

Returning to our example (in year one), we have $\mathcal{H}_1(b) = \{(y_1^3 y_2^3, 2)\}$, so that $D_2(b) = y_1^{3/2}$. We can take $\mathcal{G}_2(b) = \{(y_2^3, 3)\}$ or, equivalently, $\mathcal{G}_2(b) = \{(y_2, 1)\}$ to get a codimension 1 presentation $\big(N_1(b), \mathcal{G}_2(b), \mathcal{E}_1(b)\big)$ of $\mathrm{inv}_{11/2}$ at b.

Now, $E^2(b) = \{H_1\}$, so that $s_2(b) = 1$. We set

$$\mathcal{F}_2(b) = \mathcal{G}_2(b) \cup \big(E^2(b), 1\big) = \{(y_1, 1), (y_2, 1)\}$$

and $\mathcal{E}_2(b) = \mathcal{E}_1(b) \setminus E^2(b) = \varnothing$. Then $\big(N_1(b), \mathcal{F}_2(b), \mathcal{E}_1(b)\big)$ is a codimension 1 presentation of inv_2 at b, and we can get a codimension 2 presentation $\big(N_2(b), \mathcal{H}_2(b), \mathcal{E}_2(b)\big)$ of inv_2 at b by taking $N_2(b) = \{y_2 = y_3 = 0\}$ and $\mathcal{H}_2(b) = \{(y_1, 1)\}$.

It follows that $\nu_3(b) = 1$. Since $E^3(b) = \varnothing$, $s_3(b) = 0$. We get a codimension 3 presentation of inv_3 at b by taking

$$N_3(b) = \{y_1 = y_2 = y_3 = 0\} = \{b\}, \qquad H_3(b) = \varnothing.$$

Therefore,

$$\mathrm{inv}_X(b) = \big(2, 0; \tfrac{3}{2}, 1; 1, 0; \infty\big)$$

and $S_{\mathrm{inv}_X}(b) = S_{\mathrm{inv}_2}(b) = \{b\}$. The latter is the centre of the next blowing-up σ_2. The set $\sigma_2^{-1}(U_1)$ is covered by three coordinate charts

$$U_{1i} = \sigma_2^{-1}(U_1) \setminus \{y_i = 0\}', \qquad i = 1, 2, 3.$$

For example, U_{12} has coordinates (z_1, z_2, z_3) with respect to which σ_2 is given by

$$y_1 = z_1 z_2, \qquad y_2 = z_2, \qquad y_3 = z_2 z_3.$$

REMARK 3.5. *Zariski-semicontinuity of the invariant.* Each point of M_j, for $j = 0, 1, \ldots$, admits a coordinate neighbourhood U such that, for all $x_0 \in U$, $\{x \in U : \mathrm{inv}(x) \leq \mathrm{inv}(x_0)\}$ is Zariski-open in U (i.e., the complement of a Zariski-closed subset of U): For $\mathrm{inv}_{1/2}$, this is just Zariski-semicontinuity of the order of a regular function g (a local generator of the ideal of X). For inv_1, the result is a consequence of the following semicontinuity assertion for $E^1(x)$: There is a Zariski-open neighbourhood of x_0 in U, in which $E^1(x) = E(x) \cap E^1(x_0)$, for all $x \in S_{\mathrm{inv}_{1/2}}(x_0) = \{x \in U : \mathrm{inv}_{1/2}(x) \geq \mathrm{inv}_{1/2}(x_0)\}$. (See [Bierstone and Milman 1997, Proposition 6.6] for a simple proof.)

For $\mathrm{inv}_{11/2}$: Suppose that $\mu_k = d \in \mathbb{N}$, for all $(h, \mu_h) \in \mathcal{H}_1(x_0)$, as above. Then, in a Zariski-open neighbourhood of x_0 where $S_{\mathrm{inv}_1(x_0)} = \{x : \mathrm{inv}_1(x) = \mathrm{inv}_1(x_0)\}$, we have

$$d\nu_2(x) = \min_{\mathcal{H}_1(x_0)} \mu_x \left(\frac{h}{D_2(x_0)^d} \right), \qquad \text{for } x \in S_{\mathrm{inv}_1(x_0)}.$$

Semicontinuity of $\nu_2(x)$ is thus a consequence of semicontinuity of the order of an element $g = h/D_2(x_0)^d$ such that $\mu_{x_0}(g) = d\nu_2(x_0)$.

Likewise for inv_2, $\mathrm{inv}_{21/2}$, ...

Year two. Let X_2 denote the strict transform X_1' of X_1 by σ_2. Then $X_2 \cap U_{12} = V(g_2)$, where

$$g_2 = z_2^{-2} g_1 \circ \sigma_2 = z_3^2 - z_1^3 z_2^4.$$

Let c be the origin of U_{12}. Then $E(c) = \{H_1, H_2\}$ where

$$H_1 = \{y_1 = 0\}' = \{z_1 = 0\},$$
$$H_2 = \sigma_2^{-1}(b) = \{z_2 = 0\}.$$

We have $\nu_1(c) = 2 = \nu_1(a)$. Therefore, $E^1(c) = \varnothing$, $s_1(c) = 0$, $\mathcal{E}_1(c) = E(c)$. $\mathcal{F}_1(c) = \mathcal{G}_1(c) = \{(g_2, 2)\}$ provides a codimension 0 presentation of inv_1 at c, and we get a codimension 1 presentation by taking

$$N_1(c) = \{z_3 = 0\}, \qquad \mathcal{H}_1(c) = \{(z_1^3 z_2^4, 2)\}.$$

Therefore $\mu_2(c) = \frac{7}{2}$, $\mu_{2H_1}(c) = \frac{3}{2}$ and $\mu_{2H_2}(c) = \frac{4}{2} = 2$, so that $\nu_2(c) = 0$ and

$$\mathrm{inv}_X(c) = (2, 0; 0).$$

Moreover, $D_2(c) = z_1^{3/2} z_2^2$, and we get a codimension 1 presentation of $\mathrm{inv}_X = \mathrm{inv}_{11/2}$ at c using

$$N_1(c) = \{z_3 = 0\}, \qquad \mathcal{G}_2(c) = \{(z_1^{3/2} z_2^2, 1)\}.$$

Therefore,

$$S_{\mathrm{inv}_X}(c) = S_{\mathrm{inv}_{11/2}}(c) = \{z_1 = z_3 = 0\} \cup \{z_2 = z_3 = 0\};$$

of course, $\{z_1 = z_3 = 0\} = S_{\mathrm{inv}_X}(c) \cap H_1$ and $\{z_2 = z_3 = 0\} = S_{\mathrm{inv}_X}(c) \cap H_2$.

REMARK 3.6. In general, suppose that $\mathrm{inv}_X(c) = \mathrm{inv}_{t+1/2}(c)$ and $v_{t+1}(c) = 0$. (We assume $c \in M_j$, for some $j = 1, 2, \ldots$.) Then inv_X has a codimension t presentation at c: $N_t(c) = \{z_{n-t+1} = \cdots = z_n = 0\}$, $\mathcal{G}_{t+1}(c) = \{(D_{t+1}(c), 1)\}$, where $D_{t+1}(c)$ is a monomial with rational exponents in the exceptional divisors z_H, $H \in \mathcal{E}_t(c)$; $N_t(c)$ has coordinates $(z_1, \ldots, z_{n-t})$ in which each such $z_H = z_i$, for some $i = 1, \ldots, n - t$. It follows that each component Z of $S_{\mathrm{inv}_X}(c)$ has the form

$$Z = S_{\mathrm{inv}_X}(c) \cap \bigcap\{H \in E(c) : Z \subset H\};$$

we will write $Z = Z_I$, where $I = \{H \in E(c) : Z \subset H\}$. It follows that, if U is any open neighbourhood of c on which $\mathrm{inv}_X(c)$ is a maximum value of inv_X, then every component Z_I of $S_{\mathrm{inv}_X}(c)$ extends to a global smooth closed subset of U:

First consider any total order on $\{I : I \subset E_j\}$. For any $c \in M_j$, set

$$J(c) = \max\{I : Z_I \text{ is a component of } S_{\operatorname{inv}_X}(c)\},$$
$$\operatorname{inv}_X^e(c) = \big(\operatorname{inv}_X(c); J(c)\big).$$

Then inv_X^e is Zariski-semicontinuous (again comparing values of inv_X^e lexicographically), and its locus of maximum values on any given open subset of M_j is smooth.

Of course, given $c \in M_j$ and a component Z_I of $S_{\operatorname{inv}_X}(c)$, we can choose the ordering of $\{J : J \subset E_j\}$ so that $I = J(c) = \max\{J : J \subset E_j\}$. It follows that, if U is any open neighbourhood of c on which $\operatorname{inv}_X(c)$ is a maximum value of inv_X, then Z_I extends to a smooth closed subset of U.

To obtain an algorithm for canonical desingularization, we can choose as each successive centre of blowing up the maximum locus of

$$\operatorname{inv}_X^e(\,\cdot\,) = \big(\operatorname{inv}_X(\,\cdot\,), J(\,\cdot\,)\big),$$

where J is defined as above using the following total ordering of the subsets of E_j: Write $E_j = \{H_1^j, \ldots, H_j^j\}$, where each H_i^j is the strict transform in M_j of the exceptional hypersurface $H_i^i = \sigma_i^{-1}(C_{i-1}) \subset M_i$, $i = 1, \ldots, j$. We can order $\{I : I \subset E_j\}$ by associating to each subset I the lexicographic order of the sequence $(\delta_1, \ldots, \delta_j)$, where $\delta_i = 0$ if $H_i^j \notin I$ and $\delta_i = 1$ if $H_i^j \in I$.

In our example, year two, we have

$$S_{\operatorname{inv}_X}(c) = \big(S_{\operatorname{inv}_X}(c) \cap H_1\big) \cup \big(S_{\operatorname{inv}_X}(c) \cap H_2\big).$$

(Each H_i is H_i^2 in the notation preceding.) The order of H_1 (respectively, H_2) is $(1, 0)$ (respectively, $(0, 1)$), so that $J(c) = \{H_1\}$ and the centre of the third blowing-up σ_3 is $C_2 = S_{\operatorname{inv}_X}(c) \cap H_1 = \{z_1 = z_3 = 0\}$.

Thus $\sigma_3^{-1}(U_{12}) = U_{121} \cup U_{123}$, where $U_{12i} = \sigma_3^{-1}(U_{12}) \setminus \{z_i = 0\}'$, $i = 1, 3$. The strict transform of $X_2 \cap U_{12}$ lies in U_{121}; the latter has coordinates (w_1, w_2, w_3) in which σ_3 can be written

$$z_1 = w_1, \qquad z_2 = w_2, \qquad z_3 = w_1 w_3.$$

Year three. Let X_3 denote the strict transform of X_2 by σ_3. Then $X_3 \cap U_{121} = V(g_3)$, where $g_3(w) = w_3^2 - w_1 w_2^4$. Let $d = 0$ in U_{121}. There are three exceptional hypersurfaces $H_1 = \{z_1 = 0\}'$, $H_2 = \{z_2 = 0\}' = \{w_2 = 0\}$ and $H_3 = \sigma_3^{-1}(C_2) = \{w_1 = 0\}$; since $H_1 \not\ni d$, $E(d) = \{H_2, H_3\}$. We have $\nu_1(d) = 2 = \nu_1(a)$. Therefore, $E^1(d) = \varnothing$, $s_1(d) = 0$ and $\mathcal{E}_1(d) = E(d)$. $\mathcal{F}_1(d) = \mathcal{G}_1(d) = \{(g_3, 2)\}$ provides a codimension 0 presentation of inv_1 at d, and we get a codimension 1 presentation by taking

$$N_1(d) = \{w_3 = 0\}, \qquad \mathcal{H}_1(d) = \{(w_1 w_2^4, 2)\}.$$

Therefore, $\mu_2(c) = \frac{5}{2}$ and $D_2(d) = w_1^{1/2}w_2^2$, so that $\nu_2(d) = 0$ and

$$\mathrm{inv}_X(d) = (2,0,0) = \mathrm{inv}_X(c)!$$

However,

$$\mu_2(d) = \tfrac{5}{2} < \tfrac{7}{2} = \mu_2(c);$$

i.e., $1 \le \mu_X(d) < \mu_X(c)$, where $\mu_X = \mu_2$ (compare (2.8) and following). We get a codimension 1 presentation of $\mathrm{inv}_X = \mathrm{inv}_{11/2}$ at d by taking

$$N_1(d) = \{w_3 = 0\}, \qquad \mathcal{G}_2(d) = \big\{(D_2(d), 1)\big\}.$$

Therefore,

$$S_{\mathrm{inv}_X}(d) = S_{\mathrm{inv}_1}(d) = \{w_2 = w_3 = 0\},$$

so we let σ_4 be the blowing-up with centre $C_3 = \{w_2 = w_3 = 0\}$. Then $\sigma_4^{-1}(U_{121}) = U_{1212} \cup U_{1213}$, where $U_{121i} = \sigma_4^{-1}(U_{121}) \setminus \{w_i = 0\}'$, $i = 2,3$; U_{1212} has coordinates (v_1, v_2, v_3) in which σ_4 is given by

$$w_1 = v_1, \qquad w_2 = v_2, \qquad w_3 = v_2 v_3.$$

Year four. Let X_4 be the strict transform of X_3. Then $X_4 \cap U_{1212} = V(g_4)$, where $g_4(v) = v_3^2 - v_1 v_2^2$. Let $e = 0$ in U_{1212}. Then $E(e) = \{H_3, H_4\}$, where $H_3 = \{w_1 = 0\}' = \{v_1 = 0\}$ and $H_4 = \sigma_4^{-1}(C_3) = \{v_2 = 0\}$. Again $\nu_1(e) = 2 = \nu_1(a)$, so that $E^1(e) = \varnothing$, $s_1(e) = 0$ and $\mathcal{E}_1(e) = E(e)$. Calculating as above, we obtain $\mu_2(e) = \frac{3}{2}$ and $D_2(e) = v_1^{1/2}v_2$, so that $\nu_2(e) = 0$ and $\mathrm{inv}_X(e) = (2,0;0)$ again. But now $\mu_X(e) = \mu_2(e) = \frac{3}{2}$. Our invariant inv_X is presented at e by

$$N_1(e) = \{v_3 = 0\}, \qquad \mathcal{G}_2(e) = \{(v_1^{1/2}v_2, 1)\}.$$

Therefore, $S_{\mathrm{inv}_X}(e) = \{v_2 = v_3 = 0\}$. Taking as σ_5 the blowing-up with centre $C_4 = S_{\mathrm{inv}_X}(e)$, the strict transform X_5 becomes smooth (over U_{1212}). ($\mu_2(e) - 1 < 1$, so $\nu_1(\cdot)$ must decrease over C_4.)

Further blowings-up are still needed to obtain the stronger assertion of embedded resolution of singularities.

REMARK 3.7. The hypersurface $V(g_4)$ in year four above is called *Whitney's umbrella*. Consider the same hypersurface $X = \{x_3^2 - x_1 x_2^2 = 0\}$ but without a history of blowings-up; i.e., $E(\cdot) = \varnothing$. Let $a = 0$. In this case, $\mathrm{inv}_{11/2}(a) = (2,0;\frac{3}{2})$, and we get a codimension 1 presentation of $\mathrm{inv}_{11/2}$ at a using

$$N_1(a) = \{x_3 = 0\}, \qquad \mathcal{G}_2(a) = \{(x_1 x_2^2, 3)\}$$

or, equivalently, $\mathcal{G}_2(a) = \{(x_1, 1), (x_2, 1)\}$, as in year zero of Example 3.1. Therefore,

$$\mathrm{inv}_X(a) = \big(2,0; \tfrac{3}{2},0; 1,0; \infty\big).$$

As a centre of blowing up we would choose $C = S_{\mathrm{inv}_X}(a) = \{a\}$ — not the x_1-axis as in year four above, although the singularity is the same!

4. Key Properties of the Invariant

Our main goal in this section is to explain why $\mathrm{inv}_X(a)$ is indeed an invariant. Once we establish invariance, the Embedded Desingularization Theorem A follows directly from local properties of inv_X. The crucial properties have already been explained in Section 3 above; we summarize them in the following theorem.

THEOREM B [Bierstone and Milman 1997, Theorem 1.14]. *Consider any sequence of* inv_X*-admissible (local) blowings-up* (1.8). *Then the following properties hold:*

(1) Semicontinuity:

    (i) *For each* j, *every point of* M_j *admits a neighbourhood* U *such that* inv_X *takes only finitely many values in* U *and, for all* $a \in U$, $\{x \in U : \mathrm{inv}_X(x) \leq \mathrm{inv}_X(a)\}$ *is Zariski-open in* U.

    (ii) inv_X *is infinitesimally upper-semicontinuous in the sense that* $\mathrm{inv}_X(a) \leq \mathrm{inv}_X(\sigma_j(a))$ *for all* $a \in M_j$, $j \geq 1$.

(2) Stabilization: *Given* $a_j \in M_j$ *such that* $a_j = \sigma_{j+1}(a_{j+1})$, $j = 0, 1, 2, \ldots$, *there exists* j_0 *such that* $\mathrm{inv}_X(a_j) = \mathrm{inv}_X(a_{j+1})$ *when* $j \geq j_0$. (*In fact, any nonincreasing sequence in the value set of* inv_X *stabilizes.*)

(3) Normal crossings: *Let* $a \in M_j$. *Then* $S_{\mathrm{inv}_X}(a)$ *and* $E(a)$ *simultaneously have only normal crossings. Suppose* $\mathrm{inv}_X(a) = (\ldots; \nu_{t+1}(a))$. *If* $\nu_{t+1}(a) = \infty$, *then* $S_{\mathrm{inv}_X}(a)$ *is smooth. If* $\nu_{t+1}(a) = 0$ *and* Z *denotes any irreducible component of* $S_{\mathrm{inv}_X}(a)$, *then*

$$Z = S_{\mathrm{inv}_X}(a) \cap \bigcap \{H \in E(a) : Z \subset H\}.$$

(4) Decrease: *Let* $a \in M_j$ *and suppose* $\mathrm{inv}_X(a) = (\ldots; \nu_{t+1}(a))$. *If* $\nu_{t+1}(a) = \infty$ *and* σ *is the local blowing-up of* M_j *with centre* $S_{\mathrm{inv}_X}(a)$, *then* $\mathrm{inv}_X(a') < \mathrm{inv}_X(a)$ *for all* $a' \in \sigma^{-1}(a)$. *If* $\nu_{t+1}(a) = 0$, *then there is an additional invariant* $\mu_X(a) = \mu_{t+1}(a) \geq 1$ *such that, if* Z *is an irreducible component of* $S_{\mathrm{inv}_X}(a)$ *and* σ *is the local blowing-up with centre* Z, *then*

$$\big(\mathrm{inv}_X(a'), \mu_X(a')\big) < \big(\mathrm{inv}_X(a), \mu_X(a)\big)$$

for all $a' \in \sigma^{-1}(a)$. (*We have* $e_t! \mu_X(a) \in \mathbb{N}$, *where* e_t *is defined as in Section 1 or in the proof following.*)

PROOF. The semicontinuity property (1)(i) has been explained in Remark 3.5. Infinitesimal upper-semicontinuity (1)(ii) is immediate from the definition of the $s_r(a)$ and from infinitesimal upper-semicontinuity of the order of a function on blowing up locally with smooth centre in its equimultiple locus. (The latter property is an elementary Taylor series computation, and is also clear from the calculation in Section 2 above.)

The stabilization property (2) for $\mathrm{inv}_{1/2}$ is obvious in the hypersurface case because then $\mathrm{inv}_{1/2}(a) = \nu_1(a) \in \mathbb{N}$. (In the general case, we need to begin with stabiization of the Hilbert–Samuel function; see [Bierstone and Milman 1989, Theorem 5.2.1] for a very simple proof of this result due originally to Bennett [1970].) The stabilization assertion for inv_X follows from that for $\text{inv}_{1/2}$ and from infinitesimal semicontinuity because, although $\nu_{r+1}(a)$, for each $r > 0$, is perhaps only rational, our construction in Section 3 shows that $e_r!\nu_{r+1}(a) \in \mathbb{N}$, where $e_1 = \nu_1(a)$ and $e_{r+1} = \max\{e_r!, e_r!\nu_{r+1}(a)\}$, $r > 0$. (In the general case, the Hilbert–Samuel function $H_{X_j,a}(l)$ coincides with a polynomial if $l \geq k$, for k large enough, and we can take as e_1 the least such k.)

The normal crossings condition (3) has also been explained in Section 3; see Remark 3.6, in particular, for the case that $\nu_{t+1}(a) = 0$. The calculation in Section 2 then gives the property of decrease (4), as is evident also in the example of Section 3.   □

When our spaces satisfy a compactness assumption (so that inv_X takes maximum values), it follows from Theorem B that we can obtain the Embedded Desingularization Theorem A by simply applying the algorithm of 1.11 above, stopping when inv_X becomes (locally) constant. To be more precise, let inv_X^e denote the extended invariant for canonical desingularization introduced in Remark 3.6. Consider a sequence of blowings-up (1.8) with inv_X-admissible centres. Note that if X_j is not smooth and $a \in \text{Sing}\,X_j$, then $S_{\text{inv}_X}(a) \subset \text{Sing}\,X_j$ because ν_1 (or, in general, $H_{X_j,a}$) already distinguishes between smooth and singular points. Since $\text{Sing}\,X_j$ is Zariski-closed, it follows that if C_j denotes the locus of maximum values of inv_X^e on $\text{Sing}\,X_j$, then C_j is smooth. By Theorem B, there is a finite sequence of blowings-up with such centres, after which X_j is smooth.

On the other hand, if X_j is smooth and $a \in S_j$, where $S_j = \{x \in X_j : s_1(x) > 0\}$, then $S_{\text{inv}_X}(a) \subset S_j$. Since S_j is Zariski-closed, it follows that if C_j denotes the locus of maximum values of inv_X^e on S_j, then C_j is smooth. Therefore, after finitely many further blowings-up $\sigma_{j+1}, \ldots, \sigma_k$ with such centres, $S_k = \varnothing$. It is clear from the definition of s_1 that, if X_k is smooth and $S_k = \varnothing$, then each $H \in E_k$ which intersects X_k is the strict transform in M_k of $\sigma_{i+1}^{-1}(C_i)$, for some i such that X_i is smooth along C_i; therefore, X_k and E_k simultaneously have only normal crossings, and we have Theorem A.

We will prove invariance of inv_X using the idea of a "presentation" introduced in Section 3 above. It will be convenient to consider "presentation" in an abstract sense, rather than associated to a particular invariant: Let M denote a manifold and let $a \in M$.

DEFINITIONS 4.1. An abstract (*infinitesimal*) *presentation of codimension* p *at* a means simply a triple $(N = N_p(a), \mathcal{H}(a), \mathcal{E}(a))$ as in Section 3; namely: N is a germ of a submanifold of codimension p at a, $\mathcal{H}(a)$ is a finite collection of pairs (h, μ_h), where $h \in \mathcal{O}_{N,a}$, $\mu_h \in \mathbb{Q}$ and $\mu_a(h) \geq \mu_h$, and $\mathcal{E}(a)$ is a finite set of smooth hypersurfaces containing a, such that N and $\mathcal{E}(a)$ simultaneously have normal crossings and $N \not\subset H$, for all $H \in \mathcal{E}(a)$.

A local blowing-up σ with centre $C \ni a$ will be called *admissible* (for an infinitesimal presentation as above) if $C \subset S_{\mathcal{H}(a)} = \{x \in N : \mu_x(h) \geq \mu_h$, for all $(h, \mu_h) \in \mathcal{H}(a)\}$.

DEFINITION 4.2. We will say that two infinitesimal presentations $(N = N_p(a)$, $\mathcal{H}(a), \mathcal{E}(a))$ and $(P = P_q(a), \mathcal{F}(a), \mathcal{E}(a))$ with given $\mathcal{E}(a)$, but not necessarily of the same codimension, are *equivalent* if (in analogy with Definition 3.2):

(1) $S_{\mathcal{H}(a)} = S_{\mathcal{F}(a)}$, as germs at a in M.

(2) If σ is an admissible local blowing-up and $a' \in \sigma^{-1}(a)$, then $a' \in N'$ and $\mu_{a'}(y_{\mathrm{exc}}^{-\mu_h} h \circ \sigma) \geq \mu_h$ for all $(h, \mu_h) \in \mathcal{H}(a)$ if and only if $a' \in P'$ and

$$\mu_{a'}(y_{\mathrm{exc}}^{-\mu_f} f \circ \sigma) \geq \mu_f \qquad \text{for all } (f, \mu_f) \in \mathcal{F}(a).$$

(3) Conditions (1) and (2) continue to hold for the transforms $(N_p(a'), \mathcal{H}(a'), \mathcal{E}(a'))$ and $(P_q(a'), \mathcal{F}(a'), \mathcal{E}(a'))$ of our data by sequences of morphisms of types (i), (ii) and (iii) as in Definition 3.2.

We will, in fact, impose a further condition on the way that exceptional blowings-up (iii) are allowed to occur in a sequence of transformations in condition (3) above; see Definition 4.5 below.

Our proof of invariance of inv_X follows the constructive definition outlined in Section 3. Let X denote a hypersurface in M, and consider any sequence of blowings-up (or local blowings-up) (1.8), where we assume (at first) that the centres of blowing up are $\frac{1}{2}$-admissible. Let $a \in M_j$, for some $j = 0, 1, 2, \ldots$. Suppose that $g \in \mathcal{O}_{M_j, a}$ generates the local ideal $\mathcal{I}_{X_j, a}$ of X_j at a, and let $\mu_g = \mu_a(g)$. Then, as in Section 3, $\mathcal{G}_1(a) = \{(g, \mu_g)\}$ determines a codimension zero presentation $(N_0(a), \mathcal{G}_1(a), \mathcal{E}_0(a))$ of $\mathrm{inv}_{1/2} = \nu_1$ at a, where $N_0(a)$ is the germ of M_j at a, and $\mathcal{E}_0(a) = \varnothing$. In particular, the equivalence class of $(N_0(a), \mathcal{G}_1(a), \mathcal{E}_0(a))$ in the sense of Definition 4.2 depends only on the local isomorphism class of (M_j, X_j) at a.

We introduce $E^1(a)$ as in 1.12 above, and let $s_1(a) = \#E^1(a)$, $\mathcal{E}_1(a) = E(a) \backslash E^1(a)$. Let

$$\mathcal{F}_1(a) = \mathcal{G}_1(a) \cup (E^1(a), 1),$$

where $(E^1(a), 1)$ denotes $\{(x_H, 1) : H \in E^1(a)\}$ and x_H means a local generator of the ideal of H. Then $(N_0(a), \mathcal{F}_1(a), \mathcal{E}_1(a))$ is a codimension zero presentation of $\mathrm{inv}_1 = (\nu_1, s_1)$ at a. Clearly, the equivalence class of $(N_0(a), \mathcal{F}_1(a), \mathcal{E}_1(a))$ depends only on the local isomorphism class of $(M_j, X_j, E_j, E^1(a))$. Moreover, $(N_0(a), \mathcal{F}_1(a), \mathcal{E}_1(a))$ has an equivalent codimension one presentation $(N_1(a), \mathcal{H}_1(a), \mathcal{E}_1(a))$ as described in Section 3. For example, let $a_k = \pi_{kj}(a)$, for $k = 0, \ldots, j$, as in 1.12, and let i denote the "earliest year" k such that $\mathrm{inv}_{1/2}(a) = \mathrm{inv}_{1/2}(a_k)$. Then $\mathcal{E}_1(a_i) = \varnothing$. As in Section 3, we can take $N_1(a_i) =$ any hypersurface of maximal contact for X_i at a_i. If $(x_1, \ldots, x_n)$ are local coordinates for M_i with respect to which $N_1(a_i) = \{x_n = 0\}$, then we can take

$$\mathcal{H}_1(a_i) = \left\{ \left(\left. \frac{\partial^q f}{\partial x_n^q} \right|_{N_1(a_i)}, \mu_f - q \right) : 0 \le q < \mu_f, \ (f, \mu_f) \in \mathcal{F}_1(a_i) \right\}.$$

A codimension one presentation $(N_1(a), \mathcal{H}_1(a), \mathcal{E}_1(a))$ of inv_1 at a can be obtained by transforming $(N_1(a_i), \mathcal{H}_1(a_i), \mathcal{E}_1(a_i))$ to a. The condition that $N_1(a)$ and $\mathcal{E}_1(a)$ simultaneously have normal crossings and $N_1(a) \not\subset H$ for all $H \in \mathcal{E}_1(a)$ is a consequence of the effect of blowing with smooth centre of codimension at least 1 in $N(a_k)$, $i \le k < j$ (as in the calculation in Section 2).

Say that $\mathcal{H}_1(a) = \{(h, \mu_h)\}$; each $h \in \mathcal{O}_{N_1(a),a}$ and $\mu_h \le \mu_a(h)$. Recall that we define

$$\mu_2(a) = \min_{\mathcal{H}_1(a)} \frac{\mu_a(h)}{\mu_h},$$

$$\mu_{2H}(a) = \min_{\mathcal{H}_1(a)} \frac{\mu_{H,a}(h)}{\mu_h}, \qquad H \in \mathcal{E}_1(a),$$

and $\quad \nu_2(a) = \mu_2(a) - \sum_{H \in \mathcal{E}_1(a)} \mu_{2H}(a)$

(Definitions 3.2, 3.4). Propositions 4.4 and 4.6 below show that each of $\mu_2(a)$ and $\mu_{2H}(a)$, $H \in \mathcal{E}_1(a)$, depends only on the equivalence class of $(N_1(a), \mathcal{H}_1(a), \mathcal{E}_1(a))$, and thus only on the local isomorphism class of $(M_j, X_j, E_j, E^1(a))$.

If $\nu_2(a) = 0$ or ∞, then we set $\mathrm{inv}_X(a) = \mathrm{inv}_{11/2}(a)$. If $0 < \nu_2(a) < \infty$, then we construct a codimension one presentation $(N_1(a), \mathcal{G}_2(a), \mathcal{E}_1(a))$ of $\mathrm{inv}_{11/2}$ at a, as in Section 3. From the construction, it is not hard to see that the equivalence class of $(N_1(a), \mathcal{G}_2(a), \mathcal{E}_1(a))$ depends only on that of $(N_1(a), \mathcal{H}_1(a), \mathcal{E}_1(a))$. (See [Bierstone and Milman 1997, 4.23 and 4.24] as well as Proposition 4.6 ff. below.)

This completes a cycle in the inductive definition of inv_X. Assume now that the centres of the blowings-up in (1.8) are $1\frac{1}{2}$-admissible. We introduce $E^2(a)$ as in 1.12, and let $s_2(a) = \#E^2(a)$, $\mathcal{E}_2(a) = \mathcal{E}_1(a) \backslash E^2(a)$. If $\mathcal{F}_2(a) = \mathcal{G}_2(a) \cup (E^2(a), 1)$, where $(E^2(a), 1)$ denotes $\{(x_H|_{N_1(a)}, 1) : H \in E_2(a)\}$, then $(N_1(a), \mathcal{F}_2(a), \mathcal{E}_2(a))$ is a codimension one presentation of $\mathrm{inv}_2 = (\mathrm{inv}_{11/2}, s_2)$ at a, whose equivalence class depends only on the local isomorphism class of $(M_j, X_j, E_j, E^1(a), E^2(a))$. It is clear from the construction of $\mathcal{G}_2(a)$ that $\mu_{\mathcal{G}_2(a)} = 1$, where

$$\mu_{\mathcal{G}_2(a)} = \min_{(g, \mu_g) \in \mathcal{G}_2(a)} \frac{\mu_a(g)}{\mu_g}.$$

Therefore $(N_1(a), \mathcal{F}_2(a), \mathcal{E}_2(a))$ admits an equivalent codimension two presentation $(N_2(a), \mathcal{H}_2(a), \mathcal{E}_2(a))$, and we define $\nu_3(a) = \mu_3(a) - \sum_{H \in \mathcal{E}_2(a)} \mu_{3H}(a)$, as above. By Propositions 4.4 and 4.6, $\mu_3(a)$ and each $\mu_{3H}(a)$ depend only on the equivalence class of $(N_2(a), \mathcal{H}_2(a), \mathcal{E}_2(a))$, ... We continue until $\nu_{t+1}(a) = 0$ or ∞ for some t, and then take $\mathrm{inv}_X(a) = \mathrm{inv}_{t+1/2}(a)$.

Invariance of inv_X thus follows from Propositions 4.4 and 4.6 below, which are formulated purely in terms of an abstract infinitesimal presentation.

Let M be a manifold, and let $(N(a), \mathcal{H}(a), \mathcal{E}(a))$ be an infinitesimal presentation of codimension $r \geq 0$ at a point $a \in M$. We write $\mathcal{H}(a) = \{(h, \mu_h)\}$, where $\mu_a(h) \geq \mu_h$ for all (h, μ_h).

DEFINITIONS 4.3. We define $\mu(a) = \mu_{\mathcal{H}(a)}$ as

$$\mu_{\mathcal{H}(a)} = \min_{\mathcal{H}(a)} \frac{\mu_a(h)}{\mu_h}.$$

Thus $1 \leq \mu(a) \leq \infty$. If $\mu(a) < \infty$, then we define $\mu_H(a) = \mu_{\mathcal{H}(a),H}$, for each $H \in \mathcal{E}(a)$, as

$$\mu_{\mathcal{H}(a),H} = \min_{\mathcal{H}(a)} \frac{\mu_{H,a}(h)}{\mu_h}.$$

We will show that each of $\mu(a)$ and the $\mu_H(a)$ depends only on the equivalence class of $(N(a), \mathcal{H}(a), \mathcal{E}(a))$ (where we consider only *presentations of the same codimension r*). The main point is that $\mu(a)$ and the $\mu_H(a)$ can be detected by "test blowings-up" (test transformations of the form (i), (ii), (iii) as allowed by the definition 4.2 of equivalence).

For $\mu(a)$, we show in fact that if $(N^i(a), \mathcal{H}^i(a), \mathcal{E}(a))$, for $i = 1, 2$, are two infinitesimal presentations of the same codimension r, then $\mu_{\mathcal{H}^1(a)} = \mu_{\mathcal{H}^2(a)}$ if the presentations are equivalent with respect to transformations of types (i) and (ii) alone (i.e., where we allow only transformations of types (i) and (ii) in Definition 4.2). This ia a stronger condition than invariance under equivalence in the sense of Definition 4.2 (using all three types of transformations) because the equivalence class with respect to transformations of types (i) and (ii) alone is, of course, larger than the equivalence class with respect to transformations of all three types (i), (ii) and (iii).

PROPOSITION 4.4. [Bierstone and Milman 1997, Proposition 4.8]. $\mu(a)$ *depends only on the equivalence class of* $(N(a), \mathcal{H}(a), \mathcal{E}(a))$ *(among presentations of the same codimension r) with respect to transformations of types (i) and (ii).*

PROOF. Clearly, $\mu(a) = \infty$ if and only if $S_{\mathcal{H}(a)} = N(a)$; i.e., if and only if $S_{\mathcal{H}(a)}$ is (a germ of) a submanifold of codimension r in M.

Suppose that $\mu(a) < \infty$. We can assume that $\mathcal{H}(a) = \{(h, \mu_h)\}$ where all $\mu_h = e$, for some $e \in \mathbb{N}$. Let $\sigma_0 \colon P_0 = M \times \mathbb{K} \to M$ be the projection from the product with a line (i.e., a morphism of type (ii)) and let $(N(c_0), \mathcal{H}(c_0), \mathcal{E}(c_0))$ denote the transform of $(N(a), \mathcal{H}(a), \mathcal{E}(a))$ at $c_0 = (a, 0) \in P_0$; i.e., $N(c_0) = N(a) \times \mathbb{K}$, $\mathcal{E}(c_0) = \{H \times \mathbb{K}, \text{ for all } H \in \mathcal{E}(a), \text{ and } M \times \{0\}\}$ and $\mathcal{H}(c_0) = \{(h \circ \sigma_0, \mu_h) \colon (h, \mu_h) \in \mathcal{H}(a)\}$. We follow σ_0 by a sequence of admissible blowings-up (morphisms of type (i)),

$$\longrightarrow P_{\beta+1} \xrightarrow{\sigma_{\beta+1}} P_\beta \longrightarrow \cdots \longrightarrow P_1 \xrightarrow{\sigma_1} P_0,$$

where each $\sigma_{\beta+1}$ is a blowing-up with centre a point $c_\beta \in P_\beta$ determined as follows: Let γ_0 denote the arc in P_0 given by $\gamma_0(t) = (a, t)$. For $\beta \geq 1$, define $\gamma_{\beta+1}$ inductively as the lifting of γ_β to $P_{\beta+1}$, and set $c_{\beta+1} = \gamma_{\beta+1}(0)$.

We can choose local coordinates $(x_1, \ldots, x_n)$ for M at a, in which $a = 0$ and $N(a) = \{x_{n-r+1} = \cdots = x_n = 0\}$. Write $(x, t) = (x_1, \ldots, x_{n-r}, t)$ for the corresponding coordinate system of $N(c_0)$. In P_1, the strict transform $N(c_1)$ of $N(c_0)$ has a local coordinate system $(x, t) = (x_1, \ldots, x_{n-r}, t)$ at c_1 with respect to which $\sigma_1(x, t) = (tx, t)$, and $\gamma_1(t) = (0, t)$ in this coordinate chart; moreover, $\mathcal{H}(c_1) = \{(t^{-e}h(tx), e), \text{ for all } (h, \mu_h) = (h, e) \in \mathcal{H}(a)\}$. After β blowings-up as above, $N(c_\beta)$ has a local coordinate system $(x, t) = (x_1, \ldots, x_{n-r}, t)$ with respect to which $\sigma_1 \circ \cdots \circ \sigma_\beta$ is given by $(x, t) \mapsto (t^\beta x, t)$, $\gamma_\beta(t) = (0, t)$ and $\mathcal{H}(c_\beta) = \{(h', \mu_{h'} = e)\}$, where

$$h' = t^{-\beta e} h(t^\beta x),$$

for all $(h, \mu_h) = (h, e) \in \mathcal{H}(a)$. By the definition of $\mu(a)$, each

$$h(t^\beta x) = t^{\beta \mu(a) e} \tilde{h}'(x, t),$$

where the $\tilde{h}'(x, t)$ do not admit t as a common divisor; for each $(h, \mu_h) \in \mathcal{H}(a)$, we have

$$h' = t^{\beta(\mu(a)-1)e} \tilde{h}'.$$

We now introduce a subset S of $\mathbb{N} \times \mathbb{N}$ depending only on the equivalence class of $(N(a), \mathcal{H}(a), \mathcal{E}(a))$ (with respect to transformations of types (i) and (ii)) as follows: First, we say that $(\beta, 0) \in S$, for $\beta \geq 1$, if after β blowings-up as above, there exists (a germ of) a submanifold W_0 of codimension r in the exceptional hypersurface $H_\beta = \sigma_\beta^{-1}(c_{\beta-1})$ such that $W_0 \subset S_{\mathcal{H}(c_\beta)}$. If so, then necessarily $W_0 = H_\beta \cap N(c_\beta) = \{t = 0\}$, and the condition that $W_0 \subset S_{\mathcal{H}(c_\beta)}$ means precisely that $\mu_{W_0, c_\beta}(h') \geq e$, for all h'; i.e., that $\beta(\mu(a) - 1)e \geq e$, or $\beta(\mu(a) - 1) \geq 1$. (In particular, since $\mu(a) \geq 1$, $(\beta, 0) \notin S$ for all $\beta \geq 1$ if and only if $\mu(a) = 1$.)

Suppose that $(\beta, 0) \in S$, for some $\beta \geq 1$, as above. Then we can blow up P_β locally with centre W_0. Set $Q_0 = P_\beta$, $d_0 = c_\beta$ and $\delta_0 = \gamma_\beta$. Let τ_1: $Q_1 \to Q_0$ denote the local blowing-up with centre W_0, and let $d_1 = \delta_1(0)$, where δ_1 denotes the lifting of δ_0 to Q_1. (Then $\tau_1|N(d_1)$: $N(d_1) \to N(d_0)$ is the identity.) We say that $(\beta, 1) \in S$ if there exists a submanifold W_1 of codimension r in the hypersurface $H_1 = \tau_1^{-1}(W_0)$ such that $W_1 \subset S_{\mathcal{H}(d_1)}$. If so, then again necessarily $W_0 = H_1 \cap N(d_1) = \{t = 0\}$. Since $\mathcal{H}(d_1) = \{(h', e)\}$, where each $h' = t^{\beta(\mu(a)-1)e - e} \tilde{h}'$ and the $\tilde{h}'$ do not admit t as a common factor, it follows that $(\beta, 1) \in S$ if and only if $\beta(\mu(a) - 1)e - e \geq e$.

We continue inductively: If $\alpha \geq 1$ and $(\beta, \alpha - 1) \in S$, let τ_α: $Q_\alpha \to Q_{\alpha-1}$ denote the local blowing-up with centre $W_{\alpha-1}$, and let $d_\alpha = \delta_\alpha(0)$, where δ_α is the lifting of $\delta_{\alpha-1}$ to Q_α. We say that $(\beta, \alpha) \in S$ if there exists (a germ of) a submanifold W_α of codimension r in the exceptional hypersurface $H_\alpha =$

$\tau_\alpha^{-1}(W_{\alpha-1})$ such that $W_\alpha \subset S_{\mathcal{H}(d_\alpha)}$. Since $\mathcal{H}(d_\alpha) = \{(h', e)\}$, where each $h' = t^{\beta(\mu(a)-1)e - \alpha e}\tilde{h}'$ and the $\tilde{h}'$ do not admit t as a common factor, it follows as before that $(\beta, \alpha) \in S$ if and only if $\beta(\mu(a) - 1) - \alpha \geq 1$.

Now S, by its definition, depends only on the equivalence class of $(N(a), \mathcal{H}(a), \mathcal{E}(a))$ (with respect to transformations of types (i) and (ii)). On the other hand, we have proved that $S = \varnothing$ if and only if $\mu(a) = 1$, and, if $S \neq \varnothing$, then

$$S = \{(\beta, \alpha) \in \mathbb{N} \times \mathbb{N} : \beta(\mu(a) - 1) - \alpha \geq 1\}.$$

Our proposition follows since $\mu(a)$ is uniquely determined by S; in the case that $S \neq \varnothing$,

$$\mu(a) = 1 + \sup_{(\beta, \alpha) \in S} \frac{\alpha + 1}{\beta}. \qquad \square$$

Suppose that $\mu(a) < \infty$. Then we can also use test blowings-up to prove invariance of $\mu_H(a) = \mu_{\mathcal{H}(a), H}$, $H \in \mathcal{E}(a)$: Fix $H \in \mathcal{E}(a)$. As before we begin with the projection $\sigma_0 \colon P_0 = M \times \mathbb{K} \to M$ from the product with a line. Let $(N(a_0), \mathcal{H}(a_0), \mathcal{E}(a_0))$ denote the transform of $(N(a), \mathcal{H}(a), \mathcal{E}(a))$ at $a_0 = (a, 0) \in P_0$ by the morphism σ_0 (of type (ii)), and let $H_0^0 = M \times \{0\}$, $H_1^0 = \sigma_0^{-1}(H) = H \times \mathbb{K}$. Thus $H_0^0, H_1^0 \in \mathcal{E}(a_0)$. We follow σ_0 by a sequence of exceptional blowings-up (morphisms of type (iii)),

$$\longrightarrow P_{j+1} \xrightarrow{\sigma_{j+1}} P_j \longrightarrow \cdots \longrightarrow P_1 \xrightarrow{\sigma_1} P_0,$$

where each σ_{j+1}, for $j \geq 0$, has centre $C_j = H_0^j \cap H_1^j$ and $H_0^{j+1} = \sigma_{j+1}^{-1}(C_j)$, $H_1^{j+1} =$ the strict transform of H_1^j by σ_{j+1}. Let a_{j+1} denote the unique intersection point of C_{j+1} and $\sigma_{j+1}^{-1}(a_j)$, for $j \geq 0$. (Thus $a_{j+1} = \gamma_{j+1}(0)$, where γ_0 denotes the arc $\gamma_0(t) = (a, t)$ in P_0 and γ_{j+1} denotes the lifting of γ_j by σ_{j+1}, for $j \geq 0$.)

We can choose local coordinates $(x_1, \ldots, x_n)$ for M at a, in which $a = 0$, $N(a) = \{x_{n-r+1} = \cdots = x_n = 0\}$, and each $K \in \mathcal{E}(a)$ is given by $x_i = 0$, for some $i = 1, \ldots, n - r$. (Set $x_i = x_K$.) Write $(x, t) = (x_1, \ldots, x_m, t)$, where $m = n - r$, for the corresponding coordinate system of $N(a_0) = N(a) \times \mathbb{K}$.

We can assume that $x_H = x_1$. In P_1, the strict transform $N(a_1)$ of $N(a_0)$ has a chart with coordinates $(x, t) = (x_1, \ldots, x_m, t)$ in which σ_1 is given by $\sigma_1(x, t) = (tx_1, x_2, \ldots, x_m, t)$ and in which $a_1 = (0, 0)$, $\gamma_1(t) = (0, t)$ and $x_1 = x_H$. (x_H now means $x_{H_1^1}$.) Proceeding inductively, for each j, $N(a_j)$ has a coordinate system $(x, t) = (x_1, \ldots, x_m, t)$ in which $a_j = (0, 0)$ and $\sigma_1 \circ \cdots \circ \sigma_j \colon N(a_j) \to N(a_0)$ is given by

$$(x, t) \mapsto (t^j x_1, x_2, \ldots, x_m, t).$$

We can assume that $\mu_h = e \in \mathbb{N}$, for all $(h, \mu_h) \in \mathcal{H}(a)$. Set

$$D = \prod_{K \in \mathcal{E}(a)} x_K^{\mu_K(a)}.$$

Thus D^e is a monomial in the coordinates $(x_1, \ldots, x_m)$ of $N(a)$ with exponents in $\mathbb{N}$, and D^e is the greatest common divisor of the h in $\mathcal{H}(a)$ which is a monomial in x_K, $K \in \mathcal{E}(a)$ (by Definitions 4.3). In particular, for some $h = D^e g$ in $\mathcal{H}(a)$, $g = g_H$ is not divisible by $x_1 = x_H$. Therefore, there exists $i \geq 1$ such that

$$\mu_{a_j}(g_H \circ \pi_j) = \mu_{a_i}(g_H \circ \pi_i),$$

for all $j \geq i$, where $\pi_j = \sigma_0 \circ \sigma_1 \circ \cdots \circ \sigma_j$. (We can simply take i to be the least order of a monomial not involving x_H in the Taylor expansion of g_H.)

On the other hand, for each $h = D^e g$ in $\mathcal{H}(a)$, $\mu_{a_j}(g \circ \pi_j)$ increases as $j \to \infty$ unless g is not divisible by x_H. Therefore, we can choose $h = D^e g_H$, as above, and i large enough so that we also have $\mu(a_j) = \mu_{a_j}(h \circ \pi_j)/e$, for all $j \geq i$. Clearly, if $j \geq i$, then

$$\mu_H(a) = \mu(a_{j+1}) - \mu(a_j).$$

Since $\mu(a)$ depends only on the equivalence class of $(N(a), \mathcal{H}(a), \mathcal{E}(a))$ among presentations of the same codimension r, as defined by 4.2, the preceding argument shows that each $\mu_H(a)$, $H \in \mathcal{E}(a)$, is also an invariant of this equivalence class. But the argument shows more precisely that the $\mu_H(a)$ depend only on a larger equivalence class obtained by allowing in Definition 4.2 only certain sequences of morphisms of types (i), (ii) and (iii):

DEFINITION 4.5.   We weaken the notion of equivalence in Definition 4.2 by allowing only the transforms induced by certain sequences of morphisms of types (i), (ii) and (iii); namely,

$$\xrightarrow{\quad} \begin{matrix} M_j \\ \mathcal{E}(a_j) \end{matrix} \xrightarrow{\sigma_j} \begin{matrix} M_{j-1} \\ \mathcal{E}(a_{j-1}) \end{matrix} \xrightarrow{\quad} \cdots \xrightarrow{\sigma_{i+1}} \begin{matrix} M_i \\ \mathcal{E}(a_i) \end{matrix} \xrightarrow{\quad} \cdots \xrightarrow{\quad} \begin{matrix} M_0 = M \\ \mathcal{E}(a_0) = \mathcal{E}(a) \end{matrix}$$

where, if $\sigma_{i+1}, \ldots, \sigma_j$ are exceptional blowings-up (iii), then $i \geq 1$ and σ_i is of either type (iii) or (ii). In the latter case, $\sigma_i \colon M_i = M_{i-1} \times \mathbb{K} \to M_{i-1}$ is the projection, each σ_{k+1}, $k = i, \ldots, j-1$, is the blowing-up with centre $C_k = H_0^k \cap H_1^k$ where $H_0^k, H_1^k \in \mathcal{E}(a_k)$, $a_{k+1} = \sigma_{k+1}^{-1}(a_k) \cap H_1^{k+1}$, and we require that the H_0^k, H_1^k be determined by some fixed $H \in \mathcal{E}(a_{i-1})$ inductively in the following way: $H_0^i = M_{i-1} \times \{0\}$, $H_1^i = \sigma_i^{-1}(H)$, and, for $k = i+1, \ldots, j-1$, $H_0^k = \sigma_k^{-1}(C_{k-1})$, $H_1^k = $ the strict transform of H_1^{k-1} by σ_k.

In other words, with this notion of equivalence, we have proved:

PROPOSITION 4.6. [Bierstone and Milman 1997, Proposition 4.11]. *Each $\mu_H(a)$, $H \in \mathcal{E}(a)$, and therefore also $\nu(a) = \mu(a) - \Sigma \mu_H(a)$ depends only on the equivalence class of $(N(a), \mathcal{H}(a), \mathcal{E}(a))$ (among presentations of the same codimension).*

Recall that in the r-th cycle of our recursive definition of inv_X, we use a codimension r presentation $(N_r(a), \mathcal{H}_r(a), \mathcal{E}_r(a))$ of inv_r at a to construct a codimension r presentation $(N_r(a), \mathcal{G}_{r+1}(a), \mathcal{E}_r(a))$ of $\mathrm{inv}_{r+1/2}$ at a. The construction involved survives transformations as allowed by Definition 4.5, but perhaps not an arbitrary sequence of transformations of types (i), (ii) and (iii) (compare [Bierstone and Milman 1997, 4.23 and 4.24]; in other words, we show only that the equivalence class of $(N_r(a), \mathcal{G}_{r+1}(a), \mathcal{E}_r(a))$ as given by Definition 4.5 depends only on that of $(N_r(a), \mathcal{H}_r(a), \mathcal{E}_r(a))$. It is for this reason that we need Proposition 4.6 as stated.

Acknowledgement

We are happy to thank Paul Centore for the line drawings in this paper.

References

[Abhyankar 1966] S. S. Abhyankar, *Resolution of singularities of embedded algebraic surfaces*, Pure and Applied Mathematics **24**, Academic Press, New York, 1966.

[Abhyankar 1982] S. S. Abhyankar, *Weighted expansions for canonical desingularization*, Lecture Notes in Math. **910**, Springer, Berlin, 1982.

[Abhyankar 1988] S. S. Abhyankar, "Good points of a hypersurface", *Adv. in Math.* **68** (1988), 87–256.

[Aroca et al. 1975] J. M. Aroca, H. Hironaka, and J. L. Vicente, *The theory of the maximal contact*, Mem. Mat. Instituto "Jorge Juan" **29**, Consejo Superior de Investigaciones Científicas, Madrid, 1975.

[Aroca et al. 1977] J. M. Aroca, H. Hironaka, and J. L. Vicente, *Desingularization theorems*, Mem. Mat. Instituto "Jorge Juan" **30**, Consejo Superior de Investigaciones Científicas, Madrid, 1977.

[Bennett 1970] B. M. Bennett, "On the characteristic functions of a local ring", *Ann. of Math.* (2) **91** (1970), 25–87.

[Bierstone and Milman 1988] E. Bierstone and P. D. Milman, "Semianalytic and subanalytic sets", *Inst. Hautes Études Sci. Publ. Math.* **67** (1988), 5–42.

[Bierstone and Milman 1989] E. Bierstone and P. D. Milman, "Uniformization of analytic spaces", *J. Amer. Math. Soc.* **2** (1989), 801–836.

[Bierstone and Milman 1990] E. Bierstone and P. D. Milman, "Arc-analytic functions", *Invent. Math.* **101** (1990), 411–424.

[Bierstone and Milman 1991] E. Bierstone and P. D. Milman, "A simple constructive proof of canonical resolution of singularities", pp. 11–30 in *Effective methods in algebraic geometry* (Castiglioncello, 1990), edited by T. Mora and C. Traverso, Progress in Math. **94**, Birkhäuser, Boston, 1991.

[Bierstone and Milman 1997] E. Bierstone and P. D. Milman, "Canonical desingularization in characteristic zero by blowing up the maximum strata of a local invariant", *Invent. Math.* **128** (1997), 207–302.

[Brill and Noether 1892–93] A. Brill and M. Noether, "Die Entwicklung der Theorie der algebraischen Funktionen in älterer und neuerer Zeit", *Jahresber. der Deutsche Math. Verein.* **3** (1892–93), 111–566.

[Encinas and Villamayor 1998] S. Encinas and O. Villamayor, "Good points and constructive resolution of singularities", *Acta Math.* **181** (1998), 109–158.

[Giraud 1974] J. Giraud, "Sur la théorie du contact maximal", *Math. Z.* **137** (1974), 285–310.

[Hironaka 1964] H. Hironaka, "Resolution of singularities of an algebraic variety over a field of characteristic zero", *Ann. of Math.* (2) **79** (1964), 109–203, 205–326.

[Hironaka 1974] H. Hironaka, *Introduction to the theory of infinitely near singular points*, Mem. Mat. Instituto "Jorge Juan" **28**, Consejo Superior de Investigaciones Científicas, Madrid, 1974.

[Hironaka 1977] H. Hironaka, "Idealistic exponents of singularity", pp. 52–125 in *Algebraic geometry: the Johns Hopkins centennial lectures* (Baltimore, 1976), edited by J.-I. Igusa, Johns Hopkins Univ. Press, Baltimore, MD, 1977.

[de Jong 1996] A. J. de Jong, "Smoothness, semi-stability and alterations", *Inst. Hautes Études Sci. Publ. Math.* **83** (1996), 51–93.

[Jung 1908] H. W. E. Jung, "Darstellung der Funktionen eines algebraischen Körpers zweier unabhängigen Veränderlichen x, y in der Umgebung einer stelle $x = a, y = b$", *J. Reine Angew. Math.* **133** (1908), 289–314.

[Lipman 1975] J. Lipman, "Introduction to resolution of singularities", pp. 187–230 in *Algebraic geometry* (Arcata, CA, 1974), edited by R. Hartshorne, Proc. Sympos. Pure Math. **29**, Amer. Math. Soc., Providence, 1975.

[Moh 1992] T. T. Moh, "Quasi-canonical uniformization of hypersurface singularities of characteristic zero", *Comm. Algebra* **20** (1992), 3207–3249.

[Villamayor 1989] O. Villamayor, "Constructiveness of Hironaka's resolution", *Ann. Sci. École Norm. Sup.* (4) **22** (1989), 1–32.

[Villamayor 1992] O. Villamayor, "Patching local uniformizations", *Ann. Sci. École Norm. Sup.* (4) **25** (1992), 629–677.

[Walker 1935] R. J. Walker, "Reduction of the singularities of an algebraic surface", *Ann. of Math.* **36** (1935), 336–365.

[Youssin 1990] B. Youssin, *Newton polyhedra without coordinates. Newton polyhedra of ideals*, Mem. Amer. Math. Soc. **433**, Amer. Math. Soc., Providence, 1990.

[Zariski 1939] O. Zariski, "The reduction of the singularities of an algebraic surface", *Ann. of Math.* **40** (1939), 639–689.

[Zariski 1943] O. Zariski, "Foundations of a general theory of birational correspondences", *Trans. Amer. Math. Soc.* **53** (1943), 490–542.

[Zariski 1944] O. Zariski, "Reduction of the singularities of algebraic three dimensional varieties", *Ann. of Math.* (2) **45** (1944), 472–542.

EDWARD BIERSTONE
DEPARTMENT OF MATHEMATICS
UNIVERSITY OF TORONTO
TORONTO, ONTARIO M5S 3G3
CANADA
bierston@math.toronto.edu

PIERRE D. MILMAN
DEPARTMENT OF MATHEMATICS
UNIVERSITY OF TORONTO
TORONTO, ONTARIO M5S 3G3
CANADA
milman@math.toronto.edu

Several Complex Variables
MSRI Publications
Volume **37**, 1999

Global Regularity of the $\bar{\partial}$-Neumann Problem:
A Survey of the L^2-Sobolev Theory

HAROLD P. BOAS AND EMIL J. STRAUBE

ABSTRACT. The fundamental boundary value problem in the function theory of several complex variables is the $\bar{\partial}$-Neumann problem. The L^2 existence theory on bounded pseudoconvex domains and the C^∞ regularity of solutions up to the boundary on smooth, bounded, strongly pseudoconvex domains were proved in the 1960s. On the other hand, it was discovered quite recently that global regularity up to the boundary fails in some smooth, bounded, weakly pseudoconvex domains.

We survey the global regularity theory of the $\bar{\partial}$-Neumann problem in the setting of L^2 Sobolev spaces on bounded pseudoconvex domains, beginning with the classical results and continuing up to the frontiers of current research. We also briefly discuss the related global regularity theory of the Bergman projection.

CONTENTS

1. Introduction

The $\bar{\partial}$-Neumann problem is a natural example of a boundary-value problem with an elliptic operator but with non-coercive boundary conditions. It is also a prototype (in the case of finite-type domains) of a subelliptic boundary-value

1991 *Mathematics Subject Classification.* 32F20, 32F15, 32H10, 35N15.

Both authors were partially supported by NSF grant DMS-9500916 and, at the Mathematical Sciences Research Institute, by NSF grant DMS-9022140.

problem, in much the same way that the Dirichlet problem is the archetypal elliptic boundary-value problem. In this survey, we discuss global regularity of the $\bar{\partial}$-Neumann problem in the L^2-Sobolev spaces $W^s(\Omega)$ for all non-negative s and also in the space $C^\infty(\bar{\Omega})$. For estimates in other function spaces, such as Hölder spaces and L^p-Sobolev spaces, see [Beals et al. 1987; Berndtsson 1994; Chang et al. 1992; Cho 1995; Christ 1991; Fefferman 1995; Fefferman and Kohn 1988; Fefferman et al. 1990; Greiner and Stein 1977; Kerzman 1971; Krantz 1979; Lieb 1993; McNeal 1991; McNeal and Stein 1994; Nagel et al. 1989; Straube 1995]; for questions of real analytic regularity, see, for example, [Chen 1988; Christ 1996b; Derridj and Tartakoff 1976; Komatsu 1976; Tartakoff 1978; 1980; Tolli 1996; Treves 1978] and [Christ 1999, Section 10] in this volume.

We also discuss the closely related question of global regularity of the Bergman projection operator. This question is intimately connected with the boundary regularity of holomorphic mappings; see, for example, [Bedford 1984; Bell 1981; 1984; 1990; Bell and Ligocka 1980; Bell and Catlin 1982; Diederich and Fornæss 1982; Forstnerič 1993].

For an overview of techniques of partial differential equations in complex analysis, see [Folland and Kohn 1972; Hörmander 1965; 1990; 1994; Kohn 1977; Krantz 1992].

2. The L^2 Existence Theory

Throughout the paper, Ω denotes a bounded domain in $\mathbb{C}^n$, where $n > 1$. We say that Ω has class C^k boundary if $\Omega = \{z : \rho(z) < 0\}$, where ρ is a k times continuously differentiable real-valued function in a neighborhood of the closure $\bar{\Omega}$ whose gradient is normalized to length 1 on the boundary $b\Omega$. We denote the standard L^2-Sobolev space of order s by $W^s(\Omega)$; see, for example, [Adams 1975; Lions and Magenes 1972; Treves 1975]. The space of $(0, q)$ forms with coefficients in $W^s(\Omega)$ is written $W^s_{(0,q)}(\Omega)$, the norm being defined by

$$\left\| {\sum_J}' a_J \, d\bar{z}_J \right\|_s^2 = {\sum_J}' \|a_J\|_s^2, \qquad (2\text{--}1)$$

where $d\bar{z}_J$ means $d\bar{z}_{j_1} \wedge d\bar{z}_{j_2} \wedge \cdots \wedge d\bar{z}_{j_q}$, and the prime indicates that the sum is taken over strictly increasing q-tuples J. We will consider the coefficients a_J, originally defined only for increasing multi-indices J, to be defined for other J so as to be antisymmetric functions of the indices. For economy of notation, we restrict attention to $(0, q)$ forms; modifications for (p, q) forms are simple (because the $\bar{\partial}$ operator does not see the dz differentials).

The $\bar{\partial}$ operator acts as usual on a $(0, q)$ form via

$$\bar{\partial}\left({\sum_J}' a_J \, d\bar{z}_J \right) = \sum_{j=1}^n {\sum_J}' \frac{\partial a_J}{\partial \bar{z}_j} d\bar{z}_{jJ}. \qquad (2\text{--}2)$$

The domain of $\bar{\partial} : L^2_{(0,q)}(\Omega) \to L^2_{(0,q+1)}(\Omega)$ consists of those forms u for which $\bar{\partial}u$, defined in the sense of distributions, belongs to $L^2_{(0,q+1)}(\Omega)$. It is routine to check that $\bar{\partial}$ is a closed, densely defined operator from $L^2_{(0,q)}(\Omega)$ to $L^2_{(0,q+1)}(\Omega)$. Consequently, the Hilbert-space adjoint $\bar{\partial}^*$ also exists and defines a closed, densely defined operator from $L^2_{(0,q+1)}(\Omega)$ to $L^2_{(0,q)}(\Omega)$.

Suppose $u = \sum'_J u_J \, d\bar{z}_J$ is continuously differentiable on the closure $\bar{\Omega}$, and ψ is a smooth test form. If the boundary $b\Omega$ is sufficiently smooth, then pairing u with $\bar{\partial}\psi$ and integrating by parts gives

$$(u, \bar{\partial}\psi) = \left(-\sum_{k=1}^{n}\sum_{K}{}' \frac{\partial u_{kK}}{\partial z_k} \, d\bar{z}_K, \psi \right) + \sum_{K}{}' \int_{b\Omega} \bar{\psi}_K \sum_{k=1}^{n} u_{kK} \frac{\partial \rho}{\partial z_k} \, d\sigma. \qquad (2\text{--}3)$$

The same calculation with a compactly supported ψ shows (without any boundary smoothness hypothesis) that if u is a square-integrable form in the domain of $\bar{\partial}^*$, then $\bar{\partial}^* u = \vartheta u$, where the *formal adjoint* ϑ is given by the equation

$$\vartheta u = -\sum_{k=1}^{n}\sum_{K}{}' \frac{\partial u_{kK}}{\partial z_k} \, d\bar{z}_K. \qquad (2\text{--}4)$$

It follows that a continuously differentiable form u is in the domain of $\bar{\partial}^*$ if and only if

$$\sum_{k=1}^{n} u_{kK} \frac{\partial \rho}{\partial z_k} \bigg|_{b\Omega} = 0 \text{ for every } K. \qquad (2\text{--}5)$$

The method of Friedrichs mollifiers shows that forms which are continuously differentiable on the closure $\bar{\Omega}$ are dense in the intersection of the domains of $\bar{\partial}$ and $\bar{\partial}^*$ with respect to the graph norm

$$\left(\|u\|^2 + \|\bar{\partial}u\|^2 + \|\bar{\partial}^* u\|^2 \right)^{1/2}$$

when the boundary $b\Omega$ is sufficiently smooth; see, for instance, [Hörmander 1965, § 1.2 and Prop. 2.1.1]. Also, forms that are continuously differentiable on the closure are dense in the domain of $\bar{\partial}$ with respect to the graph norm

$$(\|u\|^2 + \|\bar{\partial}u\|^2)^{1/2}.$$

The fundamental L^2 existence theorem for the $\bar{\partial}$-Neumann problem is due to Hörmander [1965]. One version of the result is the following.

THEOREM 1. *Let Ω be a bounded pseudoconvex domain in $\mathbb{C}^n$, where $n \geq 2$. Let D denote the diameter of Ω, and suppose $1 \leq q \leq n$.*

(i) *The complex Laplacian $\square = \bar{\partial}\bar{\partial}^* + \bar{\partial}^*\bar{\partial}$ is an unbounded, self-adjoint, surjective operator from $L^2_{(0,q)}(\Omega)$ to itself having a bounded inverse N_q (the $\bar{\partial}$-Neumann operator).*

(ii) *For all u in $L^2_{(0,q)}(\Omega)$, we have the estimates*

$$\|N_q u\| \leq \left(\frac{D^2 e}{q}\right)\|u\|, \quad \|\bar{\partial}^* N_q u\| \leq \left(\frac{D^2 e}{q}\right)^{1/2}\|u\|, \quad \|\bar{\partial} N_q u\| \leq \left(\frac{D^2 e}{q}\right)^{1/2}\|u\|.$$
$$(2\text{-}6)$$

(iii) *If f is a $\bar{\partial}$-closed $(0,q)$ form, then the canonical solution of the equation $\bar{\partial} u = f$ (the solution orthogonal to the kernel of $\bar{\partial}$) is given by $u = \bar{\partial}^* N_q f$; if f is a $\bar{\partial}^*$-closed $(0,q)$ form, then the canonical solution of the equation $\bar{\partial}^* u = f$ (the solution orthogonal to the kernel of $\bar{\partial}^*$) is given by $u = \bar{\partial} N_q f$.*

The Hilbert space method for proving Theorem 1 is based on estimating the norm of a form u in terms of the norms of $\bar{\partial} u$ and $\bar{\partial}^* u$. Hörmander discovered that it is advantageous to introduce weighted spaces $L^2(\Omega, e^{-\varphi})$, even for studying the unweighted problem. We denote the norm in the weighted space by $\|u\|_\varphi = \|u e^{-\varphi/2}\|$ and the adjoint of $\bar{\partial}$ with respect to the weighted inner product by $\bar{\partial}^*_\varphi(\cdot) = e^\varphi \bar{\partial}^*(\cdot e^{-\varphi})$. More generally, one can choose different exponential weights in $L^2_{(0,q-1)}$, $L^2_{(0,q)}$, and $L^2_{(0,q+1)}$; see [Hörmander 1990] for this method and applications.

The following identity is the basic starting point. The proof involves integrating by parts and manipulating the boundary integrals with the aid of the boundary condition (2–5) for membership in the domain of $\bar{\partial}^*$. The idea of introducing a second auxiliary function a was originally used in [Ohsawa and Takegoshi 1987; Ohsawa 1988], which deal with extending square-integrable holomorphic functions from submanifolds. The formulation given below comes from [Siu 1996; McNeal 1996]. In these papers (see also [Diederich and Herbort 1992]) the freedom to manipulate both the weight factor φ and the twisting factor a is essential.

PROPOSITION 2. *Let Ω be a bounded domain in $\mathbb{C}^n$ with class C^2 boundary; let u be a $(0,q)$ form (where $1 \leq q \leq n$) that is in the domain of $\bar{\partial}^*$ and that is continuously differentiable on the closure $\bar{\Omega}$; and let a and φ be real functions that are twice continuously differentiable on $\bar{\Omega}$, with $a \geq 0$. Then*

$$\|\sqrt{a}\,\bar{\partial} u\|^2_\varphi + \|\sqrt{a}\,\bar{\partial}^*_\varphi u\|^2_\varphi$$

$$= {\sum_K}' \sum_{j,k=1}^n \int_{b\Omega} a\frac{\partial^2 \rho}{\partial z_j \partial \bar{z}_k} u_{jK} \bar{u}_{kK} e^{-\varphi}\, d\sigma$$

$$+ {\sum_J}' \sum_{j=1}^n \int_\Omega a\left|\frac{\partial u_J}{\partial \bar{z}_j}\right|^2 e^{-\varphi}\, dV + 2\,\mathrm{Re}\left({\sum_K}' \sum_{j=1}^n u_{jK}\frac{\partial a}{\partial z_j}\, d\bar{z}_K, \bar{\partial}^*_\varphi u\right)_\varphi$$

$$+ {\sum_K}' \sum_{j,k=1}^n \int_\Omega \left(a\frac{\partial^2 \varphi}{\partial z_j \partial \bar{z}_k} - \frac{\partial^2 a}{\partial z_j \partial \bar{z}_k}\right) u_{jK} \bar{u}_{kK} e^{-\varphi}\, dV. \qquad (2\text{-}7)$$

For $a \equiv 1$ see [Hörmander 1965]; the case $a \equiv 1$ and $\varphi \equiv 0$ is the classical Kohn–Morrey formula [Kohn 1963; 1964; Morrey 1958]; see also [Ash 1964]. The usual proof of the L^2 existence theorem is based on a variant of (2–7) with $a \equiv 1$ and with different exponential weights φ in the different $L^2_{(0,q)}$ spaces; see [Catlin 1984b] for an elegant implementation of this approach. Here we will give an argument that has not appeared explicitly in the literature: we take $\varphi \equiv 0$ and make a good choice of a.

Suppose that Ω is a pseudoconvex domain: this means that the complex Hessian of the defining function ρ is a non-negative form on the vectors in the complex tangent space. Consequently, the boundary integral in (2–7) is non-negative. In particular, taking a to be identically equal to 1 gives

$$\|\bar{\partial} u\|^2 + \|\bar{\partial}^* u\|^2 \geq {\sum_{J}}' \sum_{j=1}^{n} \left\| \frac{\partial u_J}{\partial \bar{z}_j} \right\|^2 , \tag{2–8}$$

so the bar derivatives of u are always under control.

If we replace a by $1 - e^b$, where b is an arbitrary twice continuously differentiable non-positive function, then after applying the Cauchy–Schwarz inequality to the term in (2–7) involving first derivatives of a, we find

$$\|\sqrt{a}\,\bar{\partial} u\|^2 + \|\sqrt{a}\,\bar{\partial}^* u\|^2 \geq {\sum_{K}}' \sum_{j,k=1}^{n} \int_{\Omega} e^b \frac{\partial^2 b}{\partial z_j \partial \bar{z}_k} u_{jK} \bar{u}_{kK}\, dV - \|e^{b/2}\bar{\partial}^* u\|^2. \tag{2–9}$$

Since $a + e^b = 1$ and $a \leq 1$, it follows that

$$\|\bar{\partial} u\|^2 + \|\bar{\partial}^* u\|^2 \geq {\sum_{K}}' \sum_{j,k=1}^{n} \int_{\Omega} e^b \frac{\partial^2 b}{\partial z_j \partial \bar{z}_k} u_{jK} \bar{u}_{kK}\, dV \tag{2–10}$$

for every twice continuously differentiable non-positive function b. Notice that this inequality becomes a strong one if there happens to exist a bounded plurisubharmonic function b whose complex Hessian has large eigenvalues. (This theme will recur later on: see the discussion after Theorem 10 and the discussion of property (P) in Section 5.)

In particular, let p be a point of Ω, and set $b(z) = -1 + |z - p|^2/D^2$, where D is the diameter of the bounded domain Ω. The preceding inequality then implies the fundamental estimate

$$\|u\|^2 \leq \frac{D^2 e}{q} \left(\|\bar{\partial} u\|^2 + \|\bar{\partial}^* u\|^2 \right). \tag{2–11}$$

Although this estimate was derived under the assumption that u is continuously differentiable on the closure $\bar{\Omega}$, it holds by density for all square-integrable forms u that are in the intersection of the domains of $\bar{\partial}$ and $\bar{\partial}^*$. We also assumed that the boundary of Ω is smooth enough to permit integration by parts. Estimate (2–11) is equivalent to every form in $L^2_{(0,q)}(\Omega)$ admitting a representation as $\bar{\partial} v + \bar{\partial}^* w$ with $\|v\|^2 + \|w\|^2 \leq (D^2 e/q)\|u\|^2$. The latter property carries over to arbitrary bounded pseudoconvex domains by exhausting a nonsmooth Ω by smooth ones, and therefore so does inequality (2–11).

Once estimate (2–11) is in hand, the proof of Theorem 1 follows from standard Hilbert space arguments; see, for example, [Catlin 1983, pp. 164–165] or [Shaw 1992, § 2]. The latter paper also shows the existence of the $\bar{\partial}$-Neumann operator N_0 on $(\ker \bar{\partial})^\perp$.

3. Regularity on General Pseudoconvex Domains

A basic question is whether one can improve Theorem 1 to get regularity estimates in Sobolev norms: $\|Nu\|_s \leq C\|u\|_s$, $\|\bar{\partial}^* Nu\|_s \leq C\|u\|_s$, $\|\bar{\partial} Nu\|_s \leq C\|u\|_s$. If such estimates were to hold for all positive s, then Sobolev's lemma would imply that the $\bar{\partial}$-Neumann operator N (together with $\bar{\partial}^* N$ and $\bar{\partial} N$) is continuous in the space $C^\infty(\bar{\Omega})$ of functions smooth up to the boundary.

At first sight, it appears that one ought to be able to generalize the fundamental L^2 estimate (2–11) directly to an estimate of the form $\|u\|_s \leq C(\|\bar{\partial} u\|_s + \|\bar{\partial}^* u\|_s)$, simply by replacing u by a derivative of u. This naive expectation is erroneous: the difficulty is that not every derivative of a form u in the domain of $\bar{\partial}^*$ is again in the domain of $\bar{\partial}^*$. The usual attempt to overcome this difficulty is to cover the boundary of Ω with special boundary charts [Folland and Kohn 1972, p. 33] in each of which one can take a frame of tangential vector fields that do preserve the domain of $\bar{\partial}^*$. Since such vector fields have variable coefficients, they do not commute with either $\bar{\partial}$ or $\bar{\partial}^*$, and so one needs to handle error terms that arise from the commutators.

In subsequent sections, we will discuss various hypotheses on the domain Ω that yield regularity estimates in Sobolev norms. In this section, we discuss firstly some completely general results on smoothly bounded pseudoconvex domains and secondly some counterexamples.

It is an observation of J. J. Kohn and his school that the $\bar{\partial}$-Neumann problem is always regular in $W^\varepsilon(\Omega)$ for a sufficiently small positive ε.

PROPOSITION 3. *Let Ω be a bounded pseudoconvex domain in $\mathbb{C}^n$ with class C^∞ boundary. There exist positive ε and C (both depending on Ω) such that $\|Nu\|_\varepsilon \leq C\|u\|_\varepsilon$, $\|\bar{\partial}^* Nu\|_\varepsilon \leq C\|u\|_\varepsilon$, and $\|\bar{\partial} Nu\|_\varepsilon \leq C\|u\|_\varepsilon$ for every $(0,q)$ form u (where $1 \leq q \leq n$).*

The idea of the proof is very simple. Since the commutator of a differential operator of order ε with $\bar{\partial}$ or $\bar{\partial}^*$ is again an operator of order ε, but with a coefficient bounded by a constant times ε, error terms can be absorbed into the main term when ε is sufficiently small.

THEOREM 4. *Let Ω be a bounded pseudoconvex domain in $\mathbb{C}^n$ with class C^∞ boundary. Fix a positive s. There exists a T (depending on s and Ω) such that for every t larger than T, the weighted $\bar{\partial}$-Neumann problem for the space $L^2_{(0,q)}(\Omega, e^{-t|z|^2} dV(z))$ is regular in $W^s(\Omega)$. In other words, N_t, $\bar{\partial}^*_t N_t$, and $\bar{\partial} N_t$ are continuous in $W^s(\Omega)$.*

Moreover, if f is a $\bar{\partial}$-closed $(0,q)$ form with coefficients in $C^\infty(\bar{\Omega})$, then there exists a form u with coefficients in $C^\infty(\bar{\Omega})$ such that $\bar{\partial}u = f$.

This fundamental result on continuity of the weighted operators is due to Kohn [1973]. It says that one can always have regularity for the $\bar{\partial}$-Neumann problem up to a certain number of derivatives if one is willing to change the measure with respect to which the problem is defined. The idea of the proof is to apply Proposition 2 with $a \equiv 1$ and $\varphi(z) = t|z|^2$ to obtain

$$\|e^{-t|z|^2/2}u\|^2 \le Ct^{-1}\big(\|e^{-t|z|^2/2}\,\bar{\partial}u\|^2 + \|e^{-t|z|^2/2}\,\bar{\partial}_t^* u\|^2\big).$$

When t is sufficiently large, the factor t^{-1} makes it possible to absorb error terms coming from commutators (see the sketch of the proof of Theorem 12 below for the ideas of the technique). The resulting *a priori* estimates are valid under the assumption that the left-hand sides of the inequalities are known to be finite; Kohn completed the proof by applying the method of elliptic regularization [Kohn and Nirenberg 1965] (see also the remarks after Theorem 7 below).

By means of a Mittag-Leffler argument ([Kohn 1977, p. 230], argument attributed to Hörmander), one can deduce solvability of the equation $\bar{\partial}u = f$ in the space $C^\infty(\bar{\Omega})$ (but the solution will not be the canonical solution orthogonal to the kernel of $\bar{\partial}$). With some extra care, the solution operator can be made linear, and also continuous from $W^{s+\varepsilon}_{(0,q+1)}(\Omega) \cap \ker\bar{\partial}$ to $W^s_{(0,q)}(\Omega)$ for every positive s and ε [Sibony 1990]. It is unknown whether or not there exists a linear solution operator for $\bar{\partial}$ that breaks even at every level in the Sobolev scale. Solvability with Sobolev estimates (with a loss of three derivatives) has recently been obtained for domains with only C^4 boundary by S. L. Yie [1995].

Given any solution of the equation $\bar{\partial}u = f$, one obtains the canonical solution by subtracting from u its projection onto the kernel of $\bar{\partial}$. In view of Kohn's result above, it is natural to study the regularity properties of the projection mapping. We denote the orthogonal projection from $L^2_{(0,q)}(\Omega)$ onto $\ker\bar{\partial}$ by P_q; when $q = 0$, this operator is the *Bergman projection*. A direct relation between the Bergman projection and the $\bar{\partial}$-Neumann operator is given by Kohn's formula $P_q = \mathrm{Id} - \bar{\partial}^* N_{q+1}\bar{\partial}$ for $0 \le q \le n$. It is evident that if the $\bar{\partial}$-Neumann operator N_{q+1} is continuous in $C^\infty(\bar{\Omega})$, then so is P_q. The exact relationship between regularity properties of the $\bar{\partial}$-Neumann operators and the Bergman projections was determined in [Boas and Straube 1990].

THEOREM 5. *Let Ω be a bounded pseudoconvex domain in $\mathbb{C}^n$ with class C^∞ boundary. Fix an integer q such that $1 \le q \le n$. Then the $\bar{\partial}$-Neumann operator N_q is continuous on $C^\infty_{(0,q)}(\bar{\Omega})$ if and only if the projection operators P_{q-1}, P_q, and P_{q+1} are continuous on the corresponding $C^\infty(\bar{\Omega})$ spaces. The analogous statement holds with the Sobolev space $W^s(\Omega)$ in place of $C^\infty(\bar{\Omega})$.*

In view of the implications for boundary regularity of biholomorphic and proper holomorphic mappings [Bedford 1984; Bell 1981; 1984; 1990; Bell and Ligocka

1980; Bell and Catlin 1982; Diederich and Fornæss 1982; Forstnerič 1993], regularity in $C^\infty(\bar\Omega)$ is a key issue.

For some years there was uncertainty over whether the Bergman projection operator P_0 of every bounded domain in $\mathbb{C}^n$ with C^∞ smooth boundary might be regular in the space $C^\infty(\bar\Omega)$. Barrett [1984] found the first counterexample, motivated by the so-called "worm domains" of Diederich and Fornæss [1977a]. In his example, for every $p > 2$ there is a smooth, compactly supported function whose Bergman projection is not in $L^p(\Omega)$. Barrett and Fornæss [1986] constructed a counterexample even more closely related to the worm domains. Although the worm domains are smoothly bounded pseudoconvex domains in $\mathbb{C}^2$, these counterexamples are not pseudoconvex. Subsequently, Kiselman [1991] showed that pseudoconvex, but nonsmooth, truncated versions of the worm domains have irregular Bergman projections.

Later Barrett [1992] (see [Barrett 1998] for a generalization) used a scaling argument together with computations on piecewise Levi-flat model domains to show that the Bergman projection of a worm domain must fail to preserve the space $W^s(\Omega)$ when s is sufficiently large. In view of Theorem 5, the $\bar\partial$-Neumann operator N_1 also fails to preserve $W_{(0,1)}^s(\Omega)$. This left open the possibility of regularity in $C^\infty(\bar\Omega)$. Finally the question was resolved by Christ [1996a], as follows.

THEOREM 6. *For every worm domain, the Bergman projection operator P_0 and the $\bar\partial$-Neumann operator N_1 fail to be continuous on $C^\infty(\bar\Omega)$ and $C_{(0,1)}^\infty(\bar\Omega)$.*

Christ's proof is delicate and indirect. Roughly speaking, he shows that the $\bar\partial$-Neumann operator does satisfy for most values of s an estimate of the form $\|N_1 u\|_s \leq C\|u\|_s$ for all u for which $N_1 u$ is known a priori to lie in $C_{(0,1)}^\infty(\bar\Omega)$. If N_1 were to preserve $C_{(0,1)}^\infty(\bar\Omega)$, then density of $C_{(0,1)}^\infty(\bar\Omega)$ in $W_{(0,1)}^s(\Omega)$ would imply continuity of N_1 in $W_{(0,1)}^s(\Omega)$, contradicting Barrett's result.

The obstruction to continuity in $W^s(\Omega)$ for every s on the worm domains is a global one: namely, the nonvanishing of a certain class in the first De Rham cohomology of the annulus of weakly pseudoconvex boundary points (this class measures the twisting of the boundary at the annulus; for details, see Theorem 15). For smoothly bounded domains Ω, it is known that for each fixed s there is no local obstruction in the boundary to continuity in $W^s(\Omega)$ [Barrett 1986; Chen 1991b].

For all domains Ω where continuity in $C^\infty(\bar\Omega)$ is known, one can actually prove continuity in $W^s(\Omega)$ for all positive s. This intriguing phenomenon is not understood at present. (The corresponding phenomenon does not hold for partial differential operators in general: see [Christ 1999, Section 3] in this volume.)

Although regularity of the $\bar\partial$-Neumann problem in $C^\infty(\bar\Omega)$ is known in large classes of pseudoconvex domains (see sections 4–6), the example of the worm domains shows that regularity sometimes fails. At present, necessary and sufficient conditions for global regularity of the $\bar\partial$-Neumann operator and of the Bergman projection are not known.

4. Domains of Finite Type

Historically, the first major development on the $\bar{\partial}$-Neumann problem was its solution by Kohn [1963; 1964] for strictly pseudoconvex domains. A strictly pseudoconvex domain can be defined by a strictly plurisubharmonic function, so by taking $a \equiv 1$ and $\varphi \equiv 0$ in (2–7) and keeping the boundary term we find that $\|\bar{\partial}u\|^2 + \|\bar{\partial}^*u\|^2 \geq C\|u\|^2_{L^2(b\Omega)}$. Roughly speaking, this inequality says that we have gained half a derivative, since the restriction map $W^{s+1/2}(\Omega) \to W^s(b\Omega)$ is continuous when $s > 0$. This gain is half of what occurs for an ordinary elliptic boundary-value problem, so we have a "subelliptic estimate."

THEOREM 7. *Let Ω be a bounded strictly pseudoconvex domain in $\mathbb{C}^n$ with class C^∞ boundary. If $1 \leq q \leq n$, then for each non-negative s there is a constant C such that the following estimates hold for every $(0, q)$ form u:*

$$\|u\|_{s+1/2} \leq C\big(\|\bar{\partial}u\|_s + \|\bar{\partial}^*u\|_s\big) \quad \text{if } u \in \operatorname{dom}\bar{\partial} \cap \operatorname{dom}\bar{\partial}^*,$$
$$\|N_q u\|_{s+1} \leq C\|u\|_s, \qquad \|\bar{\partial}N_q u\|_{s+1/2} + \|\bar{\partial}^* N_q u\|_{s+1/2} \leq C\|u\|_s. \tag{4-1}$$

The standard reference for the proof of this result is [Folland and Kohn 1972], where the theory is developed for almost complex manifolds; see also [Krantz 1992]. The estimates can be localized, as in Theorem 8 below.

A key technical point in the proof of Theorem 7 is that after establishing the estimates under the assumption that the left-hand side is a priori finite, one then has to convert the a priori estimates into genuine estimates, in the sense that the left-hand side is finite when the right-hand side is finite. Kohn's original approach was considerably simplified in [Kohn and Nirenberg 1965] in a very general framework, via the elegant device of "elliptic regularization." The idea of the method is to add to $\Box$ an elliptic operator times ε (thereby obtaining a standard elliptic problem), to prove estimates independent of ε, and to let ε go to zero. (The analysis used in [Christ 1996a] to prove Theorem 6 shows that indeed a priori estimates cannot always be converted into genuine estimates. For this phenomenon in the context of the Bergman projection, see [Boas and Straube 1992a].) Another interesting approach to the proof of Theorem 7 was indicated by Morrey [1963; 1964].

A number of authors (see [Beals et al. 1987; Greiner and Stein 1977] and their references) have refined the results for strictly pseudoconvex domains in various ways, such as estimates in other function spaces and anisotropic estimates. In particular, N gains two derivatives in complex tangential directions; this gain results from the bar derivatives always being under control (see (2–8)). Integral kernel methods have also been developed successfully on strictly pseudoconvex domains; see [Grauert and Lieb 1970; Henkin 1969; 1970; Lieb 1993; Lieb and Range 1987; Ramirez 1970; Range 1986; 1987] and their references.

The gain of one derivative for the $\bar{\partial}$-Neumann operator N_1 in Theorem 7 is sharp, and the domain is necessarily strictly pseudoconvex if this estimate holds.

For discussion of this point, see [Catlin 1983; Folland and Kohn 1972, §III.2; Hörmander 1965, §3.2; Krantz 1979, §4].

More generally, one can ask when the $\bar{\partial}$-Neumann operator gains some fractional derivative. One says that a subelliptic estimate of order ε holds for the $\bar{\partial}$-Neumann problem on $(0, q)$ forms in a neighborhood U of a boundary point z_0 of a pseudoconvex domain in $\mathbb{C}^n$ if there is a constant C such that

$$\|u\|_\varepsilon^2 \leq C(\|\bar{\partial}u\|_0^2 + \|\bar{\partial}^* u\|_0^2) \qquad (4\text{--}2)$$

for every smooth $(0, q)$ form u that is supported in $U \cap \bar{\Omega}$ and that is in the domain of $\bar{\partial}^*$. The systematic study of subelliptic estimates in [Kohn and Nirenberg 1965] provides the following "pseudolocal estimates."

THEOREM 8. *Let Ω be a bounded pseudoconvex domain in $\mathbb{C}^n$ with class C^∞ boundary. Suppose that a subelliptic estimate* (4–2) *holds in a neighborhood U of a boundary point z_0. Let χ_1 and χ_2 be smooth cutoff functions supported in U with χ_2 identically equal to 1 in a neighborhood of the support of χ_1. For every non-negative s, there is a constant C such that the $\bar{\partial}$-Neumann operator N_q and the Bergman projection P_q satisfy the estimates*

$$\|\chi_1 N_q u\|_{s+2\varepsilon} \leq C(\|\chi_2 u\|_s + \|u\|_0), \quad \text{for } 1 \leq q \leq n,$$

$$\|\chi_1 \bar{\partial}^* N_q u\|_{s+\varepsilon} + \|\chi_1 \bar{\partial} N_q u\|_{s+\varepsilon} \leq C(\|\chi_2 u\|_s + \|u\|_0), \quad \text{for } 1 \leq q \leq n, \qquad (4\text{--}3)$$

$$\|\chi_1 P_q u\|_s \leq C(\|\chi_2 u\|_s + \|u\|_0), \quad \text{for } 0 \leq q \leq n.$$

Consequently, if a subelliptic estimate (4–2) holds in a neighborhood of every boundary point of a smooth bounded pseudoconvex domain Ω in $\mathbb{C}^n$, then the Bergman projection is continuous from $W_{(0,q)}^s(\Omega)$ to itself, and the $\bar{\partial}$-Neumann operator is continuous from $W_{(0,q)}^s(\Omega)$ to $W_{(0,q)}^{s+2\varepsilon}(\Omega)$.

In a sequence of papers [D'Angelo 1979; 1980; 1982; Catlin 1983; 1984a; 1987b], Kohn's students David Catlin and John D'Angelo resolved the question of when subelliptic estimates hold in a neighborhood of a boundary point of a smooth bounded pseudoconvex domain in $\mathbb{C}^n$. The necessary and sufficient condition is that the point have "finite type" in an appropriate sense. We briefly sketch this work; for details, consult the above papers as well as [Catlin 1987a; D'Angelo 1993, 1995; Diederich and Lieb 1981; Greiner 1974; Kohn 1979a; 1979b; 1981; 1984] and the survey [D'Angelo and Kohn 1999] in this volume.

The simplest obstruction to a subelliptic estimate is the presence of a germ of an analytic variety in the boundary of a domain. Indeed, examples show that local regularity of the $\bar{\partial}$-Neumann problem fails when there are complex varieties in the boundary; see [Catlin 1981; Diederich and Pflug 1981]. If the boundary is real-analytic near a point, then the absence of germs of q-dimensional complex-analytic varieties in the boundary near the point is necessary and sufficient for the existence of a subelliptic estimate on $(0, q)$-forms [Kohn 1979b]. This was first proved by combining a sufficient condition from Kohn's theory of ideals of subelliptic multipliers [1979b] with a theorem of Diederich and Fornæss [1978] on analytic varieties. Moreover, Diederich and Fornæss showed that a compact real-analytic manifold contains no germs of complex-analytic varieties of positive dimension, so subelliptic estimates hold for every bounded pseudoconvex domain in $\mathbb{C}^n$ with real-analytic boundary.

The first positive results in the C^∞ category were established in dimension two. A boundary point of a domain in $\mathbb{C}^2$ is of finite type if the boundary has finite order of contact with complex manifolds through the point; equivalently, if some finite-order commutator of complex tangential vector fields has a component that is transverse to the complex tangent space to the boundary. If m is an upper bound for the order of contact of complex manifolds with the boundary, then a subelliptic estimate (4–2) holds with $\varepsilon = 1/m$. For these results, see [Greiner 1974; Kohn 1979b]; for the equivalence of the two notions of finite type, see [Bloom and Graham 1977]. For pseudoconvex domains of finite type in dimension two, sharp estimates for the $\bar{\partial}$-Neumann problem are now known in many function spaces; see [Chang et al. 1992; Christ 1991] and their references.

In higher dimensions, it is no longer the case that all reasonable notions of finite type agree; for relations among them, see [D'Angelo 1987a]. D'Angelo's notion of finite type has turned out to be the right one for characterizing subelliptic estimates for the $\bar{\partial}$-Neumann problem. His idea to measure the order of contact of varieties with a real hypersurface M in $\mathbb{C}^n$ at a point z_0 is to fix a defining function ρ for M and to consider the order of vanishing at the origin of $\rho \circ f$, where f is a nonconstant holomorphic mapping from a neighborhood of the origin in $\mathbb{C}$ to $\mathbb{C}^n$ with $f(0) = z_0$. Since the variety that is the image of f may be singular, it is necessary to normalize by dividing by the order of vanishing at the origin of $f(\,\cdot\,) - z_0$. The supremum over all f of this normalized order of contact of germs of varieties with M is the D'Angelo 1-type of z_0.

THEOREM 9. *The set of points of finite 1-type of a smooth real hypersurface M in $\mathbb{C}^n$ is an open subset of M, and the 1-type is a locally bounded function on M.*

This fundamental result of D'Angelo [1982] is remarkable, because the 1-type may fail to be an upper semi-continuous function (see [D'Angelo 1993, p. 136] for a simple example). The theorem implies that if every point of a bounded domain in $\mathbb{C}^n$ is of finite 1-type, then there is a global upper bound on the 1-type.

For higher-dimensional varieties, there is no canonical way that serves all purposes to define the order of contact with a hypersurface. Catlin [1987b] defined a quantity $D_q(z_0)$ that measures the order of contact of q-dimensional varieties in "generic" directions (and D_1 agrees with D'Angelo's 1-type). Catlin's fundamental result is the following.

THEOREM 10. *Let Ω be a bounded pseudoconvex domain in $\mathbb{C}^n$ with class C^∞ boundary. A subelliptic estimate for the $\bar{\partial}$-Neumann problem on $(0, q)$ forms holds in a neighborhood of a boundary point z_0 if and only if $D_q(z_0)$ is finite. The ε in the subelliptic estimate (4–2) satisfies $\varepsilon \leq 1/D_q(z_0)$.*

The necessity of finite order of contact, together with the upper bound on ε, was proved in [Catlin 1983] (see also [Catlin 1981]), and the sufficiency in [Catlin 1987b]. Catlin's proof of sufficiency has two parts. His theory of multitypes [1984a] implies the existence of a stratification of the set of weakly pseudoconvex boundary points. The stratification is used to construct families of bounded plurisubharmonic functions whose complex Hessians in neighborhoods of the boundary have eigenvalues that blow up like inverse powers of the thickness of the neighborhoods. Such powers heuristically act like derivatives, and so it should be plausible that the basic inequality (2–10) leads to a subelliptic estimate (4–2).

It is unknown in general how to determine the optimal value of ε in a subelliptic estimate in terms of boundary data. For convex domains of finite type in $\mathbb{C}^n$, the optimal ε in a subelliptic estimate for $(0,1)$ forms is the reciprocal of the D'Angelo 1-type [Fornæss and Sibony 1989; McNeal 1992]; this is shown by a direct construction of bounded plurisubharmonic functions with suitable Hessians near the boundary. McNeal [1992] proved that for convex domains, the D'Angelo 1-type can be computed simply as the maximal order of contact of the boundary with complex lines. (There is an elementary geometric proof of McNeal's result in [Boas and Straube 1992b] and an analogue for Reinhardt domains in [Fu et al. 1996].) It is clear that in general, the best ε cannot equal the reciprocal of the type, simply because the type is not necessarily upper semi-continuous. For more about this subtle issue, see [D'Angelo 1993; 1995; D'Angelo and Kohn 1999; Diederich and Herbort 1993].

5. Compactness

A subelliptic estimate (4–2) implies, in particular, that the $\bar{\partial}$-Neumann operator is compact as an operator from $L^2_{(0,q)}(\Omega)$ to itself. This follows because the embedding from $W^\varepsilon_{(0,q)}(\Omega)$ into $L^2_{(0,q)}(\Omega)$ is compact when Ω is bounded with reasonable boundary, by the Rellich–Kondrashov theorem; see, for example, [Treves 1975, Prop. 25.5]. One might think of compactness in the $\bar{\partial}$-Neumann problem as a limiting case of subellipticity as $\varepsilon \to 0$.

The following lemma reformulates the compactness condition.

LEMMA 11. *Let Ω be a bounded pseudoconvex domain in $\mathbb{C}^n$, and suppose that $1 \le q \le n$. The following statements are equivalent.*

(i) *The $\bar{\partial}$-Neumann operator N_q is compact from $L^2_{(0,q)}(\Omega)$ to itself.*

(ii) *The embedding of the space $\operatorname{dom} \bar{\partial} \cap \operatorname{dom} \bar{\partial}^*$, provided with the graph norm $u \mapsto \|\bar{\partial} u\|_0 + \|\bar{\partial}^* u\|_0$, into $L^2_{(0,q)}(\Omega)$ is compact.*

(iii) *For every positive ε there exists C_ε such that*

$$\|u\|_0^2 \le \varepsilon \left(\|\bar{\partial} u\|_0^2 + \|\bar{\partial}^* u\|_0^2 \right) + C_\varepsilon \|u\|_{-1}^2 \tag{5–1}$$

when $u \in \operatorname{dom} \bar{\partial} \cap \operatorname{dom} \bar{\partial}^*$.

Statement (c) is called a *compactness estimate* for the $\bar{\partial}$-Neumann problem. Its equivalence with statement (b) is in [Kohn and Nirenberg 1965, Lemma 1.1]. The equivalence of statement (a) with statements (b) and (c) follows easily from the L^2 theory discussed in Section 2 and the compactness of the embedding $L^2_{(0,q)}(\Omega) \to W^{-1}_{(0,q)}(\Omega)$.

In view of Theorem 8, it is a reasonable guess that compactness in the $\bar{\partial}$-Neumann problem implies global regularity of the $\bar{\partial}$-Neumann operator in the sense that N_q maps $W^s_{(0,q)}(\Omega)$ into itself. Work of Kohn and Nirenberg [1965] shows that this conjecture is correct.

THEOREM 12. *Let Ω be a bounded pseudoconvex domain in $\mathbb{C}^n$ with class C^∞ boundary, and suppose $1 \le q \le n$. If a compactness estimate (5–1) holds for the $\bar{\partial}$-Neumann problem on $(0, q)$ forms, then the $\bar{\partial}$-Neumann operator N_q is a compact (in particular, continuous) operator from $W^s_{(0,q)}(\Omega)$ into itself for every non-negative s.*

It suffices to prove the result for integral s; the intermediate cases follow from standard interpolation theorems [Bergh and Löfström 1976; Persson 1964]. We sketch the argument for $s = 1$, which illustrates the method. To prove the compactness of the $\bar{\partial}$-Neumann operator in $W^1_{(0,q)}(\Omega)$, we will establish the (a priori) estimate $\|N_q u\|_1^2 \le \varepsilon \|u\|_1^2 + C_\varepsilon \|u\|_0^2$ for arbitrary positive ε under the assumption that u and $N_q u$ are both in $C^\infty(\bar{\Omega})$.

First we show that the compactness estimate (5–1) lifts to 1-norms: namely,

$$\|u\|_1^2 \le \varepsilon \left(\|\bar{\partial} u\|_1^2 + \|\bar{\partial}^* u\|_1^2 \right) + C_\varepsilon \|u\|_{-1}^2$$

for smooth forms u in dom $\bar{\partial}^*$ (with a new constant C_ε). In a neighborhood of a boundary point, we complete $\bar{\partial}\rho$ to an orthogonal basis of $(0, 1)$ forms and choose dual vector fields. To estimate tangential derivatives of u, we apply (5–1) to these derivatives (valid since they preserve the domain of $\bar{\partial}^*$). We then commute the derivatives with $\bar{\partial}$ and $\bar{\partial}^*$, which gives an error term that is of the same order as the quantity on the left-hand side that we are trying to estimate, but multiplied by a factor of ε. We also need to estimate the normal derivative of u, but since the boundary is noncharacteristic for the elliptic complex $\bar{\partial} \oplus \bar{\partial}^*$, the normal derivative of u can be expressed in terms of $\bar{\partial} u$, $\bar{\partial}^* u$, and tangential derivatives of u. Summing over a collection of special boundary charts that cover the boundary, and using interior elliptic regularity to estimate the norm on a compact set, we obtain an inequality of the form

$$\|u\|_1^2 \le A\varepsilon \left(\|\bar{\partial} u\|_1^2 + \|\bar{\partial}^* u\|_1^2 + \|u\|_1^2 \right) + B \left(\|\bar{\partial} u\|_0^2 + \|\bar{\partial}^* u\|_0^2 + \|u\|_0^2 \right),$$

where the constants A and B are independent of ε. We can use the standard interpolation inequality $\|f\|_s \le \varepsilon \|f\|_{s+1} + C_\varepsilon \|f\|_{s-1}$ to absorb terms into the left-hand side when ε is sufficiently small.

The lifted compactness estimate together with the L^2 boundedness of the $\bar{\partial}$-Neumann operator implies

$$\|N_q u\|_1^2 \leq \varepsilon\big(\|\bar{\partial}N_q u\|_1^2 + \|\bar{\partial}^* N_q u\|_1^2\big) + C_\varepsilon \|u\|_0^2. \tag{5-2}$$

Working as before in special boundary charts, we commute derivatives and integrate by parts on the right-hand side to make $\bar{\partial}\bar{\partial}^* N_q u + \bar{\partial}^* \bar{\partial} N_q u = u$ appear; see [Kohn 1984, p. 140; Boas and Straube 1990, p. 31]. Keeping track of commutator error terms and applying the Cauchy–Schwarz inequality, we find

$$\|\bar{\partial}N_q u\|_1^2 + \|\bar{\partial}^* N_q u\|_1^2$$
$$\leq A\big(\|N_q u\|_1 \|u\|_1 + \|u\|_0^2 + (\|\bar{\partial}N_q u\|_1 + \|\bar{\partial}^* N_q u\|_1)\|N_q u\|_1\big) \tag{5-3}$$

for some constant A. Consequently $\|\bar{\partial}N_q u\|_1^2 + \|\bar{\partial}^* N_q u\|_1^2 \leq B(\|N_q u\|_1^2 + \|u\|_1^2)$ for some constant B. Combining this with (5–2) gives the required a priori estimate

$$\|N_q u\|_1^2 \leq \varepsilon \|u\|_1^2 + C_\varepsilon \|u\|_0^2. \tag{5-4}$$

Kohn and Nirenberg [1965] developed the method of elliptic regularization (described above after Theorem 7) to convert these a priori estimates into genuine ones.

There is a large class of domains for which the $\bar{\partial}$-Neumann operator is compact [Catlin 1984b; Sibony 1987]. Catlin [1984b] introduced "property (P)" and showed that it implies a compactness estimate (5–1) for the $\bar{\partial}$-Neumann problem. A domain Ω has property (P) if for every positive number M there exists a plurisubharmonic function λ in $C^\infty(\bar{\Omega})$, bounded between 0 and 1, whose complex Hessian has all its eigenvalues bounded below by M on $b\Omega$:

$$\sum_{j,k=1}^{n} \frac{\partial^2 \lambda}{\partial z_j \partial \bar{z}_k}(z) w_j \bar{w}_k \geq M|w|^2, \qquad \text{for } z \in b\Omega, \ w \in \mathbb{C}^n. \tag{5-5}$$

That property (P) implies a compactness estimate (5–1) follows directly from (2–10) and interior elliptic regularity.

It is easy to see that the existence of a strictly plurisubharmonic defining function implies property (P), so strictly pseudoconvex domains satisfy property (P). So do pseudoconvex domains of finite type: this was proved by Catlin [1984b] as a consequence of the analysis of finite type boundaries in [Catlin 1984a; D'Angelo 1982].

Property (P) is, however, much more general than the condition of finite type. For instance, it is easy to see that a domain that is strictly pseudoconvex except for one infinitely flat boundary point must have property (P). More generally, property (P) holds if the set of weakly pseudoconvex boundary points has Hausdorff two-dimensional measure equal to zero [Boas 1988; Sibony 1987]. This latter reference contains a systematic study of the property (under the name of "B-regularity"). In particular, Sibony found examples of B-regular domains whose boundary points of infinite type form a set of positive measure.

It is folklore that an analytic disc in the boundary of a pseudoconvex domain in $\mathbb{C}^2$ obstructs compactness of the $\bar{\partial}$-Neumann problem: this can be proved by an adaptation of the argument used in [Catlin 1981; Diederich and Pflug 1981] to show (in any dimension) that analytic discs in the boundary preclude hypoellipticity of $\bar{\partial}$. In higher dimensions, tamely embedded analytic discs in the boundary obstruct compactness, but the general situation seems not to be understood; see [Krantz 1988; Ligocka 1985] for a discussion of some interesting examples. Salinas found an obstruction to compactness phrased in terms of the C^*-algebra generated by the operators of multiplication by coordinate functions; see the survey [Salinas 1991] and its references.

In view of the maximum principle, property (P) excludes analytic structure from the boundary: in particular, the boundary cannot contain analytic discs. However, the absence of analytic discs in the boundary does not guarantee property (P) [Sibony 1987, p. 310], although it does in the special cases of convex domains and complete Reinhardt domains [Sibony 1987, Prop. 2.4].

It is not yet understood how much room there is between property (P) and compactness. Having necessary and sufficient conditions on the boundary of a domain for compactness of the $\bar{\partial}$-Neumann problem would shed considerable light on the interactions among complex geometry, pluripotential theory, and partial differential equations.

6. The Vector Field Method

In the preceding section, we saw that the $\bar{\partial}$-Neumann problem is globally regular in domains that support bounded plurisubharmonic functions with arbitrarily large complex Hessian at the boundary. Now we will discuss a method that applies, for example, to domains admitting defining functions that are plurisubharmonic on the boundary. The method is based on the construction of certain vector fields that almost commute with $\bar{\partial}$.

We begin with some general remarks about proving a priori estimates of the form $\|N_q u\|_s \leq C\|u\|_s$ and $\|P_q u\|_s \leq C\|u\|_s$ in Sobolev spaces for the $\bar{\partial}$-Neumann operator and the Bergman projection. Firstly, all the action is near the boundary. This is clear for the Bergman projection on functions, because the mean-value property shows that every Sobolev norm of a holomorphic function on a compact subset of a domain is dominated by a weak norm on the whole domain (for instance, the L^2 norm). The corresponding property holds for the $\bar{\partial}$-Neumann operator due to interior elliptic regularity.

Secondly, the conjugate holomorphic derivatives $\partial/\partial \bar{z}_j$ are always under control. This is obvious for the case of the Bergman projection P_0 on functions (since holomorphic functions are annihilated by anti-holomorphic derivatives), and the inequality (2–8) shows that anti-holomorphic derivatives are tame for the $\bar{\partial}$-Neumann problem.

Thirdly, differentiation by vector fields whose restrictions to the boundary lie in the complex tangent space is also innocuous. Indeed, integrating by parts turns tangential vector fields of type $(1,0)$ into vector fields of type $(0,1)$, which are tame, plus terms that can be absorbed [Boas and Straube 1991, formula (3)].

Thus, we only need to estimate derivatives in the complex normal direction near the boundary. Moreover, since the bar derivatives are free, it will do to estimate either the real part or the imaginary part of the complex normal derivative. That is, we can get by with estimating either the real normal derivative, or a tangential derivative that is transverse to the complex tangent space.

A simple application of these ideas shows, for example, that the Bergman projection P_0 on functions for every bounded Reinhardt domain Ω in $\mathbb{C}^n$ with class C^∞ boundary is continuous from $W^s(\Omega)$ to itself for every positive integer s [Boas 1984; Straube 1986]. Indeed, the domain is invariant under rotations in each variable, so the Bergman projection commutes with each angular derivative $\partial/\partial\theta_j$. At every boundary point, at least one of these derivatives is transverse to the complex tangent space, so

$$\|P_0 u\|_1 \le C \sum_{j=1}^n \|(\partial/\partial\theta_j)P_0 u\|_0 = C \sum_{j=1}^n \|P_0(\partial u/\partial\theta_j)\|_0 \le C' \|u\|_1.$$

Higher derivatives are handled analogously. A similar technique proves global regularity of the $\bar{\partial}$-Neumann operator on bounded pseudoconvex Reinhardt domains [Boas et al. 1988; Chen 1989].

Thus, the nicest situation for proving estimates in Sobolev norms for the $\bar{\partial}$-Neumann operator is to have a tangential vector field, transverse to the complex tangent space, that commutes with the $\bar{\partial}$-Neumann operator, or what is nearly the same thing, that commutes with $\bar{\partial}$ and $\bar{\partial}^*$. (This method is classical: see [Derridj 1978; Derridj and Tartakoff 1976; Komatsu 1976].) Actually, it would be enough for the commutator with each anti-holomorphic derivative $\partial/\partial\bar{z}_j$ to have vanishing $(1,0)$ component in the complex normal direction. However, [Derridj 1991, Théorème 2.6 and the remark following it] shows that no such field can exist in general.

If we have a real tangential vector field T, transverse to the complex tangent space, whose commutator with each $\partial/\partial\bar{z}_j$ has $(1,0)$ component in the complex normal direction of modulus less than ε, then we get an estimate of the form

$$\|T^s N_q u\|_0 \le A_s(\|u\|_s + \varepsilon\|N_q u\|_s) + C_{s,T}\|u\|_0.$$

If the field T is normalized so that its coefficients and its angle with the complex tangent space are bounded away from zero, then $\|T^s N_q u\|_0$ controls $\|N_q u\|_s$ (independently of ε), so we get global regularity of N_q up to a certain level in the Sobolev scale. (By making estimates uniformly on a sequence of interior approximating strongly pseudoconvex domains, we can convert the a priori estimates to genuine ones.) Moreover, it suffices if T is approximately tangential in the sense that its normal component is of order ε. (This idea comes from [Barrett 1986]; see the proof of Theorem 16.) If we can find a sequence of such normalized vector fields corresponding to progressively smaller values of ε, then the $\bar{\partial}$-Neumann problem is globally regular at every level in the Sobolev scale. Because of the local regularity at points of finite type, the vector fields need exist only in (progressively smaller) neighborhoods of the boundary points of infinite type. In other words, we have the following result (where the imaginary parts of the X_ε correspond to the vector fields described above) [Boas and Straube 1991; Boas and Straube 1993].

THEOREM 13. *Let Ω be a bounded pseudoconvex domain in $\mathbb{C}^n$ with class C^∞ boundary and defining function ρ. Suppose there is a positive constant C such that for every positive ε there exists a vector field X_ε of type $(1,0)$ whose coefficients are smooth in a neighborhood U_ε in $\mathbb{C}^n$ of the set of boundary points of Ω of infinite type and such that*

(i) *$|\arg X_\varepsilon \rho| < \varepsilon$ on U_ε, and moreover $C^{-1} < |X_\varepsilon \rho| < C$ on U_ε, and*
(ii) *when $1 \le j \le n$, the form $\partial\rho$ applied to the commutator $[X_\varepsilon, \partial/\partial\bar{z}_j]$ has modulus less than ε on U_ε.*

Then the $\bar{\partial}$-Neumann operators N_q (for $1 \le q \le n$) and the Bergman projections P_q (for $0 \le q \le n$) are continuous on the Sobolev space $W^s_{(0,q)}(\Omega)$ when $s \ge 0$.

For a simple example in which the hypothesis of this theorem can be verified, consider a ball with a cap sliced off by a real hyperplane, and the edges rounded. The normal direction to the hyperplane will serve as X_ε (the U_ε being shrinking neighborhoods of the flat part of the boundary), so the $\bar{\partial}$-Neumann operator for this domain is continuous at every level in the Sobolev scale.

Indeed, the hypothesis of Theorem 13 can be verified for all convex domains. (The regularity of the $\bar{\partial}$-Neumann problem for convex domains in dimension two was obtained independently by Chen [1991a] using related ideas.) More generally, the theorem applies to domains admitting a defining function that is plurisubharmonic on the boundary [Boas and Straube 1991]. We state this as a separate result and sketch the proof. (Continuity in $W^{1/2}(\Omega)$ in the presence of a plurisubharmonic defining function was obtained earlier by Bonami and Charpentier [1988; 1990].)

THEOREM 14. *Let Ω be a bounded pseudoconvex domain in $\mathbb{C}^n$ with class C^∞ boundary. Suppose that Ω has a C^∞ defining function ρ that is plurisubharmonic on the boundary: $\sum_{j,k=1}^n (\partial^2\rho/\partial z_j \partial\bar{z}_k) w_j \bar{w}_k \ge 0$ for all $z \in b\Omega$ and all $w \in \mathbb{C}^n$. Then for every positive s there exists a constant C such that for all $u \in W^s_{(0,q)}(\Omega)$ we have*

$$\|N_q u\|_s \le C\|u\|_s, \quad \text{for } 1 \le q \le n,$$

$$\|P_q u\|_s \le C\|u\|_s, \quad \text{for } 0 \le q \le n.$$

Pseudoconvexity says that on the boundary, $\sum_{j,k=1}^n (\partial^2\rho/\partial z_j \partial\bar{z}_k) w_j \bar{w}_k \ge 0$ for vectors w in the complex tangent space: that is, those vectors for which

$\sum_{j=1}^{n}(\partial\rho/\partial z_j)w_j = 0$. The hypothesis of the theorem is that on the boundary, the complex Hessian of ρ is non-negative on all vectors, not just complex tangent vectors. (There are examples of pseudoconvex domains, even with real-analytic boundary, that do not admit such a defining function even locally [Behrens 1984; 1985; Fornæss 1979.) We now sketch how this extra information can be used to construct the special vector fields needed to invoke Theorem 13.

The key observation is that for each j, derivatives of $\partial\rho/\partial z_j$ of type $(0,1)$ in directions that lie in the null space of the Levi form must vanish. Indeed, if $\partial/\partial\bar{z}_1$ (say) is in the null space of the Levi form at a boundary point p, then $\partial^2\rho/\partial z_1\partial\bar{z}_1(p) = 0$, but since the matrix $\partial^2\rho/\partial z_j\partial\bar{z}_k(p)$ is positive semidefinite, its whole first column must vanish. (It was earlier observed by Noell [1991] that the unit normal to the boundary of a convex domain is constant along Levi-null curves.)

To construct the required global vector field, it will suffice to construct a vector field whose commutator with each complex tangential field of type $(1,0)$ has vanishing component in the complex normal direction at a specified boundary point p. Indeed, these components will be bounded by ε in a neighborhood of p by continuity, and we can use a partition of unity to patch local fields into a global field. (Terms in the commutator coming from derivatives of the partition of unity cause no difficulty because they are complex tangential.) It is easy to extend the field from the boundary to the inside of the domain to prescribe the proper commutator with the complex normal direction.

Suppose that $\partial\rho/\partial z_n(p) \neq 0$. We want to correct the field $(\partial\rho/\partial z_n)^{-1}(\partial/\partial z_n)$ by subtracting a linear combination of complex tangential vector fields so as to adjust the commutators. Since the Levi form may have some zero eigenvalues at p, we need a compatibility condition to solve the resulting linear system. The observation above that type $(0,1)$ derivatives in Levi-null directions annihilate $\partial\rho/\partial z_n$ at p is precisely the condition needed for solvability. For details of the proof, see [Boas and Straube 1991].

Kohn [1998] has found a new proof and generalization of Theorem 14. By a theorem of Diederich and Fornæss [1977b] (see also [Range 1981]), a smooth bounded pseudoconvex domain admits a defining function such that some (small) positive power δ of its absolute value is plurisuperharmonic inside Ω. Kohn shows that the $\bar{\partial}$-Neumann problem is regular in $W^s(\Omega)$ for s up to a level depending, roughly speaking, on δ. (More generally, the power of the defining function need only be "approximately plurisuperharmonic"; compare [Kohn 1998, Remark 5.1].)

Theorem 13 applies to other situations besides the one described in Theorem 14. For instance, it is possible to construct the vector fields on pseudoconvex domains that are regular in the sense of Diederich and Fornæss [1977c] and Catlin [1984b]. (This gives no new theorem, however, since the $\bar{\partial}$-Neumann problem is known to be compact on such domains [Catlin 1984b]; nor does it give a simplified proof of global regularity in the finite type case, since the construction of the vector fields still requires Catlin's stratification [1984a] of the set of weakly pseudoconvex points.)

As mentioned in Section 3, global regularity for the $\bar{\partial}$-Neumann problem breaks down on the Diederich–Fornæss worm domains. On those domains, the set of weakly pseudoconvex boundary points is precisely an annulus, and it is possible to compute directly that the vector fields specified in Theorem 13 cannot exist on this annulus.

For domains of this kind, where the boundary points of infinite type form a nice submanifold of the boundary, there is a natural condition that guarantees the existence of the vector fields needed to apply Theorem 13. Following the notation of [D'Angelo 1987b; 1993], we let η denote a purely imaginary, non-vanishing one-form on the boundary $b\Omega$ that annihilates the complex tangent space and its conjugate. Let T denote the purely imaginary tangential vector field on $b\Omega$ orthogonal to the complex tangent space and its conjugate and such that $\eta(T) \equiv 1$. Up to sign, the Levi form of two complex tangential vector fields X and Y is $\eta([X,\overline{Y}])$. The (real) one-form α is defined to be minus the Lie derivative of η in the direction of T:

$$\alpha = -\mathcal{L}_T \eta. \tag{6-1}$$

One can show [Boas and Straube 1993, § 2] that if M is a submanifold of the boundary whose real tangent space is contained in the null space of the Levi form, then the restriction of the form α to M is closed, and hence represents a cohomology class in the first De Rham cohomology $H^1(M)$. (In the special case when M is a complex submanifold, this closedness corresponds to the pluriharmonicity of certain argument functions, as in [Barrett and Fornæss 1988; Bedford and Fornæss 1978, Prop. 3.1; Bedford and Fornæss 1981, Lemma 1; Diederich and Fornæss 1977a, p. 290].) This class is independent of the choice of η. If this cohomology class vanishes on such a submanifold M, and if M contains the points of infinite type, then the vector fields described in Theorem 13 do exist. Thus, we have the following result [Boas and Straube 1993].

THEOREM 15. *Let Ω be a bounded pseudoconvex domain in $\mathbb{C}^n$ with class C^∞ boundary. Suppose there is a smooth real submanifold M (with or without boundary) of $b\Omega$ that contains all the points of infinite type of $b\Omega$ and whose real tangent space at each point is contained in the null space of the Levi form at that point (under the usual identification of $\mathbb{R}^{2n}$ with $\mathbb{C}^n$). If the $H^1(M)$ cohomology class $[\alpha|_M]$ is zero, then the $\bar{\partial}$-Neumann operators N_q (for $1 \leq q \leq n$) and the Bergman projections P_q (for $0 \leq q \leq n$) are continuous on the Sobolev space $W^s_{(0,q)}(\Omega)$ when $s \geq 0$.*

On the worm domains, one can compute directly that the class $[\alpha|_M]$ is not zero. The appearance of this cohomology class explains, in particular, why an analytic annulus in the boundary of the worm domains is bad for Sobolev estimates, while an annulus in the boundary of other domains may be innocuous [Boas and

Straube 1992a], and an analytic disc is always benign (same reference). In the special case that $n = 2$ and M is a bordered Riemann surface, Barrett has shown that there is a pluripolar subset of $H^1(M)$ such that estimates in $W^k(\Omega)$ fail for sufficiently large k if $[\alpha|_M]$ lies outside this subset [Barrett 1998]. When M is a complex submanifold of the boundary, $[\alpha|_M]$ has a geometric interpretation as a measure of the winding of the boundary of Ω around M (equivalently, the winding of the vector normal to the boundary). For details, see [Bedford and Fornæss 1978]. In the context of Hartogs domains in $\mathbb{C}^2$, see also [Boas and Straube 1992a].

The constructions of the vector fields (needed to apply Theorem 13) in the proofs of Theorems 14 and 15 are more closely related than appears at first glance. The vector fields can be written locally in the form

$$e^h L_n + \sum_{j=1}^{n-1} a_j L_j,$$

where $L_1, \ldots, L_{n-1}$ form a local basis for the tangential vector fields of type $(1, 0)$, L_n is the normal field of type $(1, 0)$, and h and the a_j are smooth functions. The commutator conditions in Theorem 13 in directions not in the null space of the Levi form can *always* be satisfied by using the a_j to correct the commutators. Computing the commutators in the remaining directions leads to the equation $dh|_{\mathcal{N}(p)} = \alpha|_{\mathcal{N}(p)}$ at points p of infinite type (where $\mathcal{N}(p)$ is the null space of the Levi form at p). The above proof of Theorem 14 amounts to showing that $\alpha|_{\mathcal{N}(p)} = 0$ when there is a defining function that is plurisubharmonic on the boundary, whence $h \equiv 0$ gives a solution. In Theorem 15, the hypothesis of the vanishing of the cohomology class of α on M allows us to solve for h (on M).

In general, the points of infinite type need not lie in a "nice" submanifold of the boundary. It is not known what should play the role of the cohomology class $[\alpha|_M]$ in the general situation. (Note that the analogue of the property that $\alpha|_M$ is closed holds in general: $d\alpha|_{\mathcal{N}(p)} = 0$; see [Boas and Straube 1993, § 2].) Furthermore, it is not understood how to combine the ideas of this section with the pluripotential theoretic methods discussed in Section 5, such as B-regularity and property (P).

7. The Bergman Projection on General Domains

In pseudoconvex domains, global regularity of the $\bar{\partial}$-Neumann problem is essentially equivalent to global regularity of the Bergman projection [Boas and Straube 1990]. In nonpseudoconvex domains, the $\bar{\partial}$-Neumann operator may not exist, yet the Bergman projection is still well defined. Since global regularity of the Bergman projection on functions is intimately connected to the boundary regularity of biholomorphic and proper holomorphic mappings (see [Bedford 1984; Bell 1981; 1984; 1990; Bell and Ligocka 1980; Bell and Catlin 1982;

Diederich and Fornæss 1982; Forstnerič 1993]), it is interesting to analyze the Bergman projection directly, without recourse to the $\bar{\partial}$-Neumann problem. Even very weak regularity properties of the Bergman projection can be exploited in the study of biholomorphic mappings [Barrett 1986; Lempert 1986].

In this section, we survey the theory of global regularity of the Bergman projection on general (that is, not necessarily pseudoconvex) domains.

The first regularity results for the Bergman projection that were obtained without the help of the $\bar{\partial}$-Neumann theory are in [Bell and Boas 1981], where it is shown that the Bergman projection P on functions maps the space $C^\infty(\bar{\Omega})$ of functions smooth up to the boundary continuously into itself when Ω is a bounded complete Reinhardt domain with C^∞ smooth boundary.

This result was generalized in [Barrett 1982] to domains with "transverse symmetries." A domain Ω is said to have transverse symmetries if it admits a Lie group G of holomorphic automorphisms acting transversely in the sense that the map $G \times \Omega \to \Omega$ taking (g, z) to $g(z)$ extends to a smooth map $G \times \bar{\Omega} \to \bar{\Omega}$, and for each point $z_0 \in b\Omega$ the map $g \mapsto g(z_0)$ of G to $b\Omega$ induces a map on tangent spaces $T_{\mathrm{Id}}G \to T_{z_0}^{\mathbb{R}}(b\Omega)$ whose image is not contained in the complex tangent space to $b\Omega$ at z_0. In other words, there exists for each boundary point z_0 a one-parameter family of automorphisms of $\bar{\Omega}$ whose infinitesimal generator is transverse to the tangent space at z_0. This class of domains includes many Cartan domains as well as all smooth bounded Reinhardt domains; in both cases, suitable Lie groups of rotations provide the transverse symmetries [Barrett 1982]. For domains with transverse symmetries, it was observed in [Straube 1986] that the Bergman projection not only maps the space $C^\infty(\bar{\Omega})$ into itself, but actually preserves the Sobolev spaces.

More generally, one can obtain regularity results in the presence of a transverse vector field of type $(1, 0)$ with holomorphic coefficients, even if it does not come from a family of automorphisms. David Barrett obtained the following result [Barrett 1986].

THEOREM 16. *Let Ω be a bounded domain in $\mathbb{C}^n$ with class C^∞ boundary and defining function ρ. Suppose there is a vector field X of type $(1, 0)$ with holomorphic coefficients in $C^\infty(\bar{\Omega})$ that is nowhere tangent to the boundary of Ω and such that $|\arg X\rho| < \pi/4k$ for some positive integer k. Then the Bergman projection on functions maps the Sobolev space $W^k(\Omega)$ continuously into itself.*

In particular, Theorem 16 implies that there are no local obstructions to W^k regularity of the Bergman projection. In other words, any sufficiently small piece of C^∞ boundary can be a piece of the boundary of a domain G whose Bergman projection is continuous in $W^k(G)$: indeed, G can be taken to be a small perturbation of a ball, and then the radial field satisfies the hypothesis of the theorem.

Theorem 16 also applies when $k = 1/2$ and the boundary is only Lipschitz smooth. For example, the hypothesis holds for $k = 1/2$ when the domain is strictly star-shaped. Lempert [1986] has exploited this weak regularity property to prove a Hölder regularity theorem for biholomorphic mappings between star-shaped domains with real-analytic boundaries.

The first step in the proof of Theorem 16 is one we have seen before in Section 6: namely, it suffices to estimate derivatives of holomorphic functions in a direction transverse to the boundary. Thus, to bound $\|Pf\|_k$ it suffices to bound $\|X^k Pf\|_0$. However, the inner product $\langle X^k Pf, X^k Pf \rangle$ is bounded above by a constant times $|\langle \varphi^k X^k Pf, X^k Pf \rangle|$ when $\operatorname{Re} \varphi^k$ is bounded away from zero. By the hypothesis of the theorem, we can take φ to be a smooth function that equals $\overline{X}\rho/X\rho$ near the boundary. We then replace $\varphi^k X^k$ on the left-hand side of the inner product by $(\varphi X - \overline{X})^k$, making a lower-order error (since $\overline{X}$ annihilates holomorphic functions). The point is that $(\varphi X - \overline{X})$ is tangential at the boundary, so we can integrate by parts without boundary terms, obtaining $|\langle Pf, X^{2k} Pf \rangle|$ plus lower-order terms. Since X is a holomorphic field, we can remove the Bergman projection operator from the left-hand side of the inner product, integrate by parts, and apply the Cauchy–Schwarz inequality to get an upper bound of the form $C\|f\|_k \|Pf\|_k$. This gives an a priori estimate $\|Pf\|_k \leq C\|f\|_k$. The estimate can be converted into a genuine estimate via an argument involving the resolvent of the semigroup generated by the real part of X. For details of the proof, see [Barrett 1986].

It is possible to combine such methods with techniques based on pseudoconvexity. Estimates for the Bergman projection and the $\bar{\partial}$-Neumann operator on pseudoconvex domains that have transverse symmetries on the complement of a compact subset of the boundary consisting of points of finite type were obtained in [Chen 1987] and [Boas et al. 1988].

A domain in $\mathbb{C}^2$ is called a Hartogs domain if, with each of its points (z, w), it contains the circle $\{(z, \lambda w) : |\lambda| = 1\}$; it is complete if it also contains the disc $\{(z, \lambda w) : |\lambda| \leq 1\}$. The (pseudoconvex) worm domains [Diederich and Fornæss 1977a] and the (nonpseudoconvex) counterexample domains in [Barrett 1984; Barrett and Fornæss 1986] with irregular Bergman projections are incomplete Hartogs domains in $\mathbb{C}^2$. It is easy to see that when a Hartogs domain in $\mathbb{C}^2$ is complete, the obstruction to regularity identified in Section 6 cannot occur; see [Boas and Straube 1992a, §1]. Actually, completeness guarantees that the Bergman projection is regular whether or not the domain is pseudoconvex [Boas and Straube 1989]. (See [Boas and Straube 1992a] for a systematic study of the Bergman projection on Hartogs domains in $\mathbb{C}^2$.)

THEOREM 17. *Let Ω be a bounded complete Hartogs domain in $\mathbb{C}^2$ with class C^∞ boundary. The Bergman projection maps the Sobolev space $W^s(\Omega)$ continuously into itself when $s \geq 0$.*

The proof again uses different arguments on different parts of the boundary. An interesting new twist occurs in that the $\bar{\partial}$-Neumann operator of the envelope of holomorphy of the domain (which is still a complete Hartogs domain) is exploited.

The Bergman projection is known to preserve the Sobolev spaces $W^s(\Omega)$ in all cases in which it is known to preserve the space $C^\infty(\bar{\Omega})$ of functions smooth up to the boundary (as is the case for the $\bar{\partial}$-Neumann operator on pseudoconvex domains). It is an intriguing question whether this is a general phenomenon.

We now turn to the connection between the regularity theory of the Bergman projection and the duality theory of holomorphic function spaces, which originates with Bell [1982b]. When k is an integer, let $A^k(\Omega)$ denote the subspace of the Sobolev space $W^k(\Omega)$ consisting of holomorphic functions, and let $A^\infty(\Omega)$ denote the subspace of $C^\infty(\bar{\Omega})$ consisting of holomorphic functions. We may view the Fréchet space $A^\infty(\Omega)$ as the projective limit of the Hilbert spaces $A^k(\Omega)$, and we introduce the notation $A^{-\infty}(\Omega)$ for the space $\bigcup_{k=1}^\infty A^{-k}(\Omega)$, provided with the inductive limit topology.

For discussion of some of the technical properties of these spaces of holomorphic functions, see [Bell and Boas 1984; Straube 1984]. In particular, $A^{-\infty}(\Omega)$ is a Montel space, and subsets of $A^{-\infty}(\Omega)$ are bounded if and only if they are contained and bounded in some $A^{-k}(\Omega)$. The inductive limit structure on $A^{-\infty}(\Omega)$ turns out to be "nice" because the embeddings $A^{-k}(\Omega) \to A^{-k-1}(\Omega)$ are compact (as a consequence of Rellich's lemma). Functions in $A^{-\infty}(\Omega)$ can be characterized in two equivalent ways: they have growth near the boundary of Ω that is at most polynomial in the reciprocal of the distance to the boundary, and their traces on interior approximating surfaces $b\Omega_\varepsilon$ converge in the sense of distributions on $b\Omega$. See [Straube 1984] for an elementary discussion of these facts.

The L^2 inner product extends to a more general pairing. Harmonic functions are a natural setting for this extension. We use the notations $h^\infty(\Omega)$ and $h^{-\infty}(\Omega)$ for the spaces of harmonic functions analogous to $A^\infty(\Omega)$ and $A^{-\infty}(\Omega)$.

PROPOSITION 18. *Let Ω be a bounded domain in $\mathbb{C}^n$ with class C^∞ boundary. For each positive integer k there is a constant C_k such that for every square-integrable harmonic function f, and every $g \in C^\infty(\bar{\Omega})$, we have the inequality*

$$\left| \int_\Omega f\bar{g} \right| \leq C_k \|f\|_{-k} \|g\|_k. \tag{7-1}$$

The proof of Proposition 18 follows from the observation that for every $g \in C^\infty(\bar{\Omega})$, there is a function g_1 vanishing to high order at the boundary of Ω such that the difference $g - g_1$ is orthogonal to the harmonic functions. See [Bell 1982c; Boas 1987, Appendix B; and Ligocka 1986] for details; the root idea originates with Bell [1979] in the context of holomorphic functions. Alternatively, Proposition 18 can be derived from elementary facts about the Dirichlet problem for the Laplace operator [Straube 1984].

Because the square-integrable harmonic functions are dense in $h^{-\infty}(\Omega)$, it follows from (7-1) that the L^2 pairing extends by continuity to a pairing $\langle f, g \rangle$ on $h^{-\infty}(\Omega) \times C^\infty(\bar{\Omega})$. In particular, this pairing is well defined and separately continuous on $A_{\mathrm{cl}}^{-\infty}(\Omega) \times A^\infty(\Omega)$, where $A_{\mathrm{cl}}^{-\infty}(\Omega)$ denotes the closure of $A^0(\Omega)$ in $A^{-\infty}(\Omega)$.

PROPOSITION 19. *Let Ω be a bounded domain in $\mathbb{C}^n$ with class C^∞ boundary. The following statements are equivalent.*

(i) *The Bergman projection P maps the space $C^\infty(\bar{\Omega})$ continuously into itself.*

(ii) *The spaces $A_{\mathrm{cl}}^{-\infty}(\Omega)$ and $A^\infty(\Omega)$ of holomorphic functions are mutually dual via the extended pairing $\langle\,,\,\rangle$.*

Proposition 19 is from [Bell and Boas 1984; Komatsu 1984]; the case of a strictly pseudoconvex domain is in [Bell 1982b], and duality of spaces of harmonic functions is studied in [Bell 1982a; Ligocka 1986]. Once Proposition 18 is in hand, Proposition 19 is easily proved. For example, suppose that the Bergman projection is known to preserve the space $C^\infty(\bar{\Omega})$, and let τ be a continuous linear functional on the space $A_{\mathrm{cl}}^{-\infty}(\Omega)$. Because τ extends to a continuous linear functional on the inductive limit $W^{-\infty}(\Omega)$ of the ordinary Sobolev spaces, it is represented by pairing with a function g in the space $W_0^\infty(\Omega)$ of functions vanishing to infinite order at the boundary. On $A^0(\Omega)$, and hence on $A_{\mathrm{cl}}^{-\infty}(\Omega)$, pairing with g is the same as pairing with Pg since, by hypothesis, $Pg \in A^\infty(\Omega)$. Therefore τ is indeed represented by an element of $A^\infty(\Omega)$.

It is nontrivial that $A_{\mathrm{cl}}^{-\infty}(\Omega) = A^{-\infty}(\Omega)$ when Ω is pseudoconvex. Examples show that density properties fail dramatically in the nonpseudoconvex case [Barrett 1984; Barrett and Fornæss 1986]. The arguments in these papers can be adapted to show that $A_{\mathrm{cl}}^{-\infty}(\Omega) \neq A^{-\infty}(\Omega)$ for these examples.

THEOREM 20. *Let Ω be a bounded pseudoconvex domain in $\mathbb{C}^n$ with class C^∞ boundary. Then the space $A^\infty(\Omega)$ of holomorphic functions is dense both in $A^k(\Omega)$ and in $A^{-k}(\Omega)$ for each non-negative integer k.*

The first part is in [Catlin 1980], the second in [Bell and Boas 1984]. In particular, the Bergman projection is globally regular on a pseudoconvex domain Ω if and only if the spaces $A^{-\infty}(\Omega)$ and $A^\infty(\Omega)$ are mutually dual via the pairing $\langle\,,\,\rangle$.

Here is a typical application of Proposition 19 to the theory of the Bergman kernel function $K(w, z)$.

COROLLARY 21. *Let Ω be a bounded domain in $\mathbb{C}^n$ with class C^∞ boundary. Suppose that the Bergman projection maps the space $C^\infty(\bar{\Omega})$ into itself. If S is a set of determinacy for holomorphic functions on Ω, then $\{K(\,\cdot\,, z) : z \in S\}$ has dense linear span in $A^\infty(\Omega)$.*

Indeed, global regularity of the Bergman projection P implies that $K(\,\cdot\,, z) \in A^\infty(\Omega)$ for each z in Ω, since $K(\,\cdot\,, z)$ is the projection of a smooth, radially symmetric bump function (this idea originates in [Kerzman 1972]). Now if a linear functional τ on $A^\infty(\Omega)$ vanishes on each $K(\,\cdot\,, z)$ for $z \in S$, then $\tau(z) = 0$ on S, whence $\tau \equiv 0$. (Note that since $\tau \in A_{\mathrm{cl}}^{-\infty}(\Omega)$, the Bergman kernel does reproduce τ, because evaluation at an interior point is continuous in the topology of $A^{-\infty}(\Omega)$.)

Corollary 21 is due to Bell [1979; 1982b]. It is the key to certain non-vanishing properties of the Bergman kernel function that are essential in the approach to boundary regularity of holomorphic mappings developed in [Bell and Ligocka 1980; Bell 1981; 1984; Ligocka 1980; 1981; Webster 1979].

References

[Adams 1975] R. A. Adams, *Sobolev spaces*, Pure App. Math. **65**, Academic Press, 1975.

[Ash 1964] M. E. Ash, "The basic estimate of the $\bar{\partial}$-Neumann problem in the non-Kählerian case", *Amer. J. Math.* **86** (1964), 247–254.

[Barrett 1982] D. E. Barrett, "Regularity of the Bergman projection on domains with transverse symmetries", *Math. Ann.* **258**:4 (1982), 441–446.

[Barrett 1984] D. E. Barrett, "Irregularity of the Bergman projection on a smooth bounded domain in $\mathbb{C}^2$", *Ann. of Math.* (2) **119**:2 (1984), 431–436.

[Barrett 1986] D. E. Barrett, "Regularity of the Bergman projection and local geometry of domains", *Duke Math. J.* **53**:2 (1986), 333–343.

[Barrett 1992] D. E. Barrett, "Behavior of the Bergman projection on the Diederich–Fornæss worm", *Acta Math.* **168**:1-2 (1992), 1–10.

[Barrett 1998] D. E. Barrett, "The Bergman projection on sectorial domains", pp. 1–24 in *Operator theory for complex and hypercomplex analysis* (Mexico City, 1994), edited by E. R. de Arellano et al., Contemp. Math. **212**, Amer. Math. Soc., Providence, RI, 1998.

[Barrett and Fornæss 1986] D. E. Barrett and J. E. Fornæss, "Uniform approximation of holomorphic functions on bounded Hartogs domains in $\mathbb{C}^2$", *Math. Z.* **191**:1 (1986), 61–72.

[Barrett and Fornæss 1988] D. E. Barrett and J. E. Fornæss, "On the smoothness of Levi-foliations", *Publicacions Matemàtiques* **32**:2 (1988), 171–177.

[Beals et al. 1987] R. Beals, P. C. Greiner, and N. K. Stanton, "L^p and Lipschitz estimates for the $\bar{\partial}$-equation and the $\bar{\partial}$-Neumann problem", *Math. Ann.* **277**:2 (1987), 185–196.

[Bedford 1984] E. Bedford, "Proper holomorphic mappings", *Bull. Amer. Math. Soc.* **10**:2 (1984), 157–175.

[Bedford and Fornæss 1978] E. Bedford and J. E. Fornæss, "Domains with pseudoconvex neighborhood systems", *Invent. Math.* **47**:1 (1978), 1–27.

[Bedford and Fornæss 1981] E. Bedford and J. E. Fornæss, "Complex manifolds in pseudoconvex boundaries", *Duke Math. J.* **48** (1981), 279–288.

[Behrens 1984] M. Behrens, *Plurisubharmonische definierende Funktionen pseudokonvexer Gebiete*, Schriftenreihe des Math. Inst. der Univ. Münster (Ser. 2) **31**, Universität Münster, 1984.

[Behrens 1985] M. Behrens, "Plurisubharmonic defining functions of weakly pseudoconvex domains in $\mathbb{C}^2$", *Math. Ann.* **270**:2 (1985), 285–296.

[Bell 1979] S. R. Bell, "Non-vanishing of the Bergman kernel function at boundary points of certain domains in $\mathbb{C}^n$", *Math. Ann.* **244**:1 (1979), 69–74.

[Bell 1981] S. R. Bell, "Biholomorphic mappings and the $\bar{\partial}$-problem", *Ann. of Math.* (2) **114**:1 (1981), 103–113.

[Bell 1982a] S. R. Bell, "A duality theorem for harmonic functions", *Michigan Math. J.* **29** (1982), 123–128.

[Bell 1982b] S. R. Bell, "A representation theorem in strictly pseudoconvex domains", *Illinois J. Math.* **26**:1 (1982), 19–26.

[Bell 1982c] S. R. Bell, "A Sobolev inequality for pluriharmonic functions", *Proc. Amer. Math. Soc.* **85**:3 (1982), 350–352.

[Bell 1984] S. R. Bell, "Boundary behavior of proper holomorphic mappings between nonpseudoconvex domains", *Amer. J. Math.* **106**:3 (1984), 639–643.

[Bell 1990] S. Bell, "Mapping problems in complex analysis and the $\bar{\partial}$-problem", *Bull. Amer. Math. Soc.* **22**:2 (1990), 233–259.

[Bell and Boas 1981] S. R. Bell and H. P. Boas, "Regularity of the Bergman projection in weakly pseudoconvex domains", *Math. Ann.* **257**:1 (1981), 23–30.

[Bell and Boas 1984] S. R. Bell and H. P. Boas, "Regularity of the Bergman projection and duality of holomorphic function spaces", *Math. Ann.* **267**:4 (1984), 473–478.

[Bell and Catlin 1982] S. Bell and D. Catlin, "Boundary regularity of proper holomorphic mappings", *Duke Math. J.* **49**:2 (1982), 385–396.

[Bell and Ligocka 1980] S. Bell and E. Ligocka, "A simplification and extension of Fefferman's theorem on biholomorphic mappings", *Invent. Math.* **57**:3 (1980), 283–289.

[Bergh and Löfström 1976] J. Bergh and J. Löfström, *Interpolation spaces*, Grundlehren der Math. Wissenschaften **223**, Springer, Berlin, 1976.

[Berndtsson 1994] B. Berndtsson, "Some recent results on estimates for the $\bar{\partial}$-equation", pp. 27–42 in *Contributions to complex analysis and analytic geometry*, edited by H. Skoda and J.-M. Trépreau, Aspects of Mathematics **E26**, Vieweg, Wiesbaden, 1994.

[Bloom and Graham 1977] T. Bloom and I. R. Graham, "A geometric characterization of points of type m on real submanifolds of $\mathbb{C}^n$", *J. Differential Geom.* **12**:2 (1977), 171–182.

[Boas 1984] H. P. Boas, "Holomorphic reproducing kernels in Reinhardt domains", *Pacific J. Math.* **112**:2 (1984), 273–292.

[Boas 1987] H. P. Boas, "The Szegő projection: Sobolev estimates in regular domains", *Trans. Amer. Math. Soc.* **300**:1 (1987), 109–132.

[Boas 1988] H. P. Boas, "Small sets of infinite type are benign for the $\bar{\partial}$-Neumann problem", *Proc. Amer. Math. Soc.* **103**:2 (1988), 569–578.

[Boas and Straube 1989] H. P. Boas and E. J. Straube, "Complete Hartogs domains in $\mathbb{C}^2$ have regular Bergman and Szegő projections", *Math. Z.* **201**:3 (1989), 441–454.

[Boas and Straube 1990] H. P. Boas and E. J. Straube, "Equivalence of regularity for the Bergman projection and the $\bar{\partial}$-Neumann operator", *Manuscripta Math.* **67**:1 (1990), 25–33.

[Boas and Straube 1991] H. P. Boas and E. J. Straube, "Sobolev estimates for the $\bar{\partial}$-Neumann operator on domains in $\mathbb{C}^n$ admitting a defining function that is plurisubharmonic on the boundary", *Math. Z.* **206**:1 (1991), 81–88.

[Boas and Straube 1992a] H. P. Boas and E. J. Straube, "The Bergman projection on Hartogs domains in $\mathbb{C}^2$", *Trans. Amer. Math. Soc.* **331**:2 (1992), 529–540.

[Boas and Straube 1992b] H. P. Boas and E. J. Straube, "On equality of line type and variety type of real hypersurfaces in $\mathbb{C}^n$", *J. Geom. Anal.* **2**:2 (1992), 95–98.

[Boas and Straube 1993] H. P. Boas and E. J. Straube, "De Rham cohomology of manifolds containing the points of infinite type, and Sobolev estimates for the $\bar{\partial}$-Neumann problem", *J. Geom. Anal.* **3**:3 (1993), 225–235.

[Boas et al. 1988] H. P. Boas, S.-C. Chen, and E. J. Straube, "Exact regularity of the Bergman and Szegő projections on domains with partially transverse symmetries", *Manuscripta Math.* **62**:4 (1988), 467–475.

[Bonami and Charpentier 1988] A. Bonami and P. Charpentier, "Une estimation Sobolev $\frac{1}{2}$ pour le projecteur de Bergman", *C. R. Acad. Sci. Paris Sér. I Math.* **307**:5 (1988), 173–176.

[Bonami and Charpentier 1990] A. Bonami and P. Charpentier, "Boundary values for the canonical solution to $\bar{\partial}$-equation and $W^{1/2}$ estimates", preprint 9004, Centre de Recherche en Mathématiques de Bordeaux, Université Bordeaux I, April 1990.

[Catlin 1980] D. Catlin, "Boundary behavior of holomorphic functions on pseudoconvex domains", *J. Differential Geom.* **15**:4 (1980), 605–625.

[Catlin 1981] D. Catlin, "Necessary conditions for subellipticity and hypoellipticity for the $\bar{\partial}$-Neumann problem on pseudoconvex domains", pp. 93–100 in *Recent developments in several complex variables*, edited by J. E. Fornæss, Annals of Mathematics Studies **100**, Princeton Univ. Press, Princeton, NJ, 1981.

[Catlin 1983] D. Catlin, "Necessary conditions for subellipticity of the $\bar{\partial}$-Neumann problem", *Ann. of Math.* (2) **117**:1 (1983), 147–171.

[Catlin 1984a] D. Catlin, "Boundary invariants of pseudoconvex domains", *Ann. of Math.* (2) **120**:3 (1984), 529–586.

[Catlin 1984b] D. W. Catlin, "Global regularity of the $\bar{\partial}$-Neumann problem", pp. 39–49 in *Complex analysis of several variables* (Madison, WI, 1982), edited by Y.-T. Siu, Proc. Symp. Pure Math. **41**, Amer. Math. Soc., Providence, 1984.

[Catlin 1987a] D. Catlin, "Regularity of solutions to the $\bar{\partial}$-Neumann problem", pp. 708–714 in *Proceedings of the International Congress of Mathematicians* (Berkeley, 1986), vol. 1, Amer. Math. Soc., 1987.

[Catlin 1987b] D. Catlin, "Subelliptic estimates for the $\bar{\partial}$-Neumann problem on pseudoconvex domains", *Ann. of Math.* (2) **126**:1 (1987), 131–191.

[Chang et al. 1992] D.-C. Chang, A. Nagel, and E. M. Stein, "Estimates for the $\bar{\partial}$-Neumann problem in pseudoconvex domains of finite type in $\mathbb{C}^2$", *Acta Math.* **169**:3-4 (1992), 153–228.

[Chen 1987] S.-C. Chen, "Regularity of the Bergman projection on domains with partial transverse symmetries", *Math. Ann.* **277**:1 (1987), 135–140.

[Chen 1988] S.-C. Chen, "Global analytic hypoellipticity of the $\bar{\partial}$-Neumann problem on circular domains", *Invent. Math.* **92**:1 (1988), 173–185.

[Chen 1989] S.-C. Chen, "Global regularity of the $\bar{\partial}$-Neumann problem on circular domains", *Math. Ann.* **285**:1 (1989), 1–12.

[Chen 1991a] S.-C. Chen, "Global regularity of the $\bar{\partial}$-Neumann problem in dimension two", pp. 55–61 in *Several complex variables and complex geometry* (Santa Cruz, CA, 1989), vol. 3, edited by E. Bedford et al., Proc. Sympos. Pure Math. **52**, Amer. Math. Soc., Providence, RI, 1991.

[Chen 1991b] S.-C. Chen, "Regularity of the $\bar{\partial}$-Neumann problem", *Proc. Amer. Math. Soc.* **111**:3 (1991), 779–785.

[Cho 1995] S. Cho, "L^p boundedness of the Bergman projection on some pseudoconvex domains in $\mathbb{C}^n$", preprint, 1995. To appear in *Complex Variables*.

[Christ 1991] M. Christ, "Precise analysis of $\bar{\partial}_b$ and $\bar{\partial}$ on domains of finite type in $\mathbb{C}^2$", pp. 859–877 in *Proceedings of the International Congress of Mathematicians* (Kyoto, 1990), vol. 2, Math. Soc. Japan, Tokyo, 1991.

[Christ 1996a] M. Christ, "Global C^∞ irregularity of the $\bar{\partial}$-Neumann problem for worm domains", *J. Amer. Math. Soc.* **9**:4 (1996), 1171–1185.

[Christ 1996b] M. Christ, "The Szegő projection need not preserve global analyticity", *Ann. of Math.* (2) **143**:2 (1996), 301–330.

[Christ 1999] M. Christ, "Remarks on global irregularity in the $\bar{\partial}$-Neumann problem", pp. 161–198 in *Several complex variables*, edited by M. Schneider and Y.-T. Siu, Math. Sci. Res. Inst. Publ. **37**, Cambridge U. Press, New York, 1999.

[D'Angelo 1979] J. P. D'Angelo, "Finite type conditions for real hypersurfaces", *J. Differential Geom.* **14**:1 (1979), 59–66.

[D'Angelo 1980] J. P. D'Angelo, "Subelliptic estimates and failure of semicontinuity for orders of contact", *Duke Math. J.* **47**:4 (1980), 955–957.

[D'Angelo 1982] J. P. D'Angelo, "Real hypersurfaces, orders of contact, and applications", *Ann. of Math.* (2) **115**:3 (1982), 615–637.

[D'Angelo 1987a] J. P. D'Angelo, "Finite-type conditions for real hypersurfaces in $\mathbb{C}^n$", pp. 83–102 in *Complex analysis* (University Park, PA, 1986), edited by S. G. Krantz, Lecture Notes in Math. **1268**, Springer, 1987.

[D'Angelo 1987b] J. P. D'Angelo, "Iterated commutators and derivatives of the Levi form", pp. 103–110 in *Complex analysis* (University Park, PA, 1986), edited by S. G. Krantz, Lecture Notes in Math. **1268**, Springer, Berlin, 1987.

[D'Angelo 1993] J. P. D'Angelo, *Several complex variables and the geometry of real hypersurfaces*, Studies in Advanced Mathematics, CRC Press, Boca Raton, FL, 1993.

[D'Angelo 1995] J. P. D'Angelo, "Finite type conditions and subelliptic estimates", pp. 63–78 in *Modern methods in complex analysis* (Princeton, 1992), edited by T. Bloom et al., Annals of Mathematics Studies **137**, Princeton Univ. Press, Princeton, NJ, 1995.

[D'Angelo and Kohn 1999] J. P. D'Angelo and J. J. Kohn, "Subelliptic estimates and finite type", pp. 199–232 in *Several complex variables*, edited by M. Schneider and Y.-T. Siu, Math. Sci. Res. Inst. Publ. **37**, Cambridge U. Press, New York, 1999.

[Derridj 1978] M. Derridj, "Regularité pour $\bar{\partial}$ dans quelques domaines faiblement pseudoconvexes", *J. Differential Geom.* **13** (1978), 559–576.

[Derridj 1991] M. Derridj, "Domaines à estimation maximale", *Math. Z.* **208**:1 (1991), 71–88.

[Derridj and Tartakoff 1976] M. Derridj and D. S. Tartakoff, "On the global real analyticity of solutions to the $\bar{\partial}$-Neumann problem", *Comm. Partial Differential Equations* **1**:5 (1976), 401–435.

[Diederich and Fornæss 1977a] K. Diederich and J. E. Fornæss, "Pseudoconvex domains: an example with nontrivial Nebenhülle", *Math. Ann.* **225**:3 (1977), 275–292.

[Diederich and Fornæss 1977b]   K. Diederich and J. E. Fornæss, "Pseudoconvex domains: bounded strictly plurisubharmonic exhaustion functions", *Inv. Math.* **39** (1977), 129–141.

[Diederich and Fornæss 1977c]   K. Diederich and J. E. Fornæss, "Pseudoconvex domains: existence of Stein neighborhoods", *Duke Math. J.* **44**:3 (1977), 641–662.

[Diederich and Fornaess 1978]   K. Diederich and J. E. Fornaess, "Pseudoconvex domains with real-analytic boundary", *Ann. of Math.* (2) **107**:2 (1978), 371–384.

[Diederich and Fornæss 1982]   K. Diederich and J. E. Fornæss, "Boundary regularity of proper holomorphic mappings", *Invent. Math.* **67**:3 (1982), 363–384.

[Diederich and Herbort 1992]   K. Diederich and G. Herbort, "Extension of holomorphic L^2-functions with weighted growth conditions", *Nagoya Math. J.* **126** (1992), 141–157.

[Diederich and Herbort 1993]   K. Diederich and G. Herbort, "Geometric and analytic boundary invariants on pseudoconvex domains. Comparison results", *J. Geom. Anal.* **3**:3 (1993), 237–267.

[Diederich and Lieb 1981]   K. Diederich and I. Lieb, *Konvexität in der komplexen Analysis: neue Ergebnisse und Methoden*, DMV Seminar **2**, Birkhäuser, Basel, 1981.

[Diederich and Pflug 1981]   K. Diederich and P. Pflug, "Necessary conditions for hypoellipticity of the $\bar{\partial}$-problem", pp. 151–154 in *Recent developments in several complex variables*, edited by J. E. Fornæss, Annals of Mathematics Studies **100**, Princeton Univ. Press, Princeton, NJ, 1981.

[Fefferman 1995]   C. Fefferman, "On Kohn's microlocalization of $\bar{\partial}$ problems", pp. 119–133 in *Modern methods in complex analysis* (Princeton, 1992), edited by T. Bloom et al., Annals of Mathematics Studies **137**, Princeton Univ. Press, Princeton, NJ, 1995.

[Fefferman and Kohn 1988]   C. L. Fefferman and J. J. Kohn, "Hölder estimates on domains of complex dimension two and on three-dimensional CR manifolds", *Adv. in Math.* **69**:2 (1988), 223–303.

[Fefferman et al. 1990]   C. L. Fefferman, J. J. Kohn, and M. Machedon, "Hölder estimates on CR manifolds with a diagonalizable Levi form", *Adv. in Math.* **84**:1 (1990), 1–90.

[Folland and Kohn 1972]   G. B. Folland and J. J. Kohn, *The Neumann problem for the Cauchy-Riemann complex*, Annals of Mathematics Studies **75**, Princeton Univ. Press, Princeton, NJ, 1972.

[Fornæss 1979]   J. E. Fornæss, "Plurisubharmonic defining functions", *Pacific J. Math.* **80** (1979), 381–388.

[Fornæss and Sibony 1989]   J. E. Fornæss and N. Sibony, "Construction of P.S.H. functions on weakly pseudoconvex domains", *Duke Math. J.* **58**:3 (1989), 633–655.

[Forstnerič 1993]   F. Forstnerič, "Proper holomorphic mappings: a survey", pp. 297–363 in *Several complex variables* (Stockholm, 1987/1988), edited by J. E. Fornæss, Mathematical notes **38**, Princeton Univ. Press, Princeton, NJ, 1993.

[Fu et al. 1996]   S. Fu, A. V. Isaev, and S. G. Krantz, "Finite type conditions on Reinhardt domains", *Complex Variables Theory Appl.* **31**:4 (1996), 357–363.

[Grauert and Lieb 1970]   H. Grauert and I. Lieb, "Das Ramirezsche Integral und die Lösung der Gleichung $\bar{\partial}f = \alpha$ im Bereich der beschränkten Formen", *Rice Univ. Studies* **56**:2 (1970), 29–50.

[Greiner 1974] P. Greiner, "Subelliptic estimates for the $\bar\partial$-Neumann problem in $\mathbb{C}^2$", *J. Differential Geometry* **9** (1974), 239–250.

[Greiner and Stein 1977] P. C. Greiner and E. M. Stein, *Estimates for the $\bar\partial$-Neumann problem*, Mathematical Notes **19**, Princeton Univ. Press, Princeton, NJ, 1977.

[Henkin 1969] G. M. Henkin, "Integral representations of functions holomorphic in strictly pseudoconvex domains and some applications", *Mat. Sbornik* **78** (1969), 611–632. In Russian; translated in *Math. USSR Sbornik* **7** (1969), 597–616.

[Henkin 1970] G. M. Henkin, "Integral representations of functions in strictly pseudoconvex domains and applications to the $\bar\partial$-problem", *Mat. Sbornik* **82** (1970), 300–308. In Russian; translated in *Math. USSR Sbornik* **11** (1970), 273–281.

[Hörmander 1965] L. Hörmander, "L^2 estimates and existence theorems for the $\bar\partial$ operator", *Acta Math.* **113** (1965), 89–152.

[Hörmander 1990] L. Hörmander, *An introduction to complex analysis in several variables*, Third ed., North-Holland, Amsterdam, 1990.

[Hörmander 1994] L. Hörmander, *Notions of convexity*, Progress in Mathematics **127**, Birkhäuser, Basel, 1994.

[Kerzman 1971] N. Kerzman, "Hölder and L^p estimates for solutions of $\bar\partial u = f$ in strongly pseudoconvex domains", *Comm. Pure Appl. Math.* **24** (1971), 301–379.

[Kerzman 1972] N. Kerzman, "The Bergman kernel. Differentiability at the boundary", *Math. Ann.* **195** (1972), 149–158.

[Kiselman 1991] C. O. Kiselman, "A study of the Bergman projection in certain Hartogs domains", pp. 219–231 in *Several complex variables and complex geometry* (Santa Cruz, CA, 1989), vol. 3, edited by E. Bedford et al., Proc. Sympos. Pure Math. **52**, Amer. Math. Soc., Providence, RI, 1991.

[Kohn 1963] J. J. Kohn, "Harmonic integrals on strongly pseudo-convex manifolds, I", *Ann. of Math.* (2) **78** (1963), 112–148.

[Kohn 1964] J. J. Kohn, "Harmonic integrals on strongly pseudo-convex manifolds, II", *Ann. of Math.* (2) **79** (1964), 450–472.

[Kohn 1973] J. J. Kohn, "Global regularity for $\bar\partial$ on weakly pseudo-convex manifolds", *Trans. Amer. Math. Soc.* **181** (1973), 273–292.

[Kohn 1977] J. J. Kohn, "Methods of partial differential equations in complex analysis", pp. 215–237 in *Several complex variables* (Williamstown, MA, 1975), vol. 1, edited by R. O. Wells, Jr., Proc. Sympos. Pure Math. **30**, Amer. Math. Soc., 1977.

[Kohn 1979a] J. J. Kohn, "Subelliptic estimates", pp. 143–152 in *Harmonic analysis in Euclidean spaces* (Williamstown, MA, 1978), vol. 2, edited by G. Weiss and S. Wainger, Proc. Sympos. Pure Math. **35**, Amer. Math. Soc., 1979.

[Kohn 1979b] J. J. Kohn, "Subellipticity of the $\bar\partial$-Neumann problem on pseudo-convex domains: sufficient conditions", *Acta Math.* **142**:1-2 (1979), 79–122.

[Kohn 1981] J. J. Kohn, "Boundary regularity of $\bar\partial$", pp. 243–260 in *Recent developments in several complex variables*, edited by J. E. Fornæss, Annals of Mathematics Studies **100**, Princeton Univ. Press, Princeton, NJ, 1981.

[Kohn 1984] J. J. Kohn, "A survey of the $\bar\partial$-Neumann problem", pp. 137–145 in *Complex analysis of several variables* (Madison, WI, 1982), edited by Y.-T. Siu, Proc. Symp. Pure Math. **41**, Amer. Math. Soc., Providence, RI, 1984.

[Kohn 1998] J. J. Kohn, "Quantitative estimates for global regularity", preprint, 1998.

[Kohn and Nirenberg 1965] J. J. Kohn and L. Nirenberg, "Non-coercive boundary value problems", *Comm. Pure Appl. Math.* **18** (1965), 443–492.

[Komatsu 1976] G. Komatsu, "Global analytic-hypoellipticity of the $\bar{\partial}$-Neumann problem", *Tôhoku Math. J.* (2) **28**:1 (1976), 145–156.

[Komatsu 1984] G. Komatsu, "Boundedness of the Bergman projection and Bell's duality theorem", *Tôhoku Math. J.* (2) **36** (1984), 453–467.

[Krantz 1979] S. G. Krantz, "Characterizations of various domains of holomorphy via $\bar{\partial}$ estimates and applications to a problem of Kohn", *Illinois J. Math.* **23**:2 (1979), 267–285.

[Krantz 1988] S. G. Krantz, "Compactness of the $\bar{\partial}$-Neumann operator", *Proc. Amer. Math. Soc.* **103**:4 (1988), 1136–1138.

[Krantz 1992] S. G. Krantz, *Partial differential equations and complex analysis*, CRC Press, Boca Raton, FL, 1992.

[Lempert 1986] L. Lempert, "On the boundary behavior of holomorphic mappings", pp. 193–215 in *Contributions to several complex variables*, edited by A. Howard and P.-M. Wong, Aspects of Mathematics **E9**, Vieweg, Braunschweig, 1986.

[Lieb 1993] I. Lieb, "A survey of the $\bar{\partial}$-problem", pp. 457–472 in *Several complex variables* (Stockholm, 1987/1988), edited by J. E. Fornæss, Mathematical Notes **38**, Princeton Univ. Press, Princeton, NJ, 1993.

[Lieb and Range 1987] I. Lieb and R. M. Range, "The kernel of the $\bar{\partial}$-Neumann operator on strictly pseudoconvex domains", *Math. Ann.* **278**:1-4 (1987), 151–173.

[Ligocka 1980] E. Ligocka, "Some remarks on extension of biholomorphic mappings", pp. 350–363 in *Analytic functions* (Kozubnik, Poland, 1979), edited by J. Lawrynowicz, Lecture Notes in Math. **798**, Springer, Berlin, 1980.

[Ligocka 1981] E. Ligocka, "How to prove Fefferman's theorem without use of differential geometry", *Ann. Polonici Math.* **39** (1981), 117–130.

[Ligocka 1985] E. Ligocka, "The regularity of the weighted Bergman projections", pp. 197–203 in *Seminar on deformations* (Łódź and Warsaw, 1982/84), edited by J. Lawrynowicz, Lecture Notes in Math. **1165**, Springer, Berlin, 1985.

[Ligocka 1986] E. Ligocka, "The Sobolev spaces of harmonic functions", *Studia Math.* **84**:1 (1986), 79–87.

[Lions and Magenes 1972] J.-L. Lions and E. Magenes, *Non-homogeneous boundary value problems and applications*, vol. I, Grundlehren der Math. Wissenschaften **181**, Springer, Berlin, 1972.

[McNeal 1991] J. D. McNeal, "On sharp Hölder estimates for the solutions of the $\bar{\partial}$-equations", pp. 277–285 in *Several complex variables and complex geometry* (Santa Cruz, CA, 1989), vol. 3, edited by E. Bedford et al., Proc. Sympos. Pure Math. **52**, Amer. Math. Soc., Providence, RI, 1991.

[McNeal 1992] J. D. McNeal, "Convex domains of finite type", *J. Funct. Anal.* **108**:2 (1992), 361–373.

[McNeal 1996] J. D. McNeal, "On large values of L^2 holomorphic functions", *Math. Res. Lett.* **3**:2 (1996), 247–259.

[McNeal and Stein 1994] J. D. McNeal and E. M. Stein, "Mapping properties of the Bergman projection on convex domains of finite type", *Duke Math. J.* **73**:1 (1994), 177–199.

[Morrey 1958] C. B. Morrey, Jr., "The analytic embedding of abstract real-analytic manifolds", *Ann. of Math.* (2) **68** (1958), 159–201.

[Morrey 1963] C. B. Morrey, Jr., "The δ-Neumann problem on strongly pseudo-convex manifolds", pp. 171–178 in *Outlines of the joint Soviet-American symposium on partial differential equations* (Novosibirsk, 1963), Izdatel'stvo Akademii Nauk SSSR, Moscow, 1963.

[Morrey 1964] C. B. Morrey, Jr., "The $\bar{\partial}$-Neumann problem on strongly pseudo-convex manifolds", pp. 81–133 in *Differential analysis* (Bombay, 1964), Tata Inst. of Fundamental Research Studies in Math. **2**, Oxford Univ. Press, London, 1964.

[Nagel et al. 1989] A. Nagel, J.-P. Rosay, E. M. Stein, and S. Wainger, "Estimates for the Bergman and Szegő kernels in $\mathbb{C}^2$", *Ann. of Math.* (2) **129**:1 (1989), 113–149.

[Noell 1991] A. Noell, "Local versus global convexity of pseudoconvex domains", pp. 145–150 in *Several complex variables and complex geometry* (Santa Cruz, CA, 1989), vol. 3, edited by E. Bedford et al., Proc. Sympos. Pure Math. **52**, Amer. Math. Soc., Providence, RI, 1991.

[Ohsawa 1988] T. Ohsawa, "On the extension of L^2 holomorphic functions, II", *Publ. Res. Inst. Math. Sci. (Kyoto Univ.)* **24**:2 (1988), 265–275.

[Ohsawa and Takegoshi 1987] T. Ohsawa and K. Takegoshi, "On the extension of L^2 holomorphic functions", *Math. Z.* **195**:2 (1987), 197–204.

[Persson 1964] A. Persson, "Compact linear mappings between interpolation spaces", *Arkiv för Mat.* **5**:13 (1964), 215–219.

[Ramirez 1970] E. Ramirez, "Ein Divisionsproblem und Randintegraldarstellungen in der komplexen Analysis", *Math. Ann.* **184** (1970), 172–187.

[Range 1981] R. M. Range, "A remark on bounded strictly plurisubharmonic exhaustion functions", *Proc. Amer. Math. Soc.* **81**:2 (1981), 220–222.

[Range 1986] R. M. Range, *Holomorphic functions and integral representations in several complex variables*, Graduate Texts in Math. **108**, Springer, New York, 1986.

[Range 1987] R. M. Range, "Integral representations in the theory of the $\bar{\partial}$-Neumann problem", pp. 281–290 in *Complex analysis, II* (College Park, MD, 1985/1986), edited by C. A. Berenstein, Lecture Notes in Math. **1276**, Springer, New York, 1987.

[Salinas 1991] N. Salinas, "Noncompactness of the $\bar{\partial}$-Neumann problem and Toeplitz C^*-algebras", pp. 329–334 in *Several complex variables and complex geometry* (Santa Cruz, 1989), vol. 3, edited by E. Bedford et al., Proc. Symp. Pure Math. **52**, Amer. Math. Soc., Providence, RI, 1991.

[Shaw 1992] M.-C. Shaw, "Local existence theorems with estimates for $\bar{\partial}_b$ on weakly pseudo-convex CR manifolds", *Math. Ann.* **294**:4 (1992), 677–700.

[Sibony 1987] N. Sibony, "Une classe de domaines pseudoconvexes", *Duke Math. J.* **55**:2 (1987), 299–319.

[Sibony 1990] N. Sibony, personal correspondence, 1990.

[Siu 1996] Y.-T. Siu, "The Fujita conjecture and the extension theorem of Ohsawa–Takegoshi", pp. 577–592 in *Geometric complex analysis* (Hayama, 1995), edited by J. Noguchi et al., World Scientific, Singapore, 1996.

[Straube 1984] E. J. Straube, "Harmonic and analytic functions admitting a distribution boundary value", *Ann. Sc. Norm. Sup. Pisa, Cl. Sci.* (4) **11**:4 (1984), 559–591.

[Straube 1986] E. J. Straube, "Exact regularity of Bergman, Szegő and Sobolev space projections in non pseudoconvex domains", *Math. Z.* **192** (1986), 117–128.

[Straube 1995] E. J. Straube, "A remark on Hölder smoothing and subellipticity of the $\bar{\partial}$-Neumann operator", *Comm. in Partial Differential Equations* **20**:1-2 (1995), 267–275.

[Tartakoff 1978] D. S. Tartakoff, "Local analytic hypoellipticity for $\Box_b$ on nondegenerate Cauchy-Riemann manifolds", *Proc. Nat. Acad. Sci. U.S.A.* **75**:7 (1978), 3027–3028.

[Tartakoff 1980] D. Tartakoff, "On the local real analyticity of solutions to $\Box_b$ and the $\bar{\partial}$-Neumann problem", *Acta Math.* **145** (1980), 117–204.

[Tolli 1996] F. Tolli, *Analytic hypoellipticity on a convex bounded domain*, Ph.D. dissertation, UCLA, March 1996.

[Treves 1975] F. Treves, *Basic linear partial differential equations*, Pure and Applied Mathematics **62**, Academic Press, New York, 1975.

[Treves 1978] F. Treves, "Analytic hypo-ellipticity of a class of pseudodifferential operators with double characteristics and applications to the $\bar{\partial}$-Neumann problem", *Comm. Partial Differential Equations* **3**:6-7 (1978), 475–642.

[Webster 1979] S. M. Webster, "Biholomorphic mappings and the Bergman kernel off the diagonal", *Invent. Math.* **51**:2 (1979), 155–169.

[Yie 1995] S. L. Yie, *Solutions of Cauchy Riemann equations on pseudoconvex domain with nonsmooth boundary*, Ph.D. thesis, Purdue University, West Lafayette, IN, 1995.

HAROLD P. BOAS
DEPARTMENT OF MATHEMATICS
TEXAS A&M UNIVERSITY
COLLEGE STATION, TX 77843-3368
UNITED STATES
boas@math.tamu.edu

EMIL J. STRAUBE
DEPARTMENT OF MATHEMATICS
TEXAS A&M UNIVERSITY
COLLEGE STATION, TX 77843-3368
UNITED STATES
straube@math.tamu.edu

Several Complex Variables
MSRI Publications
Volume **37**, 1999

Recent Developments in the Classification Theory of Compact Kähler Manifolds

FRÉDÉRIC CAMPANA AND THOMAS PETERNELL

ABSTRACT. We review some of the major recent developments in global complex geometry, specifically:

1. Mori theory, rational curves and the structure of Fano manifolds.
2. Non-splitting families of rational curves and the structure of compact Kähler threefolds.
3. Topology of compact Kähler manifolds: topological versus analytic isomorphism.
4. Topology of compact Kähler manifolds: the fundamental group.
5. Biregular classification: curvature and manifolds with nef tangent/anti-canonical bundles.

Introduction

This article reports some of the recent developments in the classification theory of compact complex Kähler manifolds with special emphasis on manifolds of non-positive Kodaira dimension (vaguely: semipositively curved manifolds). In the introduction we want to give some general comments on classification theory concerning main principles, objectives and methods. Of course one could ask more generally for a classification theory of arbitrary compact manifolds but this seems hopeless as most of the techniques available break down in the "general" case (such as Hodge theory). Also there are a lot of pathologies which tell us to introduce some reasonable assumptions. From an algebraic point of view one will restrict to projective manifolds but from a more complex-analytic viewpoint, the Kähler condition is the most natural. Clearly manifolds which are only bimeromorphic to a projective or Kähler manifold are interesting, too, but these will be mainly ignored in this article and might occur only as intermediate products. The most basic questions in classification theory are the following.

(A) Which topological or differentiable manifolds carry a complex (algebraic or Kähler) structure? If a topological manifold carries a complex structure, try to describe them (moduli spaces, deformations, invariants).

(B) Birational or bimeromorphic classification: describe manifolds up to bimeromorphic equivalence and try to find nice models in every class.

(C) Biregular theory: try to describe manifolds up to biholomorphic equivalence;
this is only possible with additional assumptions (such as curvature), study
their properties and invariants.

There are also intermediate questions such as: What happens to the bimeromorphic class of a manifold in a deformation?

We give some more explanations to the single problems and relate then to the
content of this article.

(A) We will mainly ignore the existence problem, which has not been of central
interest in the past except for low dimensions. As to the moduli problem for
complex structures, the first thing is to look for invariants, in particular the
Kodaira dimension. For surfaces there is a big difference between topological
and differentiable isomorphy: the Kodaira dimension is a diffeomorphic invariant
but not a topological one (Donaldson). In dimension 3 the difference between
topological and differentiable equivalence vanishes and therefore 3-folds which
are diffeomorphic need not have the same Kodaira dimension. Nevertheless
one can still ask to "classify" all complex Kähler structures for a given Kähler
manifold and in particular to determine as many invariants as possible. The
strongest assertions one can look for would predict that for restricted classes
of manifolds topological equivalence already implies biholomorphic or at least
deformation equivalence — for example, for Fano manifolds with $b_2(X) = 1$.
This is a very difficult question and even in dimension 3 it is known only for a
few examples such as projective space. The problem gets still more difficult if
one looks for all complex structures; then we are far from giving the answer even
for projective 3-space. One main difficulty is the lacking of a new topological
invariant, such as the holomorphic Euler characteristic, in dimension 3. These
and related questions will be discussed in section 4. One of the most subtle
topological invariants of compact Kähler manifolds is certainly the fundamental
group which has attracted much interest in the last few years. We discuss this
in section 5.

(B) The most important birational (bimeromorphic) invariant is the Kodaira
dimension. Therefore one wants to study the structure of the particular classes of
manifolds X of a given Kodaira dimension $\kappa = \kappa(X)$. The most interesting cases
are $\kappa = -\infty, 0$ and $\dim X$, while the cases $1 \leq \kappa \leq \dim X - 1$ are "interpolations"
of these (in lower dimension) in terms of fiber spaces. We concentrate on the class
of varieties with negative κ; it is studied in detail in Section 1. In the context of
birational geometry two varieties are considered equal, if they coincide after some
birational surgery such as blow-ups. Therefore one is looking for good birational
models. The construction of such models in dimension 3 in the projective case,
the so-called Mori theory, is discussed in Section 2. It depends on a numerical

theory of the canonical bundle on X. The theory lives from projective techniques but the results should hold in the Kähler case, too. Some results in this direction are discussed in Section 3; it seems that a general theory needs a new, analytic way to construct rational curves.

(C) Here we want to study a single manifold as individual or a specific class of manifolds. A typical problem: classify manifolds with certain curvature conditions: for example, semipositive holomorphic bisectional or Ricci curvature. This is discussed in Section 6. Most of this type of problems deal with manifolds which are not of general or of non-positive Kodaira dimension because these have a richer geometry (and are fewer, hence more rigid.)

Sections 1, 2 and 5 have mainly written up by Campana, while Peternell is responsible for the rest. Both authors had the opportunity to spend a significant period at MSRI during the special year on complex analysis 1995/96. They would like to thank the institute for the support and the excellent working conditions.

1. Birational Classification. The Kodaira Dimension.

In this section we introduce the Kodaira dimension of compact (Kähler) manifolds and discuss the class of projective manifolds with negative Kodaira dimension. Furthermore a refined version of the Kodaira dimension is introduced.

Let X be a compact Kähler manifold with canonical bundle K_X and L a line bundle on X. Let $n = \dim X$.

DEFINITION 1.1. The **Iitaka dimension** of L is defined by

$$\kappa(X, L) := \varlimsup_{\substack{m > 0 \\ m \to \infty}} \left(\frac{\log h^0(X, mL)}{\log m} \right).$$

The **Kodaira dimension** of X is

$$\kappa(X) := \kappa(X, K_X).$$

This definition is short but not very illuminating. It can be more concretely described as follows:

$$\kappa(X, L) = \begin{cases} -\infty & \text{if and only if } h^0(X, mL) = 0 \text{ for all } m > 0, \\ 0 & \text{if and only if } h^0(X, mL) \leq 1 \text{ (and not always 0).} \end{cases}$$

In fact Iitaka showed that if $\kappa(X, L) \geq 0$ there exist $d \in \{0, \dots, n\}$, a number $m_0 > 0$ and constants $0 < A < B$ such that

$$Am^d \leq h^0(X, mL) \leq Am^d \quad \text{for all } m \in \mathbb{N} \text{ divisible by } m_0.$$

In geometrical terms we have

$$d = \max \dim \Phi_m(X),$$

where $\Phi_m : X \dashrightarrow \mathbb{P}(H^0(X, mL)^*)$ is defined by the linear system $|mL|$ (when $\kappa(X) \geq 0$).

EXAMPLES 1.2. (1) If $\dim X = 1$, then:

$$\kappa(X) = \begin{cases} -\infty & \text{if and only if } X = \mathbb{P}_1, \\ 0 & \text{if and only if } g(X) = 1 \quad (X \text{ is an elliptic curve}), \\ 1 & \text{if and only if } g(X) \geq 2. \end{cases}$$

(2) If $\dim X = 2$, we have the Kodaira–Enriques classification of algebraic surfaces with invariants $\kappa = \kappa(X)$ and $q = h^1(X, \mathcal{O}_X)$, which are both birational invariants. The following table gives the values of κ and q when X is bimeromorphic to a surface of the specified type:

$\kappa = -\infty$	$q = 0$	$\mathbb{P}_2$
$\kappa = -\infty$	$q \geq 1$	$\mathbb{P}_1 \times C$, where C is a smooth curve with $g(C) = q$
$\kappa = 0$	$q = 0$	a K3 or Enriques surface ($q = 0$; $K = \mathcal{O}_X$ or $2K_X = \mathcal{O}_X$)
$\kappa = 0$	$q = 1$	a bielliptic surface
$\kappa = 0$	$q = 2$	an abelian surface
$\kappa = 1$		an elliptic fibration given by $|mK_X|$
$\kappa = 2$		general type: $|mK_X|$ gives a birational map (for $m \geq 5$)

(3) $n = \dim X$ arbitrary. Assume $\kappa(X) \geq 0$. The linear system $|mK_X|$ defines a rational dominant map

$$\Phi_m : X \dashrightarrow Y$$

with connected fibers, called the **Iitaka fibration** of X [Ueno 1975] and $\dim Y = \kappa(X)$. The general fiber X_y of Φ_m has $\kappa(X_y) = 0$. This map is a birational invariant of X via the birational invariance of the plurigenera $P_m := h^0(X, mK_X)$.

The class of projective (smooth) n-folds thus falls into $n + 1$ classes, according to the value of κ. There are 3 "new" classes in each dimension n:

(a) $\kappa(X) = -\infty$: The linear systems $|mK_X|$ do not give any information.

(b) $\kappa(X) = 0$.

(c) $\kappa(X) = n$ (X is said to be of "general type").

Indeed, for the classes $1 \leq \kappa(X) \leq n - 1$, the Iitaka fibration expresses X as a fibration over a lower-dimensional manifold, and with fibers having $\kappa = 0$. This reduces largely the structure of X to lower-dimensional cases.

As we can see from the case of curves and surfaces, the 3 classes above differ completely: the special ones have $\kappa = -\infty$ or $\kappa = 0$, whereas the general ones have $\kappa = 2$ (hence the name).

We now discuss manifolds with $\kappa = -\infty$. Here we have a standard conjecture:

CONJECTURE 1.3. Let X be a projective manifold. Then $\kappa = -\infty$ if and only if X is uniruled.

Recall that X is said to be **uniruled** if there exists a dominant rational map $\psi : \mathbb{P}_1 \times T \dashrightarrow X$ with $\dim T = \dim X - 1$. In other words: there exists a rational curve going through the general point of X.

The "only if" part is an easy exercise. The converse is known for $n \leq 3$. For $n = 2$ this is the Enriques classification and results from the famous "Castelnuovo criterion"; for $n = 3$, this is a deep theorem proved by Y. Miyaoka using results of Kawamata and Mori (see Section 2).

There is however a big difference (from the birational point of view) between $\mathbb{P}_1 \times \mathbb{P}_1$ say and $\mathbb{P}_1 \times C$ where C is a curve with $g(C) \geq 1$. In fact $\mathbb{P}_1 \times C$ has much less rational curves than $\mathbb{P}_1 \times \mathbb{P}_1$. We try to make this more precise:

DEFINITION 1.4. Let X be a projective manifold.

(1) X is **rationally generated** if for any dominant map $\varphi : X \dashrightarrow Y$, the variety Y is uniruled.

(2) X is **rationally connected** if any two generic points on X can be joined by a **rational chain** C (that is, a connected curve with all its irreducible components rational).

(2′) X is **strongly rationally connected** if moreover the chain C in (2) can be chosen to be irreducible.

(3) X is **unirational** if there is a dominant rational map $\varphi : \mathbb{P}_n \dashrightarrow X$. If moreover φ is birational we say that X is **rational**.

Notice the following obvious implications:

X rational $\Rightarrow X$ unirational $\Rightarrow X$ strongly rationally connected $\Rightarrow$
$$\Rightarrow X \text{ rationally connected} \Rightarrow X \text{ rationally generated} \Rightarrow X \text{ uniruled.}$$

COMMENTS. When $n = 1$, all these properties are equivalent.

When $n = 2$, rationality is equivalent to rational generatedness but of course uniruledness is weaker then the other properties.

When $n = 3$, rational connectedness is equivalent to strong rational connectedness by [Kollár et al. 1992a], and unirationality is distinct from rationality (as shown in [Clemens and Griffiths 1972], the cubic hypersurface in $\mathbb{P}_4$ is non-rational, but unirational). It is also unknown whether rational connectedness implies unirationality; this is in fact doubtfull: the general quartic hypersurfaces in $\mathbb{P}_4$ are rationally connected but expected not to be unirational.

The only known reverse implication for arbitrary dimension is that rational connectedness implies strong rational connectedness; see [Kollár et al. 1992a] (the smoothness assumption is essential here: consider the cone over an elliptic curve!). This is based on relative deformation theory of maps.

1.5. We now discuss the difference between rational connectedness and rational generatedness in a special case: let $\varphi : X \to \mathbb{P}_1$ be a (regular) map with generic fiber X_λ rationally connected. Then X is obviously rationally generated. But it is rationally connected if and only if there is a rational curve C in X such that $\varphi(C) = \mathbb{P}_1$. It is in fact sufficient to check the equivalence of rational generatedness and rational connectedness in this special case in order to prove that the two notions coincide in general. However this equivalence might very well be a low-dimensional phenomenon.

An important example of rationally connected manifolds are the Fano manifolds (see Section 2). Conversely, we ask:

REMARK 1.6. It is unlikely that every rationally connected manifold is birational to some Fano variety. In fact, there are infinitely many birationally inequivalent families of conic bundles over surfaces, whereas it is expected that there are only finitely many families of Fano 3-folds. This last fact is known — as explained later — in the smooth case and it is also known in the singular case if $b_2(X) = 1$ (Kawamata).

THEOREM 1.7 [Campana 1992; Kollár et al. 1992a]. *Let X be a smooth projective n-fold. There exists a unique dominant rational map $\rho : X \dashrightarrow X_1$ such that, for x "**general**" in X, the fiber of ρ through x consists of the points $x' \in X$ which can be joined to x by some rational chain C. Moreover, ρ is a "**quasi-fibration**", so that its generic fiber is smooth and rationally connected. It is characterized by the following property: for any dominant $\rho' : X \dashrightarrow Y$ with generic fiber rationally connected, ρ' dominates ρ (that is, there exists $\psi : Y \dashrightarrow X_1$ such that $\psi \circ \rho' = \rho$). The map ρ is a birational invariant of X.*

Recall that $x \in X$ is **general** if it lies outside a countable union of Zariski closed subsets with empty interior, and that $\rho : X \to X_1$ is a **quasi-fibration** if there exist Zariski open nonempty subsets V of Y_1, and U of X such that the restriction of ρ to U is regular, maps U to V and $\rho : U \to V$ is proper. (In other words: the indeterminacy locus of ρ is **not** mapped **onto** X_1).

The map ρ above is called the maximal rationally connected fibration in [Kollár et al. 1992a] and the rational quotient in [Campana 1992].

Notice that this construction holds also for X compact Kähler [Campana 1992] and for X only normal. But in the normal case it is in general no longer a birational invariant.

Now it might happen that X_1 is again uniruled (this would happen precisely if the general fiber of $\rho_1 \circ \rho$ is rationally generated but not rationally connected, ρ_1 being explained in the next sentence). So X_1 has a rational quotient $\rho_1 : X_1 \to X_2$ as well. Proceeding this way, the dimension decreases by at least one at each step until it finally stops. Therefore we can state:

COROLLARY 1.8 [Campana 1995a]. *There exists a (unique) rational dominant map $\sigma : X \dashrightarrow S(X)$ to a non-uniruled variety $S(X)$ and with generic rationally generated fiber. It dominates any other $\sigma' : X \dashrightarrow Y$ with Y non-uniruled, and* *is dominated by any* $\sigma' : X \dashrightarrow Y$ *with generic rational generated fiber. This map* σ *is a quasi-fibration and a birational invariant (for X smooth).*

We call σ the LNU-quotient of X (for "largest non-uniruled"), or the MRG-fibration of X (for "maximal rationally generated").

Notice that, by convention, **a point is not uniruled** in case X itself is rationally generated, and that $X = S(X)$ if X is not uniruled.

We now introduce, after [Campana 1995a], a refined Kodaira dimension which should (at least conjecturally) calculate $\kappa(S(X))$, and plays an essential role in Section 5; it leads also to refinements of Conjecture 1.3 above.

DEFINITION 1.9 [Campana 1995a]. Let X be a compact complex manifold. We define

$$\kappa_+(X) := \max\{\kappa(Y) \mid \text{there exists } \varphi : X \dashrightarrow Y \text{ dominant}\}$$

and

$$\kappa^+(X) := \max\{\kappa(X, \det \mathcal{F}) \mid \mathcal{F} \neq 0 \text{ is a coherent subsheaf of } \Omega_X^p, \text{ for some } p > 0\}.$$

Here $\det(\mathcal{F})$ is the saturation of $\det \mathcal{F} \subset \bigwedge^r \Omega_X^p$ if $r = \mathrm{rk}(\mathcal{F})$.

We have the following easy properties of κ_+ and κ^+. Here $a(X)$ denotes the algebraic dimension; see Theorem and Definition 3.1.

PROPOSITION 1.10 [Campana 1995a]. (1) κ_+ *and* κ^+ *are birational invariants.*
(2) *If* $\varphi : X \to Y$ *is dominant, then* $\kappa^+(X) \geq \kappa^+(Y)$ *(and similarly of course for* κ_+).
(3) $\dim(X) \geq a(X) \geq \kappa^+(X) \geq \kappa_+(X) \geq \kappa(X) \geq -\infty$.
(4) *If* $\varphi : X \to Y$ *has a generic fiber which is rationally generated, then* $\kappa^+(X) = \kappa^+(Y)$ *and* $\kappa_+(X) = \kappa_+(Y)$.
(4') *If X is rationally generated, then* $\kappa^+(X) = \kappa_+(X) = -\infty$.

EXAMPLES 1.11 (curves and surfaces). (1) If $\dim X = 1$, then $\kappa^+ = \kappa_+ = \kappa$.
(2) If $\dim X = 2$, the situation is more interesting:

(a) $\kappa^+(X) = \kappa_+(X) = \kappa(X)$ if $\kappa(X) \geq 0$ (use for example the Castelnuovo–de Franchis theorem). Thus only when $\kappa(X) = -\infty$ we get more information on X from κ^+ than from κ.

(b) If $\kappa(X) = -\infty$ then $\kappa^+(X) = \kappa_+(X) = -\infty$ if and only if X is rational; and $\kappa^+(X) = \kappa_+(X) = 0$ (respectively 1) if and only if X is birational to $\mathbb{P}_1 \times B$, where B is a curve of genus $g = 1$ (respectively $g \geq 2$).

CONJECTURE 1.12 [Campana 1995a]. Let X be a projective (or compact Kähler) manifold. Then:

(a) $\kappa^+(X) = \kappa_+(X) = \kappa(X)$ if $\kappa(X) \geq 0$.
(b) $\kappa^+(X) = \kappa_+(X) = \kappa(S(X))$ if $S(X)$ is the LNU-quotient of X (see paragraph after Corollary 1.8), unless X is rationally generated (that is, $S(X)$ is a point).

(c) $\kappa^+(X) = -\infty$ if and only if X is rationally generated.

This conjecture is in fact a consequence of Conjecture 1.3 and of standard conjectures in the Minimal Model Program. More precisely: 1.11 holds if 1.3 holds and if every projective (or compact Kähler) manifold with $\kappa(X) = 0$ is bimeromorphic to a variety X' with only $\mathbb{Q}$-factorial terminal singularities and such that $K_{X'} \equiv 0$ (or $c_1(X') = 0$). See Section 2 for the terminology.

The reduction of 1.11 to these other conjectures rests in the projective case on Miyaoka's generic semipositivity theorem (Theorem 2.7), and thus on characteristic $p > 0$ methods.

Observe finally that the class of rationally generated manifolds is invariant under deformations, and that all sections of tensor bundles vanish for manifolds in that class. This makes these manifolds difficult to distinguish from rationally connected or unirational manifolds. Should the properties "rationally connected" and "rationally generated" be different, the right class to consider (characterized by $\kappa^+ = -\infty$) is the class of rationally generated manifolds.

We shall see in Section 5 that π_1 vanishes for these, too.

EXAMPLES 1.13. We give some instances in which Conjecture 1.11 holds.

(1) If $n \leq 3$ and if X is projective, the conjecture holds:

    (a) For $n = 1$, this is obvious since $\kappa^+ = \kappa$ for curves.

    (b) For $n = 2$, this is easy, too, because the only non-trivial sheaves $\mathcal{F} \subsetneq \Omega_X^p$ appearing are of rank one in Ω_X^1. The Castelnuovo–de Franchis theorem then applies and solves the problem (in the Kähler case as well).

    (c) For $n = 3$, this is a consequence of the fact that the Minimal Model Program and Abundance conjecture have been solved in that dimension by the Japanese School (Kawamata, Miyaoka, Mori). See Section 2 for more details.

(2) If $c_1(X) = 0$, the conjecture holds (that is, $\kappa^+(X) = \kappa(X) = 0$). This is proved in [Campana 1995a] using Miyaoka's generic semipositivity theorem (our Theorem 2.7) if X is projective. In the Kähler case, this is an easy consequence of the existence of Ricci-flat Kähler metrics: holomorphic tensors are parallel.

(3) If K_X is nef and $\kappa^+(X) = n$, then $\kappa(X) = n$, too. The proof involves Miyaoka's generic semipositivity theorem again, but is more involved.

2. Numerical Theory: The Minimal Model Program

This section gives a short introduction to the minimal model theory or Mori theory of projective manifolds. It consists of two parts:

(a) producing a "contraction" of X when the canonical bundle K_X of a projective manifold X is not nef;

(b) giving a structure theorem in case K_X is nef, namely that mK_X is generated by global sections.

Let X be a projective manifold, and L a line bundle on X. Recall that L is **nef** if $L.C \geq 0$ for any effective curve C in X, and ample if the linear system $|mL|$ provides an embedding for some $m > 0$. If mL is generated by global sections for some $m > 0$, then L is nef (the converse is not true in general).

There is also a relative version: if $\varphi : X \to Y$ is a morphism, then L is φ-nef if $L.C \geq 0$ for any curve C in X contained in some fiber of φ; and L is φ-ample if the natural evaluation map $\varphi^*\varphi_*(mL) \to mL$ is surjective for some $m > 0$ and defines an embedding of X over Y in $\mathbb{P}\big(\varphi^*\varphi_*(mL)\big)$.

We denote by $\equiv$ numerical equivalence of Cartier divisors.

2.1. INTRODUCTION. As already seen, projective n-folds X fall into 2 classes, according to their value of κ:

(1) $\kappa(X) = -\infty$.

(2) $\kappa(X) \geq 0$: the Iitaka fibration $I_X : X \dashrightarrow Y$ reduces the structure of X to that of $I(X)$ and its general fiber X_y, which has $\kappa(X_y) = 0$. One is thus largely reduced to lower-dimensional varieties, except in the two extreme cases: $\kappa(X) = 0$ and $\kappa(X) = n$.

Classes 1 and 2 above contain their numerical analogues (1′) and (2′) defined as follows:

(1′) X is a **Fano fibration** (that is, there exists a map $\varphi : X \to Y$ such that K_X^{-1} is φ-ample). An extreme case is when φ is constant (that is, K_X^{-1} is ample). In this case by definition X is said to be **Fano** (or **del Pezzo** when $n = 2$).

(2′) K_X is nef and mK_X is generated by global sections for some $m \gg 0$.

In this case, $I_X : X \to Y$ is a morphism defined by the linear system $|m'K_X|$, for a suitable m'. In the special case $\kappa(X) = 0$ condition (2′) means that K_X is **torsion**, and in the case $\kappa(X) = n$ it means that K_X is **ample**. Observe that K_{X_y} is torsion for the generic fiber X_y of I_X.

A natural question is whether conversely any X has a (birational) **minimal model** X' in one of the classes (1′) or (2′) above, with mild singularities.

As we shall see below, the answer is yes for $n \leq 3$ (and conjecturally for all n). The interest in dealing with varieties in classes (1′) and (2′) is twofold:

(a) a precise biregular classification of X' can be expected by the study of Fano manifolds in case (1′) and the study of the linear systems $|mK_X|$ for the class (2′).

(b) The knowledge of the numerical invariants of X' — in contrary to the birational ones — allows the use of Riemann–Roch formula and vanishing theorems.

We look first to the case of dimension 2, where the situation is classically under-stood, although not from that point of view. Later on, we shall describe what happens for $n = 3$, where new phenomena occur, discovered mainly by S. Mori in 1980–1988.

THEOREM 2.2. *Let X be a smooth projective surface. Exactly one of the follow-ing possibilities occurs*:

(1) K_X *is nef.*
(2) X *contains a* (-1)-*curve.*
(3) X *is ruled, that is, it admits a $\mathbb{P}_1$-bundle structure $\rho : X \to B$ over a curve B.*
(4) $X \cong \mathbb{P}_2$.

Recall that a (-1)-curve is a curve C such that $C \simeq \mathbb{P}_1$ and $N_{C|X} \simeq \mathcal{O}(-1)$. Such curves are numerically characterized by: $K.C < 0$ and $C^2 < 0$; see [Barth et al. 1984], for example. Every (-1)-curve is the exceptional divisor of a contraction: $\gamma : X \to X_1$, where X_1 is a smooth surface, $\gamma(C) = x_1 \in X_1$ is a point and γ is the blow-up of this point in X_1, with $C = \gamma^{-1}(x_1)$. Such a contraction decreases b_2 by one. Thus, after contracting finitely many (-1)-curves, one gets a smooth surface X', birational to X, such that either (1), (3) or (4) holds for X'. In case (1) we say that X' is a **minimal model for** X (it is in fact unique in that case, so that its numerical invariants are birational invariants for X). In case (3) and (4), we get a Fano-fibration for X' (the map ρ might be the constant map).

We say a few words about a possible proof of Theorem 2.2: Assume that (1) and (2) do not hold. Then $-K_X.C > 0$ for some curve C, and $C^2 \geq 0$ for any such C. The all point is then to show that C can be chosen to be **rational**. This can be shown easily when $q(X) := h^1(\mathcal{O}_X) = 0$ by using the arguments of ([Barth et al. 1984]), and similar ones when $q(X) > 0$ (after introducing the Albanese map whose image is a curve in that case; the point is just to show that this is the desired ruling).

The second step to conclude the program above is then:

THEOREM 2.3. *Assume that K_X is nef. Then mK_X is generated by global sections for some $m > 0$.*

Here again, the proof can be divided into cases (we know that $K_X^2 \geq 0$ since K_X is nef):

(1) $K_X^2 > 0$. This case is easy — no special property of K is needed.
(2) $K_X^2 = 0$, but $K_X \not\equiv 0$. One just has to show that $P_m := h^0(K_X^m) \geq 2$ for some $m > 0$. We thus only need to consider the special case of surfaces with $p_g := h^0(K_X) = 0, 1$.
(3) $K_X \equiv 0$. One has to show that: $h^0(mK_X) = P_m > 0$ for some m. Again one has to consider the special case $p_g = 0$.

By the Noether formula, $\chi(\mathcal{O}_X) \geq 0$ and $q \leq 2$ in all the special cases—with equality only if the Albanese image is an abelian surface. The situations in 2 and 3 can then be classified (one can use the arguments of [Beauville 1978, VI and VII], for example).

Theorem 2.2 generalizes to the 3-dimensional case as follows:

THEOREM 2.4 [Mori 1982]. *Let X be a smooth projective 3-fold. Exactly one of the following situations occurs*:

(1) K_X *is nef*.

(2) X *contains a rational curve C such that $-4 \leq K_X.C \leq -1$, and there exists a unique morphism $\varphi : X \to Y$ with connected fibers to a projective normal variety Y such that K_X^{-1} is φ-ample, $\rho(X) = 1 + \rho(Y)$ and φ maps to points exactly the curves C' which are numerically proportional to C. The map φ, called an* **extremal contraction**, *is of one of the following types*:

    (2a) φ *is birational. It then contracts an irreducible divisor E to either a point of a curve. There are five possible situations*:

        (2a1) $\varphi(E)$ *is a smooth curve of Y blown-up by φ*.

        (2a2) $E \simeq \mathbb{P}_2$; $N_{E|X} \simeq \mathcal{O}(-1)$; $y = \varphi(E)$ *is a smooth point blown-up by φ*.

        (2a3) $E = \mathbb{P}_2$; $N_{E|X} \simeq \mathcal{O}(-2)$; $y = \varphi(E)$ *is singular*.

        (2a4) $E \simeq \mathbb{P}_1 \times \mathbb{P}_1$; $N_{E|X} \simeq \mathcal{O}(-1,-1)$; $\varphi(E) = y$ *is a ordinary double point*.

        (2a5) E *is a quadric cone in $\mathbb{P}_3$*; $N_{E|X} \simeq \mathcal{O}_{\mathbb{P}_3(-1)}$; $\varphi(E) = y$ *is analytically* $u^2 + v^2 + w^2 + t^3 = 0$.

    (2b) Y *is a surface; then K_X^{-1} is φ-ample and φ is a conic bundle*.

    (2c) Y *is a curve; then K_X^{-1} is φ-ample and one says φ is a del Pezzo fibration*.

    (2d) Y *is a point; then X is Fano with $\rho(X) = 1$*.

This result also shows the non-apparent relationship between the cases (2), (3), (4) of Theorem 2.2.

We say a few words of the proof of Theorem 2.4: it is very different from the proof of Theorem 2.2, which proceeds by classical methods using linear systems and Riemann–Roch. These methods are not available in higher dimensions since the curves are no longer divisors. (The proof of S. Mori about the existence of a rational curve C such that $0 < K_X^{-1}.C < n + 1$ and about the existence of φ works in every dimension $n = \dim X$.) Instead, the proof of Theorem 2.4 is based on deformation theory (of maps) and uses in an essential way the Frobenius morphisms in characteristic $p > 0$. The curve C is first constructed by reduction (mod p) in characteristic p and then lifted in characteristic zero. The existence (and list) of the extremal contraction φ is deduced from a detailed study of the deformations of C (with $0 < K_X^{-1}.C \leq 4$ taken as small as possible).

There is another approach, cohomological, to the existence of extremal contractions. It has been developed essentially by Kawamata, and works for varieties with only $\mathbb{Q}$-factorial, terminal singularities (see below). It does not give in general the existence of a rational curve C as above.

One of the main differences between Theorems 2.2 and 2.4 is that $Y = X_1$ is no longer smooth in general, so that the operation can a priori not be iterated.

So the next step is whether there is a reasonable class of singularities to allow for which the elementary contractions can be defined without leaving that class.

There are two guiding principles for conditions to be imposed on the singularities:

(1) K_X being nef (or φ-ample) should have a meaning in terms of intersection numbers, that is, $K_X \cdot C$ must have a meaning for every curve $C \subset X$. This is true if K_X is not only a Weil, but a Cartier divisor. However K_Y is not Cartier in case (2a3) of Theorem 2.4. But at least, some multiple mK_Y of K_Y becomes Cartier; so that one can define $K_Y.C := (1/m)(mK_Y.C)$ with the usual properties. One therefore says that Y has only $\mathbb{Q}$-**factorial singularities** if any Weil divisor D on Y is $\mathbb{Q}$-**Cartier** (that is, mY is Cartier for some $m \neq 0$). See [Reid 1983; 1987] for a detailed introduction to these questions.

(2) The second property one can ask is that the singularities do not effect the plurigenera. In other words: if $\tilde{Y} \overset{\delta}{\to} Y$ is any resolution of Y and mK_Y is Cartier, then $H^0(\tilde{Y}, mK_{\tilde{Y}}) = \delta^* H^0(Y, mK_Y)$. This does not depend on the resolution and is certainly guaranteed if

$$K_Y = \delta_* K_{\tilde{Y}}.$$

To understand this condition, write

$$K_{\tilde{Y}} = \delta^* K_Y + \Sigma \delta_i E_i,$$

where the E_i's are the divisors contracted by δ, where $\delta_i \in \mathbb{Z}$. Then the equality above holds precisely when $\delta_i \geq 0$ for any i. Such singularities are called **canonical**. Since however we are mostly interested in birational contractions $\varphi : X \to Y$ for which K_X^{-1} is φ-ample, it is natural to impose a more restrictive condition, namely that $\delta_i > 0$ for all i. We then say that Y has only **terminal singularities** if this is the case.

It turns out that the class of normal varieties Y "with only $\mathbb{Q}$-factorial terminal singularities" seems to be precisely the right one to consider: extremal contractions still exist and if $\varphi : X \to Y$ is such a contraction which is birational and contracting some divisor, then Y is again in the same class. Moreover terminal surface singularities are smooth points and, more generally, terminal singularities occur only in codimension at least 3. In particular, they are isolated in dimension 3.

Fortunately, if Y is a projective 3-fold with at most $\mathbb{Q}$-factorial terminal singularities with K_Y not nef, then a rational curve with $-K_X.C > 0$ still exists, and also an extremal contraction $\varphi : X \to Y$ still exists. However, unlike in the case X is smooth, this contraction may be **small**. This means that φ is birational, but contracts only finitely many curves (and not divisors) C_i's necessarily rational.

In this case, Y acquires a bad singularity at the image points since K_Y is no longer $\mathbb{Q}$-Cartier. Indeed, if K_Y is $\mathbb{Q}$-Cartier, then $K_X = \varphi^* K_Y$ since the exceptional locus has codimension 2. But $0 > K_X.C_i = \varphi^* K_Y.C_i = 0$.

To proceed with the construction of a minimal model, another birational transformation called a **flip** has been introduced; see [Kawamata et al. 1987; Mori 1988], for example. A flip is a commutative diagram

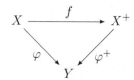

where

(1) X^+ is $\mathbb{Q}$-factorial with only terminal singularities,
(2) f is isomorphic outside the indeterminacy locus of φ and φ^+ which are both at least 2-codimensional, and
(3) $K_{X^+}.C^+ > 0$ for any curve $C^+ \subset X^+$ such that $\dim \varphi^+(C^+) = 0$.

In particular, $\rho(X) = \rho(X^+)$ and $K_{X^+} = f_* K_X$. (Notice that small contractions do not exist in dimension 2).

The existence of flips was established in dimension 3 by S. Mori [1988]; this is the deepest part of the minimal model program in dimension 3. It uses classification of all "extremal neighborhoods" of irreducible curves (necessarily smooth rational) contracted by a small contraction. The existence of flips is unknown when $n \geq 4$; a proof would presumably require new methods since a similar classification does not seem to be possible in higher dimension.

The final result is:

THEOREM 2.5. *Let X be a projective $\mathbb{Q}$-factorial 3-fold with only terminal singularities. Then there exists a birational map $\psi : X \dashrightarrow X'$ which is a finite sequence of extremal contractions and flips, and such that X' is again $\mathbb{Q}$-factorial with only terminal singularities, and either*

(1) $K_{X'}$ *is nef, or*
(2) *there exists an extremal contraction $\varphi : X' \to Y'$ such that $\dim Y' \leq 2$, and of course $-K_{X'}$ is φ-ample (similar to cases (2), (3) or (4) of Theorem 2.2).*

We call X' a **minimal model of X** in case 1.

The second part of the program — the so-called "Abundance Conjecture" — has been solved by Y. Kawamata and Y. Miyaoka (and by E. Viehweg when $q(X) > 0$ and $\kappa(X) \geq 0$ using the solution of Iitaka's conjecture in dimension 3).

THEOREM 2.6. *Let X' be $\mathbb{Q}$-factorial with only terminal singularities and assume that $K_{X'}$ is nef. Then $mK_{X'}$ is generated by global sections for suitable $m > 0$ such that $mK_{X'}$ is Cartier. Hence the induced map is "the" Iitaka reduction of X'.*

This result says that $\kappa(X') = \nu(K_{X'})$, where $\nu(K_{X'})$ is the **numerical Kodaira dimension**, defined as $\min\{0 \leq d \leq n \mid K_X^d \not\equiv 0\}$ if $\dim(X') = n$.

As for $n = 2$, the proof distinguishes the 4 cases:

(1) $K_X^3 > 0$ ($\nu = 3$). This case is easy (even if $n > 3$).
(2) $K_X^3 = 0$ but $K_X^2 \not\equiv 0$ ($\nu = 2$).
(3) $K_X^2 \equiv 0$; $K_X \not\equiv 0$ ($\nu = 1$).
(4) $K_X \equiv 0$ ($\nu = 0$).

To deal with the remaining cases (2), (3) and (4), a very delicate analysis of the elements in $|mK_{X'}|$ (which is assumed to be non-empty) is necessary. However, the very first step is to prove non-emptyness for some $m > 0$. In other words, one has to show that $\kappa(X') \geq 0$ if $K_{X'}$ is nef. This is especially hard when $q(X') = 0$, otherwise the Albanese map can be used. This step is easy when $n = 2$, because if X is a smooth surface with $q = p_g = 0$, then

$$h^0(2K) + h^0(-K) \geq \chi(2K) = K^2 + \chi(\mathcal{O}_X) \geq \chi(\mathcal{O}_X) = 1 - q + p_g = 1.$$

But K_X being nef also implies $h^0(-K) = 0$, so $P_2 > 0$.

But in dimension 3, a new approach has to be found. It was discovered by Y. Miyaoka, who gave criteria for uniruledness in arbitrary dimension n. As in S. Mori's approach, the method of reduction to characteristic $p > 0$ is used in an essential way. But this time, not only the numerical properties of K_X^{-1}, but also those of T_X, come into the game.

THEOREM 2.7 (Y. Miyaoka). *Let X be a smooth projective n-fold, H an ample divisor on X and C a complete intersection curve cut out by general elements in $|mH|$ for $m \gg 0$. If X is not uniruled, then $\Omega_X^1|C$ is semi-positive (or, equivalently, nef); that is, all rank-one quotient sheaves have non-negative degree.*

This result is known as the generic semi-positivity theorem.

COROLLARY 2.8 (Y. Miyaoka). *Let X be a normal projective n-fold with singularities in codimension at least 3 (this is the case if X is $\mathbb{Q}$-factorial with only terminal singularities). Assume that X is not uniruled. Then*

$$K_X^2.H^{n-2} \leq 3c_2(T_X).H^{n-2}$$

for H nef (and $c_2(T_X)$ being the direct image of the corresponding c_2 of any smooth model of X).

In dimension 3, we have the following important consequence:

THEOREM 2.9. *Let X' be a $\mathbb{Q}$-factorial projective 3-fold with at most terminal singularities. Assume $K_{X'}$ nef. Then X' is uniruled if and only if $\kappa(X') = -\infty$.*

We give the argument in case X' is Gorenstein; in the non-Gorenstein case one needs additional arguments: see [Miyaoka 1988]. First notice that if X' is Gorenstein we have

$$\chi(X', \mathcal{O}_{X'}) = -\frac{1}{24} K_{X'} \cdot c_2(X').$$

In general, this is false; see [Reid 1987]. From Corollary 2.8 we thus get the inequality $\chi(X', \mathcal{O}) \leq 0$. If $q(X') = 0$, we deduce $h^{3,0} > 0$, so that $\kappa(X') \geq 0$. (Recall that canonical (and so terminal) singularities are rational, so $h^{3,0}(X') = h^{3,0}(X)$ if X is any smooth model of X'.) If $q(X') > 0$, we can use the Albanese map and various versions of $C_{n,m}$ to conclude, using results of Viehweg.

Notice that X' is uniruled if $\chi(\mathcal{O}_X) > 0$.

2.10. We now turn to **Fano manifolds**, the building blocks for the Fano fibrations. Recall that a $\mathbb{Q}$-factorial projective variety X with only terminal singularities is said to be Fano (in full, $\mathbb{Q}$-Fano) if $-K_X$ is ample in an obvious sense (replace $-K_X$ by $-mK_X$).

Consider first the cases $n = 1, 2, 3$:

- $n = 1$: There is a single one: $\mathbb{P}_1$.

- $n = 2$: There are 10 deformation families: $\mathbb{P}_2$, $\mathbb{P}_1 \times \mathbb{P}_1$ and $\mathbb{P}_2$ blown-up in $1 \leq d \leq 8$ points in general position (not 3 on a line; 6 on a conic or 7 on a singular cubic, this singular point being one of them). Recall that terminal means smooth for surfaces.

- $n = 3$ and X is smooth: There are then 104 deformation families. Seventeen of them have $b_2 = 1$ and were classified by Iskovskih and Shokurov using linear systems. The basic invariant in this classification is the **index of X**, defined as $r(X) := \max\{S > 0 \mid -K_X = S.H \text{ for some } H \text{ in } \operatorname{Pic}(X)\}$. One has $1 \leq r(X) \leq n + 1$. If $r = n + 1$ then $X = \mathbb{P}_n$, and if $r = n$ then $X = Q_n$ the n-dimensional quadric [Kobayashi and Ochiai 1973]. The other 87 families have been classified by Mori and Mukai using Theorem 2.4 (see Section 2.14 below). Note that since $b_2 \geq 2$ the variety X has at least 2 different extremal contractions.

With one exception, all these families are obtained by standard methods, which we now describe (in any dimension n):

- Take $X = \mathbb{P}_n$, $X = Q_n$ (the n-dimensional quadric in $\mathbb{P}_{n+1}$), or X rational homogeneous (it is easy to see that these are Fano).

- Take smooth blow-ups of X along submanifolds Y. (Y has to be of small anticanonical degree. It may happen that no such Y can be blown-up so that the result is Fano: if X is $\mathbb{P}_2$ already blown-up in 8 points for example).

- Take complete intersections of hypersurfaces of small (anticanonical) degree: this works (by adjunction formula) and gives examples if the index r of X is large. For example, if $X = \mathbb{P}_{n+1}$, its smooth hypersurfaces of degree $d \leq n+1$ are Fano (of index $n + 2 - d$).

- Take double (or cyclic) coverings of X, branched along smooth hypersurfaces of small degree. Again this gives examples (by adjunction formula) if r is large. Double coverings of $\mathbb{P}_n$ branched along hypersurfaces of degree $2d \leq 2n$ are Fano, of index $(n + 1 - d)$.

In all these constructions it is easy by counting dimensions to see the existence of many rational curves on the manifolds obtained. This is a general phenomenon, moreover the study of rational curves on Fano manifolds leads to essential results concerning their structure. This is illustrated by the following result:

THEOREM 2.11 [Campana 1992; Kollár et al. 1992a]. *Let X be a smooth Fano n-fold. Then X is rationally connected.*

It was implicitly shown in [Mori 1979] that X is uniruled. The proof rests on ideas similar to those found in that reference.

Theorem 2.11 implies that $\pi_1(X) = \{1\}$ if X is Fano; see Section 6.

The study of the birational structure of Fano n folds is very difficult, but interesting, already in dimension 3:

- Cubic hypersurfaces in $\mathbb{P}_4$ are non-rational, but unirational (by a result of Clemens and Griffiths) and birationally distinct if non-isomorphic. (So the set of birational classes of Fano threefolds is not countable).

- Quadric hypersurfaces in $\mathbb{P}_4$ are non-rational; some are unirational (of degree 24) (Iskovskih–Manin). It is unknown whether or not the general one is unirational.

QUESTION 2.12. Assume X is rationally connected. Under which conditions is X birational to some $\mathbb{Q}$-Fano n-fold ($\mathbb{Q}$-factorial with at most terminal singularities)?

THEOREM 2.13 [Kollár et al. 1992b]. *There exists an explicit constant $A(n)$ such that every smooth Fano n-fold X satisfies*

$$c_1(X)^n \leq A(n)^n.$$

The big Matsusaka theorem then implies that the family of smooth Fano n-folds is bounded (that is, they can all be embedded in $\mathbb{P}_{N(n)}$ with degree less than $d(n)$, where $d(n)$ and $N(n)$ are explicit constants). In particular, there are only finitely many deformation families (and diffeomorphism types) of Fano n-folds.

We emphasize that the existence of a bound in Theorem 2.13 rests in an essential way on Theorem 2.11, and hence on the study of rational curves. The idea is in fact to join two general points on a Fano n-fold X by an **irreducible** (rational; this is not essential, but these are the objects that one can produce) curve of **anticanonical** degree $\delta \leq A(n)$. The "gluing Lemma" of [Kollár et al. 1992a] is used here. An easy ingenious argument (due to F. Fano) then gives the bound as in Theorem 2.13.

2.14. REFERENCES. We now give references for the proofs, and further studies. We try to cite books and introductory papers; references to the original proofs may be found there.

(1) Canonical and terminal singularities: [Reid 1983; 1987; Clemens et al. 1988].
(2) Minimal model program (cone theorem): [Kawamata et al. 1987; Clemens et al. 1988; Kollár 1989; Miyaoka and Peternell 1997].
(3) Flips: [Mori 1988; Clemens et al. 1988; Kollár 1992; Miyaoka and Peternell 1997].
(4) Abundance conjecture: [Kollár 1992; Miyaoka and Peternell 1997].
(5) Fano manifolds: [Iskovskih 1977; 1978; 1989; Shokurov 1979; Mori and Mukai 1983 (for 3-folds); Campana 1992; Kollár et al. 1992a; Kollár 1996].

3. Compact Kähler Manifolds

In this section we want to discuss the global structure of (connected) compact Kähler manifolds. First we measure how far a compact manifold is from being algebraic.

THEOREM AND DEFINITION 3.1. *Let X be a compact complex manifold of dimension n. Let $\mathcal{M}(X)$ denote its field of meromorphic functions. Let $a(X)$ be the transcendence degree of $\mathcal{M}(X)$. Then*

$$0 \leq a(X) \leq n.$$

Moreover $\mathcal{M}(X)$ is an algebraic function field, that is, there is a projective manifold Y with $\dim Y = a(X)$, such that $\mathcal{M}(X) \simeq \mathcal{M}(Y)$. The number $a(X)$ is called the algebraic dimension of X.

*If $a(X) = n$, the manifold X is called **Moishezon**.*

This theorem is due to Siegel. For this and for the elementary theory of the algebraic dimension as well as algebraic reductions which we are going to define next, we refer to [Ueno 1975], or, for a less detailed and shorter presentation, [Grauert et al. 1994]. Most prominent examples of non-algebraic compact (Kähler) manifolds are of course general tori and general K3-surfaces. To define and construct algebraic reductions, fix a compact manifold X and take Y as in Theorem 3.1. Then there is a meromorphic map $f : X \dashrightarrow Y$ such that $f^*(\mathcal{M}(Y)) = \mathcal{M}(X)$. Of course there is no unique choice of Y (unless $a(X) = 0, 1$ and unless we agree to choose Y normal). Every Y' bimeromorphic to Y does the same job.

DEFINITION 3.2. Let X be a compact manifold (or irreducible reduced compact complex space). Let Y be a normal projective variety. A meromorphic map $f : X \dashrightarrow Y$ is called an **algebraic reduction** of X if

$$\mathcal{M}(X) = f^*(\mathcal{M}(Y)).$$

The extreme, and often the most difficult, case is $a(X) = 0$. Then one knows that there are only finitely many irreducible hypersurfaces in X and possibly none. One can say that the more algebraic X is the more compact subvarieties it has. This can be made precise in the following way.

DEFINITION 3.3. A compact manifold X is **algebraically connected** if

(a) every irreducible component of $\mathcal{C}_1(X)$, the cycle space or Barlet space of 1-cycles, is compact, and

(b) every two general points in X can be joined by a connected compact complex curve.

In a moment we will comment on the cycle space or Barlet space (Chow scheme in the algebraic case); the compactness is fulfilled if X is Kähler, for example. Compactness allows one to take limits of families of cycles. The importance of the notion of algebraic connectedness is demonstrated by the following result [Campana 1981].

THEOREM 3.4. *Let X be an algebraically connected compact (Kähler) manifold. Then X is Moishezon.*

The converse of Theorem 3.4 is obvious. There are many counterexamples (twistor spaces) to Theorem 3.4 if one drops condition (a) in Definition 3.3.

Instead of assuming the existence of many curves one might think of supposing the existence of a "big" submanifold forcing X to be algebraic. For example, if $Y \subset X$ is a hypersurface with ample normal bundle, then X is Moishezon.

PROBLEM 3.5. Let X be a compact Kähler manifold and $Y \subset X$ a compact submanifold with ample normal bundle. Is X Moishezon (hence projective)?

3.6. One cannot expect a reasonable structure theory for arbitrary compact complex manifolds. Pathologies will be given in Section 4. The most reasonable assumption (without assuming projectivity) is the Kähler assumption, which we will choose, or, slightly more generally, the assumption that manifolds should be bimeromorphic to a Kähler manifold; such manifolds form the so-called class $\mathcal{C}$.

Here we collect some major tools for the investigation of compact Kähler manifolds.

(1) The Albanese map $X \to \mathrm{Alb}(X)$ to the Albanese torus $\mathrm{Alb}(X)$ given by integration of d-closed 1-forms. This map exists in general for compact manifolds, however in the Kähler case every 1-form is d-closed, hence contributes to the Albanese, which in general is false. See Section 6 for some application of the Albanese in classification theory.

(2) Hodge decomposition (or better Hodge theory). This is completely false for general compact manifolds.

(3) The cycle space or Barlet space $\mathcal{C}(X)$. This is the analogue of the Chow scheme in complex geometry. $\mathcal{C}_q(X)$ parametrises q-cycles Z, that is,

$$Z = \sum n_i Z_i,$$

where $n_i \geq 0$ and Z_i are irreducible reduced compact subspaces of dimension q, the sum of course being finite. One of the most basic results is, as already mentioned, the compactness of every irreducible component of the cycle space, if X is compact Kähler (or in class $\mathcal{C}$). In algebraic geometry varieties are usually studied via ample line bundles, vanishing theorems, linear systems etc. These concepts do not work on general compact Kähler manifolds. In some sense the substitute should be cycles, in particular curves, as we shall see later in this section. For an overview of the theory of cycle spaces and applications as well as references, see [Grauert et al. 1994, Chapter 8].

How far is a compact Kähler manifold from being projective? There is a basic criterion for projectivity, due to Kodaira (see [Morrow and Kodaira 1971], for example):

THEOREM 3.7. *Let X be a compact Kähler manifold with $H^2(X, \mathcal{O}_X) = 0$. Then X is projective.*

This indicates that 2-forms should play an important role in the theory of non-algebraic compact Kähler. There is a "conjecture", due to Kodaira (or Andreotti), concerning the vague question posed above.

DEFINITION 3.8. Let X be a compact n-dimensional Kähler manifold. We say that X can be **approximated algebraically** if the following condition holds. There is a complex manifold $\mathcal{X}$ and a proper surjective submersion

$$\pi : \mathcal{X} \to \Delta = \{z \in \mathbb{C}^n \mid \|z\| < 1\},$$

such that, putting $X_t = \pi^{-1}(t)$, we have:

(a) $X_0 \simeq X$;

(b) there is a sequence (t_ν) converging to 0 and such that all the X_{t_ν} are projective.

We call $\pi : \mathcal{X} \to \Delta$ a family of compact manifolds and we often denote it by (X_t).

PROBLEM 3.9. Can every compact Kähler manifold be approximated algebraically?

This is true for surfaces, but it is proved in a rather indirect way, via the Kodaira–Enriques classification. And this is the only evidence we have. Certainly it would be very interesting to find a conceptual proof for surfaces.

3.10. Our main intention is now to try to understand compact Kähler manifolds according to their Kodaira dimension. More precisely we ask whether there is a Mori theory in the Kähler case. This means:

(1) proving that a compact Kähler manifold X has $\kappa(X) = -\infty$ if and only if it is uniruled (one direction being clear) and trying to find a birational model which has a Fano fibration as described in Section 2;

(2) if $\kappa(X) \geq 0$, finding a minimal model X', that is, X' has only terminal singularities and $K_{X'}$ is nef;

(3) if K_X is nef, then mK_X is generated by global sections for some m (abundance).

But Mori theory is somewhat more: it predicts how to find a minimal model and a Fano fibration. Namely, if K_X is not nef, then there should be a "canonical" contraction, the contraction of an extremal ray in the algebraic category. So we first have to explain what "nef" means in the Kähler case.

DEFINITION 3.11. Let X be a compact complex manifold and L a line bundle on X. Fix a positive (not necessarily closed) $(1,1)$-form ω on X. Then L is **nef** if for every $\varepsilon > 0$ there exists a hermitian metric h_ε on L with curvature

$$\Theta_{h_\varepsilon} \geq -\varepsilon\omega.$$

REMARKS 3.12. (1) Obviously the definition is independent on the choice of ω.

(2) If X is projective, then L is nef if and only if $L \cdot C \geq 0$ for all curves $C \subset X$. For this and many more basic properties of nef line bundles we refer to [Demailly et al. 1994].

(3) Call L **algebraically nef** if $L \cdot C \geq 0$ for all curves C. Then in general "algebraically nef" does not imply "nef". For an example take any compact Kähler surface X with $a(X) = 1$. Then the algebraic reduction is an elliptic fibration $f : X \to B$. Any curve in X is contained in some fiber of f. Now take an ample line bundle G on B and put $L = f^*(G^*)$. Then clearly L is not nef (because its dual has a section with zeroes!) but it is algebraically nef.

(4) There are examples of nef line bundles which do not admit a metric of semipositive curvature [Demailly et al. 1994]. So we really need to work with ε in the definition 3.11.

(5) Assume X is Kähler. Let $KC(X)$ denote the (closed) Kähler cone of X. This is the closed cone inside $H^{1,1}(X) \cap H^2(X, \mathbb{R})$ generated by the classes of the Kähler forms. Let L be a line bundle on X. Then L is nef if and only if $c_1(L) \in KC(X)$. See [Peternell 1998b].

3.13. The first basic question for a Mori theory in the Kähler case is therefore the following: given a compact Kähler manifold X with K_X not nef, is there a curve C such that $K_X \cdot C < 0$? If yes, can we choose C rational? We have seen that for general L the answer is no, but K_X of course has special properties. We will give a positive answer in some cases below. A general method to attack the problem would be to deform the complex structure to a generic almost complex structure and then to try to use the theory of J-holomorphic curves. However for the approach one would need the following openness property (we state it only in the holomorphic category).

PROBLEM 3.14. Let $\mathfrak{X} \to \Delta$ be a family of compact Kähler manifolds. Assume that K_{X_0} is not nef. Is then K_{X_t} not nef for all (small) t?

This is unknown even in the projective case. See [Andreatta and Peternell 1997] for some results in this direction.

The standard approach to Mori theory in the projective case is as follows. Assume K_X not nef. Fix an ample line bundle H. Let r be the uniquely determined positive number such that $K_X + rH$ is on the boundary of the ample cone, which is to say it is nef but not ample. Then r is rational. Now $m(K_X + rH)$ is generated by global sections and the associated morphism gives a contraction we are looking for. Needless to say that the approach completely breaks down in the Kähler case. The substitute should be the theory of non-splitting families of rational curves, this allows to avoid thoroughly to speak about line bundles (except the canonical bundle, of course), sections and linear systems. It was Kollár [1991a] who reconstructed contractions for smooth threefolds using this method. We are now going to explain the geometry of non-splitting families of rational curves.

DEFINITION 3.15. A **non-splitting family** $(C_t)_{t \in T}$ of (rational) curves is a family of curves (C_t) such that the parameter space T is compact irreducible and C_t is an irreducible reduced (rational) curve for every $t \in T$. It is described by its graph $\mathcal{C}$ with projections $p : \mathcal{C} \to X$ and $q : \mathcal{C} \to T$ such that $C_t = p(q^{-1}(t))$.

3.16.   Mori's breaking lemma [1979] is an indispensable tool in dealing with families of rational curves. It holds on every compact complex manifold X for which condition 3.3(a) holds and states that if (C_t) is a family of rational curves (with compact and irreducible T as usual, of course) and if there are points $p, q \in X, p \neq q$, such that $p, q \in C_t$ for all $t \in T$, then (C_t) has to split (ig $\dim T > 0$). Moreover Mori proved that if $\dim X = n$ and if (C_t) is non-splitting, then $K_X \cdot C_t \geq -n - 1$. Equality holds for the family of lines on the projective space $\mathbb{P}_n$ and it is conjectured that this is the only example.

Now we describe the structure of non-splitting families of rational curves in compact Kähler threefolds as given in [Campana and Peternell 1997].

THEOREM 3.17. *Let X be a compact Kähler threefold and $(C_t)_{t \in T}$ a non-splitting family of rational curves.*

(1) *If $K_X \cdot C_t = -4$ and if $\dim T = 4$, then $X \simeq \mathbb{P}_3$.*
(2) *If $K_X \cdot C_t = -3$, and if $\dim T = 3$, then either $X \simeq Q_3$, the 3-dimensional quadric, or X is a $\mathbb{P}_2$-bundle over a smooth curve.*
(3) *Assume $K_X \cdot C_t = -2$ and $\dim T = 2$.*

(3a) *If X is non-algebraic and if the C_t fill up a surface $S \subset X$, then $S \simeq \mathbb{P}_2$ with normal bundle $N_{S|X} = \mathcal{O}(-1)$. The same holds for X projective if S is normal.*

(3b) *If X is covered by the C_t, we are in one of the following cases.*

  (3b1) *X is Fano with $b_2(X) = 1$ and index 2.*

  (3b2) *X is a quadric bundle over a smooth curve with C_t contained in fibers.*

  (3b3) *X is a $\mathbb{P}_1$-bundle over a surface, the C_t being the fibers.*

  (3b4) *X is the blow-up of a $\mathbb{P}_2$-bundle over a curve along a section. Here the C_t are the strict transforms of the lines in the $\mathbb{P}_2'$s meeting the section.*

(4) *Let $K_X \cdot C_t = -1$ and $\dim T = 1$. Then the C_t fill up a surface S. Assume that S is non-algebraic.*

(4a) *If S is normal, we are in one of the following cases.*

  (4a1) *$S = \mathbb{P}_2$ with $N_S = \mathcal{O}(-2)$.*

  (4a2) *$S = \mathbb{P}_1 \times \mathbb{P}_1$ with $N_S = \mathcal{O}(-1,-1)$.*

  (4a3) *$S = Q_0$, the quadric cone, with $N_S = \mathcal{O}(-1)$.*

  (4a4) *S is a ruled surface over a smooth curve and X is the blow-up of a smooth threefold along C such that S is the exceptional divisor.*

(4b) *Let S be non-normal. Then $\kappa(X) = -\infty$. If moreover X can be approximated algebraically, then we have $a(X) = 1$, and under some further (necessary and sufficient) condition [Peternell 1998b, 5.2], X is a conic bundle over a surface Y with $a(Y) = 1$. The surface S consists of the reducible conics and the C_t are the irreducible components of the reducible conics.*

The essential content of the theorem can be rephrased as follows. Assume that C is a rational curve with $K_X \cdot C = k$, where $-1 \geq k \geq 4$. If no deformation of C splits, the conclusion of the theorem states that the C_t give rise to a special geometric situation. There are basically two different situations in the theorem. Either the C_t fill up X, then one can consider the "rational quotient" with respect to that family [Campana 1992; Kollár et al. 1992a], which is a priori only meromorphic, and investigate its structure. The results are just the fibrations one has in the algebraic case in the Mori theory. Or the C_t fill up a surface S. Now one has to study in great detail the structure of S. As result, in the normal case and at least if X is non-algebraic, one can blow down S to obtain a birational contraction $X \to Y$ of the same type as in the Mori theory, however in general Y will not be Kähler. We will come to this point later. In the normal case, with some extra assumption we get conic bundles. For all details of proof we refer to [Campana and Peternell 1997]. Of course it would be interesting to prove something along the lines of Theorem 3.17 also in the higher-dimensional or singular case.

Theorem 3.17 was used in [Peternell 1998b] to prove the following result:

THEOREM 3.18. *Let X be a non-algebraic compact Kähler threefold satisfying one of the following conditions.*

(I) *X can be approximated algebraically.*

(II) *$\kappa(X) = 2$.*

(III) *X has a good minimal model (that is, $mK_{X'}$ is generated by global sections).*

Assume that K_X is not nef. Then

(1) *X contains a rational curve C with $K_X \cdot C < 0$;*

(2) *there exists a surjective holomorphic map $\varphi : X \to Y$ to a normal complex space Y with $\varphi_*(\mathcal{O}_X) = \mathcal{O}_Y$ of one of the following types.*

  (2a) *φ is a $\mathbb{P}_1$- bundle or a conic bundle over a non-algebraic surface. (This can only happen in case (1).)*

  (2b) *φ is bimeromorphic contracting an irreducible divisor E to a point, and E together with its normal bundle N is one of*

  $$(\mathbb{P}_2, \mathcal{O}(-1)), \quad (\mathbb{P}_2, \mathcal{O}(-2)), \quad (\mathbb{P}_1 \times \mathbb{P}_1, \mathcal{O}(-1,-1)), \quad (Q_0, \mathcal{O}(-1)),$$

  where Q_0 is the quadric cone.

  (2c) *Y is smooth and φ is the blow-up of Y along a smooth curve.*

φ is called an extremal contraction.

Y is (a possibly singular) Kähler space in all cases except possibly (2c). Moreover in all cases but possibly (2c), φ is the contraction of an extremal ray in the cone of curves $\overline{NE}(X)$.

A normal complex space is **Kähler** if there is a Kähler metric h on the regular part of X with the following property. Every singular point has a neighborhood U and a closed embedding $U \subset V$ where V is an open subset of some $\mathbb{C}^n$ such that there is Kähler metric h' on V with $h'|U \setminus \mathrm{Sing}(X) = h$.

REMARK. Theorem 3.18 has been proved for all smooth compact Kähler threefolds X unless X is simple with $\kappa(X) = -\infty$; see [Peternell 1998c]. The same paper also proves abundance for minimal Kähler threefolds which are not both simple and non-Kummer; see 3.20 below.

ABOUT THE PROOF OF THEOREM 3.18. In order to apply Theorem 3.17 one needs a non-splitting family (C_t) of rational curves with $-4 \leq K_X \cdot C_t < 0$. For this it is sufficient to have one rational curve C with $-4 \leq K_X \cdot C < 0$. Then one can apply deformation theory to obtain a family; if this family splits, take an irreducible part C' of a splitting member with $K_X \cdot C' < 0$ and deform again. This procedure must terminate since X is Kähler.

In case (I) one shows that $K_{X_{t_\nu}}$ is not nef in terms of the algebraic approximation (this is of course a major step) and then apply Mori theory in the algebraic case to obtain a rational $C_{t_\nu} \subset X_{t_\nu}$ for a fixed t_ν with $K_{X_{t_\nu}} \cdot C_{t_\nu} < 0$. This curve can then be deformed into X_0 to obtain a rational curve C_0 with $K_{X_0} \cdot C_0 < 0$.

In case (II) we consider the linear system $|mK_X|$ defining a meromorphic map $f : X \dashrightarrow Y$ to a projective surface. Then we choose a general element $D_0 \in |mK_X|$. Now our linear system must have fixed components A_i and has a movable part B. Examining carefully the structure of B and A_i we first obtain some curve C with $K_X \cdot C < 0$ and then in a second step a rational one. A similar thing can be done if $\kappa(X) = 1$ to find at least some curve C with $K_X \cdot C < 0$. $\square$

3.19. Let X be a compact Kähler threefold with K_X not nef. We have seen that at least with some additional assumptions we can construct a rational curve C with $K_X \cdot C < 0$. Hence we can construct a map $\phi : X \to Y$ as described in Theorem 3.18. In order to continue the process in case $\dim Y = 3$ it is now very important that ϕ can be chosen in such a way that Y is again Kähler. Let E denote that exceptional locus of ϕ. If $\dim \phi(E) = 0$ then it turns out that Y is *always* Kähler. But if $\dim \phi(E) = 1$, that is, if ϕ is the blow-up of a smooth curve in the manifold Y, this is not necessarily the case, even in the projective case (Y could be Moishezon). Instead one has — in the projective case — to choose ϕ carefully: it has to be the contraction of an extremal ray in $\overline{NE}(X)$. In the Kähler case we can introduce the dual cone $\overline{NA}(X)$ to the Kähler cone in $H^{2,2}(X)$ and can prove that Y is Kähler if and only if the ray $R = \mathbb{R}_+[l]$ is extremal in $\overline{NA}(X)$, where l is a fiber of ϕ.

PROBLEMS. (1) Is Y Kähler if and only if R is extremal in $\overline{NE}(X)$?
(2) How can one find extremal rays in $\overline{NA}(X)$ or in $\overline{NE}(X)$? Is there a "Cone Theorem"?

3.20. Even if one has shown the existence of contraction $\phi : X \to Y$ for a compact Kähler threefold X with K_X not nef such that Y is Kähler, it is still necessary to do the same also for normal projective $\mathbb{Q}$-factorial Kähler threefolds X with at most terminal singularities in order to be able to repeat the process. Of course then one will run into the same trouble as in the algebraic case, namely that sometimes a small contraction will appear so that Y has bad singularities and we have to flip. However the existence and termination of flips are basically analytically local and have been settled in [Kawamata 1988; Mori 1988].

We next indicate how the expected answer to the problems (1) and (2) in Section 3.10 would give a new insight into the structure of non-algebraic Kähler threefolds far away from the "usual" algebraic applications of Mori theory.

A compact Kähler manifold X is **simple** if there is no covering family of positive dimensional subvarieties (hence through a very general point of X there is no positive dimensional compact subvariety). Note that using cycle space methods, the classification of compact Kähler manifolds can be reduced to a large extent to the classification of the simple manifolds; see [Grauert et al. 1994] and the references given there. In dimension three, simple compact Kähler threefolds are conjectured to be "Kummer" in the following sense. X is called Kummer if X is bimeromorphic to a variety T/G, with T a torus and G a finite group acting on T.

Observe that the set of points of T having non-trivial isotropy is finite in this situation.

THEOREM 3.21. *If (3.10(1)) and (3.10(2)) have positive answers in dimension three, every simple smooth compact Kähler threefold is simple.*

INDICATION OF PROOF. If X is simple, it cannot be uniruled, hence it has a minimal model by 3.10(1). By 3.10(2), $mK_{X'}$ is generated by global sections for $m \gg 0$. Again by the simplicity it follows $\kappa(X') = 0$, hence $mK_{X'} = \mathcal{O}_{X'}$. Assume X' Gorenstein and $m = 1$ for the sake of simplicity (in general one has to pass to a covering $\tilde{X} \to X'$ which is étale over the smooth part of X'). Then one can apply Riemann–Roch and obtain

$$\chi(X', \mathcal{O}_{X'}) = 0.$$

Since $\dim H^3(X', \mathcal{O}_{X'}) = 1$ by Serre duality, and since

$$H^2(X', \mathcal{O}_{X'}) \neq 0$$

(pass to a desingularisation, apply Theorem 3.7 and come back to X' using the rationality of the singularities of X') we deduce

$$H^1(X', \mathcal{O}_{X'}) \neq 0.$$

Therefore we have an Albanese map $X' \to \mathrm{Alb}(X')$. Now the structure of X' allows one to prove that the Albanese map is an isomorphism. □

We close the section with the following recent structure theorem from [Campana and Peternell 1998]:

THEOREM 3.22. *Let X be a smooth compact Kähler threefold which is not both simple and non-Kummer. Then*

(i) *If $\kappa(X) = -\infty$, X is uniruled.*
(ii) *If $\kappa(X) = 0$ and if X carries a holomorphic 2-form (for example, if X is not projective), then X is bimeromorphic to some threefold X' (possibly with quotient singularities) which has a finite cover $\tilde{X}'$ étale in codimension 1 such that $\tilde{X}'$ is either a torus or a product of an elliptic curve and a K3-surface.*

4. Topological Classification

In this section we mainly discuss the following question: given a compact complex manifolds X, can one describe all complex structures on the underlying topological (differentiable) manifold, if X has some nice properties (Fano etc.). In other words, we consider a topological manifold and ask for all complex structures if there is any. A typical question: if X is "nice" and Y homeomorphic to X, is $X \simeq Y$ biholomorphically? And: what are the analytically defined topological invariants?

4.1. In dimension 1 everything is clear: there is one (in fact analytically defined) topological invariant, the genus, and $X \simeq Y$ if and only if $g(X) = g(Y)$. Moreover every compact topological 2-dimensional real manifold carries a complex structure. The structure is unique if and only if $g = 0$, i.e., $X \simeq \mathbb{P}_1$. This already gives a hint that we should look for in higher dimension to those manifolds which are "natural" generalisations of $\mathbb{P}_1$. If $g \geq 2$, or, in higher dimensions, if X is of general type, then the task is to describe moduli spaces. This is a completely different topic and therefore systematically omitted.

4.2. In dimension 2 there has been spectacular progress in the last fifteen years due to the work of Freedman, the Donaldson theory and the Seiberg–Witten invariants. It is now known that the Kodaira dimension is a $\mathcal{C}^\infty$-invariant of compact Kähler surfaces but not a topological invariant. We will completely ignore this vast area and refer to [Donaldson and Kronheimer 1990; Okonek and Van de Ven 1990; Friedman and Morgan 1994; Okonek and Teleman 1999]. In the topological case there are still open problems, for example whether there is a surface of general type homeomorphic to $\mathbb{P}_1 \times \mathbb{P}_1$ (although the answer is known to be negative for $\mathbb{P}_2$.

The surface results imply that the Kodaira dimension is not a differentiable invariant of compact Kähler threefolds: let S be the Barlow surface, a minimal surface of general type homeomorphic to $\mathbb{P}_2$ blown up in 8 points. Then take an elliptic curve C and let $X_1 = C \times S$ and $X_2 = C \times \mathbb{P}_2(x_1, \ldots, x_8)$. Then $\kappa(X_1) = 2$ whereas $\kappa(X_2) = -\infty$. Note that X_1 and X_2 are even diffeomorphic since topological and differentiable equivalences are the same here. If we take C to have genus ≥ 2, then we even find a threefold with K_X ample diffeomorphic to a threefolds with negative Kodaira dimension.

We will now go to higher dimensions and will see that only few things are known. The most basic question is certainly the following:

QUESTION 4.3. What are the complex structures on the complex projective space $\mathbb{P}_n$?

A first answer was given by Hirzebruch and Kodaira [1957]:

THEOREM 4.4. Let X be a compact Kähler manifold homeomorphic to $\mathbb{P}_n$. Then $X \simeq \mathbb{P}_n$ biholomorphically unless n is even and K_X is ample.

The proof makes essential use of the fact that the Pontrjagin classes $p_i(X) \in H^{4i}(X, \mathbb{R})$ are topological invariants. Actually in 1957 it was only known that the $p_i(X)$ were differentiable invariants, so Hirzebruch and Kodaira could formulate only a differentiable version of Theorem 4.4, but afterwards Novikov [1965] proved that the Pontrjagin classes are actually topological invariants. Hirzebruch and Kodaira could determine only the sign of $c_1(X)^n$, so that in even dimension the case K_X ample (and divisible by 4) remained open until Yau proved the Calabi conjectures. Using the latter one can rule out the case of K_X ample as follows. By calculating invariants one finds the following Chern class equality

$$nc_1(X)^n = 20 = 2(n+1)c_2(X)c_1(X)^{n-2}.$$

This is just the borderline for the Yau inequality and by the existence of a Kähler–Einstein metric, a classical differential-geometric argument shows that the universal cover of X is the unit ball in $\mathbb{C}^n$. On the other hand X is simply connected, contradiction. This is the only known argument to rule out the existence of complex structures of general type on projective space.

THEOREM 4.5. *Let X be a compact Kähler manifold homeomorphic to $\mathbb{P}_n$ Then $X \simeq \mathbb{P}_n$ analytically.*

One can ask the same question for, e.g., the n-dimensional quadric Q_n, $n \geq 3$. If $n = 2$, one has to admit the Hirzebruch surfaces $\mathbb{P}(\mathcal{O} \oplus \mathcal{O}(-2n))$; this case is special because $b_2 = 2$. In this context Brieskorn [1964] proved a result analogous to Theorem 4.4, with the same exception, namely that there could be a projective manifold of even dimension n with K_X ample homeomorphic to Q_n. Since there are, for example, surfaces with $c_1^2 = c_2$ which are simply connected, one would need here a completely different argument from Yau's.

In Theorem 4.4 the Kähler assumption, which is obviously equivalent to projectivity, is important. On one hand it allows to compute $H^q(X, \mathcal{O}_X)$ by Hodge decomposition, on the other hand one can use the Kodaira vanishing theorem to calculate $\chi(X, \mathcal{O}_X(k))$ for the ample generator $\mathcal{O}_X(1)$. If no Kähler assumption is made, then the problem gets very complicated and is essentially unsolved. Here is a possibly tractable subproblem:

PROBLEM 4.6. Let X be a compact manifold homeomorphic to $\mathbb{P}_n$. Assume that $\dim X \geq 3$ and $a(X) > 0$. Is $X \simeq \mathbb{P}_n$?

If $n \geq 4$ nothing is known in this regard except for a result of Nakamura [1992] for $n = 4$, which gives a positive answer if $a(X) = 4$ and X not of general type. If $n = 3$ and $a(X) = 3$, the problem is completely solved [Kollár 1991b; Nakamura 1987; Peternell 1986; 1998a]:

THEOREM 4.7. *Every Moishezon threefold homeomorphic to $\mathbb{P}_3$ is $\mathbb{P}_3$.*

The same holds for the quadric and some Fano threefolds (V_5 and the cubic); see [Kollár 1991b; Nakamura 1988; 1996].

If $a(X) < \dim X$, virtually nothing is known. There is an interesting relation to the existence problem on complex structures on 6-spheres, we will come to this in Section 4.18. If $a(X) = 0$ the problem seems hopeless at the moment, but at least for threefolds Theorem 4.6 seems not to be unsolvable.

4.8. It is conjectured that in the situation of Theorem 4.4 it is not actually necessary to assume that X and $\mathbb{P}_n$ are homeomorphic. It should be sufficient to assume that the cohomology rings $H^*(X, \mathbb{Z})$ and $H^*(\mathbb{P}_n, \mathbb{Z})$ are isomorphic (as graded rings). This is proven by Van de Ven and Fujita up to dimension

6 [van de Ven 1962; Fujita 1980]. It should be mentioned that Mumford has constructed surfaces of general type with $b_1 = 0$ and $b_2 = 1$ so that in Theorem 4.4 it is not sufficient to assume equality of the Betti numbers.

There is another weakening of the problem of complex structures on projective space: one considers only complex structures near to the standard one. This has been solved by Siu [1989] (see also [Hwang 1996]):

THEOREM 4.9. *Let* $\mathfrak{X} = (X_t)_{t \in \Delta}$ *be a family of compact complex manifolds* (Definition 3.8), *parametrised by the unit disc* $\Delta \subset \mathbb{C}$. *Assume that* $X_t \simeq \mathbb{P}_n$ *for all* $t \neq 0$. *Then* $X_0 \simeq \mathbb{P}_n$.

In other words, $\mathbb{P}_n$ is stable under *global* deformations. Note that automatically all X_t are Moishezon and X_0 has a lot of vector fields. The analogous problem for the quadric was solved by Hwang [1995] and for hermitian symmetric manifolds with $b_2 = 1$ by Hwang and Mok [1998]. It should also be true for rational-homogeneous manifolds. More generally, one can ask:

PROBLEM 4.10. *Let* X_0 *be a rational-homogeneous manifold with* $b_2 = 1$. *Let* X *be a compact manifold homeomorphic to* X_0. *Does it follow that* $X \simeq X_0$?

PROBLEM 4.11. *Let* X_0 *be a Fano manifold with* $b_2(X_0) = 1$ *and* X *a projective manifold homeomorphic to* X_0. *What is the structure of* X? *Is* $\kappa(X) = -\infty$? *What happens for* $b_2(X_0) \geq 2$?

We shall restrict the discussion now to $\dim X = 3$. First we discuss the case $b_2 = 1$. We should expect that X is again Fano. This, however, is unknown even in very simple cases. For example:

PROBLEM 4.12. *Let* $X_0 \subset \mathbb{P}_4$ *be a cubic hypersurface. Is there a projective threefold* X *with* K_X *ample such that* X_0 *and* X *are homeomorphic?*

The difficulty is the lack of topological invariants, compared to surfaces we do not know any new topological invariant; however it might be possible to solve Problem 4.12 by carefully examining the linear system $|L|$ or $|2L|$, where L is the ample generator of $\mathrm{Pic}(X) = \mathbb{Z}$. Maybe it is now time to loose some words on topological invariants.

4.13. Here are the known topological invariants — by which we mean analytic invariants which a posteriori turn out to be topological invariants. First we have the Chern class $c_n(X)$, which by Hopf's theorem is nothing that the topological Euler characteristic $\chi_{\mathrm{top}}(X)$. By Hodge decomposition $q(X) = h^1(X, \mathcal{O}_X)$ is a topological invariant. Next we have the second Stiefel–Whitney class

$$w_2(X) = c_1(X)/ \mod 2 \in H^2(X, \mathbb{Z}_2).$$

Finally there are the Pontrjagin classes

$$p_i(X) \in H^{4i}(X, \mathbb{R}).$$

We have $p_1(X) = 2c_2 - c_1^2$ and $p_2(X) = 2c_4 - 2c_1 c_3 + c_2^2$.

PROBLEM 4.14. Are $h^q(X, \mathcal{O}_X)$ topological invariants of compact Kähler manifolds? Is at least $\chi(X, \mathcal{O}_X)$ a topological invariant of compact Kähler threefolds?

4.15. We now look at Fano threefolds with higher b_2. So let X_0 be a Fano threefold with $b_2 = 2$. Let X be a projective threefold homeomorphic to X_0. By Hodge decomposition we have

$$H^2(X_0, \mathcal{O}) \simeq H^2(X, \mathcal{O}).$$

In order to make progress we need to assume that $b_3 = 0$. Then we conclude that $H^3(X_0, \mathcal{O}) \simeq H^3(X, \mathcal{O})$. Thus $\chi(X, \mathcal{O}_X) = 1$. Now a fundamental theorem of Miyoka [1987] says that a threefold X with K_X nef has

$$\chi(X, \mathcal{O}_X) \leq 0.$$

This is a consequence of his inequality $c_1^2 \leq 3c_2$ for minimal threefolds. Hence K_X cannot be nef and therefore by Mori theory there must be an extremal contraction $\phi : X \to Y$. This gives us a tool to investigate the structure of X and one can prove:

THEOREM 4.16. *Let X_0 be a Fano threefold with $b_2 = 2$ and $b_3 = 0$. Let X be a projective threefold homeomorphic to X. Then $X_0 \simeq X$ or there is an explicit description for X.*

An example for "an explicit" description as mentioned in the theorem is the following. Let $X_0 = \mathbb{P}(T_{\mathbb{P}_2})$ and take for X a vector bundle E on $\mathbb{P}_2$ with the same Chern classes and let $X = \mathbb{P}(E)$.

The theorem is proved in [Campana and Peternell 1994] in the case that X_0 is not the blow-up of another Fano threefold along a smooth curve and in [Freitag 1994] in the remaining cases. One also might ask whether one can release the projectivity assumption. In this context the paper [Summerer 1997] proves that the flag manifold $\mathbb{P}(T_{\mathbb{P}_2})$ is rigid under global deformation and that $\mathbb{P}_1 \times \mathbb{P}_2$ has only "the obvious" (projective) deformations.

We saw in (4.2) that the statement "K_X is not nef" is not topologically invariant. However, if we start with a threefold X_0 such that $\chi(X_0, \mathcal{O}_{X_0}) > 0$, any projective threefold X homeomorphic to X_0 is not minimal, that is, carries an extremal contraction, *once* we know that $\chi(X_0, \mathcal{O}_{X_0}) = \chi(X, \mathcal{O}_X)$. This means that the problem of projective complex structures for threefolds with $b_2 > 1$ is most tractable in the case of positive holomorphic Euler characteristic, for example Fano threefolds.

4.17. A somehow related result of [Campana and Peternell 1994] is the following. Some non-projective Moishezon twistor space X_0 is constructed with the property that there is not projective threefold homeomorphic to X_0.

4.18. We next mention the fundamental problem asking which topological manifolds admit a complex structure. We concentrate on simply connected manifolds of dimension 6. The topological 6-manifolds which have torsion-free homology are classified by the work of Wall [1966] and Jupp [1973]. They can be completely described by a system of invariants: $H^2(X, \mathbb{Z})$, the Betti number $b_3(X)$, the cup product on $H^2(X, \mathbb{Z})$, the Pontrjagin class $p_1(X)$, the Stiefel–Whitney class $w_2(X)$ and the triangulation class $\tau(X) \in H^4(X, \mathbb{Z})$ with a certain relation. Now several fundamental questions arise:

(a) Which complex cubics can be realised as cup form of a compact complex threefold (up to equivalence)?

(b) Which systems of invariants can be realised by almost complex manifold?

(c) Which systems of invariants can be realised by complex manifolds (by Kähler manifolds)?

Instead of describing results we refer to the papers [Okonek and Van de Ven 1995; Schmitt 1995; 1997; 1996].

4.19. One of the most natural questions in the context of Section 4.18 is certainly the problem of complex structures on spheres. The situation is as follows.

(a) The only complex structure on S^2 is of course the complex structure $\mathbb{P}^1$.

(b) The spheres S^{2n} do not admit almost complex structures for $n \geq 2, n \neq 3$. If S^{2n} is equipped with the standard differentiable structure, this is due to Kirchhoff [1948], in general to Borel and Serre [1953].

(c) There remains the question of complex structures on S^6. Here almost complex structures do exist; one induced by the Cayley numbers: see [Steenrod 1951]. Hence there is the question of integrability. This is still unsolved. It is clear that a complex structure on S^6 is far from being Kähler.

(d) The only result is the following [Campana et al. 1998a]: If X is a compact manifold homeomorphic to S^6, then X does not admit a non-constant meromorphic function. More generally one can show:

THEOREM 4.20. *Let X be a smooth compact threefold with $b_2(X) = 0$. If $a(X) \geq 1$, then either $b_1(X) = 1$ and $b_3(X) = 0$ or $b_1(X) = 0$ and $b_3(X) = 2$.*

The first alternative is realised by Hopf manifolds, and the second by Calabi–Eckmann manifolds, which are complex structures on $S^3 \times S^3$. Note finally the relation between complex structures on S^6 and $\mathbb{P}_3$: let X be a complex structure on S^6 and $\hat{X} \to X$ the blow-up of a point $p \in X$. Then $\hat{X}$ is a complex structure on $\mathbb{P}_3$. In [Huckleberry et al. 1999] it is shown that there is no complex Lie group acting on X with an open orbit, in particular X can have at most two independent vector fields. A consequence: If S^6 has a complex structure, then $\mathbb{P}_3$ has a 1-dimensional family of exotic complex structures.

5. The Fundamental Group

A very interesting topological invariant is the fundamental group of a compact Kähler or projective manifold. We survey here some results concerning the following questions:

(a) Which groups are Kähler (that is, of the form $\pi_1(X)$, for some adequate compact Kähler manifold X)? Many restrictions are known.

(b) How does $\kappa(X)$ influence $\pi_1(X)$ or $\widetilde{X}$, the universal cover of X (any compact Kähler manifold)?

(c) Do the classes of groups of the form $\pi_1(X)$ for X compact Kähler and X projective differ?

We shall concentrate on question (b), and to some extend on (c), which is closely related to (b).

5.1. Restrictions on Kähler groups. We shall only give here some very brief indications, refering to [Amorós et al. 1996] for more details, where the known obstructions and examples are systematically surveyed.

There are three main types of known restrictions:

5.1.1. Restrictions on the lower central series of $\pi_1(X)$. These are deduced from classical Hodge theory (the $\partial\bar\partial$-Lemma). The basic two restrictions are that (up to torsion) this lower central series is determined by its first 2 terms (that is, by the natural map $\bigwedge^2 H^1(X, \mathbb{Q}) \to H^2(X, \mathbb{Q})$). This was shown with $\mathbb{R}$-coefficients in [Deligne et al. 1975], and was later related to the Albanese map in [Campana 1995b].

Notice, however, that even for nilpotent groups, it is not known which ones are Kähler. Only recently were non-trivial examples given [Campana 1995b; Sommese and Van de Ven 1986]). Only for the very special case of Heisenberg groups is the situation more or less understood [Campana 1995b; Carlson and Toledo 1995]. But no example is known of torsion-free nilpotent Kähler groups of nilpotency class 3 or more, although no obstruction to their existence is known.

5.1.2. The second type of restriction is that $\overline{H}^1\big(\pi_1(X)\big), \ell^2\big(\pi_1(X)\big) \neq 0$ implies that $\pi_1(X)$ is commensurable to a surface group (proved by M. Gromov, using L^2-methods). These methods show that a Kähler group has at most one end [Arapura et al. 1992]. See [Amorós et al. 1996, Sections 1 and 4] for more details.

5.1.3. Obstructions for lattices in semisimple Lie groups to be Kähler. These are derived from the theory of harmonic maps to negatively curved manifolds. Its extension to the case of Bruhat–Tits buildings and negatively curved metric spaces, which appears in papers by Gromov and Schoen and by Korevaar and Schoen, seems to be a very promising new tool. See [Amorós et al. 1996, Sections 5, 6, 7].

5.1.4. Remark. The Kähler assumption seems very essential (and minimal) to obtain restrictions on $\pi_1(X)$. Indeed: any finitely presented group is the fundamental group of a compact complex 3-fold, which can be choosen symplectic, and a twistor space on some appropriate self-dual Riemannian 4-fold after a deep result of C. Taubes. Notice that a twistor space which is Kähler (Hitchin) or even bimeromorphic to Kähler [Campana 1991] is simply-connected, so that the twistor construction does not produce any non-trivial fundamental group in the Kähler case.

5.2. Kodaira dimension and fundamental group.

We denote by X a compact Kähler manifold. In Riemannian or Kähler geometry, positivity assumptions on the Ricci curvature imply restrictions on the fundamental group (compare Section 6). For example:

5.2.1. If $\mathrm{Ricci}(X) > 0$, then $\pi_1(X)$ is finite (denoted: $|\pi_1(X)| < +\infty$).

5.2.2. If $\mathrm{Ricci}(X) \geq 0$, then $\pi_1(X)$ is almost abelian (that is, has a finite index subgroup which is abelian).

The analogous numerical assumptions read:

1. $c_1(X) > 0$ and

2. $c_1(X) \geq 0$ respectively.

In fact the analogous statements turn out to be true (due to the existence of Kähler–Einstein metrics in the case of 5.2.2'):

5.2.1'. If X is Fano, then $\pi_1(X) = 1$.

5.2.2'. If $c_1(X) = 0$, then $\pi_1(X)$ is almost abelian. (We assume that X is Kähler!)

As in Section 2, however, one expects this kind of result to be true under weaker assumptions, since π_1 is a birational invariant, and the results above should remain true for "minimal models". The questions then become:

QUESTION 5.2.1''. Assume $\kappa^+(X) = -\infty$. Is then $\pi_1(X) = 1$? (We shall see below that this is true).

Observe that here, the condition $\kappa^+(X) = -\infty$ a priori is weaker than assuming that X is birational to some Fano manifold. So that a vanishing theorem for π_1 in that case is the best one can expect by using a hypothesis on Kodaira dimensions.

QUESTION 5.2.2''. Assume $\kappa(X) = 0$. Is then $\pi_1(X)$ almost abelian?

This is unknown, but is conjectured to be true. We shall see an important special case below. It holds for surfaces and also for projective threefolds by a result of Y. Namikawa and J. Steenbrink [1995].

In fact a relative version of 5.2.2'' can reasonably be expected, too:

QUESTION 5.2.3. Let X be a projective manifold with $\kappa(X) \geq 0$; let $\Phi : X \to Y$ be its Iitaka fibration, and $\Phi_* : \pi_1(X) \to \pi_1(Y)$ the induced map (it is well defined if we assume, as we can, that X and Y are smooth -moreover it is surjective since Φ is connected). Let $K := \operatorname{Ker} \Phi_*$. Is then K almost abelian — at least after replacing X by some suitable finite étale cover $\overline{X} \to X$?

Observe that the generic fiber X_y of Φ has $\kappa = 0$. (However, the natural map $\pi_1(X_y) \mapsto K$ is not surjective in general, so that Question 5.2.3 does not reduce to 5.2.2''.)

Notice that this question is empty when X is of general type. Thus the only really new fundamental groups are to be found in this class — for X a surface, by Lefschetz theorem in the projective case.

A similar question can be asked for the algebraic reduction.

QUESTION 5.2.4. Let X be a compact Kähler manifold; let $r : X \to A$ be its algebraic reduction, and $r_* : \pi_1(X) \to \pi_1(A)$ be the induced map. (As in Question 5.2.3 this is well-defined and onto). Let $R := \operatorname{Ker} r_*$. Is then R almost-abelian? Here the generic fiber X_a of r has $\kappa(X_a) \leq 0$.

Some special cases are known, which shall be discussed below.

5.3. Γ-reduction

THEOREM 5.3.1 [Campana 1994]. *Let X be a compact Kähler manifold. There exists a quasi-fibration $\gamma_X : X \to \Gamma(X)$ such that for a general in X, the fiber X_a of γ_X passing through a is the largest among the **connected** compact analytic subsets A of X containing a such that the natural map : $i_* : \pi_1(\widehat{A}) \to \pi_1(X)$ has finite image, where i_* is induced by the inclusion of A in X composed with the normalisation map $\nu : \widehat{A} \to A$.*

*The map γ_X, called the Γ-**reduction** of X, is bimeromorphically invariant; its generic fiber is smooth. We denote by $\gamma d(X) := \dim \Gamma(X)$ its γ-**dimension**.*

The special case where X is projective has been shown independently by J. Kollár [1993], who named the map above the **Shafarevich map** of X.

The result above has been shown in [Campana 1994] with another (trivially equivalent) formulation for the universal cover $\widetilde{X}$ (or any Galois cover) of X. See also [Campana 1994] or [Kollár 1993] for the relation ship of Theorem 5.3.1 with Shafarevich's conjecture. (The Shafarevich conjecture implies in particular that γ_X is regular, and that $a \in X$ can be any point.)

Note that $\gamma d(X) = 0$ is equivalent to $A = X$ and also to $|\pi_1(X)| < +\infty$; on the other extreme: $\gamma d(X) = \dim X$ means that for any positive dimensional A through general a, the map $\pi_1(\widehat{A}) \to \pi_1(X)$ has infinite image. Obviously, X is not uniruled in that case (in fact: if $\widehat{A} = \mathbb{P}_1$, the map above has trivial image). This remark will be generalized below.

EXAMPLES. (a) Curves: $\gamma d(X) = 0$ if $g(X) = 0$; and $\gamma d(X) = 1$ if $g(X) \geq 1$.
(b) Surfaces: If $\kappa(X) < 2$, then $\gamma d(X) = q'(X) + \chi'(\mathcal{O}_X)$, where

$$q'(X) = \inf\left(q(X), 1\right) \quad \text{and} \quad \chi'(\mathcal{O}_X) = \begin{cases} 0 & \text{if } \chi(\mathcal{O}_X) \neq 0, \\ 1 & \text{if } \chi(\mathcal{O}_X) = 0. \end{cases}$$

This formula can be checked directly from the classification of Enriques–Kodaira if $\kappa(X) \leq 0$; in the elliptic case with $\kappa(X) = 1$, it can be shown that $\text{Ker}\big(\pi_1(X) \to \pi_1(B)\big)$ is infinite precisely when the singular fibers are all multiple elliptic, that is, when $\chi(\mathcal{O}_X) = 0$. (See [Gurjar and Shastri 1985], for example). For surfaces of general type, there does not seem to be any simple relationship between c_1^2 and c_2 and $\gamma d(X)$. The values attained by $\gamma d(X)$ are all possible (0, 1 or 2).

(c) Tori: We have $\gamma d(X) = \dim X$ if X is a complex torus, since its universal cover is Stein. Notice that we also have $\chi(\mathcal{O}_X) = 0$.

THEOREM 5.3.2 [Campana 1994; Kollár et al. 1992a]. *If X is rationally connected, $\pi_1(X) = \{1\}$. In particular, if X is Fano, $\pi_1(X) = \{1\}$.*

PROOF. This is an easy consequence of Theorem 5.3.1. Let a, b be general in X. They can be joined by a connected rational chain A. Then $\pi_1(\widehat{A})$ maps trivially to $\pi_1(X)$. So a and b are in the some fiber of γ_X, which is thus constant. □

This proof is the one given in [Campana 1994] (except that one works on $\widetilde{X}$ there). The proof given in [Kollár et al. 1992a] is more difficult, since it uses first that rational connectedness implies strong rational connectedness, and then uses this stronger property to conclude. (In the case of strong rational connectedness a simple argument does exist; see [Campana 1991].)

Actually, 5.3.2 holds in the relative version as well:

THEOREM 5.3.3 [Kollár 1993]. *Let $f : X \to Y$ be a dominant rational map with X, Y smooth. Assume the generic fiber of f is rationally connected. Then $f_* : \pi_1(X) \to \pi_1(Y)$ is an isomorphism.*

The proof rests on Theorem 5.3.2 and an analysis of the π_1 of fibers of f in codimension 1 on Y, to show they are simply connected.

Notice that the result above is no longer valid if one only assumes that the generic (smooth) fibers of X are simply connected:

EXAMPLE 5.3.4. Let S be an Enriques surface; $u : \widetilde{S} \mapsto S$ its universal cover (a K3 surface) and E be an elliptic curve. Let $\mathbb{Z}_2$ act on $\overline{X} := E \times \widetilde{S}$ by $i(z, \tilde{s}) = \big(-z; j(\tilde{s})\big)$, where i is a generator of $\mathbb{Z}_2$ and $j : \widetilde{S} \to \widetilde{S}$ the "Enriques Involution" $\big(\widetilde{S}/(j) = S\big)$. Let $\pi : X := \overline{X}/(i) \to E/\pm \simeq \mathbb{P}_1$ be induced by the first projection of $\overline{X}$. Then $\pi_* : \pi_1(X) \to \pi_1(\mathbb{P}_1) = \{1\}$ has infinite kernel (observe that the singular fibers are not simply connected, and that $\chi(\mathcal{O}_{\tilde{S}}) = 2 > 1$).

As a consequence of Theorem 5.3.3, we have:

COROLLARY 5.3.5. *If X is rationally generated, then $\pi_1(X) = \{1\}$.*

PROOF. We just have to iterate the MRC fibrations to eventually arrive at a point. □

This corollary will be strengthened below (with rational generatedness replaced by $\kappa^+(X) = -\infty$).

The drawback in Theorem 5.3.1 is that $a \in X$ has to be choosen to be general, so that it does not solve the following problem:

QUESTION 5.3.6 (M. Nori). *Let X be a smooth projective surface. Assume that X contains a rational curve C (singular possibly), such that $C^2 > 0$. Is then $\pi_1(X)$ finite?*

If $a \in X$ could be choosen to be any point, the answer would be yes. In particular, if the Shafarevich conjecture holds this is the case (as observed first by Gurjar). It is easy to see, using the Albanese map, that $q(X) = 0$.

The best results obtained are that linear representations of $\pi_1(X)$ have finite image. The results below also show easily that $\kappa(X) = 2$ if $\pi_1(X)$ is infinite (this was first shown by Gurjar–Shastri using classification of surfaces and showing that the Shafarevich's conjecture holds for surfaces with $\kappa \leq 1$).

5.4. The comparison theorem

THEOREM 5.4.1 [Campana 1995a]. *Let X be a compact Kähler manifold with $\chi(\mathcal{O}_X) \neq 0$. Then either $\kappa^+(X) \geq \gamma d(X)$ or $\kappa^+(X) = -\infty$ and $\pi_1(X) = \{1\}$.*

REMARKS 5.4.2. (1) The condition $\chi(\mathcal{O}_X) \neq 0$ cannot be dropped: tori X present the maximum failure to the inequality ($\kappa^+ = 0, \gamma d = n$). They might be characterized (birationally up to finite etale covers) by that property. See below.

(2) Theorem 5.4.1 extends an earlier result of M. Gromov [1991]: "If the universal cover $\tilde{X}$ of X does not contain any positive dimensional compact subvariety and $\chi(\mathcal{O}_X) \neq 0$, then X is projective". The generalisation of this result lead to the introduction of the invariants γd, κ^+ and the construction of the Γ-reduction $\gamma_X : X \to \Gamma(X)$.

(3) The proof of Gromov (and of Theorem 5.4.1) rests on L_2-methods, and especially the Atiyah's L_2-index theorem.

When $\kappa^+(X) \leq 0$, Theorem 5.4.1 gives finiteness criteria for $\pi_1(X)$, as follows:

COROLLARY 5.4.3. *Let $\kappa^+(X) = -\infty$. Then $\pi_1(X) = \{1\}$.*

PROOF. It is easy to show that $h^0(X, \Omega_X^p) = 0$ ($p > 0$) if $\kappa^+(X) = -\infty$. Thus $\chi(\mathcal{O}_X) = 1 \neq 0$, and Theorem 5.4.1 applies. □

Notice that $\kappa^+(X) = -\infty$ if X is rationally generated. So we get Corollary 5.3.5 again by a different method. Conjecturally if X is rationally generated, then $\kappa^+(X) = -\infty$; if it is false, Corollary 5.4.3 is strictly stronger than 5.3.5.

COROLLARY 5.4.4. *Let $\kappa^+(X) = 0$, and let $\chi(\mathcal{O}_X) \neq 0$. Then $|\pi_1(X)| \leq 2^{n-1}/|\chi(\mathcal{O}_X)|$, where $n = \dim X$ and $|\pi_1(X)|$ is the cardinality of $\pi_1(X)$.*

PROOF. By Theorem 5.4.1, only the inequality has to be shown (finiteness results from 5.4.1). So we are reduced to bounding π_1^{alg}, the algebraic fundamental group instead of π_1. This follows from the usual covering trick, plus the following easy inequality:

LEMMA 5.4.5. *Assume $\kappa^+(X) = 0$. Then*

$$h^0(X, \Omega_X^p) \leq \binom{n}{p} \quad and \quad |\chi(\mathcal{O}_X)| \leq 2^{n-1}. \qquad \square$$

Conjecturally, Corollary 5.4.4 should hold with $\kappa^+(X) = 0$ replaced by $\kappa(X) = 0$. If $\kappa(X) = 0$, Lemma 5.4.5 is a conjecture of K. Ueno (proved by Y. Kawamata if $p = 1$).

COROLLARY 5.4.6. *Let $\chi(\mathcal{O}_X) \neq 0$ and assume $c_1(X) = 0$. Then $|\pi_1(X)| \leq 2^{n-1}$ if X is projective.*

Indeed: $\kappa^+(X) = 0$ in that case. Of course, by the existence of Ricci-flat metrics, this is known if X is Kähler. But the proof given here is more elementary.

A special case is:

COROLLARY 5.4.7. *Let X be a K3 surface (so that $q(X) = 0$ and $K_X = \mathcal{O}_X$). Then $\pi_1(X) = \{1\}$.*

The proof of 5.4.1 shows this (even without assuming X to be Kähler). This is for sure the simplest proof of this result, not requiring any knowledge of either deformation theory or Ricci-flat metrics.

In a similar vein:

COROLLARY 5.4.8. *Let X be a compact Kähler manifold with $a(X) = 0$ (that is, X has no non-constant meromorphic function). Asume that $\chi(\mathcal{O}_X) \neq 0$, too. Then $|\pi_1(X)| \leq 2^{n-1}$.*

Indeed, $a(X) \geq \kappa^+(X)$.

Notice that general tori (with $a(X) = 0$) again show that the assumption $\chi(\mathcal{O}_X) \neq 0$ cannot be dropped.

When $n = 3$, the assumption $\chi(\mathcal{O}_X) \neq 0$ can be weakened and the bound improved:

COROLLARY 5.4.9. *Let X a compact Kähler 3-fold with $a(X) = q(X) = 0$. Then $|\pi_1(X)| \leq 3$ (hence $\pi_1(X) = \{1\}$, $\mathbb{Z}_2$ or $\mathbb{Z}_3$; the last two possibilities are probably impossible).*

PROOF. We only need to show that $0 \neq \chi(\mathcal{O}_X) (= 1 - q + n^{2,0} - h^{3,0} \geq h^{2,0} > 0)$; the first inequality holds because $h^{3,0} \leq 1$, the second because $h^{2,0} = 0$ implies X is projective (Kodaira). $\qquad \square$

The only known compact Kähler 3-folds with $a(X) = q(X) = 0$ are bimeromorphically ruled fibrations $\pi : X \to S$ where S is a K3-surface with $a(S) = 0$. Conjecturally, these are the only ones. If one assumes the Kähler version of the minimal model program and abundance conjecture in dimension 3, this conjecture is true (see [Peternell 1998b] and Section 3; the main point is to give a meaning to the statement "K_X is nef" in this situation).

The method of proof of Theorem 5.4.1 gives also a part of the relative versions of Questions 5.2.3 and 5.2.4:

THEOREM 5.4.10 [Campana 1994]. *Let* $\Psi : X \to Y$ *be either the Iitaka fibration or the algebraic reduction of the compact Kähler manifold* X. *Let* $L := \mathrm{Ker}\,\big(\psi_* : \pi_1(X) \to \pi_1(Y)\big)$ *be the kernel of the induced map. Let* X_y *be a smooth fiber of* ψ, *and* $j_* : \pi_1(X_y) \to K$ *be the morphism induced by the natural inclusion* $j : X_y \hookrightarrow X$. *Then the image of* j_* *is finite if* $\chi(\mathcal{O}_X) \neq 0$ *(and if moreover* $\kappa^+(X_y) = \kappa(X_y) = 0$ *in case* ψ *is the Iitaka fibration).*

This motivates the following question:

QUESTION 5.4.11. Let X be compact Kähler with $\chi(\mathcal{O}_X) \neq 0$; let $r : X \to A$ be its algebraic reduction. Is then $\mathrm{Ker}\big(r_* : \pi_1(\overline{X}) \to \pi_1(\overline{A})\big)$ a finite group for some suitable finite étale cover $\overline{X}$ of X?

For the general case of Questions 5.2.3 and 5.2.4, there is another partial positive answer. In order to state it, we introduce some notation. For any group Γ, set $\Gamma^{\mathrm{nilp}} = \Gamma/\Gamma'_\infty$, where $\Gamma'_\infty := \bigcap_{n \geq 2} \Gamma'_n$ with

$$\Gamma'_n := \mathrm{Ker}\,\big(\Gamma \to \Gamma/\Gamma_n \to (\Gamma/\Gamma_n)/\,\mathrm{Torsion}\,\big).$$

THEOREM 5.4.12 [Campana 1995b]. *Let* $\Psi : X \to Y$ *be either the algebraic reduction, or the Iitaka fibration of the compact Kähler manifold* X. *Let* $\psi_*^{\mathrm{nilp}} : \pi_1(X)^{\mathrm{nilp}} \to \pi_1(Y)^{\mathrm{nilp}}$ *be the natural morphism (see [Campana 1995b] for the precise definition). Then* $K := \mathrm{Ker}(\psi_*^{\mathrm{nilp}}) \cong \mathbb{Z}^{\oplus 2s}$, *where* $s = q(X) - q(Y)$, *and the exact sequence*

$$1 \to K \to \pi_1(X)^{\mathrm{nilp}} \to \pi_1(Y)^{\mathrm{nilp}} \to 1$$

splits (non-canonically).

The proof is given in [Campana 1995b] only for the algebraic reduction, but the sume proof applies for the Iitaka fibration.

We conclude this section with some conjectures concerning n-dimensional compact Kähler manifolds X:

CONJECTURES 5.4.13. (1) Let X be such that $\kappa^+(X) = 0$ (or $\kappa(X) = 0$), $\gamma d(X) = n$. Then X is bimeromorphic to some X_0 which is covered by a torus.

(2) More generally: assume that $\kappa^+(X) = 0$ (or $\kappa(X) = 0$), and that $\gamma d(X) = d$. Then: some finite étale cover $\overline{X}$ of X is bimeromorphic to a product $\overline{Y} \times T$, where T is a torus and $\kappa^+(\overline{Y}) = \kappa(\overline{Y}) = 0$ with $\pi_1(\overline{Y}) = \{1\}$.

Conjecture 5.4.12(2) should also have a relative version (for the Iitaka fibration).

Finally, we refer to [Kollár 1993] to see some other aspects of the application of Theorem 5.4.1 in the projective setting.

6. Biregular Classification

By "biregular classification" we mean a more or less explicit description of varieties of a certain type. Of course this is only possible under very restrictive circumstances. In differential geometry one classifies roughly in terms of curvature conditions: positive, negative and zero curvature. A curvature condition is suitable for biregular classification rather than birational classification because the sign of the curvature makes the variety more or less rigid in the birational category: blow-ups destroy the curvature condition. In the context of complex geometry a slightly more general notion than the sign of curvature will be useful as we shall see in this section. However we can still, cum grano salis, say that the aim of this section is to understand projective or Kähler manifolds with semipositive (bisectional or Ricci) curvature. The class of negatively curved manifolds is much larger and it is hopeless to get a biregular classification. We begin with a very short review of the situation in dimension 1.

6.0. Let X be a compact Riemann surface. If $-K_X$ is ample, then $X \simeq \mathbb{P}_1$, if $K_X = \mathcal{O}_X$, then X is a torus and if K_X is ample, then X has genus ≥ 2. The same classification holds in terms of positive, zero and negative curvature. In this case of course the holomorphic bisectional and Ricci curvatures are equivalent. In higher dimensions the tangent bundle T_X and the anticanonical bundle $-K_X$ are no longer the same, that is, we have to distinguish between bisectional and Ricci curvature; we will first look at the tangent bundle.

The classification theory in higher dimensions starts with this result:

THEOREM 6.1 (Mori). *Let X be a compact manifold with ample tangent bundle. Then $X \simeq \mathbb{P}_n$.*

X is automatically projective and Mori's proof is to rediscover the lines (through a given point). A priori however it is not at all clear whether there is any rational curve; these are constructed by Mori's reduction to characteristic p. Given a projective manifold X with $K_X \cdot C < 0$ for some curve C, Mori constructs a rational curve with the same property. The point is that in characteristic p, the inequality $K_X \cdot C < 0$ allow one to deform C, at least after having applied a suitable Frobenius. The rational curve appears since a certain rational map is not a morphism.

For a proof of Theorem 6.1 not using characteristic p, see [Peternell 1996].

In the same year (1979) Siu and Yau proved Theorem 6.1 with a weaker assumption, namely that X has a Kähler metric with positive holomorphic bisectional curvature.

In the spirit of Siu and Yau, but using characteristic p, Mok [1988] proved the following result:

THEOREM 6.2. *Let X be a compact Kähler manifold of semi-positive holomorphic bisectional curvature. Then, after taking a finite étale cover, X is of the* *form*

$$X \simeq T \times \prod Y_j,$$

where T is a torus and the Y_j are hermitian symmetric manifolds with $b_2 = 1$.

In Mori's theorem no assumption on curvature is made; ampleness is "just" an algebraic property. To check it, it is not necessary to construct a metric. We are looking for an equivalent result in the semipositive case. Note by the way that in Mok's theorem one needs a **Kähler** metric of semipositive curvature which is much stronger than just assuming the existence of some hermitian metric with the same curvature. In the case of line bundles on projective manifolds it is clear how to get rid of curvature conditions: one assumes L to be nef, that is, $L \cdot C \geq 0$ for all curve $C \subset X$. In the Kähler case however this definition clearly fails. The substitute is Definition 3.11. In the vector bundle case we define:

DEFINITION 6.3. Let X be a compact manifold. A vector bundle E on X is nef, if $\mathcal{O}_{\mathbb{P}(E)}(1)$ is *nef* on $\mathbb{P}(E)$.

Now the problem is: *Determine the structure of compact Kähler manifolds X such that T_X is nef, or, alternatively, $-K_X$ nef.*

For T_X nef we have a structure theorem, proved in [Demailly et al. 1994]. In order to state it we introduce the following "irregularity":

$$\tilde{q}(X) = \sup\{q(\tilde{X}) \mid \tilde{X} \to X \text{ is finite étale }\}.$$

THEOREM 6.4. *Let X be a compact Kähler manifold with T_X nef.*

(1) *The Albanese map $\alpha : X \to \mathrm{Alb}(X)$ is a surjective submersion with nef relative tangent bundle.*
(2) *If $\tilde{q}(X) = q(X)$, then the fibers of α are Fano manifolds.*
(3) *X is Fano if and only if $c_1(X)^n \neq 0$.*
(4) *$\pi_1(X)$ is almost abelian, that is, an extension of $\mathbb{Z}^m$ by a finite group.*

Note that the structure of α is not arbitrary: it has a flat nature in the sense that the bundles $\alpha_*(-mK_X)$ are numerically flat (nef and with nef dual). An important step in the proof of Theorem 6.4 is the study of numerically flat vector bundles:

THEOREM 6.5. *Let X be a compact Kähler manifold. Let E be a numerically flat vector bundle on X. Then E admits a filtration*

$$0 = E_0 \subset E_1 \subset \cdots \subset E_p = E$$

by subbundles such that the quotients E_i/E_{i+1} are hermitian flat, that is, defined by a representation $\pi_1(X) \to U(r)$.

For proofs see [Demailly et al. 1994]. For ideas and background relevant to all of this section see also [Peternell 1996].

Theorem 6.5 is used in the proof of 6.4 to show the existence of a 1-form after finite étale cover, if there is a p-form for p odd.

Theorem 6.4 reduces the structure problem for manifolds with nef tangent bundles to that of Fano manifolds with nef tangent bundles. If X is Fano with T_X nef, then consider the contraction of an extremal ray, say $\varphi : X \to Y$. One can prove that φ is a surjective submersion with nef relative tangent bundle, so that the main difficulty is provided by Fano manifolds with $b_2(X) = 1$. Here is the main conjecture about these varieties.

CONJECTURE 6.6. Let X be a Fano manifold with T_X nef. Then X is rational homogeneous.

As already said, the main difficulty arises when $b_2(X) = 1$. If this case is settled, then one has to study Mori fibrations over rational homogeneous manifolds whose fibers are rational homogeneous and need to lift vector fields. The evidence for Conjecture 6.6 is the validity in dimensions 2 and 3 and that X behaves as if it is homogeneous: every effective divisor is nef, the deformations of a rational curve fill up all of X etc. The classification in dimension 3 uses however classification theory and therefore does not shed any light on the higher-dimensional case. One is tempted to prove the existence at least of some vector fields by proving

$$\chi(X, T_X) > 0 \qquad (*)$$

which together with the vanishing $H^q(X, T_X) = 0, q \geq 2$ would give us some vector field. The nefness of T_X yields inequalities for the Chern classes of X. Unfortunately these inequalities are not strong enough to give $(*)$ via Riemann–Roch. Instead one should study the family of rational curves of minimal degree in X. They already cover X and experience shows that they dictate the geometry of X. For more comments see [Peternell 1996].

We now turn to compact Kähler manifolds X with $-K_X$ nef. The building blocks of these varieties are Fano manifolds, that is, $-K_X$ is ample, and manifolds with $K_X \equiv 0$, i.e. tori, Calabi–Yau manifolds and symplectic manifolds up to finite étale cover. We want to see how manifolds with $-K_X$ nef are constructed from these "prototypes". The starting point is to separate the torus part by considering the Albanese map.

THEOREM 6.7. Let X be a compact Kähler manifold with $-K_X$ hermitian semi-positive (that is, there is a metric on $-K_X$ with semi-positive curvature). Then the Albanese map is a surjective submersion.

The proof goes by constructing for every holomorphic 1-form ω a differentiable vector field v such that the contraction gives $\|v\|^2$. Now the curvature condition implies that v is holomorphic, since ω is holomorphic and therefore $\|v\|$ is a constant. Hence ω has no zeroes which proves the claim. See [Demailly et al. 1993] for details. If $-K_X$ is merely nef, the proof apparently does not work. At least the surjectivity was proved by Qi Zhang [1996] in the algebraic case:

THEOREM 6.8. Let X be a projective manifold with $-K_X$ nef. Then the Albanese map is surjective.

PROOF. If not, X would admit a map onto a variety Y of general type, which can be ruled by cutting down to a curve in Y and applying the results of [Miyaoka 1993]. □

This last paper relies on characteristic p, so the Kähler case remains unsettled in general. However in [Campana et al. 1998b] it is shown that a compact Kähler n-fold with $-K_X$ nef cannot have a map onto a variety of general type of dimension 1 (this case is proved in [Demailly et al. 1993]), $n-2$ or $n-1$. This settles in particular Theorem 6.8 in the Kähler case up to dimension 4.

Concerning smoothness, the following theorem settles the threefold case:

THEOREM 6.9 [Peternell and Serrano 1998]. *Let X be a smooth projective three-fold with $-K_X$ nef. Then the Albanese is smooth.*

The proof relies on a careful analysis of the Mori contractions on X. The Kähler case is settled in [Demailly et al. 1998].

In the hermitian semi-positive case one can prove much more [Demailly et al. 1996]:

THEOREM 6.10. *Let X be a compact Kähler manifold with $-K_X$ hermitian semi-positive. Then:*

(1) *The universal cover $\tilde{X}$ admits a holomorphic and isometric splitting*

$$\tilde{X} \simeq \mathbb{C}^q \times \prod X_i,$$

where the X_i are Calabi–Yau manifolds or symplectic manifolds or manifolds having the property that

$$H^0(X_i, \Omega_{X_i}^{\otimes m}) = 0$$

for all $m > 0$.

(2) *There exists a finite étale Galois cover $\hat{X} \to X$ such that the Albanese map is a locally trivial fiber bundle to the q-dimensional torus A whose fibers are all simply connected and of types descirbed in (1).*

(3) *We have $\pi_1(\hat{X}) \simeq \mathbb{Z}^{2q}$.*

In the nef case it is at least known that $\pi_1(X)$ has subexponential growth. In [Campana 1995a] (see also Definiton 1.9) a refined version of Kodaira dimension is defined:

DEFINITION 6.11. Let X be a compact manifold. Then

$$\kappa^+(X) = \max\{\kappa(\det \mathcal{F}) \mid \mathcal{F} \subset \Omega_X^p \text{ for some } p > 0\}.$$

Replacing Ω_X^p by $\Omega_X^{\otimes m}$ we obtain an invariant $\kappa^{++}(X)$. In these terms the varieties X in Theorem 6.10 which are neither Calabi–Yau nor symplectic satisfy $\kappa^{++}(X) = -\infty$. Moreover we have $\kappa_+(X) = \kappa^{++}(X)$ if $-K_X$ is hermitian semi-positive.

To conclude we collect problems on varieties with nef anticanonical bundles as well as problems on the new Kodaira type invariants.

PROBLEMS 6.12. Let X be compact Kähler.

(1) What is the relation between $\kappa^+(X)$ and $\kappa^{++}(X)$? Of course $\kappa^+(X) \leq \kappa^{++}(X)$.

(2) Suppose $\kappa^+(X) = -\infty$. Is X rationally generated or even rationally connected, at least if $-K_X$ is nef?

(3) Is the structure theorem 6.10 true for compact Kähler manifolds with $-K_X$ nef?

(4) Assume $-K_X$ nef. Is $\kappa^+(X) \leq 0$?

References

[Amorós et al. 1996] J. Amorós, M. Burger, K. Corlette, D. Kotschick, and D. Toledo, *Fundamental groups of compact Kähler manifolds*, Math. Surveys and Monographs **44**, Amer. Math. Soc., Providence, RI, 1996.

[Andreatta and Peternell 1997] M. Andreatta and T. Peternell, "On the limits of manifolds with nef canonical bundles", pp. 1–6 in *Complex analysis and geometry* (Trento, 1995), edited by V. Ancona et al., Pitman Res. Notes Math. **366**, Longman, Harlow, 1997.

[Arapura et al. 1992] D. Arapura, P. Bressler, and M. Ramachandran, "On the fundamental group of a compact Kähler manifold", *Duke Math. J.* **68**:3 (1992), 477–488.

[Barth et al. 1984] W. Barth, C. Peters, and A. van de Ven, *Compact complex surfaces*, Ergebnisse der Mathematik und ihrer Grenzgebiete (3. Folge) **4**, Springer, Berlin, 1984.

[Beauville 1978] A. Beauville, *Surfaces algébriques complexes*, Astérisque **54**, Soc. math. France, Paris, 1978.

[Borel and Serre 1953] A. Borel and J.-P. Serre, "Groupes de Lie et puissances réduites de Steenrod", *Amer. J. Math.* **75** (1953), 409–448.

[Brieskorn 1964] E. Brieskorn, "Ein Satz über die komplexen Quadriken", *Math. Ann.* **155** (1964), 184–193.

[Campana 1981] F. Campana, "Coréduction algébrique d'un espace analytique faiblement kählérien compact", *Invent. Math.* **63**:2 (1981), 187–223.

[Campana 1991] F. Campana, "On twistor spaces of the class $\mathcal{C}$", *J. Differential Geom.* **33**:2 (1991), 541–549.

[Campana 1992] F. Campana, "Connexité rationnelle des variétés de Fano", *Ann. Sci. École Norm. Sup.* (4) **25**:5 (1992), 539–545.

[Campana 1994] F. Campana, "Remarques sur le revêtement universel des variétés kählériennes compactes", *Bull. Soc. Math. France* **122**:2 (1994), 255–284.

[Campana 1995a] F. Campana, "Fundamental group and positivity of cotangent bundles of compact Kähler manifolds", *J. Algebraic Geom.* **4**:3 (1995), 487–502.

[Campana 1995b]   F. Campana, "Remarques sur les groupes de Kähler nilpotents", *Ann. Sci. École Norm. Sup.* (4) **28**:3 (1995), 307–316.

[Campana and Peternell 1994]   F. Campana and T. Peternell, "Rigidity theorems for primitive Fano 3-folds", *Comm. Anal. Geom.* **2**:2 (1994), 173–201.

[Campana and Peternell 1997]   F. Campana and T. Peternell, "Towards a Mori theory on compact Kähler threefolds, I", *Math. Nachr.* **187** (1997), 29–59.

[Campana and Peternell 1998]   F. Campana and T. Peternell, "Holomorphic 2-forms on complex threefolds", preprint, 1998. To appear in *J. Alg. Geom.*

[Campana et al. 1998a]   F. Campana, J.-P. Demailly, and T. Peternell, "The algebraic dimension of compact complex threefolds with vanishing second Betti number", *Compositio Math.* **112**:1 (1998), 77–91.

[Campana et al. 1998b]   F. Campana, T. Peternell, and Q. Zhang, "On the Albanese maps of compact Kähler manifolds with nef anticanonical bundles", preprint, 1998.

[Carlson and Toledo 1995]   J. A. Carlson and D. Toledo, "Quadratic presentations and nilpotent Kähler groups", *J. Geom. Anal.* **5**:3 (1995), 359–377.

[Clemens and Griffiths 1972]   C. H. Clemens and P. A. Griffiths, "The intermediate Jacobian of the cubic threefold", *Ann. of Math.* (2) **95** (1972), 281–356.

[Clemens et al. 1988]   H. Clemens, J. Kollár, and S. Mori, *Higher-dimensional complex geometry*, Astérisque **166**, Soc. math. France, Paris, 1988.

[Deligne et al. 1975]   P. Deligne, P. Griffiths, J. Morgan, and D. Sullivan, "Real homotopy theory of Kähler manifolds", *Invent. Math.* **29**:3 (1975), 245–274.

[Demailly et al. 1993]   J.-P. Demailly, T. Peternell, and M. Schneider, "Kähler manifolds with numerically effective Ricci class", *Compositio Math.* **89**:2 (1993), 217–240.

[Demailly et al. 1994]   J.-P. Demailly, T. Peternell, and M. Schneider, "Compact complex manifolds with numerically effective tangent bundles", *J. Algebraic Geom.* **3**:2 (1994), 295–345.

[Demailly et al. 1996]   J.-P. Demailly, T. Peternell, and M. Schneider, "Compact Kähler manifolds with Hermitian semipositive anticanonical bundle", *Compositio Math.* **101**:2 (1996), 217–224.

[Demailly et al. 1998]   J. P. Demailly, T. Peternell, and M. Schneider, "Pseudo-effective line bundles on projective varieties", preprint, 1998.

[Donaldson and Kronheimer 1990]   S. K. Donaldson and P. B. Kronheimer, *The geometry of four-manifolds*, The Clarendon Press Oxford University Press, New York, 1990. Oxford Science Publications.

[Freitag 1994]   P. Freitag, *Homöomorphietyp von imprimitiven Fano 3-Mannigfaltigkeiten mit $b_2 = 2$ und $b_3 = 0$*, Thesis, Univ. Bayreuth, 1994.

[Friedman and Morgan 1994]   R. Friedman and J. W. Morgan, *Smooth four-manifolds and complex surfaces*, Ergebnisse der Mathematik und ihrer Grenzgebiete (3. Folge) **27**, Springer, Berlin, 1994.

[Fujita 1980]   T. Fujita, "On topological characterizations of complex projective spaces and affine linear spaces", *Proc. Japan Acad. Ser. A Math. Sci.* **56**:5 (1980), 231–234.

[Grauert et al. 1994]   H. Grauert, T. Peternell, and R. Remmert (editors), *Several complex variables, VII: Sheaf-theoretical methods in complex analysis*, edited by H. Grauert et al., Encycl. Math. Sci. **74**, Springer, Berlin, 1994.

[Gromov 1991] M. Gromov, "Kähler hyperbolicity and L_2-Hodge theory", *J. Differential Geom.* **33**:1 (1991), 263–292.

[Gurjar and Shastri 1985] R. V. Gurjar and A. R. Shastri, "Covering spaces of an elliptic surface", *Compositio Math.* **54**:1 (1985), 95–104.

[Hirzebruch and Kodaira 1957] F. Hirzebruch and K. Kodaira, "On the complex projective spaces", *J. Math. Pures Appl. (9)* **36** (1957), 201–216.

[Huckleberry et al. 1999] A. T. Huckleberry, S. Kebekus, and T. Peternell, "Group actions on S^6 and complex structures on $\mathbb{P}_3$", *Duke Math. J.* (1999). To appear.

[Hwang 1995] J.-M. Hwang, "Nondeformability of the complex hyperquadric", *Invent. Math.* **120**:2 (1995), 317–338.

[Hwang 1996] J.-M. Hwang, "Characterization of the complex projective space by holomorphic vector fields", *Math. Z.* **221**:3 (1996), 513–519.

[Hwang and Mok 1998] J.-M. Hwang and N. Mok, "Rigidity of irreducible Hermitian symmetric spaces of the compact type under Kähler deformation", *Invent. Math.* **131**:2 (1998), 393–418.

[Iskovskih 1977] V. A. Iskovskih, "Fano threefolds, I", *Izv. Akad. Nauk SSSR Ser. Mat.* **41**:3 (1977), 516–562. In Russian; translated in *Math. USSR Izv.* **11** (1977), 485–527.

[Iskovskih 1978] V. A. Iskovskih, "Fano threefolds, II", *Izv. Akad. Nauk SSSR Ser. Mat.* **42**:3 (1978), 506–549. In Russian; translated in *Math. USSR Izv.* **12** (1978), 469–506.

[Iskovskih 1989] V. A. Iskovskih, "Double projection from a line onto Fano 3-folds of the first kind", *Mat. Sb.* **180**:2 (1989), 260–278. In Russian; translated in *Math. USSR Sb.* **66** (1990), 265-284.

[Jupp 1973] P. E. Jupp, "Classification of certain 6-manifolds", *Proc. Cambridge Philos. Soc.* **73** (1973), 293–300.

[Kawamata 1988] Y. Kawamata, "Crepant blowing-up of 3-dimensional canonical singularities and its application to degenerations of surfaces", *Ann. of Math. (2)* **127**:1 (1988), 93–163.

[Kawamata et al. 1987] Y. Kawamata, K. Matsuda, and K. Matsuki, "Introduction to the minimal model problem", pp. 283–360 in *Algebraic geometry* (Sendai, 1985), edited by T. Oda, Adv. Studies in Pure Math. **10**, Kinokuniya, Tokyo and North-Holland, Amsterdam, 1987.

[Kirchoff 1948] A. Kirchoff, "Sur l'existence de certains champs tensoriels sur les sphères à n dimensions", *C. R. Acad. Sci. Paris* **223** (1948), 1258–1260.

[Kobayashi and Ochiai 1973] S. Kobayashi and T. Ochiai, "Characterizations of complex projective spaces and hyperquadrics", *J. Math. Kyoto Univ.* **13** (1973), 31–47.

[Kollár 1989] J. Kollár, "Minimal models of algebraic threefolds: Mori's program", pp. Exp. No. 712, 303–326 in *Séminaire Bourbaki, 1988/89*, Astérisque **177-178**, Soc. math. France, Paris, 1989.

[Kollár 1991a] J. Kollár, "Extremal rays on smooth threefolds", *Ann. Sci. École Norm. Sup. (4)* **24**:3 (1991), 339–361.

[Kollár 1991b] J. Kollár, "Flips, flops, minimal models, etc.", pp. 113–199 in *Surveys in differential geometry* (Cambridge, MA, 1990), Lehigh Univ., Bethlehem, PA, and Amer. Math. Soc., 1991.

[Kollár 1992] J. Kollár, *Flips and abundance for algebraic threefolds* (Salt Lake City, UT, 1991), Astérisque **211**, Soc. math. France, Montrouge, 1992.

[Kollár 1993] J. Kollár, "Shafarevich maps and plurigenera of algebraic varieties", *Invent. Math.* **113**:1 (1993), 177–215.

[Kollár 1996] J. Kollár, *Rational curves on algebraic varieties*, Ergebnisse der Mathematik und ihrer Grenzgebiete (3. Folge) **32**, Springer, Berlin, 1996.

[Kollár et al. 1992a] J. Kollár, Y. Miyaoka, and S. Mori, "Rationally connected varieties", *J. Algebraic Geom.* **1**:3 (1992), 429–448.

[Kollár et al. 1992b] J. Kollár, Y. Miyaoka, and S. Mori, "Rational connectedness and boundedness of Fano manifolds", *J. Differential Geom.* **36**:3 (1992), 765–779.

[Miyaoka 1987] Y. Miyaoka, "The Chern classes and Kodaira dimension of a minimal variety", pp. 449–476 in *Algebraic geometry* (Sendai, 1985), edited by T. Oda, Adv. Studies in Pure Math. **10**, Kinokuniya, Tokyo and North-Holland, Amsterdam, 1987.

[Miyaoka 1988] Y. Miyaoka, "On the Kodaira dimension of minimal threefolds", *Math. Ann.* **281**:2 (1988), 325–332.

[Miyaoka 1993] Y. Miyaoka, "Relative deformations of morphisms and applications to fibre spaces", *Comment. Math. Univ. St. Paul.* **42**:1 (1993), 1–7.

[Miyaoka and Peternell 1997] Y. Miyaoka and T. Peternell, *Geometry of higher-dimensional algebraic varieties*, DMV seminar **26**, Birkhäuser, Basel, 1997.

[Mok 1988] N. Mok, "The uniformization theorem for compact Kähler manifolds of nonnegative holomorphic bisectional curvature", *J. Differential Geom.* **27**:2 (1988), 179–214.

[Mori 1979] S. Mori, "Projective manifolds with ample tangent bundles", *Ann. of Math.* (2) **110**:3 (1979), 593–606.

[Mori 1982] S. Mori, "Threefolds whose canonical bundles are not numerically effective", *Ann. of Math.* (2) **116**:1 (1982), 133–176.

[Mori 1988] S. Mori, "Flip theorem and the existence of minimal models for 3-folds", *J. Amer. Math. Soc.* **1**:1 (1988), 117–253.

[Mori and Mukai 1983] S. Mori and S. Mukai, "On Fano 3-folds with $B_2 \geq 2$", pp. 101–129 in *Algebraic varieties and analytic varieties* (Tokyo, 1981), edited by S. Iitaka, Adv. Studies in Pure Math. **1**, North-Holland, Amsterdam, 1983.

[Morrow and Kodaira 1971] J. Morrow and K. Kodaira, *Complex manifolds*, Holt Rinehart and Winston, New York, 1971.

[Nakamura 1987] I. Nakamura, "Moishezon threefolds homeomorphic to $\mathbb{P}^3$", *J. Math. Soc. Japan* **39**:3 (1987), 521–535.

[Nakamura 1988] I. Nakamura, "Threefolds homeomorphic to a hyperquadric in $\mathbb{P}^4$", pp. 379–404 in *Algebraic geometry and commutative algebra: in honor of Masayoshi Nagata*, edited by H. Hijikata et al., Kinokuniya, Tokyo, 1988.

[Nakamura 1992] I. Nakamura, "On Moishezon manifolds homeomorphic to $\mathbb{P}^n_{\mathbb{C}}$", *J. Math. Soc. Japan* **44**:4 (1992), 667–692.

[Nakamura 1996]   I. Nakamura, "Moishezon threefolds homeomorphic to a cubic hypersurface in $\mathbb{P}^{4}$", *J. Algebraic Geom.* **5**:3 (1996), 537–569.

[Namikawa and Steenbrink 1995]   Y. Namikawa and J. H. M. Steenbrink, "Global smoothing of Calabi-Yau threefolds", *Invent. Math.* **122**:2 (1995), 403–419.

[Novikov 1965]   S. P. Novikov, "Topological invariance of rational classes of Pontrjagin", *Dokl. Akad. Nauk SSSR* **163** (1965), 298–300.

[Okonek and Teleman 1999]   C. Okonek and A. Teleman, "Recent developments in Seiberg–Witten theory and complex geometry", pp. 391–427 in *Several complex variables*, edited by M. Schneider and Y.-T. Siu, Math. Sci. Res. Inst. Publ. **37**, Cambridge U. Press, New York, 1999.

[Okonek and Van de Ven 1990]   C. Okonek and A. Van de Ven, "Stable bundles, instantons and C^{∞}-structures on algebraic surfaces", pp. 197–249 in *Several complex variables, VI*, edited by W. Barth and R. Narasimhan, Encycl. Math. Sci. **69**, Springer, Berlin, 1990.

[Okonek and Van de Ven 1995]   C. Okonek and A. Van de Ven, "Cubic forms and complex 3-folds", *Enseign. Math.* (2) **41**:3-4 (1995), 297–333.

[Peternell 1986]   T. Peternell, "Algebraic structures on certain 3-folds", *Math. Ann.* **274**:1 (1986), 133–156.

[Peternell 1996]   T. Peternell, "Manifolds of semi-positive curvature", pp. 98–142 in *Transcendental methods in algebraic geometry* (Cetraro, 1994), edited by J.-P. Demailly et al., Lecture Notes in Math. **1646**, Springer, Berlin, 1996.

[Peternell 1998a]   T. Peternell, "Moishezon manifolds and rigidity theorems", *Bayreuth. Math. Schr.* **54** (1998), 1–108.

[Peternell 1998b]   T. Peternell, "Towards a Mori theory on compact Kähler threefolds, II", *Math. Ann.* **311**:4 (1998), 729–764.

[Peternell 1998c]   T. Peternell, "Towards a Mori theory on compact Kähler threefolds, III", preprint, Universität Bayreuth, 1998.

[Peternell and Serrano 1998]   T. Peternell and F. Serrano, "Threefolds with nef anti-canonical bundles", *Collect. Math. (Barcelona)* **49**:2-3 (1998), 465–517. Dedicated to the memory of Fernando Serrano.

[Reid 1983]   M. Reid, "Minimal models of canonical 3-folds", pp. 131–180 in *Algebraic varieties and analytic varieties* (Tokyo, 1981), edited by S. Iitaka, Adv. Studies in Pure Math. **1**, North-Holland, Amsterdam, 1983.

[Reid 1987]   M. Reid, "Young person's guide to canonical singularities", pp. 345–414 in *Algebraic geometry* (Brunswick, ME, 1985), edited by S. J. Bloch, Proc. Symp. Pure Math. **46**, Amer. Math. Soc., Providence, RI, 1987.

[Schmitt 1995]   A. Schmitt, *Zur Topologie dreidimensionaler komplexer Mannig-faltigkeiten*, Thesis, Universität Zürich, 1995.

[Schmitt 1996]   A. Schmitt, "On the non-existence of Kähler structures on certain closed and oriented differentiable 6-manifolds", *J. Reine Angew. Math.* **479** (1996), 205–216.

[Schmitt 1997]   A. Schmitt, "On the classification of certain 6-manifolds and applications to algebraic geometry", *Topology* **36**:6 (1997), 1291–1315.

[Shokurov 1979] V. V. Shokurov, "Existence of a straight line on Fano 3-folds", *Izv. Akad. Nauk SSSR Ser. Mat.* **43**:4 (1979), 922–964, 968. In Russian; translated in *Math. USSR Izv.* **15** (1980), 173–209.

[Siu 1989] Y. T. Siu, "Nondeformability of the complex projective space", *J. Reine Angew. Math.* **399** (1989), 208–219. Errata in **431** (1992), 65–74.

[Sommese and Van de Ven 1986] A. J. Sommese and A. Van de Ven, "Homotopy groups of pullbacks of varieties", *Nagoya Math. J.* **102** (1986), 79–90.

[Steenrod 1951] N. Steenrod, *The topology of fibre bundles*, Princeton Mathematical Series **14**, Princeton Univ. Press, Princeton, NJ, 1951.

[Summerer 1997] A. Summerer, *Globale Deformationen der Mannigfaltigkeiten* $\mathbb{P}(T_{\mathbb{P}_2})$ *und* $\mathbb{P}_1 \times \mathbb{P}_2$, Bayreuth. Math. Schr. **52**, Universität Bayreuth, 1997.

[Ueno 1975] K. Ueno, *Classification theory of algebraic varieties and compact complex spaces*, Lecture Notes in Math. **439**, Springer, Berlin, 1975. Notes written in collaboration with P. Cherenack.

[van de Ven 1962] A. van de Ven, "Analytic compactifications of complex homology cells", *Math. Ann.* **147** (1962), 189–204.

[Wall 1966] C. T. C. Wall, "Classification problems in differential topology, V: On certain 6-manifolds", *Invent. Math.* **1** (1966), 355–374. Corrigendum in **2**, 306.

[Zhang 1996] Q. Zhang, "On projective manifolds with nef anticanonical bundles", *J. Reine Angew. Math.* **478** (1996), 57–60.

FRÉDÉRIC CAMPANA
INSTITUT ÉLIE CARTAN (NANCY)
UNIVERSITÉ HENRI POINCARÉ (NANCY I)
FACULTÉ DES SCIENCES, B.P. 239
54506 VANDOEUVRE-LES-NANCY
FRANCE
Frederic.Campana@iecn.u-nancy.fr

THOMAS PETERNELL
MATHEMATISCHES INSTITUT
UNIVERSITÄT BAYREUTH
D-95440 BAYREUTH
GERMANY
Thomas.Peternell@uni-bayreuth.de

Several Complex Variables
MSRI Publications
Volume **37**, 1999

Remarks on Global Irregularity
in the $\bar{\partial}$–Neumann Problem

MICHAEL CHRIST

ABSTRACT. The Bergman projection on a general bounded, smooth pseudo-convex domain in two complex variables need not be globally regular, that is, need not preserve the class of all functions that are smooth up to the boundary. In this article the construction of the worm domains is reviewed, with emphasis on those features relevant to their role as counterexamples to global regularity. Prior results, and related issues such as the commutation method and compactness estimates, are discussed. A model in two real variables for global irregularity is discussed in detail. Related work on real analytic regularity, both local and global, is summarized. Several open questions are posed.

CONTENTS

Research supported by NSF grant DMS 96-23007.

1. Introduction

Let $n > 1$, and let $\Omega \subset \mathbb{C}^n$ be a bounded domain with C^∞ boundary. The $\bar{\partial}$–Neumann problem for $(0,1)$-forms on Ω is a boundary value problem

$$\square u = f \qquad \text{on } \Omega, \tag{1-1}$$

$$u \lrcorner \bar{\partial}\rho = 0 \qquad \text{on } \partial\Omega, \tag{1-2}$$

$$\bar{\partial}u \lrcorner \bar{\partial}\rho = 0 \qquad \text{on } \partial\Omega. \tag{1-3}$$

where u, f are $(0,1)$-forms, ρ is any defining function for Ω, $\square = \bar{\partial}\bar{\partial}^* + \bar{\partial}^*\bar{\partial}$ and $\lrcorner$ denotes the interior product of forms. $\mathbb{C}^n$ is regarded as being equipped with its canonical Hermitian metric, and $\bar{\partial}^*$ denotes the formal adjoint of $\bar{\partial}$ with respect to that metric.

The boundary conditions may be reformulated so as to apply to functions that are not very regular at the boundary: $u \in \text{Domain}(\bar{\partial}^*)$ and $\bar{\partial}u \in \text{Domain}(\bar{\partial}^*)$ [Folland and Kohn 1972]. In the L^2 setting there is then a satisfactory global theory [Folland and Kohn 1972; Catlin 1984]; if Ω is pseudoconvex, then for every $f \in L^2(\Omega)$ there exists a unique solution $u \in L^2(\Omega)$. Moreover, if $\bar{\partial}f = 0$, then $\bar{\partial}u = f$, and u is the solution with smallest L^2 norm. The Neumann operator N is the bounded linear operator on $L^2(\Omega)$ that maps datum f to solution u.

The $\bar{\partial}$–Neumann problem is useful as a tool for solving the primary equation $\bar{\partial}u = f$ because it often leads to a solution having good regularity properties at the boundary. For large classes of domains, in particular for all strictly pseudoconvex domains, it is a hypoelliptic boundary value problem, that is, the solution u is smooth[1] in any relatively open subset of $\bar{\Omega}$ in which the datum f is smooth. Whereas the main goal of the theory is regularity in spaces and norms such as C^∞, C^k, Sobolev or Hölder, basic estimates and existence and uniqueness theory are most naturally expressed in L^2.

For some time it remained an open question whether the global C^∞ theory was as satisfactory as the L^2 theory.

THEOREM 1.1 [Christ 1996b]. *There exist a smoothly bounded, pseudoconvex domain in $\mathbb{C}^2$ and a datum $f \in C^\infty(\bar{\Omega})$ such that the unique solution $u \in L^2(\Omega)$ of the $\bar{\partial}$–Neumann problem does not belong to $C^\infty(\bar{\Omega})$.*

There were antecedents. Barrett [1984] gave an example of a smoothly bounded, nonpseudoconvex domain for which the Bergman projection B fails to preserve $C^\infty(\bar{\Omega})$. Kiselman [1991] showed that B fails to preserve $C^\infty(\bar{\Omega})$ for certain bounded but nonsmooth pseudoconvex Hartogs domains. Barrett [1992] added a fundamental insight and deduced that for the so-called worm domains, which are smoothly bounded and pseudoconvex, B fails to map the Sobolev space H^s to itself, for large s. Finally Christ [1996b] proved an *a priori* H^s estimate

---

[1] "Smooth" and "C^∞" are synonymous throughout this article.

for smooth solutions on worm domains, and observed that this estimate would contradict Barrett's result if global C^∞ regularity were valid.

This article discusses background, related results, the proof of global irregularity, and open questions. It is an expanded version of lectures given at MSRI in the Fall of 1995. A brief report on analytic hypoellipticity is also included.

I am indebted to Emil Straube for useful comments on a preliminary draft.

2. Background

The equation $\Box u = f$ is a linear system of n equations. The operator $\Box$ is simply a constant multiple of the Euclidean Laplacian, acting diagonally with respect to the standard basis $\{d\bar{z}_j\}$, so is elliptic. However, the boundary conditions are not coercive; that is, the *a priori* inequality

$$\sum_{|\alpha| \le 2} \|\partial^\alpha u\|_{L^2(\Omega)} \le C\|f\|_{L^2(\Omega)}$$

for all $u \in C^\infty(\bar{\Omega})$ satisfying the boundary conditions $(1\text{–}1), (1\text{–}1)$ is not valid for any nonempty Ω. For strictly pseudoconvex domains one has a weaker *a priori* inequality: the H^1 norm of u is bounded by a constant multiple of the $L^2 = H^0$ norm of f, provided that $u \in C^\infty(\bar{\Omega})$ satisfies the boundary conditions [Kohn 1963; 1964]. Even this inequality breaks down for domains that are pseudoconvex but not strictly pseudoconvex; the regularity of solutions is governed by geometric properties of the boundary.

There are two different fundamental notions of regularity in the C^∞ category, hypoellipticity and global regularity.[2] Hypoellipticity means that for every open set $V \subset \mathbb{C}^n$ and every $f \in L^2(\Omega) \cap C^\infty(V \cap \bar{\Omega})$, the $\bar{\partial}$–Neumann solution u belongs to $C^\infty(V \cap \bar{\Omega})$. Global regularity in C^∞ means that for every $f \in C^\infty(\bar{\Omega})$, the $\bar{\partial}$–Neumann solution u also belongs to $C^\infty(\bar{\Omega})$. Hypoellipticity thus implies global regularity.

Consider any linear partial differential operator L, with C^∞ coefficients, defined on a smooth compact manifold M without boundary. Such an operator is said to be hypoelliptic if for any open set $V \subset M$ and any $u \in \mathcal{D}'(V)$ such that $Lu \in C^\infty(V)$, necessarily $u \in C^\infty(V)$. It is globally regular in C^∞ if for all $u \in \mathcal{D}'(M)$ such that $Lu \in C^\infty(M)$, necessarily $u \in C^\infty(M)$. The definitions given for the $\bar{\partial}$–Neumann problem in the preceding paragraph are the natural analogues of these notions for boundary value problems, with minor modifications.

In general, global C^∞ regularity is a far weaker property than hypoellipticity. As a first example, consider the two dimensional torus $\mathbb{T}^2 = \mathbb{R}^2/\mathbb{Z}^2$, equipped with coordinates (x_1, x_2). Let $L = \partial_{x_1} + \alpha \partial_{x_2}$, where α is a real constant. The vector field L defines a foliation of $\mathbb{T}^2$, and any function u defined in some

---

[2]The latter is sometimes called global hypoellipticity.

open subset and locally constant along each leaf is annihilated by L. From the relationship $\widehat{Lu}(k) = (2\pi i)(k_1 + \alpha k_2)\hat{u}(k)$ it follows that L is globally regular in C^∞ if and only if α satisfies a Diophantine inequality $|k_1 + \alpha k_2| \geq |k|^{-N}$ for some $N < \infty$ as $|k| \to \infty$. Thus global regularity holds for almost every α. No such Diophantine behavior has been encountered for the $\bar{\partial}$–Neumann problem for domains in $\mathbb{C}^n$; irregularity for that problem has a rather different source.[3]

As a second example, consider any torus $\mathbb{T}^n$ and distribution $K \in \mathcal{D}'(\mathbb{T}^n)$. Denote by 0 the identity element of the group $\mathbb{T}^n$. The convolution operator $Tf = f * K$ then preserves $C^\infty(\mathbb{T}^n)$. On the other hand, T is pseudolocal[4] if and only if $K \in C^\infty(\mathbb{T}^n \backslash \{0\})$.

The principal results known, in the positive direction, concerning hypoellipticity in the $\bar{\partial}$–Neumann problem for smoothly bounded pseudoconvex domains in $\mathbb{C}^n$ are as follows. For all strictly pseudoconvex domains, the $\bar{\partial}$–Neumann problem is hypoelliptic [Kohn 1963; 1964]. For all $s \geq 0$, the solution belongs to the Sobolev class H^{s+1} in every relatively compact subset of any relatively open subset of $\bar{\Omega}$ in which the datum belongs to H^s. Precise results describe the gain in regularity in various function spaces and the singularities of objects such as the Bergman kernel.[5]

Hypoellipticity holds more generally, for all domains of finite type in the sense of [D'Angelo 1982]. (The defining property of such domains is that at any $p \in \partial\Omega$, no complex subvariety of $\mathbb{C}^n$ has infinite order of contact with $\partial\Omega$.) The $\bar{\partial}$–Neumann problem satisfies subelliptic estimates up to the boundary: there exists $\varepsilon > 0$ such that for every $s \geq 0$, every relatively open subset U of $\bar{\Omega}$ and every datum $f \in L^2(\Omega) \cap C^\infty(U)$, the $\bar{\partial}$–Neumann solution u belongs to $H^{s+\varepsilon}$ on every relatively compact subset of U [Catlin 1987]. Conversely, subellipticity implies finite type. No characterization of the optimal ε is known in general.

The case of domains of finite type in $\mathbb{C}^2$ is far simpler than that in higher dimensions, and is well understood. Finite type in $\mathbb{C}^2$ is characterized by Lie brackets of vector fields in $T^{1,0} \oplus T^{0,1}(\partial\Omega)$, and the optimal exponent ε equals $2/m$ where m is the type as defined by Lie brackets or by the maximal order of contact of complex submanifolds with $\partial\Omega$.[6] Closely related to the $\bar{\partial}$–Neumann problem for domains of finite type in $\mathbb{C}^2$ is the theory of sums of squares of smooth real vector fields satisfying the bracket condition of Hörmander [1967].

---

[3]Somewhat artificial examples of operators with variable coefficients exhibiting similar behavior are analyzed in [Himonas 1995].

[4]An operator T is said to be pseudolocal if it preserves $\mathcal{D}'(\mathbb{T}^n) \cap C^\infty(V)$ for every open subset V of $\mathbb{T}^n$.

[5]There is likewise a gain of one derivative in the Hölder and L^p-Sobolev scales [Greiner and Stein 1977]. Moreover, there is a gain of two derivatives in the so-called "good" directions; for any smooth vector fields V_1, V_2 defined on $\bar{\Omega}$ such that V_i and JV_i are tangent to $\partial\Omega$, $V_1 V_2 u \in H^s$ wherever $f \in H^s$ [Greiner and Stein 1977].

[6]There is still a gain of two derivatives in good directions, and a gain of $2/m$ derivatives in the Hölder and L^p-Sobolev scales, for the $\bar{\partial}$–Neumann problem as well as for a related equation on $\partial\Omega$. See [Chang et al. 1992; Christ 1991a; 1991b] and the many references cited there.

So far as this author is aware, little has been established in the positive direction concerning hypoellipticity for domains of infinite type. There are however several interesting theorems guaranteeing global C^∞ regularity, or a closely related property, for classes of domains for which hypoellipticity need not hold. The first result of this type [Kohn 1973] concerned the weighted $\bar{\partial}$–Neumann problem, associated to any plurisubharmonic function $\varphi \in C^\infty(\bar\Omega)$. In this problem $\Box$ is replaced by $\Box_\varphi = \bar{\partial}\bar{\partial}_\varphi^* + \bar{\partial}_\varphi^*\bar{\partial}$, where $\bar{\partial}_\varphi^*$ is the formal adjoint of $\bar{\partial}$ in the Hilbert space $L^2(\Omega, e^{-\varphi}dzd\bar{z})$, and the boundary conditions are that $u, \bar{\partial}u$ should belong to the domain of $\bar{\partial}_\varphi^*$ (on forms of degrees one and two, respectively). Kohn showed that given any Ω and any exponent $s \geq 0$, there exists φ such that for every $f \in H^s(\Omega)$, the solution u of the $\bar{\partial}$–Neumann problem with weight $\exp(-\varphi)$ also belongs to $H^s(\Omega)$. Work of Bell and Ligocka [1980], however, demonstrated that the problem for $\varphi \equiv 0$ has a special significance.

Consider the quadratic form

$$Q(u, u) = \|\bar{\partial}u\|^2_{H^0(\Omega)} + \|\bar{\partial}^* u\|^2_{H^0(\Omega)}.$$

Compactness of the Neumann operator is equivalent to an inequality

$$\|u\|^2_{H^0} \leq \varepsilon Q(u, u) + C_\varepsilon \|u\|^2_{H^{-1}} \tag{2–1}$$

for all $u \in C^1(\bar\Omega)$ satisfying the first boundary condition (1–1), for all $\varepsilon > 0$. Subellipticity implies compactness, which in turn implies [Kohn and Nirenberg 1965] global regularity. See [Catlin 1984; Sibony 1987] for compactness criteria in terms of auxiliary plurisubharmonic functions having suitable growth properties.

A second type of result asserts that global C^∞ regularity holds for all domains enjoying suitable symmetries, in particular, for any Reinhardt domain, or more generally, for any circular or Hartogs domain for which the orbit of the symmetry group is transverse to the complex tangent space to Ω at every boundary point.[7] Such results are essentially special cases of a general principle to the effect that global regularity always holds in the presence of suitable global symmetries, one version of which is formulated in the real analytic category in [Christ 1994a].

More general results in the positive direction have been obtained by Boas and Straube [Boas and Straube 1991a; 1991b; 1993], after earlier work of Bonami and Charpentier [1988]. Denote by $W_\infty \subset \partial\Omega$ the set of points at which the boundary has infinite type. A sufficient condition for global C^∞ regularity is that there exist a smooth real vector field V defined on some neighborhood of W_∞ in $\partial\Omega$ and transverse to $[T^{1,0} \oplus T^{0,1}](\partial\Omega)$ at every point of W_∞, such that

$$[V, T^{1,0} \oplus T^{0,1}] \subset T^{1,0} \oplus T^{0,1}. \tag{2–2}$$

In fact, it suffices that for each $\varepsilon > 0$ there exist V_ε, defined on some neighborhood $U = U_\varepsilon$ of W_∞ in $\partial\Omega$ and transverse to $T^{1,0} \oplus T^{0,1}$ at every point of W_∞,

---

[7]Results of this genre have been obtained by numerous authors including So-Chin Chen, Cordaro and Himonas [Cordaro and Himonas 1994], Derridj [1997], Barrett, and Straube.

such that[8]

$$[V_\varepsilon, T^{1,0} \oplus T^{0,1}] \subset T^{1,0} \oplus T^{0,1} \text{ modulo } O(\varepsilon) \text{ on } U. \tag{2-3}$$

For Hartogs or circular domains having transverse symmetries, the action of the symmetry group S^1 gives rise to a single vector field V having the stronger commutation property $[V, \bar{\partial}] = 0$, $[V, \bar{\partial}^*] = 0$.

One corollary of the theorem of Boas and Straube is $\bar{\partial}$–Neumann global regularity for all *convex* domains [Boas and Straube 1991b; Chen 1991]. To formulate a second special case, consider any Ω for which the set $W_\infty \subset \partial\Omega$ of all boundary points of infinite type consists of a smoothly bounded, compact complex submanifold $\mathcal{V}$ of $\mathbb{C}^n$ with boundary, of positive dimension. A second corollary is global C^∞ regularity for Ω whenever $\mathcal{V}$ is simply connected. A third case where the required vector field exists is when there exists a defining function $\rho \in C^\infty(\bar{\Omega})$ that is plurisubharmonic at the boundary[9] [Boas and Straube 1991b].[10]

3. Exact Regularity and Positivity

Consider any smoothly bounded, pseudoconvex domain $\Omega \subset \mathbb{C}^n$. Denote by $L^2(\Omega)$ the space of square integrable $(0,1)$-forms defined on Ω, and for each $s \geq 0$ denote by $H^s = H^s(\Omega)$ the space of $(0,1)$-form valued functions Ω possessing s derivatives in L^2 in the usual sense of Sobolev theory.

The Neumann operator N (for $(0,1)$-forms) is the unique bounded linear operator on $L^2(\Omega)$ that maps any f to the unique solution u of the $\bar{\partial}$–Neumann problem with datum f. Existence and uniqueness stem from the fundamental inequality

$$\|u\|_{L^2(\Omega)}^2 \leq C\|\bar{\partial}u\|_{L^2(\Omega)}^2 + C\|\bar{\partial}^* u\|_{L^2(\Omega)}^2, \tag{3-1}$$

valid for all $u \in C^1(\bar{\Omega})$ satisfying the first boundary condition (1–1). A proof may be found in [Catlin 1984].

DEFINITION. For each $s \geq 0$, the $\bar{\partial}$–Neumann problem for Ω is exactly regular in H^s if the Neumann operator N maps $H^s(\Omega)$ into $H^s(\Omega)$.

Corresponding notions may be defined for an operator L on a compact manifold without boundary. By virtue of the Sobolev embedding theorem, exact regularity implies global C^∞ regularity in either setting. There is of course no converse in general, as illustrated by the operators $\partial_{x_1} + \alpha\partial_{x_2}$ on $\mathbb{T}^2$. If $|k_1 + \alpha k_2| \geq c|k|^{-N}$ as $|k| \to \infty$, then L^{-1} exists modulo a finite dimensional kernel and cokernel,

---

[8]Fix finitely many coordinate patches $O_\alpha \subset \partial\Omega$ whose union contains W_∞ and fix, for each α, a basis of sections $X_{\alpha,j}$ of $T^{1,0} \oplus T^{0,1}(O_\alpha)$. It is required that for each ε and each $N < \infty$ there exist V_ε such that for all α and all j, $[V_\varepsilon, X_{\alpha,j}]$ may be decomposed in $U_\varepsilon \cap O_\alpha$ as a section of $T^{1,0} \oplus T^{0,1}(O_\alpha)$ plus a vector field whose C^N norm is at most ε.

[9]The complex Hessian of ρ is required to be positive semidefinite at each point of $\partial\Omega$.

[10]This result has been reproved and refined by Kohn [$\geq$ 1999].

and maps $H^s(\mathbb{T}^2)$ to H^{s-N} for all s, but since the limit infimum of $|k_1 + \alpha k_2|$ always equals zero, L^{-1} cannot preserve any class H^s.[11]

Why is exact regularity of such importance? The theory begins with an H^0 estimate, $\|u\|_{H^0} \leq C\|\Box u\|_{H^0}$. For the very degenerate boundary conditions arising at boundary points of infinite type, there is no hope of any parametrix formula that will express u in terms of $\Box u$, modulo a smoothing term. Attempts to exploit the H^0 inequality to majorize derivatives of u lead to error terms, for instance from the commutation of $\Box$ with partial derivatives, which appear on the right hand side of an inequality. One arrives at an estimate of the form

$$\|u\|_{H^t} \leq C\|\Box u\|_{H^s} + C'\|u\|_{H^s}. \tag{3-2}$$

Such an inequality is useful only if (i) $t > s$, (ii) both $t = s$ and $C' < 1$, or (iii) $C' = 0$ because all commutator terms vanish identically.

For general pseudoconvex domains whose boundaries contain points of infinite type, there is no smoothing effect to make $t > s$. Estimates with $t \leq s$ are highly unstable, potentially being destroyed by perturbations by operators of order zero. In practice, (ii) requires that C' be made arbitrarily small, to ensure that it is < 1. Commutator terms can be expected to vanish identically only for domains with symmetries.

For any smoothly bounded, pseudoconvex domain Ω there exists $\delta > 0$ such that the $\bar{\partial}$–Neumann problem is exactly regular in H^s for all $0 \leq s < \delta$. This holds essentially because $C' = O(s)$ in (3–2) for small $s \geq 0$.

All proofs of exact regularity have relied on $Q(u, u)$ being sufficiently large relative to commutator terms. Consider first the compact case. The H^0 inequality (2–1) leads for each s and each $\varepsilon > 0$ to an inequality

$$\varepsilon^{-1}\|u\|_{H^s} \leq \left[\|\Box u\|_{H^s} + C\|u\|_{H^s}\right] + C'_{\varepsilon, s}\|u\|_{H^0},$$

where C depends only on s. The factor ε^{-1} on the left hand side permits absorption of the term $C\|u\|_{H^s}$, whence the H^s norm of u is majorized by the H^s norm of $\Box u$.

Consider next the weighted theory. Fix Ω and a strictly plurisubharmonic function $\varphi \in C^\infty(\bar{\Omega})$. Denote by $\bar{\partial}^*_\lambda$ the adjoint of $\bar{\partial}$ in $\mathcal{H}_\lambda = L^2(\Omega, \exp(-\lambda\varphi))$, and set $Q_\lambda(u, u) = \|\bar{\partial}u\|^2_{\mathcal{H}_\lambda} + \|\bar{\partial}^*_\lambda u\|^2_{\mathcal{H}_\lambda}$. Then for all $u \in C^1(\bar{\Omega})$ satisfying the first boundary condition (1–1), $\|u\|^2_{\mathcal{H}_\lambda} \leq C\lambda^{-1}Q_\lambda(u, u)$. This inequality is intermediate between the basic unweighted majorization $\|u\|^2_{L^2} \leq CQ(u, u)$ and the compactness inequality (2–1). The norms of $\mathcal{H}_\lambda$ and L^2 are equivalent, though not uniformly in λ, so the weighted inequality implies [Kohn 1973] that for all sufficiently large λ, for all $s \leq c\lambda^{1/2}$ and all $u \in C^\infty(\bar{\Omega})$, $\|u\|_{H^s} \leq C_\lambda\|\Box_\lambda u\|_{H^s}$. It is possible to pass from this *a priori* majorization to the conclusion that the

---

[11]No analogous example is known to this author for the $\bar{\partial}$–Neumann problem for domains in $\mathbb{C}^n$; global C^∞ regularity has always been been proved via exact regularity. Kohn has asked whether they are in fact equivalent.

$\bar{\partial}$–Neumann problem for Ω with weight $\exp(-\lambda\varphi)$ is exactly regular in H^s, in the range $s \leq c\lambda^{1/2}$.

Finally, in the results of Boas and Straube, one begins with a weaker inequality $\|u\|^2 \leq CQ(u, u)$ for a fixed constant C. Outside any neighborhood of the set W_∞ of boundary points of infinite type, this is supplemented by a compactness estimate. By exploiting the special vector field V it can be arranged that for each s, the commutator terms leading to the potentially harmful term $C'\|u\|_{H^s}$ on the right hand side of (3–2) are of three types. Those of the first type are supported outside a neighborhood of W_∞, hence are harmless by virtue of the compactness inequality. Those of the second type are majorized by arbitrarily small multiples of $\|u\|_{H^s}$. Those of the third type, arising from the $T^{1,0} \oplus T^{0,1}(\partial\Omega)$ components of commutators of V with sections of $T^{1,0} \oplus T^{0,1}$, are majorized by lower order Sobolev norms of u.

The common theme is that a successful analysis is possible because the basic L^2 inequality is stronger than the harmful commutator terms. In the first situation, the L^2 inequality is arbitrarily strong; in the third, the error terms are arbitrarily weak near W_∞, and in the second, the weight $\exp(-\lambda\varphi)$ is chosen so as to make the L^2 inequality sufficiently strong relative to the error terms.

4. Worm Domains

The worm domains, invented by Diederich and Fornæss [1977], are examples of pseudoconvex domains whose closures have no Stein neighborhood bases. This means that there exists $\delta > 0$ such that there exists no pseudoconvex domain containing $\bar{\Omega}$, and contained in $\{z : \mathrm{distance}(z, \Omega) < \delta\}$.

DEFINITION. A worm domain in $\mathbb{C}^2$ is a bounded open set of the form

$$W = \left\{ z : |z_1 + e^{i \log |z_2|^2}|^2 < 1 - \phi(\log |z_2|^2) \right\} \tag{4–1}$$

having the following properties:

(i) W has smooth boundary and is pseudoconvex.
(ii) $\phi \in C^\infty$ takes values in $[0, 1]$, vanishes identically on $[-r, r]$ for some $r > 0$, and vanishes nowhere else.
(iii) W is strictly pseudoconvex at every boundary point where $\left|\log |z_2|^2\right| > r$.

We will sometimes write $W_r = W$.

Diederich and Fornæss proved that ϕ can be chosen so that these properties hold; beyond this the choice of ϕ is of no consequence. Important properties of worm domains include:

(i) ∂W_r contains the annular complex manifold with boundary

$$A_r = \left\{ z : z_1 = 0 \text{ and } \left|\log |z_2|^2\right| \leq r \right\}. \tag{4–2}$$

(ii) W is strictly pseudoconvex at every boundary point not in A_r.

If $r \geq \pi$ then $\partial \mathcal{W}_r$ contains the annulus $\mathcal{A}_\pi$ as well as the two circles

$$\left\{ z : |z_1 + e^{i\pi}| = 1 \text{ and } \log |z_2|^2 = \pm \pi \right\}.$$

Applying the standard extension argument, one finds that any function holomorphic in any neighborhood of the union of $\mathcal{A}_\pi$ and the two circles must extend holomorphically to a fixed such neighborhood, which thus is contained in every pseudoconvex neighborhood of $\overline{\mathcal{W}_r}$. But if $r < \pi$ then $\overline{\mathcal{W}_r}$ does have a basis of pseudoconvex neighborhoods [Fornæss and Stensønes 1987; Bedford and Fornæss 1978].

The worm domains had long been regarded as important test cases for global regularity when Barrett [1992] achieved a breakthrough.

THEOREM 4.1. *For each $r > 0$ there exists $t \in \mathbb{R}^+$ such that for any worm domain $\mathcal{W}_r$ and any $s \geq t$, the $\bar{\partial}$–Neumann problem fails to be exactly regular in H^s. Moreover $t \to 0$ as $r \to \infty$.*

The proof focused on the Bergman projection B rather than on the Neumann operator. B is the orthogonal projection mapping scalar valued functions in $L^2(\Omega)$ onto the closed subspace of all holomorphic square integrable functions. It is related to the $\bar{\partial}$–Neumann problem via the formula [Kohn 1963; 1964].

$$B = I - \bar{\partial}^* N \bar{\partial}, \tag{4-3}$$

where I denotes the identity operator. In $\mathbb{C}^2$, for any exponent s, B preserves (scalar valued) H^s if and only if N preserves $((0,1)$-form valued) H^s; B preserves $C^\infty(\overline{\Omega})$ if and only if N does so [Boas and Straube 1990][12].

The proof had two parts, of which the first was an elegant direct analysis of the nonsmooth domains

$$\mathcal{W}_r' = \left\{ z : \left| z_1 + e^{i \log |z_2|^2} \right| < 1 \text{ and } -r < \log |z_2|^2 < r \right\}.$$

B not only fails to preserve H^t, but even fails to map $C^\infty(\overline{\mathcal{W}_r'})$ to H^t.

This step has much in common with the contemporaneous proof by Christ and Geller [1992] that the Szegő projection for certain real analytic domains of finite type fails to be analytic pseudolocal. In both analyses, separation of variables leads to a synthesis of the projection operator in terms of explicitly realizable projections onto one dimensional subspaces.[13] The expression for such a rank one projection carries a factor of the reciprocal of the norm squared of a basis element. Analytic continuation of this reciprocal with respect to a natural Fourier parameter leads to poles off of the real axis, which are the source of irregularity.

The second part was a proof by contradiction. It was shown that if the Bergman projection for $\mathcal{W}_r$ preserves some H^s, then the Bergman projection for

---

[12]There exists a generalization valid for all dimensions [Boas and Straube 1990].

[13]This decomposition and synthesis in [Christ and Geller 1992] was taken from work of Nagel [1986].

$\mathcal{W}'_r$ must also preserve H^s. The reasoning relied on a scaling argument, in which it was essential that the norms H^s on the left and right hand sides of the *a priori* inclusion inequality have identical scaling properties. Consequently this indirect method did not exclude the possibility that B might map H^s to $H^{s-\varepsilon}$, for all $\varepsilon > 0$, for all s.

5. A Cohomology Class

The worm domains have another property of vital importance for any discussion of global regularity, whose significance in this context was recognized by Boas and Straube [1993]. Consider any smoothly bounded domain Ω for which the set W_∞ of all boundary points of infinite type forms a smooth, compact complex submanifold R, with boundary. The worm domains are examples.

The embedding of R into the Cauchy–Riemann manifold $\partial\Omega$ induces an element of the de Rham cohomology group $H^1(R)$, defined as follows. Fix any purely real, nowhere vanishing one-form η, defined in a neighborhood in $\partial\Omega$ of R, that annihilates $T^{0,1} \oplus T^{1,0}(\partial\Omega)$. Fix likewise a smooth real vector field V, transverse to $T^{0,1} \oplus T^{1,0}$, satisfying $\eta(V) \equiv 1$. Consider the one-form $\alpha = -\mathcal{L}_V \eta\big|_R$, the Lie derivative of $-\eta$ with respect to V, restricted to R.[14] Moreover, if Ω is pseudoconvex, then α is a closed form [Boas and Straube 1993], hence represents an element $[\alpha]$ of the cohomology group $H^1(R)$. This element is independent of the choices of η and of V.

DEFINITION. The winding class $w(R, \partial\Omega)$ of $\partial\Omega$ about R is the cohomology class $[\alpha] \in H^1(R)$.

This class is determined by the first-order jet of the CR structure of $\partial\Omega$ along R. A fundamental property of worm domains is that

$$\text{For every worm domain, } w(\mathcal{A}_r, \partial\mathcal{W}_r) \neq 0. \tag{5--1}$$

A theorem of Boas and Straube [1993] asserts that if $w(R, \partial\Omega) = 0$, then there exist vector fields V satisfying the approximate commutation relation (2–3). Consequently the $\bar{\partial}$–Neumann problem is globally regular in C^∞.

To understand $w(R, \partial\Omega)$ in concrete terms [15], suppose that $\Omega \subset \mathbb{C}^2$ and R is a smooth Riemann surface with boundary, embeddable in $\mathbb{C}^1$. Choose coordinates $(x + iy, t) \in \mathbb{C} \times \mathbb{R}$ in a neighborhood of R in $\partial\Omega$ such that $R \subset \{t = 0\}$; identify R with $\{x + iy : (x + iy, 0) \in R\}$. A Cauchy–Riemann operator has the form $\bar{\partial}_b = X + iY$ where X, Y are real vector fields of the form $X = \partial_x + a\partial_t$, $Y = \partial_y + b\partial_t$, where a, b are smooth real valued functions of

---

[14]α is a section over R of the tangent bundle TR, not of $T\partial\Omega$.

[15]Bedford and Fornæss [1978] gave a geometric interpretation of $w(R, \partial\Omega)$, and had shown that whenever it is smaller than a certain threshold value in a natural norm on cohomology, $\bar{\Omega}$ has a pseudoconvex neighborhood basis. α had appeared earlier in work of D'Angelo [1979; 1987].

(x, y, t) and $a(x+iy, 0) \equiv 0 \equiv b(x+iy, 0)$. The Levi form may be identified with the function $\lambda(x + iy, t) = (b_x + ab_t) - (a_y + ba_t)$, where the subscripts indicate partial differentiation. By hypothesis, $R = \{(x + iy, t) : \lambda(x + iy, t) = 0\}$.

By choosing $\eta = dt - a\,dx - b\,dy$ and $V = \partial_t$, we obtain $-\mathcal{L}_V \eta = a_t\,dx + b_t\,dy$ and hence $\alpha(x + iy) = a_t(x, y, 0)\,dx + b_t(x, y, 0)\,dy$, for $x + iy \in R$. Note that

$$d\alpha = (a_{t,y} - b_{t,x})(x + iy, 0)\,dx\,dy = (\partial_t \lambda)(x + iy, 0)\,dx\,dy.$$

Pseudoconvexity of $\partial\Omega$ means that $\lambda(x + iy, t) \geq 0$ everywhere, which forces $\partial_t \lambda(x + iy, 0) \equiv 0$ since $\lambda(x + iy, 0) \equiv 0$. Thus α is indeed closed.

To what extent does the CR structure of $\partial\Omega$ coincide with the Levi flat CR structure $R \times \mathbb{R}$ near R? More precisely, do there exist coordinates $(x + iy, t)$ in which $R \subset \{t = 0\}$ and $\bar{\partial}_b$ takes the form $(\partial_x + \tilde{a}\partial_t) + i(\partial_y + \tilde{b}\partial_t)$ with $\tilde{a}(x+iy, t), \tilde{b}(x+iy, t) = O(t^2)$ as $t \to 0$ for every $x+iy \in R$? By an elementary analysis, the answer is affirmative if and only if $w(R, \partial\Omega) = 0$. Thus the theorem of Boas and Straube asserts rather paradoxically that global regularity holds whenever the CR structure near R is sufficiently degenerate.

In the absence of any pseudoconvexity hypothesis, α need not be closed, but exactness of the form α remains the criterion for existence of the desired coordinate system. There exists a hierarchy of invariants $w_k(R, \partial\Omega)$, with $w(R, \partial\Omega) = w_1$. Each w_k is defined if $w_{k-1} = 0$, and represents the obstruction to existence of coordinates in which $\tilde{a}, \tilde{b} = O(t^{k+1})$. Each w_k is an equivalence class of forms modulo exact forms; in the pseudoconvex case, w_k is represented by a closed form for even k. These invariants have no relevance to C^∞ regularity, but we believe that they may play a role in the theory of global regularity in Gevrey classes, partially but not completely analogous to the role of w_1 in the C^∞ case.

6. Special Vector Fields and Commutation

The use of auxiliary vector fields V satisfying the commutation equations

$$[V, X_j] \in \text{span}\{X_i\} \qquad \text{for all } j \tag{6–1}$$

together with a transversality condition, for sums of squares operators $L = \sum_i X_i^2$, and of analogous commutation equations in related situations such as the $\bar{\partial}$–Neumann problem, has not been restricted to the question of global C^∞ regularity. Real analytic vector fields satisfying (6–1) globally on a compact manifold have been used by Derridj and Tartakoff [1976], Komatsu [1975; 1976] and later authors to prove global regularity in C^ω [Tartakoff 1996]. This work has depended also on what are known as maximal estimates and their generalizations.[16]

---

[16]Maximal estimates and their connection with representations of nilpotent Lie groups are the subject of a deep theory initiated by Rothschild and Stein [1976] and developed by Helffer and Nourrigat in a series of works including [Helffer and Nourrigat 1985] and leading up to [Nourrigat 1990] and related work of Nourrigat.

For sums of squares operators, maximal estimates take the form

$$\sum_{i,j} \|X_i X_j u\|_{L^2} \le C\|Lu\|_{L^2} + C\|u\|_{L^2}.$$

They are used to absorb certain error terms that arise from commutators $[V, X_j]$ in a bootstrapping argument in which successively higher derivatives are estimated. Chen [1988; 1989], Cordaro and Himonas [1994], Derridj [1997] and Christ [1994a] have obtained cruder results based on the existence of vector fields for which the commutators vanish identically. Such results require far weaker bounds than maximal estimates.

Auxiliary vector fields with this commutation property have also been used to establish analytic hypoellipticity in certain cases. In the method of Tartakoff [1980], this requires the modification of V by cutoff functions having appropriate regularity properties, to take into account the possible lack of global regularity or even global definition of the data. Sjöstrand [1982; 1983] has developed a microlocal analogue, in which a vector field corresponds to a one parameter deformation of the operator being studied.

The use of auxiliary vector fields having this commutation property should be regarded not as a special trick, but rather as the most natural approach to exact regularity. The remainder of this section is devoted to a justification of this assertion. For simplicity we restrict the discussion to any sum of squares operator $L = \sum_j X_j^2$, on a compact manifold M without boundary.[17] We assume $\|u\|_{L^2} \le C\|Lu\|_{L^2}$, for all $u \in C^2$.

Consider any first order elliptic, self adjoint, strictly positive pseudodifferential operator Λ on M. Then the powers Λ^s are well defined for all $s \in \mathbb{C}$, and Λ^s maps $H^r(M)$ bijectively to H^{r-s} for all $s, r \in \mathbb{R}$. Define $L_s = \Lambda^s \circ L \circ \Lambda^{-s}$. Then for each $0 \le s \in \mathbb{R}$, L is exactly regular in H^s if and only if for all $u \in H^{-s}(M)$,

$$L_s u \in H^0 \text{ implies } u \in H^0. \tag{6–2}$$

Thus one seeks an *a priori* inequality for all $u \in C^\infty(M)$ of the form[18]

$$\|u\|_{L^2} \le C_s \|L_s u\|_{L^2} + C_s \|u\|_{H^{-1}}. \tag{6–3}$$

Since such an inequality holds for $L_s = L$, it is natural to ask whether L_s may be analyzed as a perturbation of L. Now $L_s = \sum_j (\Lambda^s X_j \Lambda^{-s})^2$. Moreover,

$$\Lambda^s [X_j, \Lambda^{-s}] = -s\Lambda^{-1}[X_j, \Lambda]$$

---

[17]The same analysis applies equally well to the $\bar{\partial}$–Neumann problem on any pseudoconvex domain in $\mathbb{C}^2$, by the method of reduction to the boundary as explained in §9.

[18]From an inequality of this type for all $s \ge 0$, with C_s bounded uniformly on compact sets, it is possible to deduce (6–2) for all $s \in \mathbb{R}$ by a continuity argument, using approximations to the identity and pseudodifferential calculus.

modulo a pseudodifferential operator of order ≤ -1; the contribution of any such operator can be shown always to be negligible for our discussion, by exploiting the L^2 inequality

$$\|X_j u\| \leq C\|Lu\| + C\|u\|. \tag{6-4}$$

Therefore modulo harmless error terms,

$$L_s \approx L - s\sum_j (X_j B_j + B_j X_j) + s^2 \sum_j B_j^2, \tag{6-5}$$

where $B_j = \Lambda^{-1}[X_j, \Lambda]$ has order ≤ 0. Since each factor $\Lambda^{-1}[X_j, \Lambda]$ has order ≤ 0, (6-4) implies

$$\|(L_s - L)u\| \leq C(|s| + s^2)(\|Lu\| + \|u\|) + C\|u\|_{H^{-1}} + C\|Lu\|_{H^{-1}}.$$

Thus (6-3) holds, and L is exactly regular in H^s, for all sufficiently small s.

Moreover, for any pseudodifferential operator E of strictly negative order, any perturbation term of the form $EX_i X_j$ is harmless, even if multiplied by an arbitrarily large coefficient, since it ultimately leads to an estimate in terms of some negative order Sobolev norm of u after exploiting (6-4) in evaluating the quadratic form $\langle L_s u, u \rangle$. Thus in order to establish (6-3), it would suffice for there to exist Λ such that each commutator $[X_j, \Lambda]$ can be expressed as $\sum_i B_{i,j} X_i$ modulo an operator of order 0, where each $B_{i,j}$ is some pseudodifferential operator of order 0. This is a property of the principal symbol of Λ alone. Moreover, by virtue of standard microlocal regularity estimates, it suffices to have such a commutation relation microlocally in a conic neighborhood of the characteristic variety $\Sigma \subset T^* M$ defined by the vanishing of the principal symbol of L.

Let us now specialize the discussion to the case where at every point of M, $\{X_j\}$ are linearly independent and span a subspace of the tangent space having codimension one. Then Σ is a line bundle. We suppose this bundle to be orientable. Thus Σ splits as the union of two half line bundles, and there exists a globally defined vector field T transverse at every point to span$\{X_j\}$.

In a conic neighborhood of either half, Λ may be expressed as a smooth real vector field V, plus a perturbation expressible as a finite sum of terms $E_{i,j} X_i X_j$ where $E_{i,j}$ has order ≤ -1, plus a negligible term of order 0. V is transverse to span$\{X_j\}$, because Λ is elliptic. The commutator of any X_j with any of these perturbation terms has already the desired form. Thus if there exists Λ for which each commutator $[X_j, \Lambda]$ takes the desired form, then there must exist V satisfying $[V, X_j] \in$ span$\{X_i\}$ for all j.

7. A Model

Global C^∞ irregularity for the worm domains was discovered by analyzing the simplest instance of a more general problem. Consider a finite collection of smooth real vector fields X_j on a compact manifold M without boundary, and an operator $L = -\sum_j X_j^2 + \sum_j b_j X_j + a$ where $a, b_j \in C^\infty$. Under what circumstances is L globally regular in C^∞?

Denote by $\|\cdot\|$ the norm in $L^2(M)$, with respect to some smooth measure. A Lipschitz path $\gamma : [0,1] \mapsto M$ is said to be admissible if $\frac{d}{ds}\gamma(s) \in \text{span}\{X_j(\gamma(s))\}$ for almost every s. A collection of vector fields X_j is said to satisfy the bracket hypothesis if the Lie algebra generated by them spans the tangent space to M.

We impose three hypotheses in order to preclude various pathologies and to mimic features present in the $\bar{\partial}$–Neumann problem for arbitrary smoothly bounded, pseudoconvex domains in $\mathbb{C}^2$.

- There exists $C < \infty$ such that for all $u \in C^2(M)$, $\|u\| \le C\|Lu\|$.
- For every $x, y \in M$ there exists an admissible path γ satisfying[19] $\gamma(0) = x$ and $\gamma(1) = y$.
- $\{X_j\}$ satisfies the bracket hypothesis on some nonempty subset $U \subset M$.

Under these hypotheses, must L be globally regular in C^∞?

This is not a true generalization of the $\bar{\partial}$–Neumann problem. But as will be explained in §9, the latter may be reduced (in $\mathbb{C}^2$) to a very similar situation, where the vector fields are the real and imaginary parts of $\bar{\partial}_b$ on $\partial\Omega$.

The first hypothesis mimics the existence of an L^2 estimate for the $\bar{\partial}$–Neumann problem. The second and third mimic respectively the absence of compact complex submanifolds without boundary in boundaries of domains in $\mathbb{C}^n$, and the presence of strictly pseudoconvex points in boundaries of all such domains, respectively. Each hypothesis excludes the constant coefficient examples on $\mathbb{T}^2$ discussed in §2. The first may be achieved, for any collection of vector fields and coefficients b_j, by adding a sufficiently large positive constant to a. These assumptions complement one another. The third builds in a certain smoothing effect, while the second provides a mechanism for that effect to propagate to all of M.

Global C^∞ regularity does not necessarily hold in this situation. As an example,[20] let $M = \mathbb{T}^2$ and fix a coordinate patch $V_0 \subset M$ along with an identification of V_0 with $\{(x,t) \in (-2,2) \times (-2\delta, 2\delta)\} \subset \mathbb{R}^2$. Set $J = [-1,1] \times \{0\}$. Let X, Y be any two smooth, real vector fields defined on M satisfying the following hypotheses.

(i) $X, Y, [X, Y]$ span the tangent space to M at every point of $M \backslash J$.
(ii) In V_0, $X \equiv \partial_x$ and $Y \equiv b(x,t)\partial_t$.
(iii) For all $|x| \le 1$ and $|t| \le \delta$, $b(x,t) = \alpha(x)t + O(t^2)$, where $\alpha(x)$ vanishes nowhere.

The collection of vector fields $\{X, Y\}$ then satisfies the second and third hypotheses imposed above.

---

[19]This property is called reachability by some authors [Sussmann 1973].

[20]The global structure of M is of no importance in this example.

The role of the Riemann surface R in the discussion in §5 is taken here by J, even though $H^1(J) = 0$. Although there appears to be no direct analogue of the one-forms η, α of that discussion, there exists no vector field V transverse to span$\{X, Y\}$ such that $[V, X]$ and $[V, Y]$ belong to span$\{X, Y\}$; nor does a family of such vector fields exist with the slightly weaker approximate commutation property (2–3).

THEOREM 7.1 [Christ 1995a]. *Let X, Y, M be as above. Let L be any operator on M of the form $L = -X^2 - Y^2 + a$, such that $a \in C^\infty$ and $\|u\|^2 \leq C\langle Lu, u\rangle$ for all $u \in C^2(M)$. Then L is not globally regular in C^∞.*

The close analogy between this result and the $\bar{\partial}$–Neumann problem for worm domains will be explained in §9. A variant of Theorem 7.1 is actually proved in [Christ 1995a], but the same proof applies.

Before discussing the proof, we will formulate more precise conclusions giving some insight into the nature of the problem and the singularities of solutions. For $|x| \leq 1$, write $a(x, t) = \beta(x) + O(t)$. Consider the one parameter family of ordinary differential operators

$$\mathcal{H}_\sigma = -\partial_x^2 + \sigma\alpha(x)^2 + \beta(x).$$

Define Σ_0 to be the set of all $\sigma \in \mathbb{C}$ for which the Dirichlet problem

$$\begin{cases} \mathcal{H}_\sigma f = 0 & \text{on } [-1, 1], \\ f(\pm 1) = 0 \end{cases}$$

has a nonzero solution. Then Σ_0 consists of a discrete sequence of real numbers $\lambda_0 < \lambda_1 < \ldots$ tending to $+\infty$. Define

$$\Sigma = \left\{ s \in [0, \infty) : (s - 1/2)^2 \in \Sigma_0 \right\}.$$

Write $\Sigma = \{s_0 < s_1 < \ldots\}$. It can be shown [Christ 1995a] that $s_0 > 0$.

Under our hypotheses, L^{-1} is a well defined bounded linear operator on $L^2(M)$.

THEOREM 7.2. *L has the following global regularity properties.*

- *For every $s < s_0$, L^{-1} preserves $H^s(M)$.*
- *For each $s > s_0$, L^{-1} fails to map $C^\infty(M)$ to H^s.*
- *Suppose that $0 \leq s < r < s_0$, or $s_j < s < r < s_{j+1}$ for some $j \geq 0$. Then any $u \in H^s(M)$ satisfying $Lu \in H^r(M)$ must belong to H^r.*
- *For each $s \notin \Sigma$ an a priori inequality is valid: There exists $C < \infty$ such that for every $u \in H^s(M)$ such that $Lu \in H^s$,*

$$\|u\|_{H^s} \leq C\|Lu\|_{H^s}. \tag{7–1}$$

- *For each $s \notin \Sigma$, $\{f \in H^s(M) : L^{-1}f \in H^s\}$ is a closed subspace of H^s with finite codimension.*

To guess the nature of the singularities of solutions, consider the following simpler problem. Define

$$\mathcal{L} = -\partial_x^2 - \alpha^2(x)(t\partial_t)^2 + \beta(x). \tag{7-2}$$

Consider the Dirichlet problem

$$\begin{cases} \mathcal{L}u = g & \text{on } [-1,1] \times \mathbb{R}, \\ u(x,t) \equiv 0 & \text{on } \{\pm 1\} \times \mathbb{R}. \end{cases} \tag{7-3}$$

To construct a singular solution for this Dirichlet problem, fix $s \in \Sigma$, set $\sigma = (s - 1/2)(s + 1/2)$, and fix a nonzero solution of $\mathcal{H}_\sigma f = 0$ with $f(\pm 1) = 0$. Fix $\eta \in C_0^\infty(\mathbb{R})$, identically equal to one in some neighborhood of 0. Set $u(x,t) = \eta(t)f(x)t^{s-1/2}$ for $t > 0$, and $u \equiv 0$ for $t < 0$. Then $u \in L^2([-1,1] \times \mathbb{R})$ is a solution of (7–3) for a certain $g \in C_0^\infty([-1,1] \times \mathbb{R})$. Thus the Dirichlet problem (7–3) for $\mathcal{L}$ on the strip is globally irregular.

The proof of Theorem 7.1 consists in reducing the global analysis of L on M to the Dirichlet problem for $\mathcal{L}$. Unfortunately, we know of no direct construction of nonsmooth solutions for L on M that uses the singular solution of the preceding paragraph as an Ansatz.

Instead, the proof[21] consists in two parts [Christ 1995a]. First, the *a priori* inequality (7–1) is established. Second, emulating Barrett [1992], we prove that for any $s \geq s_0$, L cannot be exactly regular in H^s.

With these two facts in hand, suppose that L were globally regular in C^∞. Fix any $s_0 < s \notin \Sigma$. Given any $f \in H^s$, fix a sequence $\{f_j\} \subset C^\infty$ converging to f in H^s. Then $\{L^{-1}f_j\}$ is Cauchy in H^s, by the *a priori* inequality, since $L^{-1}f_j \in C^\infty$ by hypothesis. On the other hand, since L^{-1} is bounded on L^2, $L^{-1}f_j \to L^{-1}f$ in L^2 norm. Consequently $L^{-1}f \in H^s$. This contradicts the result that L fails to be exactly regular in H^s.

8. A Tale of Three Regions

The main part of the analysis is the proof of the *a priori* estimate (7–1) for $0 < s \notin \Sigma$. The main difficulty is as follows.

Associated to the operator L is a sub-Riemannian structure on the manifold M. Define a metric $\rho(x,y)$ to equal the minimal length of any Lipschitz path γ joining x to y, such that the tangent vector to γ is almost everywhere of the form $s_1 X + s_2 Y$ with $s_1^2 + s_2^2 \leq 1$. Points having coordinates (x, ε) with $|x| \leq 1/2$ are at distance $> 1/2$ from J in this degenerate metric, no matter how small $\varepsilon > 0$ may be; paths approaching J "from above" have infinite length, but paths such as $s \mapsto (s, 0)$ approaching J "from the side" have finite length.

For the purpose of analyzing L, M is divided naturally into three regions. Region I is $M \backslash J$; L satisfies the bracket hypothesis on any compact subset of

---

[21]We have subsequently found a reformulation of the proof that eliminates the second part of the argument and has a less paradoxical structure. But this reformulation involves essentially the same ingredients, and is no simpler.

region I, so a very satisfactory regularity theory is known: L is hypoelliptic and gains at least one derivative. Region II is an infinitesimal tubular neighborhood $\{|x| \leq 1, 0 \neq t \sim 0\}$. Here L is an elliptic polynomial in $\partial_x, t\partial_t$, so a natural tool for its analysis is the partial Mellin transform in the variable t. The subregions $t > 0$, $t < 0$ are locally decoupled where $|x| < 1$; the relationship between $u(x,0^+)$ and $u(x,0^-)$ is determined by global considerations. Region III is another infinitesimal region, lying to both sides of J, where $t \sim 0$ and $1 < |x| \sim 1$. In this transitional region, if Y is expanded as a linear combination $c(x)\partial_t + O(t)\partial_t$, the coefficient $c(x)$ vanishes to infinite order as $|x| \to 1^+$. In such a situation no parametrix construction can be hoped for. The only tool available appears to be *a priori* L^2 estimation stemming from integration by parts.

One needs not only an analysis for each region, but three compatible analyses. No attack by decomposing M into three parts by a partition of unity has succeeded; error terms resulting from commutation of L with the partition functions are too severe to be absorbed.

The proof of the *a priori* estimate proceeds in several steps. For simplicity we assume $u \in C^\infty$. The following discussion is occasionally imprecise; correct statements may be found in [Christ 1995a].

First step. For any $\varepsilon > 0$, u may be assumed to be supported where $|x| < 1+\varepsilon$ and $|t| < \varepsilon$. Indeed, since $X, Y, [X, Y]$ span the tangent space outside J, the H^{s+1} norm of u is controlled on any compact subset of $M \backslash J$ by $\|Lu\|_{H^s} + \|u\|_{H^0}$.

Second step. Fix a globally defined, self adjoint, strictly positive elliptic first order pseudodifferential operator Λ on M, and set $L_s = \Lambda^s \circ L \circ \Lambda^{-s}$. Then L satisfies an *a priori* exact regularity estimate in H^s if and only if there exist ε, C such that

$$\|u\| \leq C\|L_s u\| + C\|u\|_{H^{-1}}$$

for all $u \in C^\infty$ supported where $|x| < 1+\varepsilon$ and $|t| < \varepsilon$, where all norms without subscripts are L^2 norms. In particular, we may work henceforth on $\mathbb{R}^2$ rather than on M.

Denote by $\Gamma \subset T^*M$ the line bundle $\{(x,t;\xi,\tau) : (x,t) \in J \text{ and } \xi = 0\}$. Microlocally on the complement of Γ, the H^1 norm of u is controlled by the H^0 norm of $L_s u$ plus the H^{-1} norm of u, for every s.

Third step.

$$L_s = -\partial_x^2 - (Y_s + A_1)(Y_s + A_2) + \beta(x) + A_3,$$

where $\beta(x) = a(x,0)$ and Y_s is a real vector field which, where $|x| \leq 1$, takes the form

$$Y_s = \alpha(x)(t\partial_t + s) + O(t^2)\partial_t.$$

The principal symbol σ_0 of each $A_j \in S_{1,0}^0$ vanishes identically on Γ.

Fourth step. Integration by parts yields

$$\|\partial_x u\| \leq C\|L_s u\| + C\|u\| \qquad \text{for all } u \in C^2. \tag{8-1}$$

By itself this inequality is of limited value, since $\|u\|$ appears on the right hand side rather than on the left.

Fifth step. The fundamental theorem of calculus together with the vanishing of $u(x,t)$ for all $|x| > 1 + \varepsilon$ yield

$$\|u\|_{L^2(\{|x|>1\})} + \|u\|_{L^2(\{-1,1\}\times\mathbb{R})} \leq C\varepsilon^{1/2}\|\partial_x u\| \leq C\varepsilon^{1/2}\left[\|L_s u\| + C\|u\|\right].$$

Combining this with (8–1) and absorbing certain terms into the left hand side gives the best information attainable without a close analysis of the degenerate region II.

$$\|u\| + \|\partial_x u\| + \varepsilon^{-1/2}\|u\|_{L^2(\{-1,1\}\times\mathbb{R})} \leq C\|L_s u\| + C\|u\|_{L^2([-1,1]\times\mathbb{R})}. \quad (8\text{–}2)$$

By choosing ε to be sufficiently small, we may absorb the last term on the right into the left hand side of the inequality.

It remains to control the L^2 norm of u in $[-1, 1] \times \mathbb{R}$. In the next step we prepare the machinery that will be used to achieve this control in step seven.

Sixth step. Define $\mathcal{L}_s = -\partial_x^2 - \alpha(x)^2(t\partial_t + s)^2 + \beta(x)$; note that $\mathcal{L}_s$ is an elliptic polynomial in $\partial_x, t\partial_t$. Conjugation with the Mellin transform[22] in the variable t reduces the analysis of $\mathcal{L}_s$ on $L^2([-1,1] \times \mathbb{R})$ to that of the one parameter family[23] of ordinary differential operators

$$\mathcal{H}_{(s+i\tau-\frac{1}{2})^2} = -\partial_x^2 - \alpha(x)^2(s + i\tau - \tfrac{1}{2})^2 + \beta(x), \qquad \tau \in \mathbb{R}.$$

The assumption that $s \notin \Sigma$ is equivalent to the assertion that for each $\tau \in \mathbb{R}$, the nullspace of $\mathcal{H}_{(s+i\tau-\frac{1}{2})^2}$ on $L^2([-1,1])$ with Dirichlet boundary conditions is $\{0\}$. Thus $\mathcal{H}_{(s+i\tau-\frac{1}{2})^2}g = f$ may be solved in $L^2([-1,1])$, with arbitrarily prescribed boundary values, and the solution is unique. On the other hand, because $\mathcal{H}_{(s+i\tau-\frac{1}{2})^2}$ is an elliptic polynomial in ∂_x and $i\tau$, the same holds automatically for all sufficiently large $|\tau|$. Quantifying all this and invoking the Plancherel and inversion properties of the Mellin transform, one deduces that the Dirichlet problem for $\mathcal{L}_s$ is uniquely solvable in $L^2([-1,1] \times \mathbb{R})$. Moreover, if $u \in C^2([-1,1] \times \mathbb{R})$ has compact support and $\mathcal{L}_s u = f_1 + t\partial_t f_2 + (t\partial_t)^2 f_3$ in $[-1,1] \times \mathbb{R}$, then[24]

$$\|u\|_{L^2([-1,1]\times\mathbb{R})} + \|\partial_x u\|_{L^2([-1,1]\times\mathbb{R})} \leq C\sum_j \|f_j\|_{L^2([-1,1]\times\mathbb{R})} + C\|u\|_{L^2(\{\pm1\}\times\mathbb{R})}.$$

$$(8\text{–}3)$$

---

[22]This applies for $t > 0$; the region $t < 0$ is handled by substituting $t \mapsto -t$ and repeating the same analysis.

[23]s is shifted to $s - \frac{1}{2}$ in order to take into account the difference between the measures dt and $t^{-1} dt$; the latter appears in the usual Plancherel formula for the Mellin transform.

[24]Up to two factors of $t\partial_t$ are permitted on the right hand side of the equation for $\mathcal{L}_s u$, because $\mathcal{H}_{(s+i\tau)^2}$ is an elliptic polynomial of degree two in $\partial_x, i\tau$ for each s.

Seventh step. On $[-1, 1] \times \mathbb{R}$, $\mathcal{L}_s u = L_s u + (\mathcal{L}_s - L_s) u$. The remainder term $(\mathcal{L}_s - L_s) u$ may be expressed as $(t \partial_t)^2 A_1 u + t \partial_t A_2 u + A_3 u$, where $\sigma_0(A_j) \equiv 0$ on Γ. Thus by (8–3),

$$\|u\|_{L^2([-1,1] \times \mathbb{R})} \le C \|u\|_{L^2(\{\pm 1\} \times \mathbb{R})} + C \sum_j \|A_j u\|.$$

Since the H^0 norm of u is controlled microlocally by the H^{-1} norms of $L_s u$ and of u on the complement of Γ, and since $\sigma_0(A_j) \equiv 0$ on Γ,

$$\|A_j u\| \le C_\eta \|L_s u\|_{H^{-1}} + C_\eta \|u\|_{H^{-1}} + \eta \|u\|_{H^0}$$

for every $\eta > 0$. By inserting this into the preceding inequality and combining the result with the conclusion of the fifth step, we arrive at the desired *a priori* inequality majorizing $\|u\|$ by $\|L_s u\| + \|u\|_{H^{-1}}$.   □

The simpler half of the proof is the demonstration that L is not exactly regular in H^{s_0}. The operator $\mathcal{L} = -\partial_x^2 - (\alpha(x) t \partial_t)^2 + \beta(x)$ is obtained from L, in the region $|x| \le 1$, by substituting $t = \varepsilon \tilde{t}$, and letting $\varepsilon \to 0$. At typical points where $|x| > 1$, the coefficient of ∂_t^2 in $\mathcal{L}$ will be nonzero, and this scaling will lead to $\varepsilon^{-1} \partial_{\tilde{t}}$, hence in the limit to an infinite coefficient.

First step. There exists $f \in C_0^\infty((-1, 1) \times \mathbb{R})$ for which the unique solution $u \in L^2([-1, 1] \times \mathbb{R})$ of $L_s u = f$ with boundary condition $u(\pm 1, t) \equiv 0$ is singular, in the sense that $|\partial_t|^{s_0} u \notin L^2([-1, 1] \times \mathbb{R})$. This follows from a Mellin transform analysis, in the spirit of the sixth step above.

The remainder of the proof consists in showing that for any s, if L is exactly regular in $H^s(M)$, then there exists $C < \infty$ such that for every $f \in C_0^\infty((-1, 1) \times \mathbb{R})$, there exists a solution $u \in L^2([-1, 1] \times \mathbb{R})$ satisfying $\mathcal{L} u = f$ and the boundary condition $u \equiv 0$ on $\{\pm 1\} \times \mathbb{R}$, such that $|\partial_t|^s u \in L^2([-1, 1] \times \mathbb{R})$ and

$$\||\partial_t|^s u\|_{L^2} \le C \|f\|_{H^s}. \tag{8–4}$$

Second step. Fix $s > 0$, and suppose L to be exactly regular in $H^s(M)$. Fix $f \in C_0^\infty((-1, 1) \times \mathbb{R})$. To produce the desired solution u, recall that L^{-1} is a well defined bounded operator on $L^2(M)$. For each small $\varepsilon > 0$, for (x, t) in a fixed small open neighborhood in M of J, set

$$u_\varepsilon(x, t) = (L^{-1} f_\varepsilon)(x, \varepsilon t) \qquad \text{where } f_\varepsilon(x, t) = f(x, \varepsilon^{-1} t).$$

f_ε is supported where $|x| < 1 - \eta$ and $|t| < C\varepsilon$ for some $C, \eta \in \mathbb{R}^+$; we extend it to be identically zero outside this set, so that it is globally defined on M. The hypothesis that L^{-1} is bounded on $H^s(M)$ implies that in a neighborhood of J, u_ε and $\partial_x u_\varepsilon$ satisfy (8–4); the essential point is that the highest order derivative with respect to t on both sides of (8–4) is $|\partial_t|^s$, hence both sides scale in the same way under dilation with respect to t, as $\varepsilon \to 0$.

Since L^{-1} is bounded on H^0, the same reasoning leads to the conclusion that $u_\varepsilon, \partial_x u_\varepsilon$ are uniformly bounded in $L^2(M)$.

Third step. Define u to be a weak $*$ limit of some weakly convergent sequence u_{ε_j}. Then $u, \partial_x u, |\partial_t|^s u \in L^2([-1,1] \times \mathbb{R})$, with norms bounded by $\|f\|_{H^s}$. Passing to the limit in the equation defining u_ε, and exploiting the *a priori* bounds, we obtain $\mathcal{L}u = f$ in $[-1,1] \times \mathbb{R}$.

The scaling and limiting procedure of steps two and three is due to Barrett [1992], who carried it out for the Bergman projection, rather than for a differential equation.[25]

Fourth step. It remains to show that $u(\pm 1, t) = 0$ for almost every $t \in \mathbb{R}$. Because $\partial_x u \in L^2$ and $u \in L^2$, $u(\pm 1, t)$ is well defined as a function in $L^2(\mathbb{R})$.

For $|x| > 1$, the differential operator obtained from this limiting procedure has infinite coefficients, and no equation for u is obtained. Instead, recall that for any neighborhood U of J, L^{-1} maps $H^0(M)$ boundedly to $H^1(M \backslash U)$. Now the space H^1 scales differently from H^0. From this it can be deduced that $u_\varepsilon \to 0$ in L^2 norm in $M \backslash U$. Coupling this with the uniform bound on $\partial_x u_\varepsilon$ in L^2, it follows that $u_\varepsilon(\pm 1, t) \to 0$ in $L^2(\mathbb{R})$. Therefore u satisfies the Dirichlet boundary condition. □

Paradoxically, then, the Dirichlet boundary condition arises from the failure for $|x| > 1$ of the same scaling procedure that gives rise to the differential operator $\mathcal{L}$ for $|x| < 1$. Global singularities arise from the interaction between the degenerate region J and the nondegenerate region $|x| > 1$ that borders it.

This analysis is objectionable on several grounds. First, it is indirect. Second, it yields little information concerning the nature of singularities, despite strong heuristic indications that for $|x| < 1$ and $t > 0$, singular solutions behave like $g(x)t^{s_j - \frac{1}{2}}$ modulo higher powers of t. Third, it relies on the ellipticity of $\mathcal{L}$ with respect to $t\partial_t$ in order to absorb terms that are $O(t^2 \partial_t)$. No such ellipticity is present in analogues on three dimensional CR manifolds, such as the boundary of the worm domain.

In §5 we pointed out another paradox: the regularity theorem of Boas and Straube guarantees global regularity whenever the CR structure near a Riemann surface R embedded in $\partial\Omega$ is sufficiently degenerate. It is interesting to reexamine this paradox from the point of view of the preceding analysis. Consider the Dirichlet problem on $[-1,1] \times \mathbb{R}$ for the operator

$$\mathcal{L} = -\partial_x^2 - \alpha^2(x)(t^m \partial_t)^2 + \beta(x).$$

The case $m = 1$ has already been analyzed; exponents $m > 1$ give rise to more degenerate situations. When $m > 1$, separation of variables leads to solutions

$$f_\lambda(x, t) = g_\lambda(x) e^{-\lambda t^{1-m}} \chi_{t>0}$$

---

[25]The Dirichlet boundary condition was not discussed in [Barrett 1992]. Instead, the limiting operator was identified as a Bergman projection by examining its actions on the space of square integrable holomorphic functions and on its orthocomplement.

where λ is a nonlinear eigenvalue parameter, χ is the characteristic function of $\mathbb{R}^+$, and g_λ satisfies the ordinary differential equation

$$-g'' - \alpha^2(x)(m-1)^2\lambda^2 g + \beta(x)g = 0$$

on $[-1, 1]$ with boundary conditions $g(\pm 1) = 0$. When $\lambda > 0$, these solutions are C^∞ at $t = 0$. The larger m becomes, the more rapidly f vanishes at $t = 0$, and hence the milder is its singularity (in the sense of Gevrey classes, for instance).

9. More on Worm Domains

We next explain how analysis of the $\bar{\partial}$–Neumann problem on worm domains may be reduced to a variant of the two dimensional model discussed in the preceding section. Assume $\Omega \Subset \mathbb{C}^2$ to have smooth defining function ρ.

The $\bar{\partial}$–Neumann problem is a boundary value problem for an elliptic partial differential equation, and as such is amenable to treatment by the method of reduction to a pseudodifferential equation on the boundary.[26] This reduction is achieved by solving instead the elliptic boundary value problem

$$\begin{cases} \Box u = f & \text{on } \Omega, \\ \quad u = v & \text{on } \partial\Omega, \end{cases} \tag{9-1}$$

where v is a section of a certain complex line bundle $\mathcal{B}^{0,1}$ on $\partial\Omega$. The section v depends on f and is to be chosen so that the unique solution u satisfies the $\bar{\partial}$–Neumann boundary conditions; The problem (9–1) is explicitly solvable via pseudodifferential operator calculus, modulo a smoothing term, and there is a precise connection between the regularity of the solution and of the data.

The section v has in principle two components, but the first $\bar{\partial}$–Neumann boundary condition says that one component vanishes identically. The second boundary condition may be expressed as an equation $\Box^+ v = g$ on $\partial\Omega$, where $\Box^+$ is a certain pseudodifferential operator of order 1, and $g = (\bar{\partial}Gf \lrcorner \bar{\partial}\rho)$ restricted to $\partial\Omega$, where Gf is the unique solution of the elliptic boundary value problem $\Box(Gf) = f$ on Ω and $Gf \equiv 0$ on $\partial\Omega$.

On $\partial\Omega$ a Cauchy–Riemann operator is the complex vector field

$$\bar{\partial}_b = (\partial_{\bar{z}_1}\rho)\partial_{\bar{z}_2} - (\partial_{\bar{z}_2}\rho)\partial_{\bar{z}_1}.$$

Define $\bar{L} = \bar{\partial}_b$, $L = \bar{\partial}_b^*$. The principal symbol of $\Box^+$ vanishes only on a line bundle Σ^+ that is one half of the characteristic variety defined by the vanishing of the principal symbol of $\bar{\partial}_b$. After composing $\Box^+$ with an elliptic pseudodifferential operator of order $+1$, $\Box^+$ takes the form

$$\mathcal{L} = \bar{L}L + B_1\bar{L} + B_2L + B_3 \tag{9-2}$$

microlocally in a conic neighborhood of Σ^+, where each B_j is a pseudodifferential operator of order less than or equal to 0. For each $s > 0$, if $t = s - 1/2$ then

---

[26]A detailed presentation is in [Chang et al. 1992].

the Neumann operator preserves $H^s(\Omega)$ if and only if whenever $v \in H^{-1/2}(\partial\Omega)$ and $\mathcal{L}v \in H^t(\partial\Omega)$, necessarily $v \in H^t(\partial\Omega)$. Since $\square^+$ is elliptic on the complement of Σ^+, all the analysis may henceforth be microlocalized to a small conic neighborhood of Σ^+.

For worm domains, the circle group acts as a group of automorphisms by $z \mapsto R_\theta z = (z_1, e^{i\theta}z_2)$, inducing corresponding actions on functions and forms. The Hilbert space of square integrable $(0, k)$-forms decomposes as the orthogonal direct sum $\bigoplus_{j \in \mathbf{Z}} \mathcal{H}_j^k$ where $\mathcal{H}_j^k$ is the set of all $(0, k)$-forms f satisfying $R_\theta f \equiv e^{ij\theta}f$. The Bergman projection and Neumann operator preserve $\mathcal{H}_j^0$ and $\mathcal{H}_j^1$, respectively.

PROPOSITION 9.1. *For each worm domain there exists a discrete subset $S \subset \mathbf{R}^+$ such that for each $s \notin S$ and each $j \in \mathbf{Z}$ there exists $C_{s,j} < \infty$ such that for every $(0, 1)$-form $u \in \mathcal{H}_j^1 \cap C^\infty(\overline{\mathcal{W}})$ such that $Nu \in C^\infty(\overline{\mathcal{W}})$,*

$$\|Nu\|_{H^s(\mathcal{W})} \leq C_{s,j}\|u\|_{H^s(\mathcal{W})}.$$

We do not know whether $C_{s,j}$ may be taken to be independent of j. The proof does imply that it is bounded by $C_s(1+|j|)^N$, for some exponent N independent of s. Thus our *a priori* inequalities can be formulated for all $u \in C^\infty$, rather than for each $\mathcal{H}_j$, but in such a formulation the norm on the right hand side should be changed to H^{s+N}.[27]

The Hilbert space $L^2(\partial\mathcal{W})$ decomposes into an orthogonal direct sum of subspaces $\mathcal{H}_j$, consisting of functions automorphic of degree j with respect to the action of the rotation group S^1 in the variable z_2. $\mathcal{H}_j$ may be identified with $L^2(\partial\mathcal{W}/S^1)$. The operators $\mathcal{L}, \bar{L}, L, B_j$ in (9–2) may be constructed so as to commute with the action of S^1, hence to preserve each $\mathcal{H}_j$. Thus for each j, the action of $\mathcal{L}$ on $\mathcal{H}_j(\partial\mathcal{W})$ may be identified with the action of an operator $\mathcal{L}_j$ on $L^2(\partial\mathcal{W}/S^1)$.

The quotient $\partial\mathcal{W}/S^1$ is a two dimensional real manifold. Coordinatizing $\partial\mathcal{W}$ by (x, θ, t) in such a way that $z_2 = \exp(x + i\theta)$ and $z_1 = \exp(i2x)(e^{it} - 1)$ where $\left|\log|z_2|^2\right| \leq r$, $\mathcal{L}_j$ takes the form $\bar{L}L + B_1\bar{L} + B_2L + B_3$ where $\bar{L}$ is a complex vector field which takes the form $\bar{L} = \partial_x + it\alpha(t)\partial_t$ where $|x| \leq r/2$, $\alpha(0) \neq 0$, and each B_k is a classical pseudodifferential operator of order ≤ 0, which depends on the parameter j in a nonuniform manner.

Setting $J = \{(x, t) : |x| \leq r/2 \text{ and } t = 0\}$, and writing $\bar{L} = X + iY$, the vector fields $X, Y, [X, Y]$ span the tangent space to $\partial\mathcal{W}/S^1$ at every point in the complement of J, and are tangent to J at each of its points. Thus the operator $\mathcal{L}_j$ on $\partial\mathcal{W}/S^1$ is quite similar to the two dimensional model discussed in §7, with two added complications: There are pseudodifferential factors, and the reduction of the $\bar{\partial}$–Neumann problem to $\mathcal{L}$, and thence to $\mathcal{L}_j$, requires only a microlocal

---

[27]The extra N derivatives are tangent to the Riemann surface $R = \mathcal{A}$ in $\partial\mathcal{W}$ along $\mathcal{A}$, and hence are essentially invariant under scaling in the direction orthogonal to $\mathcal{A}$, just as was $t\partial_t$ in the discussion in §7.

a priori estimate for $\mathcal{L}_j$ in a certain conic subset of phase space. The proof of Theorem 7.1 can be adapted to this situation.

The lower order terms $B_1\bar{L}, B_2L, B_3$ are not negligible in this analysis; indeed, they determine the values of the exceptional Sobolev exponents $s \in \Sigma$, but the analysis carries through for any such lower order terms. The set Σ turns out to be independent of j.

At the end of § 8 we remarked that the two dimensional analysis relies on a certain ellipticity absent in three dimensions. For the worm domain, the global rotation symmetry makes possible a reduction to two dimensions; the lack of ellipticity results in a lack of uniformity of estimates with respect to j, but has no effect on the analysis for fixed j.

10. Analytic Regularity

This section is a brief report on recent progress on analytic hypoellipticity and global analytic regularity not only for the $\bar{\partial}$–Neumann problem, but also for related operators such as sums of squares of vector fields, emphasizing the author's contributions. More information, including references, can be found in the expository articles [Christ 1995b; 1996c]. Throughout the discussion, all domains and all coefficients of operators are assumed to be C^ω.

It has been known since about 1978, through the fundamental work of Tartakoff [1978; 1980] and Treves [1978], that the $\bar{\partial}$–Neumann problem is analytic hypoelliptic (that is, the solution is real analytic up to the boundary wherever the datum is) for all strictly pseudoconvex domains. Other results and methods in this direction have subsequently been introduced by Geller, Métivier and Sjöstrand.

On the other hand, Baouendi and Goulaouic discovered that

$$\partial_x^2 + \partial_y^2 + x^2\partial_t^2$$

is not analytic hypoelliptic, despite satisfying the bracket hypothesis. Métivier generalized this by showing that for sums of squares of d linearly independent real vector fields in $\mathbb{R}^{d+1}$, analytic hypoellipticity fails to hold if an associated quadratic form, analogous to the Levi form, is degenerate at every point of an open set. Nondegeneracy of this form is equivalent to the characteristic variety defined by the vanishing of the principal symbol being a symplectic submanifold of $T^*\mathbb{R}^{d+1}$.

There remained the intermediate case, which arises in the study of the $\bar{\partial}$–Neumann problem for bounded, pseudoconvex, real analytic domains in $\mathbb{C}^n$. Subsequent investigations have fallen into three categories.

(i) Analytic hypoellipticity has been proved in certain weakly pseudoconvex and nonsymplectic cases, by extending the methods known for the strictly pseudoconvex and symplectic case. Much work in this direction has been done, in particular, by Derridj and Tartakoff [1988; 1991; 1993; 1995]; perhaps the furthest advance is [Grigis and Sjöstrand 1985]. All this work has required that the degeneration from strict pseudoconvexity to weak pseudoconvexity have a very special algebraic form; the methods seem to be decidedly limited in scope.

(ii) Global C^ω regularity has been proved for certain very special domains and operators possessing global symmetries [Chen 1988; Christ 1994a; Derridj 1997; Cordaro and Himonas 1994].

(iii) Various counterexamples and negative results have been devised. Some of these will be described below. Despite progress, there still exist few theorems of much generality; one of those few is in [Christ 1994b].

At present a wide gap separates the positive results from the known negative results. However, through the development of these negative results it has become increasingly evident that analytic hypoellipticity, and even global regularity in C^ω, are valid only rarely in the weakly pseudoconvex/nonsymplectic setting. While analytic hypoellipticity remains an open question for most weakly pseudoconvex domains, we believe that it fails to hold in the vast majority of cases.[28] Thus any method for proving analytic hypoellipticity must necessarily be very limited in scope.

An interesting conjecture has recently been formulated by Treves [1999], concerning the relationship between analytic hypoellipticity of a sum of squares operator, and the symplectic geometry of certain strata of the characteristic variety defined by the vanishing of its principal symbol.

Another proposed connection between hypoellipticity, in the real analytic, Gevrey, and C^∞ categories, and symplectic geometry is explored in [Christ 1998].

10.1. Global Counterexamples. It had been hoped that in both the C^∞ and the C^ω categories, at least global regularity would hold in great generality.

THEOREM 10.1. *There exist a bounded, pseudoconvex domain $\Omega \subset \mathbb{C}^2$ with C^ω boundary and a function $f \in C^\omega(\partial\Omega)$, whose Szegő projection does not belong to $C^\omega(\partial\Omega)$.*

The analysis [Christ 1996d] is related in certain broad aspects to the proof of global C^∞ irregularity for worm domains. Symmetry permits a reduction in dimension; more sophisticated analysis permits a reduction to one real dimension modulo certain error terms; existence of nonlinear eigenvalues for certain associated operators is at the core of the analysis; a deformation is introduced to evade the nonlinear eigenvalues; *a priori* estimates are proved for certain deformations; coupling these with singularities at the nonlinear eigenvalue parameters leads to a contradiction.

---

[28]This is another context in which second order equations are less well behaved than are those of first order. For operators of principal type, such as $\bar\partial_b$, there is a very satisfactory theory, and many such operators are analytic hypoelliptic, microlocally in appropriate regions.

This example has been refined by Tolli [1998]: there exists a convex domain having the same property, which is weakly pseudoconvex at only a single boundary point.

10.2. Victory in $\mathbb{R}^2$. For a relatively simple test class of operators with no artificial symmetry assumptions, analytic hypoellipticity has essentially been characterized. Consider any two real, C^ω vector fields X, Y, satisfying the bracket condition in an open subset of $\mathbb{R}^2$.

THEOREM 10.2 [Christ 1995a]. *For generic[29] pairs of vector fields, $L = X^2 + Y^2$ is analytic hypoelliptic at a point $p \in \mathbb{R}^2$ if and only if there exist an exponent $m \geq 1$ and coordinates with origin at p in which*

$$\operatorname{span}\{X, Y\} = \operatorname{span}\{\partial_x, x^{m-1}\partial_t\}. \tag{10-1}$$

Equality of these spans is to be understood in the sense of C^ω modules, not pointwise.

Sufficiency of the condition stated was proved long ago by Grušin; what is new is the necessity. The principal corollary is that analytic hypoellipticity holds quite rarely indeed. We believe that the same happens in higher dimensions and for other operators.

The main step is to show that L is analytic hypoelliptic if and only if a certain nonlinear eigenvalue problem has no solution. This problem takes the following form. To L is associated a one parameter family of ordinary differential operators $\mathcal{L}_z = -\partial_x^2 + Q(x, z)^2$, with parameter $z \in \mathbb{C}^1$, where Q is a homogeneous polynomial in $(x, z) \in \mathbb{R} \times \mathbb{C}$ that is monic with respect to x, and has degree $m - 1$ where m is the "type" at p; that is, the bracket hypothesis holds to order exactly m at p. The polynomial Q, modulo a simple equivalence relation, and a numerical quantity $q \in \mathbb{Q}^+$ used to define it, are apparently new geometric invariants of a pair of vector fields, satisfying the bracket condition, in $\mathbb{R}^2$. These invariants are not defined in terms of Lie brackets; q is related to a sort of directed order of contact at p between different branches of the complexified variety in $\mathbb{C}^2$ defined by the vanishing of the determinant of X, Y. The analytic hypoelliptic case arises precisely when this variety is nonsingular at p, that is, has only one branch. The pair $\{X, Y\}$ satisfies (10-1) if and only if $Q(x, z) \equiv x^{m-1}$ (modulo the equivalence relation).

A parameter z is said to be a nonlinear eigenvalue if $\mathcal{L}_z$ has nonzero nullspace in $L^2(\mathbb{R})$.

THEOREM 10.3 [Christ 1996a]. *If there exists at least one nonlinear eigenvalue for $\{\mathcal{L}_z\}$, then L fails to be analytic hypoelliptic in any neighborhood of p.*

---

[29]The meaning of "generic" will not be explained here; the set of all nongeneric pairs has been proved to be small, and may conceivably be empty. There is a corresponding microlocal theorem for $(X + iY) \circ (X - iY)$, a model for $\bar{\partial}_b^* \bar{\partial}_b$, under a pseudoconvexity hypothesis, in which no assumption of genericity is needed.

For generic[30] polynomials Q, there exist infinitely many nonlinear eigenvalues.

The restriction to $\mathbb{R}^2$ is essential to the analysis. However, the restriction to two vector fields is inessential and has been made only for the sake of simplicity.

For operators $(X + iY) \circ (X - iY)$ under a suitable "pseudoconvexity" hypothesis, there is an analogous but complete theory [Christ 1996a]: analytic hypoellipticity microlocally in the appropriate conic subset of $T^*\mathbb{R}^2$, the geometric condition (10–1), and nonexistence of nonlinear eigenvalues for the associated family of ordinary differential operators are all equivalent. Moreover, nonlinear eigenvalues fail to exist if and only if Q is equivalent to x^{m-1}.

For analyses of two classes of nonlinear eigenvalue problems for ordinary differential operators see [Christ 1993; 1996a].

10.3. Gevrey Hypoellipticity. Consider any sum of squares operator L in any dimension. Assume that the bracket hypothesis holds to order exactly m at a point p. Then, by [Derridj and Zuily 1973], L is hypoelliptic in the Gevrey class G^s for all $s \geq m$. Until about 1994, for every example known to this author, either L was analytic hypoelliptic, or it was Gevrey hypoelliptic for no $s < m$. The proof of Theorem 10.2 led to detailed information on Gevrey regularity, and in particular to the discovery of a whole range of intermediate behavior.

A simplified analysis applies to the following examples. They are of limited interest in themselves, but serve to demonstrate the intricacy of the Gevrey theory, and the fact that subtler geometric invariants than m come into play. Let $1 \leq p \leq q \in \mathbb{N}$, let (x, t) be coordinates in $\mathbb{R} \times \mathbb{R}^2$, and define

$$L = \partial_x^2 + x^{2(p-1)}\partial_{t_1}^2 + x^{2(q-1)}\partial_{t_2}^2.$$

Through work of Grušin, Oleĭnik and Radkevič, these are known to be analytic hypoelliptic if and only if $p = q$. The bracket condition is satisfied to order $m = q$ at 0.

THEOREM 10.4 [Christ 1997b]. *L is G^s hypoelliptic in some neighborhood of 0 if and only if $s \geq q/p$.*

This result has been reproved from another point of view by Bove and Tartakoff [1997], who obtained a still more refined result in terms of certain nonisotropic Gevrey classes.

An example in the opposite direction has been developed by Yu [1998]. In $\mathbb{R}^5$ with coordinates $(x, y, t) \in \mathbb{R}^{2+2+1}$ consider the examples

$$L_m = \partial_{x_1}^2 + (\partial_{y_1} + x_1^{m-1}\partial_t)^2 + \partial_{x_2}^2 + (\partial_{y_2} + x_2\partial_t)^2 .$$

---

[30]The set of nongeneric polynomials has Hausdorff codimension at least two, in a natural parameter space. We do not know whether it is empty; this question is analogous to one raised by Barrett [1995].

L_m is analytic hypoelliptic when $m = 2$. For $m > 2$ it is Gevrey hypoelliptic of all orders $s \geq 2$ [Derridj and Zuily 1973]. Clearly it becomes more degenerate as m increases; brackets of length m in ∂_{x_1} and $\partial_{y_1} + x_1^{m-1}\partial_t$ are required to span the direction ∂_t. What is less clear is that increasing degeneracy should have no effect on Gevrey hypoellipticity.

THEOREM 10.5 [Yu 1998]. *For any even $m \geq 4$, L_m fails to be analytic hypoelliptic. More precisely, L_m is G^s hypoelliptic only if $s \geq 2$.*

The proof relies on the asymptotic behavior of nonlinear eigenvalues ζ_j as $j \to \infty$, not merely on the existence of one eigenvalue. It is quite a bit more intricate than the treatment of examples like $\partial_{x_1}^2 + (\partial_{y_1} + x_1^{m-1}\partial_t)^2$ in $\mathbb{R}^3$.

10.4. Speculation.

PREDICTION. In nonsymplectic and weakly pseudoconvex situations, analytic hypoellipticity holds very rarely, and only for special types of degeneracies. The algebraic structure of a degeneracy is decisive.

One instance in which this deliberately vague principle can be made precise is the theory for operators $X^2 + Y^2$ in $\mathbb{R}^2$. According to Theorem 10.2, for generic vector fields, analytic hypoellipticity holds at $p \in \mathbb{R}^2$ if and only if the complex variety $W \subset \mathbb{C}^2$ defined by the vanishing of $\det(X, Y)$ has a single branch at p.

For operators $X^2 + Y^2$ in $\mathbb{R}^3$, and for the $\bar{\partial}$–Neumann problem for weakly pseudoconvex, real analytic domains in $\mathbb{C}^2$, we believe that the following examples are the key to understanding what condition might characterize analytic hypoellipticity. With coordinates $(x, y, t) \in \mathbb{R}^3$ consider vector fields $X = \partial_x$, $Y = \partial_y + a(x, y)\partial_t$, which correspond to so called "rigid" CR structures. The fundamental invariant is the Levi form $\lambda(x, y, t) = \lambda(x, y) = \partial a(x, y)/\partial x$.

Let $(x, y, t; \xi, \eta, \tau)$ be coordinates in $T^*\mathbb{R}^3$. Consider examples

$$\lambda_1(x, y) = x^{2p} + y^{2p}$$
$$\lambda_2(x, y) = x^p y^p + x^{2q} + y^{2q}$$

where $0 < p < q$ and p is even. In each case, the variety in $T^*\mathbb{R}^3$ defined by the vanishing of the principal symbols of X, Y and $[X, Y]$ is the symplectic submanifold $V = \{\xi = \eta = x = y = 0\}$. The Poisson stratifications conjectured by Treves [1999] to govern analytic hypoellipticity do not distinguish between λ_1 and λ_2. Operators with $\partial a/\partial x = \lambda_1$ are known to be analytic hypoelliptic [Grigis and Sjöstrand 1985]. There is an algebraic obstruction to the application of existing methods to Levi forms λ_2, and analytic hypoellipticity remains an open question in this case.

QUESTION 10.1. Are operators $X^2 + Y^2$ in $\mathbb{R}^3$ with Levi forms $[X, Y] = \lambda_2(x, y)\partial_t$ analytic hypoelliptic?

Further remarks explaining the difference between λ_2 and λ_1 can be found in [Christ 1998].

11. Questions

We conclude with speculations and possible directions for further investigation. Many of the questions posed here have been raised by earlier authors and are of long standing. Throughout the discussion we assume that $\Omega \Subset \mathbb{C}^n$ is smoothly bounded and pseudoconvex. Denote by Λ^δ and G^s the usual Hölder and Gevrey classes, respectively.

QUESTION 11.1. Let Ω be a domain of finite type in $\mathbb{C}^n$, $n \geq 3$. Does the Neumann operator map L^∞ to $\Lambda^\delta(\overline{\Omega})$ for some $\delta > 0$?

Convex domains behave better than general pseudoconvex domains, in several respects: (i) Global C^∞ regularity always holds.[31] (ii) The Bergman and Szegő projections and associated kernels for smoothly bounded convex domains of finite type are reasonably well understood in the C^∞, L^p Sobolev and Hölder categories, through work of McNeal [1994] and McNeal and Stein [1994], whereas much less is known for general pseudoconvex domains of finite type in $\mathbb{C}^n$, $n > 2$. (iii) For any convex domain in $\mathbb{C}^2$, the equation $\bar{\partial} u = f$ has an L^p solution for any L^p datum, for all $1 < p < \infty$ [Polking 1991].

QUESTION 11.2. Is the equation $\bar{\partial} u = f$ solvable in L^p and Hölder classes, for all smoothly bounded convex domains in $\mathbb{C}^n$, for all n?

A basic example of a nonconvex, pseudoconvex domain of finite type is the cross of iron in $\mathbb{C}^3$:
$$\Omega^\dagger : y_0 > |z_1|^6 + |z_1 z_2|^2 + |z_2|^6 \,,$$
where $z_j = x_j + i y_j$. Separation of variables leads to formulae for the Bergman and Szegő kernels, analogous to but more complicated than the formula of Nagel [1986] for certain domains in $\mathbb{C}^2$. So far as this author is aware, all questions beyond the existence of subelliptic estimates are open, including pointwise bounds for the Szegő and Bergman kernels, L^p and Hölder class mapping properties, analyticity, and analytic pseudolocality.[32] It might be possible to extract some information from the kernel formulae.

For further information concerning Hölder, supremum and L^p norm estimates, see [Sibony 1980/81; 1993; Fornæss and Sibony 1991; 1993]. A survey concerning weakly pseudoconvex domains is [Sibony 1991].

PROBLEM 11.3. Analyze $\Omega^\dagger$.

Work of Morimoto [1987b] and of Bell and Mohammed [1995] suggests the following conjecture concerning hypoellipticity (in C^∞) for domains of infinite type. Denote by $\lambda(z)$ the smallest eigenvalue of the Levi form at a point $z \in \partial\Omega$, and by W_∞ the set of all boundary points at which Ω is not of finite type.

---

[31] On the other hand, Tolli [1998] has proved that for a certain convex real analytic domain in $\mathbb{C}^2$ having only a single weakly pseudoconvex boundary point, the Szegő projection fails to preserve the class of functions globally real analytic on the boundary.

[32] I am indebted to J. McNeal for useful conversations concerning $\Omega^\dagger$.

CONJECTURE 11.4. Suppose that W_∞ is contained in a smooth real hypersurface M of $\partial\Omega$. Suppose that there exist $c > 0$ and $0 < \delta < 1$ such that for all $z \in \partial\Omega$,

$$\lambda(z) \geq c \exp(-\operatorname{distance}(z, M)^{-\delta}).$$

Then the $\bar{\partial}$-Neumann problem for Ω is hypoelliptic.

Conversely, there exist domains for which $\lambda(z) \geq c \exp(-C \operatorname{distance}(z, M))$, yet the $\bar{\partial}$-Neumann problem is not hypoelliptic.

The hypothesis that W_∞ is contained in a smooth hypersurface is unnatural; if this conjecture can be proved then a further generalization more in the spirit of Kohn's work [1979] on subellipticity and finite ideal type should be sought.

Now that global C^∞ regularity is known not to hold in general, it is natural to seek sufficient conditions. Compactness of the $\bar{\partial}$-Neumann problem is a more robust property that may prove more amenable to a satisfactory analysis. It is a purely local property; Diophantine inequalities and related pathology should not intervene in discussions of compactness.

PROBLEM 11.5. Characterize compactness of the Neumann operator N for pseudoconvex domains in $\mathbb{C}^2$.

At the least, this should be feasible for restricted classes of domains. Compactness is equivalent to the absence of complex discs in the boundary for Reinhardt domains, and presence of complex discs precludes compactness for arbitrary domains at least in $\mathbb{C}^2$, but the equivalence breaks down for Hartogs domains [Matheos 1998]. This problem and the next question appear to be related to the existence of nowhere dense compact subsets of $\mathbb{C}^1$ with positive logarithmic capacity.

A satisfactory characterization of global C^∞ regularity appears not to be a reasonable goal, but at least two natural questions beckon.

QUESTION 11.6. Does there exist a smoothly bounded, pseudoconvex domain $\Omega \subset \mathbb{C}^2$ whose boundary contains no analytic discs, yet the $\bar{\partial}$-Neumann problem for Ω is not globally regular in C^∞?

QUESTION 11.7. For the $\bar{\partial}$-Neumann problem on smoothly bounded pseudoconvex domains, does global C^∞ regularity always imply exact regularity in H^s for all s?

I suspect the answer to be negative. Barrett [1995; 1998] has studied exact regularity for domains in $\mathbb{C}^2$ for which W_∞ is a smoothly bounded Riemann surface, and shown that (i) exact regularity is violated whenever a certain nonlinear eigenvalue problem on the Riemann surface has a positive solution, and (ii) for generic Riemann surfaces, the nonlinear eigenvalue problem has indeed a positive solution. If there exist exceptional Riemann surfaces without nonlinear eigenvalues, some of those would be candidates for examples of global regularity without exact regularity. Whether such domains exist remains an open question.

It is also conceivable that the very instability of estimates with loss of derivatives could be exploited to show that for some one parameter family of domains Ω_t, global regularity holds for generic t even though exact regularity does not, in the same way that the Diophantine condition $|k_1 + \alpha k_2| \geq c|k|^{-N}$ holds for generic α, without establishing global regularity for any particular value of t.[33]

QUESTION 11.8. Does global regularity fail to hold for every domain in $\mathbb{C}^2$ for which W_∞ is a smoothly bounded Riemann surface R, satisfying $w(R, \partial\Omega) \neq 0$?

There is some interesting intuition for global C^∞ irregularity for the operators described in § 7, based on the connection between degenerate elliptic second order operators with real coefficients and stochastic processes. For information on this connection see [Bell 1995]. This intuition, together with conditional expectation arguments applied to random paths, predicts global irregularity for the models discussed in § 7, and explains in more geometric terms the seemingly paradoxical regularity for the more degenerate cases $m > 1$ discussed at the end of that section.

PROBLEM 11.9. Understand global C^∞ irregularity, for second order degenerate elliptic operators with real coefficients, from the point of view of Malliavin calculus and related stochastic techniques.

The next two problems and next question concern gaps in our understanding of global C^∞ irregularity for worm domains, and are of lesser importance.

PROBLEM 11.10. Prove global C^∞ irregularity for worm domains by working directly on the domain, rather than by reducing to the boundary.

PROBLEM 11.11. Generalize the analysis of the worm domains to higher dimensional analogues.

QUESTION 11.12. For worm domains, for Sobolev exponents s not belonging to the discrete exceptional set, is there an *a priori* estimate for the Neumann operator in H^s, with no loss of derivatives?

This amounts to asking whether bounds are uniform in the parameter j.

Much of the interest in global regularity for the $\bar{\partial}$–Neumann problem stems from a theorem of Bell and Ligocka [1980]: If $\Omega_1, \Omega_2 \subset \mathbb{C}^n$ are bounded, pseudoconvex domains with C^∞ boundaries, if $f : \Omega_1 \mapsto \Omega_2$ is a biholomorphism, and if the Bergman projection for each domain preserves $C^\infty(\bar{\Omega}_j)$, then f extends to a C^∞ diffeomorphism of their closures. For worm domains, the Bergman projection fails to preserve smoothness up to the boundary; this property is equivalent to global regularity for the $\bar{\partial}$–Neumann problem [Boas and Straube 1990]. But the proof leads to no counterexample for the mapping problem. Chen [1993] has shown that every automorphism of any worm domain is a rotation $(z_1, z_2) \mapsto (z_1, e^{i\theta} z_2)$, and hence certainly extends smoothly to the boundary.

---

[33] For a slightly related problem in which estimates with loss of derivatives have been established by exploiting such instability see [Christ and Karadzhov $\geq$ 1999; Christ et al. 1996].

QUESTION 11.13. Does every biholomorphic mapping between two smoothly bounded, pseudoconvex domains extend to a diffeomorphism of their closures?

Our next set of questions concerns the real analytic theory.

QUESTION 11.14. Suppose that $\Omega \Subset \mathbb{C}^2$ is pseudoconvex and has a real analytic boundary. For which Ω is the $\bar{\partial}$–Neumann problem analytic hypoelliptic? For which is it globally regular in C^ω?

Relatively recent examples [Christ 1997a; 1997b; Bove and Tartakoff 1997; Yu 1998] have demonstrated that for analytic nonhypoelliptic operators, determination of the optimal exponent for Gevrey hypoellipticity is a subtle matter. It is not at all apparent how geometric properties of the domain determine this exponent.[34]

QUESTION 11.15. If Ω has a real analytic boundary but the $\bar{\partial}$–Neumann problem is not analytic hypoelliptic, for which exponents s is it hypoelliptic in the Gevrey class G^s? For which exponents is it globally regular in G^s?

There appear to exist wormlike domains whose defining functions belong to every Gevrey class G^s with $s > 1$, and for which the higher invariants w_k are nonzero. Both the examples discussed at the end of § 7 and formal analysis of commutators suggest that for $m > 1$, nonvanishing of w_m may be related to global irregularity in Gevrey classes G^s for $s < m/(m-1)$.

QUESTION 11.16. Do the higher invariants w_k introduced in § 5 play a role in the theory of global Gevrey class hypoellipticity?

Another fundamental issue pertaining to singularities is their propagation. Consider the operator $\bar{\partial}_b^* \circ \bar{\partial}_b$, on the boundary of any real analytic, pseudoconvex domain $\Omega \Subset \mathbb{C}^2$. Suppose there exists a smooth, nonconstant curve $\gamma \subset \partial\Omega$ whose tangent vector lies everywhere in the span of the real and imaginary parts of $\bar{\partial}_b$, and which is contained in the set of all weakly pseudoconvex points of $\partial\Omega$.[35] Consider only functions u whose analytic wave front sets are contained in the subset of phase space in which $\bar{\partial}_b^*\bar{\partial}_b$ is C^∞ hypoelliptic.

QUESTION 11.17. If γ intersects the analytic singular support of u, must γ be contained in its analytic singular support?

The same may be asked for operators $X^2 + Y^2$, where X, Y are equal to, or analogous to, the real and imaginary parts of $\bar{\partial}_b$. For these, the question has been answered affirmatively by Grigis and Sjöstrand [1985] in the special case where the type is 3 at every "weakly pseudoconvex" point.

The $\bar{\partial}$–Neumann problem is a method for reducing the overdetermined first order system $\bar{\partial}u = f$ to a determined second order equation, analogous to Hodge

---

[34]A conjecture in this direction has been formulated by Bove and Tartakoff [1997].

[35]It has been shown [Christ 1994b] that whenever such a curve exists, $\bar{\partial}_b^*\bar{\partial}_b$ fails to be analytic hypoelliptic, microlocally in the region of phase space where it is C^∞ hypoelliptic.

theory. At the root of the counterexamples discovered in the last few years for global C^∞ regularity, for analytic hypoellipticity, and for global C^ω regularity are certain nonlinear eigenvalue problems that are associated to second order equations, but seem to have no counterparts for first order equations. Other methods for solving $\bar\partial u = f$ do exist, including solution by integral operators, and generalization of the $\bar\partial$–Neumann method to a twisted $\bar\partial$ complex [Ohsawa and Takegoshi 1987; McNeal 1996; Siu 1996]. Another method, which solves the $\bar\partial$ and $\bar\partial_b$ equations globally in the C^ω category, has been described by Christ and Li [1997].

QUESTION 11.18. Do these counterexamples represent limitations inherent in the nature of second order equations, or can the method of reduction of the $\bar\partial$ system to a determined second order equation be modified so as to avoid them? Do they have analogues for the $\bar\partial$ system itself?

Addenda. After this paper was written the first part of Conjecture 11.4 was proved for $\mathbb{C}^2$ by the author.

(Added in proof.) Since this paper was written, the author has learned of additional references concerning hypoellipticity of infinite type sums of squares of vector fields. They include [Kajitani and Wakabayashi 1991; Morimoto 1987a; Morimoto and Morioka 1997]. Further speculation may be found in [Christ 1998].

References

[Barrett 1984] D. E. Barrett, "Irregularity of the Bergman projection on a smooth bounded domain in $\mathbb{C}^2$", *Ann. of Math.* (2) **119**:2 (1984), 431–436.

[Barrett 1992] D. E. Barrett, "Behavior of the Bergman projection on the Diederich–Fornæss worm", *Acta Math.* **168**:1-2 (1992), 1–10.

[Barrett 1995] D. E. Barrett, "Duality between A^∞ and $A^{-\infty}$ on domains with nondegenerate corners", pp. 77–87 in *Multivariable operator theory* (Seattle, 1993), edited by R. E. Curto et al., Contemp. Math. **185**, Amer. Math. Soc., Providence, RI, 1995.

[Barrett 1998] D. E. Barrett, "The Bergman projection on sectorial domains", pp. 1–24 in *Operator theory for complex and hypercomplex analysis* (Mexico City, 1994), edited by E. R. de Arellano et al., Contemp. Math. **212**, Amer. Math. Soc., Providence, RI, 1998.

[Bedford and Fornæss 1978] E. Bedford and J. E. Fornæss, "Domains with pseudoconvex neighborhood systems", *Invent. Math.* **47**:1 (1978), 1–27.

[Bell 1995] D. R. Bell, *Degenerate stochastic differential equations and hypoellipticity*, Longman, Harlow (UK), 1995.

[Bell and Ligocka 1980] S. Bell and E. Ligocka, "A simplification and extension of Fefferman's theorem on biholomorphic mappings", *Invent. Math.* **57**:3 (1980), 283–289.

[Bell and Mohammed 1995] D. R. Bell and S. E. A. Mohammed, "An extension of Hörmander's theorem for infinitely degenerate second-order operators", *Duke Math. J.* **78**:3 (1995), 453–475.

[Boas and Straube 1990] H. P. Boas and E. J. Straube, "Equivalence of regularity for the Bergman projection and the $\bar{\partial}$-Neumann operator", *Manuscripta Math.* **67**:1 (1990), 25–33.

[Boas and Straube 1991a] H. P. Boas and E. J. Straube, "Sobolev estimates for the complex Green operator on a class of weakly pseudoconvex boundaries", *Comm. Partial Differential Equations* **16**:10 (1991), 1573–1582.

[Boas and Straube 1991b] H. P. Boas and E. J. Straube, "Sobolev estimates for the $\bar{\partial}$-Neumann operator on domains in $\mathbb{C}^n$ admitting a defining function that is plurisubharmonic on the boundary", *Math. Z.* **206**:1 (1991), 81–88.

[Boas and Straube 1993] H. P. Boas and E. J. Straube, "De Rham cohomology of manifolds containing the points of infinite type, and Sobolev estimates for the $\bar{\partial}$-Neumann problem", *J. Geom. Anal.* **3**:3 (1993), 225–235.

[Bonami and Charpentier 1988] A. Bonami and P. Charpentier, "Une estimation Sobolev $\frac{1}{2}$ pour le projecteur de Bergman", *C. R. Acad. Sci. Paris Sér. I Math.* **307**:5 (1988), 173–176.

[Bove and Tartakoff 1997] A. Bove and D. Tartakoff, "Optimal non-isotropic Gevrey exponents for sums of squares of vector fields", *Comm. Partial Differential Equations* **22**:7-8 (1997), 1263–1282.

[Catlin 1984] D. W. Catlin, "Global regularity of the $\bar{\partial}$-Neumann problem", pp. 39–49 in *Complex analysis of several variables* (Madison, WI, 1982), edited by Y.-T. Siu, Proc. Symp. Pure Math. **41**, Amer. Math. Soc., Providence, 1984.

[Catlin 1987] D. Catlin, "Subelliptic estimates for the $\bar{\partial}$-Neumann problem on pseudoconvex domains", *Ann. of Math.* (2) **126**:1 (1987), 131–191.

[Chang et al. 1992] D.-C. Chang, A. Nagel, and E. M. Stein, "Estimates for the $\bar{\partial}$-Neumann problem in pseudoconvex domains of finite type in $\mathbb{C}^2$", *Acta Math.* **169**:3-4 (1992), 153–228.

[Chen 1988] S.-C. Chen, "Global analytic hypoellipticity of the $\bar{\partial}$-Neumann problem on circular domains", *Invent. Math.* **92**:1 (1988), 173–185.

[Chen 1989] S.-C. Chen, "Global regularity of the $\bar{\partial}$-Neumann problem on circular domains", *Math. Ann.* **285**:1 (1989), 1–12.

[Chen 1991] S.-C. Chen, "Global regularity of the $\bar{\partial}$-Neumann problem in dimension two", pp. 55–61 in *Several complex variables and complex geometry* (Santa Cruz, CA, 1989), vol. 3, edited by E. Bedford et al., Proc. Sympos. Pure Math. **52**, Amer. Math. Soc., Providence, RI, 1991.

[Chen 1993] S.-C. Chen, "Characterization of automorphisms on the Barrett and the Diederich-Fornæss worm domains", *Trans. Amer. Math. Soc.* **338**:1 (1993), 431–440.

[Christ 1991a] M. Christ, "On the $\bar{\partial}_b$ equation for three-dimensional CR manifolds", pp. 63–82 in *Several complex variables and complex geometry* (Santa Cruz, CA, 1989), vol. 3, edited by E. Bedford et al., Proc. Sympos. Pure Math. **52**, Amer. Math. Soc., Providence, RI, 1991.

[Christ 1991b] M. Christ, "Precise analysis of $\bar{\partial}_b$ and $\bar{\partial}$ on domains of finite type in
$\mathbb{C}^2$", pp. 859–877 in *Proceedings of the International Congress of Mathematicians*
(Kyoto, 1990), vol. 2, Math. Soc. Japan, Tokyo, 1991.

[Christ 1993] M. Christ, "Analytic hypoellipticity, representations of nilpotent groups,
and a nonlinear eigenvalue problem", *Duke Math. J.* **72**:3 (1993), 595–639.

[Christ 1994a] M. Christ, "Global analytic hypoellipticity in the presence of symmetry", *Math. Res. Lett.* **1**:5 (1994), 559–563.

[Christ 1994b] M. Christ, "A necessary condition for analytic hypoellipticity", *Math.
Res. Lett.* **1**:2 (1994), 241–248.

[Christ 1995a] M. Christ, "Global irregularity for mildly degenerate elliptic operators",
preprint, Math. Sci. Res. Inst., Berkeley, 1995. Available at http://www.msri.org/
publications/preprints/online/1995-097.html.

[Christ 1995b] M. Christ, "On local and global analytic and Gevrey hypoellipticity",
Exp. No. IX, 7 in *Journées "Équations aux Dérivées Partielles"* (Saint-Jean-de-
Monts, 1995), École Polytech., Palaiseau, 1995.

[Christ 1996a] M. Christ, "Analytic hypoellipticity in dimension two", preprint, Math.
Sci. Res. Inst., Berkeley, 1996. Available at http://www.msri.org/publications/
preprints/online/1996-009.html.

[Christ 1996b] M. Christ, "Global C^∞ irregularity of the $\bar{\partial}$-Neumann problem for
worm domains", *J. Amer. Math. Soc.* **9**:4 (1996), 1171–1185.

[Christ 1996c] M. Christ, "A progress report on analytic hypoellipticity", pp. 123–146
in *Geometric complex analysis* (Hayama, 1995), edited by J. Noguchi et al., World
Sci. Publishing, Singapore, 1996.

[Christ 1996d] M. Christ, "The Szegő projection need not preserve global analyticity",
Ann. of Math. (2) **143**:2 (1996), 301–330.

[Christ 1997a] M. Christ, "Examples pertaining to Gevrey hypoellipticity", *Math. Res.
Lett.* **4** (1997), 725–733.

[Christ 1997b] M. Christ, "Intermediate optimal Gevrey exponents occur", *Comm.
Partial Differential Equations* **22**:3-4 (1997), 359–379.

[Christ 1998] M. Christ, "Hypoellipticity: geometrization and speculation", preprint,
Math. Sci. Res. Inst., Berkeley, 1998. Available at http://www.msri.org/publications/
preprints/online/1998-003.html. To appear in *Proceedings of Conference on Complex
Analysis and Geometry in honor of P. Lelong*, Progress in Mathematics, Birkhaüser.

[Christ and Geller 1992] M. Christ and D. Geller, "Counterexamples to analytic
hypoellipticity for domains of finite type", *Ann. of Math.* (2) **135**:3 (1992), 551–
566.

[Christ and Karadzhov ≥ 1999] M. Christ and G. Karadzhov, "Local solvability for
a class of partial differential operators with double characteristics". To appear in
Revista Mat. Iberoamericana.

[Christ and Li 1997] M. Christ and S.-Y. Li, "On analytic solvability and hypoellipticity
for $\bar{\partial}$ and $\bar{\partial}_b$", *Math. Z.* **226**:4 (1997), 599–606.

[Christ et al. 1996] M. Christ, G. E. Karadzhov, and D. Müller, "Infinite-dimensional
families of locally nonsolvable partial differential operators", *Math. Res. Lett.* **3**:4
(1996), 511–526.

[Cordaro and Himonas 1994] P. D. Cordaro and A. A. Himonas, "Global analytic hypoellipticity of a class of degenerate elliptic operators on the torus", *Math. Res. Lett.* **1**:4 (1994), 501–510.

[D'Angelo 1979] J. P. D'Angelo, "Finite type conditions for real hypersurfaces", *J. Differential Geom.* **14**:1 (1979), 59–66.

[D'Angelo 1982] J. P. D'Angelo, "Real hypersurfaces, orders of contact, and applications", *Ann. of Math.* (2) **115**:3 (1982), 615–637.

[D'Angelo 1987] J. P. D'Angelo, "Iterated commutators and derivatives of the Levi form", pp. 103–110 in *Complex analysis* (University Park, PA, 1986), edited by S. G. Krantz, Lecture Notes in Math. **1268**, Springer, Berlin, 1987.

[Derridj 1997] M. Derridj, "Analyticité globale de la solution canonique de $\bar{\partial}_b$ pour une classe d'hypersurfaces compactes pseudoconvexes de $\mathbb{C}^2$", *Math. Res. Lett.* **4**:5 (1997), 667–677.

[Derridj and Tartakoff 1976] M. Derridj and D. S. Tartakoff, "On the global real analyticity of solutions to the $\bar{\partial}$-Neumann problem", *Comm. Partial Differential Equations* **1**:5 (1976), 401–435.

[Derridj and Tartakoff 1988] M. Derridj and D. S. Tartakoff, "Local analyticity for $\Box_b$ and the $\bar{\partial}$-Neumann problem at certain weakly pseudoconvex points", *Comm. Partial Differential Equations* **13**:12 (1988), 1521–1600.

[Derridj and Tartakoff 1991] M. Derridj and D. S. Tartakoff, "Local analyticity for the $\bar{\partial}$-Neumann problem and $\Box_b$—some model domains without maximal estimates", *Duke Math. J.* **64**:2 (1991), 377–402.

[Derridj and Tartakoff 1993] M. Derridj and D. S. Tartakoff, "Local analyticity for the $\bar{\partial}$-Neumann problem in completely decoupled pseudoconvex domains", *J. Geom. Anal.* **3**:2 (1993), 141–151.

[Derridj and Tartakoff 1995] M. Derridj and D. S. Tartakoff, "Microlocal analyticity for the canonical solution $\bar{\partial}_b$ on some rigid weakly pseudoconvex hypersurfaces in $\mathbb{C}^2$", *Comm. Partial Differential Equations* **20**:9-10 (1995), 1647–1667.

[Derridj and Zuily 1973] M. Derridj and C. Zuily, "Régularité analytique et Gevrey pour des classes d'opérateurs elliptiques paraboliques dégénérés du second ordre", pp. 371–381 in *Colloque International CNRS sur les équations aux dérivées partielles linéaires* (Orsay, 1972), Astérisque **2-3**, Soc. Math. France, Paris, 1973.

[Diederich and Fornæss 1977] K. Diederich and J. E. Fornæss, "Pseudoconvex domains: an example with nontrivial Nebenhülle", *Math. Ann.* **225**:3 (1977), 275–292.

[Folland and Kohn 1972] G. B. Folland and J. J. Kohn, *The Neumann problem for the Cauchy-Riemann complex*, Annals of Mathematics Studies **75**, Princeton Univ. Press, Princeton, NJ, 1972.

[Fornæss and Sibony 1991] J. E. Fornæss and N. Sibony, "On L^p Estimates for $\bar{\partial}$", pp. 129–163 in *Several complex variables and complex geometry* (Santa Cruz, CA, 1989), vol. 3, edited by E. Bedford et al., Proc. Sympos. Pure Math. **52**, Amer. Math. Soc., Providence, RI, 1991.

[Fornæss and Sibony 1993] J. E. Fornæss and N. Sibony, "Smooth pseudoconvex domains in C^2 for which the corona theorem and L^p estimates for $\bar{\partial}$ fail", pp. 209–222 in *Complex analysis and geometry*, edited by V. Ancona and A. Silva, Univ. Ser. Math., Plenum, New York, 1993.

[Fornæss and Stensønes 1987] J. E. Fornæss and B. Stensønes, *Lectures on counter-examples in several complex variables*, Mathematical Notes **33**, Princeton Univ. Press, Princeton, NJ, 1987.

[Greiner and Stein 1977] P. C. Greiner and E. M. Stein, *Estimates for the $\bar{\partial}$-Neumann problem*, Mathematical Notes **19**, Princeton Univ. Press, Princeton, NJ, 1977.

[Grigis and Sjöstrand 1985] A. Grigis and J. Sjöstrand, "Front d'onde analytique et sommes de carrés de champs de vecteurs", *Duke Math. J.* **52**:1 (1985), 35–51.

[Helffer and Nourrigat 1985] B. Helffer and J. Nourrigat, *Hypoellipticité maximale pour des opérateurs polynômes de champs de vecteurs*, vol. 58, Progr. Math., Birkhäuser, Boston, 1985.

[Himonas 1995] A. A. Himonas, "On degenerate elliptic operators of infinite type", *Math. Z.* **220**:3 (1995), 449–460.

[Hörmander 1967] L. Hörmander, "Hypoelliptic second order differential equations", *Acta Math.* **119** (1967), 147–171.

[Kajitani and Wakabayashi 1991] K. Kajitani and S. Wakabayashi, "Propagation of singularities for several classes of pseudodifferential operators", *Bull. Sc. Math. Ser. 2* **115** (1991), 397–449.

[Kiselman 1991] C. O. Kiselman, "A study of the Bergman projection in certain Hartogs domains", pp. 219–231 in *Several complex variables and complex geometry* (Santa Cruz, CA, 1989), vol. 3, edited by E. Bedford et al., Proc. Sympos. Pure Math. **52**, Amer. Math. Soc., Providence, RI, 1991.

[Kohn 1963] J. J. Kohn, "Harmonic integrals on strongly pseudo-convex manifolds, I", *Ann. of Math.* (2) **78** (1963), 112–148.

[Kohn 1964] J. Kohn, "Harmonic integrals on strongly pseudo-convex manifolds, II", *Ann. of Math.* (2) **79** (1964), 450–472.

[Kohn 1973] J. J. Kohn, "Global regularity for $\bar{\partial}$ on weakly pseudo-convex manifolds", *Trans. Amer. Math. Soc.* **181** (1973), 273–292.

[Kohn 1979] J. J. Kohn, "Subellipticity of the $\bar{\partial}$-Neumann problem on pseudo-convex domains: sufficient conditions", *Acta Math.* **142**:1-2 (1979), 79–122.

[Kohn ≥ 1999] J. J. Kohn, Personal communication.

[Kohn and Nirenberg 1965] J. J. Kohn and L. Nirenberg, "Non-coercive boundary value problems", *Comm. Pure Appl. Math.* **18** (1965), 443–492.

[Komatsu 1975] G. Komatsu, "Global analytic-hypoellipticity of the $\bar{\partial}$-Neumann problem", *Proc. Japan Acad.* **51** (1975), suppl., 818–820.

[Komatsu 1976] G. Komatsu, "Global analytic-hypoellipticity of the $\bar{\partial}$-Neumann problem", *Tôhoku Math. J.* (2) **28**:1 (1976), 145–156.

[Matheos 1998] P. Matheos, *Failure of compactness for the $\bar{\partial}$ Neumann problem for two complex dimensional Hartogs domains with no analytic disks in the boundary*, Ph.d. dissertation, UCLA, Los Angeles, June 1998. To appear in *J. Geom. Anal.*

[McNeal 1994] J. D. McNeal, "Estimates on the Bergman kernels of convex domains", *Adv. Math.* **109**:1 (1994), 108–139.

[McNeal 1996] J. D. McNeal, "On large values of L^2 holomorphic functions", *Math. Res. Lett.* **3**:2 (1996), 247–259.

[McNeal and Stein 1994] J. D. McNeal and E. M. Stein, "Mapping properties of the Bergman projection on convex domains of finite type", *Duke Math. J.* **73**:1 (1994), 177–199.

[Morimoto 1987a] Y. Morimoto, "A criterion for hypoellipticity of second order differential operators", *Osaka J. Math.* **24** (1987), 651–675.

[Morimoto 1987b] Y. Morimoto, "Hypoellipticity for infinitely degenerate elliptic operators", *Osaka J. Math.* **24**:1 (1987), 13–35.

[Morimoto and Morioka 1997] Y. Morimoto and T. Morioka, "The positivity of Schrödinger operators and the hypoellipticity of second order degenerate elliptic operators", *Bull. Sc. Math.* **121** (1997), 507–547.

[Nagel 1986] A. Nagel, "Vector fields and nonisotropic metrics", pp. 241–306 in *Beijing lectures in harmonic analysis* (Beijing, 1984), edited by E. M. Stein, Annals of Mathematics Studies **112**, Princeton Univ. Press, Princeton, NJ, 1986.

[Nourrigat 1990] J. Nourrigat, "Subelliptic systems", *Comm. Partial Differential Equations* **15**:3 (1990), 341–405.

[Ohsawa and Takegoshi 1987] T. Ohsawa and K. Takegoshi, "On the extension of L^2 holomorphic functions", *Math. Z.* **195**:2 (1987), 197–204.

[Polking 1991] J. C. Polking, "The Cauchy-Riemann equations in convex domains", pp. 309–322 in *Several complex variables and complex geometry* (Santa Cruz, CA, 1989), vol. 3, edited by E. Bedford et al., Proc. Sympos. Pure Math. **52**, Amer. Math. Soc., Providence, RI, 1991.

[Rothschild and Stein 1976] L. P. Rothschild and E. M. Stein, "Hypoelliptic differential operators and nilpotent groups", *Acta Math.* **137**:3-4 (1976), 247–320.

[Sibony 1980/81] N. Sibony, "Un exemple de domaine pseudoconvexe régulier où l'équation $\bar{\partial}u = f$ n'admet pas de solution bornée pour f bornée", *Invent. Math.* **62**:2 (1980/81), 235–242.

[Sibony 1987] N. Sibony, "Une classe de domaines pseudoconvexes", *Duke Math. J.* **55**:2 (1987), 299–319.

[Sibony 1991] N. Sibony, "Some aspects of weakly pseudoconvex domains", pp. 199–231 in *Several complex variables and complex geometry* (Santa Cruz, CA, 1989), vol. 1, edited by E. Bedford et al., Proc. Sympos. Pure Math. **52**, Amer. Math. Soc., Providence, RI, 1991.

[Sibony 1993] N. Sibony, "On Hölder estimates for $\bar{\partial}$", pp. 587–599 in *Several complex variables* (Stockholm, 1987/1988), edited by J. E. Fornæss, Mathematical Notes **38**, Princeton Univ. Press, Princeton, NJ, 1993.

[Siu 1996] Y.-T. Siu, "The Fujita conjecture and the extension theorem of Ohsawa–Takegoshi", pp. 577–592 in *Geometric complex analysis* (Hayama, 1995), edited by J. Noguchi et al., World Scientific, Singapore, 1996.

[Sjöstrand 1982] J. Sjöstrand, "Singularités analytiques microlocales", pp. 1–166 in *Astérisque* **95**, Soc. Math. France, Paris, 1982.

[Sjöstrand 1983] J. Sjöstrand, "Analytic wavefront sets and operators with multiple characteristics", *Hokkaido Math. J.* **12**:3, part 2 (1983), 392–433.

[Sussmann 1973] H. J. Sussmann, "Orbits of families of vector fields and integrability of distributions", *Trans. Amer. Math. Soc.* **180** (1973), 171–188.

[Tartakoff 1978] D. S. Tartakoff, "Local analytic hypoellipticity for $\Box_b$ on nondegenerate Cauchy-Riemann manifolds", *Proc. Nat. Acad. Sci. U.S.A.* **75**:7 (1978), 3027–3028.

[Tartakoff 1980] D. S. Tartakoff, "The local real analyticity of solutions to $\Box_b$ and the $\bar{\partial}$-Neumann problem", *Acta Math.* **145**:3-4 (1980), 177–204.

[Tartakoff 1996] D. S. Tartakoff, "Global (and local) analyticity for second order operators constructed from rigid vector fields on products of tori", *Trans. Amer. Math. Soc.* **348**:7 (1996), 2577–2583.

[Tolli 1998] F. Tolli, "Failure of global regularity of $\bar{\partial}_b$ on a convex domain with only one flat point", *Pacific J. Math.* **185**:2 (1998), 363–398.

[Treves 1978] F. Treves, "Analytic hypo-ellipticity of a class of pseudodifferential operators with double characteristics and applications to the $\bar{\partial}$-Neumann problem", *Comm. Partial Differential Equations* **3**:6-7 (1978), 475–642.

[Treves 1999] F. Treves, "Symplectic geometry and analytic hypo-ellipticity", pp. 201–219 in *Differential equations* (La Pietra, 1996), edited by M. Giaquinta et al., Proc. Sympos. Pure Math. **65**, Amer. Math. Soc., Providence, RI, 1999.

[Yu 1998] C.-C. Yu, *Nonlinear eigenvalues and analytic-hypoellipticity*, Mem. Amer. Math. Soc. **636**, Amer. Math. Soc., 1998.

MICHAEL CHRIST
DEPARTMENT OF MATHEMATICS
UNIVERSITY OF CALIFORNIA
BERKELEY, CA 94720
UNITED STATES
christ@math.berkeley.edu

Several Complex Variables
MSRI Publications
Volume **37**, 1999

Subelliptic Estimates and Finite Type

JOHN P. D'ANGELO AND JOSEPH J. KOHN

ABSTRACT. This paper surveys work in partial differential equations and several complex variables that revolves around subelliptic estimates in the $\bar{\partial}$-Neumann problem. The paper begins with a discussion of the question of *local regularity*; one is given a bounded pseudoconvex domain with smooth boundary, and hopes to solve the inhomogeneous system of Cauchy–Riemann equation $\bar{\partial}u = \alpha$, where α is a differential form with square integrable coefficients and satisfying necessary compatibility conditions. Can one find a solution u that is smooth wherever α is smooth? According to a fundamental result of Kohn and Nirenberg, the answer is yes when there is a *subelliptic estimate*. The paper sketches the proof of this result, and goes on to discuss the history of various finite-type conditions on the boundary and their relationships to subelliptic estimates. This includes finite-type conditions involving iterated commutators of vector fields, subelliptic multipliers, finite type conditions measuring the order of contact of complex analytic varieties with the boundary, and Catlin's multitype.

The paper also discusses additional topics such as nonpseudoconvex domains, Holder and L^p estimates for $\bar{\partial}$, and finite-type conditions that arise when studying holomorphic extension, convexity, and the Bergman kernel function. The paper contains a few new examples and some new calculations on CR manifolds. The paper ends with a list of nine open problems.

CONTENTS

1. Introduction

The solution of the Levi problem during the 1950's established the fundamental result in function theory characterizing domains of holomorphy. Suppose that Ω is a domain in complex Euclidean space $\mathbb{C}^n$. The solution establishes that three conditions on Ω are identical: Ω is a domain of holomorphy, Ω is pseudoconvex, and the sheaf cohomology groups $H^q(\Omega, O)$ are trivial for each $q \geq 1$. The first property is a global function-theoretic property, the second is a local property of the boundary, and the third tells us that certain overdetermined systems of linear partial differential equations (the inhomogeneous Cauchy–Riemann equations) always have smooth solutions.

After the solution of the Levi problem, research focused upon domains with smooth boundaries and mathematicians hoped to establish deeper connections between partial differential equations and complex analysis. This led to the study of the Cauchy–Riemann equations on the closed domain and to many questions relating the boundary behavior of the Cauchy–Riemann operator $\bar{\partial}$ to the function theory on Ω. We continue the introduction by describing the question of local regularity for $\bar{\partial}$, and how its study motivated various geometric notions of "finite type".

Suppose that Ω is a bounded domain and that its boundary $b\Omega$ is a smooth manifold. We define $\bar{\partial}$ in the sense of distributions. Let α be a differential $(0, q)$ form with square-integrable coefficients and satisfying the compatibility condition $\bar{\partial}\alpha = 0$. What geometric conditions on $b\Omega$ guarantee that we can solve the Cauchy–Riemann equation $\bar{\partial}u = \alpha$ so that the $(0, q-1)$ form u must be smooth wherever α is? Here smoothness up to the boundary is the issue.

One approach to regularity results is the $\bar{\partial}$-Neumann problem. See [Folland and Kohn 1972; Kohn 1977; 1984] for extensive discussion. Let $L^2_{(0,q)}(\Omega)$ denote the space of $(0, q)$ forms with square-integrable coefficients. The $\bar{\partial}$-Neumann problem generalizes Hodge theory; careful attention to boundary conditions is now required. Under certain geometric conditions on $b\Omega$, Kohn constructed an operator N on $L^2_{(0,q)}(\Omega)$ such that $u = \bar{\partial}^* N\alpha$ gives the unique solution to $\bar{\partial}u = \alpha$ that is orthogonal to the null space of $\bar{\partial}$ on Ω. This is called the *canonical solution* or the $\bar{\partial}$-Neumann solution. In particular the Neumann operator N exists on bounded pseudoconvex domains. What additional geometric conditions on $b\Omega$ guarantee that N is a pseudo-local operator, and hence yield *local regularity* for the canonical solution u? By local regularity we mean that u is smooth wherever α is smooth. We shall see that pseudolocality for N follows from subelliptic estimates.

Kohn [1963; 1964] solved the $\bar{\partial}$-Neumann problem on strongly pseudoconvex domains in 1962. Subsequent work by Kohn and Nirenberg [1965] exposed clearly the subelliptic nature of the problem. Local regularity holds on strongly pseudoconvex domains because there is a subelliptic estimate; in this case one can take ε equal to $\frac{1}{2}$ in Definition 3.4 of this paper. Local regularity follows from a subelliptic estimate for any positive ε (see Theorem 3.5); this led Kohn to seek geometric conditions for subelliptic estimates. For domains in two dimensions he introduced in [1972] a finite-type condition (called "finite commutator-type" in this paper) enabling him to prove a subelliptic estimate. Greiner [1974] established the necessity of finite commutator-type in two dimensions. These theorems generated much work concerned with intermediate conditions between pseudoconvexity and strong pseudoconvexity. Different analytic problems lead to different intermediate, or finite-type, conditions on $b\Omega$. After contributions by many authors, Catlin [1983; 1984; 1987] completely solved one major problem of this kind. He proved that a certain finite-type condition is both necessary and sufficient for subelliptic estimates on $(0, q)$ forms for $q \geq 1$ for the $\bar{\partial}$-Neumann problem on smoothly bounded pseudoconvex domains. The finite-type condition is that the maximum order of contact of q-dimensional complex-analytic varieties with the boundary be finite at each point.

In this paper we survey those finite-type conditions arising from subelliptic estimates for the $\bar{\partial}$-Neumann problem and we indicate their relationship to function theory, geometry, and partial differential equations. We provide greater detail when we discuss subelliptic multipliers; we consider their use both on domains that are not pseudoconvex and on domains in CR manifolds. We indicate directions for further research and end the paper with a list of open problems.

2. The Levi Form

We begin by considering the geometry of the boundary of a domain in complex Euclidean space and its relationship to the function theory on the domain, using especially the Cauchy–Riemann operator $\bar{\partial}$ and the $\bar{\partial}$-Neumann problem. Let Ω denote a domain in $\mathbb{C}^n$ whose boundary is a smooth manifold denoted by $b\Omega$ or by M. Pseudoconvexity is a geometric property of $b\Omega$ that is necessary and sufficient for Ω to be a domain of holomorphy; for domains with smooth boundaries, pseudoconvexity is determined by the Levi form.

We recall an invariant definition of the Levi form that makes sense also for CR manifolds of hypersurface type. Thus we suppose that M is a smooth real manifold of dimension $2n - 1$ and that $\mathbb{C}TM$ denotes its complexified tangent bundle.

We say that M is a CR manifold of hypersurface type if there is a subbundle $T^{1,0}M \subset \mathbb{C}TM$ such that the following conditions hold:

1. $T^{1,0}M$ is integrable (closed under the Lie bracket operation).
2. $T^{1,0}M \cap \overline{T^{1,0}M} = \{0\}$.
3. The bundle $T^{1,0}M \oplus \overline{T^{1,0}M}$ has codimension one in $\mathbb{C}TM$.

For real submanifolds in $\mathbb{C}^n$ the bundle $T^{1,0}M$ is defined by $\mathbb{C}TM \cap T^{1,0}\mathbb{C}^n$, and thus local sections of $T^{1,0}M$ are complex $(1,0)$ vector fields tangent to M.

The bundle $T^{1,0}M$ is closed under the Lie bracket, or commutator, $[\,,\,]$. For CR manifolds this integrability condition is part of the definition.

On the other hand, the bundle $T^{1,0}M \oplus \overline{T^{1,0}M}$ is generally not integrable. The Levi form measures the failure of integrability. To define it, we denote by η a purely imaginary non-vanishing 1-form that annihilates $T^{1,0}M \oplus \overline{T^{1,0}M}$. When M is a hypersurface, and r is a local defining function, we may put $\eta = \frac{1}{2}(\partial - \bar{\partial})r$. We write $\langle\,,\,\rangle$ for the contraction of a one-form and a vector field.

DEFINITION 2.1. The *Levi form* λ is the Hermitian form on $T^{1,0}M$ defined (up to a multiple) by

$$\lambda(L, \overline{K}) = \langle \eta, [L, \overline{K}] \rangle. \tag{1}$$

The CR manifold M is called strongly pseudoconvex when λ is definite, and is called weakly pseudoconvex when λ is semi-definite but not definite. We say that the domain lying on one side of a real hypersurface is pseudoconvex when λ is positive semi-definite on the hypersurface.

We can also interpret the Levi form as the restriction of the complex Hessian of a defining function to the space $T^{1,0}M$. To see this we use the Cartan formula for the exterior derivative of η. Because $L, \overline{K}$ are annihilated by η and are tangent to M, we can write

$$\langle \partial\bar{\partial}r, L \wedge \overline{K} \rangle = \langle -d\eta, L \wedge \overline{K} \rangle = -L\langle \eta, \overline{K} \rangle + \overline{K}\langle \eta, L \rangle + \langle \eta, [L, \overline{K}] \rangle = \lambda(L, \overline{K}).$$

It is also useful to express the entries of the matrix λ with respect to a special local basis of the $(1,0)$ vector fields. Suppose that r is a defining function, and that we are in a neighborhood where $r_{z_n} \neq 0$. We put

$$T = \frac{1}{r_{z_n}} \frac{\partial}{\partial z_n} - \frac{1}{r_{\bar{z}_n}} \frac{\partial}{\partial \bar{z}_n}.$$

Then $\langle \eta, T \rangle = 1$. For $i = 1, 2, \ldots, n-1$ we define L_i by

$$L_i = \frac{\partial}{\partial z_i} - \frac{r_{z_i}}{r_{z_n}} \frac{\partial}{\partial z_n}.$$

Then the L_i, for $i = 1, 2, \ldots n-1$, form a commuting local basis for sections of $T^{1,0}M$. Furthermore $[L_i, \overline{L}_j] = \lambda_{ij}T$. Using subscripts for partial derivatives we have

$$\lambda_{ij} = \frac{r_{i\bar{j}}|r_n|^2 - r_{i\bar{n}}r_n r_{\bar{j}} - r_{n\bar{j}}r_i r_{\bar{n}} + r_{n\bar{n}}r_i r_{\bar{j}}}{|r_n|^2}.$$

Strong pseudoconvexity is a non-degeneracy condition: if λ is positive-definite at a point $p \in M$, then it is positive-definite in a neighborhood. Furthermore strong pseudoconvexity is "finitely determined": if M' is another hypersurface containing p and osculating M to second order there, then M' is also strongly pseudoconvex at p. In seeking generalizations of strong pseudoconvexity that have applications in analytic problems we expect that generalizations will be both open and finitely determined conditions.

As a simple example we compute the Levi form for domains in $\mathbb{C}^n$ defined locally by the equation

$$r(z, \bar{z}) = 2\operatorname{Re}(z_n) + \sum_{k=1}^{N} |f^k(z)|^2 < 0. \tag{2}$$

Here the functions f^k are holomorphic near the origin, vanish there, and depend only on the variables $z_1, z_2, \ldots, z_{n-1}$. The domain defined by (2) is pseudoconvex. Its Levi form near the origin has the nice expression

$$(\lambda_{ij}) = \left(\sum_{k=1}^{N} f_{z_i}^k \overline{f_{z_j}^k} \right) = (\partial f)^* (\partial f). \tag{3}$$

It follows immediately from (3) that the origin will be a weakly pseudoconvex point if and only if the rank of ∂f (as a mapping on $\mathbb{C}^{n-1}$) is less than full there. It is instructive to consider finite-type notions in this case and compare them with standard notions of singularities from algebraic and analytic geometry. For example, we will see that the origin is a point of finite D_1-type if and only if the germs of the functions f^k define a trivial variety, and more generally a point of finite D_q-type if and only if the functions define a variety of dimension less than q. The origin is a point of finite commutator-type if and only if some f^k is not identically zero; we see that this is the same as being finite D_{n-1}-type. This simple example allows us to glimpse the role of commutative algebra in later discussions, and it illustrates why different finiteness conditions arise.

It will be important to understand the *determinant* of the Levi form. To do so we make some remarks about restricting a linear map to a subspace. Suppose that $A : \mathbb{C}^n \to \mathbb{C}^n$ is a self-adjoint linear map, and that $\zeta \in \mathbb{C}^n$ is a unit vector. We form two new linear transformations using this information.

First we extend A to a map $(E_\zeta A) : \mathbb{C}^n \times \mathbb{C} \to \mathbb{C}^n \times \mathbb{C}$ given by

$$(E_\zeta A)(z, t) = (Az + t\zeta, \langle z, \zeta \rangle). \tag{4}$$

Second we restrict A to a map on the orthogonal complement of the span of ζ, and identify this with a map $R_\zeta A : \mathbb{C}^{n-1} \to \mathbb{C}^{n-1}$, by composing with an isometry in the range. Then, assuming $n \geq 2$, we have

$$\det(R_\zeta A) = \det(E_\zeta A).$$

One way to see this is to choose coordinates so that $\zeta = (0, 0, \ldots, 0, 1)$ and the matrix of the map $E_\zeta A$ has lots of zeroes. Expanding by cofactors (twice) shows that the determinant equals the determinant of the $n-1$ by $n-1$ principal minor of A, which equals the determinant of R_ζ by the same computation that one does to write the Levi form as an $n-1$ by $n-1$ matrix.

According to this result we may express the determinant of the Levi form as the determinant of the $n+1$ by $n+1$ bordered Hessian matrix

$$E = \begin{pmatrix} r_{z_1 \bar{z}_1} & r_{z_2 \bar{z}_1} & \cdots & r_{z_n \bar{z}_1} & r_{\bar{z}_1} \\ r_{z_1 \bar{z}_2} & r_{z_2 \bar{z}_2} & \cdots & r_{z_n \bar{z}_2} & r_{\bar{z}_2} \\ \vdots & \vdots & \ddots & \vdots & \vdots \\ r_{z_1 \bar{z}_n} & r_{z_2 \bar{z}_n} & \cdots & r_{z_n \bar{z}_n} & r_{\bar{z}_n} \\ r_{z_1} & r_{z_2} & \cdots & r_{z_n} & 0 \end{pmatrix}. \tag{5}$$

It will be convenient later to write (5) in simpler notation. To do so, we imagine ∂r as a row, and $\bar{\partial} r$ as a column. We get

$$E = \begin{pmatrix} \partial \bar{\partial} r & \bar{\partial} r \\ \partial r & 0 \end{pmatrix}. \tag{6}$$

Finally we remark that when the defining equation is given by (3), there is a simple formula for the determinant of the Levi form. We have

$$\det(\lambda) = \sum |J(f_{i_1}, \ldots, f_{i_{n-1}})|^2,$$

where the sum is taken over all choices of $n-1$ of the functions f_k, and J denotes the Jacobian determinant in $n-1$ dimensions. Thus the determinant of the Levi form is the squared norm of a holomorphic mapping in this case.

3. Subelliptic Estimates for the $\bar{\partial}$-Neumann Problem

From the introduction we have seen that the $\bar{\partial}$-Neumann problem constructs a particular solution to the inhomogeneous Cauchy–Riemann equations. The $\bar{\partial}$-Neumann problem is a boundary value problem; the equation is elliptic, but the boundary conditions are not elliptic. One of the most important results, due to Kohn and Nirenberg, states that local regularity for the canonical solution to the inhomogeneous Cauchy–Riemann equations follows from a subelliptic estimate. In this section we define subelliptic estimates, and sketch a proof of the Kohn–Nirenberg result.

We begin by recalling the definition of the tangential Sobolev norms. We write $\mathbb{R}^m_-$ for the subset of $\mathbb{R}^m$ whose last coordinate is negative. For convenience we denote the first $m-1$ components by t and the last component by r.

DEFINITION 3.1 (PARTIAL FOURIER TRANSFORM). Suppose that $u \in C_0^\infty(\mathbb{R}^m_-)$. The partial Fourier transform of u is given by

$$\tilde{u}(\xi, r) = \int_{\mathbb{R}^{m-1}} e^{-it \cdot \xi} u(t, r) \, dt.$$

DEFINITION 3.2. Suppose that $u \in C_0^\infty(\mathbb{R}^m_-)$. We define the tangential pseudo-differential operator Λ^s and the tangential Sobolev norm $|||u|||_s$ by

$$(\widetilde{\Lambda^s u})(\xi, r) = (1 + |\xi|^2)^{s/2} \tilde{u}(\xi, r), \qquad |||u|||_s = \|\Lambda^s u\|.$$

Note that the L^2 norm is computed over $\mathbb{R}^m_-$. Suppose that Ω is a smoothly bounded domain in $\mathbb{C}^n$, and $p \in b\Omega$. On a sufficiently small neighborhood U of p we introduce coordinates $(t_1, \ldots, t_{2n-1}, r)$ where r is a defining function for Ω. We may also assume that $\omega_1, \ldots \omega_n$ form an orthonormal basis for the $(1,0)$ forms on U and that $\omega_n = (\partial r)/|\partial r|$.

Thus a $(0,1)$ form ϕ defined on U may be written

$$\phi = \sum_1^n \phi_j \bar{\omega}_j$$

We write

$$|||\phi|||_s^2 = \sum_1^n |||\phi_j|||_s^2.$$

We denote by $\bar{\partial}^*$ the L^2-adjoint of (the maximal extension) of $\bar{\partial}$ and let $\mathcal{D}(\bar{\partial}^*)$ denote its domain. In terms of the ω_j there is a simple expression for the boundary condition required for a form to be in $\mathcal{D}(\bar{\partial}^*)$. If $\phi \in C^\infty(U \cap \bar{\Omega})$, then ϕ is in $\mathcal{D}(\bar{\partial}^*)$ if and only if $\phi_n = 0$ on $U \cap b\Omega$.

We define (in terms of the $L^2(\Omega)$ inner product) the quadratic form Q by

$$Q(\phi, \psi) = (\bar{\partial}\phi, \bar{\partial}\psi) + (\bar{\partial}^*\phi, \bar{\partial}^*\psi).$$

Integration by parts yields the following formula for $Q(\phi, \phi)$ on $(0,1)$-forms, where r is a local defining function for $b\Omega$.

LEMMA 3.3. *The quadratic form Q satisfies*

$$Q(\phi, \phi) = \sum_{i,j=1}^n \int_\Omega |(\phi_i)_{\bar{z}_j}|^2 dV + \sum_{i,j=1}^n \int_{b\Omega} r_{z_i \bar{z}_j} \phi_i \bar{\phi}_j dS = \|\phi\|_{\bar{z}}^2 + \int_{b\Omega} \lambda(\phi, \phi)\, dS. \quad (7)$$

This formula reveals an asymmetry between the barred and unbarred derivatives; this is a consequence of the boundary conditions. Observe also that the integral of the Levi form appears. This term is non-negative when Ω is pseudoconvex. The *basic estimate* asserts that the terms on the right of (7) are dominated by a constant times $Q(\phi, \phi)$. For pseudoconvex domains in $\mathbb{C}^n$ we also have the estimate

$$\|\phi\|^2 \leq CQ(\phi, \phi). \quad (8)$$

This estimate does not hold generally for domains in manifolds, unless the manifold admits a strongly plurisubharmonic exhaustion function.

In order to prove local regularity for the $\bar{\partial}$-Neumann solution to the inhomogeneous Cauchy–Riemann equations, we use a stronger estimate, called a subelliptic estimate.

DEFINITION 3.4. Suppose that $\Omega \Subset \mathbb{C}^n$ is smoothly bounded and pseudoconvex. Let $p \in \bar{\Omega}$ be any point in the closure of the domain. The $\bar{\partial}$-Neumann problem satisfies a subelliptic estimate at p on $(0,1)$ forms if there exist positive constants C, ε and a neighborhood $U \ni p$ such that

$$\|\phi\|_\varepsilon^2 \leq C(\|\bar\partial\phi\|^2 + \|\bar\partial^*\phi\|^2) \tag{9}$$

for every $(0,1)$-form ϕ that is smooth, compactly supported in U, and in $\mathcal{D}(\bar\partial^*)$.

We usually say simply *a subelliptic estimate holds* when the definition applies. Although the definition of $\bar\partial^*$ (and hence that of Q) depends on the Hermitian metric used, whether a subelliptic estimate holds is independent of the metric [Sweeney 1972].

We begin with the connection to local regularity. Suppose that α is a $(0,1)$ form in $L^2(\Omega)$ and that $\alpha|_{U \cap \bar\Omega}$ is smooth. Let ϕ in $\mathcal{D}(\bar\partial^*)$ be the unique form that satisfies

$$Q(\phi, \psi) = (\alpha, \psi)$$

for all ψ in $\mathcal{D}(\bar\partial) \cap \mathcal{D}(\bar\partial^*)$. Then $\phi = N\alpha$ and we have $\bar\partial(\bar\partial^*\phi) = \alpha$. A subelliptic estimate implies that $\phi|_{U \cap \bar\Omega} \in C^\infty(U \cap \bar\Omega)$. The basic theorem of Kohn and Nirenberg [1965] gives this and additional consequences of a subelliptic estimate.

THEOREM 3.5 (KOHN AND NIRENBERG). *Suppose that a subelliptic estimate holds. Then ϕ restricted to $U \cap \bar\Omega$ is smooth. More generally the Neumann operator N is pseudolocal. We also have, in terms of local Sobolev norms H_s,*

$$\begin{aligned} \alpha \in H_s &\Rightarrow N\alpha \in H_{s+2\varepsilon}, \\ \alpha \in H_s &\Rightarrow \bar\partial^* N\alpha \in H_{s+\varepsilon}. \end{aligned} \tag{10}$$

SKETCH OF PROOF. Suppose that a subelliptic estimate holds, and that D is an arbitrary first order partial differential operator. The first step is to prove the estimate

$$\||D\phi\||_{\varepsilon-1}^2 \leq Q(\phi, \phi) \tag{11}$$

for all $\phi \in C_0^\infty(U \cap \bar\Omega) \cap \mathcal{D}(\bar\partial^*)$. This is clear when D is tangential, so it suffices to consider $D = \frac{\partial}{\partial r}$. Observe that $b\Omega$ is non-characteristic for the quadratic form Q (in fact Q is elliptic, although the boundary conditions are not). Therefore we have an estimate

$$\left\|\frac{\partial\phi}{\partial r}\right\|^2 \leq C\left(Q(\phi, \phi) + \||\phi\||_1^2\right). \tag{12}$$

After using cut-off functions to give a meaning to $Q(\Lambda^{\varepsilon-1}\phi, \Lambda^{\varepsilon-1}\phi)$, we replace ϕ by $\Lambda^{\varepsilon-1}\phi$ in (12). This yields

$$\||\frac{\partial\phi}{\partial r}\||_{\varepsilon-1}^2 \leq C\left(Q(\Lambda^{\varepsilon-1}\phi, \Lambda^{\varepsilon-1}\phi) + \||\phi\||_\varepsilon^2\right). \tag{13}$$

We next require some calculations involving the commutators $[\bar\partial, \Lambda^{\varepsilon-1}]$ and $[\bar\partial^*, \Lambda^{\varepsilon-1}]$. We omit the proofs, but both $\|[\bar\partial, \Lambda^{\varepsilon-1}]\phi\|$ and $\|[\bar\partial^*, \Lambda^{\varepsilon-1}]\phi\|$ can be estimated in terms of a constant times $\||\phi\||_{\varepsilon-1}$. Given this we can estimate

$$Q(\Lambda^{\varepsilon-1}\phi, \Lambda^{\varepsilon-1}\phi) \leq cQ(\phi, \phi). \tag{14}$$

Combining (13) and (14) with the subelliptic estimate proves (11) when $D = \partial/\partial r$.

Assume that ϕ is smooth. Let ζ and ζ' be cutoff functions with $\text{supp}(\zeta) \Subset \text{supp}(\zeta')$ and suppose that $\zeta' = 1$ on a neighborhood of the support of ζ. We need an estimate involving higher derivatives:

$$\sum_{|\gamma| \leq m+2} \|D^\gamma \zeta \phi\|_{(k+2)\varepsilon - |\gamma|} \leq C_{mk} \Big(\sum_{|\gamma| \leq m} \|D^\gamma \zeta' \alpha\|_{m\varepsilon - |\gamma|} + \|\phi\| \Big).$$

The proof of this is complicated, and we omit it.

The next step is to introduce elliptic regularization. For $\delta > 0$ we consider the quadratic form Q_δ defined by

$$Q_\delta(\phi, \psi) = Q(\phi, \psi) + \delta \sum_{|\gamma| \leq 1} (D^\gamma \phi, D^\gamma \psi).$$

The form Q_δ is elliptic. We can solve

$$Q_\delta(\phi_\delta, \psi) = (\alpha, \psi)$$

so that ϕ_δ is smooth wherever α is smooth. From estimate (8) we obtain $\|\phi_\delta\| \leq C\|\alpha\|$ where C is independent of δ. One then proves that a subsequence of the ϕ_δ converges in the C^∞ topology to a solution ϕ of the original problem. $\qquad \square$

We close the section by making a few remarks about the definition of a subelliptic estimate. Observe that the set of points for which a subelliptic estimate holds must be an open subset of the closed domain. For interior points, the estimate (9) is elliptic, and holds with $\varepsilon = 1$. At strongly pseudoconvex boundary points, the estimate holds for $\varepsilon = \frac{1}{2}$. Catlin has found necessary and sufficient conditions for a subelliptic estimate for some $\varepsilon > 0$ to hold. See Theorem 7.1. In the weakly pseudoconvex case there is no general result giving the largest possible value of the parameter ε in terms of the geometry of $b\Omega$ at the boundary point p.

4. Ideals of Subelliptic Multipliers

We assume that Ω is a smoothly bounded pseudoconvex domain. The estimate (9) holds at interior points; we next let x be a boundary point of Ω. For a neighborhood U containing x, consider the set of all functions $f \in C_0^\infty(U \cap \Omega)$ such that there are $C, \varepsilon > 0$ for which

$$\|f\phi\|_\varepsilon^2 \leq C(\|\bar{\partial}\phi\|^2 + \|\bar{\partial}^* \phi\|^2) \tag{15}$$

for all $\phi \in C_0^\infty(U \cap \Omega) \cap \mathcal{D}(\bar{\partial}^*)$. Here both constants may depend on f. Let $\mathcal{J}_x$ denote the collection of all germs of such functions at x; its elements are called subelliptic multipliers. We see immediately that a subelliptic estimate holds precisely when the constant function 1 is a subelliptic multiplier.

LEMMA 4.1. *Suppose that* λ_{ij} *are the components of the Levi matrix with respect to the local basis* $\{L_1, \ldots, L_{n-1}\}$ *of* $T^{10}(b\Omega)$. *Then there is a constant* C *so that*

$$\sum_{i,j=1}^{n-1} (\lambda_{ij}\Lambda^{1/2}\phi_i, \Lambda^{1/2}\phi_j) \leq CQ(\phi, \phi). \tag{16}$$

We omit the proof, which uses the expression

$$\det \begin{pmatrix} \partial\bar{\partial}r & \bar{\partial}r \\ \partial r & 0 \end{pmatrix}$$

for the determinant of the Levi form (see the discussion between (4) *and* (6)) *and also requires properties of commutators of tangential pseudodifferential operators.*

PROPOSITION 4.2. *Suppose that* Ω *is pseudoconvex. The defining function* r *is a subelliptic multiplier, with* $\varepsilon = 1$. *The determinant of the Levi form is a subelliptic multiplier, with* $\varepsilon = \frac{1}{2}$.

PROOF. To show that r is a subelliptic multiplier with $\varepsilon = 1$ is easy. It follows from integration by parts that $\|(r\phi_k)_{z_i}\|^2 = \|(r\phi_k)_{\bar{z}_i}\|^2$. Therefore it suffices to estimate the first order barred derivatives. To do so we replace ϕ by $r\phi$ in Lemma 3.3 and observe that $Q(r\phi, r\phi) \leq CQ(\phi, \phi)$.

That $\det(\lambda_{ij})$ is a subelliptic multiplier with $\varepsilon = \frac{1}{2}$ follows from Lemma 4.1.
□

Starting with Proposition 4.2, Kohn [1979] developed an algorithmic procedure for constructing new multipliers, for which the corresponding value of epsilon is typically smaller. We now discuss a slight reformulation of this procedure.

PROPOSITION 4.3. *Let* x *be a boundary point of the pseudoconvex domain* Ω. *Then the collection of subelliptic multipliers* $\mathfrak{J}_x$ *on* (0,1) *forms is a radical ideal. In particular,*

$$f \in \mathfrak{J}_x, |g|^N \leq f \implies g \in \mathfrak{J}_x. \tag{17}$$

When $m\varepsilon \leq 1$, we also have the estimate

$$\||g\phi|\|_\varepsilon^2 \leq c\||g^m\phi|\|_{m\varepsilon}^2 + c\|\phi\|^2 \tag{18}$$

PROPOSITION 4.4. *Suppose that* f *is a subelliptic multiplier, and that*

$$\||f\phi|\|_\varepsilon^2 \leq cQ(\phi, \phi) \tag{19}$$

for all appropriate ϕ *and for* $0 < \varepsilon \leq 1$. *Then there is a constant* $c > 0$ *so that*

$$\|\sum_{j=1}^n \frac{\partial f}{\partial z_j}\phi_j\|_{\varepsilon/2}^2 \leq cQ(\phi, \phi) \tag{20}$$

We will use Proposition 4.4 by augmenting the Levi matrix by adding the rows ∂f and the column $\bar{\partial} f$ in the same way we did this for ∂r and $\bar{\partial} r$. More precisely, suppose that $f_1, \ldots, f_N$ are subelliptic multipliers. We define the $n + 1 + N$ by $n + 1 + N$ matrix $A(f)$ by

$$A(f) = \begin{pmatrix} \partial\bar{\partial} r & \bar{\partial} r & \bar{\partial} f_1 & \cdots & \bar{\partial} f_N \\ \partial r & 0 & 0 & \cdots & 0 \\ \partial f_1 & 0 & 0 & \cdots & 0 \\ \vdots & \vdots & \vdots & \ddots & \vdots \\ \partial f_N & 0 & 0 & \cdots & 0 \end{pmatrix}. \tag{21}$$

PROPOSITION 4.5. *Suppose that f_i are subelliptic multipliers. Then the determinant of $A(f)$ is a subelliptic multiplier.*

Define I_0 to be the real radical of the ideal generated by r and the determinant of the Levi form $\det(\lambda)$. For $k \geq 1$, define I_k to be the real radical of the ideal generated by I_{k-1} and all determinants $\det(A(f))$ for $f_j \in I_{k-1}$.

By Proposition 4.2 we know that r and $\det(\lambda)$ are subelliptic multipliers. By Proposition 4.3 all the elements in I_0 are subelliptic multipliers. By Propositions 4.4 and 4.5, and induction, for each k all the elements of I_k are subelliptic multipliers. Thus a subelliptic estimate holds whenever 1 lies in some I_k.

DEFINITION 4.6. The point p in a pseudoconvex real hypersurface M is of *finite ideal-type* if there is an integer k such that $1 \in I_k$. (Equivalently I_k is the ring of germs of smooth functions at p.)

As for the subelliptic estimate, whether p is of finite ideal-type is independent of the Hermitian metric used. Next we prove directly that the existence of a complex analytic variety V in $b\Omega$ prevents points on V from being of finite ideal-type. This theorem motivates Section 6.

THEOREM 4.7. *Suppose that Ω is pseudoconvex and that there is a complex analytic variety V lying in $b\Omega$. Then points of V cannot be of finite ideal-type.*

PROOF. The condition of finite ideal-type is an open condition, so we may assume that p is a smooth point of V. We may find a non-zero vector field L that is tangent to V and is a holomorphic combination of the usual L_i. Then L is in the kernel of the Levi form along V, so $\det(\lambda)$ vanishes along V. Therefore all elements of I_0 vanish on V. We proceed by induction. Suppose that all elements of I_{k-1} vanish along V. Choosing $f_j \in I_{k-1}$ we have $L(f_j) = 0$ because L is tangent. Therefore the matrix whose entries are $L_i(f_j)$ must have a non-trivial kernel, and hence $\det(A(f))$ must vanish on V, and thus all elements of I_k vanish on V also. $\qquad\square$

For real-analytic pseudoconvex domains, the sequence of ideals stabilizes after finitely many steps [Kohn 1979]. Either $1 \in I_k$ for some k, or the process uncovers a real-analytic real subvariety in the boundary of "positive holomorphic dimension". A CR submanifold of M has positive holomorphic dimension when it has a non-zero tangent vector field annihilated by the Levi form of M. Diederich and Fornaess [1978] then proved (assuming M is pseudoconvex and real-analytic) that a variety with positive holomorphic dimension can lie in M and pass through p only when there are complex-analytic varieties in the boundary passing through points arbitrarily close by. This is equivalent to the statement that there are no complex-analytic varieties in the boundary passing through p. See [D'Angelo 1993; 1991] for a proof of this last equivalence that applies without the hypothesis of pseudoconvexity. Conversely by Theorem 4.7 the estimate cannot hold when there is a complex-analytic variety passing through p and lying in the boundary.

This gives the result in the pseudoconvex real-analytic case.

THEOREM 4.8. *Let $\Omega \Subset \mathbb{C}^n$ be pseudoconvex, and suppose that its boundary is real-analytic near p. Then there is a subelliptic estimate at p on $(0, q)$ forms if and only if there is no germ of a complex-analytic variety of dimension q lying in $b\Omega$ and passing through p. (and thus, in the language of Section 6, if and only if $\Delta_q(M, p)$ is finite).*

5. Finite Commutator-Type

The definition of finite commutator-type for a point p on a CR manifold involves only the CR structure. For imbedded hypersurfaces finite commutator-type is equivalent to regular $(n-1)$-type, namely, the order of tangency of every complex hypersurface with M at p is finite. See Section 6. For domains in $\mathbb{C}^2$, finite commutator-type, finite ideal-type, and finite D_1-type are equivalent conditions.

Suppose that $p \in M$, and that L is a local section of $T^{1,0}M$. We define the type of L at p by

$$t(L, p) = \min\{k : \text{there is a commutator } X = [\ldots [L_1, L_2], \ldots L_k]$$
$$\text{such that} \langle X, \eta \rangle(p) \neq 0\}.$$

In this definition each L_i equals either L or $\bar{L}$. Thus the type of a vector field at p equals two precisely when the Levi form $\lambda(L, \bar{L})(p)$ is non-zero. Taking higher commutators is closely related to but not precisely the same as taking higher derivatives of $\lambda(L, \bar{L})$ in the directions of L and $\bar{L}$. Because of the distinction it is worth introducing a related number. We define

$$c(L, p) = \min\{k : Y \langle [L, \bar{L}], \eta \rangle(p) \neq 0\},$$

where Y is a monomial differential operator $Y = \prod_{j=1}^{k-2} L_j$ and again each L_j equals either L or $\bar{L}$. Thus $c(L, p) = 2$ precisely when the Levi form $\lambda(L, \bar{L})(p)$ is non-zero. By computing higher commutators, we observe that some but not all of the terms arising are those in the definition of $c(L, p)$. For points in a CR manifold where the Levi form has eigenvalues of opposite sign, there are vector fields for which these numbers are different. It is believed to be true, but not proved in the literature, that these two numbers are the same for all vector fields in the pseudoconvex case. See [Bloom 1981; D'Angelo 1993] for what is known.

Next we define the commutator-type of a point on a CR manifold of hypersurface type.

DEFINITION 5.1. The point p on a CR manifold M of hypersurface type is a point of finite commutator-type if $t(L, p)$ is finite for some local section L of $T^{1,0} M$. The commutator-type of p is the minimum of the types of all such $(1, 0)$ vector fields L.

We next discuss some geometric aspects of this notion. For a 3-dimensional CR manifold such as a hypersurface in $\mathbb{C}^2$, the space $T_p^{1,0}(M)$ is 1-dimensional, so the types of all vector fields non-zero at p are the same. In this case we also have $t(L, p) = c(L, p)$ for all L and p. When the Levi form has $n - 2 > 0$ positive eigenvalues and one vanishing eigenvalue on a pseudoconvex CR manifold of dimension $2n - 1$, the minimum value of $t(L, p)$ is two, but furthermore there can be only one possible value for $t(L, p)$ other than 2 and again $t(L, p) = c(L, p)$ for all L and p. For real hypersurfaces, the commutator-type equals the maximum order of tangency of a complex hypersurface. The geometry becomes more complicated when the Levi form has several vanishing eigenvalues, and the types of vector fields give incomplete information. In particular the condition that all $(1, 0)$ vector fields L satisfy $t(L, p) < \infty$ does not prevent complex-analytic varieties from lying in a hypersurface.

REMARK. We discuss the geometric interpretation of the type of a single vector field. Suppose that M is a real hypersurface in $\mathbb{C}^n$ and that V is a complex manifold osculating M to order N at p. Then there is a $(1, 0)$ vector field L with $L(p) \neq 0$ and $t(L, p) \geq N$. We may take L to be tangent to V. The converse is not generally true, but the first author believes that it may be true in the pseudoconvex case. We give an example due to Bloom [1981].

EXAMPLE 5.2. Put $r(z, \bar{z}) = 2 \operatorname{Re}(z_3) + (z_2 + \bar{z}_2 + |z_1|^2)^2$, and let M denote the zero set of r. Let p be the origin. Put $L_j = \partial/\partial z_j - r_{z_j} \partial/\partial z_3$ for $j = 1, 2$. In this case L_1 and L_2 form a global basis for sections of $T^{1,0} M$. We put $L = L_1 - \bar{z}_1 L_2$. Then $\lambda(L, \bar{L})(0) = 0$, and the iterated bracket $[[L, \bar{L}], L]$ vanishes identically. Consequently $t(L, p) = \infty$. On the other hand, it is easy to check that the maximum order of contact of a complex-analytic curve (whether singular or not) with M at p is 4; in the notation of the next section, $\Delta_1^{\operatorname{Reg}}(M, p) = \Delta_1(M, p) = 4$.

Singularities create a new difficulty. Suppose that $t(L, p)$ is finite for every local vector field that is non-zero at p. There may nevertheless be a complex *variety* lying in M and passing through p. Thus the notion of type of a vector field does not detect singularities.

EXAMPLE 5.3. Put $r(z, \bar{z}) = 2 \operatorname{Re}(z_3) + |f(z_1, z_2)|^2$ and let M denote its zero set. Here f is a holomorphic polynomial with $f(0, 0) = 0$. The complex subvariety of $\mathbb{C}^3$ defined by the vanishing of z_3 and f lies in M and passes through the origin. Depending on f we can exhibit several phenomena. Rather than giving a complete discussion, we choose several different f to illustrate the possibilities:

Consider the real hypersurfaces in $\mathbb{C}^3$ defined by $r(z, \bar{z}) = 2\,\mathrm{Re}(z_3) + |f(z_1, z_2)|^2$ when f is as follows:

1. $f(z_1, z_2) = z_1^m$
2. $f(z_1, z_2) = z_1^2 - z_2^3$
3. $f(z_1, z_2) = z_1 z_2$

The first hypersurface contains the complex manifold defined by $z_1 = z_3 = 0$. We detect it by commutators because the type of $L = \partial/\partial z_2$ is infinity. The second hypersurface contains an irreducible complex variety V that has a singularity at the origin. (The variety is not normal). All non-zero vector fields have type either 4 or 6 there. Consider the $(1, 0)$ vector field defined by

$$L = 3z_1 L_1 + 2z_2 L_2 = 3z_1 \frac{\partial}{\partial z_1} + 2z_2 \frac{\partial}{\partial z_2} - 6|f(z_1, z_2)|^2 \frac{\partial}{\partial z_3}.$$

A simple calculation shows that L is tangent to V, has infinite type along V except at 0, but vanishes at 0. The third hypersurface contains a reducible complex variety W. Commutators detect this, because each irreducible branch is a complex manifold. These examples motivated the first author to express notions of finite-type directly in terms of orders of contact and the resulting commutative algebra.

6. Orders of Contact and Finite D_q-type

D'Angelo defined several numerical functions measuring the order of contact of possibly singular complex varieties of dimension q with a real hypersurface M. For each q with $1 \le q \le n - 1$, we have the functions $\Delta_q(M, p)$ and $\Delta_q^{\mathrm{Reg}}(M, p)$. The first measures the maximum order of contact of all q-dimensional complex-analytic varieties, and the second measures the maximum order of contact of all q-dimensional complex manifolds. Catlin's necessary and sufficient condition for subellipticity for $(0, q)$ forms on a pseudoconvex domain is equivalent to $\Delta_q(M, p)$ being finite. Understanding these functions defining orders of contact requires some elementary commutative algebra. The idea is first to consider Taylor polynomials of the defining function to reduce to the algebraic case. The methods of [D'Angelo 1993; 1982] show how to express everything using numerical invariants of families of ideals of holomorphic polynomials. In this section we give the definition of these functions and state some of the geometric results known.

Suppose first that J is an ideal in the ring of germs of smooth functions at $p \in \mathbb{C}^n$. We wish to assign a numerical invariant called the order of contact to J that mixes the real and complex categories. Often J will be $I(M, p)$, the germs of smooth functions vanishing on a hypersurface M near p. A local defining function r for M at p then generates the principal ideal $I(M,p)$.

It is convenient for the definition to write $(\mathbb{C}^k, x)$ for the germ of $\mathbb{C}^k$ at the point x, and to write $z : (\mathbb{C}, 0) \to (\mathbb{C}^n, p)$ when z is the germ of a holomorphic mapping with $z(0) = p$. To define the order of contact of J with such a z, we pull back the ideal J to one dimension. We write $\nu(z) = \nu_p(z)$ for the multiplicity of z; this is the minimum of the orders of vanishing of the mappings $t \to z_j(t) - p_j$. We write $\nu(z^*r)$ for the order of vanishing of the function $t \to r(z(t))$ at the origin. The ratio $\Delta(J, z) = \inf_{r \in J} \nu(z^*r)/\nu_p(z)$ is called the *order of contact* of J with the holomorphic curve z. Note that the germ of a curve z is non-singular if $\nu(z) = 1$. The crucial point is that we allow the curves to be singular. For a hypersurface we have the following definition.

DEFINITION 6.1. The order of contact of (the germ at 0 of) a holomorphic curve z with the real hypersurface M at p is the number

$$\Delta(M, p, z) = \inf_{r \in I(M,p)} \frac{\nu(z^*r)}{\nu_p(z)}.$$

We can compute $\Delta(M, p, z)$ by letting r in the definition be a defining function; this gives the infimum.

There are several ways to generalize to singular complex varieties of higher dimension. Below we do this by pulling back to holomorphic curves after we have restricted to subspaces of the appropriate dimension. Thus we let $\phi : \mathbb{C}^{n-q+1} \to \mathbb{C}^n$ be a linear embedding, and we consider the subset $\phi^* M \subset \mathbb{C}^{n-q+1}$. For generic choices of ϕ this will be a hypersurface; when it is not we work with ideals. We are now prepared to define the numbers $\Delta_q(M, p)$ and $\Delta_q^{\text{Reg}}(M, p)$.

DEFINITION 6.2. Let M be a smooth real hypersurface in $\mathbb{C}^n$. For each integer q with $1 \le q \le n$ we define $\Delta_q(M, p)$ and $\Delta_q^{\text{Reg}}(M, p)$ as follows:

$$\Delta_1(M, p) = \sup_z \Delta(M, p, z),$$

where the supremum is taken over non-constant germs of holomorphic curves;

$$\Delta_q(M, p) = \inf_\phi \Delta_1(\phi^* M, p),$$

where the infimum is taken over linear imbeddings $\phi : \mathbb{C}^{n-q+1} \to \mathbb{C}^n$; and

$$\Delta_1^{\text{Reg}}(M, p) = \sup_{z:\nu(z)=1} \Delta(M, p, z),$$

where and the supremum is taken over the non-singular germs of holomorphic curves. The last expression is called the *regular order of contact*.

For $q = 1, \ldots, n-1$ we take the supremum over all germs $z : (\mathbb{C}^q, 0) \to (\mathbb{C}^n, p)$ for which $dz(0)$ is injective:

$$\Delta_q^{\text{Reg}}(M, p) = \sup_z \inf_{r \in I(M,p)} \nu(z^*r)$$

We also put $\Delta_n(M,p) = \Delta_n^{\text{Reg}}(M,p) = 1$ for convenience.

EXAMPLE 6.3. Put $r(z,\bar{z}) = \text{Re}(z_4) + |z_1 z_2 - z_3^5|^2$, and let p be the origin. Note that the image of the map $(s,t) \to (s^5, t^5, st, 0)$ lies in M, but that its derivative is not injective at 0. This shows that $\Delta_2(M,p) = \infty$. On the other hand $\Delta_2^{\text{Reg}}(M,0) = 10$; the map $(s,t) \to (s,0,t,0)$ for example gives the supremum. We have

$$(\Delta_4(M,0), \Delta_3(M,0), \Delta_2(M,0), \Delta_1(M,0)) = (1, 4, \infty, \infty),$$
$$(\Delta_4^{\text{Reg}}(M,0), \Delta_3^{\text{Reg}}(M,0), \Delta_2^{\text{Reg}}(M,0), \Delta_1^{\text{Reg}}(M,0)) = (1, 4, 10, \infty).$$

DEFINITION 6.4. Let M be a smooth real hypersurface in $\mathbb{C}^n$. The point $p \in M$ is of *finite D_q- type* if $\Delta_q(M,p)$ is finite. It is of *finite regular D_q-type* is $\Delta_q^{\text{Reg}}(M,p)$ is finite.

One of the main geometric results is local boundedness for the function $p \to \Delta_q(M,p)$. This shows that finite D_q-type is an open non-degeneracy condition. The condition is also finitely determined. See [D'Angelo 1993] for a complete discussion of these functions.

THEOREM 6.5. *Let M be a smooth real hypersurface in $\mathbb{C}^n$. The function $p \to \Delta_q(M,p)$ is locally bounded; if p is near p_o then*

$$\Delta_q(M,p) \le 2(\Delta_q(M,p_o))^{n-q}$$

Suppose additionally that M is pseudoconvex. For each q with $1 \le q \le n-1$ the function $p \to \Delta_q(M,p)$ satisfies the following sharp bounds: if p is near p_o then

$$\Delta_q(M,p) \le \Delta_q(M,p_o)^{n-q}/2^{n-1-q}.$$

COROLLARY. *For each $q \ge 1$, the set of points of finite D_q-type is an open subset of M.*

The set of points of finite *regular* D_q-type is not generally open when $q < n-1$. See Example 5.3.2.

We remark also on additional information available in the real-analytic case [D'Angelo 1993; 1991; Diederich and Fornaess 1978] and sharper information in the algebraic case (when there is a polynomial defining equation) [D'Angelo 1983].

THEOREM 6.6. *Let M be a real-analytic real hypersurface in $\mathbb{C}^n$. Then either $\Delta_1(M,p)$ is finite or there is a 1-dimensional complex-analytic variety contained in M and passing through p. If M is compact, then the first alternative must hold.*

When the defining equation is a polynomial there is quantitative information depending only on the dimension and the degree of the polynomial [D'Angelo 1983].

THEOREM 6.7. *Let M be a real hypersurface in $\mathbb{C}^n$ defined by a polynomial equation of degree d. Then either $\Delta_1(M, p) \leq 2d(d-1)^{n-1}$ or there is a complex-analytic 1-dimensional variety contained in M and passing through p. Furthermore, there is an explicit way to find the defining equations of the complex variety directly from the defining equation for M.*

Theorems 6.5 and 6.7 rely upon writing real-valued polynomials as differences of squared norms of holomorphic mappings; it is easy to decide when the zero sets of such expressions contain complex analytic varieties. The method enables one to work in the category of holomorphic polynomials and to use elementary commutative algebra.

We mention briefly what this entails. We consider the ring of germs of holomorphic functions at a point and its maximal ideal $\mathcal{M}$. Saying that a proper ideal I of germs of holomorphic functions is primary to the maximal ideal $\mathcal{M}$ is equivalent to saying that its elements vanish simultaneously only at the origin (Nullstellensatz). It is then possible to assign numerical invariants that measure the singularity defined by the primary ideal, such as the order of contact, the smallest power of $\mathcal{M}$ contained in I, the codimension of I, etc. Inequalities among these invariants are crucial to the proofs of Theorems 6.5 and 6.7. Consider again the domains defined by (2); the origin is of finite D_1-type if and only if the ideal $(f_1, \ldots f_N, z_n)$ is primary to $\mathcal{M}$. One sees that the passage from strongly pseudoconvex points to points of finite D_1-type precisely parallels the passage from the maximal ideal $\mathcal{M}$ to ideals primary to it.

7. Catlin's Multitype and Sufficient Conditions for Subelliptic Estimates

Catlin generalized Theorem 4.8 to the smooth case. In [Catlin 1983; 1984; 1987] he established that finite type is a necessary and sufficient condition for subellipticity on pseudoconvex domains. In most of this section we consider the results for $(0, 1)$ forms.

THEOREM 7.1. *Let $\Omega \Subset \mathbb{C}^n$ be a pseudoconvex domain with smooth boundary. Then there is a subelliptic estimate at p if and only if $\Delta^1(b\Omega, p) < \infty$. The parameter epsilon from Definition 3.4 must satisfy $\varepsilon \leq \frac{1}{\Delta^1(b\Omega, p)}$.*

We start by discussing the proof that finite type implies that subelliptic estimates hold. Catlin applies the method of weight functions used earlier by Hörmander [1966]. Rather than working with respect to Lebesgue measure dV, consider the measure $e^{-\Phi} dV$ where Φ will be chosen according to the needs of the problem. After this choice is properly made, one employs, as a substitute for Lemma 3.3, the inequality

$$\int_\Omega \sum_{i,j=1}^n \Phi_{z_i \bar{z}_j} a_i \bar{a}_j \, dV + \sum_{j,k=1}^n \|\bar{L}_j a_k\|^2 \leq CQ(a, a), \qquad (22)$$

where $|\Phi| \leq 1$. Here $\bar{L}_j$ are $(0,1)$ vector fields on $\mathbb{C}^n$. There could be also a term on the left side involving the boundary integral of the Levi form, but such a term does not need to be used in this approach to the estimates. Instead, one needs to choose Φ with a large Hessian. One step in Catlin's proof is the following reduction:

THEOREM 7.2. *Suppose that* $\Omega \in \mathbb{C}^n$ *is a pseudoconvex domain defined by* $\Omega = \{r < 0\}$, *and that* $p \in b\Omega$. *Let* U *be a neighborhood of* p. *Suppose that for all* $\delta > 0$ *there is a smooth real-valued function* Φ_δ *satisfying the properties:*

$$|\Phi_\delta| \leq 1 \qquad on \ U,$$

$$(\Phi_\delta)_{z_i \bar{z}_j} \geq 0 \qquad on \ U,$$

$$\sum_{i,j=1}^n (\Phi_\delta)_{z_i \bar{z}_j} a_i \bar{a}_j \geq c \frac{|a|^2}{\delta^{2\varepsilon}} \qquad on \ U \cap \{-\delta < r \leq 0\}. \tag{23}$$

Then there is a subelliptic estimate of order ε *at* p.

Theorem 7.2 reduces the problem to constructing such bounded smooth plurisubharmonic functions whose Hessians are at least as large as $\delta^{-2\varepsilon}$. One of the crucial ingredients is the use of an n-tuple of rational numbers ($+\infty$ is also allowed) called the multitype. This n-tuple differs from both the n-tuples of orders of contact or of regular orders of contact. There are inequalities in one direction; in simple geometric situations there may be equality. Later we mention the work of Yu in this direction.

We give some motivation for the use of the multitype. Suppose that $W \subset M$ is a manifold of holomorphic dimension zero. Recall that W is a CR submanifold of M, and that the Levi form for M does not annihilate any $(1,0)$ vector fields tangent to W. It follows from the discussion in Section 3 that the distance d_W is a subelliptic multiplier. Suppose that we have a subelliptic estimate away from W. We then obtain a subelliptic estimate (with a smaller epsilon) on W as well, because d_W is a subelliptic multiplier. Hence manifolds of holomorphic dimension zero are small sets as far as the estimates are concerned. This suggests a stratification of M.

Suppose now that p is a point of finite D_q-type, and that U is a neighborhood of p in $b\Omega$ where $\Delta_q(M,p) \leq 2(\Delta_q(M,p_o))^{n-q}$. Catlin defines the multitype as an n-tuple of rational numbers, and shows that it assumes only finitely many values in U. The stratification is then given by the level sets of the multitype function. Catlin proves that each such level set is locally contained in a manifold of holomorphic dimension at most $q - 1$. Establishing the properties of the multitype is difficult, and involves showing that the multitype equals another n-tuple called the commutator type. The commutator type generalizes the notions of Section 3. See [Catlin 1984] for this material.

We next define the multitype. Let $\mu = (\mu_1, \ldots, \mu_n)$ be an n-tuple of numbers (or plus infinity) with $1 \leq \mu_j \leq \infty$ and such that $\mu_1 \leq \mu_2 \leq \ldots \leq \mu_n$. We demand that, whenever μ_k is finite, we can find integers n_j so that

$$\sum_{j=1}^{k} \frac{n_j}{\mu_j} = 1.$$

We call such n-tuples weights, and order them lexicographically. Thus, for example, $(1, 2, \infty)$ is considered smaller than $(1, 4, 6)$. A weight is called distinguished if we can find local coordinates so that p is the origin and such that

$$\sum_{j=1}^{n} \frac{a_j + b_j}{\mu_j} < 1 \ \Rightarrow \ D^a \overline{D}^b r(0) = 0, \tag{24}$$

where $a = (a_1, \ldots, a_n)$ and $b = (b_1, \ldots, b_n)$ are multi-indices. The multitype $m(p)$ is the smallest weight that dominates (in the lexicographical ordering) every distinguished weight μ. In some sense we are assigning weights m_j to the coordinate direction z_j and measuring orders of vanishing. The following statements are automatic from the definition. If the Levi form has rank $q - 1$ at p, then $m_j(p) = 2$ for $2 \leq j \leq q$. In general $m_1(p) = 1$, and $m_2(p) = \Delta_{n-1}(M, p) = \Delta_{n-1}^{\text{Reg}}(M, p)$.

EXAMPLE 7.3. Let M be the hypersurface in $\mathbb{C}^3$ defined by

$$r(z, \bar{z}) = 2 \operatorname{Re}(z_3) + |z_1^2 - z_2^3|^2.$$

The multitype at the origin is $(1, 4, 6)$ and $(\Delta_3(M, 0), \Delta_2(M, 0), \Delta_1(M, 0)) = (1, 4, \infty)$. Thus a finite multitype at p does not guarantee that p is of finite D_1-type. At points of the form $(t^3, t^2, 0)$ for a non-zero complex number t, the multitype will be $(1, 2, \infty)$. This illustrates the upper semicontinuity of the multitype in the lexicographical sense, because $(1, 2, \infty)$ is smaller than $(1, 4, 6)$.

Catlin proved that the multitype on a pseudoconvex hypersurface is upper semicontinuous in this lexicographical sense. He also proved the collection of inequalities given, for $1 \leq q \leq n$, by

$$m_{n+1-q}(p) \leq \Delta_q(M, p). \tag{25}$$

Yu [1994] defined a point to be h-$extendible$ if equality in (25) holds for each q. This class of boundary points exhibits simpler geometry than the general case. Yu proved that convex domains of finite D_1-type are h-extendible, after McNeal [1992] had proved for convex domains with boundary M that $\Delta_1(M, p) = \Delta_1^{\text{Reg}}(M, p)$. Yu then gave a nice application, that h-extendible boundary points must be peak points for the algebra of functions holomorphic on the domain and continuous up to the boundary. McNeal applied his result to the boundary behavior of the Bergman kernel function on convex domains.

Sibony has studied the existence of strongly plurisubharmonic functions with large Hessians as in Theorem 7.2. He introduced the notion that a compact subset $X \subset \mathbb{C}^n$ be *B-regular*. The intuitive idea is that such a subset is B-regular when it contains no analytic structure in a certain strong sense. Sibony has given several equivalent formulations of this notion; one is that the algebra of continuous functions on X is the same as the closure of the algebra of continuous plurisubharmonic functions defined near the set. Another equivalence is the existence, given a real number s, of a plurisubharmonic function, defined near X and bounded by unity, whose Hessian has minimum eigenvalue at least s everywhere on X. Catlin proved for example that a submanifold of holomorphic dimension 0 in a pseudoconvex hypersurface is necessarily B-regular. See [Sibony 1991] for considerable discussion of B-regularity and additional applications.

8. Necessary Conditions and Sharp Subelliptic Estimates

We next discuss necessity results for subelliptic estimates. Greiner [1974] proved that finite commutator-type is necessary for subelliptic estimates in two dimensions. Rothschild and Stein [1976] proved in two dimensions that the largest possible value for ε is the reciprocal of $t(L,p)$, where L is any $(1,0)$ vector field that doesn't vanish at p. In higher dimensions finite commutator-type does not guarantee a subelliptic estimate on $(0,1)$ forms. Furthermore, Example 5.3.2 shows that $t(L,p)$ can be finite for every $(1,0)$ vector field L while subelliptic estimates fail.

Although finite D_1-type is necessary and sufficient, an example of D'Angelo shows that one cannot in general choose epsilon as large as the reciprocal of the order of contact [D'Angelo 1982; 1980]. The result is very simple. The function $p \to \Delta^1(b\Omega, p)$ is not in general upper semicontinuous, so its reciprocal is not lower semicontinuous. Definition 3.4 reveals that, if there is a subelliptic estimate of order epsilon at one point, then there also is one at nearby points. Catlin has shown that the parameter value cannot be determined by information based at one point alone [Catlin 1983]. Nevertheless Theorem 6.5 shows that the condition of finite type does propagate to nearby points. This suggests that one can always choose epsilon as large as $\varepsilon = 2^{n-2}/(\Delta^1(b\Omega, p))^{n-1}$. A more precise conjecture is that we may always choose epsilon as large as $\varepsilon = 1/B(b\Omega, p)$. The denominator is the "multiplicity" of the point, defined in [D'Angelo 1993], where it is proved that the function $p \to B(M, p)$ is upper semicontinuous.

Determining the precise largest value for ε seems to be difficult. See Example 8.1 and Proposition 8.3 below. An example from [D'Angelo 1995] considers domains of the form (2), where for $j = 1, \ldots, n-1$ the functions f_j are arbitrary Weierstrass polynomials of degree m_j in z_j that depend only on $(z_1, \ldots, z_j)$. The multiplicity in this case is $B(b\Omega, p) = 2 \prod m_j$. It is possible, using the method of subelliptic multipliers, to obtain a value of epsilon that works uniformly over all such choices of Weierstrass polynomials and depends only upon these exponents. The result is much smaller than the reciprocal of the multiplicity.

We next illustrate the difficulty in obtaining sharp subelliptic estimates. The existence of the estimate at a point can be decided by examining a finite Taylor polynomial of the defining function there, because finite D_1-type is a finitely determined condition. This Taylor polynomial does not determine the sharp value of epsilon. Suppose that l, m are integers with $m \geq l \geq 2$.

EXAMPLE 8.1. Consider the pseudoconvex domain, defined near the origin, by the function r, where

$$r(z, \bar{z}) = 2\operatorname{Re}(z_3) + |z_1^2 - z_2 z_3^l|^2 + |z_2|^4 + |z_1 z_3^m|^2.$$

We have $\Delta_1(b\Omega, 0) = 4$ and $B(b\Omega, 0) = 8$. Catlin [1983] proved that the largest ε for which there is a subelliptic estimate in a neighborhood of the origin equals $\frac{m+2l}{4(2m+l)}$. This number takes on values between $\frac{1}{4}$ (when $m = l$) and $\frac{1}{8}$. This information supports the conjecture that the value of the largest ε satisfies

$$\frac{1}{B(b\Omega, 0)} \leq \varepsilon \leq \frac{1}{\Delta_1(b\Omega, 0)}.$$

In order to avoid singularities and obtain precise results, Catlin [1983] considers families of complex manifolds. Suppose that T is a collection of positive numbers whose limit is 0. For each $t \in T$ we consider a biholomorphic image $M_t = g_t(B_t)$ of the ball of radius t about 0 in $\mathbb{C}^q$. We suppose that the derivatives dg_t satisfy appropriate uniformity conditions. In particular we need certain q by q minor determinants of dg_t to be uniformly bounded away from 0. We may then pull back r to g_t and define the phrase "The order of contact of the family M_t with $b\Omega$ at the origin is at least η" by decreeing that $\sup |g_t^* r| \leq Ct^\eta$.

Catlin proved the following precise necessity result.

THEOREM 8.2. *Suppose that* $b\Omega$ *is smooth and pseudoconvex and that there is a family* M_t *of q-dimensional complex manifolds whose order of contact with* $b\Omega$ *at a boundary point p is at least η. If there is a subelliptic estimate at p on $(0, q)$ forms for some ε, then $\varepsilon \leq \frac{1}{\eta}$.*

Catlin has also proved the following unpublished result.

PROPOSITION 8.3. *Suppose that ε is a real number with $0 < \varepsilon \leq \frac{1}{4}$. Then there is a smooth pseudoconvex domain in $\mathbb{C}^3$ such that a subelliptic estimate holds with parameter ε but for no larger number. If ε is a rational number in this interval, then there is a pseudoconvex domain in $\mathbb{C}^3$ with a polynomial defining equation such that a subelliptic estimate holds with parameter ε but for no larger number.*

SKETCH OF PROOF. Consider the pseudoconvex domain Ω defined by the following generalization of Example 8.1. We suppose that f, g are holomorphic functions, vanishing at the origin, and satisfying $|g(z)| \leq |f(z)|$. We put

$$r(z, \bar{z}) = 2\operatorname{Re}(z_3) + |z_1^m - z_2 f(z_3)|^2 + |z_2|^{2m} + |g(z_3)z_2|^2.$$

It is easy to see that $\Delta_1(b\Omega, p) = 2m$ and that $B(b\Omega, p) = 2m^2$ no matter what the choices of f, g are. It is possible to explicitly compute the largest possible value of the parameter ε in many cases. By putting $f(z_3) = z_3^p$ and $g(z_3) = z_3^q$ one can show that $\varepsilon = (q + p(m-1))/(2m^2 q)$, exhibiting the entire range of rational numbers between the reciprocals of the the the D_1-type $2m$ and the multiplicity $2m^2$. To see that ε is at most this number one considers the family of complex one-dimensional manifolds M_t defined by the parametric curves $\zeta \to (\zeta, \zeta^m/(it)^p, it)$ for $\zeta \in \mathbb{C}$ satisfying $|\zeta| \leq |t|^\alpha$ for some exponent α. By choosing α appropriately one can compute the contact of this family of complex manifolds, and obtain $\varepsilon \leq (q + p(m-1))/(2m^2 q)$. To show that equality holds one must construct explicitly the functions needed in Theorem 7.2. More complicated choices of f, g enable us to obtain any real number in this range. $\qquad\square$

REMARK. Catlin has made other choices of f, g in the examples from Proposition 8.3 to draw a remarkable conclusion. For any η with $0 < \eta \leq \frac{1}{4}$, there is a smoothly bounded pseudoconvex domain in $\mathbb{C}^3$ such that a subelliptic estimate holds for all ε less than η, but not for η.

9. Domains in Manifolds

Suppose now that X is a complex manifold with Hermitian metric g_{ij}. Let Ω be a pseudoconvex domain in X with compact closure and smooth boundary. We still have the notions of defining function, vector fields and forms of type (1,0) as before. We have $|\partial r|^2 = \sum g_{ij} r_{z_i} r_{\bar{z}_j}$ in a local coordinate system. In a small neighborhood U of a point $p \in b\Omega$ we suppose that $\omega_1, \ldots \omega_n$ form an orthonormal basis for the (1,0) forms. We may suppose that $\omega_n = (\partial r)/|\partial r|$. Let $\{L_i\}$ be a basis of (1,0) vector fields dual to $\{\omega_j\}$ in U. We can write a (0,1)-form ϕ as $\phi = \sum \phi_i \bar{\omega}_i$. When u is a function we have $\bar{\partial} u = \sum \bar{L}_i(u) \bar{\omega}_i$. Applying $\bar{\partial}$ to ϕ we have

$$\bar{\partial}\phi = \sum_{i<j} \bar{L}_i(\phi_j) \bar{\omega}_i \wedge \bar{\omega}_i + \sum \phi_i \bar{\partial}\bar{\omega}_i.$$

Suppose now that ϕ is supported in U, and that $\phi \in \mathcal{D}(\bar{\partial}^*)$. We can write

$$(\bar{\partial}^* \phi, u) = \sum \bar{L}_i{}^* \phi_i, u + \sum \int_{b\Omega} L_i(r) \phi_i \bar{u} \, dS. \tag{26}$$

We have $L_i(r) = 0$ unless $i = n$. From (26) we see that the boundary condition is given by $\phi_n = 0$, and that

$$\bar{\partial}^* \phi = -\sum L_i \phi_i + \sum a_i \phi_i$$

for smooth functions a_i.

Following the proof of Lemma 3.3, and absorbing terms appropriately we obtain the *basic estimate*. Note that we require $\|\phi\|^2$ on the right hand side:

$$\sum_{i,j} \|\bar{L}_i \phi_j\|^2 + \sum_{i,j=1}^n \int_{b\Omega} \lambda_{ij} \phi_i \bar{\phi}_j dS \le C(Q(\phi,\phi) + \|\phi\|^2)$$

Recall our earlier remark that, when $X = \mathbb{C}^n$, we can estimate $\|\phi\|^2 \le CQ(\phi,\phi)$. This implies that the space of harmonic $(0,1)$ forms $\mathcal{H}^{0,1}$ consists of 0 alone. For a general Hermitian manifold X, this will not be true.

The definition of the tangential Sobolev norms in the manifold setting uses partitions of unity. Assuming this definition, suppose that Ω is a domain in a Hermitian complex manifold X. We say that the $\bar{\partial}$-Neumann problem satisfies a subelliptic estimate at $p \in b\Omega$ if, for a sufficiently small neighborhood U of p, there are positive constants C, ε so that

$$\|\|\phi\|\|_\varepsilon^2 \le C(\|\bar{\partial}\phi\|^2 + \|\bar{\partial}^*\phi\|^2) \tag{27}$$

for every $(0,1)$-form ϕ that is smooth, compactly supported in U, and in $\mathcal{D}(\bar{\partial})$. Note that (8) holds for ϕ supported in sufficiently small neighborhoods, so we do not require putting $\|\phi\|^2$ on the right side of (27).

Suppose that ϕ^ν is a bounded sequence in the $\|\|\phi\|\|_\varepsilon$ norm. Then there is a convergent subsequence in L^2. In other words, the inclusion mapping is a compact operator. Hence the harmonic space $\mathcal{H}^{0,1}$ is finite-dimensional. Furthermore harmonic forms are smooth on $\bar{\Omega}$. Finally we have the usual Hodge decomposition. See [Kohn and Nirenberg 1965] for the details.

10. Domains That Are Not Pseudoconvex

Suppose now that Ω is a domain in $\mathbb{C}^n$ with smooth but not pseudoconvex boundary. Let λ denote the Levi form, considered at each boundary point p as a linear transformation from $T_p^{10}(M)$ to itself. We write $\mathrm{Tr}(\lambda)$ for the trace of this linear mapping. (Since the Levi form is defined up to a multiple, the trace is also defined up to a multiple.) We write Id for the identity operator on $T_p^{10}(M)$.

For a point $p \in b\Omega$, we consider two possible positivity conditions.

Condition 1 (Pseudoconvexity). There is a neighborhood of p on which $\lambda \ge 0$.
Condition 2. There is a neighborhood of p on which $\lambda \ge \mathrm{Tr}(\lambda)\,\mathrm{Id}$.

In case 1 holds we have already defined finite ideal-type and seen that finite ideal-type implies that a subelliptic estimate holds. We now define ideals of subelliptic multipliers in case condition 2 holds. (See [Kohn 1985]).

We let J_0 be the real radical of the ideal generated by a defining function r and by the determinant of the mapping $\lambda - \mathrm{Tr}(\lambda)\,\mathrm{Id}$. Given a collection of functions $f = f_1, \ldots, f_N$ we define a linear transformation $B(f)$ on $T^{10}(M)$ and corresponding Hermitian form by

$$\langle B(f)\zeta, \zeta \rangle = \langle (\lambda - \mathrm{Tr}(\lambda)\,\mathrm{Id})\zeta, \zeta \rangle + \sum_{j=1}^{N} (|\bar{\partial}_b f_j \otimes \zeta|^2 - |\langle \bar{\partial}_b f_j, \bar{\zeta} \rangle|^2).$$

In coordinates we have

$$\sum_{m,l} B_{ml}(f)\zeta_m \bar{\zeta}_l = \sum_{i,j} \lambda_{ij}\zeta_i \bar{\zeta}_j - \mathrm{Tr}(\lambda) \sum_i |\zeta_i|^2 + \sum_{i,j,k} |\bar{L}_j(f_k)\zeta_i - \bar{L}_i(f_k)\zeta_j|^2.$$

When condition 2 holds we define the ideals J_k inductively. We let J_k be the real radical of the ideal generated by J_{k-1} and the determinants of all matrices $B(f)$ for $f_j \in J_{k-1}$. When condition 2 holds we say that p is of *finite ideal-type* if there is an integer k for which $1 \in J_k$, that is, the ideal J_k is the full ring of germs of smooth functions at p.

PROPOSITION 10.1. *Suppose that condition 2 holds, and that p is of finite ideal-type. Then there is a subelliptic estimate in the $\bar{\partial}$-Neumann problem on $(0,1)$ forms.*

PROOF. We begin with some lemmas.

LEMMA 10.2. *Suppose that i, j are less than n. Then*

$$\|\bar{L}_i(\phi_j)\|^2 = \|L_i(\phi_j)\|^2 - \int_{b\Omega} \lambda_{ii}|\phi_j|^2 dS + 0(\|\phi_j\| \sum_{k<n} \|L_k \phi_j\| + \|\bar{L}_n \phi_j\|^2) + \|\phi_j\|^2)$$

$$(28)$$

SKETCH OF PROOF. Begin with $\|\bar{L}_i(\phi_j)\|^2 = (\bar{L}_i(\phi_j), \bar{L}_i(\phi_j))$ and integrate by parts twice using Stokes's theorem. At one point write $\bar{L}_i L_i = L_i \bar{L}_i - [L_i, \bar{L}_i]$. Then note that the T component of $[L_i, \bar{L}_i]$ equals λ_{ii}, and integrate the term containing this by parts again to get a boundary integral of $\lambda_{ii}|\phi_j|^2$. The other terms get estimated by the Schwartz inequality. $\square$

LEMMA 10.3. *There is a positive constant C such that, for smooth $\phi \in \mathcal{D}(\bar{\partial}^*)$,*

$$\sum_{i,j<n} \|L_i(\phi_j)\|^2 + \sum_j \|\bar{L}_n \phi_j\|^2 + \sum_{i,j=1}^n \int_{b\Omega} \lambda_{ij}\phi_i \bar{\phi}_j \, dS$$

$$- \int_{b\Omega} \mathrm{Tr}(\lambda)|\phi|^2 \, dS \le C(Q(\phi, \phi) + \|\phi\|^2).$$

PROOF. Take the sum over i, j in (28) and substitute in the basic estimate. Estimate the other terms by the small constant large constant trick. $\square$

To finish the proof of Proposition 10.1, we suppose that f is a subelliptic multiplier, so that $\||f\phi\||_\varepsilon^2 \le CQ(\phi, \phi)$. Next we verify that

$$\sum_{i,j<n} \||\bar{L}_i(f)\phi_j - \bar{L}_j(f)\phi_i\||_{\frac{\varepsilon}{2}}^2 \le CQ(\phi, \phi).$$

This inequality is *dual* to the estimate of Proposition 4.4. Suppose that f is a subelliptic multiplier. Given any Hermitian form $W(\phi, \phi)$ whose determinant is a subelliptic multiplier, we form a new form W' defined by

$$W'(\phi, \phi) = W(\phi, \phi) + \sum_{i,j<n} \|\bar{L}_i(f)\phi_j - \bar{L}_j(f)\phi_i\|^2.$$

As before we see that the determinant of the coefficient matrix of W' is also a subelliptic multiplier.

Proposition 10.1 follows by iterating this operation. $\qquad\square$

Ho [1991] has proved sharp subelliptic estimates on $(0, n-1)$ forms at p for domains that are not pseudoconvex, under the following assumption: there is a $(1,0)$ vector field L for which $t(L, p)$ is finite and for which $\lambda(L, \bar{L}) \geq 0$ near p.

11. A Result for CR Manifolds

We next study subellipticity on a pseudoconvex CR manifold M of hypersurface type and of dimension $2n - 1$. (See [Kohn 1985].) We replace $\bar{\partial}$ by the tangential Cauchy–Riemann operator $\bar{\partial}_b$ and the quadratic form Q by Q_b defined by

$$Q_b(\phi, \phi) = (\bar{\partial}_b\phi, \bar{\partial}_b\phi) + (\bar{\partial}_b^*\phi, \bar{\partial}_b^*\phi).$$

We say that Q_b is subelliptic at p if there is a neighborhood U of p and positive constants C, ε such that

$$\|\phi\|_\varepsilon^2 \leq CQ_b(\phi, \phi)$$

for all smooth forms supported in U.

As before we suppose that the L_i, for $i = 1, \ldots, n-1$, form a local basis for $T^{1,0}M$ and that $L_1, \ldots, L_{n-1}, \bar{L}_1, \ldots, \bar{L}_{n-1}, T$ form a local basis for $\mathbb{C}TM$. We also assume that $\bar{T} = -T$. Definition 2.1 of the Levi form shows that its components λ_{ij} with respect to this local basis are equal to the T coefficient of $[L_i, \bar{L}_j]$. Since M is assumed to be pseudoconvex we may choose the signs so that λ is positive semi-definite. We also define the matrix $\beta = (\beta_{ij})$ by $\beta = \mathrm{Tr}(\lambda)\,\mathrm{Id} - \lambda$. Recall that condition 2 from Section 10 is that β is negative semi-definite. Both λ and β are size $n-1$ by $n-1$. A simple inequality holds when $n > 2$.

LEMMA 11.1. *Suppose that $n > 2$. Then $\det(\beta) \geq \det(\lambda)$.*

PROOF. For completeness we first observe that $\beta = 0$ when $n = 2$, and the inequality fails. When $n = 3$ the two determinants are equal. Otherwise we suppose that we are working at one point, and that λ is diagonal. We may suppose that the eigenvalues of λ satisfy $0 \leq \lambda_1 \leq \ldots \leq \lambda_{n-1}$. We then have

$$\det(\lambda) = \prod \lambda_j \leq \lambda_{n-2}(\lambda_{n-1})^{n-2}.$$

We have $\beta_j = \text{Tr}(\lambda) - \lambda_j = \sum_{k \neq j} \lambda_k$. Since all the λ_j are non-negative, we can drop terms and easily obtain

$$\det(\beta) = \prod \beta_j \geq \lambda_{n-2}(\lambda_{n-1})^{n-2},$$

and the result follows.   □

Proceeding as in Section 10 we obtain the two basic estimates

$$\sum \|\bar{L}_i\phi_j\|^2 + \sum (\lambda_{ij}T\phi_i, \phi_j) \leq Q_b(\phi, \phi) + 0(\|\phi\|^2 + \|\phi\| \sum \|\bar{L}_k\phi_j\|),$$

$$\sum \|L_i\phi_j\|^2 - \sum (\beta_{ij}T\phi_i, \phi_j) \leq Q_b(\phi, \phi) + 0(\|\phi\|^2 + \|\phi\| \sum \|L_k\phi_j\|).$$

PROPOSITION 11.2. *Suppose that $n > 2$. Let $\lambda = (\lambda_{ij})$ be the Levi matrix with respect to the local basis $\{L_1, \ldots, L_{n-1}\}$ of $T^{1,0}(M)$. Then there is a constant C so that, for all smooth ϕ supported in U,*

$$(\det(\lambda)\Lambda^{1/2}\phi, \Lambda^{1/2}\phi) \leq CQ_b(\phi, \phi).$$

PROOF. We need to microlocalize the two basic estimates. We suppose that we are working in a coordinate neighborhood U of a point p, where our coordinates are denoted by $x_1, \ldots, x_{2n-2}, t$. We may assume that these coordinates have been chosen so that, at p, we have $T = (1/\sqrt{-1})(\partial/\partial t)$ and $L_j = \frac{1}{2}(\partial/\partial x_{2j-1} - \sqrt{-1}\partial/\partial x_{2j})$.

Let $\xi_1, \ldots, \xi_{2n-2}, \tau$ denote the dual coordinates in the Fourier transform space. We may also assume that $T = (1/\sqrt{-1})\partial/\partial t$ in the full neighborhood.

Suppose that u is smooth and supported in U. Write

$$u = u^+ + u^- + u^0,$$

where $\hat{u}^+$ is supported in a conical neighborhood of 0 with $\tau > 0$, $\hat{u}^-$ is supported in a conical neighborhood of 0 with $\tau < 0$, and u^0 is supported outside of such neighborhoods.

Since Q_b is elliptic on the support of $\hat{u}^0$, we have the estimate

$$\|\det(\lambda)\phi^0\|_{\frac{1}{2}}^2 \leq C\|\phi^0\|_1^2 \leq CQ_b(\phi, \phi)$$

By Gårding's inequality we have the estimates

$$(\det(\lambda)T\phi^+, \phi^+) \geq -c\|\phi\|^2, \qquad (\det(\beta)T\phi^-, \phi^-) \geq -c\|\phi\|^2.$$

We also have

$$(\det(\lambda)T\phi^+, \phi^+) = (\det(\lambda)\Lambda^{1/2}\phi^+, \Lambda^{1/2}\phi^+) + \cdots,$$

$$(\det(\beta)T\phi^-, \phi^-) = (\det(\beta)\Lambda^{1/2}\phi^-, \Lambda^{1/2}\phi^-) + \cdots,$$

Here the dots denote error terms. Using the basic estimates we obtain

$$\sum_{i,j=1}^{n-1} (\lambda_{ij}\Lambda^{1/2}\phi_i^+, \Lambda^{1/2}\phi_j^+) \leq CQ_b(\phi, \phi)$$

and

$$\sum_{i,j=1}^{n-1} (b_{ij}\Lambda^{1/2}\phi_i^-, \Lambda^{1/2}\phi_j^-) \leq CQ_b(\phi, \phi).$$

Combining the separate estimates for ϕ^0, ϕ^+, and ϕ^- and adding gives Proposition 11.2. $\qquad\square$

As before we augment the Levi form. Suppose that $f_1, \ldots, f_N$ are subelliptic multipliers. We form the matrix

$$A(f) = \begin{pmatrix} \lambda & \bar{\partial}_b f_1 & \ldots & \bar{\partial}_b f_N \\ \partial_b f_1 & 0 & \ldots & 0 \\ \vdots & \vdots & \ddots & \vdots \\ \partial_b f_N & 0 & \ldots & 0 \end{pmatrix}.$$

Similarly we form matrices $B(f)$ as in Section 10. This gives us sequences of ideals I_k and J_k of germs of smooth functions. Note that the inequality from Lemma 11.1 gives $I_0 \subset J_0$. A simple induction then shows that

$$\det(A(f_1, \ldots, f_n)) \leq \det(B(f_1, \ldots, f_n))$$

Therefore if $n > 2$ and if $1 \in I_k$ for some k, we also have $1 \in J_k$. We obtain the following result.

THEOREM 11.3. *Suppose that M is a pseudoconvex CR manifold of dimension $2n - 1$ and of hypersurface type. If $n > 2$, and $1 \in I_k$ for some k, a subelliptic estimate holds.*

REMARK. We mentioned earlier the asymmetry between the L's and the $\bar{L}$'s in Lemma 3.3. For a CR manifold we eliminate this asymmetry by obtaining two basic estimates, one for the $\bar{L}$'s using λ and one for the L's using β.

12. Hölder and L^p estimates for $\bar{\partial}$

Optimal Hölder estimates for $\bar{\partial}$ and estimates for the Bergman projection and kernel function are known only in two dimensions and in some special cases. See for example [Christ 1988; Chang et al. 1992; McNeal 1989; Nagel et al. 1989; McNeal and Stein 1994]. In the elliptic case Hölder estimates are equivalent to elliptic estimates, but Hölder estimates do not necessarily hold for subelliptic operators. See [Guan 1990] for examples of second order subelliptic operators for which Hölder regularity fails completely.

As before we wish to solve the equation $\bar{\partial}u = \alpha$, where u and α are in $L^2(\Omega)$. We set Lip(0) to be the set of bounded functions and, for $0 < s < 1$ we let Lip(s) denote the space of functions u satisfying a Hölder estimate $|u(x) - u(y)| \leq C|x - y|^s$. We extend the definition to all real s inductively by applying the definition to first derivatives.

Let p be a boundary point of commutator-type m. Suppose that $\zeta\alpha \in \mathrm{Lip}(s)$ for all smooth cut-off functions ζ for some s. Let u denote the $\bar{\partial}$-Neumann solution to $\bar{\partial}u = \alpha$, so u is orthogonal to the holomorphic functions. Fefferman and Kohn showed that $\zeta u \in \mathrm{Lip}(s + \frac{1}{m})$ for all smooth cut-off functions ζ. This result requires that $s + \frac{1}{m}$ not be an integer, although they obtain the corresponding result in that case as well, by giving a different definition of the Lipschitz spaces for integer values of s. Write $\mathrm{LIP}(s)$ for this class of spaces; for s not an integer $\mathrm{Lip}(s) = \mathrm{LIP}(s)$. They proved also that both the Bergman and Szegő projection preserve Lipschitz spaces.

For bounded pseudoconvex domains in $\mathbb{C}^n$ the range of $\bar{\partial}_b$ in L^2 is closed; see [Kohn 1986; Shaw 1985]. For general CR manifolds (even in the strongly pseudo-convex 3-dimensional case) it is required to assume this. Given the assumption of closed range, all these results follow from the analysis of a second-order pseudodifferential operator A on $\mathbb{R}^3$. Fefferman and Kohn [1988] prove that solutions u They prove that solutions u to the equation $Au = f$ lie in $\mathrm{LIP}(s + \frac{2}{m})$ when $f \in \mathrm{LIP}(s)$ near a point of commutator-type m. See [Fefferman 1995] for a discussion of the operator A and the techniques of microlocal analysis needed. The techniques also work [Fefferman et al. 1990] in the restricted case in higher dimensions where the Levi form is smoothly diagonalizable.

There are many special cases where estimates for the Bergman and Szegő projections and L^p estimates for $\bar{\partial}$ have been proved by other methods. Fornaess and Sibony [1991] construct a smoothly bounded pseudoconvex domain such that, for certain $\alpha \in L^p$ with $p > 2$, the equation $\bar{\partial}u = \alpha$ has no solution in $L^{p'}$ for all p' in a certain range of values $\leq p$. They also prove a positive result for Runge domains. Chang, Nagel and Stein [Chang et al. 1992] give precise estimates in various function spaces for solutions of $\bar{\partial}u = \alpha$ on domains of finite commutator-type in $\mathbb{C}^2$. See [McNeal and Stein 1994] for estimates (Sobolev, Lipschitz, anisotropic Lipschitz) on convex domains of finite type in arbitrary dimensions. We do not discuss these results here.

13. Brief Discussion of Related Topics

Many different finite-type conditions arise in complex analysis. Here we briefly describe some situations where precise theorems are known in various finite-type settings. The reader should consult the bibliographies in the papers we mention for a complete overview.

Hans Lewy [1956] first studied the extension of CR functions from a strongly pseudoconvex real hypersurface. After work by many authors, usually involving commutator finite type, Trepreau [1986] established that every germ of a CR function at a point p extends to one side of the hypersurface M if and only if there is no germ of a complex hypersurface passing through p and lying in M. Tumanov [1988] introduced the concept of minimality that gives necessary and

sufficient conditions for the holomorphic extendability to wedges of CR functions
defined on generic CR manifolds of higher codimension.

Baouendi, Treves and Jacobowitz [Baouendi et al. 1985] introduced the notion
of *essentially finite* for a point on a real-analytic hypersurface. It is a sufficient
condition in order that the germ of a CR diffeomorphism between real analytic
real hypersurfaces must be a real-analytic mapping. Using elementary commutative algebra, one can extend the definition of essential finiteness to points on
smooth hypersurfaces [D'Angelo 1987] and show that the set of such points is an
open set. Furthermore, if p is of finite D_1-type, then p is essentially finite. The
converse does not hold. It is possible to measure the extent of essential finiteness
by computing the multiplicity (codimension) of an ideal of formal power series.
This number is called the *essential type*. Baouendi and Rothschild developed
the notion of essential type and used it to prove some beautiful results about
extension of mappings between real analytic hypersurfaces; see the bibliography
in [Baouendi and Rothschild 1991]. We mention one of these results. Let M, M'
be real analytic hypersurfaces in $\mathbb{C}^n$ containing 0. Suppose that $f : M \to M'$ is
smooth with $f(0) = 0$, and that f extends to be holomorphic on the intersection
of a neighborhood of 0 with one side of M. If M' is essentially finite at 0, and
f is of *finite multiplicity*, then f extends to be holomorphic on a full neighborhood of 0. In this case the essential type of the point in the domain equals the
multiplicity of the mapping times the essential type of the point in the target.
The notion of finite multiplicity also comes from commutative algebra; again an
appropriate ideal must be of finite codimension.

The essential type also arises when considering infinitesimal CR automorphisms of real analytic hypersurfaces. Stanton [1996] introduced the notion of
holomorphic nondegeneracy for a real hypersurface at a point p. A real hypersurface is called *holomorphically nondegenerate* at p if there is no nontrivial
ambient holomorphic vector field (a vector field of type $(1, 0)$ on $\mathbb{C}^n$ with holomorphic coefficients) tangent to M near p. If this condition holds at one point
on a real analytic hypersurface, it holds at all points. Stanton proves that M is
holomorphically nondegenerate if and only if the set of points of finite essential-
type is both open and dense. The condition at one point is different from any of
the finiteness notions we have discussed so far. Holomorphic nondegeneracy is
important because it provides a necessary and sufficient condition for the finite
dimensionality of the distinguished subspace of the infinitesimal CR automorphisms consisting of real parts of holomorphic vector fields. We mention this here
to emphasize again that different finite-type notions arise in different problems.

A smoothly bounded domain Ω is strongly pseudoconvex if and only if it
is locally biholomorphically equivalent to a strictly (linearly) convex domain.
We say that the boundary is locally convexifiable. A necessary condition for
local convexifiability at a boundary point p is that there is a local holomorphic
support function at p. An example of Kohn and Nirenberg [1973] gives a weakly
pseudoconvex domain with polynomial boundary for which there is no local

holomorphic support function. This means that there is no holomorphic function f on Ω that vanishes at p, but is non-zero for all nearby points of the domain Ω. The existence of a (strict) holomorphic support function at p implies that there is a holomorphic function peaking at p. We have mentioned the result of Yu about peak points in certain finite-type cases. Earlier Bedford and Fornaess [1978] proved that every boundary point of finite type in a pseudoconvex domain in $\mathbb{C}^2$ is a peak point.

One fascinating question we do not consider in this paper is the behavior of the Bergman and Szegő kernels near points of finite type. Only in a few cases are exact formulas for these kernels known, and estimates from above and below are not known in general.

14. Open Problems

1. *Finite ideal-type.* Finite D_1-type is equivalent to subelliptic estimates on $(0,1)$ forms (Theorem 7.1). Finite ideal-type implies subelliptic estimates (Section 4), and the existence of a complex variety in the boundary prevents points along it from being of finite ideal-type. The circle is not complete; does finite D_1-type imply finite ideal-type? This would give a simpler proof of the sufficiency in Theorem 7.1.

2. *Global regularity.* Global regularity for the $\bar{\partial}$-Neumann problem means that the $\bar{\partial}$-Neumann solution u to $\bar{\partial}u = \alpha$ is smooth on the closed domain when α is. Global regularity follows of course from subelliptic estimates, but global regularity holds in some cases when subellipticity does not. Also global regularity fails in general smoothly bounded pseudoconvex domains. (See Christ's article in these proceedings). The necessary and sufficient condition is unknown.

3. *Peak points.* Let Ω be pseudoconvex. Is every point of finite D_1-type a peak point for the algebra of functions holomorphic on Ω and continuous on the closure?

4. *Type conditions for vector fields.* Does $c(L,p)$ equal $t(L,p)$ for each vector field on a pseudoconvex CR manifold of hypersurface type?

5. *Contact of complex manifolds.* Suppose that M is a pseudoconvex real hypersurface, and that $t(L,p) = N$ for some local $(1,0)$ vector field L. Must there be a complex-analytic 1-dimensional manifold tangent to M at p of order m, that is, is it necessarily true that $\Delta_1^{\mathrm{Reg}}(M,p) \geq N$?

6. *Sharp subelliptic estimates.* Suppose that a subelliptic estimate holds at p. Can one express the largest possible ε in terms of the geometry? If this isn't possible, can we always choose ε to be the reciprocal of the multiplicity, as defined in [D'Angelo 1993]?

7. *Hölder estimates.* Extend the results of [Fefferman et al. 1990] to domains of finite type.

8. *Bergman kernel.* Describe precisely the boundary behavior of the Bergman kernel function at a point of finite type.

9. *Hölder continuous CR structures.* Suppose that M is a smooth manifold with a Hölder continuous pseudoconvex CR structure. Discuss the Hölder regularity of solutions to the equation $\bar{\partial}_b u = \alpha$. (Results here would help understand non-linear problems involving $\bar{\partial}$ and $\bar{\partial}_b$.)

References

[Baouendi and Rothschild 1991] M. S. Baouendi and L. P. Rothschild, "Holomorphic mappings of real analytic hypersurfaces", pp. 15–25 in *Several complex variables and complex geometry* (Santa Cruz, CA, 1989), vol. 3, edited by E. Bedford et al., Proc. Sympos. Pure Math. **52**, Amer. Math. Soc., Providence, RI, 1991.

[Baouendi et al. 1985] M. S. Baouendi, H. Jacobowitz, and F. Treves, "On the analyticity of CR mappings", *Ann. of Math.* (2) **122**:2 (1985), 365–400.

[Bedford and Fornaess 1978] E. Bedford and J. E. Fornaess, "A construction of peak functions on weakly pseudoconvex domains", *Ann. of Math.* (2) **107**:3 (1978), 555–568.

[Bloom 1981] T. Bloom, "On the contact between complex manifolds and real hypersurfaces in $\mathbb{C}^3$", *Trans. Amer. Math. Soc.* **263**:2 (1981), 515–529.

[Catlin 1983] D. Catlin, "Necessary conditions for subellipticity of the $\bar{\partial}$-Neumann problem", *Ann. of Math.* (2) **117**:1 (1983), 147–171.

[Catlin 1984] D. Catlin, "Boundary invariants of pseudoconvex domains", *Ann. of Math.* (2) **120**:3 (1984), 529–586.

[Catlin 1987] D. Catlin, "Subelliptic estimates for the $\bar{\partial}$-Neumann problem on pseudoconvex domains", *Ann. of Math.* (2) **126**:1 (1987), 131–191.

[Chang et al. 1992] D.-C. Chang, A. Nagel, and E. M. Stein, "Estimates for the $\bar{\partial}$-Neumann problem in pseudoconvex domains of finite type in $\mathbb{C}^2$", *Acta Math.* **169**:3-4 (1992), 153–228.

[Christ 1988] M. Christ, "Regularity properties of the $\bar{\partial}_b$ equation on weakly pseudoconvex CR manifolds of dimension 3", *J. Amer. Math. Soc.* **1**:3 (1988), 587–646.

[D'Angelo 1980] J. P. D'Angelo, "Subelliptic estimates and failure of semicontinuity for orders of contact", *Duke Math. J.* **47**:4 (1980), 955–957.

[D'Angelo 1982] J. P. D'Angelo, "Real hypersurfaces, orders of contact, and applications", *Ann. of Math.* (2) **115**:3 (1982), 615–637.

[D'Angelo 1983] J. P. D'Angelo, "A Bézout type theorem for points of finite type on real hypersurfaces", *Duke Math. J.* **50**:1 (1983), 197–201.

[D'Angelo 1987] J. P. D'Angelo, "The notion of formal essential finiteness for smooth real hypersurfaces", *Indiana Univ. Math. J.* **36**:4 (1987), 897–903.

[D'Angelo 1991] J. P. D'Angelo, "Finite type and the intersection of real and complex subvarieties", pp. 103–117 in *Several complex variables and complex geometry* (Santa Cruz, CA, 1989), vol. 3, edited by E. Bedford et al., Proc. Sympos. Pure Math. **52**, Amer. Math. Soc., Providence, RI, 1991.

[D'Angelo 1993] J. P. D'Angelo, *Several complex variables and the geometry of real hypersurfaces*, Studies in Advanced Mathematics, CRC Press, Boca Raton, FL, 1993.

[D'Angelo 1995] J. P. D'Angelo, "Finite type conditions and subelliptic estimates", pp. 63–78 in *Modern methods in complex analysis* (Princeton, 1992), edited by T. Bloom et al., Annals of Mathematics Studies **137**, Princeton Univ. Press, Princeton, NJ, 1995.

[Diederich and Fornaess 1978] K. Diederich and J. E. Fornaess, "Pseudoconvex domains with real-analytic boundary", *Ann. of Math.* (2) **107**:2 (1978), 371–384.

[Fefferman 1995] C. Fefferman, "On Kohn's microlocalization of $\bar{\partial}$ problems", pp. 119–133 in *Modern methods in complex analysis* (Princeton, 1992), edited by T. Bloom et al., Annals of Mathematics Studies **137**, Princeton Univ. Press, Princeton, NJ, 1995.

[Fefferman and Kohn 1988] C. L. Fefferman and J. J. Kohn, "Hölder estimates on domains of complex dimension two and on three-dimensional CR manifolds", *Adv. in Math.* **69**:2 (1988), 223–303.

[Fefferman et al. 1990] C. L. Fefferman, J. J. Kohn, and M. Machedon, "Hölder estimates on CR manifolds with a diagonalizable Levi form", *Adv. in Math.* **84**:1 (1990), 1–90.

[Folland and Kohn 1972] G. B. Folland and J. J. Kohn, *The Neumann problem for the Cauchy-Riemann complex*, Annals of Mathematics Studies **75**, Princeton Univ. Press, Princeton, NJ, 1972.

[Fornæss and Sibony 1991] J. E. Fornæss and N. Sibony, "On L^p Estimates for $\bar{\partial}$", pp. 129–163 in *Several complex variables and complex geometry* (Santa Cruz, CA, 1989), vol. 3, edited by E. Bedford et al., Proc. Sympos. Pure Math. **52**, Amer. Math. Soc., Providence, RI, 1991.

[Greiner 1974] P. Greiner, "Subelliptic estimates for the $\bar{\delta}$-Neumann problem in $\mathbb{C}^2$", *J. Differential Geometry* **9** (1974), 239–250.

[Guan 1990] P. Guan, "Hölder regularity of subelliptic pseudodifferential operators", *Duke Math. J.* **60**:3 (1990), 563–598.

[Ho 1991] L.-H. Ho, "Subelliptic estimates for the $\bar{\partial}$-Neumann problem for $n-1$ forms", *Trans. Amer. Math. Soc.* **325**:1 (1991), 171–185.

[Hörmander 1966] L. Hörmander, *An introduction to complex analysis in several variables*, Van Nostrand, Princeton, NJ, 1966.

[Kohn 1963] J. J. Kohn, "Harmonic integrals on strongly pseudo-convex manifolds, I", *Ann. of Math.* (2) **78** (1963), 112–148.

[Kohn 1964] J. J. Kohn, "Harmonic integrals on strongly pseudo-convex manifolds, II", *Ann. of Math.* (2) **79** (1964), 450–472.

[Kohn 1972] J. J. Kohn, "Boundary behavior of δ on weakly pseudo-convex manifolds of dimension two", *J. Differential Geometry* **6** (1972), 523–542.

[Kohn 1977] J. J. Kohn, "Methods of partial differential equations in complex analysis", pp. 215–237 in *Several complex variables* (Williamstown, MA, 1975), vol. 1, edited by R. O. Wells, Jr., Proc. Sympos. Pure Math. **30**, Amer. Math. Soc., 1977.

[Kohn 1979] J. J. Kohn, "Subellipticity of the $\bar{\partial}$-Neumann problem on pseudo-convex domains: sufficient conditions", *Acta Math.* **142**:1-2 (1979), 79–122.

[Kohn 1984]   J. J. Kohn, "A survey of the $\bar{\partial}$-Neumann problem", pp. 137–145 in *Complex analysis of several variables* (Madison, WI, 1982), edited by Y.-T. Siu, Proc. Symp. Pure Math. **41**, Amer. Math. Soc., Providence, RI, 1984.

[Kohn 1985]   J. J. Kohn, "Estimates for $\bar{\partial}_b$ on pseudoconvex CR manifolds", pp. 207–217 in *Pseudodifferential operators and applications* (Notre Dame, IN, 1984), edited by F. Treves, Proc. Symp. Pure Math. **43**, Amer. Math. Soc., Providence, 1985.

[Kohn 1986]   J. J. Kohn, "The range of the tangential Cauchy–Riemann operator", *Duke Math. J.* **53**:2 (1986), 525–545.

[Kohn and Nirenberg 1965]   J. J. Kohn and L. Nirenberg, "Non-coercive boundary value problems", *Comm. Pure Appl. Math.* **18** (1965), 443–492.

[Kohn and Nirenberg 1973]   J. J. Kohn and L. Nirenberg, "A pseudo-convex domain not admitting a holomorphic support function", *Math. Ann.* **201** (1973), 265–268.

[Lewy 1956]   H. Lewy, "On the local character of the solutions of an atypical linear differential equation in three variables and a related theorem for regular functions of two complex variables", *Ann. of Math.* (2) **64** (1956), 514–522.

[McNeal 1989]   J. D. McNeal, "Boundary behavior of the Bergman kernel function in C^{2}", *Duke Math. J.* **58**:2 (1989), 499–512.

[McNeal 1992]   J. D. McNeal, "Convex domains of finite type", *J. Funct. Anal.* **108**:2 (1992), 361–373.

[McNeal and Stein 1994]   J. D. McNeal and E. M. Stein, "Mapping properties of the Bergman projection on convex domains of finite type", *Duke Math. J.* **73**:1 (1994), 177–199.

[Nagel et al. 1989]   A. Nagel, J.-P. Rosay, E. M. Stein, and S. Wainger, "Estimates for the Bergman and Szegő kernels in $\mathbb{C}^{2}$", *Ann. of Math.* (2) **129**:1 (1989), 113–149.

[Rothschild and Stein 1976]   L. P. Rothschild and E. M. Stein, "Hypoelliptic differential operators and nilpotent groups", *Acta Math.* **137**:3-4 (1976), 247–320.

[Shaw 1985]   M.-C. Shaw, "L^{2}-estimates and existence theorems for the tangential Cauchy–Riemann complex", *Invent. Math.* **82**:1 (1985), 133–150.

[Sibony 1991]   N. Sibony, "Some aspects of weakly pseudoconvex domains", pp. 199–231 in *Several complex variables and complex geometry* (Santa Cruz, CA, 1989), vol. 1, edited by E. Bedford et al., Proc. Sympos. Pure Math. **52**, Amer. Math. Soc., Providence, RI, 1991.

[Stanton 1996]   N. K. Stanton, "Infinitesimal CR automorphisms of real hypersurfaces", *Amer. J. Math.* **118**:1 (1996), 209–233.

[Sweeney 1972]   W. J. Sweeney, "Coerciveness in the Neumann problem", *J. Differential Geometry* **6** (1972), 375–393.

[Trépreau 1986]   J.-M. Trépreau, "Sur le prolongement holomorphe des fonctions C–R défines sur une hypersurface réelle de classe C^{2} dans $\mathbb{C}^{n}$", *Invent. Math.* **83**:3 (1986), 583–592.

[Tumanov 1988]   A. E. Tumanov, "Extending CR functions on manifolds of finite type to a wedge", *Mat. Sb.* (*N.S.*) **136(178)**:1 (1988), 128–139. In Russian; translated in *Math. USSR Sb.* **64**:1 (1989), 129–140.

[Yu 1994]   J. Y. Yu, "Peak functions on weakly pseudoconvex domains", *Indiana Univ. Math. J.* **43**:4 (1994), 1271–1295.

JOHN P. D'ANGELO
DEPTARTMENT OF MATHEMATICS
UNIVERSITY OF ILLINOIS
URBANA, IL 61801
UNITED STATES
  jpda@math.uiuc.edu

JOSEPH J. KOHN
DEPTARTMENT OF MATHEMATICS
PRINCETON UNIVERSITY
PRINCETON, NJ 08540
UNITED STATES
  kohn@math.princeton.edu

Several Complex Variables
MSRI Publications
Volume **37**, 1999

Pseudoconvex-Concave Duality and Regularization of Currents

JEAN-PIERRE DEMAILLY

To the memory of Michael Schneider

ABSTRACT. We investigate some basic properties of Finsler metrics on holomorphic vector bundles, in the perspective of obtaining geometric versions of the Serre duality theorem. We establish a duality framework under which pseudoconvexity and pseudoconcavity properties get exchanged — up to some technical restrictions. These duality properties are shown to be related to several geometric problems, such as the conjecture of Hartshorne and Schneider, asserting that the complement of a q-codimensional algebraic subvariety with ample normal bundle is q-convex. In full generality, a functorial construction of Finsler metrics on symmetric powers of a Finslerian vector bundle is obtained. The construction preserves positivity of curvature, as expected from the fact that tensor products of ample vector bundles are ample. From this, a new shorter and more geometric proof of a basic regularization theorem for closed $(1, 1)$ currents is derived. The technique is based on the construction of a mollifier operator for plurisubharmonic functions, depending on the choice of a Finsler metric on the cotangent bundle and its symmetric powers.

CONTENTS

1991 *Mathematics Subject Classification.* 14B05, 14J45, 32C17, 32J25, 32S05.

Key words and phrases. Serre duality, pseudoconvexity, pseudoconcavity, Finsler metric, symmetric power, Chern curvature, Hartshorne–Schneider conjecture, plurisubharmonic function, positive current, regularization of currents, Legendre transform, Lelong number, Ohsawa–Takegoshi theorem, Skoda L^2 estimates.

Introduction

The goal of the present paper is to investigate some duality properties connecting pseudoconvexity and pseudoconcavity. Our ultimate perspective would be a geometric duality theory parallel to Serre duality, in the sense that Serre duality would be the underlying cohomological theory. Although similar ideas have already been used by several authors in various contexts — for example, for the study of direct images of sheaves [Ramis et al. 1971], or in connection with the study of Fantappie transforms and lineal convexity [Kiselman 1997], or in the study of Monge–Ampère equations [Lempert 1985] — we feel that the "convex-concave" duality theory still suffers from a severe lack of understanding.

Our main concern is about *Finsler metrics* on holomorphic vector bundles. As is well known, a holomorphic vector bundle E on a compact complex manifold is ample in the sense of [Hartshorne 1966] if and only if its dual $E^\star$ admits a strictly pseudoconvex tubular neighborhood of 0, that is, if and only if $E^\star$ has a strictly plurisubharmonic smooth Finsler metric. In that case, we expect E itself to have a tubular neighborhood of the zero section such that the Levi form of the boundary has everywhere signature $(r-1, n)$, where r is the rank of E and $n = \dim X$; in other words, E has a Finsler metric whose Levi form has signature (r, n). This is indeed the case if E is positive in the sense of Griffiths, that is, if the above plurisuharmonic Finsler metric on $E^\star$ can be chosen to be *hermitian*; more generally, Sommese [1978; 1979; 1982] has observed that everything works well if the Finsler metric is fiberwise convex (in the ordinary sense). The Kodaira–Serre vanishing theorem tells us that strict pseudoconvexity of $E^\star$ implies that the cohomology of high symmetric powers $S^m E$ is concentrated in degree 0, while the Andreotti–Grauert vanishing theorem tells us that (r, n) convexity-concavity of E implies that the cohomology of $S^m E^\star$ is concentrated in degree n. Of course, both properties are connected from a cohomological view point by the Serre duality theorem, but the related geometric picture seems to be far more involved. A still deeper unsolved question is Griffiths' conjecture [1969] on the equivalence of ampleness and positivity of curvature for hermitian metrics.

One of the difficulties is that "linear" duality between E and $E^\star$ is not sufficient to produce the expected biduality properties relating convexity on one side and concavity on the other side. What seems to be needed rather, is a duality between large symmetric powers $S^m E$ and $S^m E^\star$, asymptotically as m goes to infinity ("polynomial duality"). Although we have not been able to find a completely satisfactory framework for such a theory, one of our results is that there is a functorial and natural construction which assigns Finsler metrics on all symmetric powers $S^m E$, whenever a Finsler metric on E is given. The assignment has the desired property that the Finsler metrics on $S^m E$ are plurisubharmonic if the Finsler metric on E was. The construction uses "polynomial duality" in an essential way, although it does not produce good metrics on the dual bundles $S^m E^\star$.

Several interesting questions depend on the solution to these problems. Robin Hartshorne [1970] raised the question whether the complement of an algebraic subvariety Y with ample normal bundle N_Y in a projective algebraic variety X is q-convex in the sense of Andreotti–Grauert, with $q = \operatorname{codim} Y$. Michael Schneider [1973] proved the result in the case the normal bundle is positive is the sense of Griffiths, thus yielding strong support for Hartshorne's conjecture. As a consequence of Sommese's observation, Schneider's result extends the case if $N_Y^\star$ has a strictly pseudoconvex and fiberwise convex neighborhood of the zero section, which is the case for instance if N_Y is ample and globally generated.

Other related questions which we treat in detail are the approximation of closed positive $(1,1)$-currents and the attenuation of their singularities. In general, a closed positive current T cannot be approximated (even in the weak topology) by smooth closed positive currents. A cohomological obstruction lies in the fact that T may have negative intersection numbergs $\{T\}^p \cdot Y$ with some subvarieties $Y \subset X$. This is the case for instance if $T = [E]$ is the current of integration on a the exceptional curve of a blown-up surface and $Y = E$. However, as we showed in [Demailly 1982; 1992; 1994], the approximation is possible if we allow the regularization T_ε to have a small negative part. The main point is to control this negative part accurately, in term of the global geometry of the ambient geometry X. It turns out that more or less optimal bounds can be described in terms of the convexity of a Finsler metric on the tangent bundle T_X. Again, a relatively easy proof can be obtained for the case of a hermitian metric [Demailly 1982; 1994], but the general Finsler case, as solved in [Demailly 1992], still required very tricky analytic techniques. We give here an easier and more natural method based on the use of "symmetric products" of Finsler metrics.

Many of the ideas presented here have matured over a long period of time, for a large part through discussion and joint research with Thomas Peternell and Michael Schneider. Especially, the earlier results [Demailly 1992] concerning smoothing of currents were strongly motivated by techniques needed in our joint work [Demailly et al. 1994]. I would like here to express my deep memory of Michael Schneider, and my gratitude for his very beneficial mathematical influence.

1. Pseudoconvex Finsler Metrics and Ample Vector Bundles

Let X be a complex manifold and E a holomorphic vector bundle over X. We set $n = \dim_{\mathbb{C}} X$ and $r = \operatorname{rank} E$.

DEFINITION 1.1 [Kobayashi 1975]. A (positive definite) Finsler metric on E is a positive complex homogeneous function $\xi \mapsto \|\xi\|_x$ defined on each fiber E_x, that is, such that $\|\lambda \xi\|_x = |\lambda| \|\xi\|_x$ for each $\lambda \in \mathbb{C}$ and $\xi \in E_x$, and $\|\xi\|_x > 0$ for $\xi \neq 0$.

We will in general assume some regularity, e.g. continuity of the function $(x, \xi) \mapsto \|\xi\|_x$ on the total space E of the bundle. We say that the metric is smooth if it is smooth on $E \smallsetminus \{0\}$. The logarithmic indicatrix of the Finsler metric is by definition the function

$$(1.2) \qquad\qquad \chi(x, \xi) = \log \|\xi\|_x.$$

We will say in addition that the Finsler metric is *convex* if the function $\xi \mapsto \|\xi\|_x$ is convex on each fiber E_x (viewed as a real vector space). A Finsler metric is convex if and only if it derives from a norm (hermitian norms are of course of a special interest in this respect); however, we will have to deal as well with non convex Finsler metrics.

The interest in Finsler metrics essentially arises from the following well-known characterization of ample vector bundles [Kodaira 1954; Grauert 1958; Kobayashi 1975].

THEOREM 1.3. *Let E be a vector bundle on a compact complex manifold X. The following properties are equivalent.*

(1) E *is ample in the sense of* [Hartshorne 1966].

(2) $\mathbb{O}_{P(E)}(1)$ *is an ample line bundle on the projectivized bundle $P(E)$ (of hyperplanes of E).*

(3) $\mathbb{O}_{P(E)}(1)$ *carries a smooth hermitian metric of positive Chern curvature form.*

(4) $E^\star$ *carries a smooth Finsler metric which is strictly plurisubharmonic on the total space $E^\star \smallsetminus \{0\}$.*

(5) $E^\star$ *admits a smoothly bounded strictly pseudoconvex tubular neighborhood U of the zero section.*

Actually, the equivalence of (1), (2) is a purely algebraic fact, while the equivalence of (2) and (3) is a consequence of the Kodaira embedding theorem. The equivalence of (3) and (4) just comes from the observation that a Finsler metric on $E^\star$ can be viewed also as a *hermitian metric $h^\star$* on the line bundle $\mathbb{O}_{P(E)}(-1)$ (as the total space of $\mathbb{O}_{P(E)}(-1)$ coincides with the blow-up of $E^\star$ along the zero section), and from the obvious identity

$$(\pi_{P(E)})^\star \Theta_{h^\star}(\mathbb{O}_{P(E)}(-1)) = -\frac{i}{2\pi} \partial \bar{\partial} \chi^\star,$$

where $\Theta_{h^\star}(\mathbb{O}_{P(E)}(-1))$ denotes the Chern curvature form of $h^\star = e^{\chi^\star}$, and $\pi_{P(E)} : E^\star \smallsetminus \{0\} \to P(E)$ the canonical projection. Finally, if we have a Finsler metric as in (4), then $U_\varepsilon = \{\xi^\star : \|\xi^\star\|^\star < \varepsilon\}$ is a fundamental system of strictly pseudoconvex neighborhood of the zero section of $E^\star$. Conversely, given such a neighborhood U, we can make it complex homogeneous by replacing U with $U^\star = \bigcap_{|\lambda| \geq 1} \lambda U$. Then $U^\star$ is the unit ball bundle of a continuous strictly plurisubharmonic Finsler metric on $E^\star$ (which can further be made smooth thanks to Richberg's regularization theorem [1968], or by the much more precise results of [Demailly 1992], which will be reproved in a simpler way in Section 9).

REMARK 1.4. It is unknown whether the ampleness of E implies the existence of a *convex* strictly plurisubharmonic Finsler metric on $E^\star$. Sommese [1978] observed that this is the case if E is ample and generated by sections. In fact, if there are sections $\sigma_j \in H^0(X, E)$ generating E, then

$$h_0(\xi^\star) = \left(\sum_j |\sigma_j(x) \cdot \xi^\star|^2 \right)^{1/2}$$

defines a weakly plurisubharmonic and strictly convex (actually hermitian) metric on $E^\star$. On the other hand, the ampleness implies the existence of a strictly plurisubharmonic Finsler metric h_1, thus $(1 - \varepsilon)h_0 + \varepsilon h_1$ is *strictly plurisubharmonic* and *strictly convex* for ε small enough. Griffiths conjectured that ampleness of E might even be equivalent to the existence of a *hermitian metric* with positive curvature, thus to the existence of a hermitian strictly plurisubharmonic metric on $E^\star$. Not much is known about this conjecture, except that it holds true if $r = \mathrm{rank}\,E = 1$ (Kodaira) and $n = \dim X = 1$ ([Umemura 1980]; see also [Campana and Flenner 1990]). Our feeling is that the general case should depend on deep facts of gauge theory (some sort of vector bundle version of the Calabi–Yau theorem would be needed).

2. Linearly Dual Finsler Metrics

Given a Finsler metric $\| \ \|$ on a holomorphic vector bundle E, one gets a dual (or rather *linearly dual*) Finsler metric $\| \ \|^\star$ on $E^\star$ by putting

$$(2.1) \qquad \|\xi^\star\|_x^\star = \sup_{\xi \in E_x \smallsetminus \{0\}} \frac{|\xi \cdot \xi^\star|}{\|\xi\|_x}, \qquad \xi^\star \in E_x^\star.$$

Equivalently, in terms of the logarithmic indicatrix, we have

$$(2.2) \qquad \chi^\star(x, \xi^\star) = \sup_{\xi \in E_x \smallsetminus \{0\}} \log|\xi \cdot \xi^\star| - \chi(x, \xi), \qquad \xi^\star \in E_x^\star.$$

It is clear that the linearly dual metric $\| \ \|^\star$ is always convex, and therefore the biduality formula $\| \ \|^{\star\star} = \| \ \|$ holds true if and only if $\| \ \|$ is convex.

A basic observation made in [Sommese 1978] is that the pseudoconvexity of a Finsler metric is related to some sort of pseudoconcavity of the dual metric, provided that the given metric is fiberwise convex. We will reprove it briefly in order to prepare the reader to the general case (which requires *polynomial* duality, and not only *linear* duality). We first need a definition.

DEFINITION 2.3. Let E be equipped with a smooth Finsler metric of logarithmic indicatrix $\chi(x, \xi) = \log \|\xi\|_x$. We say that $\| \ \|$ has transversal Levi signature (r, n) (where $r = \mathrm{rank}\,E$ and $n = \dim X$) if, at every point $(x, \xi) \in E \smallsetminus \{0\}$, the

Levi form $i\partial\bar{\partial}(e^{\chi})$ is positive definite along the fiber E_x and negative definite on some n-dimensional subspace $W \subset T_{E,(x,\xi)}$ which is transversal to the fiber E_x.

This property can also be described geometrically as follows.

PROPOSITION 2.4. *The Finsler metric $\| \ \|$ on E has transversal Levi signature (r, n) if and only if it is fiberwise strictly pseudoconvex, and through every point (x_0, ξ_0) of the unit sphere bundle $\|\xi\| = 1$ passes a germ of complex n-dimensional submanifold M_0 which is entirely contained in the unit ball bundle $\{\|\xi\| \leq 1\}$ and has (strict) contact order 2 at (x_0, ξ_0).*

SKETCH OF PROOF. If the geometric property (2.4) is satisfied, we simply take $W = T_{M_0,(x_0,\xi_0)}$. Conversely, if $i\partial\bar{\partial}e^{\chi}$ has signature (r, n) as in 2.3, then $i\partial\bar{\partial}\chi$ has signature $(r-1, n)$ (with one zero eigenvalue in the radial direction, since χ is log homogeneous). The Levi form of the hypersurface $\chi = 0$ thus has signature $(r-1, n)$ as well, and we can take the negative eigenspace $W \subset T_{E,(x_0,\xi_0)}$ to be tangent to that hypersurface. The germ M_0 is then taken to be the graph of a germ of holomorphic section $\sigma : (X, x_0) \to E$ tangent to W, with the second order jet of σ adjusted in such a way that $\chi(x, \sigma(x)) \leq -\varepsilon|x - x_0|^2$ (as $\partial\chi(x_0) \neq 0$, one can eliminate the holomorphic and antiholomorphic parts in the second order jet of $\chi(x, \sigma(x))$). $\qquad\square$

The main part, (a), of the following basic result is due to A. Sommese [1978].

THEOREM 2.5. *Let E be equipped with a smooth Finsler metric of logarithmic indicatrix $\chi(x, \xi) = \log \|\xi\|_x$. Assume that the metric is (fiberwise) strictly convex.*

(a) *If the metric $\| \ \|$ is strictly plurisubharmonic on $E \setminus \{0\}$, then the dual metric $\| \ \|^* = e^{\chi^*}$ has transversal Levi signature (r, n) on $E^* \setminus \{0\}$.*

(b) *In the opposite direction, if $\| \ \|$ has transversal Levi signature (r, n), then $\| \ \|^*$ is strictly plurisubharmonic on $E^* \setminus \{0\}$.*

REMARK 2.6. Theorem 2.5 still holds under the following more general, but more technical hypothesis, in place of the strict convexity hypothesis:

(H) *For every point $(x, [\xi^*]) \in P(E_x)$, the supremum*

$$\chi^*(x, \xi^*) = \sup_{\xi \in E_x \setminus \{0\}} \log |\xi \cdot \xi^*| - \chi(x, \xi), \qquad \xi^* \in E_x^*$$

is reached on a unique line $[\xi] = f(x, [\xi^*]) \in P(E_x)$, where $[\xi]$ is a non critical maximum point along $P(E_x)$.

Notice that the supremum is always reached in at least one element $[\xi] \in P(E_x)$, just by compactness. The assumption that there is a unique such point $[\xi] = f(x, [\xi^*])$ which is non critical ensures that f is smooth by the implicit function theorem, hence χ^* will be also smooth.

The uniqueness assumption is indeed satisfied if the Finsler metric of E is *strictly convex*. Indeed, if the maximum is reached for two non colinear vectors

ξ_0, ξ_1 and if we adjust ξ_0 and ξ_1 by taking multiples such that $\xi_0 \cdot \xi^\star = \xi_1 \cdot \xi^\star = 1$, then again $\xi_t \cdot \xi^\star = 1$ for all $\xi_t = (1 - t)\xi_0 + t\xi_1 \in\ ]\xi_0, \xi_1[$, while the strict convexity implies $\chi(x, \xi_t) < \chi(x, \xi_0) = \chi(x, \xi_1)$, contradiction. We see as well that the maximum must be a non critical point, and that the Finsler metric $\| \ \|^\star$ is strictly convex. Thus, in this case, $\| \ \|$ is strictly plurisubharmonic if and only if $\| \ \|^\star$ has transversal Levi signature (r, n).

REMARK 2.7. In Theorem 2.5, the extra convexity assumption — or its weaker counterpart (H) — is certainly needed. In fact, if the conclusions were true without any further assumption, the linear bidual of a continuous plurisubharmonic Finsler metric would still be plurisubharmonic (since we can approximate locally such metrics by smooth strictly plurisubharmonic ones). This would imply that the convex hull of a pseudoconvex circled tubular neighborhood is pseudoconvex. However, if we equip the trivial rank two vector bundle $\mathbb{C} \times \mathbb{C}^2$ over $\mathbb{C}$ with the plurisubharmonic Finsler metric

$$\|\xi\|_x = \max \left(|\xi_1|,\ |\xi_2|,\ |x|\sqrt{|\xi_1|\, |\xi_2|} \right),$$

a trivial computation shows that the convex hull is associated with the metric

$$\|\xi\|'_x = \max \left(|\xi_1|,\ |\xi_2|,\ \frac{|x|^2}{1 + |x|^2}(|\xi_1| + |\xi_2|) \right)$$

which is not plurisubharmonic in x.

PROOF OF THEOREM 2.5. (a) First observe that $\exp(\chi^\star) = \| \ \|^\star$ is convex, and even strictly convex since the assumptions are not affected by small smooth C^∞ or C^2 perturbations on χ. Thus $i\partial\bar{\partial}\exp(\chi^\star)$ has at least r positive eigenvalues eigenvalues along the vertical directions of $E^\star \to X$.

Let $f : P(E^\star) \to P(E)$ be defined as in condition 2.5 (H), and let $\tilde{f} : E \setminus \{0\}^\star \to E \setminus \{0\}$ be a lifting of f. One can get such a global lifting $\tilde{f}$ by setting e.g. $\tilde{f}(x, \xi^\star) \cdot \xi^\star = 1$, so that $\tilde{f}$ is uniquely defined. By definition of $\chi^\star$ and f, we have

$$\chi^\star(x, \xi^\star) = \log\left|\tilde{f}(x, \xi^\star) \cdot \xi^\star\right| - \chi(x, \tilde{f}(x, \xi^\star))$$

in a neighborhood of $(x_0, \xi_0^\star)$. Fix a local trivialization $E_{|U} \simeq U \times V$ xhere $V \simeq \mathbb{C}^r$ and view $\tilde{f}$ as a map $\tilde{f} : E_{|U}^\star \simeq U \times V^\star \to V$ defined in a neighborhood of $(x_0, \xi_0^\star)$. As $\dim E = n + r$ and $\dim V = r$, the kernel of the $\bar{\partial}$-differential

$$\bar{\partial}\tilde{f}_{(x_0, \xi_0^\star)} : T_{E^\star, (x_0, \xi_0^\star)} \to V$$

is a complex subspace $W_0 \subset T_{E^\star, (x_0, \xi_0^\star)}$ of dimension $p \geq n$. By definition of W_0, there is a germ of p-dimensional submanifold $M \subset E^\star$ with $T_{M, (x_0, \xi_0^\star)} = W_0$, and a germ of holomorphic function $g : M \to V$ such that

$$\tilde{f}(x, \xi^\star) = g(x, \xi^\star) + O(|x - x_0|^2 + |\xi^\star - \xi_0^\star|^2) \quad \text{on } M.$$

This implies

$$\chi^\star(x, \xi^\star) = \log |g(x, \xi^\star) \cdot \xi^\star| - \chi(x, g(x, \xi^\star)) + O(|x - x_0|^3 + |\xi^\star - \xi_0^\star|^3) \quad \text{on } M.$$

In fact, since $\xi_0 = \widetilde{f}(x_0, \xi_0^\star)$ is a stationary point for $\xi \mapsto \log|\xi \cdot \xi^\star| - \chi(x, \xi)$, the partial derivative in ξ is $O(|x - x_0| + |\xi - \xi_0|)$, and a substitution of $\xi = \widetilde{f}(x, \xi^\star)$ by $\xi_1 = g(x, \xi^\star)$ introduces an error

$$O(|x - x_0| + |\xi - \xi_0| + |\xi_1 - \xi_0|) \, |\xi - \xi_1| = O(|x - x_0|^3 + |\xi^\star - \xi_0^\star|^3)$$

at most. Therefore

$$i\partial\bar\partial\chi^\star(x, \xi^\star) = -i\partial\bar\partial\chi(x, g(x, \xi^\star)) < 0 \quad \text{in restriction to } W_0 = T_{M,(x_0, \xi_0^\star)}.$$

This shows that $i\partial\bar\partial\chi^\star$ has at least $p \geq n$ negative eigenvalues. As there are already r negative eigenvalues, the only possibility is that $p = n$.

(b) The assumption on (E, χ) means that for every $(x_0, \xi_0) \in E \smallsetminus \{0\}$, there is a germ of holomorphic section $\sigma : X \to E$ such that $-\chi(x, \sigma(x))$ is strictly plurisubharmonic and $\sigma(x_0) = \xi_0$. Fix $\xi_0^\star \in E_{x_0}^\star \smallsetminus \{0\}$ and take $\xi_0 \in E_{x_0} \smallsetminus \{0\}$ to be the unique point where the maximum defining $\chi^\star$ is reached. Then we infer that $\chi^\star(x, \xi^\star) \geq \log|\xi^\star \cdot \sigma(x)| - \chi(x, \sigma(x))$, with equality at $(x_0, \xi_0^\star)$. An obvious application of the mean value inequality then shows that $\chi^\star$ is plurisubharmonic and that $i\partial\bar\partial\chi^\star$ is strictly positive in all directions of $T_{E^\star}$, except the radial vertical direction.   □

3. A Characterization of Signature (r, n) Concavity

Let E be a holomorphic vector bundle equipped with a smooth Finsler metric which satisfies the concavity properties exhibited by Theorem 2.5. We then have the following results about supremum of plurisubharmonic functions.

THEOREM 3.1. *Assume that the Finsler metric $\|\ \|_E$ on E has transversal Levi signature (r, n). Then, for every plurisubharmonic function $(x, \xi) \mapsto u(x, \xi)$ on the total space E, the function*

$$M_u(x, t) = \sup_{\|\xi\|_E \leq |e^t|} u(x, \xi)$$

is plurisubharmonic on $X \times \mathbb{C}$.

PROOF. First consider the restriction $x \mapsto M_u(x, 0)$, and pick a point x_0 in X. Let $\xi_0 \in E$, $\|\xi_0\|_E = 1$ be a point such that $M_u(x_0, 0) = u(x_0, \xi_0)$. By Proposition 2.4, there a germ of holomorphic section $\sigma : (X, x_0) \to E$ such that $\sigma(x_0) = \xi_0$, whose graph is contained in the unit ball bundle $\|\xi\|_E \leq 1$. Thus $M_u(x, 0) \geq u(x, \sigma(x))$ and $u(x_0, \sigma(x_0)) = M_u(x_0, 0)$. This implies that $M_u(x, 0)$ satisfies the mean value inequality at x_0. As x_0 is arbitrary, we conclude that $x \mapsto M_u(x, 0)$ is plurisubharmonic. The plurisubharmonicity in (x, t) follows by considering the pull-back of $E^\star$ to $X \times \mathbb{C}$ by the projection $(x, t) \mapsto x$, equipped with the Finsler metric $|e^{-t}| \|\xi\|_E$ at point (x, t). We again have osculating holomorphic sections contained in the unit ball bundle $\|\xi\|_E \leq |e^t|$, and the conclusion follows as before.   □

We now turn ourselves to the "converse" result:

THEOREM 3.2. *Let $\| \ \|_E$ be a smooth Finsler metric on E which is fiberwise strictly plurisubharmonic on all fibers E_x. Assume that X is Stein and that*

$$M_u(x,t) = \sup_{\|\xi\|_E \leq |e^t|} u(x,\xi)$$

is plurisubharmonic on $X \times \mathbb{C}$ for every plurisubharmonic function $(x,\xi) \mapsto u(x,\xi)$ on the total space E. Then the Levi form of $\| \ \|_E$ has at least n seminegative eigenvalues, in other words $\| \ \|_E$ is, locally over X, a limit of smooth Finsler metrics of transversal Levi signature (r,n).

PROOF. Once we know that there are at least n seminegative eigenvalues, we can produce metrics of signature (r,n) by considering

$$(x,\xi) \mapsto \|\xi\|_E \, e^{-\varepsilon|x|^2}, \qquad \varepsilon > 0$$

in any coordinate patch, whence the final assertion. Now, assume that the Levi form of $\| \ \|_E$ has at least $(r+1)$ positive eigenvalues at some point $(x_0,\xi_0) \in E$. Then the direct sum of positive eigenspaces in $T_{E,(x_0,\xi_0)}$ projects to a positive dimensional subspace in T_{X,x_0}. Consider a germ of smooth complex curve $\Gamma \subset X$ passing through x_0, such that its tangent at x_0 is contained in that subspace. Then (after shrinking Γ if necessary) the restriction of the metric $\| \ \|_E$ to $E_{|\Gamma}$ is strictly plurisubharmonic. By the well-known properties of strictly pseudoconvex domains the unit ball bundle $\|\xi\|_E < 1$ admits a peak function u at (x_0,ξ_0), that is, there is a smooth strictly plurisuharmonic function u on $E_{|\Gamma}$ which is equal to 0 at (x_0,ξ_0) and strictly negative on the set $\{(x,\xi) \neq (x_0,\xi_0) : \|\xi\| \leq 1\}$. As u is smooth, we can extend it to $E_{|B}$, where $B = B(x_0,\delta)$ is a small ball centered at x_0. As X is Stein, we can even extend it to E, possibly after shrinking B. Now $M_u(x,0)$ is equal to 0 at x_0 and strictly negative elsewhere on the curve Γ. This contradicts the maximum principle and shows that M_u cannot be plurisubharmonic. Hence the assumption was absurd and the Levi form of $\| \ \|_E$ has at least n seminegative eigenvalues.   □

4. A Conjecture of Hartshorne and Schneider on Complements of Algebraic Subvarieties

Our study is closely connected to the following interesting (and unsolved) conjecture of R. Hartshorne, which was first partially confirmed by Michael Schneider [1973] in the case of a Griffiths positive normal bundle.

CONJECTURE 4.1. *If X is a projective n-dimensional manifold and $Y \subset X$ is a complex submanifold of codimension q with ample normal bundle N_Y, then $X \smallsetminus Y$ is q-convex in the sense of Andreotti–Grauert. In other words, $X \smallsetminus Y$ has a smooth exhaustion function whose Levi form has at least $n - q + 1$ positive eigenvalues on a neighborhood of Y.*

Using Sommese's result 2.5(a), one can settle the following special case of the conjecture.

PROPOSITION 4.2 (Sommese). *In addition to the hypotheses in the conjecture, assume that $N_Y^\star$ has a strictly convex plurisubharmonic Finsler metric (this is the case for instance if N_Y is generated by global sections). Then $X \smallsetminus Y$ is q-convex.*

PROOF. By adding ε times a strictly plurisubharmonic Finsler metric on $N_Y^\star$ (which exists thanks to the assumption that N_Y is ample), we can even assume that the metric on $N_Y^\star$ is strictly convex *and* strictly plurisubharmonic. Then the dual metric on N_Y has a Levi form of signature $(q, n - q)$. Let $\widetilde{X} \to X$ be the blow-up of X with center Y, and $\widetilde{Y} = P(N_Y^\star)$ the exceptional divisor. Then, by Theorem 2.5, the Finsler metric on N_Y corresponds to a hermitian metric on

$$\mathcal{O}_{P(N_Y^\star)}(-1) \simeq N_{\widetilde{Y}} = \mathcal{O}_{\widetilde{X}}(\widetilde{Y})_{|\widetilde{Y}},$$

whose curvature form has signature $(q-1, n-q)$ on $\widetilde{Y}$. Take an arbitrary smooth extension of that metric to a metric of $\mathcal{O}_{\widetilde{X}}(\widetilde{Y})$ on $\widetilde{X}$. After multiplying the metric by a factor of the form $\exp(C\, d(z, \widetilde{Y})^2)$ in a neighborhood of $\widetilde{Y}$ (where $C \gg 0$ and $d(z, \widetilde{Y})$ is the riemannian distance to $\widetilde{Y}$ with respect to some metric), we can achieve that the curvature of $\mathcal{O}_{\widetilde{X}}(\widetilde{Y})$ acquires an additional negative eigenvalue in the normal direction to $\widetilde{Y}$. In this way, the curvature form of $\mathcal{O}_{\widetilde{X}}(\widetilde{Y})$ has signature $(q - 1, n - q + 1)$ in a neighborhood of $\widetilde{Y}$. We let $\sigma_{\widetilde{Y}} \in H^0(\widetilde{X}, \mathcal{O}_{\widetilde{X}}(\widetilde{Y}))$ be the canonical section of divisor $\widetilde{Y}$. An exhaustion of $X \smallsetminus Y = \widetilde{X} \smallsetminus \widetilde{Y}$ with the required properties is obtained by putting $\psi(z) = -\log \|\sigma_{\widetilde{Y}}(z)\|$. $\qquad\square$

5. Symmetric and Tensor Products of Finsler Metrics

Let E be a holomorphic vector bundle of rank r. In the sequel, we consider the m-th symmetric product $S^{m_1} E \times S^{m_2} E \to S^{m_1 + m_2} E$ and the m-th symmetric power $E \to S^m E$, $\xi \mapsto \xi^m$, which we view as the result of taking products of polynomials on $E^\star$. We also use the duality pairing $S^m E^\star \times S^m E \to \mathbb{C}$, denoted by $(\theta_1, \theta_2^\star) \mapsto \theta_1 \cdot \theta_2^\star$. In multi-index notation, we have

$$(e)^\alpha \cdot (e^\star)^\beta = \delta_{\alpha\beta} \frac{\alpha!}{(|\alpha|)!},$$

where $(e_j)_{1 \le j \le r}$ is a basis of E, $(e_j^\star)_{1 \le j \le r}$ the dual basis in $E^\star$, $1 \le j \le r$, and $(e)^\alpha = e_1^{\alpha_1} \ldots e_r^{\alpha_r}$ [Caution: this formula implies that $\theta_1^p \cdot \theta_2^{\star p} \ne (\theta_1 \cdot \theta_2^\star)^p$ for general elements $\theta_1 \in S^m E$, $\theta_2 \in S^m E^\star$, although this is true if $m = 1$.]

Whilst the linear dual $\| \ \|_{E^\star}$ of a Finsler metric $\| \ \|_E$ is not well behaved if $\| \ \|_E$ is not convex, we will see that (positive) symmetric powers and tensor powers can always be equipped with natural well behaved Finsler metrics. For an element $\theta^\star \in S^m E^\star$, viewed as a homogeneous polynomial of degree m on E, we set

$$(5.1) \qquad \|\theta^\star\|_{S^m E^\star, L^\infty_{1,m}} = \sup_{\xi \in E \setminus \{0\}} \frac{|\theta^\star \cdot \xi^m|}{\|\xi\|^m_E} = \sup_{\|\xi\|_E \leq 1} |\theta^\star \cdot \xi^m|.$$

[In the notation $L^\infty_{1,m}$, the upper index ∞ refers to the fact that we use sup norms, while the lower indices 1 refers to the fact that $\theta^\star$ appears with exponent 1, and ξ^m with exponent m.] This definition just reduces to the definition of the dual metric in the case $m = 1$, and thus need not be better behaved than the dual metric from the view point of curvature. On the other hand, for all $\theta_i^\star \in S^{m_i} E^\star$, $i = 1, 2$, it satisfies the submultiplicative law

$$\|\theta_1^\star \theta_2^\star\|_{S^{m_1+m_2} E^\star, L^\infty_{1,m_1+m_2}} \leq \|\theta_1^\star\|_{S^{m_1} E^\star, L^\infty_{1,m_1}} \|\theta_2^\star\|_{S^{m_2} E^\star, L^\infty_{1,m_2}}.$$

On the "positive side", i.e. for $\tau \in S^m E$, we define a sequence of metrics $\| \ \|_{S^m E, L^\infty_{p,1}}$ on $S^m E$, $p \geq 1$, and their "limit" $\| \ \|_{S^m E, L^\infty_{\infty,1}}$ by putting

$$(5.2) \qquad \|\tau\|_{S^m E, L^\infty_{p,1}} = \sup_{\theta^\star \in S^{mp} E^\star \setminus \{0\}} \left(\frac{|\tau^p \cdot \theta^\star|}{\|\theta^\star\|_{S^{pm} E^\star, L^\infty_{1,pm}}} \right)^{1/p}$$

$$= \sup_{\|\theta^\star\|_{S^{mp} E^\star, L^\infty_{1,pm}} \leq 1} |\tau^p \cdot \theta^\star|^{1/p},$$

$$(5.3) \qquad \|\tau\|_{S^m E, L^\infty_{\infty,1}} = \limsup_{p \to +\infty} \|\tau\|_{S^m E, L^\infty_{p,1}}.$$

In the case $m = 1$, we have of course $S^1 E = E$, but neither $\| \ \|_{S^1 E, L^\infty_{p,1}}$ nor $\| \ \|_{S^1 E, L^\infty_{\infty,1}}$ necessarily coincide with the original metric $\| \ \|_E$. In fact, by definition, it is easily seen that the unit ball bundle $\|\xi\|_{S^1 E, L^\infty_{\infty,1}} \leq 1$ is just the (fiberwise) *polynomial hull* of the ball bundle $\|\xi\|_E \leq 1$. In particular, $\| \ \|_{S^1 E, L^\infty_{\infty,1}}$ and $\| \ \|_E$ do coincide if and only if $\| \ \|_E$ is plurisubharmonic on all fibers E_x, which is certainly the case if $\| \ \|_E$ is globally plurisubharmonic on E. [By contrast, the unit ball bundle $\|\xi\|_{S^1 E, L^\infty_{1,1}} \leq 1$ is the convex hull of $\|\xi\|_E \leq 1$, and need not be pseudoconvex even if the latter is; see Remark 2.7.] Our first observation is this:

PROPOSITION 5.4. *The L^∞ metric $\| \ \|_{S^m E, L^\infty_{\infty,1}}$ is always well defined and non degenerate (in the sense that the $\limsup$ is finite and non zero for $\tau \neq 0$), and it defines a continuous Finsler metric on $S^m E$.*

PROOF. If in (5.2) we restrict $\theta^\star$ to be of the form $\theta^\star = (\xi^\star)^{mp}$, then

$$\tau^p \cdot \theta^\star = (\tau \cdot \xi^{\star m})^p, \qquad \|(\xi^\star)^{mp}\|_{S^{mp} E^\star, L^\infty_{1,mp}} = \|\xi^\star\|^{mp}_{E^\star},$$

where $\| \ \|_{E^\star}$ is the linear dual of $\| \ \|_E$. From this we infer

$$\|\tau\|_{S^m E, L^\infty_{p,1}} \geq \sup_{\|\xi^\star\|_{E^\star} \leq 1} |\tau \cdot \xi^{\star m}|$$

for all $p = 1, 2, \ldots, \infty$, in particular $\|\tau\|_{S^m E, L^\infty_{\infty,1}}$ is non degenerate. In the other direction, we have to show that $\|\tau\|_{S^m E, L^\infty_{\infty,1}}$ is finite. We first make an explicit calculation when $\| \ \|_E$ is a hermitian norm. We may assume $E = \mathbb{C}^r$ with its standard hermitian norm. Then, writing $\theta^\star \cdot \xi^m = \sum_{|\alpha|=m} c_\alpha \xi^\alpha$ in multi-index notation, we get

$$\|\theta^\star\|^2_{S^m E^\star, L^\infty_{1,m}} = \sup_{\|\xi\|=1} \left| \sum_{|\alpha|=m} c_\alpha \xi^\alpha \right|^2 \geq \sup_{t_1+\cdots+t_r=1} \sum_{|\alpha|=m} |c_\alpha|^2 t^\alpha.$$

This is obtained by integrating over the n-torus $\xi_j = t_j^{1/2} e^{iu_j}$, $0 \leq u_j < 2\pi$ (with $t = (t_j)$ fixed, $\sum t_j = 1$), and applying Parseval's formula. We can now replace the right hand supremum by the average over the $(n-1)$-simplex $\sum t_j = 1$. A short computation yields

$$\|\theta^\star\|^2_{S^m E^\star, L^\infty_{1,m}} \geq \frac{(r-1)!}{(m+1)(m+2)\ldots(m+r-1)} \sum_{|\alpha|=m} |c_\alpha|^2 \frac{\alpha!}{(|\alpha|)!}.$$

However $\sum_{|\alpha|=m} |c_\alpha|^2 \frac{\alpha!}{(|\alpha|)!}$ is just the hermitian norm on $S^m E^\star$ induced by the inclusion $S^m E^\star \subset (E^\star)^{\otimes m}$. The dual norm is the hermitian norm on $S^m E$. From this, we infer

$$\|\tau\|_{S^m E, L^\infty_{p,1}} \leq \left(\frac{(mp+1)(mp+2)\ldots(mp+r-1)}{(r-1)!} \right)^{1/2p} \|\tau^p\|^{1/p}_{S^{mp} E, \text{herm}},$$

$$\|\tau\|_{S^m E, L^\infty_{\infty,1}} \leq \|\tau\|_{S^m E, \text{herm}}$$

[using the obvious fact that hermitian norms are submultiplicative], whence the finiteness of $\|\tau\|_{S^m E, L^\infty_{\infty,1}}$. Finally, given any two Finsler metrics $_1\| \ \|_E$ and $_2\| \ \|_E$ such that $_1\| \ \|_E \leq {}_2\| \ \|_E$, it is clear that $_1\|\tau\|_{S^m E, L^\infty_{\infty,1}} \leq {}_2\|\tau\|_{S^m E, L^\infty_{\infty,1}}$. By comparing a given Finsler norm $\| \ \|_E = {}_1\| \ \|_E$ with a hermitian norm $_2\| \ \|_E$, we conclude that the metric $\| \ \|_{S^m E, L^\infty_{\infty,1}}$ must be finite. Moreover, comparing the metrics $\| \ \|_E$ at nearby points, we see that $\| \ \|_{S^m E, L^\infty_{\infty,1}}$ varies continuously (and that it depends continuously on $\| \ \|_E$). □

Our next observation is that the L^∞ metrics on the negative symmetric powers $S^m E^\star$ can be replaced by L^2 metrics without changing the final metric $\| \ \|_{S^m E, L^\infty_{\infty,1}}$ on $S^m E$. To see this, fix an arbitrary smooth positive volume form dV on $P(E^\star_x)$, with (say) $\int_{P(E^\star_x)} dV = 1$. We can view any element $\theta^\star \in S^m E^\star$ as a section of $H^0(P(E^\star_x), \mathcal{O}_{P(E^\star_x)}(m))$. Let $\|\theta^\star\|^2_{\mathcal{O}(m)}$ be the pointwise norm on $\mathcal{O}_{P(E^\star)}(m)$ induced by $\| \ \|_E$, and let $d\sigma$ be the area measure on the unit sphere bundle $\Sigma(E)$ induced by dV. We then set

$$\|\theta^\star\|^2_{S^m E^\star, L^2_{1,m}} = \int_{P(E^\star_x)} \|\theta^\star\|^2_{\mathcal{O}(m)} dV = \int_{\xi \in \Sigma(E^\star_x)} |\theta^\star \cdot \xi^m|^2 d\sigma(\xi).$$

Clearly

$$\|\theta^\star\|_{S^m E^\star, L^2_{1,m}} \leq \|\theta^\star\|_{S^m E^\star, L^\infty_{1,m}}.$$

On the other hand, there exists a constant C such that

$$(5.5) \qquad \|\theta^\star\|^2_{S^m E^\star, L^\infty_{1,m}} \leq C\, m^{r-1} \|\theta^\star\|^2_{S^m E^\star, L^2_{1,m}}.$$

This is seen by applying the mean value inequality for subharmonic functions, on balls of radius $\sim 1/\sqrt{m}$ centered at arbitrary points in $P(E_x)$. In fact, in a suitable local trivialization of $\mathcal{O}_{P(E^\star)}$ near a point $[\xi_0] \in P(E_x^\star)$, we can write $\|\theta^\star\|_{\mathcal{O}(m)}^2 = |\theta_0^\star|^2 e^{-m\psi}$ where $\theta_0^\star$ is the holomorphic function representing $\theta^\star$, and ψ is the weight of the metric on $\mathcal{O}_{P(E^\star)}(1)$. We let l be the holomorphic part in the first jet of ψ at $[\xi_0]$, and apply the mean value inequality to

$$|\theta_0^\star e^{-ml}|^2 e^{-m(\psi - 2\operatorname{Re} l)}.$$

As $\psi - 2\operatorname{Re} l$ vanishes at second order at $[\xi_0]$, its maximum on a ball of radius $1/\sqrt{m}$ is $O(1/m)$. Hence, up to a constant independent of m, we can replace $|\theta_0^\star e^{-ml}|^2 e^{-m(\psi - 2\operatorname{Re} l)}$ by the subharmonic function $|\theta_0^\star e^{-ml}|^2$. Inequality (5.5) then follows from the mean value inequality on the ball $B([\xi_0], 1/\sqrt{m})$ [noticing that the volume of this ball is $\sim 1/m^{r-1}$]. Now (5.5) shows that the replacement of $\|\theta^\star\|_{S^{pm}E^\star, L_{1,pm}^\infty}$ by $\|\theta^\star\|_{S^{pm}E^\star, L_{1,pm}^2}$ in (5.2) and (5.3) does not affect the limit as p tends to $+\infty$.

If $\|\ \|_E$ is (globally) plurisubharmonic, we can even use more global L^2 metrics without changing the limit. Take a small Stein open subset $U \Subset X$ and fix a Kähler metric ω on $P(E_{|U}^\star)$. To any section $\sigma \in H^0(\pi^{-1}(U), \mathcal{O}_{P(E^\star)}(m)) = H^0(U, S^m E^\star)$, we associate the L^2 norm

$$\|\sigma\|_{S^m E^\star, L_{1,m}^2(U)}^2 = \int_{\pi^{-1}(U)} |\sigma|^2 dV_\omega,$$

where $\pi : P(E^\star) \to X$ is the canonical projection. In this way, we obtain a Hilbert space

$$\mathcal{H}_{E,m}(U) = \left\{\sigma : \|\sigma\|_{S^m E^\star, L_1^2(U)}^2 < +\infty\right\} \subset H^0(\pi^{-1}(U), \mathcal{O}_{P(E^\star)}(m)),$$

and associated (non hermitian!) metrics

$$\|\tau\|_{S^m E_x, L_{p,1}^2(U)} = \sup_{\sigma \in \mathcal{H}_{E,mp}(U), \|\sigma\| \leq 1} |\sigma(x) \cdot \tau^p|^{1/p},$$

$$\|\tau\|_{S^m E_x, L_{\infty,1}^2(U)} = \limsup_{p \to +\infty} \|\tau\|_{S^m E_x, L_{p,1}^2(U)}, \qquad \tau \in S^m E_x^\star.$$

As these metrics are obtained by taking sups of plurisubharmonic functions $((x, \tau) \mapsto \sigma(x) \cdot \tau^p$ is holomorphic on the total space of $S^m E)$, it is clear that the corresponding metrics are plurisubharmonic on $S^m E$. Furthermore, an argument entirely similar to the one used for (5.5) shows that

$$\|\sigma(x)\|_{S^m E_x^\star, L_{1,m}^2}^2 \leq C\, m^n \|\sigma\|_{S^m E^\star, L_{1,m}^2(U)}^2 \qquad \text{for all } x \in U' \Subset U.$$

In order to get this, we apply the mean value inequality on balls of radius $1/\sqrt{m}$ centered at points of the fiber $P(E_x^\star)$ and transversal to that fiber, in combination with Fubini's theorem. In the other direction, the Ohsawa–Takegoshi L^2 extension theorem [Ohsawa and Takegoshi 1987; Ohsawa 1988; Manivel 1993] shows that every element $\theta^\star \in S^m E_x^\star$, viewed as a section of $\mathcal{O}_{P(E_x^\star)}(m)$, can be extended to a section $\sigma \in H^0(\pi^{-1}(U), \mathcal{O}_{P(E^\star)}(m))$ such that

$$\|\sigma\|_{S^m E^\star, L^2_{1,m}(U)} \leq C' \|\theta^\star\|_{S^m E_x^\star, L^2_{1,m}},$$

where C' does not depend on $x \in U$. For this, we use the fundamental assumption that $\|\ \|_E$ is plurisubharmonic (and take profit of the fact that $\mathcal{O}_{P(E^\star)}(1)$ is relatively ample to get enough positivity in the curvature estimates: write e.g. $\mathcal{O}_{P(E^\star)}(m) = \mathcal{O}_{P(E^\star)}(m - m_0) \otimes \mathcal{O}_{P(E^\star)}(m_0)$, keep the original metric on the first factor $\mathcal{O}_{P(E^\star)}(m - m_0)$, and put a metric with uniformly positive curvature on the second factor). From this, we conclude that $\|\tau\|_{S^m E_x, L^\infty_{\infty,1}(U)}$ coincides with the metric defined in (5.3). Since this metric depends *in fine* only on $\|\ \|_E$, we will simply denote it by $\|\ \|_{S^m E}$. We have thus proven:

THEOREM 5.6. *If $\|\ \|_E$ is (strictly) plurisubharmonic on E, then $\|\ \|_{S^m E}$ is (strictly) plurisubharmonic on $S^m E$.*

The case of strict plurisubharmonicity can be handled by more or less obvious perturbation arguments and will not be detailed here. As a consequence, we get the Finsler metric analogue of the fact that a direct sum or tensor product of ample vector bundles is ample.

COROLLARY 5.7. *If E, F are holomorphic vector bundles, and $\|\ \|_E$, $\|\ \|_F$ are (strictly) plurisubharmonic Finsler metrics on E, F, there exist naturally defined (strictly) plurisubharmonic Finsler metrics $\|\ \|_{E\oplus F}$, $\|\ \|_{E\otimes F}$ on $E \oplus F$, $E \otimes F$ respectively.*

PROOF. In the case of the direct sum, we simply set $\|\xi \oplus \eta\|_{E\oplus F} = \|\xi\|_E + \|\eta\|_F$. The logarithmic indicatrix is given by

$$\chi_{E\oplus F}(x, \xi, \eta) = \log\big(\exp(\chi_E(x, \xi)) + \exp(\chi_F(x, \eta))\big),$$

and it is clear from there that $\chi_{E\oplus F}$ is plurisubharmonic. Now, we observe that $S^2(E \oplus F) = S^2 E \oplus S^2 F \oplus (E \otimes F)$. Hence $E \otimes F$ can be viewed as a subbundle of $S^2(E \oplus F)$. To get the required Finsler metric on $E \otimes F$, we just apply Theorem 5.5 to $S^2(E \oplus F)$ and take the induced metric on $E \otimes F$. $\qquad\square$

REMARK 5.8. It would be interesting to know whether good Finsler metrics could be defined as well on the dual symmetric powers $S^m E^\star$. One natural candidate would be to use the already defined metrics $\|\ \|_{S^m E}$ and to set

$$\|\tau^\star\|_{S^m E^\star, L^\infty_{p,1}} = \sup_{\|\theta\|_{S^{pm} E} \leq 1} |\tau^{\star p} \cdot \theta|^{1/p},$$

$$\|\tau^\star\|_{S^m E^\star, L^\infty_{\infty,1}} = \limsup_{p \to +\infty} \|\tau^\star\|_{S^m E^\star, L^\infty_{p,1}}.$$

However, we do not know how to handle these "bidually defined" Finsler metrics, and the natural question whether $\|\ \|_{S^m E^\star, L^\infty_{\infty,1}}$ has transversal signature $(\dim S^m E^\star, n)$ probably has a negative answer if $\|\ \|_E$ is not convex (although this might be "asymptotically true" as m tends to $+\infty$).

5.9. RELATION TO COHOMOLOGY VANISHING AND DUALITY THEOREMS. If $\| \ \|_{E^*}$ is smooth and strictly plurisubharmonic, then E is ample, thus its symmetric powers $S^m E$ have a lot of sections and the Kodaira–Serre vanishing theorem holds true, i.e.

$$H^q(X, S^m E \otimes \mathcal{F}) = 0, \qquad q \neq 0,$$

for every coherent sheaf $\mathcal{F}$ and $m \geq m_0(\mathcal{F})$ large enough. In a parallel way, if $\| \ \|_E$ has a metric of signature (r, n), then the line bundle $\mathcal{O}_{P(E^*)}(1)$ has a hermitian metric such that the curvature has signature $(r-1, n)$ over $P(E^*)$. From this, by the standard Bochner technique, we conclude that

$$H^q(P(E^*), \mathcal{O}_{P(E^*)}(m) \otimes \mathcal{G}) = 0, \qquad q \neq n,$$

for every locally free sheaf $\mathcal{G}$ on $P(E^*)$ and $m \geq m_0(\mathcal{G})$. The Leray spectral sequence shows that

$$H^q(X, S^m E^* \otimes \mathcal{F}) = H^q(P(E^*), \mathcal{O}_{P(E^*)}(m) \otimes \pi^* \mathcal{F}),$$

thus we have vanishing of this group as well is $\mathcal{F}$ is locally free and $q \neq n$, $m \geq m_0(\mathcal{F})$. The Serre duality theorem connects the two facts via an isomorphism

$$H^q(X, S^m E^* \otimes \mathcal{F})^* = H^{n-q}(X, S^m E \otimes \mathcal{F}^* \otimes K_X).$$

What we are looking for, in some sense, is a "Finsler metric version" of the Serre duality theorem. Up to our knowledge, the duality works well only for convex Finsler metrics (and also asymptotically, for high symmetric powers $S^m E$ which carry positively curved hermitian metrics).

6. A Trick on Taylor Series

Let $\pi : E \to X$ be a holomorphic vector bundle, such that E^* is equipped with a continuous plurisubharmonic Finsler metric $\|\xi^*\|_{E^*} = \exp(\chi^*(x, \xi^*))$. Thanks to Section 5, we are able to define plurisubharmonic Finsler metrics $\| \ \|_{S^m E^*}$ on all symmetric powers of E^*. Our goal is to use these metrics in order to define plurisubharmonic sup functionals for holomorphic or plurisubharmonic functions. We first start with the simpler case when $\| \ \|_{E^*}$ is convex.

THEOREM 6.1. *Assume that* $\| \ \|_{E^*}$ *is plurisubharmonic and convex, and let* $\|\xi\|_E = \exp(\chi(x, \xi))$ *be the (linearly) dual metric. Then, for every plurisubharmonic function* $(x, \xi) \mapsto u(x, \xi)$ *on the total space* E, *the function*

$$M_u^\chi(x, t) = \sup_{\|\xi\|_E \leq |e^t|} u(x, \xi)$$

is plurisubharmonic on $X \times \mathbb{C}$.

PROOF. This is a local result on X, so we can assume that X is an open set $\Omega \subset \mathbb{C}^n$ and that $E = \Omega \times \mathbb{C}^r$ is trivial. By the standard approximation techniques, we can approximate $\| \ \|_{E^\star}$ by smooth strictly convex and strictly plurisubharmonic metrics $_\varepsilon\| \ \|_{E^\star} \geq \| \ \|_{E^\star}$ which decrease to $\| \ \|_{E^\star}$ as ε decreases to 0. We then get a decreasing family $\lim \downarrow_{(\varepsilon \to 0)} {}_\varepsilon M_u(x,t) = M_u(x,t)$. It is thus enough to treat the case of smooth strictly convex and strictly plurisubharmonic metrics $\| \ \|_{E^\star}$. In that case, $\| \ \|_E$ has a Levi form of signature (r,n) and we conclude by Theorem 3.1. □

Unfortunately, in the general case when $\| \ \|_{E^\star}$ is not convex, this simple approach does not work [in the sense that M_u is not always plurisubharmonic]. We circumvent this difficulty by using instead the well-known trick of Taylor expansions, and replacing the sup with a more sophisticated evaluation of norms. If f is a holomorphic function on the total space of E, the Taylor expansion of f along the fibers of E can be written as

$$f(x,\xi) = \sum_{m=0}^{+\infty} a_m(x) \cdot \xi^m, \qquad \xi \in E_x,$$

where a_m is a section in $H^0(X, S^m E^\star)$. In that case, we set

(6.2) $$\widehat{M}_f^\chi(x,t) = \sum_{m=0}^{+\infty} \|a_m(x)\|_{S^m E^\star} |e^{mt}|.$$

This is by definition a plurisubharmonic function on $X \times \mathbb{C}$. In fact, $\log \widehat{M}_f^\chi(x,t)$ is a plurisubharmonic function as well. As we will see in the following lemma, $\widehat{M}_f^\chi$ will play essentially the same role as $M_{|f|}$ could have played.

LEMMA 6.3. *Fix a hermitian metric* $\| \ \|_{E^\star,\mathrm{herm}} \geq \| \ \|_{E^\star}$, *and let* $\| \ \|_{E,\mathrm{herm}}$ *be the dual metric. Then there is an inequality*

$$\sup_{\|\xi\|_E \leq |e^t|} |f(x,\xi)| \leq \widehat{M}_f^\chi(x,t) \leq \left(1 + \frac{1}{\varepsilon}\right)^r \sup_{\|\xi\|_{E,\mathrm{herm}} \leq (1+\varepsilon)|e^t|} |f(x,\xi)|.$$

PROOF. The left hand inequality is obtained by expanding

$$f(x,\xi) \leq \sum_{m=0}^{+\infty} |a_m(x) \cdot \xi^m| \leq \sum_{m=0}^{+\infty} \|a_m(x)\|_{S^m E^\star} \|\xi\|_E^m,$$

thanks to the fact that $a_m(x)^p \cdot \xi^{mp} = (a_m(x) \cdot \xi^m)^p$. In the other direction, we have

$$\|a_m(x)\|_{S^m E^\star} \leq \|a_m(x)\|_{S^m E^\star,\mathrm{herm}}$$
$$\leq \frac{(m+1)\dots(m+r-1)}{(r-1)!} \sup_{\|\xi\|_{E,\mathrm{herm}} \leq 1} |a_m(x) \cdot \xi^m|$$

thanks to the inequalities obtained in the proof of Proposition 5.4. Now, the standard Cauchy inequalities imply

$$\sup_{\|\xi\|_{E,\mathrm{herm}}=1} |a_m(x) \cdot \xi^m| \leq \frac{1}{R^m} \sup_{\|\xi\|_{E,\mathrm{herm}}=R} |f(x,\xi)|.$$

Combining all the above with $R > |e^t|$, we get

$$\widehat{M}_f^\chi(x,t) \leq \sup_{\|\xi\|_{E,\mathrm{herm}}\leq R} |f(x,\xi)| \sum_{m=0}^{+\infty} \frac{|e^{mt}|}{R^m} \frac{(m+1)\ldots(m+r-1)}{(r-1)!}$$

$$\leq \frac{1}{\left(1-\frac{|e^t|}{R}\right)^r} \sup_{\|\xi\|_{E,\mathrm{herm}}\leq R} |f(x,\xi)| \leq \left(1+\frac{1}{\varepsilon}\right)^r \sup_{\|\xi\|_{E,\mathrm{herm}}\leq(1+\varepsilon)|e^t|} |f(x,\xi)|.$$

The lemma is proved.   □

REMARK 6.4. It is clear that the sup functional $M_{|f|}^\chi$ is submultiplicative, i.e.

$$M_{|fg|}^\chi(x,t) \leq M_{|f|}^\chi(x,t) \, M_{|g|}^\chi(x,t).$$

However, the analogous property for $\widehat{M}_f^\chi$ would require to know whether

$$\|a \cdot b\|_{S^{m_1+m_2}E^\star} \leq \|a\|_{S^{m_1}E^\star} \|b\|_{S^{m_2}E^\star},$$

(or a similar inequality with a constant C independent of m_1, m_2). It is not clear whether such a property is true, since the precise asymptotic behaviour of the metrics $\| \ \|_{S^m E^\star}$ is hard to understand. In order to circumvent this problem, we select a non increasing sequence of real numbers $\rho_m \in \,]0,1]$ with $\rho_0 = 1$, such that

$$(6.5) \qquad \rho_{m_1+m_2}\|a \cdot b\|_{S^{m_1+m_2}E^\star} \leq \rho_{m_1}\rho_{m_2}\|a\|_{S^{m_1}E^\star}\|b\|_{S^{m_2}E^\star}$$

for all m_1, m_2. One can easily find such a sequence $\rho = (\rho_m)$ by induction on m, taking ρ_m/ρ_{m-1} small enough. Then

$$(6.6) \qquad \widehat{M}_f^{\chi,\rho}(x,t) := \sum_{m=0}^{+\infty} \rho_m\|a_m(x)\|_{S^m E^\star}|e^{mt}|$$

obviously satisfies the submultiplicative property. On the other hand, we lose the left hand inequality in Lemma 6.3. This unsatisfactory feature will create additional difficulties which we can only solve at the expense of using deeper analytic techniques.

We are mostly interested in the case when $E = T_X$ is the tangent bundle, and assume that a plurisubharmonic Finsler metric $\|\xi^\star\|_{T_X^\star} = \exp(\chi^\star(x,\xi^\star))$ is given. Locally, on a small coordinate open set $U \Subset U_0 \subset X$ associated with a holomorphic chart

$$\tau : U_0 \to \tau(U_0) \subset \mathbb{C}^n,$$

we have a corresponding trivialization $\tau' : T_{X|U} \simeq \tau(U) \times \mathbb{C}^n$. Given a holomorphic function f in a neighborhood of $\bar{U}$, we consider the holomorphic function such that $F(x,\xi) = f(\alpha(x,\xi))$ where $\alpha(x,\xi) = \tau^{-1}(\tau(x) + \tau'(x)\xi)$. It is defined on a sufficiently small ball bundle $B_\varepsilon(T_{X|U}) = \{(x,\xi) \in T_{X|U} : \|\xi\| < \varepsilon\}$, $\varepsilon > 0$. Thus

$$(6.7) \qquad \widetilde{M}_f^{\chi,\rho}(x,t) := \widehat{M}_F^{\chi,\rho}(x,t)$$

makes sense for $|e^t| < c\varepsilon$, $c > 0$. Again, by construction, this is a plurisubharmonic function of (x,t) on $U \times \{|e^t| < c\varepsilon\}$. This function will be used as a replacement of the sup of f on the Finsler ball $\alpha(x, B(0, |e^t|)) \subset X$ (which we unfortunately know nothing about). However, the definition is not coordinate invariant, and we have to investigate the effect of coordinate changes.

LEMMA 6.8. *Consider two holomorphic coordinate coordinate charts τ_j on a neighborhood of $\overline{U}$, for $j = 1,2$, and the corresponding maps*

$$\alpha_j : B_\varepsilon(T_{X|U}) \to U_0, \qquad \alpha_j(x,\xi) = \tau_j^{-1}(\tau_j(x) + \tau_j'(x)\xi).$$

Let $F_j = f \circ \alpha_j$, $j = 1,2$, and let $\delta > 0$ be fixed. Then there is a choice of a decreasing sequence $\rho = (\rho_m)$ such that

$$\widehat{M}_{F_2}^{\chi,\rho}(x,t) \le (1+\delta)\widehat{M}_{F_1}^{\chi,\rho}(x,t),$$

where ρ depends on U, τ_1, τ_2, but not on f (here $|e^t|$ is suppose to be chosen small enough so that both sides are defined). Any sequence ρ with ρ_m/ρ_{m-1} smaller that a given suitable sequence of small numbers works.

In other words, if the sequence (ρ_m) decays sufficiently fast, the functional $\widetilde{M}_f^{\chi,\rho}(x,t)$ defined above can be chosen to be "almost" coordinate invariant.

PROOF. It is easy to check by the implicit function theorem that there exists a (uniquely) defined map $w : T_X \to T_X$, defined near the zero section and tangent to the identity at 0, such that

$$\alpha_2(x,\xi) = \alpha_1(x, w(x,\xi)), \qquad w(x,\xi) = \xi + O(\xi^2).$$

Hence, if we write

$$f \circ \alpha_j(x,\xi) = \sum_{m=0}^{+\infty} a_{m,j}(x) \cdot \xi^m, \qquad j = 1, 2,$$

the series corresponding to index $j = 2$ is obtained from the $j = 1$ series by substituting $\xi \mapsto w(x,\xi)$. It follows that

$$a_{m,2}(x) = a_{m,1}(x) + \sum_{\mu < m} L_{m,\mu}(x) \cdot a_{\mu,1}(x)$$

where $L_{m,\mu} : S^\mu T_X^\star \to S^m T_X^\star$ are certain holomorphic linear maps depending only on the chart mappings τ_1, τ_2. If ρ_m/ρ_{m-1} is small enough, the contribution given by $\rho_m \sum_{\mu < m} L_{m,\mu}(x) \cdot a_{\mu,1}(x)$ is negligible compared to the $\rho_\mu \|a_{\mu,1}(x)\|_{S^\mu T_X^\star}$. The lemma follows. $\qquad \square$

7. Approximation of Plurisubharmonic Functions by Logarithms of Holomorphic Functions

The next step is to extend the $\widetilde{M}_f^{\chi,\rho}$ functional to plurisubharmonic functions defined on a complex manifold, when the cotangent bundle T_X^* is equipped with a Finsler metric. The simplest way to do this is to approximate such functions by logarithms of holomorphic functions, by means of the Ohsawa–Takegoshi L^2 extension theorem [Ohsawa and Takegoshi 1987; Ohsawa 1988; Manivel 1993]. We reproduce here some of the techniques introduced in [Demailly 1992], but with substantial improvements. The procedure is still local and not completely canonical, so we will have later to apply a gluing procedure.

THEOREM 7.1. *Let φ be a plurisubharmonic function on a bounded pseudoconvex open set $U \subset \mathbb{C}^n$. For every $p > 0$, let $\mathcal{H}_{p\varphi}(U)$ be the Hilbert space of holomorphic functions f on U such that $\int_U |f|^2 e^{-2p\varphi} d\lambda < +\infty$ and let*

$$\varphi_p = \frac{1}{2p} \log \sum |\sigma_l|^2,$$

where (σ_l) is an orthonormal basis of $\mathcal{H}_{p\varphi}(U)$. Then there are constants $C_1 > 0$ and $C_2 > 0$ independent of p such that

(i) $\varphi(z) - \dfrac{C_1}{p} \leq \varphi_p(z) \leq \sup\limits_{|\zeta-z|<r} \varphi(\zeta) + \dfrac{1}{p} \log \dfrac{C_2}{r^n}$

   for every $z \in U$ and $r < d(z, \partial U)$. In particular, φ_p converges to φ pointwise and in L^1_{loc} topology on U when $p \to +\infty$ and

(ii) $\nu(\varphi, z) - \dfrac{n}{p} \leq \nu(\varphi_p, z) \leq \nu(\varphi, z)$ *for every $z \in U$.*

PROOF. Note that $\sum |\sigma_l(z)|^2$ is the square of the norm of the evaluation linear form $f \mapsto f(z)$ on $\mathcal{H}_{p\varphi}(U)$. As φ is locally bounded above, the L^2 topology is actually stronger than the topology of uniform convergence on compact subsets of U. It follows that the series $\sum |\sigma_l|^2$ converges uniformly on U and that its sum is real analytic. Moreover, we have

(7.2) $$\varphi_p(z) = \sup_{f \in B_p(1)} \frac{1}{p} \log |f(z)|$$

where $B_p(1)$ is the unit ball of $\mathcal{H}_{p\varphi}(U)$. For $r < d(z, \partial U)$, the mean value inequality applied to the plurisubharmonic function $|f|^2$ implies

$$|f(z)|^2 \leq \frac{1}{\pi^n r^{2n}/n!} \int_{|\zeta-z|<r} |f(\zeta)|^2 d\lambda(\zeta)$$

$$\leq \frac{1}{\pi^n r^{2n}/n!} \exp\left(2p \sup_{|\zeta-z|<r} \varphi(\zeta)\right) \int_U |f|^2 e^{-2p\varphi} d\lambda.$$

If we take the supremum over all $f \in B_p(1)$ we get

$$\varphi_p(z) \leq \sup_{|\zeta-z|<r} \varphi(\zeta) + \frac{1}{2p} \log \frac{1}{\pi^n r^{2n}/n!}$$

and the second inequality in (i) is proved. Conversely, the Ohsawa–Takegoshi extension theorem [Ohsawa and Takegoshi 1987; Ohsawa 1988; Manivel 1993] applied to the 0-dimensional subvariety $\{z\} \subset U$ shows that for any $a \in \mathbb{C}$ there is a holomorphic function f on U such that $f(z) = a$ and

$$\int_U |f|^2 e^{-2p\varphi} d\lambda \leq C_3 |a|^2 e^{-2p\varphi(z)},$$

where C_3 only depends on n and diam U. We fix a such that the right hand side is 1. This gives the other inequality

$$\varphi_p(z) \geq \frac{1}{p} \log |a| = \varphi(z) - \frac{\log C_3}{2p}.$$

The above inequality implies $\nu(\varphi_p, z) \leq \nu(\varphi, z)$. In the opposite direction, we find

$$\sup_{|x-z|<r} \varphi_p(x) \leq \sup_{|\zeta-z|<2r} \varphi(\zeta) + \frac{1}{p} \log \frac{C_2}{r^n}.$$

Divide by $\log r$ and take the limit as r tends to 0. The quotient by $\log r$ of the supremum of a plurisubharmonic function over $B(x, r)$ tends to the Lelong number at x. Thus we obtain

$$\nu(\varphi_p, x) \geq \nu(\varphi, x) - \frac{n}{p}. \qquad \square$$

Another important fact is that the approximations φ_p do no depend much on the open set U, and they have a good dependence on φ under small perturbations. In fact, let $U', U'' \subset U$ be Stein open subsets, and let φ', φ'' be plurisubharmonic functions on U', U'' such that $|\varphi' - \varphi''| \leq \varepsilon$ on $U' \cap U''$. If f' is a function in the unit ball of $\mathcal{H}_{p\varphi}(U')$, then

$$\int_{U' \cap U''} |f'|^2 e^{-2p\varphi''} d\lambda \leq e^{2p\varepsilon}$$

by the hypothesis on $\varphi' - \varphi''$. For every $x_0 \in U'$, we can find a function $f'' \in \mathcal{H}_{p\varphi''}(U)$ such that $f''(x_0) = f'(x_0)$ and

$$\int_{U''} |f''|^2 e^{-2p\varphi''} d\lambda \leq \frac{C}{(d(x_0, \complement U))^{2n+2}} e^{2p\varepsilon} \int_{U'} |f'|^2 e^{-2p\varphi'} d\lambda.$$

This is done as usual, by solving the equation $\bar{\partial} g = \bar{\partial}(\theta f')$ with a cut-off function θ supported in the ball $B(x_0, \delta/2)$ and equal to 1 on $B(x_0, \delta/4)$, $\delta = d(x_0, \complement U)$, with the weight $2p\varphi(z) + 2n \log |z - x_0|$; the desired function is then $f'' = \theta f' - g$. By readjusting f'' by a constant so that f'' is in the unit sphere, and by taking the sup of $\log |f'(x_0)|$ and $\log |f''(x_0)|$ for all f' and f'' in the unit ball of their respective Hilbert spaces, we conclude that

$$\varphi_p'(x) \leq \varphi_p''(x) + \varepsilon + \frac{1}{2p} \log \frac{C}{d(x, \complement(U' \cap U''))^{2n+2}} \qquad \text{on } U' \cap U'',$$

with some constant $C > 0$ depending only on the pair (U', U''). By symmetry, we get

$$(7.3) \qquad |\varphi'_p(x) - \varphi''_p(x)| \le \varepsilon + \frac{1}{2p} \log \frac{C}{d(x, \complement(U' \cap U''))^{2n+2}} \qquad \text{on } U' \cap U''.$$

The next idea would be to take Taylor series much in the same way as we did in § 6, and look e.g. at

$$\Phi^{\chi,\rho,p}(x,t) = \sup_{f \in B_p(1)} \frac{1}{p} \log \widetilde{M}_f^{\chi,\rho}(x,t).$$

The main problem with this approach occurs when we want to check the effect of a change of coordinate patch. We then want to compare the jets with those of the functions f obtained on another coordinate patch, say up to an order Cp for $C \gg 0$ large. The comparison would be easy (by the usual Hörmander–Bombieri $\bar{\partial}$-technique, as we did for the 0-jets in (7.3)) for jets of small order in comparison to p, but going to such high orders introduces intolerable distortion in the required bounds. A solution to this problem is to introduce further approximations of φ_p for which we have better control on the jets. This can be done by using Skoda's L^2-estimates for surjective bundle morphisms [Skoda 1972a; 1978]. This approach was already used in [Demailly 1992], but in a less effective fashion.

Let $K_p^U : \varphi \mapsto \varphi_p$ be the transformation defined above. This transformation has the effect of converting the singularities of φ, which are a priori arbitrary, into logarithmic analytic singularities (and, as a side effect, the multiplicities get discretized, with values in $\frac{1}{p}\mathbb{N}$). We simply iterate the process twice, and look at

$$(7.4) \qquad \varphi_{p,q} = K_{pq}^U(K_p^U(\varphi))$$

for large integers $q \gg p \gg 1$. In other words,

$$\varphi_p(z) = \frac{1}{2p} \log \sum |\sigma_l(z)|^2, \qquad \varphi_{p,q}(z) = \frac{1}{2pq} \log \sum |\tilde{\sigma}_l(z)|^2$$

where $\sigma = (\sigma_l)_{l \in \mathbb{N}}$ and $\tilde{\sigma} = (\tilde{\sigma}_l)_{l \in \mathbb{N}}$ are Hilbert bases of the L^2 spaces

$$\mathcal{H}_{p\varphi}(U) = \left\{ \int_U |f|^2 e^{-2p\varphi} d\lambda < +\infty \right\}, \qquad \mathcal{H}_{pq\varphi_p}(U) = \left\{ \int_U |f|^2 |\sigma|^{-2q} d\lambda < +\infty \right\}.$$

Theorem 7.1 shows that we still have essentially the same estimates for $\varphi_{p,q}$ as we had for φ_p, namely

$$(7.5 \text{ i}) \qquad \varphi(z) - \frac{C_1}{p} \le \varphi_{p,q}(z) \le \sup_{|\zeta - z| < r} \varphi(\zeta) + \left(\frac{1}{p} + \frac{1}{pq} \right) \log \frac{C_2}{r^n}$$

$$(7.5 \text{ ii}) \qquad \nu(\varphi, z) - n\left(\frac{1}{p} + \frac{1}{pq} \right) \le \nu(\varphi_{p,q}, z) \le \nu(\varphi, z).$$

The major improvement is that we can now compare the jets when U varies, even when we allow a small perturbation on φ as well.

PROPOSITION 7.6. *Suppose that we have plurisubharmonic functions φ', φ'' defined on bounded Stein open sets U', $U'' \Subset \mathbb{C}^n$, with $|\varphi' - \varphi''| \leq \varepsilon$ on $U' \cap U''$. Let $\sigma' = (\sigma'_l)_{l \in \mathbb{N}}$, $\sigma'' = (\sigma''_l)_{l \in \mathbb{N}}$ be the associated Hilbert bases of $\mathcal{H}_{p\varphi'}(U')$, $\mathcal{H}_{p\varphi''}(U'')$, and $\tilde{\sigma}' = (\tilde{\sigma}'_l)_{l \in \mathbb{N}}$, $\tilde{\sigma}'' = (\tilde{\sigma}''_l)_{l \in \mathbb{N}}$ the bases of $\mathcal{H}_{pq\varphi'_p}(U')$, $\mathcal{H}_{pq\varphi''_p}(U'')$. Fix a Stein open set $W \Subset U' \cap U''$ and a holomorphic function f'' on U'' such that*

$$\int_{U''} |f''|^2 |\sigma''|^{-2q} d\lambda \leq 1, \qquad q > n + 1.$$

(i) *One can write $f'' = \sum_{L \in \mathbb{N}^m} g_L(\sigma')^L$ on W with $m = q - n - 1$ and*

$$\int_W \sum_L |g_L|^2 |\sigma'|^{-2(n+1)} d\lambda \leq C^{2q} e^{2pq\varepsilon}$$

with a constant $C > 1$ depending only on $d(W, \complement(U' \cap U''))$.

(ii) *There are holomorphic functions h_l on W such that $f'' = \sum h_l \hat{\sigma}'_l$ on W, and*

$$\sup_W \sum_l |h_l|^2 \leq C_1(p) C^{2q} e^{2pq\varepsilon}$$

where C is as in (i) and $C_1(p)$ depends on p (and U', U'', W as well).

PROOF. (i) Thanks to (7.3), we have $|\sigma'| \geq C^{-1} e^{-p\varepsilon} |\sigma''|$ on W for some constant $C > 1$ depending only only on $d(W, \complement(U' \cap U''))$. Therefore

$$\int_W |f|^2 |\sigma'|^{-2q} d\lambda \leq C^{2q} e^{2pq\varepsilon}.$$

We apply Skoda's L^2 division theorem (Corollary 10.6) with $r = n$, $m = q - n - 1$, $\alpha = 1$, on the Stein open set W. Our assertion (i) follows, after absorbing the extra constant $(q - n)$ in C^{2q}.

(ii) We first apply (i) on a Stein open set W_1 such that $W \Subset W_1 \Subset U' \cap U''$, and write in this way $f'' = \sum_L g_L(\sigma')^L$ with the L^2 estimate as in (i). By [Nadel 1990] (see also [Demailly 1993]), the ideal sheaf $\mathcal{J}$ of holomorphic functions v on U' such that

$$\int_{U'} |v|^2 |\sigma'|^{-2(n+1)} d\lambda < +\infty$$

is coherent and locally generated by its global L^2 sections (of course, this ideal depends on the σ'_l, hence on p and φ'). It follows that we can find finitely many holomorphic functions $v_1, \ldots, v_N$, $N = N(p)$, such that

$$\int_{U'} |v_j|^2 |\sigma'|^{-2(n+1)} d\lambda = 1$$

and $\mathcal{J}(W_1) = \sum v_j \mathcal{O}(W_1)$. As the topology given by the L^2 norm on the L^2 sections of $\mathcal{J}(W_1)$ is stronger than the Fréchet topology of uniform convergence on compact subsets, and as we have a Fréchet epimorphism $\mathcal{O}(W_1)^{\oplus N} \to \mathcal{J}(W_1)$,

$(a_1, \ldots, a_N) \mapsto \sum a_j v_j$, the open mapping theorem shows that we can write $g = \sum a_j v_j$ with

$$\sup_W \sum_{j=1}^N |a_j|^2 \le A(p) \int_{W_1} |g|^2 |\sigma'|^{-2(n+1)} d\lambda$$

for every holomorphic function g on W_1 for which the right hand side is finite [the constant $A(p)$ depends on $\mathfrak{I}$, hence on p]. In particular, we can write $g_L = \sum_j a_{j,L} v_j$ with

$$\sup_W \sum_{j,L} |a_{j,L}|^2 \le A(p) \int_{W_1} \sum_L |g_L|^2 |\sigma'|^{-2(n+1)} d\lambda \le A(p) C^{2q} e^{2pq\varepsilon}.$$

We find

$$f'' = \sum_L g_L (\sigma')^L = \sum_{j,L} a_{j,L} g_j (\sigma')^L,$$

and as L runs over all multiindices of length $m = q - n - 1$ we get

$$\int_{U'} \sum_{j,L} |g_j (\sigma')^L|^2 |\sigma'|^{-2q} d\lambda = \int_{U'} \sum_j |g_j|^2 |\sigma'|^{-2(n+1)} d\lambda = N = N(p).$$

We can therefore express the function $g_j (\sigma')^L$ in terms of the Hilbert basis $(\hat{\sigma}'_l)$

$$g_j (\sigma')^L = \sum_l b_{j,L,l} \hat{\sigma}'_l, \qquad b_{j,L,l} \in \mathbb{C}, \qquad \sum_{j,L,l} |b_{j,L,l}|^2 = N(p).$$

Summing up everything, we obtain

$$f'' = \sum_{j,L,l} a_{j,L} b_{j,L,l} \hat{\sigma}'_l = \sum_l h_l \hat{\sigma}'_l, \qquad h_l = \sum_{j,L} a_{j,L} b_{j,L,l}.$$

The Cauchy–Schwarz inequality implies

$$\sup_W \sum_l |h_l|^2 \le \sup_W \sum_{j,L} |a_{j,L}|^2 \sum_{j,L,l} |b_{j,L,l}|^2 \le N(p) A(p) C^{2q} e^{2pq\varepsilon},$$

as desired. $\qquad\qquad\qquad\qquad\qquad\qquad\qquad\qquad\qquad\qquad\qquad\qquad\qquad\square$

Now, assume that X is a complex manifold such that $T^\star_X$ is equipped with a plurisubharmonic Finsler metric. As all constructions to be used are local, we may suppose that we are in a small coordinate open subset $U_0 \Subset X$ or, equivalently, in a Stein open set $U_0 \Subset \mathbb{C}^n$, with φ being defined on U_0. We fix Stein open sets $U \Subset U_1 \Subset U_0$ and select a sequence $\rho = (\rho_m)$ satisfying property (6.5) on each fiber $T^\star_{U,x}$. Finally, for $(x,t) \in U \times \mathbb{C}$, we set

$$(7.7) \qquad \Phi^{\chi,\rho,p,q}(x,t) := \sup_{f \in B_{p,q}(1)} \frac{1}{pq} \log \widehat{M}^{\chi,\rho}_F(x,t) + \frac{C_0}{p}, \qquad C_0 \gg 0,$$

where f runs over the unit ball $B_{p,q}(1)$ of $\mathcal{H}_{pq\varphi_p}(U_1)$ and $F(x,\xi) = f(x+\xi)$. Then $\Phi^{\chi,\rho,p,q}(x,t)$ is well defined on $U \times \{\operatorname{Re} t < -A\}$ for $A \ge 0$ sufficiently large. Thanks to Lemma 6.8, the choice of coordinates on U_0 is essentially irrelevant when we compute $\widehat{M}_F^{\chi,\rho}(x,t)$, provided that ρ decays fast enough. Moreover, a change of coordinate $\tau : U_0 \mapsto \tau(U_0)$ has the effect of replacing φ by $\varphi^\tau = \varphi \circ \tau^{-1}$ and φ_p by $\varphi_p^\tau = \varphi_p \circ \tau^{-1} + O(1/p)$, since the only change occurring in the definition of $\mathcal{H}_{p\varphi^\tau}(\tau(U_1))$ is the replacement of the Lebesgue volume form $d\lambda$ by $\tau^\star d\lambda$, which affects the L^2 norm by at most a constant. Similarly, the L^2 norm of $\mathcal{H}_{pq\varphi_p^\tau}(\tau(U_1))$ gets modified by an irrelevant multiplicative factor $\exp(O(q))$, inducing a negligible error term $O(1/p)$ in (7.7). If $C_0 \geq 0$ is large enough, (7.2) combined with 7.5 (i) implies that

$$\varphi(z) \leq \varphi_{p,q}(z) + \frac{C_0}{p} = \sup_f \frac{1}{pq} \log |f(x)| + \frac{C_0}{p} \leq \sup_{|\zeta-z|<r} \varphi(\zeta) + \left(\frac{1}{p} + \frac{1}{pq}\right) \log \frac{C}{r^n}$$

for some $C > 0$, where f runs over $B_{p,q}(1) \subset \mathcal{H}_{pq\varphi_p}(U_1)$. Lemma 6.3 applied with $r = |e^t|$ then gives

(7.8 i)
$$\Phi^{\chi,\rho,p,q}(x,t) \leq \sup_{|z-x|\leq C'|e^t|} \varphi_{p,q}(z) + \frac{C_0}{p}$$
$$\leq \sup_{|z-x|\leq C|e^t|} \varphi(z) - n\left(\frac{1}{p} + \frac{1}{pq}\right) \operatorname{Re} t + \frac{C}{p},$$

(7.8 ii)
$$\Phi^{\chi,\rho,p,q}(x,t) \geq \sup_{|z-x|\leq c_{\chi,\rho,p,q,\varphi,U}|e^t|} \varphi_{p,q}(z) + \frac{C_0}{p}$$
$$\geq \sup_{|z-x|\leq c_{\chi,\rho,p,q,\varphi,U}|e^t|} \varphi(z) \geq \varphi(x)$$

where C, C' are universal constants, and $c_{\chi,\rho,p,q,\varphi,U}$ depends on all given data, but is independent of x. The last inequality is a simple consequence of the fact that the Taylor series $\widehat{M}_f^{\chi,\rho}(x,t) = \sum_{m\geq 0} \rho_m \|a_m(x)\| \, |e^{mt}|$ are never identically zero, hence their behavior as $|e^t| \to 0$ is the same as for the series $\sum_{m\leq N} \rho_m \|a_m(x)\| \, |e^{mt}|$, truncated at some rank $N = N_{p,q,\varphi,U}$. The constant $c_{\chi,\rho,p,q,\varphi,U}$ then essentially depends only on $\inf_{m\leq N} \rho_m$. The upper and lower bound provided by (7.8) imply in particular

(7.9 i)
$$\lim_{\operatorname{Re} t \to -\infty} \frac{\Phi^{\chi,\rho,p,q}(x,t)}{\operatorname{Re} t} = \lim_{\operatorname{Re} t \to -\infty} \frac{\sup_{|z-x|\leq|e^t|} \varphi_{p,q}(z)}{\operatorname{Re} t} = \nu(\varphi_{p,q}, x),$$

(7.9 ii)
$$\lim_{\operatorname{Re} t \to -\infty} \left| \frac{\Phi^{\chi,\rho,p,q}(x,t)}{\operatorname{Re} t} - \frac{\sup_{|z-x|\leq|e^t|} \varphi_{p,q}(z)}{\operatorname{Re} t} \right| = 0,$$

where the second limit is uniform on U [For this, we use the convexity of $\operatorname{Re} t \mapsto \sup_{|z-t|\leq|e^t|} \varphi(z)$ to check that the constants C in $\sup_{|z-t|\leq C|e^t|} \varphi(z)$ are irrelevant.] For future reference, we also note

(7.10) The functions $\Phi^{\chi,\rho,p,q}(x,t)$ are continuous on $U \times \{\operatorname{Re} t < -A\}$.

This is an immediate consequence of the fact that the unit ball $B_{p,q}(1)$ of $\mathcal{H}_{pq\varphi_p}(U_1)$ is a normal family of holomorphic functions. We now investigate the effect of a perturbation on φ.

PROPOSITION 7.11. *Let* $U' \Subset U_1' \Subset U_0$, $U'' \Subset U_1'' \Subset U_0$ *and let* φ', φ'' *be plurisubharmonic functions on* U_1', U_1'' *such that* $|\varphi'' - \varphi' - \operatorname{Re} g| \le \varepsilon$ *on* $U_1' \cap U_1''$, *for some holomorphic function* $g \in \mathcal{O}(U_1' \cap U_1'')$. *There are constants* $C_2(p)$ *and* C_3 (*depending also on* φ', φ'', g, U', U'') *such that*

$$\left| \Phi''^{\,\chi,\rho,p,q}(x,t) - \Phi'^{\,\chi,\rho,p,q}(x,t) - \operatorname{Re} g(x) \right| \le 2\varepsilon + \frac{C_2(p)}{q} + \frac{C_3}{p}$$

for all $x \in U' \cap U''$ *and* $|e^t| < r_0(\varepsilon)$ *small enough.*

PROOF. We first treat the simpler case when $g = 0$. By 7.6 (ii), every function $f'' \in B_{p,q}''(1) \subset \mathcal{H}_{pq\varphi_p}(U_1'')$ can be written

$$f'' = \sum_{l \in \mathbb{N}} h_l \hat{\sigma}_l', \qquad \sup_W \sum_l |h_l|^2 \le C_1(p) C^{2q} e^{2pq\varepsilon}$$

on any relatively compact neighborhood W of $\overline{U' \cap U''}$ in $U_1' \cap U_1''$. Fix a small polydisk $\overline{D}(r) \subset \mathbb{C}^n$ such that $\overline{U' \cap U''} + \overline{D}(r) \subset U_1' \cap U_1''$, and expand

$$h_l(x) = \sum_{\alpha \in \mathbb{N}^n} a_{l,\alpha} (x - x_0)^\alpha$$

as a power series at each point $x_0 \in U' \cap U''$. By integrating $\sum |h_l|^2$ over the polycircle $\prod \partial D(x_{0,j}, r_j)$, we find

(7.12) $$\sum_{l \in \mathbb{N}, \alpha \in \mathbb{N}^n} |a_{l,\alpha}|^2 |r^\alpha|^2 \le C_1(p) C^{2q} e^{2pq\varepsilon}.$$

By substituting h_l with its Taylor expansion in the definition of f'', we find

$$f''(x) = \sum_{\alpha \in \mathbb{N}^n} p_\alpha(x)\, w_\alpha(x)$$

where

$$p_\alpha(x) = (x - x_0)^\alpha, \qquad w_\alpha(x) = \sum_{l \in \mathbb{N}} a_{l,\alpha} \hat{\sigma}_l'.$$

The L^2 norm of w_α in $\mathcal{H}_{pq\varphi_p'}(U_1')$ is $\left(\sum_l |a_{l,\alpha}|^2 \right)^{1/2}$, hence by definition

$$\frac{C_0}{p} + \frac{1}{pq} \log \frac{\widehat{M}_{w_\alpha}^{\chi,\rho}(x_0,t)}{\left(\sum_l |a_{l,\alpha}|^2 \right)^{1/2}} \le \Phi_{p,q}'(x_0,t).$$

On the other hand, if $|e^t| \ll \|r\|$, Lemma 6.3 implies that

$$\widehat{M}_{p_\alpha}^{\chi,\rho}(x_0,t) \le \sup_{D(x_0,r/3)} |(x - x_0)^\alpha| \le 2^{-\alpha} r^\alpha$$

From this, we infer

$$\hat{M}_{f''}^{\chi,\rho}(x_0,t) \leq \sum_{\alpha \in \mathbb{N}^n} \hat{M}_{p_\alpha}^{\chi,\rho}(x_0,t)\hat{M}_{w_\alpha}^{\chi,\rho}(x_0,t)$$

$$\leq \sum_{\alpha \in \mathbb{N}^n} 2^{-\alpha} r^\alpha \left(\sum_l |a_{l,\alpha}|^2\right)^{1/2} \exp\left(pq(\Phi'_{p,q}(x_0,t) - C_0/p)\right)$$

$$\leq \sum_{\alpha \in \mathbb{N}^n} 2^{-\alpha} \left(C_1(p)C^{2q}e^{2pq\varepsilon}\right)^{1/2} \exp\left(pq(\Phi'_{p,q}(x_0,t) - C_0/p)\right)$$

thanks to (7.12). By taking $\frac{1}{pq}\log(\dots)$ and passing to the sup over all f'', we get

$$\Phi''_{p,q}(x_0,t) \leq \frac{1}{pq}\log\left(2^n \left(C_1(p)C^{2q}e^{2pq\varepsilon}\right)^{1/2}\right) + \Phi'_{p,q}(x_0,t).$$

Proposition 7.11 is thus proved for the case $g = 0$, even with ε instead of 2ε in the final estimate. In case g is non zero, we observe that the replacement of φ' by $\varphi' + \operatorname{Re} g$ yields isomorphisms of Hilbert spaces

$$\mathcal{H}_{p\varphi'}(U'_1) \to \mathcal{H}_{p(\varphi'+\operatorname{Re} g)}(U'_1), \qquad f \mapsto e^{pg}f,$$

$$\mathcal{H}_{pq\varphi'_p}(U'_1) \to \mathcal{H}_{pq(\varphi'_p+\operatorname{Re} g)}(U'_1), \qquad f \mapsto e^{pqg}f.$$

The only difference occurring in the proof is that we get

$$f'' = e^{pqg} \sum_{l \in \mathbb{N}} h_l \hat{\sigma}'_l$$

instead of $f'' = \sum_{l \in \mathbb{N}} h_l \hat{\sigma}'_l$. In the upper bound for $\hat{M}_{f''}^{\chi,\rho}(x_0,t)$, this introduces an extra term $\hat{M}_{e^{pqg}}^{\chi,\rho}(x_0,t)$, which we evaluate as $\exp(pq(\operatorname{Re} g(x_0) + O(|e^t|)))$ thanks to Lemma 6.3. The general estimate follows, possibly with an additional ε error when $|e^t|$ is small enough. □

The final step in the construction is to "glue" together the functions $\Phi^{\chi,\rho,p,q}(x,t)$, $(p,q) \in \mathbb{N}^2$. We choose a fast increasing sequence $p \mapsto q(p)$, in such a way that $C_2(p)/q(p) \leq 1/p$, where $C_2(p)$ is the constant occurring in Proposition 7.11. We now define

$$\tilde{M}_\varphi^{\chi,\rho,s}(x,t) := \Phi^{\chi,\rho,s}(x,t)$$

$$(7.13) \qquad := \sup_{p \geq s} \left(\Phi^{\chi,\rho,p,q(p)}(x,t - \log p) + \frac{\log p}{p} + n\left(\frac{1}{p} + \frac{1}{pq(p)}\right)\operatorname{Re} t\right).$$

[The terms in $\log p$ are there only for a minor technical reason, to make sure that $\tilde{M}_\varphi^{\chi,\rho,s}(x,t)$ is a continuous function.] In this way, we achieve the expected goals, namely:

PROPOSITION 7.14. *Let φ be a plurisbharmonic function defined on a bounded Stein open set $U_0 \Subset \mathbb{C}^n$ such that $T^\star_{U_0}$ is equipped with a plurisubharmonic smooth Finsler metric, and let $U \Subset U_0$. Then there is a functional $\tilde{M}^{\chi,\rho,s}$ (associated with the choice of a sequence $q(p)$ which may have to be adjusted when φ varies,* *but can be taken fixed if φ remains in a bounded set of $L^1(U_0)$), such that the functions $\Phi^{\chi,\rho,s}(x,t) = \widetilde{M}^{\chi,\rho,s}_\varphi(x,t)$ satisfy the following properties:*

(i) *The functions $\Phi^{\chi,\rho,s}(x,t)$ are defined on $U \times \{\operatorname{Re} t < -A\}$ for $A > 0$ large, and are locally bounded continuous plurisubharmonic functions depending only on $\operatorname{Re} t$; moreover, $p \mapsto \Phi^{\chi,\rho,s}(x,t)$ is a decreasing family of functions.*

(ii) $\varphi(x) \le \Phi^{\chi,\rho,s}(x,t) \le \displaystyle\sup_{\|z-x\| \le Cs^{-1}|e^t|} \varphi(z) + C\,\dfrac{\log s}{s}$ *for some $C \gg 0$.*

(iii) $\displaystyle\lim_{\operatorname{Re} t \to -\infty} \left| \dfrac{\Phi^{\chi,\rho,s}(x,t)}{\operatorname{Re} t} - \dfrac{\sup_{\|z-x\| \le |e^t|} \varphi(z)}{\operatorname{Re} t} \right| = 0$

   uniformly on every compact subset of U, in particular

$$\lim_{\operatorname{Re} t \to -\infty} \frac{\Phi^{\chi,\rho,s}(x,t)}{\operatorname{Re} t} = \nu(\varphi,x)$$

   for every $x \in U$.

(iv) *For every holomorphic change of coordinates $\tau : U_0 \to \tau(U_0)$, the sequence $\rho = (\rho_m)$ can be chosen (depending only on τ) such that for some constant $C > 0$ we have*

$$\left| \widetilde{M}^{\chi,\rho,s}_{\varphi \circ \tau^{-1}}(\tau(x),t) - \widetilde{M}^{\chi,\rho,s}_\varphi(x,t) \right| \le \frac{C}{s} \qquad \text{for all } x \in U,$$

   when $T^\star_{\tau(U_0)}$ is equipped with the induced Finsler metric.

(v) *Let $U' \Subset U'_1 \Subset U_0$, $U'' \Subset U''_1 \Subset U_0$ be Stein open subsets, and let φ', φ'' be plurisubharmonic functions on U'_1, U''_1 such that $|\varphi' - \varphi| \le 1$ on U'_1, $|\varphi'' - \varphi| \le 1$ on U''_1 and $|\varphi'' - \varphi' - \operatorname{Re} g| \le \varepsilon$ on $U'_1 \cap U''_1$ for some holomorphic function $g \in \mathcal{O}(U'_1 \cap U''_1)$. Then*

$$\left| \widetilde{M}^{\chi,\rho,s}_{\varphi''}(x,t) - \widetilde{M}^{\chi,\rho,s}_{\varphi'}(x,t) \right| \le 2\varepsilon + \frac{C}{s} \qquad \text{for all } x \in U' \cap U'' \text{ and } |e^t| < r_0(\varepsilon),$$

   where $C = C(\varphi, U', U'')$.

PROOF. All properties are almost immediate consequences of the properties already obtained for $\Phi^{\chi,\rho,p,q}$, simply by taking the supremum. We check e.g. the continuity of $\Phi^{\chi,\rho,s}$, inequality (ii) and the second statement of (iii). In fact, (7.8 i,ii) imply

$$\varphi(x) + \frac{\log p}{p} + n\left(\frac{1}{p} + \frac{1}{p\,q(p)}\right)\operatorname{Re} t,$$

$$\le \Phi^{\chi,\rho,p,q(p)}(x, t - \log p) + \frac{\log p}{p} + n\left(\frac{1}{p} + \frac{1}{p\,q(p)}\right)\operatorname{Re} t$$

$$\le \sup_{\|z-x\| \le Cp^{-1}|e^t|} \varphi(z) + C\,\frac{\log p}{p}$$

and (7.14 ii) follows from this. Moreover, the function $\Phi^{\chi,\rho,p,q(p)}(x, t - \log p) + \cdots$ converges to $\varphi(x)$ as $p \to +\infty$, while its terms get $> \varphi(x)$ for p large, thanks to the lower bound. It follows that the sup in (7.13) is locally finite, therefore $\Phi^{\chi,\rho,s}$ is continuous. To prove (iii), we first observe that the right hand inequality in (i) gives

$$\lim_{\operatorname{Re} t \to -\infty} \frac{\Phi^{\chi,\rho,s}(x,t)}{\operatorname{Re} t} \geq \lim_{\operatorname{Re} t \to -\infty} \frac{\sup_{|z-x| \leq |e^t|} \varphi(z)}{\operatorname{Re} t} = \nu(\varphi,x).$$

In the other direction, the definition of $\Phi^{\chi,\rho,s}(x,t)$ combined with (7.8 ii) implies

$$\Phi^{\chi,\rho,s}(x,t) \geq \Phi^{\rho,p,q(p)}(x,t-\log p) + n\left(\frac{1}{p} + \frac{1}{p\,q(p)}\right)\operatorname{Re} t$$

$$\geq \sup_{|z-x| \leq p^{-1}c_{\rho,p,q(p),\varphi,U}|e^t|} \varphi(z) + n\left(\frac{1}{p} + \frac{1}{p\,q(p)}\right)\operatorname{Re} t$$

for all $p \geq s$, hence

$$\lim_{\operatorname{Re} t \to -\infty} \frac{\Phi^{\chi,\rho,s}(x,t)}{\operatorname{Re} t} \leq \lim_{\operatorname{Re} t \to -\infty} \frac{\sup_{|z-x| \leq |e^t|} \varphi(z)}{\operatorname{Re} t} + n\left(\frac{1}{p} + \frac{1}{p\,q(p)}\right).$$

We get the desired conclusion by letting $p \to +\infty$. $\qquad\square$

8. A Variant of Kiselman's Legendre Transform

To begin with, let φ be a plurisubharmonic function on a bounded pseudoconvex open set $U \Subset \mathbb{C}^n$. Consider the trivial vector bundle $T_U = U \times \mathbb{C}^n$, and assume that $T_U^\star$ is equipped with a smooth Finsler metric $_{\chi^\star}\|\xi^\star\|^\star = \exp(\chi^\star(z,\xi^\star))$ for $\xi^\star \in T_{U,z}^\star$. We assume that the curvature of the Finsler metric $\|\xi^\star\|_z^\star = e^{\chi^\star(z,\xi^\star)}$ on $T_U^\star$ satisfies

$$(8.1) \qquad \frac{i}{\pi}\partial\bar{\partial}\chi^\star(z,\xi^\star) + \pi_U^\star u(z) \geq 0$$

for some nonnegative continuous $(1,1)$-form u on U, where $\pi_U : T_U^\star \to U$ is the projection. If $\chi^\star = \log h^\star$ is a hermitian metric on $T_X^\star$, we let h be the dual metric on T_X and set

$$(8.2)^h \qquad \Phi_\infty^h(z,w) = \sup_{_h\|\xi\|_z \leq |e^w|} \varphi(z+\xi).$$

By Theorem 6.1, this definition works equally well when $h^\star$ is a *fiberwise convex* Finsler metric. Clearly $\Phi_\infty^h(z,w)$ depends only on the real part $\operatorname{Re} w$ of w and is defined on the open set Ω of points $(z,w) \in U \times \mathbb{C}$ such that $\operatorname{Re} w < \log d_z(z,\partial U)$, where d_z denotes euclidean distance with respect to $_h\|\ \|_z$. Now, we would like to extend this to the case of a general Finsler metric, without any convexity assumption. As a replacement for the "sup formula" $(8.2)^h$, we set

$$(8.2)^\chi \qquad \Phi_\infty^{\chi,\rho,s}(z,w) = \widetilde{M}_\varphi^{\chi,\rho,s}(z,w)$$

where $\widetilde{M}^{\chi,\rho,s}$ denotes the functional associated with χ, as in § 7. Here, however, $\chi(z,\xi^*)$ need not be plurisubharmonic. This is not a real difficulty, since the definition of the $\widehat{M}^{\chi,\rho}$ functional in (6.6) shows that

$$\widetilde{M}_{\varphi}^{\chi,\rho,s}(z,w) = \widetilde{M}_{\varphi}^{\chi_v,\rho,s}(z,w - v(z))$$

for any smooth function v on U such that $\chi_v(z,\xi^*) := \chi(z,\xi^*)+v(z)$ is plurisubharmonic on U (and such a function always exists by our assumption (8.1)). One of our main concern is to investigate singularities of φ and these singularities are reflected in the way $\Phi_{\infty}^h(z,w)$ and $\Phi_{\infty}^{\chi,\rho,s}$ decay to $-\infty$ as $\operatorname{Re} w$ goes to $-\infty$. In this perspective, Proposition 7.14 (iii) shows that considering $\Phi_{\infty}^{\chi,\rho,s}(z,w)$ instead of $\Phi_{\infty}^h(z,w)$ does not make any difference. Moreover $\Phi_{\infty}^{\chi,\rho,s}(z,w)$ and $\Phi_{\infty}^h(z,w)$ are both convex increasing function of $\operatorname{Re} w$. For $(z,w) \in \Omega$ and $c > 0$, we introduce the (generalized) *Legendre transform*

$$(8.3)^h \qquad \Phi_c^h(z,w) = \inf_{t \leq 0} \Phi_{\infty}^h(z,w+t) - ct,$$

$$(8.3)^\chi \qquad \Phi_c^{\chi,\rho,s}(z,w) = \inf_{t \leq 0} \Phi_{\infty}^{\chi,\rho,s}(z,w+t) - ct.$$

It is easy to see that these functions are increasing in c and that

$$(8.4)^h \qquad \lim_{c \to 0} \Phi_c^h(z,w) = \varphi(z), \qquad \lim_{c \to +\infty} \Phi_c^h = \Phi_{\infty}^h.$$

The analogue for $\Phi_c^{\chi,\rho,s}$ is

$$(8.4)^\chi \quad \lim_{c \to 0} \Phi_c^{\chi,\rho,s}(z,w) = \lim_{\operatorname{Re} t \to -\infty} \Phi_{\infty}^{\chi,\rho,s}(z,\operatorname{Re} t) \in \left[\varphi(z),\, \varphi(z) + C\log s/s\right],$$

$$\lim_{c \to +\infty} \Phi_c^{\chi,\rho,s} = \Phi_{\infty}^{\chi,\rho,s}.$$

When $_h\|\xi\|_z$ is taken to be a constant metric, we know by [Kiselman 1978] that Φ_{∞}^h and Φ_c^h are plurisubharmonic functions of the pair (z,w), and that the Lelong numbers of $\Phi_c^h(\,\cdot\,,w)$ are given by

$$(8.5)^h \qquad \nu\big(\Phi_c^h(\,\cdot\,,w),z\big) = \big(\nu(\varphi,z) - c\big)_+, \qquad \text{for all } (z,w) \in \Omega.$$

Since $(8.5)^h$ depends only on the maps $z \mapsto \Phi_{\infty}^h(z,w)$ with w fixed, the equality is still valid when h is a variable hermitian metric, and Proposition 7.14 (iii) even shows that the analogous property for $\Phi_c^{\chi,\rho,s}$ is true:

$$(8.5)^\chi \qquad \nu\big(\Phi_c^{\chi,\rho,s}(\,\cdot\,,w),z\big) = \big(\nu(\varphi,z) - c\big)_+, \qquad \text{for all } (z,w) \in \Omega.$$

As usual we denote by

$$(8.6) \qquad E_c(\varphi) = \{z \in U : \nu(\varphi,x) \geq c\}$$

the Lelong sublevel sets of φ. From now on, we omit the superscripts in the notation Φ_c^h or $\Phi_c^{\chi,\rho,s}$ since all properties are the same in both cases. In general, Φ_{∞} is continuous on Ω and its right derivative $\partial\Phi_{\infty}(z,w)/\partial \operatorname{Re} w_+$ is upper semicontinuous; indeed, this partial derivative is the decreasing limit of $\big(\Phi_{\infty}(z,w+t) - \Phi_{\infty}(z,w)\big)/t$ as $t \downarrow 0_+$. It follows that Φ_c is continuous on

$\Omega \smallsetminus (E_c(\varphi) \times \mathbb{C})$: in fact, we have $\nu(\varphi, z) = \lim_{t \to -\infty} \partial \Phi_\infty(z, t)/\partial t_+ < c$ on every compact set $K \subset \Omega \smallsetminus (E_c(\varphi) \times \mathbb{C})$, so by the upper semicontinuity there is a constant t_0 such that $\partial \Phi_\infty(z, w + t)/\partial t_+ < c$ for $(z, w) \in K$ and $t < t_0$. Therefore

$$\Phi_c(z, w) = \inf_{t_0 \leq t \leq 0} \Phi_\infty(z, w + t) - ct \qquad \text{on } K,$$

and this infimum with compact range is continuous. Our next goal is to investigate the plurisubharmonicity of Φ_c.

PROPOSITION 8.7. *Assume the curvature of the Finsler metric* $\|\xi^\star\|_z^\star = e^{\chi^\star(z, \xi^\star)}$ *on* $E^\star$ *satisfies*

$$\frac{i}{\pi} \partial \bar{\partial} \chi^\star(z, \xi^\star) + \pi_X^\star u(z) \geq 0$$

for some nonnegative continuous $(1,1)$*-form* u *on* X, *where* $\pi_X : E^\star \to X$ *is the projection. Then* $\Phi_c = \Phi_c^{\chi, \rho, s}$ *[and likewise* $\Phi_c = \Phi_c^h$*] enjoys the following properties.*

(i) *For all* $\eta \geq 0$, *we have*

$$\Phi_c(z, w - \eta) \geq \Phi_c(z, w) - \min \left\{ \frac{\partial \Phi_\infty(z, w)}{\partial \operatorname{Re} w_-}, c \right\} \eta;$$

(ii) *For* $(\zeta, \eta) \in T_U \times \mathbb{C}$ *and* $c \in \,]0, +\infty]$, *the Hessian of* Φ_c *satisfies*

$$\frac{i}{\pi} \partial \bar{\partial} (\Phi_c)_{(z,w)}(\zeta, \eta) \geq - \min \left\{ \frac{\partial \Phi_\infty(z, w)}{\partial \operatorname{Re} w_+}, c \right\} u_z(\zeta).$$

PROOF. (i) For $\eta \geq 0$ and $t \leq 0$, the convexity of $\Phi_\infty(z, w)$ in $\operatorname{Re} w$ implies

$$\Phi_\infty(z, w + t - \eta) \geq \Phi_\infty(z, w + t) - \eta \frac{\partial \Phi_\infty(z, w + t)}{\partial \operatorname{Re} w_-}$$

As $\partial \Phi_\infty(z, w)/\partial \operatorname{Re} w_-$ is increasing in $\operatorname{Re} w$, the infimum of both sides minus ct gives

$$\Phi_c(z, w - \eta) \geq \Phi_c(z, w) - \eta \frac{\partial \Phi_\infty(z, w)}{\partial \operatorname{Re} w_-}.$$

On the other hand, the change of variables $t = t' + \eta$ yields

$$\Phi_c(z, w - \eta) \geq \inf_{t' \leq -\eta} \Phi_\infty(z, w + t') - c(t' + \eta) \geq \Phi_c(z, w) - c\eta.$$

Property (i) follows.

(ii) Fix $(z_0, w_0) \in \Omega$ and a semipositive quadratic function $v(z)$ on $\mathbb{C}^n$ such that $\frac{i}{\pi} \partial \bar{\partial} v(0) > u_{z_0}$. Then the inequality $\frac{i}{\pi} \partial \bar{\partial} v(z - z_0) > u(z)$ still holds on a neighborhood U_0 of z_0, and the Finsler metric $\|\xi^\star\|_z^\star e^{v(z - z_0)}$ is plurisubharmonic on this neighborhood. From this, we conclude by Lemma 7.7 that the associated function

$$h(z, w) := \Phi_\infty(z, w + v(z - z_0))$$

is plurisubharmonic on U_0. Its Legendre transform

$$h_c(z, w) = \inf_{t \leq 0} h(z, w - t) = \Phi_c(z, w + v(z - z_0))$$

is again plurisubharmonic. For small $(\zeta, \eta) \in T_U \times \mathbb{C}$, the mean value inequality yields

$$\int_0^{2\pi} \Phi_c(z_0 + e^{i\theta}\zeta, w_0 + e^{i\theta}\eta) \frac{d\theta}{2\pi} = \int_0^{2\pi} h_c(z_0 + e^{i\theta}\zeta, w_0 + e^{i\theta}\eta - v(\zeta)) \frac{d\theta}{2\pi}$$

$$\geq h_c(z_0, w_0 - v(\zeta)) = \Phi_c(z_0, w_0 - v(\zeta))$$

$$\geq \Phi_c(z_0, w_0) - \min\left\{\frac{\partial \Phi_\infty(z_0, w_0)}{\partial \operatorname{Re} w_+}, c\right\} v(\zeta)$$

[the last inequality follows from (i)]. For $A > \partial \Phi_\infty(z_0, w_0)/\partial \operatorname{Re} w_+$, we still have $A > \partial \Phi_\infty(z, w)/\partial \operatorname{Re} w_+$ in a neighborhood of (z_0, w_0) by the upper semicontinuity, and we conclude that the function $\Phi_c(z, w) + \min\{A, c\}v(z)$ satisfies the mean value inequality near (z_0, w_0). Hence $\Phi_c(z, w) + \min\{A, c\}v(z)$ is plurisubharmonic near (z_0, w_0). Since this is still true as A tends to $\partial \Phi_\infty(z, w)/\partial \operatorname{Re} w_+$ and $\frac{i}{\pi}\partial\bar\partial v$ tends to u_{z_0}, the proof of (ii) is complete. $\qquad\square$

9. Regularization of Closed Positive $(1, 1)$-Currents

The next step is to describe a gluing process for the construction of global regularizations of almost plurisubharmonic functions. We suppose that $T_X^\star$ is equipped with a Finsler metric $\|\xi^\star\|_x^\star = e^{\chi^\star(x, \xi^\star)}$ satisfying

$$\frac{i}{\pi}\partial\bar\partial\chi^\star(x, \xi^\star) + \pi_X^\star u(x) \geq 0,$$

where u is a smooth semipositive $(1, 1)$-form on X. Notice that $\frac{i}{\pi}\partial\bar\partial\chi^\star(z, \xi^\star)$ is just the Chern curvature of the induced hermitian metric on $\mathcal{O}_{TX}(1)$. An *almost positive* $(1, 1)$-current is by definition a real $(1, 1)$-current such that $T \geq \gamma$ for some real $(1, 1)$-form γ with locally bounded coefficients. An *almost psh function* is a function ψ which can be written locally as $\psi = \varphi + w$ where φ is plurisubharmonic and w smooth. With these definitions, $\frac{i}{\pi}\partial\bar\partial\psi$ is almost positive if and only if ψ is almost psh.

The following thereom was proved in [Demailly 1992] with a rather long and tricky proof. We present here a shorter and better approach using our modified Kiselman–Legendre transforms.

THEOREM 9.1. *Let T be a closed almost positive $(1, 1)$-current and let α be a smooth real $(1, 1)$-form in the same $\partial\bar\partial$-cohomology class as T, i.e. $T = \alpha + \frac{i}{\pi}\partial\bar\partial\psi$ where ψ is an almost psh function. Let γ be a continuous real $(1, 1)$-form such that $T \geq \gamma$. Suppose that $\mathcal{O}_{TX}(1)$ is equipped with a smooth hermitian metric such that the Chern curvature form satisfies*

$$\Theta(\mathcal{O}_{TX}(1)) + \pi_X^\star u \geq 0$$

with $\pi_X : P(T^*X) \to X$ and with some nonnegative smooth $(1,1)$-form u on X. Fix a hermitian metric ω on X. Then for every $c > 0$, there is a sequence of closed almost positive $(1,1)$-currents $T_{c,k} = \alpha + \frac{i}{\pi}\partial\bar\partial\psi_{c,k}$ such that $\psi_{c,k}$ is smooth on $X \smallsetminus E_c(T)$ and decreases to ψ as k tends to $+\infty$ (in particular, the current $T_{c,k}$ is smooth on $X \smallsetminus E_c(T)$ and converges weakly to T on X), and such that

(i) $T_{c,k} \geq \gamma - \min\{\lambda_k, c\}u - \varepsilon_k\omega$, where

(ii) $\lambda_k(x)$ is a decreasing sequence of continuous functions on X such that the limit $\lim_{k\to+\infty} \lambda_k(x)$ equals $\nu(T, x)$ at every point,

(iii) ε_k is positive decreasing and $\lim_{k\to+\infty} \varepsilon_k = 0$,

(iv) $\nu(T_{c,k}, x) = \big(\nu(T, x) - c\big)_+$ at every point $x \in X$.

PROOF. We first show that we indeed can write $T = \alpha + \frac{i}{\pi}\partial\bar\partial\psi$ with α smooth. Let (U_j^0) be a finite covering of X by coordinate balls and (θ_j) a partition of unity subordinate to (U_j^0). If T is written locally $T = \frac{i}{\pi}\partial\bar\partial\psi_j$ with ψ_j defined on U_j^0, then $\psi = \sum \theta_j\psi_j$ has the property that $\alpha := T - \frac{i}{\pi}\partial\bar\partial\psi$ is smooth. This is an easy consequence of the fact that $\psi_k - \psi_j$ is plurisubharmonic, hence smooth, on $U_j^0 \cap U_k^0$, writing T as $\frac{i}{\pi}\partial\bar\partial\psi_k$ over U_k^0. By replacing T with $T - \alpha$ and γ with $\gamma - \alpha$, we can assume that $\alpha = 0$ (in other words, Theorem 9.1 essentially deals only with the singular part of T).

We can therefore assume that $T = \frac{i}{\pi}\partial\bar\partial\psi$, where ψ is an almost plurisubharmonic function on X such that $T \geq \gamma$ for some continuous $(1,1)$-form γ. We select a finite covering $\mathcal{W} = (W_\nu)$ of X by open coordinate charts. Given $\delta > 0$, we take in each W_ν a maximal family of points with (coordinate) distance to the boundary $\geq 3\delta$ and mutual distance $\geq \delta$. In this way, we get for $\delta > 0$ small a finite covering of X by open balls U_j of radius δ, such that the concentric ball U_j^0 of radius 2δ is relatively compact in the corresponding chart W_ν. Let $\tau_j : U_j^0 \to B_j^0 := B(a_j, 2\delta)$ be the isomorphism given by the coordinates of W_ν and

$$B_j \Subset B_j^1 \Subset B_j^0, \qquad B_j = B(a_j, \delta), \qquad B_j^1 = B(a_j, \sqrt{2}\,\delta), \qquad B_j^0 = B(a_j, 2\delta),$$
$$U_j \Subset U_j^1 \Subset U_j^0, \qquad U_j = \tau_j^{-1}(B_j), \qquad U_j^1 = \tau_j^{-1}(B_j^1), \qquad U_j^0 = \tau_j^{-1}(B_j^0).$$

Let $\varepsilon(\delta)$ be a modulus of continuity for γ on the sets U_j^0, such that $\lim_{\delta\to 0} \varepsilon(\delta) = 0$ and $\gamma_x - \gamma_{x'} \leq \frac{1}{2}\varepsilon(\delta)\,\omega_x$ for all $x, x' \in U_j^0$. We denote by γ_j the $(1,1)$-form with constant coefficients on B_j^0 such that $\tau_j^*\gamma_j$ coincides with $\gamma - \varepsilon(\delta)\,\omega$ at $\tau_j^{-1}(a_j)$. Then we have

$$(9.2) \qquad\qquad 0 \leq \gamma - \tau_j^*\gamma_j \leq 2\varepsilon(\delta)\,\omega \qquad \text{on } U_j$$

for $\delta > 0$ small. We set $\psi_j = \psi \circ \tau_j^{-1}$ on B_j^0 and let $\tilde\gamma_j$ be the homogeneous quadratic function in $z - a_j$ such that $\frac{i}{\pi}\partial\bar\partial\tilde\gamma_j = \gamma_j$ on B_j^0. Finally, we set

$$(9.3) \qquad\qquad \varphi_j(z) = \psi_j(z) - \tilde\gamma_j(z) \qquad \text{on } B_j^0.$$

It is clear that φ_j is plurisubharmonic, since

$$\frac{i}{\pi}\partial\bar{\partial}(\varphi_j \circ \tau_j) = T - \tau_j^\star \gamma_j \geq \gamma - \tau_j^\star \gamma_j \geq 0.$$

We combine $(8.2)^\chi$ and $(8.3)^\chi$ to define "regularized" functions

(9.4) $\quad \Phi_{j,c}^{\chi,\rho,s}(z,w) = \inf_{t \leq 0} \widetilde{M}_{\varphi_j}^{\chi,\rho,s}(z,w), \qquad z \in B_j^1,$

(9.5) $\quad \Psi_{j,c}^{\chi,\rho,s}(z,w) = \Phi_{j,c}^{\chi,\rho,s}(z,w) + \tilde{\gamma}_j(z) - \varepsilon(\delta)^{1/2}|z - a_j|^2, \qquad z \in B_j^1,$

(9.6) $\quad \Psi_c^{\chi,\rho,s}(x,w) = \sup_{U_j^1 \ni x} \Psi_{j,c}^{\chi,\rho,s}(\tau_j(x),w), \qquad x \in X,$

for $\mathrm{Re}\, w < -A$, with $A \gg 0$. We have to check that the gluing procedure used in the definition of $\Psi_c^{\chi,\rho,s}$ does not introduce discontinuities when x passes through a boundary ∂U_j^1. For this, we must compare $\Psi_{j,c}^{\chi,\rho,s}(\tau_j(x),w)$ and $\Psi_{k,c}^{\chi,\rho,s}(\tau_k(x),w)$ on overlapping open sets U_j^1, U_k^1. The comparison involves two points:

- effect of replacing ψ_j with $\psi_j - \tilde{\gamma}_j$, and
- effect of coordinate changes.

First assume for simplicity that U_j^1 and U_k^1 are contained in the same coordinate patch W_ν (in such a way that $\tau_j = \tau_k$ on $U_j^1 \cap U_k^1$, therefore in this case, we do not have to worry about coordinate changes). Then $\psi_j = \psi_k$ on $B_j^1 \cap B_k^1$, and therefore $\varphi_k - \varphi_j = \tilde{\gamma}_j - \tilde{\gamma}_k$ is a quadratic function whose Levi form is $O(\varepsilon(\delta))$, by the assumption on the modulus of continuity of γ. This quadratic function can be written as

$$\tilde{\gamma}_j(z) - \tilde{\gamma}_k(z) = \mathrm{Re}\, g_{jk}(z) + q_{jk}(z - z_{jk}^0),$$

the sum of an affine pluriharmonic part $\mathrm{Re}\, g_{jk}$ and a quadratic term $q_{jk}(z - z_{jk}^0)$ which takes $O(\varepsilon(\delta)\delta^2)$ values (since $\mathrm{diam}\, B_j^1 \cap B_k^1 \leq \delta$). Therefore we have

$$|\varphi_k - \varphi_j - \mathrm{Re}\, g_{jk}| \leq C\varepsilon(\delta)\delta^2.$$

By 7.14 (v), we conclude that

$$\left| \Phi_{k,c}^{\chi,\rho,s}(z,w) - \Phi_{j,c}^{\chi,\rho,s}(z,w) - \mathrm{Re}\, g_{jk}(z) \right| \leq 2C\varepsilon(\delta)\delta^2 + \frac{C'}{s}$$

for some constants C, C', hence

$$\left| \left(\Phi_{k,c}^{\chi,\rho,s}(z,w) + \tilde{\gamma}_k(z) \right) - \left(\Phi_{j,c}^{\chi,\rho,s}(z,w) + \tilde{\gamma}_j(z) \right) \right| \leq 3C\varepsilon(\delta)\delta^2 + \frac{C'}{s}.$$

Now, in case U_j^1 and U_k^1 are not equipped with the same coordinates, 7.14 (iv) shows that an extra error term C/s is introduced by the change of coordinates $\tau_{jk} = \tau_j \circ \tau_k^{-1}$, and also possibly a further $O(\delta^3)$ term due to the fact that

$\tilde{\gamma}_j \circ \tau_{jk}$ differs from a quadratic function by terms of order 3 or more in the τ_k-coordinates. Combining everything together, we get

$$\left| \left(\Phi_{k,c}^{\chi,\rho,s}(\tau_k(z), w) + \tilde{\gamma}_k(\tau_k(z)) \right) - \left(\Phi_{j,c}^{\chi,\rho,s}(\tau_j(z), w) + \tilde{\gamma}_j(\tau_j(z)) \right) \right|$$

$$\leq C'' \left(\varepsilon(\delta)\delta^2 + \delta^3 + \frac{1}{s} \right) \leq C''' \varepsilon(\delta)\delta^2$$

if we choose $s \geq 1/(\varepsilon(\delta)\delta^2)$. We assume from now on that s is chosen in this way. For $x \in \partial U_j^1 = \tau_j^{-1}(S(a_j, \sqrt{2}\delta))$, formula (9.5) yields

$$\Psi_{j,c}^{\chi,\rho,s}(\tau_j(z), w) = \Phi_{j,c}^{\chi,\rho,s}(\tau_j(z), w) + \tilde{\gamma}_j(\tau_j(z)) - \varepsilon(\delta)^{1/2} 2\delta^2,$$

whereas there exists k such that $x \in U_k = \tau_k^{-1}(B(a_k, \delta))$, hence

$$\Psi_{k,c}^{\chi,\rho,s}(\tau_k(z), w) \geq \Phi_{k,c}^{\chi,\rho,s}(\tau_j(z), w) + \tilde{\gamma}_k(\tau_k(z)) - \varepsilon(\delta)^{1/2}\delta^2.$$

We infer from this

$$\Psi_{k,c}^{\chi,\rho,s}(\tau_k(z), w) - \Psi_{j,c}^{\chi,\rho,s}(\tau_j(z), w) \geq \varepsilon(\delta)^{1/2}\delta^2 - C''' \varepsilon(\delta)\delta^2 > 0$$

for δ small enough. This shows that formula (9.6) makes sense for δ small. Formulas (9.2) and (9.5) show that

$$(9.7) \qquad \frac{i}{\pi} \partial \bar{\partial}_z \Psi_{j,c}^{\chi,\rho,s}(\tau_j(z), w) \geq \frac{i}{\pi} \partial \bar{\partial}_z \Phi_{j,c}^{\chi,\rho,s}(\tau_j(z), w) + \gamma - C\varepsilon(\delta)^{1/2}\omega$$

for some constant $C > 0$. The sequence of approximations $\psi_{c,k}$ needed in the theorem is obtained by taking sequences $\delta_k \downarrow 0$, $s_k \geq 1/(\varepsilon(\delta_k)\delta_k^2)$ and $A_k \uparrow +\infty$, and putting

$$\tilde{\psi}_{c,k}(z) = \Psi_c^{\chi,\rho,s_k}(z, -A_k) + \frac{1}{k}$$

where Ψ_c^{χ,ρ,s_k} is constructed as above by means of an open covering $\mathcal{U}_k$ of X with balls of radii $\sim \delta_k$. By (9.7) and Proposition 8.7 ii), we find

$$\frac{i}{\pi} \partial \bar{\partial} \tilde{\psi}_{c,k} \geq - \min \left(\frac{\partial \Phi_\infty^{\chi,\rho,s_k}}{\partial \operatorname{Re} w_-}(z, -A_k), c \right) u - C\varepsilon(\delta_k)^{1/2}\omega.$$

As $\displaystyle \lim_{\operatorname{Re} w \to -\infty} \frac{\partial \Phi^{\chi,\rho,s}}{\partial \operatorname{Re} w_-}(z, w) = \nu(\varphi, z) = \nu(\psi, z)$, a suitable choice of A_k ensures that

$$\tilde{\lambda}_k(z) := \frac{\partial \Phi_\infty^{\chi,\rho,s_k}}{\partial \operatorname{Re} w_-}(z, -A_k) \to \nu(\psi, z) \qquad \text{as } k \to +\infty.$$

Furthermore, an appropriate choice of the sequences δ_k, s_k, A_k guarantees that the sequence $\tilde{\psi}_{c,k}$ is non increasing. [The only point we have to mind about is the effect of a change of the open covering, as the radius δ_k of the covering balls decreases to 0. However, Proposition 7.14 (iv, v) shows that the effect can be made negligible with respect to $\frac{1}{k} - \frac{1}{k+1}$, and then everything is ok.] We can ensure as well that λ_k is decreasing, by replacing if necessary $\tilde{\lambda}_k$ with

$$\lambda_k(z) = \sup_{l \geq k} \tilde{\lambda}_l(z).$$

Finally, the functions $\tilde{\psi}_{c,k}$ that we got are (a priori) just known to be continuous on $X \setminus E_c(T)$, thanks to Proposition 7.14 (i) and the discussion before Proposition 8.7. Again, Richberg's approximation theorem [1968] shows that we can replace $\tilde{\psi}_{c,k}$ with a smooth approximation $\psi_{c,k}$ on $X \setminus E_c(T)$, with $|\tilde{\psi}_{c,k} - \psi_{c,k}|$ arbitrarily small in uniform norm, and at the expense of losing an extra error term $\varepsilon_k \omega$ in the lower bound for $\frac{i}{\pi} \partial \bar{\partial} \psi_{c,k}$. Theorem 9.1 is proved. $\square$

10. Appendix: Basic Results on L^2 Estimates

We state here the basic L^2 existence theorems used in the above sections, concerning $\bar{\partial}$ equations or holomorphic functions. The first of these is the intrinsic manifold version of Hörmander's L^2 estimates [Hörmander 1965; 1966], based on the Bochner–Kodaira–Nakano technique. See also [Andreotti and Vesentini 1965].

THEOREM 10.1. *Let L be a holomorphic line bundle on a weakly pseudoconvex n-dimensional manifold X equipped with a Kähler metric ω. Suppose that L has a smooth hermitian metric whose curvature form satisfies*

$$2\pi \, \Theta(L) + i\partial\bar{\partial}\varphi \geq A\omega$$

where φ is an almost psh function and A a positive continuous function on X. Then for every form v of type (n,q), $q \geq 1$, with values in L, such that $\bar{\partial}v = 0$ and

$$\int_X \frac{1}{A}|v|^2 e^{-\varphi} dV_\omega < +\infty,$$

there exists a form u of type $(n, q-1)$ with values in L such that $\bar{\partial}u = v$ and

$$\int_X |u|^2 e^{-\varphi} dV_\omega \leq \frac{1}{q} \int_X \frac{1}{A}|v|^2 e^{-\varphi} dV_\omega.$$

A *weakly pseudoconvex manifold* is by definition a complex manifold possessing a smooth weakly pseudoconvex exhaustion function (examples: Stein manifolds, compact manifolds, the total space of a Griffiths weakly negative vector bundle, and so on). Suppose that φ has Lelong number $\nu(\varphi, x) = 0$ at a given point x. Then for every m the weight $e^{-m\varphi}$ is integrable in a small neighborhood V of x [Skoda 1972b]. Let θ be a cut-off function equal to 1 near x, with support in V. Let z be coordinates and let e be a local frame of L on V. For ε small enough, the curvature form

$$2\pi \, \Theta(L) + i\partial\bar{\partial}\big(\varphi(z) + 2\varepsilon\theta(z) \log|z - x|\big)$$

is still positive definite. We apply A.1 to the bundle L^m equipped with the corresponding weight $m(\varphi(z) + 2\varepsilon\theta(z)\log|z - x|)$, and solve the equation $\bar{\partial}u = v$ for the $(n,1)$-form $v = \bar{\partial}(\theta(z)P(z)dz_1 \wedge \cdots \wedge dz_n \otimes e^m)$ associated to an arbitrary polynomial P. The L^2 estimate shows that the solution u has to vanish at order $\geq q + 1$ at x where $q = [m\varepsilon] - n$, hence

$$\theta(z)P(z)dz_1 \wedge \cdots \wedge dz_n \otimes e^m - u(z)$$

is a holomorphic section of $K_X \otimes L^m$ with prescribed jet of order q at x.

COROLLARY 10.2. *Suppose that* $2\pi\,\Theta(L) + i\partial\bar{\partial}\varphi \geq \delta\,\omega$ *for some* $\delta > 0$. *Let* $x \in X$ *be such that* $\nu(\varphi, x) = 0$. *Then there exists* $\varepsilon > 0$ *such that the sections in* $H^0(X, K_X \otimes L^m)$ *generate all jets of order* $\leq m\varepsilon$ *at* x *for* m *large.* □

We now state the basic L^2 extension theorem which was needed in several occasions. A detailed proof can be found in [Ohsawa and Takegoshi 1987; Ohsawa 1988; Manivel 1993]; see also [Demailly 1996, Theorem 13.6]. Only the case $q = 0$ (dealing with holomorphic sections) does play a role in this work.

THEOREM 10.3 (OHSAWA–TAKEGOSHI). *Let* X *be a weakly pseudoconvex* n-*dimensional complex manifold equipped with a Kähler metric* ω, *let* L *be a hermitian holomorphic line bundle,* E *a hermitian holomorphic vector bundle of rank* r *over* X, *and* s *a global holomorphic section of* E. *Assume that* s *is generically transverse to the zero section, and let*

$$Y = \big\{x \in X : s(x) = 0,\ \Lambda^r ds(x) \neq 0\big\}, \qquad p = \dim Y = n - r.$$

Moreover, assume that the $(1,1)$-*form* $i\Theta(L) + r\,i\,\partial\bar{\partial}\log|s|^2$ *is semipositive and that there is a continuous function* $\alpha \geq 1$ *such that the following two inequalities hold everywhere on* X :

(a) $i\Theta(L) + r\,i\,\partial\bar{\partial}\log|s|^2 \geq \alpha^{-1}\dfrac{\{i\Theta(E)s, s\}}{|s|^2}$,

(b) $|s| \leq e^{-\alpha}$.

Then for every smooth $\bar{\partial}$-*closed* $(0,q)$-*form* f *over* Y *with values in the line bundle* $\Lambda^n T_X^\star \otimes L$ *(restricted to* Y*), such that* $\int_Y |f|^2 |\Lambda^r(ds)|^{-2} dV_\omega < +\infty$, *there exists a* $\bar{\partial}$-*closed* $(0,q)$-*form* F *over* X *with values in* $\Lambda^n T_X^\star \otimes L$, *such that* F *is smooth over* $X \smallsetminus \{s = \Lambda^r(ds) = 0\}$, *satisfies* $F_{|Y} = f$ *and*

$$\int_X \frac{|F|^2}{|s|^{2r}(-\log|s|)^2}\,dV_{X,\omega} \leq C_r \int_Y \frac{|f|^2}{|\Lambda^r(ds)|^2}dV_{Y,\omega}\,,$$

where C_r *is a numerical constant depending only on* r.

COROLLARY 10.4. *Let* Y *be a pure dimensional closed complex submanifold of* $\mathbb{C}^n$, *let* Ω *be a bounded pseudoconvex open set and let* φ *be a plurisubharmonic function on* Ω. *Then for any holomorphic function* f *on* $Y \cap \Omega$ *with*

$$\int_{Y \cap \Omega} |f|^2 e^{-\varphi} dV_Y < +\infty,$$

there exists a holomorphic extension F to Ω such that

$$\int_\Omega |F|^2 e^{-\varphi} dV \le A \int_{Y \cap \Omega} |f|^2 e^{-\varphi} dV_Y < +\infty.$$

Here A depends only on Y and on the diameter of Ω.

Finally, a crucial application of Skoda's L^2 estimates [1972a; 1978] for ideals of holomorphic functions was made in Section 5:

THEOREM 10.5. *Let φ be a plurisubharmonic function on a pseudoconvex open set $\Omega \subset \mathbb{C}^n$ and let $\sigma_1, \ldots, \sigma_N$ be holomorphic functions on Ω (the sequence σ_j can be infinite). Set $r = \min\{N-1, n\}$ and $|\sigma|^2 = \sum |\sigma_j|^2$. Then, for every holomorphic function f on Ω such that*

$$\int_\Omega |f|^2 |\sigma|^{-2(r+1+\alpha)} e^{-\varphi} dV < +\infty, \ \alpha > 0,$$

there exist holomorphic functions $g_1, \ldots, g_N$ on Ω such that $f = \sum_{1 \le j \le N} g_j \sigma_j$ and

$$\int_\Omega |g|^2 |\sigma|^{-2(r+\alpha)} e^{-\varphi} dV \le \frac{\alpha+1}{\alpha} \int_\Omega |f|^2 |\sigma|^{-2(r+1+\alpha)} e^{-\varphi} dV < +\infty.$$

COROLLARY 10.6. *With the same notations, suppose that*

$$\int_\Omega |f|^2 |\sigma|^{-2(r+m+\alpha)} e^{-\varphi} dV < +\infty$$

for some $\alpha > 0$ and some integer $m \ge 1$. Then there exist holomorphic functions g_L for all $L = (l_1, \ldots, l_m) \in \{1, \ldots, N\}^m$ such that

$$f = \sum_L g_L \sigma^L \ \text{with} \ \sigma^L = \sigma_{l_1} \sigma_{l_2} \ldots \sigma_{l_m},$$

$$\int_\Omega \sum_L |g_L|^2 |\sigma|^{-2(r+\alpha)} e^{-\varphi} dV \le \frac{\alpha+m}{\alpha} \int_\Omega |f|^2 |\sigma|^{-2(r+m+\alpha)} e^{-\varphi} dV < +\infty.$$

PROOF. Use induction on m: if the result is true for $(m - 1, \alpha + 1)$ then $f = \sum_\Lambda g_\Lambda \sigma^\Lambda$ with Λ of length $m - 1$, and each function g_Λ can be written $g_\Lambda = \sum_{l_m} g_L \sigma_{l_m}$ with $L = (\Lambda, l_m)$ and

$$\int_\Omega \sum_{l_m} |g_L|^2 |\sigma|^{-2(r+\alpha)} e^{-\varphi} dV \le \frac{\alpha+1}{\alpha} \int_\Omega |G_\Lambda|^2 |\sigma|^{-2(r+1+\alpha)} e^{-\varphi} dV < +\infty,$$

$$\int_\Omega \sum_\Lambda |g_\Lambda|^2 |\sigma|^{-2(r+1+\alpha)} e^{-\varphi} dV \le \frac{\alpha+m}{\alpha+1} \int_\Omega |f|^2 |\sigma|^{-2(r+m+\alpha)} e^{-\varphi} dV < +\infty. \ \square$$

References

[Andreotti and Vesentini 1965] A. Andreotti and E. Vesentini, "Carleman estimates for the Laplace–Beltrami equation on complex manifolds", *Inst. Hautes Études Sci. Publ. Math.* **25** (1965), 81–130.

[Campana and Flenner 1990] F. Campana and H. Flenner, "A characterization of ample vector bundles on a curve", *Math. Ann.* **287**:4 (1990), 571–575.

[Demailly 1982] J.-P. Demailly, "Estimations L^2 pour l'opérateur $\bar{\partial}$ d'un fibré vectoriel holomorphe semi-positif au-dessus d'une variété kählérienne complète", *Ann. Sci. École Norm. Sup.* (4) **15**:3 (1982), 457–511.

[Demailly 1992] J.-P. Demailly, "Regularization of closed positive currents and inter-section theory", *J. Algebraic Geom.* **1**:3 (1992), 361–409.

[Demailly 1993] J.-P. Demailly, "A numerical criterion for very ample line bundles", *J. Differential Geom.* **37**:2 (1993), 323–374.

[Demailly 1994] J.-P. Demailly, "Regularization of closed positive currents of type $(1, 1)$ by the flow of a Chern connection", pp. 105–126 in *Contributions to complex analysis and analytic geometry*, edited by H. Skoda and J.-M. Trépreau, Aspects of Mathematics **E26**, Vieweg, Wiesbaden, 1994.

[Demailly 1996] J.-P. Demailly, "L^2 estimates for the $\bar{\partial}$-operator on complex mani-folds", Lecture notes, summer school on complex analysis, Institut Fourier, Grenoble, 1996.

[Demailly et al. 1994] J.-P. Demailly, T. Peternell, and M. Schneider, "Compact complex manifolds with numerically effective tangent bundles", *J. Algebraic Geom.* **3**:2 (1994), 295–345.

[Grauert 1958] H. Grauert, "On Levi's problem and the imbedding of real-analytic manifolds", *Ann. of Math.* (2) **68** (1958), 460–472.

[Griffiths 1969] P. A. Griffiths, "Hermitian differential geometry, Chern classes, and positive vector bundles", pp. 185–251 in *Global Analysis: Papers in Honor of K. Kodaira*, edited by D. C. Spencer and S. Iyanaga, Princeton mathematical series **29**, Princeton Univ. Press, Princeton, NJ, and Univ. Tokyo Press, Tokyo, 1969.

[Hartshorne 1966] R. Hartshorne, "Ample vector bundles", *Inst. Hautes Études Sci. Publ. Math.* **29** (1966), 63–94.

[Hartshorne 1970] R. Hartshorne, *Ample subvarieties of algebraic varieties*, Lecture Notes in Math. **156**, Springer, Berlin, 1970. Notes written in collaboration with C. Musili.

[Hörmander 1965] L. Hörmander, "L^2 estimates and existence theorems for the $\bar{\partial}$ operator", *Acta Math.* **113** (1965), 89–152.

[Hörmander 1966] L. Hörmander, *An introduction to complex analysis in several variables*, Van Nostrand, Princeton, NJ, 1966.

[Kiselman 1978] C. O. Kiselman, "The partial Legendre transformation for plurisub-harmonic functions", *Invent. Math.* **49**:2 (1978), 137–148.

[Kiselman 1997] C. O. Kiselman, "Duality of functions defined in lineally convex sets", *Univ. Iagel. Acta Math.* **35** (1997), 7–36.

[Kobayashi 1975] S. Kobayashi, "Negative vector bundles and complex Finsler struc-tures", *Nagoya Math. J.* **57** (1975), 153–166.

[Kodaira 1954] K. Kodaira, "Some results in the transcendental theory of algebraic varieties", *Ann. of Math.* (2) **59** (1954), 86–134.

[Lempert 1985] L. Lempert, "Symmetries and other transformations of the complex Monge–Ampère equation", *Duke Math. J.* **52**:4 (1985), 869–885.

[Manivel 1993] L. Manivel, "Un théorème de prolongement L^2 de sections holomorphes d'un fibré hermitien", *Math. Z.* **212**:1 (1993), 107–122.

[Nadel 1990] A. M. Nadel, "Multiplier ideal sheaves and Kähler–Einstein metrics of positive scalar curvature", *Ann. of Math.* (2) **132**:3 (1990), 549–596.

[Ohsawa 1988] T. Ohsawa, "On the extension of L^2 holomorphic functions, II", *Publ. Res. Inst. Math. Sci. (Kyoto Univ.)* **24**:2 (1988), 265–275.

[Ohsawa and Takegoshi 1987] T. Ohsawa and K. Takegoshi, "On the extension of L^2 holomorphic functions", *Math. Z.* **195**:2 (1987), 197–204.

[Ramis et al. 1971] J. P. Ramis, G. Ruget, and J. L. Verdier, "Dualité relative en géométrie analytique complexe", *Invent. Math.* **13** (1971), 261–283.

[Richberg 1968] R. Richberg, "Stetige streng pseudokonvexe Funktionen", *Math. Ann.* **175** (1968), 257–286.

[Schneider 1973] M. Schneider, "Über eine Vermutung von Hartshorne", *Math. Ann.* **201** (1973), 221–229.

[Skoda 1972a] H. Skoda, "Application des techniques L^2 à la théorie des idéaux d'une algèbre de fonctions holomorphes avec poids", *Ann. Sci. École Norm. Sup.* (4) **5** (1972), 545–579.

[Skoda 1972b] H. Skoda, "Sous-ensembles analytiques d'ordre fini ou infini dans $\mathbb{C}^{n}$", *Bull. Soc. Math. France* **100** (1972), 353–408.

[Skoda 1978] H. Skoda, "Morphismes surjectifs de fibrés vectoriels semi-positifs", *Ann. Sci. École Norm. Sup.* (4) **11**:4 (1978), 577–611.

[Sommese 1978] A. J. Sommese, "Concavity theorems", *Math. Ann.* **235**:1 (1978), 37–53.

[Sommese 1979] A. J. Sommese, "Complex subspaces of homogeneous complex manifolds, I: Transplanting theorems", *Duke Math. J.* **46**:3 (1979), 527–548.

[Sommese 1982] A. J. Sommese, "Complex subspaces of homogeneous complex manifolds, II: Homotopy results", *Nagoya Math. J.* **86** (1982), 101–129.

[Umemura 1980] H. Umemura, "Moduli spaces of the stable vector bundles over abelian surfaces", *Nagoya Math. J.* **77** (1980), 47–60.

JEAN-PIERRE DEMAILLY
UNIVERSITÉ DE GRENOBLE I
DÉPARTEMENT DE MATHÉMATIQUES
INSTITUT FOURIER, UMR 5582 DU CNRS
38402 SAINT-MARTIN D'HÈRES CEDEX
FRANCE
demailly@ujf-grenoble.fr

Several Complex Variables
MSRI Publications
Volume **37**, 1999

Complex Dynamics in Higher Dimension

JOHN ERIK FORNÆSS AND NESSIM SIBONY

Dedicated to the memory of Michael Schneider

ABSTRACT. We discuss a few new results in the area of complex dynamics in higher dimension. We investigate generic properties of orbits of biholomorphic symplectomorphisms of $\mathbb{C}^n$. In particular we show (Corollary 3.4) that for a dense G_δ set of maps, the set of points with bounded orbit has empty interior while the set of points with recurrent orbits nevertheless has full measure. We also investigate the space of real symplectomorphisms of $\mathbb{R}^n$ which extend to $\mathbb{C}^n$. For this space we show (Theorem 3.10) that for a dense G_δ set of maps, the set of points with bounded orbit is an F_σ with empty interior.

CONTENTS

1. Introduction

In this paper we discuss low-dimensional dynamical systems described by complex numbers. There is a parallel theory for real numbers. The real numbers have the advantage of being more directly tuned to describing real-life systems. However, complex numbers offer additional regularity and besides, real systems usually complexify in a way that makes phenomena more clear: for example, periodic points disappear under parameter changes in the real case, but remain in the complex case.

1991 *Mathematics Subject Classification.* 32H50.

Fornæss is supported by an NFS grant.

In the case of the solar system and other complicated systems, one has to resign oneself to studying the time evolution of a small number of variables, since if one wants to precisely predict long-term evolution one runs into unsurmountable computer problems. One cannot forget unavoidable errors that are just necessary limits of knowledge. And some knowledge is hence limited to a phenomenological type.

Here we give a brief overview of some of the open questions in the area of complex dynamics in dimension 2 or more. We also discuss some new results by the authors about symplectic geometry and Hamiltonian mechanics, belonging to higher-dimensional complex dynamics.

2. Questions in Higher-Dimensional Complex Dynamics

Complex dynamics in one complex dimension arose in the end of the last century as an outgrowth of studies of Newton's method and the three body problem in celestial mechanics. See [Alexander 1994] for a historical treatment.

2.1. Local Theory. In the local theory one studies the behavior near a fixed point, $f(x) = x$. This was the beginning of the theory in one complex variable: see [Schröder 1871]. Schröder discussed the case when the derivative $f'(x)$ of the map at the fixed point had absolute value less than one. He asked whether after a change of coordinates — that is, a conjugation — the map could be made linear in a small neighborhood (if the derivative was non-zero). This gave rise to the Schröder equation, which was later solved by Farkas [1884]. The case when the derivative was 1 was discussed by Fatou [1919; 1920a; 1920b] and Julia [1918], who proved the so-called flower theorem, describing the shape of the set of points whose orbit converges to the fixed point (the basin of attraction). The more general neutral case, i.e. when $|f'(x)| = 1$ is still not completely understood. The first result in this direction was proved by Siegel [1942]. He showed that f is conjugate to $f = e^{i\theta}z$ in case θ is sufficiently far from being rational. This was shown later to be valid for a larger class of angles by Brjuno [1965; 1971; 1972] and the question whether this was a necessary and sufficient condition was discussed by Yoccoz [1992].

The same problem arises for fixed points in higher dimension. In the case of a sufficiently irrational indifferent fixed point, Sternberg [1961] showed that the Theorem of Siegel is still valid. See [1987] for a more detailed history.

In general, let $f : (\mathbb{C}^n, 0) \to (\mathbb{C}^n, 0)$ be a germ of a holomorphic map with $f(0) = 0$. The objective is to describe the local nature of the set of points converging to the fixed point. There is as yet no systematic study of this, and the work that has been done is more of a global nature. See the next section.

2.2. Global Theory. The case when $f'(x)$ has one eigenvalue 1 and the other is λ was studied by Ueda [1986], who showed that in the case of $|\lambda| < 1$, and when f is an automorphism of $\mathbb{C}^2$, the basin of attraction of the fixed point is a biholomorphic copy of $\mathbb{C}^2$ — what is called a Fatou–Bieberbach domain. A similar result when both eigenvalues are 1 and under suitable conditions on the higher order terms was proved recently by Weickert [1998]. A general result in any dimension was proved subsequently by Hakim [1998; 1997]. These two works are also of a local nature.

The global analogues of the distinction between attracting, repelling and indifferent behaviour at fixed points, are the distinction between Fatou sets, Julia sets and borderline cases, like Siegel domains. This started after the Montel Theorem was proved in one dimension, or with the equivalent notion of Kobayashi hyperbolicity in higher dimension.

In the theory of iteration of polynomials or rational functions in one variable, the Fatou sets are completely classified into 5 types [Sullivan 1985]. In higher dimension one has the same kinds of Fatou sets but there are others as well. There is as yet no complete classification of Fatou components for holomorphic maps on $\mathbb{P}^2$ say; see [Fornæss and Sibony 1995a]. For example, one knows that there are no wandering Fatou components Ω (that is, components Ω such that $f^n(\Omega) \cap \Omega = \varnothing$ for all n) in one dimension, but this is unknown in higher dimension. Another simple open problem in the case of polynomial automorphisms of $\mathbb{C}^2$ is whether a Fatou component can be biholomorphic to an annulus cross $\mathbb{C}$.

As far as the Julia set is concerned, one has a basic tool available, pluripotential theory. This is based on the fact that for example if one lifts a holomorphic map on $\mathbb{P}^n$ to a homogenous polynomial F on $\mathbb{C}^{n+1}$ of degree d, then the limit $G = \lim_{n\to\infty} d^{-n} \log \|F^n\|$ exists and the $(1,1)$ current $T = dd^c G$ has support precisely over the Julia set and therefore is an invariant object measuring the dynamics. This tool lies behind much of what is known about complex dynamics in higher dimension. See [Fornæss and Sibony 1994; Fornæss 1996].

The function G and the current T are naturally restricted to the case of iteration of polynomial and rational maps. In the case of entire maps in one and several variables, it seems one must get by without pluripotential theory. It is also appropriate to mention that one of the successful tools in iteration in one dimension, quasiconformal maps, have so far no higher-dimensional complex analogue. This is perhaps the reason why one hasn't so far been able to decide whether wandering Fatou components exist in higher dimension for polynomial automorphisms say. We should also say that although one doesn't have pluripotential theory in the study of entire maps, one has instead much more freedom to work with holomorphic functions, so in the case of holomorphic automorphisms of higher dimension and holomorphic endomorphisms in one dimension, one can show that wandering components exist [Fornaess and Sibony 1998a]. See [Bergweiler 1993] for the 1-dimensional case.

Several classes of maps on $\mathbb{C}^n$ have been studied. One can divide into two major classes, biholomorphic maps and endomorphisms.

The class that has been studied the most are the Hénon maps in $\mathbb{C}^2$, these are the polynomial automorphism with nontrivial dynamics. See [Bedford and Smillie 1991a; 1991b; 1992; 1998; ≥ 1999a; ≥ 1999b; Bedford et al. 1993; Hubbard and Oberste-Vorth 1994; 1995; Fornæss and Sibony 1992].

However, when it comes to the next step, extending the theory to $\mathbb{C}^3$ or higher, there are very few results at present (but see [Bedford and Pambuccian 1998]). A first step is to classify the polynomial automorphisms hopefully in a manner analogous to the Friedland–Milnor classification [1989] in $\mathbb{C}^2$. This has so far been done only for degree-2 maps in $\mathbb{C}^3$ [Fornæss and Wu 1998].

In the case of entire automorphisms, one can study for example the behaviour of orbits in general, asking whether they usually tend to infinity. This has been studied in [Fornæss and Sibony 1995b; [1996]; [1996]]. But we don't know for example, in the case of symplectomorphisms of $\mathbb{C}^{2n}$, whether there can exist a set of positive measure of bounded orbits which persist under all small perturbations. We discuss this type of question in the next section.

If we go to polynomial endomorphisms of $\mathbb{C}^2$ — that is, beyond the case of Hénon maps, which are invertible — the study is wide open. There is a classification [Alcarez and de Medrano ≥ 1999] of endomorphisms which are polynomial maps of degree 2 in $\mathbb{R}^2$ up to composition with linear automorphisms (not up to conjugation), but so far no systematic study exists of these classes.

2.3. Flows of Holomorphic Vector Fields. The iteration of maps refers to discrete dynamics. That is, one describes how a system changes in one unit of time. One can make an analogous study of continuous dynamics, where the maps are given by flows of holomorphic vector fields. Some work has been done on this [Fornæss and Sibony 1995b], but less than in the discrete case. Again one can ask questions about the local flow near a fixed point for the flow (a zero of the vector field) and about the global flow. Some work has been done in [Forstneric 1996] in the case of complete vector fields (those for which the flow is defined for all time), and in [Fornæss and Grellier 1996] on the question of the size of the set of points with exploding orbits (orbits which reach infinity in finite time).

2.4. Holomorphic Foliations and Laminations. Holomorphic vector fields foliate space by integral curves. In general, one can study foliations or more general, laminations. A compact set K is said to be laminated if there exist through every point $p \in K$, a complex manifold $M_p \subset K$ and these are either analytic continuations of each other or disjoint. One doesn't for example know if there can exist a lamination of some compact set in $\mathbb{P}^2$ so that no leaf is a compact complex manifold. See [Brunella 1994; 1996; Brunella and Ghys 1995; Camacho 1991; Camacho et al. 1992; Gómez-Mont 1988; 1987] for some work on foliations and further references.

In dealing with holomorphic endomorphisms of $\mathbb{P}^2$, one studies for example the unstable set of the saddle set of hyperbolic maps [Fornæss and Sibony 1998b]. This can locally be written as a union of graphs of local unstable manifolds. However, since these might not be pairwise disjoint in the endomorphism case (the unstable manifold through a point depends in general on the prehistory of the point), one gets to study a more general concept than laminations.

3. Symplectic Geometry and Hamiltonian Mechanics

We discuss here three new results. The first concerns the abundance of recurrent points for complex symplectomorphims of $\mathbb{C}^{2k}$. The second deals with real symplectomorphisms which can be complexified. Finally the third topic concerns the estimates of decompositions of Hamiltonians into sums of Hamiltonians whose associated Hamiltonian vector fields give arise to globally defined symplectomorphisms.

3.1. Recurrent points. In this section we discuss biholomorphic symplectomorphisms of $\mathbb{C}^{2k}$, that is, maps $f : \mathbb{C}^{2k} \to \mathbb{C}^{2k}$ which preserve the symplectic form $\omega := \sum_{j=1}^{k} dz_j \wedge dw_j$. We denote the class of all symplectomorphisms by $\mathcal{S}$. We put the topology of uniform convergence on compact sets on $\mathcal{S}$.

We are interested in the generic, long-term behavior of orbits. See [Fornæss and Sibony 1996].

DEFINITION 3.1. Let $f \in \mathcal{S}$ and $p \in \mathbb{C}^{2k}$. We say that the point p is recurrent if for every neighborhood U of p there is a point $q \in U$ and an integer $n > 1$ so that $f^n(q) \in U$.

THEOREM 3.2. *The set of recurrent points R_f is of full measure for a G_δ dense set of symplectic maps f.*

This can be contrasted with a previous result from [Fornæss and Sibony 1996]:

THEOREM 3.3. *There is a dense G_δ set $\mathcal{S}' \subset \mathcal{S}$ so that for each $f \in \mathcal{S}'$, the set $K_f \subset \mathbb{C}^{2k}$ of points whose orbit is bounded has empty interior.*

Combining these two results, one gets:

COROLLARY 3.4. *There is a dense G_δ set $\mathcal{S}'' \subset \mathcal{S}$ so that for each $f \in \mathcal{S}''$, the set R_f has full measure, while K_f has no interior.*

PROOF OF THEOREM 3.2. Let $f \in \mathcal{S}$. Let B_m denote the closed ball of center 0 and radius m, and set

$$U_m^f = \{x : x \in B_m \text{ and } |f^n(x) - x| < 1/m \text{ for some } n > 0\},$$
$$\mathcal{S}_m = \{f \in \mathcal{S} : |B_m \setminus U_m^f| < 2^{-m}\}.$$

The set $\mathcal{S}_m$ is open for the compact open topology.

CLAIM 3.5. $\mathcal{S}_m$ is dense in $\mathcal{S}$.

Assuming the claim, the set $\mathcal{S}' = \cap \mathcal{S}_m$ is a G_δ dense set. Let $f \in \mathcal{S}'$. Define $R = \bigcup_N \left(\bigcap_{m>N} U_m^f \right)$. Then R is of full measure in any ball; indeed,

$$\left| \bigcup_{m>N} (B_m \setminus U_m^f) \right| < 2^{-N}.$$

Every point x in R is recurrent.

We want now to prove the claim, thus completing the proof of the theorem. Let $f_0 \in \mathcal{S}$. We want to approximate f_0 on a given compact X by maps in $\mathcal{S}_m$. We can assume $X \subset B_m$. Set $\Omega_m := U_m^{f_0}$.

If Ω_m is nonempty we choose a compact set $K_m \subset \Omega_m$ such that $|\Omega_m \setminus K_m| < 2^{-2m}$. There is N_0 such that for $x \in K_m$ there is $n \leq N_0$ with $|f^n(x) - x| < 1/m$. We want to enlarge the set of recurrent points (up to order $1/m$) by perturbing slightly f_0 on B_m.

We choose finitely many disjoint compact rectangles $\widehat{S}_m^j{}_{j \leq L} \subset B_m \setminus K_m$ with $\mathrm{diam}(\widehat{S}_m^j) < 1/(3m)$, and such that

$$\left| B_m \setminus \left(K_m \cup \bigcup_j \widehat{S}_m^j \right) \right| < 2^{-(m+2)}.$$

Set $S_m^j := \widehat{S}_m^j \setminus \Omega_m$. Since every S_m^j is disjoint from Ω_m we have for every $l \geq 1$

$$f_0^l(S_m^j) \cap S_m^j = \varnothing$$

and consequently for $r, l \in \mathbb{Z}$ with $r \neq l$ we have

$$f_0^l(S_m^j) \cap f_0^r(S_m^j) = \varnothing. \tag{3-1}$$

Fix k such that $\bigcup_{|n| \leq N_0} f_0^n(B_m) \subset \mathring{B}_k$. Since f_0 is volume preserving, condition (3–1) implies the existence of an $l_0 \geq N_0$ such that

$$\left\{ x \in \bigcup_j S_m^j : \text{there is } l \text{ satisfying } |l| \geq l_0 \text{ and } f^l(x) \in B_k \right\}$$

is of measure less than $2^{-(m+1)}$. Let B_r, for $r > k$, be a ball containing $\bigcup_{|n| \leq l_0} f^n(B_m)$. Similarly there is an $l_1 \in \mathbb{N}$ such that $l_1 \geq l_0$ and

$$\left\{ x \in \bigcup_j S_m^j : \text{there is } l \text{ satisfying } |l| \geq l_1 \text{ and } f^l(x) \in B_r \right\}$$

is of measure less than $2^{-(m+1)}$. Shrinking the sets S_m^j we can assume that there are finitely many disjoint compact sets $\left(\widetilde{S}_m^j \right)_{j \leq L'}$ in $B_m \setminus \Omega_m$ such that

$$\left| B_m \setminus (K_m \cup \widetilde{S}_m^j) \right| < 2^{-m}$$

and for $|l| \geq l_1$, the set $f^l(\widetilde{S}_m^j) \cap B_r$ is empty.

For any $x \in \widetilde{S}_m^j$ consider the complete orbit $\mathcal{O}(x) = \{f^n(x)\}_{n \in \mathbb{Z}}$. Let $n(x)$ be the first exit time of the orbit from B_r, and $-n'(x)$ the last entry time of the orbit into B_r. More precisely, $f^{n(x)}(x) \notin B_r$ but $f^{n(x)-p}(x) \in B_r$ for $0 < p < n(x)$; similarly $f^{-n'(x)}(x) \notin B_r$ but $f^{-n'(x)+p}(x) \in B_r$, if $n'(x) > p > 0$. Define

$$\Omega^+ = \left\{ f^{n(x)}(x) : x \in \bigcup_j \widetilde{S}_m^j \right\},$$
$$\Omega^- = \left\{ f^{-n'(x)}(x) : x \in \bigcup_j \widetilde{S}_m^j \right\}.$$

We can remove from $\bigcup_j \widetilde{S}_m^j$ a set of arbitrarily small measure such that the maps $x \to n(x)$ and $x \to n'(x)$ are locally constant.

The map $\phi : \Omega^+ \to \Omega^-$ defined by $\phi(f^{n(x)}(x)) = f^{-n'(x)}(x)$ is a bijection.

We can assume that we can cover $\bigcup_j \tilde{S}_m^j$ by a finite union of small pairwise disjoint rectangles K^+, with $|\Omega^+ \setminus K^+|$ and $|K^+ \setminus \Omega^+| \ll 1$, and such that ϕ is locally given by $y \to f^{-s}(y)$ with $s \in \mathbb{N}$ fixed in a neighborhood.

We want to approximate $\phi_{|K^+}$ on most of K^+ by a global symplectomorphism σ such that σ is close to the identity on B_r. Then the map $f := f_0 \circ \sigma$ will be close to f_0 on X, and the orbit of most $f^{n(x)}(x) \in K^+$ will pass near x. Observe that we control the orbit of x under f in B_r, it stays close to the orbit of x under f_0. Hence $B_m \setminus U_f$ will be of arbitrarily small measure. We first need a lemma.

LEMMA 3.6. *Let K be a compact set disjoint from the closed ball $\overline{B}$. Assume that K is a finite union of disjoint polynomially convex sets. Then for every $\varepsilon > 0$ there is a compact set $K_\varepsilon \subset K$, such that $\overline{B} \cup K_\varepsilon$ is polynomially convex and $|K \setminus K_\varepsilon| < \varepsilon$.*

PROOF OF THE LEMMA. We first consider the case when $B = \varnothing$. Let l be a complex linear form. Fix a real number α. For any $0 < \delta \ll 1$ define

$$K_\delta^- = \{x \in K : \operatorname{Re} l(x) \leq \alpha - \delta\},$$
$$K_\delta^+ = \{x \in K : \operatorname{Re} l(x) \geq \alpha + \delta\}.$$

For δ small enough the measure of $K \setminus (K_\delta^+ \cup K_\delta^-)$ is arbitrarily small. It is easy to verify that $\widehat{K_\delta^- \cup K_\delta^+} = \widehat{K_\delta^-} \cup \widehat{K_\delta^+}$. Repeating the process with finitely many hyperplanes one gets easily the set K_ε such that $\overline{B} \cup K_\varepsilon$ is polynomially convex.

Next consider the general case. Let l be a again complex linear form. Assume the real hyperplane $H = \{x : \operatorname{Re} l(x) = \alpha\}$ does not intersect $\overline{B}$, but possibly intersects K. For any $0 < \delta \ll 1$ define

$$Y = \overline{B} \cup K,$$
$$Y_\delta^- = \{x \in Y : \operatorname{Re} l(x) \leq \alpha - \delta\},$$
$$Y_\delta^+ = \{x \in Y : \operatorname{Re} l(x) \geq \alpha + \delta\}.$$

For δ small enough the measure of $Y \setminus (Y_\delta^+ \cup Y_\delta^-)$ is arbitrarily small. It is easy to verify that $\widehat{Y_\delta^- \cup Y_\delta^+} = \widehat{Y_\delta^-} \cup \widehat{Y_\delta^+}$. Repeating the process with finitely many hyperplanes $l_j = \alpha_j$ one can choose things so that $\overline{B} \subset \bigcap_j \{l_j < \alpha_j\}$ while $K \subset \bigcup_j \{l_j > \alpha_j\}$, and one gets easily a set K_ε such that $\overline{B} \cup K_\varepsilon$ is polynomially convex. □

The second result we need is due to Forstneric.

PROPOSITION 3.7 [Forstneric 1996]. *Let U be a simply connected Runge domain in $\mathbb{C}^{2k}$ such that $H^1(U, \mathbb{C}) = 0$. Let Φ_t be a biholomorphic map from U into $\mathbb{C}^{2k}$ of class $\mathcal{C}^2$ in $(t, z) \in [0, 1] \times U$. Assume each domain $U_t = \Phi_t(U)$ is Runge, and that every Φ_t is a symplectomorphism and Φ_1 can be approximated on U by global symplectomorphisms. Then Φ_0 can be approximated on U by global symplectomorphisms. A similar result holds for volume preserving maps in $\mathbb{C}^k$ (if one assumes that $H^{k-1}(U, \mathbb{C}) = 0$).*

We now finish the proof of the claim. Let K_ε be a compact obtained from K^+ by applying Lemma 3.6 to $\bar{B}_k \cup K^+$. We have to construct the family Φ_t such that $\phi = \Phi_1$ on a small neighborhood of $\bar{B} \cup K_\varepsilon$, which is topologically trivial. We are going to use the real hyperplanes as in the proof of the lemma. Suppose that the first hyperplane is just $H = \{(z, w) : \operatorname{Re} z_1 = \alpha\}$ and that $\bar{B} \subset \{\operatorname{Re} z_1 < \alpha\}$. Let $\chi(s)$ be a smooth approximation to $(s - \alpha)^+$. Define

$$\tau_s^t(z, w) = (z_1 + t\chi(s)w_1, w_1, z_2, \ldots, w_k)$$

for $(z, w) \in \{\operatorname{Re} z_1 > \alpha\}$; here t is a large constant. Let $\psi_s := \tau_s^1 \circ \phi$. Then we have left the part in $\{\operatorname{Re} z_1 < \alpha\}$ unchanged and the part of K^+ in $\{\operatorname{Re} z_1 > \alpha + \varepsilon\}$, for $0 < \varepsilon \ll 1$, has slid far away.

Using sliding away from finitely many hyperplanes, removing possibly a set of very small measure from K_ε, we can construct $\Phi_s := \phi \circ \tau_s^l \circ \cdots \circ \tau_s^1 \circ \phi$ such that $\Phi_0 = \phi$ and Φ_1 is defined on the ball and on finitely many rectangles.

The image of $\tilde{B} \cup K^+$ under ϕ_1 is contained in a union of balls $(B_j)_{j \leq L}$ which are very far apart, in particular all their projections on the coordinate axes are disjoint. We can choose balls $(\tilde{B})_j$ with $B_j \Subset \tilde{B}_j$ and still the balls $\{\tilde{B}_j\}$ are very far apart. We connect ϕ_1^{-1} to the identity. Composing ϕ_1^{-1} with a finite number of shears (s_j) of type $(z_1 + h(w_1), w_1, z_2, \ldots, w_k)$ with h entire we can achieve that the $\Theta_1 := s_p \circ \cdots \circ s_1 \circ \phi_1^{-1}$ satisfies $\Theta_1(B_j) \Subset \tilde{B}_j$ and Θ_1 has a fixed point in each B_j, we can then write a homotopy to the identity. $\square$

REMARKS 3.8. 1. The authors proved in [Fornæss and Sibony 1996] the existence of a G_δ dense set $S' \subset S$ such that for any $f \in S'$ the set of recurrent points R_f is a G_δ dense set. It follows from the previous theorem that generically, in the Baire sense, R_f is a G_δ dense set of full measure.

2. The same results hold for the group $\mathcal{V}$ of volume preserving biholomorphisms in $\mathbb{C}^k$. We can just start with a neighborhood U of K_ε which satisfies $H^{k-1}(U, \mathbb{C}) = 0$.

3.2. Real Symplectomorphisms. Let $S_\mathbb{R} := \{f \in S : f : \mathbb{C}^{2k} \to \mathbb{C}^{2k}$ such that $f(\mathbb{R}^{2k}) = \mathbb{R}^{2k}\}$. More precisely, let (z_j, w_j) be complex coordinates in $\mathbb{C}^{2k}$. Assume that $z_j = p_j + ip_j'$, $w_j = q_j + iq_j'$, the coordinates on $\mathbb{R}^{2k}$ are (p_j, q_j). The restriction of the form $\omega = \sum dz_j \wedge dw_j$ to $\mathbb{R}^{2k}$ is the standard symplectic form $\omega_0 = \sum dp_j \wedge dq_j$.

PROPOSITION 3.9. *The group $S_\mathbb{R}$ consists of diffeomorphisms f_0 of $\mathbb{R}^{2k}$ such that $f_0^* \omega_0 = \omega_0$, which extend biholomorphically to $\mathbb{C}^{2k}$.*

PROOF. Left to the reader. $\square$

A family $(f_i)_{i \in I}$ in $S_\mathbb{R}$ converges to $f \in S_\mathbb{R}$ if and only if (f_i) converge to f uniformly on compact sets of $\mathbb{C}^{2k}$ and the restriction of f_i to $\mathbb{R}^{2k}$ converges to $f_{|\mathbb{R}^{2k}}$ in the fine topology in $\mathbb{R}^{2k}$, which means that given any continuous function $\eta > 0$ on $\mathbb{R}^{2k}$ and given any n, $\sup_{|\alpha| \leq n} |D^\alpha f_i - D^\alpha f|(p, q) < \eta(p, q)$ for i in a cofinal set.

It is easy to verify that $\mathcal{S}_\mathbb{R}$ with this topology is a Baire space.

THEOREM 3.10. *Let $k = 1, 2$. There is a G_δ dense set $\mathcal{S}'_\mathbb{R} \subset \mathcal{S}_\mathbb{R}$ such that for $f \in \mathcal{S}'_\mathbb{R}$ the set $K_f := \{(z, w) \in \mathbb{C}^{2k} : f^n(z, w) \text{ is bounded}\}$ is an F_σ of empty interior.*

PROOF. Let $H = \mathbb{C}^{2k} \times \mathcal{S}_\mathbb{R}$ with the product topology. Set

$$K := \{(z, w, f) \text{ with bounded forward orbit}\}.$$

Let Δ_n be a basis for the topology of $\mathbb{C}^{2k}$. For each n, define

$$\mathcal{S}_n := \{f : |f^m(z, w)| < n \text{ for every } m \text{ and every}(z, w) \in \bar{\Delta}_n\}.$$

If K has interior then some $\mathcal{S}_n$ has nonempty interior. Assume $n = R$. Let U_R be the interior of $\{(z, w, f) : |f^m(z, w)| \leq R \text{ for every } m\}$. Let U be the projection of U_R in $\mathcal{S}_R$.

For $(z, w, f) \in U_R$ let V_f be the slice of U_R for fixed f. The open set V_f is Runge for every f. Moreover it is clearly invariant under the map $(z, w) \to (\bar{z}, \bar{w})$. From the Schwarz Lemma, given ε_0 there is $\alpha > 0$ so that

$$|x - x'| < \alpha \Rightarrow |f^m(x) - f^m(x')| < \varepsilon_0 \tag{3-2}$$

for every $f \in U$, close enough to a given $f_0 \in U$.

Since the maps f are volume preserving on each slice, V_f is also backward invariant and each connected component of V_f is periodic. It follows from the Cartan Theorem and from the fact that the maps are volume preserving that for every such f and any component U_f of V_f the closure of the subgroup generated by f restricted to $\bigcup_n f^n(U_f)$ is a compact Abelian Lie group G_f. Consequently $G_f = T^l \times A$, where T is the unit circle, $l \in \mathbb{N}$, and A is a finite group. For $a \in U$, $a = (z, w)$, let $\bar{a} = (\bar{z}, \bar{w})$. Let $X_a = G_f(a)$, $Y_a = X_a \cup X_{\bar{a}}$.

LEMMA 3.11. *Let V be a Runge, bounded open set in $\mathbb{C}^{2k}$, for $k = 1, 2$, stable under $(z, w) \to (\bar{z}, \bar{w})$. Assume that V is invariant under a symplectic map $f \in \mathcal{S}_\mathbb{R}$. Let $G = \overline{(f^n_{|V})_n}$ and assume G is not discrete. There is a point a such that Y_a is polynomially convex and in every neighborhood of Y_a we can find $Y_{a'}$ such that $Y_a \cup Y_{a'}$ is polynomially convex and $Y_a \cap Y_{a'} = \varnothing$.*

PROOF. For any $x \in V$, the set Y_x is a union of disjoint tori (possibly points). The polynomially convex hull $\hat{Y}_x$ of Y_x is stable under f. As in [Fornæss and Sibony 1996, Lemma 4.3], we can find an a such that Y_a is polynomially convex. If Y_a is a finite set hence a union of periodic orbits the map f is then linearizable near each periodic orbit. Since the map is symplectic (holomorphic) it follows that the map is conjugate to a matrix with blocks

$$\begin{pmatrix} e^{i\theta_1} & 0 \\ 0 & e^{-i\theta_1} \end{pmatrix}.$$

One then computes easily that *most* orbits near the periodic points are polynomially convex. And even the union of finitely many of them is normally polynomially convex. More precisely: if, for example,

$$f(z_1, z_2, w_1, w_2) = (e^{i\theta_1} z_1, \, e^{-i\theta_1} w_1, \, e^{i\theta_2} z_2, \, e^{-i\theta_2} w_2),$$

the orbits that avoid the axes are polynomially convex.

Assume Y_a is a finite union of tori. Let $\tilde{Y}_a$ be the local complexification. If G acts effectively on Y_a, the dimension of Y_a equals that of G; let this dimension be l. Generically the orbits are of dimension l. Orbits close to Y_a are also polynomially convex. If $Y_{a'}$ is polynomially convex and the complexifications are disjoint then $Y_a \cup Y_{a'}$ is also polynomially convex. This finishes the lemma if $k = 1$.

We next consider the case where there is a non discrete isotropy group for Y_a and Y_a is not a finite union of periodic orbits. It suffices to consider only the identity component of G.

Let G_0 be the isotropy group. Then Y_a and G_0 are both a finite union of tori. Then G_0 is generated by a holomorphic Hamiltonian vector field with Hamiltonian H. Necessarily $\nabla H \equiv 0$ on Y_a, so H is constant on Y_a. Let V_a be a Runge neighborhood of Y_a, stable under f.

Next consider a point p close to a where $H(p)$ is nonzero. Let Y be the H orbit of p. Changing p a little, we may assume that Y is polynomially convex and still lies in the same level set of H. Let G' in G be a T^1 subgroup transverse to G_0. Then T^1 is generated by a holomorphic Hamiltonian vector field with Hamiltonian K.

Consider the K orbit Z of p. There is a projection of Z to Y_a given by mapping p to a and following the vectorfield of K. This can be extended to a holomorphic projection of a neighborhood of the full G orbit of p by letting the projection be constant on H orbits. Here we use the fact that the group G acts real analytically. Then it follows that Y_a together with the orbit of p is polynomially convex: Indeed Y_a is a totally real torus [Fornæss and Sibony 1996] and continuous functions on Y_a are uniform limits of polynomials [Wermer 1976]. To check polynomial convexity of a set X which projects on Y_a, it is enough by [Wermer 1976] to check the polynomial convexity of the fibers under π.   □

We continue with the proof of Theorem 3.10. Let $Y_a, Y_{a'}$ be orbits under $G = G_{f_0}$ with $f_0 \in U$. Suppose $f_0^{n_0}$ is in the identity component of G. Assume $Y_a \cup Y_{a'}$ is polynomially convex as in the Lemma. Let f^t be a one parameter subgroup in G such that $f^1 = f_0^{n_0}$. Define $\xi = (df^t/dt)_{|t=0}$. It is proved in [Fornæss and Sibony 1996] Lemma 4.4 that ξ is a Hamiltonian vector field. We define $\tilde{\xi} = \xi$ on a neighborhood of Y_a and $\tilde{\xi} = -\xi$ in a neighborhood of $Y_{a'}$.

Let h be a Hamiltonian for $\tilde{\xi}$ defined on a Runge neighborhood of $Y_a \cup Y_{a'}$. We can assume that $h(\bar{z}, \bar{w}) = \bar{h}(z, w)$. We approximate h by a polynomial P, real on $\mathbb{R}^{2k}$, uniformly on a Runge neighborhood V_d containing a $d-$neighborhood of $Y_a \cup Y_{a'}$, we need the Hamiltonian vector field generated by P, has a small angle ε, with $\tilde{\xi}$ on $Y_a \cup Y_{a'}$.

We can write $P = \sum_{j=1}^{N} P_j$ where each P_j is a polynomial such that the associated Hamiltonian vector field is complete and the flow is a shear s_j. We can assume that the P_j are real on $\mathbb{R}^{2k}$. For $\delta > 0$, write $\delta P = \sum_{j=1}^{N} \delta P_j$, S_j^{δ} the shear associated to δP_j and $S^{\delta} = S_N^{\delta} \circ \cdots \circ S_1^{\delta}$.

Fix a large ball $B(0, r)$ in $\mathbb{C}^{2k}$. Assume $f(B(0,r)) \subset B(0, r')$. Choose $\delta \ll 1$ so that $|S^{\delta} - \mathrm{Id}|_{B(0, r')} \leq \varepsilon$.

We have $|S^{\delta} \circ f - f|_{B(0,r)} \leq \varepsilon$.

We will modify $S^{\delta} \circ f_0$ to bring it inside the open set U.

We show first that $(S^{\delta} \circ f_0)$ do not satisfy condition (3–2). Indeed, $(S^{\delta} \circ f_0)^{(m)}(x)$ is in V_d as soon as $m \leq d/(100\varepsilon\delta)$ for $x \in Y_a$, $x' \in Y_{a'}$. But S^{δ} push points in Y_a and $Y_{a'}$ in opposite directions so for $m \sim 1/\delta$ we have

$$\left|(S^{\delta} \circ f_0)^m(a) - (S^{\delta} \circ f_0)^m(a')\right| \sim 1 \geq \varepsilon_0,$$

contradicting (3–2).

However, we have to modify $S^{\delta} \circ f_0$ to keep the previous estimates and to put in the given neighborhood of f_0 in the fine topology.

Fix ε_j and $r_j \to \infty$ so that if $|f - f_0|_{B(r)} < \varepsilon_0$ and

$$|f - f_0|_{(B(r_{j+1}) \setminus B(r_j)) \cap \mathbb{R}^{2k}} < \varepsilon_j$$

then f is in the given neighborhood of f_0 in the fine topology where (3–2) is valid.

We need just to use inductively the following lemma.

LEMMA 3.12. *Let $f \in \mathcal{S}_{\mathbb{R}}$ and fix $R_1 < R_2 < R_3$. Assume $f_{|B(0,R_2) \cap \mathbb{R}^{2k}}$ is close to the identity. Then there is $f_1 \in \mathcal{S}_{\mathbb{R}}$ such that f_1 is close to the identity on $B(0, R_1)$ and close to f on $(B(0, R_3) \setminus B(0, R_2)) \cap \mathbb{R}^{2k}$.*

PROOF. We can write f as a time-1 map of a time-dependent Hamiltonian vector field $X(t)$, which is close to zero on $B(0, R_2)$. For every t the Hamiltonian is close to zero on $B(0, R_2)$. Multiply each Hamiltonian by a cut-off function equal to zero on $B(0, R_1 - \varepsilon)$ and 1 out of $B(0, R_1)$. We approximate each Hamiltonian on $B(0, R_1) \cup (B(0, R_3) \cap \mathbb{R}^{2k})$ by entire functions real on reals. Then we consider the associated symplectomorphism and approximate by their composition, see [Fornæss and Sibony 1996, p. 316; Forstneric 1996]. □

This concludes the proof of the theorem. □

3.3. Decomposition of Homogeneous Polynomials. Let X denote a homogeneous polynomial in $\mathbb{C}^2$ of degree $d = m + 1$. In this section we will discuss how to decompose X into a finite sum of powers of linear functions. This problem arises in the study of symplectomorphisms when one wants to do computer calculations. More precisely, the problem is to truncate a power series for a symplectomorphism, and then to symplectify the truncation without "loosing too much". But that doesn't seem to be the case according to our computations. The problem arises also in approximation of symplectomorphisms with compositions of shears.

A step in this procedure is to consider symplectomorphisms of the form $F = \mathrm{Id} + (A_m, B_m) + O(\|z\|^{m+1})$.

LEMMA 3.13. *Let $f = \mathrm{Id} + Q_m + O(|z|^{m+1})$ be a germ of holomorphic map in $\mathbb{C}^{2p}$, where Q_m is a polynomial mapping homogeneous of degree m. Then Q_m is a Hamiltonian vector field if $f^*(\omega) - \omega = O(|z|^m)$.*

PROOF. To check that a holomorphic vector field X in $\mathbb{C}^{2p}$ is Hamiltonian we have to show that $X \rfloor \omega$ is a closed 1–form. Assume $X = (A_1, B_1, \ldots, A_p, B_p)$. Then

$$\omega(X, \cdot) = \sum_{j=1}^{p} (A_j dw_j - B_j dz_j),$$

and

$$d(X \rfloor \omega) = \sum (dA_j \wedge dw_j + dz_j \wedge dB_j).$$

On the other hand

$$f^*\omega = \sum \left(dz_j + dA_j + dO(|z|^{m+1}) \right) \wedge (dw_j + dB_j + \cdots)$$
$$= \omega + d(X \rfloor \omega) + O(|z|^m).$$

Hence $d(X \rfloor \omega) = 0$. $\square$

Therefore one can write

$$(A_m, B_m) = \left(-\frac{\partial X}{\partial w}, \frac{\partial X}{\partial z} \right),$$

where X is a uniquely determined homogeneous polynomial of degree $m+1 = d$. It is easy to decompose $X = \sum_{j=0}^{d} c_j Q_j$ where the Q_j are powers of linear functions, forming a basis for the homogeneous polynomials.

Hence, letting $(C_j, D_j) = c_j(-\partial Q_j/\partial w, \partial Q_j/\partial z)$, we can write

$$F = \tilde{F} + O(\|z\|^{m+1}),$$
$$\tilde{F} = (\mathrm{Id} + (C_0, D_0)) \circ \cdots \circ (\mathrm{Id} + (C_d, D_d)).$$

We are concerned here with the magnitude of the $c_j Q_j$. We will see below that the basis $\{Q_j\}$ can be chosen to be essentially as good as an orthonormal basis.

Note that if the (A_m, B_m) are less than some small ε, then we will show that the (C_j, D_j) are bounded by $c\varepsilon$ and hence the terms of $\tilde{F}$ of order at least $m+1$ are at most $c_m \varepsilon^2$. Hence, we get that if we start with a symplectomorphism close to the identity, this process can be repeated a few times, to approximate the original map to higher and higher order and the resulting symplectomorphism remains close to the identity.

In the rest of this paper we will deal with the estimates on the $c_j Q_j$. We leave it for a later paper to carry this project further for arbitrary dimension and for the estimates on the $\tilde{F}$.

THEOREM 3.14. *Let $d \geq 2$ be an integer. There exist $d + 1$ homogeneous polynomials P_d^j of degree d of the form*

$$P_d^j = \frac{\sqrt{(d+1)(d+2)}}{\pi} (\alpha_j z + \beta_j w)^d \qquad for \ j = 0, \ldots, d$$

such that $\|(\alpha_j, \beta_j)\| = \sqrt{|\alpha_j|^2 + |\beta_j|^2} = 1$, $\|P_d^j\|_{L^2(\mathbb{B})} = 1$, and the following properties are satisfied:

1. *The P_d^j is a basis for the space $\mathcal{P}_d$ of homogeneous holomorphic polynomials of degree d.*
2. *If $P \in \mathcal{P}_d$ is of the form $P = \sum_j c_j P_d^j$, then $|c_j| \leq C\sqrt{d}\|P\|_{L^2(B)}$, with C independent of P and d.*

REMARK 3.15. We can probably drop the coefficient $\sqrt{d}$ from the estimate in condition 2. See the end of the proof.

The main difficulty is that the powers of linear functions is only a $\mathbb{P}^1$ in the high-dimensional space $\mathbb{P}^d$ of all homogeneous polynomials (up to multiples). Hence it is somewhat remarkable that one can choose a basis practically as good as an orthonormal basis.

We are mainly interested for the moment in the asymptotic estimate when $d \to \infty$, rather than an optimal value for C. Hence we can restrict ourselves to large d.

Recall the following fact about Vandermonde determinants:

$$\begin{vmatrix} 1 & x_1 & x_1^2 & \cdots & x_1^n \\ 1 & x_2 & x_2^2 & \cdots & x_2^n \\ \vdots & \vdots & \vdots & \ddots & \vdots \\ 1 & x_{n+1} & x_{n+1}^2 & \cdots & x_{n+1}^n \end{vmatrix} = \prod_{j>i}(x_j - x_i).$$

We also need a slightly more general classical formula:

$$\begin{vmatrix} 1 & x_1 & \cdots & x_1^{j-1} & x_1^{j+1} & \cdots & x_1^n \\ 1 & x_2 & \cdots & x_2^{j-1} & x_2^{j+1} & \cdots & x_2^n \\ \vdots & \vdots & \ddots & \vdots & \vdots & \ddots & \vdots \\ 1 & x_n & \cdots & x_n^{j-1} & x_n^{j+1} & \cdots & x_{n+1}^n \end{vmatrix} = \prod_{j>i}(x_j - x_i)S_{n-j},$$

where S_k is the sum of all distinct products of k of the x_i, $S_0 = 1$.

Our next step is to choose, for a given degree $d \geq 2$, a set of $d+1$ points in $\mathbb{P}^1$ that are evenly distributed in the spherical metric: say $p_{i,d} = [\alpha_{i,d} : \beta_{i,d}] = p_i = [\alpha_i : \beta_i]$, with $i = 0, \ldots d$. Moreover we assume that $|\alpha_i|^2 + |\beta_i^2| = 1$. For this, we will first describe the points p_i on a sphere and then project to the complex plane.

We assume that the points (a_i, b_i) are distributed on a 2-sphere of radius $\frac{1}{2}$ (so that the area is π, the area of $\mathbb{P}^1$ in the sperical metric) centered at $(0,0,0)$ in bands with angle from the negative z-axis between $(2j\sqrt{\pi/d})$ and $2(j+1)\sqrt{\pi/d}$, where $j = 0, \ldots, \lfloor \sqrt{d\pi}/2 \rfloor - 1$. Set $\theta_j = (2j+1)\sqrt{\pi/d}$.

We want the points to be about $\sqrt{\pi/d}$ apart from each other. This is approximately what you can do if you put d points in a square lattice in $\mathbb{R}^2$ with width $\sqrt{\pi/d}$, covering an area π.

Each band has circumference $\pi \sin(\theta_j)$ and hence contains n_j evenly spread points, where n_j equals either $\lfloor \pi \sin(\theta_j)/\sqrt{\pi/d} \rfloor$ or one more than this value (the latter case is allowed if $(\sqrt{\pi d})/8 < 2j + 1 < (7\sqrt{\pi d})/8$).

Next we move the sphere up to have center at $(0, 0, 1/2)$ and use stereographic projection to map the points to the z-plane. The points in the j-th band have $|z| = \tan(\theta_j/2) = (\sin\theta_j)/(1 + \cos\theta_j)$ and are equally spaced. In projective coordinates they are

$$[\frac{\sin\theta_j}{1 + \cos\theta_j} : 1] = [\frac{\sin\theta_j}{\sqrt{2(1 + \cos\theta_j)}} : \sqrt{\frac{1 + \cos\theta_j}{2}}] = [\alpha_j : \beta_j]$$

with $|\alpha_j|^2 + |\beta_j|^2 = 1$.

In fact, we let the n_j points have arguments ω^k in the first coordinate, where $k = 1, \ldots, n_j$ and ω a primitive n_j-th root of unity. So the points are of the form

$$\left[\frac{\omega^k \sin\theta_j}{\sqrt{1(1 + \cos(\theta_j))}} : \sqrt{\frac{1 + \cos(\theta_j)}{2}} \right].$$

LEMMA 3.16.
$$d(p_i, p_k) \geq \sqrt{\pi/d} - O(1/d), \qquad \text{for } i \neq k.$$

PROOF. The distance can only be smaller than $\sqrt{\pi/d}$ if the two points are on the same circle for $\sqrt{\pi d}/8 < 2j + 1 < 7\sqrt{\pi d}/8$. Hence we get

$$d(p_i, p_k) \geq \frac{\pi \sin((2j+1)\sqrt{\pi/d})}{\dfrac{\pi \sin((2j+1)\sqrt{\pi/d})}{\sqrt{\pi/d}} + 1},$$

$$d(p_i, p_k) \geq \sqrt{\frac{\pi}{d}} \frac{\pi \sin(\pi/8)}{\pi \sin(\pi/8) + \sqrt{\pi/d}},$$

$$d(p_i, p_k) \geq \sqrt{\frac{\pi}{d}} - \frac{1}{d \sin(\pi/8)}.$$

$\square$

LEMMA 3.17. Let $i = 0, \ldots, d$. Then

$$\int_{\mathbb{B}} |z|^{2i} |w|^{2d-2i} = \frac{\pi^2}{(d+2)(d+1)} \frac{1}{\binom{d}{i}}.$$

PROOF. We calculate the integral in the w-direction and introduce polar coordinates to reduce the integral to calculation of

$$\frac{\pi^2}{d+1-i} \int_0^1 r^i (1-r)^{d+1-i} \, dr.$$

Now use induction to get the result. $\qquad\square$

LEMMA 3.18. *The functions* $Q_{i,d} = Q_i := (1/\pi)\sqrt{(d+1)(d+2)}P_i$ *have* L^2 *norm* 1.

PROOF. This follows from the previous lemma applied to the function z^d and using rotational invariance. $\qquad\square$

We will next estimate the deviation of the set Q_i from being an orthonormal basis. Observe first that the vectors

$$e_i := \frac{\sqrt{(d+1)(d+2)}}{\pi}\sqrt{\binom{d}{i}}z^i w^{d-i}$$

is an orthonormal basis by the lemma above.

We express first Q_i in terms of the e_j. The following formula is immediate. We are abusing notation from now on by writing a, b instead of α, β.

LEMMA 3.19. $Q_i = \sum_{j=0}^{j=d} a_i^j b_i^{d-j} \sqrt{\binom{d}{j}} e_j =: \sum c_i^j e_j.$

The basic estimate we need is the determinant of the transition matrix from the e_j to the Q_i. The basic idea is that this determinant measures the failure of the $\{Q_i\}$ to be orthonormal.

Set $x_i = a_i/b_i$. The next lemma shows in particular that any $d+1$ distinct points in $\mathbb{P}^1$ gives rise to a basis.

LEMMA 3.20.

$$\begin{vmatrix} c_0^0 & c_0^1 & c_0^2 & \cdots & c_0^d \\ c_1^0 & c_1^1 & c_1^2 & \cdots & c_1^d \\ \vdots & \vdots & \vdots & \ddots & \vdots \\ c_d^0 & c_d^1 & c_d^2 & \cdots & c_d^d \end{vmatrix} = \prod_{i=0}^d \left(\sqrt{\binom{d}{i}} b_i^d \right) \begin{vmatrix} 1 & x_0 & x_0^2 & \cdots & x_0^d \\ 1 & x_1 & x_1^2 & \cdots & x_1^d \\ \vdots & \vdots & \vdots & \ddots & \vdots \\ 1 & x_d & x_d^2 & \cdots & x_d^d \end{vmatrix}$$

$$= \left(\prod_{i=0}^d \sqrt{\binom{d}{i}} b_i^d \right) \prod_{j>i} \left(\frac{a_j}{b_j} - \frac{a_i}{b_i} \right)$$

$$= \left(\prod_{i=0}^d \sqrt{\binom{d}{i}} \right) (\pm 1) \sqrt{\prod_{j\neq i} b_i b_j} \sqrt{\prod_{j\neq i}(a_j/b_j - a_i/b_i)}$$

$$= \left(\prod_{i=0}^d \sqrt{\binom{d}{i}} \right) (\pm 1) \sqrt{\prod_{j\neq i}(a_j b_i - a_i b_j)}.$$

PROOF. Immediate, using Vandermonde determinants. $\qquad\square$

Next, let $\Sigma_j := \mathrm{Span}(Q_0, \ldots, \widehat{Q}_j, \ldots, Q_d)$. Write

$$Q_{j+1} = Q'_{j+1} + Q''_{j+1},$$

where Q'_{j+1} is the component of Q_{j+1} perpendicular to Σ_j. Let $d_j := \|Q'_j\|_{L^2(B)}$.

Let $G_d := \det(\langle Q_i, Q_j \rangle)$ be the Gram determinant of the vectors $Q_0, \ldots, Q_d$, and let $G_d^j := \det(\langle Q_i, Q_l \rangle)$ be the Gram determinant of $(Q_0, \ldots, \widehat{Q}_j, \ldots, Q_d)$. We will use the following classical fact:

Let $x_1, \ldots, x_n$ be linearly independent vectors in a Hilbert space, spanning a subspace Σ. Suppose $y \notin \Sigma$ and let $Py = \sum_{i=1}^{n} a_i x_i$ be the orthogonal projection of y onto Σ, so that $(y - Py) \perp x_i$ for each i. Then

$$\|y - Py\|^2 = \|y\|^2 - \sum_{i=1}^{n} a_i \langle y, x_i \rangle = \frac{G^+}{G}, \tag{3-3}$$

where $G = \det(\langle x_i, x_j \rangle)$ is the Gram determinant of $(x_1, \ldots, x_n)$ and G^+ is the Gram determinant of $(y, x_1, \ldots, x_n)$. To see this, expand the determinant G^+ by its first row and use repeatedly the equalities

$$\langle y, x_j \rangle = \sum_{i=1}^{n} a_i \langle x_i, x_j \rangle.$$

Applying (3-3) to the situation at hand, we obtain

$$d_j = \frac{\sqrt{G_d}}{\sqrt{G_d^j}}.$$

With the notation of Lemma 4.5, set $A = (c_i^j)$. Then

$$G_d = \det(A^t \bar{A}) = \left(\prod_{i=0}^{d} \binom{d}{i} \right) \left| \prod_{j \neq i} (a_j b_i - a_i b_j) \right|.$$

Now consider G_d^j. For $0 \leq j \leq d$, let A_j be the matrix obtained from A by removing row j. Then

$$G_d^j = \det(A_j^t \bar{A}_j) = \sum_{0 \leq l \leq d} \left| \det(A_j^l) \right|^2,$$

where A_j^l is obtained from A_j by removing the l-th column. Hence

$$d_j^2 = \frac{G_d}{G_d^j} = \frac{|\det A^t \bar{A}|}{|\det A_j^t \bar{A}_j|} = \frac{|\det A|^2}{\sum_{l=0}^{d} |\det A_j^l|^2}.$$

It follows that

$$d_j^2 = \cfrac{\left(\prod_{i=0}^d \binom{d}{i}\right) \prod_{k \neq i} |a_k b_i - a_i b_k|}{\sum_{0 \leq l \leq d} \left| \det \begin{bmatrix} c_0^0 & \cdots & c_0^{l-1} & c_0^{l+1} & \cdots & c_0^d \\ \vdots & \ddots & \vdots & \vdots & \ddots & \vdots \\ c_{j-1}^0 & \cdots & c_{j-1}^{l-1} & c_{j-1}^{l+1} & \cdots & c_{j-1}^d \\ c_{j+1}^0 & \cdots & c_{j+1}^{l-1} & c_{j+1}^{l+1} & \cdots & c_{j+1}^d \\ \vdots & \ddots & \vdots & \vdots & \ddots & \vdots \\ c_d^0 & \cdots & c_d^{l-1} & c_d^{l+1} & \cdots & c_d^d \end{bmatrix} \right|^2}.$$

Therefore,

$$d_j^2 = \cfrac{\left(\prod_{i=0}^d \binom{d}{i}\right) \prod_{k \neq i} |a_k b_i - a_i b_k|}{\sum_{0 \leq l \leq d} \left(\prod_{i \neq l} \binom{d}{i}\right)\left(\prod_{i \neq j} |b_i|^{2d}\right) \left| \det \begin{bmatrix} 1 & \cdots & \left(\frac{a_0}{b_0}\right)^{d-1} \\ \vdots & \ddots & \vdots \\ 1 & \cdots & \left(\frac{a_{j-1}}{b_{j-1}}\right)^{d-1} \\ 1 & \cdots & \left(\frac{a_{j+1}}{b_{j+1}}\right)^{d-1} \\ \vdots & \ddots & \vdots \\ 1 & \cdots & \left(\frac{a_d}{b_d}\right)^{d-1} \end{bmatrix} \right|^2 |S_{d-l}^j|^2}.$$

where S_{d-l}^j is the sum of all distinct products of $d - l$ of the terms $a_0/b_0, \ldots,$ $a_{j-1}/b_{j-1}, a_{j+1}/b_{j+1}, \ldots, a_d/b_d$ with $S_0^j = 1$.

We obtain

$$d_j^2 = \cfrac{\left(\prod_{i=0}^d \binom{d}{i}\right) \prod_i |b_i|^{2d} \prod_{k \neq i} \left|\frac{a_k}{b_k} - \frac{a_i}{b_i}\right|}{\sum_{0 \leq l \leq d} \left(\prod_{i \neq l} \binom{d}{i}\right)\left(\prod_{i \neq j} |b_i|^{2d}\right)\left(\prod_{\substack{k \neq i \\ k, i \neq j}} \left|\frac{a_k}{b_k} - \frac{a_i}{b_i}\right|\right)|S_{d-l}^j|^2}.$$

Simplification leads to the following equality:

LEMMA 3.21.

$$d_j^2 = \cfrac{|b_j|^{2d}\left(\prod_{k \neq j} \left|\frac{a_k}{b_k} - \frac{a_j}{b_j}\right|\right)^2}{\sum_{0 \leq l \leq d} \dfrac{|S_{d-l}^j|^2}{\binom{d}{l}}}.$$

We can expand the numerator.

LEMMA 3.22.

$$d_j^2 = \cfrac{|b_j|^{2d}\left(\left|\sum_{l=0}^d (-1)^{d-l}\left(\frac{a_j}{b_j}\right)^l S_{d-l}^j\right|\right)^2}{\sum_{0 \leq l \leq d} \dfrac{|S_{d-l}^j|^2}{\binom{d}{l}}}.$$

We now try to write down a general formula. Let us number the circles C_k and suppose that the circle C_k contains n_k points $\frac{a_k}{b_k}\omega_k^i$. We choose our point $p = (a_j, b_j)$ on circle C_r. Fix any m. We want to find a formula for S_m^j. For any integer i such that $0 \le i < n_r$, consider any possible way of writing $i + \sum \delta_k n_k = m$ where each δ_k equals 0 or 1, and $\delta_r = 0$. So we are selecting those contributions coming from i points on C_r and all the points on the circles C_k, with $\delta_k = 1$. This procedure captures all non zero contributions to S_m^j. A general formula is $(a_j/b_j)S_m^j + S_{m+1}^j = S_{m+1}$, where S_m denotes all symmetric combinations of order m.

LEMMA 3.23.

$$S_m^j = \sum_{i+\sum \delta_k n_k = m} \left(-\frac{a_j}{b_j}\right)^i \prod_{\{\delta_k=1\}} \left(\frac{a_k}{b_k}\right)^{n_k}.$$

LEMMA 3.24.

$$d_j^2 = \frac{|b_j|^{2d}\left(\left|\sum_{l=0}^{d}\sum_{i=0}^{n_r-1}(-1)^{d-l-i}\left(\frac{a_j}{b_j}\right)^{l+i}\sum_{\delta_k n_k = d-l-i}\prod\left(\frac{a_k}{b_k}\right)^{n_k}\right|\right)^2}{\sum_{0\le l\le d}\frac{|S_{d-l}^j|^2}{\binom{d}{l}}}.$$

Since the circle C_r is excluded from contributing to the last product, we have the restriction $d - l - i \le d - n_r$, so $l + i \ge n_r$.

LEMMA 3.25.

$$d_j^2 = \frac{|b_j|^{2d}\left(\left|\sum_{s=n_r}^{d} n_r\left(-\frac{a_j}{b_j}\right)^s\sum_{\delta_k n_k = d-s}\prod\left(\frac{a_k}{b_k}\right)^{n_k}\right|\right)^2}{\sum_{0\le l\le d}\frac{|S_{d-l}^j|^2}{\binom{d}{l}}}.$$

The expression in the numerator has an obvious largest term. Namely, take $\delta_k = 1$ for those circles closer to the north pole than p. Set $\delta_k = 0$ otherwise. The term is

$$n_r\left(-\frac{a_j}{b_j}\right)^s \prod_{|b_k|<|b_j|}\left(\frac{a_k}{b_k}\right)^{n_k}, \qquad s = d - \sum \delta_k n_k.$$

This term is larger than the sum of all the others.

Before we proceed, we do some point counting. Let $t(r)$ denote the number of points on and the circle C_r and south of it. Put

$$S = \widetilde{S}_{t(r)}^j = \prod_{|b_k|<|b_r|}\left(\frac{a_k}{b_k}\right)^{n_k},$$

$$S = \prod_{k>r}(\tan(\theta_k/2))^{\sqrt{\pi d}}\sin(\theta_k),$$

$$t(r) = \sum_{i=0}^{r} n_i = \sum_{i=0}^{r} \sqrt{\pi d}\sin\left((2i+1)\sqrt{\pi/d}\right).$$

Write $\theta_k = 2y$ to get $dk = \sqrt{d/\pi}\,dy$,

$$\log S = \int_{r+1}^{\sqrt{d\pi}/2-1} \sqrt{\pi d}\,\sin(\theta_k)\log(\tan(\theta_k/2))\,dx$$

$$\sim d\int_{r\sqrt{\pi/d}}^{\pi/2-\varepsilon} \sin(2y)\log(\tan(y))\,dy$$

$$= -d\cos(2y)/2\log(\tan y)\big|_{r\sqrt{\pi/d}}^{\pi/2} + d\int_{r\sqrt{\pi/d}}^{\pi/2-\varepsilon} \frac{\cos(2y)}{\sin(2y)}\,dy$$

$$= d/2\cos\theta_r\log(\tan(\theta_r/2)) - d/2\cos(\pi-2\varepsilon)\log(\tan(\pi/2-\varepsilon))$$

$$+ d/2\log(\sin(\pi-2\varepsilon)) - d/2\log(\sin\theta_r)$$

$$\sim d/2\cos\theta_r\log(\tan(\theta_r)) - d/2\log(\sin\theta_r) + (d\log 2)/2.$$

Hence we get:

LEMMA 3.26.

$$\widetilde{S}^j_{t(r)} \sim 2^{d/2}\frac{(\tan(\theta_r/2))^{d\cos\theta_r/2}}{(\sin\theta_r)^{d/2}}.$$

LEMMA 3.27. $t(r) \sim d/2 - d/2\cos\theta_r$.

PROOF.

$$t(r) \sim \int_0^r \sqrt{\pi d}\,\sin((2x+1)\sqrt{\pi/d})\,dx = d/2\cos(\sqrt{\pi/d}) - d/2\cos\theta_r. \qquad \square$$

So we get, for the numerator in d_j^2,

$$\left(\sqrt{\frac{1+\cos\theta_r}{2}}\right)^{2d} \left(\frac{\pi(\sin\theta_r)}{\sqrt{\pi/d}}\right)^2 \left(\frac{\sin\theta_r}{1+\cos\theta_r}\right)^{d-d\cos\theta_r} 2^d\frac{(\tan(\theta_r/2))^{d\cos\theta_r}}{(\sin\theta_r)^d}$$

$$= \pi d\sin^2\theta_r.$$

We have, more or less independently of which point is removed, the equality

$$S^j_{d-l} = \sqrt{\tbinom{d}{l}}.$$

But we want a lower bound for the numerator, hence cancellations are crucial.
We next estimate $S = S_r^j$.

LEMMA 3.28.

$$(S_m^j)^2 \lesssim 2^d\frac{(\tan(\theta_r/2))^{d\cos\theta_r}}{(\sin\theta_r)^d}.$$

PROOF. Recall that

$$S_j^m = \sum_{\{i+\Sigma\delta_k n_k=m\}} \left(-\frac{a_j}{b_j}\right)^i \prod_{\{\delta_k=1\}} \left(\frac{a_k}{b_k}\right)^{n_k}.$$

We should have

$$|S_m^j| \lesssim \prod\left|\frac{a_k}{b_k}\right|,$$

where the product is taken over the m points closest to the north pole.

Let C_r be the northernmost circle under those m points. By Lemma 4.14,

$$t(r) = \frac{d}{2} - \frac{d}{2}\cos\theta_r = d - m.$$

Let $\tilde{j}$ be an index for a point on C_r. Then $|S_m^j| \lesssim \widetilde{S}_{t(r)}^{\tilde{j}}$. Hence we have the estimate

$$|S_m^j| \lesssim 2^{d/2}\frac{(\tan\frac{\theta_r}{2})^d\cos(\frac{\theta_r}{2})}{(\sin\theta_r)^{d/2}}. \qquad \square$$

LEMMA 3.29.

$$|S_m^j|^2 \lesssim \binom{d}{m}.$$

PROOF. Since $m = d - t(r)$, we have $\binom{d}{m} = \binom{d}{t(r)}$. Then

$$\frac{(S_m^j)^2}{\binom{d}{m}} \lesssim \frac{2^d(\tan\frac{\theta_r}{2})^d\cos\theta_r}{(\sin\theta_r)^d}\frac{(\frac{d}{2}-\frac{d}{2}\cos\theta_r^{d/2-d/2\cos\theta_r})}{d^d}\left(d-\frac{d}{2}+\frac{d}{2}\cos\theta_r\right)^{d-d/2+d/2\cos\theta_r}$$

$$= \frac{2^d(\tan\frac{\theta_r}{2})^d\cos\theta_r}{(\sin\theta_r)^d}\frac{(\frac{d}{2}-\frac{d}{2}\cos\theta_r^{d/2-d/2\cos\theta_r})}{d^d}\left(\frac{d}{2}+\frac{d}{2}\cos\theta_r\right)^{d/2+d/2\cos\theta_r}$$

$$= \frac{2^d(\tan\frac{\theta_r}{2})^d\cos\theta_r}{(\sin\theta_r)^d}\frac{(1-\cos\theta_r^{d/2-d/2\cos\theta_r})}{2^d}(1+\cos\theta_r)^{d/2+d/2\cos\theta_r}$$

$$= \frac{2^d(\tan\frac{\theta_r}{2})^d\cos\theta_r}{(\sin\theta_r)^d}\frac{(\sin\theta_r)^d}{2^d}\frac{(1+\cos\theta_r)^{d/2\cos\theta_r}}{(1-\cos\theta_r)^{d/2\cos\theta_r}}$$

$$= \frac{(\tan\frac{\theta_r}{2})^d\cos\theta_r(1+\cos\theta_r)^{d/2\cos\theta_r}}{(1-\cos\theta_r)^{d/2\cos\theta_r}}$$

$$= 1. \qquad \square$$

Hence:

LEMMA 3.30.

$$d_j^2 \gtrsim (\sin\theta_r)^2 \gtrsim \frac{1}{d}.$$

Remark that the estimate for d_j is most degenerate near the poles. Away from the poles, we have $d^j \sim 1$. However, we have not made optimal estimates and it is likely that the optimal lower bound is independent of whether or not we are close to the poles. So probably, we should have $d_j \sim 1$ everywhere.

Hence we get $d_j \sim 1/\sqrt{d}$.

Next, take any vector X of L^2 norm 1. We can write $X = X_j + c_jQ_j$ where $X_j \in \Sigma_j$. Then $X = (X_j + c_jQ_j'') + c_jQ_j'$. We get:

If X is homogenous of degree d, then $X = \sum c_j Q_j$, with $|c_j| \lesssim \|X\|_{L^2(\mathbb{B})} \sqrt{d}$.

References

[Alcarez and de Medrano ≥ 1999] G. G. Alcarez and S. L. de Medrano, "Iterations of quadratic maps of the plane", Preprint.

[Alexander 1994] D. S. Alexander, *A history of complex dynamics: From Schröder to Fatou and Julia*, Vieweg, Braunschweig, 1994.

[Bedford and Pambuccian 1998] E. Bedford and V. Pambuccian, "Dynamics of shift-like polynomial diffeomorphisms of $\mathbb{C}^N$", *Conform. Geom. Dyn.* **2** (1998), 45–55.

[Bedford and Smillie 1991a] E. Bedford and J. Smillie, "Polynomial diffeomorphisms of $\mathbb{C}^2$: currents, equilibrium measure and hyperbolicity", *Invent. Math.* **103**:1 (1991), 69–99.

[Bedford and Smillie 1991b] E. Bedford and J. Smillie, "Polynomial diffeomorphisms of $\mathbb{C}^2$, II: Stable manifolds and recurrence", *J. Amer. Math. Soc.* **4**:4 (1991), 657–679.

[Bedford and Smillie 1992] E. Bedford and J. Smillie, "Polynomial diffeomorphisms of $\mathbb{C}^2$, III: Ergodicity, exponents and entropy of the equilibrium measure", *Math. Ann.* **294**:3 (1992), 395–420.

[Bedford and Smillie 1998] E. Bedford and J. Smillie, "Polynomial diffeomorphisms of $\mathbb{C}^2$, VI: Connectivity of J", *Ann. of Math.* (2) **148**:2 (1998), 695–735.

[Bedford and Smillie ≥ 1999a] E. Bedford and J. Smillie, "Polynomial diffeomorphisms of $\mathbb{C}^2$, V: Critical points and Lyapunov exponents". To appear in *J. Geom. Anal.*

[Bedford and Smillie ≥ 1999b] E. Bedford and J. Smillie, "Polynomial diffeomorphisms of $\mathbb{C}^2$, VII: Hyperbolicity and external rays", Preprint.

[Bedford et al. 1993] E. Bedford, M. Lyubich, and J. Smillie, "Polynomial diffeomorphisms of $\mathbb{C}^2$, IV: The measure of maximal entropy and laminar currents", *Invent. Math.* **112**:1 (1993), 77–125.

[Bergweiler 1993] W. Bergweiler, "Iteration of meromorphic functions", *Bull. Amer. Math. Soc.* (*N.S.*) **29**:2 (1993), 151–188.

[Brjuno 1965] A. D. Brjuno, "On convergence of transforms of differential equations to the normal form", *Dokl. Akad. Nauk SSSR* **165** (1965), 987–989.

[Brjuno 1971] A. D. Brjuno, "Analytic form of differential equations, I", *Trudy Moskov. Mat. Obšč.* **25** (1971), 119–262.

[Brjuno 1972] A. D. Brjuno, "Analytic form of differential equations, II", *Trudy Moskov. Mat. Obšč.* **26** (1972), 199–239.

[Brunella 1994] M. Brunella, "Vanishing holonomy and monodromy of certain centres and foci", pp. 37–48 in *Complex analytic methods in dynamical systems* (Rio de Janeiro, 1992), edited by C. Camacho, Astérisque **222**, Soc. math. France, Paris, 1994.

[Brunella 1996] M. Brunella, "On transversely holomorphic flows, I", *Invent. Math.* **126**:2 (1996), 265–279.

[Brunella and Ghys 1995] M. Brunella and É. Ghys, "Umbilical foliations and transversely holomorphic flows", *J. Differential Geom.* **41**:1 (1995), 1–19.

[Camacho 1991]   C. Camacho, "Problems on limit sets of foliations on complex projective spaces", pp. 1235–1239 in *Proceedings of the International Congress of Mathematicians* (Kyoto, 1990), vol. II, Math. Soc. Japan, Tokyo, 1991.

[Camacho et al. 1992] C. Camacho, A. Lins Neto, and P. Sad, "Foliations with algebraic limit sets", *Ann. of Math.* (2) **136**:2 (1992), 429–446.

[Farkas 1884]   J. Farkas, "Sur les fonctions itératives", *J. Math. Pure Appl.* (3) **10** (1884), 101–108.

[Fatou 1919]   P. Fatou, "Sur les équations fonctionnelles", *Bull. Soc. Math. France* **47** (1919), 161–271.

[Fatou 1920a]   P. Fatou, "Sur les équations fonctionnelles", *Bull. Soc. Math. France* **48** (1920), 33–94.

[Fatou 1920b]   P. Fatou, "Sur les équations fonctionnelles", *Bull. Soc. Math. France* **48** (1920), 208–314.

[Fornæss 1996]   J. E. Fornæss, *Dynamics in several complex variables*, CBMS Regional Conference Series in Mathematics **87**, Amer. Math. Soc., Providence, RI, 1996.

[Fornæss and Grellier 1996]   J. E. Fornæss and S. Grellier, "Exploding orbits of Hamiltonians and contact structures", pp. 155–172 in *Complex analysis and geometry* (Trento, 1993), edited by V. Ancona et al., Lecture Notes in Pure and Applied Mathematics **173**, Marcel Dekker, New York, 1996.

[Fornæss and Sibony 1992]   J. E. Fornæss and N. Sibony, "Complex Hénon mappings in $\mathbb{C}^2$ and Fatou-Bieberbach domains", *Duke Math. J.* **65**:2 (1992), 345–380.

[Fornæss and Sibony 1994]   J. E. Fornæss and N. Sibony, "Complex dynamics in higher dimensions", pp. 131–186 in *Complex potential theory* (Montreal, 1993), edited by P. M. Gauthier and G. Sabidussi, NATO Adv. Sci. Inst. Ser. C Math. Phys. Sci. **439**, Kluwer, Dordrecht, 1994. Notes partially written by Estela A. Gavosto.

[Fornæss and Sibony 1995a]   J. E. Fornæss and N. Sibony, "Classification of recurrent domains for some holomorphic maps", *Math. Ann.* **301**:4 (1995), 813–820.

[Fornæss and Sibony 1995b]   J. E. Fornæss and N. Sibony, "Holomorphic symplectomorphisms in $\mathbb{C}^{2}$", pp. 239–262 in *Dynamical systems and applications*, edited by R. P. Agarwal, World Sci. Ser. Appl. Anal. **4**, World Scientific, River Edge, NJ, 1995.

[Fornæss and Sibony 1996]   J. E. Fornæss and N. Sibony, "Holomorphic symplectomorphisms in $\mathbb{C}^{2p}$", *Duke Math. J.* **82**:2 (1996), 309–317.

[Fornaess and Sibony 1998a]   J. E. Fornaess and N. Sibony, "Fatou and Julia sets for entire mappings in $\mathbb{C}^k$", *Math. Ann.* **311**:1 (1998), 27–40.

[Fornæss and Sibony 1998b]   J. E. Fornæss and N. Sibony, "Hyperbolic maps on $\mathbb{P}^2$", *Math. Ann.* **311**:2 (1998), 305–333.

[Fornæss and Wu 1998]   J. E. Fornæss and H. Wu, "Classification of degree 2 polynomial automorphisms of $\mathbf{C}^3$", *Publ. Mat.* **42**:1 (1998), 195–210.

[Forstneric 1996]   F. Forstneric, "Actions of $(\mathbb{R}, +)$ and $(\mathbb{C}, +)$ on complex manifolds", *Math. Z.* **223**:1 (1996), 123–153.

[Friedland and Milnor 1989]   S. Friedland and J. Milnor, "Dynamical properties of plane polynomial automorphisms", *Ergodic Theory Dynamical Systems* **9**:1 (1989), 67–99.

[Gómez-Mont 1987] X. Gómez-Mont, "Universal families of foliations by curves", pp. 109–129 in *Singularités d'équations différentielles* (Dijon, 1985), Astérisque **150-151**, Soc. math. France, Paris, 1987.

[Gómez-Mont 1988] X. Gómez-Mont, "The transverse dynamics of a holomorphic flow", *Ann. of Math.* (2) **127**:1 (1988), 49–92.

[Hakim 1997] M. Hakim, "Transformations tangent to the identity: Stable pieces of manifolds", Preprint, Orsay, 1997.

[Hakim 1998] M. Hakim, "Analytic transformations of $(\mathbb{C}^p, 0)$ tangent to the identity", *Duke Math. J.* **92**:2 (1998), 403–428.

[Herman 1987] M.-R. Herman, "Recent results and some open questions on Siegel's linearization theorem of germs of complex analytic diffeomorphisms of $\mathbb{C}^n$ near a fixed point", pp. 138–184 in *VIIIth International Congress on Mathematical Physics* (Marseille, 1986), edited by M. Mebkhout and R. Sénéor, World Sci. Publishing, Singapore, 1987.

[Hubbard and Oberste-Vorth 1994] J. H. Hubbard and R. W. Oberste-Vorth, "Hénon mappings in the complex domain, I: The global topology of dynamical space", *Inst. Hautes Études Sci. Publ. Math.* **79** (1994), 5–46.

[Hubbard and Oberste-Vorth 1995] J. H. Hubbard and R. W. Oberste-Vorth, "Hénon mappings in the complex domain, II: Projective and inductive limits of polynomials", pp. 89–132 in *Real and complex dynamical systems* (Hillerød, 1993), edited by B. Branner and P. Hjorth, NATO Adv. Sci. Inst. Ser. C Math. Phys. Sci. **464**, Kluwer, Dordrecht, 1995.

[Julia 1918] G. Julia, "Mémoire sur l'itération des fonctions rationnelles", *J. Math. Pure Appl.* **8** (1918), 47–245.

[Schröder 1871] E. Schröder, "Über iterierte Functionen", *Math. Ann.* **3** (1871), 296–322.

[Siegel 1942] C. L. Siegel, "Iteration of analytic functions", *Ann. of Math.* (2) **43** (1942), 607–612.

[Sternberg 1961] S. Sternberg, "Infinite Lie groups and the formal aspects of dynamical systems", *J. Math. Mech.* **10** (1961), 451–474.

[Sullivan 1985] D. Sullivan, "Quasiconformal homeomorphisms and dynamics, I: Solution of the Fatou–Julia problem on wandering domains", *Ann. of Math.* (2) **122**:3 (1985), 401–418.

[Ueda 1986] T. Ueda, "Local structure of analytic transformations of two complex variables, I", *J. Math. Kyoto Univ.* **26**:2 (1986), 233–261.

[Weickert 1998] B. J. Weickert, "Attracting basins for automorphisms of $\mathbb{C}^2$", *Invent. Math.* **132**:3 (1998), 581–605.

[Wermer 1976] J. Wermer, *Banach algebras and several complex variables*, 2nd ed., Graduate Texts in Mathematics **35**, Springer, New York, 1976. 3rd ed. by Herbert Alexander and John Wermer, Springer, 1998.

[Yoccoz 1992] J.-C. Yoccoz, "An introduction to small divisors problems", pp. 659–679 in *From number theory to physics* (Les Houches, 1989), edited by M. Waldschmidt et al., Springer, Berlin, 1992.

JOHN ERIK FORNÆSS
MATHEMATICS DEPARTMENT
UNIVERSITY OF MICHIGAN
ANN ARBOR, MICHIGAN 48109
UNITED STATES
  fornaess@umich.edu

NESSIM SIBONY
UNIVERSITÉ PARIS SUD
URA757
BÂT 425, MATHÉMATIQUES
91405 ORSAY
FRANCE
  sibony@math.u-psud.fr

Several Complex Variables
MSRI Publications
Volume **37**, 1999

Attractors in $\mathbb{P}^2$

JOHN ERIK FORNÆSS AND BRENDAN WEICKERT

ABSTRACT. We investigate attractors for holomorphic maps from $\mathbb{P}^k \to \mathbb{P}^k$, emphasizing the case $k = 2$. The interest in attractors stems from the fact that when a map is subject to small random perturbations, the long-term dynamics of the resulting system live near the map's attractors. In the case $k = 1$, that is, the case of rational functions on the Riemann sphere, the attractors are either periodic orbits or the whole sphere. In higher dimensions, however, there are other possibilities, which we call nontrivial. In addition to giving some examples of nontrivial attractors, we prove some general results about such attractors in $\mathbb{P}^2$, among them that a given map can have at most one nontrivial attractor K, that K is then connected, has pseudoconvex complement, and contains a nonconstant entire image of $\mathbb{C}$, and that an attractor for a map f is also an attractor for any iterate f^n.

CONTENTS

1. Introduction

We recall first some general notions from the theory of dynamical systems. See [Ruelle 1989] for background.

Let (X, d) be a compact metric space and f a continuous map from X to X. The sequence $(x_j)_{1 \le j \le n}$ is an ε-pseudo-orbit if $d(f(x_j), x_{j+1}) < \varepsilon$ for $j = 1, \ldots, n-1$. For $a, b \in X$, we write $a \succ b$ if for every $\varepsilon > 0$ there is an ε-pseudo-orbit from a to b. We also write $a \succ a$. We write $a \sim b$ if $a \succ b$ and $b \succ a$, and

---

1991 *Mathematics Subject Classification.* 32H50, 32H20.

Fornæss is supported by an NSF grant. Weickert is supported by an NSF postdoctoral fellowship.

denote by $[a]$ the equivalence class of a under this relation. Define an *attractor* to be a minimal equivalence class for $\sim$. The following proposition is an easy consequence of Zorn's lemma.

PROPOSITION 1.1. *Let* $f : X \to X$ *be a continuous map on a compact metric space* X. *Then given any* $x \in X$, *there is an attractor* $[a]$ *such that* $x \succ a$.

It is also easy to show that an attractor K is compact and satisfies $f(K) = K$. See [Ruelle 1989].

We have also the notion of an *attracting set*. A nonempty compact subset $K \subset X$ is an attracting set if it satisfies these conditions:

(i) There exists an open neighborhood $U \supset K$ such that $f(U) \Subset U$.

(ii) $K = \bigcap f^n(U)$.

LEMMA 1.2. *Suppose* $\varnothing \neq U \subset X$ *is an open set such that* $f(U) \Subset U$. *Then* U *contains an attracting set* $\bigcap f^n(U)$.

PROOF. See [Ruelle 1989, Proposition 8.2]. $\qquad\qquad\qquad\qquad\qquad\square$

LEMMA 1.3. *Let* K *be an attractor. Then* K *is a decreasing limit of countably many attracting sets.*

PROOF. Let U be any open neighborhood of K. Then there exists $\rho > 0$ such that no ρ-pseudo-orbit from K leaves U. For $\varepsilon < \rho$, let V be the set of points which can be reached by an ε-pseudo-orbit starting at K. Then V is an open subset of U, and, for each $x \in V$, we have $d(f(x), \partial V) \geq \varepsilon$; otherwise points in V^c could be reached from K by an ε-pseudo-orbit, contradicting the definition of V. Thus $f(V) \Subset V$, and

$$K' := \bigcap f^n(V)$$

is an attracting set, by Lemma 1.2. Since $f(K) = K$, we have $K' \supset K$. Since U was arbitrary, we are done. $\qquad\qquad\qquad\qquad\qquad\qquad\qquad\qquad\square$

2. Size of Attractors

THEOREM 2.1. *Let* $f : \mathbb{P}^k \to \mathbb{P}^k$ *be a holomorphic map of degree at least two. Suppose that* K *is an attractor for* f. *Then either* K *is an attracting periodic orbit for* f, *or* K *contains a nonconstant, entire image of* $\mathbb{C}$.

PROOF. By the previous lemma, K is a decreasing limit of attracting sets. So we can put $K = \bigcap_{i=1}^{\infty} K_i$ where the K_i are attracting sets, $K_{i+1} \subset K_i$. We can also find open sets U_i, $U_{i+1} \Subset U_i$ and $f(U_i) \Subset U_i$, $K_i = \bigcap f^n(U_i)$, $K = \bigcap U_i$. Fix i. Let $\mathcal{A}_i$ denote the affine automorphisms of $\mathbb{P}^k$ close enough to Id. More precisely we want all $A \in \mathcal{A}_i$ to have the property that $A \circ f(U_i) \Subset U_i$. Let $\delta > 0$ be so small that if $\mathrm{dist}(p, q) < \delta$ then there exists an $A_{p,q} \in \mathcal{A}_i$ so that $A_{p,q}(p) = q$.

Let $t = t_i < 1$ be fixed so that if $p \in \mathbb{P}^k$, there exists an $A_p \in \mathcal{A}_i$ which fixes p and for which the derivative at p is scaling by the factor $1/t$. Suppose next that $p \in K_i$ and that $q := f^n(p)$ is closer to p than δ. (In fact, we can take any point $w \in K_i$ and let $p := f^m(w)$ for large m.) Let

$$ B := A_{q,p} \circ f \circ A_{f^{n-1}(p)} \circ f \circ \cdots \circ A_{f(p)} \circ f. $$

Then $B(U_i) \Subset U_i$ and $B(p) = p$.

There are two cases:

Suppose first that for each i we can always find at least one such B with some eigenvalue of $B'(p)$ strictly larger than 1 in modulus. In that case, let ξ be a corresponding eigenvector. Let $\phi : \Delta \to U$ be a holomorphic map with $\phi(0) = p$ and $\phi'(p)$ a nonzero multiple of ξ. Using the sequence $\phi_n := B^n \circ \phi_1$ we get a map ψ_i from the unit disc into U_i with $\psi_i(0) = p$ and $|\psi_i'(0)| > i$. By Brody's theorem there must be a nonconstant entire image X of $\mathbb{C}$ in $\bigcap U_i = K$.

The second case is that for some i one never can have some eigenvalue of some such $B'(p)$ larger than one. In that case, it follows that

$$ A_{f^n(p),p} \circ f^n(p) $$

has derivative bounded by t^{n-1} whenever $\text{dist}(f^n(p), p) < \delta$. We cover K_i by a finite number $\{\Delta_j\}_{j=1}^k$ of discs of radius δ. Consider any finite orbit $\{f^n(p)\}_{j=1}^N$, where $p \in K$. We can always break the orbit up in at most k blocks. The first and last point of each block of consecutive iterates are in the same disc. To define the first block, take all the iterates up to and including the last one in the same disc as the first. To get the second block, take the first iterate after the block and take all interates up to and including the last one in that disc, etc. It follows from the above estimates that for some $C > 0$, we have $\|f^n(p)'\| \leq Ct^n$ for any $p \in K$ and any $n \geq 1$. Hence $f^M|_K$ is contracting for large M. Since $f(K) = K$, it follows that the attractor is just an attracting periodic orbit. $\quad\square$

COROLLARY 2.2. *Let $f : \mathbb{P}^k \to \mathbb{P}^k$ be a holomorphic map of degree at least two, and let K be an attractor for f. Then either K is an attracting periodic orbit for f, or $K \cap J \neq \varnothing$, where J is the Julia set for f.*

PROOF. It is a result of Ueda [1994] that the Fatou set for f is Kobayashi hyperbolic. By the theorem, if K is not an attracting periodic orbit, it contains an entire nonconstant image of $\mathbb{C}$, in which case the hyperbolicity of the Fatou set implies that $K \cap J \neq \varnothing$. $\quad\square$

EXAMPLE 2.3. Let $f : \mathbb{P}^2 \to \mathbb{P}^2$ be a holomorphic map which restricts to a polynomial self map of $\mathbb{C}^2$ and preserves the line L at infinity. Then the line at infinity is an attracting set. The map $f : L \to L$ can be chosen to have a Siegel disc or a parabolic basin and no other Fatou components (except preimages). In that case L is an attractor which lies partly in the Fatou set and partly in the Julia set.

DEFINITION 2.4. We say that an attractor is trivial if it consists of a finite periodic attracting orbit or the whole space. Otherwise we say that the attractor is nontrivial.

We want to analyze nontrivial attractors. From the proof of the above theorem, we get in particular:

THEOREM 2.5. *Suppose that K is a nontrivial attractor and that U is an open set containing K. Then there exist an open set W with $K \subset W \subset U$, a positive number $t < 1$ such that if A_p is the linear map expanding by a factor $1/t$ at p, then $A_p \circ f(W) \Subset W$, and a $\delta > 0$ so that if $p, q \in K$ with $\mathrm{dist}(p, q) < \delta$, then there exists a linear map $A_{p,q}$ close to Id so that $A_{p,q}(p) = q$ and $A \circ f(W) \Subset W$. Moreover, there exists a point $p \in K$ and an integer n such that $\mathrm{dist}(f^n(p), p) < \delta$ and $B := A_{f^n(p),p} \circ f \circ A_{f^{n-1}(p)} \circ \cdots \circ A_{f(p)} \circ f$ satisfies $B(p) = p$ and for some nonzero tangent vector ξ we have $B'(p)(\xi) = \lambda \xi$, with $|\lambda| > 1$.*

COROLLARY 2.6. *Let $q \in K$, a nontrivial attractor and W, U as above. Then there exists a map $g : \mathbb{P}^2 \to \mathbb{P}^2$ with $g(W) \Subset W$ and $g(q) = q$, and also some vector $\xi \neq 0$ such that $g'(q)\xi = \lambda \xi$ with $|\lambda| > 1$.*

PROOF. Let $f_1, \ldots f_m, \tilde{f}_1, \ldots, \tilde{f}_k$ be small perturbations of f mapping W relatively compact to W, with $f_m \circ \cdots f_1(q) = p$ and $\tilde{f}_k(p) \circ \cdots \circ \tilde{f}_1(p) = q$. Wiggling a little more, we may assume that q and p are not critical points for $f_n \circ \cdots \circ f_1$ and $\tilde{f}_k \circ \cdots \circ \tilde{f}_1$, respectively. Then the composition $\tilde{f}_k \circ \cdots \circ \tilde{f}_1 \circ B^N \circ f_n \circ \cdots \circ f_1$ works for large N.   □

COROLLARY 2.7. *Let U be any neighborhood of a nontrivial attractor K. Then for every point $p \in K$, and any $R > 0$ there exists a holomorphic map $\Phi : \Delta \to U$ with $\Phi(0) = p$, $\|\Phi'(0)\| = R$.*

THEOREM 2.8. *A nontrivial attractor K is connected.*

PROOF. Suppose not. Then there exists two open sets U, V with $K \subset U \cup V$, $\bar{U} \cap \bar{V} = \varnothing$. Define $K_1 := K \cap U \neq \varnothing$ and $K_2 := K \cap V \neq \varnothing$. By the above construction, there exist entire images $\Phi_i(\mathbb{C}) \subset K_i$. The theorem follows then from the following two results.   □

3. Pseudoconvexity of the Complement of an Attractor

LEMMA 3.1. $\mathbb{P}^2 \setminus \overline{\Phi_i(\mathbb{C})}$ *is pseudoconvex.*

PROOF. If not, there is a Hartogs figure H in $\mathbb{P}^2 \setminus \overline{\Phi_i(\mathbb{C})}$, so that part of $\Phi_i(\mathbb{C})$ is in $\tilde{H} \setminus H$. But then one can find a bounded subharmonic non-constant function on $\Phi_i(\mathbb{C})$, hence on $\mathbb{C}$, which is impossible.   □

PROPOSITION 3.2. *A pseudoconvex set in $\mathbb{P}^2$ has connected complement.*

COROLLARY 3.3. *f can have at most one nontrivial attractor.*

COROLLARY 3.4. *A nontrivial attractor A for f is also an attractor for any iterate f^n.*

PROOF. Since A is a countable decreasing intersection of attracting sets for f, A is also a countable intersection of attracting sets for f^n. Hence A contains an attractor B for f^n. Since B is an attractor for f^n, we have $f^n(B) = B$. For any $x \in A$, we have $x \succ [a]$ under f^n for some attractor $[a]$. But since any pseudo-orbit for f^n is a pseudo-orbit for f, we must have $[a] \subset A$. Since A contains no attracting periodic orbits, $[a]$ must be nontrivial, and thus by Corollary 3.3 $[a] = B$. For $i < n$, let O be an ε-pseudo-orbit for f^n from $f^i(B)$ to B. Then $f^{n-i}(O)$ is an $L^{n-i}\varepsilon$- pseudo-orbit from $f^n(B) = B$ to $f^{n-i}(B)$, where L is a Lipschitz constant for f^n. Since ε was arbitrary, we have $B \succ f^{n-i}(B)$. Since B is an attractor for f^n, we must have $f^{n-i}(B) \subset B$. This holds for each $i < n$. Applying f^i to this inclusion, we obtain $B \subset f^i(B)$ for each $i < n$. Thus $B = f^i(B)$ for each $i < n$. In particular, $f(B) = B$.

Now let $a \in A$ and $b \in B$. Given $\varepsilon > 0$, there is an ε-pseudo-orbit for f from b to a. We may write

$$a = \tau_k \circ f \circ \ldots \circ \tau_1 \circ f \circ \tau_0(b),$$

where each τ_i is a translation by a vector in $B(0, \varepsilon)$. Let $j = k \bmod n$. Write

$$\tau_j \circ f \circ \ldots \circ \tau_1 \circ f \circ \tau_0 = \sigma_0 \circ f^j$$

if $j \geq 1$, where σ_0 is a translation by a vector in $B(0, \varepsilon')$, and where $\varepsilon' \to 0$ as $\varepsilon \to 0$. If $j = 0$, just take $\sigma_1 = \tau_0$. Similarly, write

$$\tau_{i+n-1} \circ f \circ \ldots \circ \tau_i \circ f = \sigma_{(i-j-1+n)/n} \circ f^n$$

for $i \in \mathbb{N}$ with $i = j + 1 \bmod n$, where again each σ is translation by a vector of modulus ε'', where $\varepsilon'' \to 0$ as $\varepsilon \to 0$. We may assume that $\varepsilon'' > \varepsilon' > \varepsilon$.

We have constructed an ε''-pseudo-orbit for f^n from $f^j(b)$ to a. But since $f^j(b) \in B$, we may also find an ε''-pseudo-orbit for f^n from b to $f^j(b)$. Putting them together, we have a $2\varepsilon''$-pseudo-orbit from b to a. Since we may make ε'' as small as we like, we have $b \succ a$ for f^n. But then $a \in B$ by the definition of B. $\qquad\square$

LEMMA 3.5. $C \cap A \neq \varnothing$.

PROOF. Obvious since the complement of the critical set is pseudoconvex. $\qquad\square$

In fact, we get for the same reason:

LEMMA 3.6. *Let X be any algebraic curve. Then $X \cap A \neq \varnothing$.*

PROPOSITION 3.7. *There is an open neighborhood $U \supset A$ so that if $p \in A$, then there exists a map $g : U \to U$ such that $g(p) = p$ and p is a saddle point.*

PROOF. First, there is a g with $g(p) = p$ and at least one eigenvalue is expanding. If the other is not attracting, we insert a detour from p close to $C \cap A$ to make the other eigenvalue small.   □

COROLLARY 3.8. *The attracting eigenvalue of g at p might be taken to be 0.*

This is obvious from the previous proof.

THEOREM 3.9. *The complement of an attractor is pseudoconvex.*

PROOF. Suppose not. Pick a point $p \in K$ with a Hartogs figure H, with $H \cap K = \varnothing$ and $p \in \tilde{H}$. We may also assume that there exists an open set $U \supset K$ such that $U \cap K = \varnothing$. Then the unstable manifold for p for g as in the previous proposition is parametrized by $\mathbf{C}$ and hence the proof of Lemma 3 applies.   □

DEFINITION 3.10. A compact set $L \subset A$ is minimal if the orbit of any point of L is dense in L.

REMARK 3.11. By Zorn's Lemma, the closure of the forward orbit of any point in A contains a minimal L.

LEMMA 3.12. *An attracting set $S \supset A$ contains A and a possibly infinite collection of attracting periodic orbits.*

This is clear.

4. Description of Fatou Components That Intersect A

To fix notation, let U_n be a sequence of neighborhoods of A,

$$V_{n+1} := f(U_{n+1}) \Subset U_{n+1} \Subset U_n,$$

such that $A = \bigcap U_n$.

THEOREM 4.1. *Let Ω be a Fatou component, $\bar{\Omega} \cap A \neq \varnothing$. Then $f^n_{|\Omega} \to A$ u.c.c.*

PROOF. It suffices to prove that if $\gamma : [0,1] \to \Omega$ is a continuous curve and $\gamma(0) \in U_n$, then $f^m(\gamma([0,1])) \subset U_n$ for all large enough m. Let $[0, r_m]$ be the largest interval for which $f^m(\gamma([0,1])) \subset \bar{U}_n$. Notice then that r_m is an increasing sequence. However, since $f^{m+1}(\gamma(r_m)) \in \bar{V}_{n+1}$, it follows by uniform continuity that there is a fixed $\varepsilon > 0$ such that $r_{m+1} \geq \min\{1, r_m + \varepsilon\}$. Hence we are done.   □

The proof shows, a little more generally:

COROLLARY 4.2. *If some Fatou component Ω intersects U_n, then $f^n_{|\Omega} \to U_n$ u.c.c.*

5. Simple Cases of Attractors

In this section we try to gain insight into attractors by working our way through examples which are gradually more complicated.

THEOREM 5.1. *Suppose that* $f : \mathbb{P}^2 \to \mathbb{P}^2$ *is a holomorphic map which restricts to a polynomial on* $\mathbb{C}^2$. *If A has a nontrivial attractor, then this is the line at infinity and f there is a rational map without attracting basins. The converse is also true.*

This is clear.

THEOREM 5.2. *Let A be a totally invariant attractor for* $f : \mathbb{P}^2 \to \mathbb{P}^2$. *Then either $A = \mathbb{P}^2$ or A is contained in a pluripolar set, and $\mathbb{P}^2 \setminus A$ is not hyperbolic. If in addition A is algebraic, then A is a hyperplane or a nonsingular quadratic curve.*

PROOF. By [Fornæss and Sibony 1994, Theorem 4.5] and the following discussion, if a proper subvariety V of $\mathbb{P}^2$ satisfies $f^{-1}(V) = V$, then either V is a nonsingular quadratic curve and f must then have odd degree, or V is a union of hyperplanes, and f has one of the forms

$$[z : w : t] \mapsto \left[f_0([z : w : t]) : f_1([z : w : t]) : t^d \right]$$
$$[z : w : t] \mapsto \left[f_0([z : w : t]) : w^d : t^d \right]$$
$$[z : w : t] \mapsto \left[z^d : w^d : t^d \right],$$

depending on whether V consists of one, two, or three hyperplanes. It is easy to verify directly that in the last two cases V is not an attractor. Thus the only possibility for a totally invariant algebraic attractor is the first case, where the attractor is a hyperplane and f is a suspension of a holomorphic map on $\mathbb{P}^1$ with empty Fatou set, or a nonsingular quadratic curve. This proves the second statement.

To prove the first, we note that by a special case of a result of Russakovskii and Shiffman [1997], given a holomorphic map $f : \mathbb{P}^k \to \mathbb{P}^k$, there exists a pluripolar set $\mathcal{E}$ such that, for any probability measure ν which gives no mass to $\mathcal{E}$,

$$((f^n)^* \nu)/d^{nk} \to \mu.$$

Taking ν to be the mass of a single point, we see that given any $p \notin \mathcal{E}$, the successive inverse images of p cluster all over supp μ. Since the complement of $\mathcal{E}$ is dense in $\mathbb{P}^k$, for any such p and any $q \in$ supp μ, we have $q \succ p$. Thus if an attractor A contains any point of supp μ, then $A = \mathbb{P}^k$. If $A \not\subset \mathcal{E}$, then there exists $p \in A$ whose inverse images cluster all over supp μ. Since A is totally invariant and closed, supp $\mu \subset A$. Thus $A = \mathbb{P}^k$.

To prove that $\Omega := \mathbb{P}^2 \setminus A$ is not hyperbolic, since supp $\mu \subset \Omega$, by a result of Briend [1996] there exists a repelling periodic point $p \in \Omega$. Since $f(\Omega) = \Omega$, there exist arbitrarily large analytic disks in Ω through p. $\qquad \square$

Next we turn to attractors which are not totally invariant.

EXAMPLE 5.3. The map $[(z - 2w)^2 : t^2 + z^2 : zt/2]$ has the line at infinity as an attractor whose preimage also containts the line $(z = 0)$.

Details: The line at infinity is forward invariant and the map restricts to a critically finite preperiodic map on the line at infinity. We need to show that $(t = 0)$ is attracting. We cover the line at infinity with two sets, $U_1 = \{|z| < |w|\}$ and $U_2 = \{|w| < |z|\}$. We introduce a metric equal the Euclidean metric in each of the two coordinates Then we show that the normal derivative if the map is at most $\frac{1}{2}$ at any point. Let $Z = (z - 2w)^2$, $W = z^2 + t^2$, $T = zt/2$.

On U_1 we have $|Z|/|W| = (1 - 2|w|/|z|)^2 > 1$. Hence U_1 is mapped into U_2. Hence the map takes the form $(z : 1 : t) \rightarrow (1 : W : T)$, or $(z, t) \rightarrow (z^2/(z - 2)^2, zt/(2(z - 2)^2))$. Hence the t derivative of the second coordinate is when $t = 0$, $|z|/(|2(z - 2)|^2) < \frac{1}{4}$ since $|z| < 1$.

On U_2 there are two cases. If $|Z| < |W|$, the map takes the form

$$(w, t) \rightarrow ((1 - 2w)^2, t/2)$$

and the normal derivative is $\frac{1}{2}$.

If $|W| < |Z|$, the map takes the form

$$(w, t) \rightarrow \left(1/(1 - 2w)^2, \, t/(2(1 - 2w))^2\right);$$

hence the normal derivative is $1/|2(1 - 2w)|^2$. But in this set $|1/(1 - 2w)^2| < 1$, so we are done again.

A general technique for generating examples of this kind for maps of degree $d \geq 3$: Take any rational map $[P(z, w) : Q(z, w)]$ of degree $d \geq 3$ without attracting basin. Then we can define $[P + t^d : Q : zt^{d-1}]$ or if necessary put the t^d on the second term.

Next we give an example of an attractor which is a smooth rational curve, but not a line. The attractor is the set $V = (zw = t^2)$. Consider, for small $\delta \neq 0$, the map

$$F_\delta = [X : Y : Z] = \left[(z + 4w - 4t)^2 : z^2 : z(z + 4w - 4t) + \delta(t^2 - zw)\right].$$

First consider $F_0 : T^2 - XY \equiv 0$ and the point of indeterminacy is $[0 : 1 : 1] \notin V$. Hence V is mapped holomorphically into itself and the map is holomorphic in a neighborhood of V. Also F can be calculated on V, parametrized by $\tau \rightarrow (\tau, 1/\tau, 1)$, which is mapped to $[(\tau - 2)^4/\tau^2 : \tau^2 : (\tau - 2)^2] = [(\tau - 2)^2/\tau^2 : \tau^2/(\tau - 2)^2 : 1]$; hence the map reduces to $x \rightarrow (x - 2)^2/x^2$ which is a critically finite maps whose Julia set is all of $\mathbb{P}^1$. For small $\delta \neq 0$, it follows that V is an attractor.

Another example, similar to the previous one: Use the map $z \rightarrow \lambda(1 - 2/z)^3$, where $\lambda \in \mathbb{C}$ is chosen to make the map critically finite, with Julia set equal to

$\mathbb{P}^1$, again realized as $zw = t^2$.

$$f : \mathbb{P}^2 \to \mathbb{P}^2,$$

$$[z : w : t] \mapsto \left[\lambda(z + 4w - 4t)^3 : z^3/\lambda : z(z - 2t)(z + 4w - 4t) + 2(z - 2t)(zw - t^2)\right].$$

In this case one calculates that $ZW - T^2 = 4(zw - t^2)^2\left(3(z - 2t)^2 + 16(zw - t^2)\right)$ and hence the variety $zw = t^2$ is contained in the critical set. Hence it is an attractor.

Given $f : \mathbb{P}^2 \to \mathbb{P}^2$, let C_1 denote the critical locus of f. Let

$$D_1 = \bigcup_{n \geq 1} f^n(C_1) \quad \text{and} \quad E_1 = \bigcap_{n \geq 0} f^n(D_1).$$

Using the terminology of Jonsson [1998], we call f *1-critically finite* if D_1 is algebraic (or, equivalently, if the union defining it is finite) and if E_1 and C_1 have no common irreducible component. Note that E_1 is algebraic if D_1 is. If f is 1-critically finite, define

$$C_2 = C_1 \cap E_1, \quad D_2 = \bigcup_{n \geq 1} f^n(C_2), \quad E_2 = \bigcap_{n \geq 0} f^n(D_2).$$

Ueda [1998] has proved that these are finite sets. Call f *2-critically finite* if $C_2 \cap E_2 = \varnothing$. It has been proved by Fornaess and Sibony [1992] and by Ueda [1998] that if f is 2-critically finite its Fatou set is empty. Further work of Jonsson [1998] and Briend [1996] has shown that for such f, supp $\mu = \mathbb{P}^2$. For 2-critically finite maps, therefore, $\mathbb{P}^2$ is an attractor. We wish to study maps which are 1-critically finite, but not necessarily 2-critically finite.

THEOREM 5.4. *Suppose that $f : \mathbb{P}^2 \to \mathbb{P}^2$ is 1-critically finite. Let A be a nontrivial attractor for f. Then either $A = \mathbb{P}^2$, or A contains a periodic cycle whose multiplier has one zero eigenvalue.*

PROOF. From Lemma 3.6, we have $C_1 \cap A \neq \varnothing$ and $E_1 \cap A \neq \varnothing$. By assumption, E_1 is algebraic, and by its definition its irreducible components are periodic. We may assume that f is not 2-critically finite; otherwise $A = \mathbb{P}^2$. Thus E_2 contains a critical point p. Since all the points in E_2 are periodic, p is periodic.

Let V be an irreducible component of E_1 containing p. Since by Corollary 3.4 an attractor for f is also an attractor for f^n, we may replace f by an iterate without loss of generality. Thus we may assume that V is invariant. If $V \subset A$, we are done. Otherwise, let U be an open set containing A with $f(U) \Subset U$. Assume that U was chosen small enough that $V \not\subset U$, and let $U' = V \cap U$. We can assume that there are only finitely many irreducible components of $V \cap U$, and these are mapped to each other. Replacing f by an iterate if necessary, we may assume that there is a component V' which is mapped into itself and which intersects A. Let $\pi : \tilde{V} \to V$ be a normalization of V, and let $\tilde{f}$ be a lift of $f|_V$ to $\tilde{V}$. Let $\tilde{V}' = \pi^{-1}(V')$. Since $\tilde{f}(\tilde{V}') \Subset \tilde{V}'$, there is a fixed point $\tilde{q}$ for $\tilde{f}$ in $\tilde{V}'$. If $\tilde{V}$ is hyperbolic, $\tilde{V}$ has only finitely many nonconstant holomorphic self maps, so this is impossible. If $\tilde{V} = \mathbb{P}^1$, then $\tilde{f}$ is a rational function with an attracting fixed point at $\tilde{q}$. If $\tilde{q}$ is not critical, then there is a critical point in its basin with infinite forward orbit. Then the image under π of this point is a critical point of f in E_1 with infinite forward orbit. This is impossible, since D_1 is a finite set. Thus again $\tilde{q}$ is critical. The final possibility is that $\tilde{V}$ is a torus. But then, since $\tilde{f}$ is not injective, every periodic point of $\tilde{f}$ is repelling, contradicting the existence of $\tilde{q}$. Thus $\tilde{q}$ is a fixed critical point for $\tilde{f}$, and $q := \pi(\tilde{q})$ is a fixed critical point for f. Since V' intersects A, and $f^n(z) \to q$ for all $z \in V'$, we must have $q \in A$. Since we have replaced f, possibly, with higher iterates, we conclude that the map we started with had a critical periodic orbit in A.   □

EXAMPLE 5.5. Consider the map $[z : w : t] \mapsto [(z - 2w)^2 : z^2 : t^2]$. The line $(t = 0)$ is an attractor, and $[1 : 1 : 0]$ is a fixed critical point in the attractor.

PROPOSITION 5.6. *If an attractor A contains a repelling periodic point p or a Siegel domain, then any path connecting p to the complement of A must intersect $\overline{D}_1$.*

PROOF. In the case of a Siegel domain Ω, we have already that $\partial\Omega \subset D_1$. In the case of a repelling periodic point p, which we may assume to be fixed, take a neighborhood of p on which branches of inverses of f^n. are defined and converge to the constant map p. Let q outside A be connected to p by a path which doesn't intersect $\overline{D}_1$. We may extend all the branches of inverses of f^n previously defined along that path, and they form a normal family in the resulting open set, by a result of Ueda. Any convergent subsequence must converge to the constant map p. Thus inverse images of p cluster on q. Thus $p \succ q$. But this is a contradiction.
□

COROLLARY 5.7. *If D_1 is algebraic, there are no repelling periodic points in A unless $A = \mathbb{P}^2$. There are no Siegel domains anywhere.*

References

[Briend 1996] J.-Y. Briend, "Exposants de Liapounoff et points périodiques d'endomorphismes holomorphes de $\mathbb{CP}^k$", *C. R. Acad. Sci. Paris Sér. I Math.* **323**:7 (1996), 805–808.

[Fornæss and Sibony 1992] J. E. Fornæss and N. Sibony, "Critically finite rational maps on $\mathbb{P}^2$", pp. 245–260 in *The Madison Symposium on Complex Analysis* (Madison, WI, 1991), edited by A. Nagel and E. L. Stout, Contemp. Math. **137**, Amer. Math. Soc., Providence, RI, 1992.

[Fornæss and Sibony 1994] J. E. Fornæss and N. Sibony, "Complex dynamics in higher dimension, I", pp. 201–231 in *Complex analytic methods in dynamical systems* (Rio de Janeiro, 1992), edited by C. Camacho, Astérisque **222**, Soc. math. France, Paris, 1994.

[Jonsson 1998] M. Jonsson, "Some properties of 2-critically finite holomorphic maps of $\mathbb{P}^2$", *Ergodic Theory Dynam. Systems* **18**:1 (1998), 171–187.

[Ruelle 1989]   D. Ruelle, *Elements of differentiable dynamics and bifurcation theory*, Academic Press, Boston, 1989.

[Russakovskii and Shiffman 1997]   A. Russakovskii and B. Shiffman, "Value distribution for sequences of rational mappings and complex dynamics", *Indiana Univ. Math. J.* **46**:3 (1997), 897–932.

[Ueda 1994]   T. Ueda, "Fatou sets in complex dynamics on projective spaces", *J. Math. Soc. Japan* **46**:3 (1994), 545–555.

[Ueda 1998]   T. Ueda, "Critical orbits of holomorphic maps on projective spaces", *J. Geom. Anal.* **8**:2 (1998), 319–334.

JOHN ERIK FORNÆSS
MATHEMATICS DEPARTMENT
UNIVERSITY OF MICHIGAN
ANN ARBOR, MICHIGAN 48109
UNITED STATES
    fornaess@umich.edu

BRENDAN WEICKERT
DEPARTMENT OF MATHEMATICS
THE UNIVERSITY OF CHICAGO
5734 UNIVERSITY AVENUE
CHICAGO, IL 60637
UNITED STATES
    brendan@math.uchicago.edu

Several Complex Variables
MSRI Publications
Volume **37**, 1999

Analytic Hilbert Quotients

PETER HEINZNER AND ALAN HUCKLEBERRY

ABSTRACT. We give a systematic treatment of the quotient theory for a holomorphic action of a reductive group $G = K^{\mathbb{C}}$ on a not necessarily compact Kählerian space X. This is carried out via the complex geometry of Hamiltonian actions and in particular uses strong exhaustion properties of K-invariant plurisubharmonic potential functions.

The open subset $X(\mu)$ of momentum semistable points is covered by analytic Luna slice neighborhoods which are constructed along the Kempf–Ness set $\mu^{-1}\{0\}$. The analytic Hilbert quotient $X(\mu) \to X(\mu)//G$ is defined on these Stein neighborhoods by complex analytic invariant theory. If X is projective algebraic, then these quotients are those given by geometric invariant theory.

The main results here appear in various contexts in the literature. However, a number of proofs are new and we hope that the systematic treatment will provide the nonspecialist with basic background information as well as details of recent developments.

CONTENTS

1. Introduction

As the title indicates, we focus here on a certain quotient construction for group actions on complex spaces. Our attention is primarily devoted to actions of (linear) reductive complex Lie groups, i.e., complex matrix groups which are complexifications $G = K^{\mathbb{C}}$ of their maximal compact subgroups.

The building blocks for these groups are $\mathbb{C}^* = (S^1)^{\mathbb{C}}$ and the simple complex Lie groups, e.g., $\mathrm{SL}_n(\mathbb{C}) = (\mathrm{SU}_n)^{\mathbb{C}}$. In fact, after a finite central extension, a connected reductive Lie group is just a product of such groups.

A holomorphic action of a complex Lie group G on a complex space X is a holomorphic map $G \times X \to X$, $(g, x) \mapsto g(x)$, which is defined by a homomorphism $G \to Aut_{\mathcal{O}}X$. Although the reductive groups themselves are easily listed, understanding their actions is an entirely different matter.

As a first step one often considers invariants of a given action. At the geometric level this can mean the orbit space construction: Two points $x_1, x_2 \in X$ are deemed equivalent if there exists $g \in G$ with $g(x_1) = x_2$.

For a compact group a quotient $\pi : X \to X/\sim =: X/K$ of this type is quite reasonable. For example, the base is Hausdorff, has a stratified manifold structure and can be dealt with methods from semi-algebraic geometry. Furthermore, this quotient has a natural invariant theory interpretation: $x_1 \sim x_2$ if and only if $f(x_1) = f(x_2)$ for every invariant continuous function $f \in \mathcal{C}(X)^K$.

On the other hand, one leaves the complex analytic category, e.g., the quotient of $\mathbb{C}$ by the standard S^1-action in $\mathbb{R}^{\geq 0}$.

From a complex analytic viewpoint an orbit space construction involving the complex Lie group G would be preferable. However, not to even mention the difficulties of constructing a complex quotient structure, due to the non-compactness of G, the orbit space X/G is often not even Hausdorff.

At least in situations where there are plenty of functions, the appropriate first steps in an analytic theory would seem to be of an invariant theoretic nature.

For example, in a case where algebraic methods are of great use, if $G \times V \to V$ is a linear action of a reductive group on a complex vector space, then the ring $R = \mathbb{C}[V]^G$ of invariant polynomials in finitely generated [Kraft 1984].

If $f_1, \ldots, f_m \in R$ is a choice for such generators, then the image of the map $F = (f_1, \ldots, f_m)$ is a good model for the quotient. Due to Hilbert's original impact on the finite generatedness of R, these are quite often referred to as Hilbert quotients. The base is the affine variety $\mathrm{Spec}(R)$ and the notation $\pi : V \to \mathrm{Spec}(R) =: V//G$ serves as a reminder that this is not necessarily an orbit space construction.

For example, for the action of $G = \mathbb{C}^*$ on $\mathbb{C}^2$ by $\lambda(z, w) = (\lambda z, \lambda^{-1}w)$, the ring R is generated by $f := zw$ and the fiber of $\pi : \mathbb{C}^2 \to \mathbb{C}^2//G = \mathbb{C}$ over $0 \in \mathbb{C}$ consists of three orbits.

Using arguments of affine algebraic geometry along with the identity principle $\mathbb{C}[V]^G = \mathbb{C}[V]^K$ one shows that the Hilbert quotient is indeed of a geometric nature. The quotient π is surjective, and $x_1 \sim x_2$ if and only if $\overline{Gx} \cap \overline{Gy} \neq \varnothing$ (see Section 3). For more general spaces and actions this may not be an equivalence relation.

Since polynomials are dense in $\mathcal{O}(V)^G$, the Hilbert quotient $\pi : V \to V//G$ for a linear action serves as the appropriate quotient if one considers $G \times V \to V$ as a holomorphic action.

Now, using the existence of a closed, equivariant embedding $X \hookrightarrow V$ in a representation space, the algebraic quotient theory is immediately extended to the category of affine varieties. For Stein spaces, even in the smooth case, there do not in general exist such equivariant embeddings [Heinzner 1988]. Nevertheless, the algebraic invariant theory can be applied, but in a certain sense only locally.

The invariant theoretic quotient $\pi : X \to X//G$ for a holomorphic action of a reductive group on a normal Stein space was constructed by D. Snow [1982]. Later the normality condition was removed and the quotient was also constructed for action of compact groups on Stein spaces [Heinzner 1989; 1991].

The key to Snow's construction is his adaptation of Luna's slice theorem (see Section 4) to the complex analytic setting. From our point of view the controlling tool for the well-definedness of the quotient is a K-invariant, strictly plurisubharmonic exhaustion function.

In the development of the complex analytic quotient theory since that time, the convexity and exhaustion properties of invariant plurisubharmonic functions have played a decisive role.

The Stein context and its rich function theory is however a bit misleading. The key geometric information is provided by K-invariant Kählerian structures and the availability of invariant plurisubharmonic potential functions which turn out to be exhaustions along the fibers of the quotients that are to be constructed.

The conceptual underlying factor is that of symplectic reduction. In order to introduce the notation we now recall this construction.

If (M, ω) is a symplectic manifold equipped with a smooth action $K \times M \to M$ of a Lie group of symplectic diffeomorphisms, a vector field $\xi_M \in \mathrm{Vect}_\omega(M)$ coming from the action of a 1-parameter group given by $\xi \in \mathrm{Lie}\, K^*$ is said to be Hamiltonian if it is associated to a function $\mu_\xi \in C^\infty(M)$ by $d\mu_\xi = i_{\xi_M}\omega$.

If these functions can be bundled together to an equivariant map $\mu : M \to \mathrm{Lie}(K)^*$ with coordinates $\xi \circ \mu = \mu_\xi$, then μ is said to be an equivariant moment map.

Vector fields $V_H \in \mathrm{Ham}(M)$ associated to K-invariant Hamiltonians $H \in C^\infty(M)$ by the rule $dH = i_{V_H}\omega$ have the full moment map as a constant of motion. Thus one is led to study the μ-fibers and their induced geometry.

It is most often sufficient to analyze a fiber of the type $M_0 := \mu^{-1}(0)$ and its embedding $i : M_0 \hookrightarrow M$. Of course, unless something is known about the group action so that, e.g., normal form theorems are applicable, M_0 may be quite singular, and in any case the form $\omega_0 := i^*\omega$ will be degenerate in the orbit directions.

If K is acting properly, e.g., for K compact, $M_{\mathrm{red}} := M_0/K$ has only mild singularities and ω_0 can be pushed down to a stratified symplectic structure. See [Sjamaar and Lerman 1991], and also [Heinzner et al. 1994] for an embedding in the Stein space setting.

The technique of symplectic reduction has been used in numerous situations to better understand existing complex group quotients; see, for example, [Mumford et al. 1994; Kirwan 1984; Neeman 1985; Guillemin and Sternberg 1984]. Here we construct quotients in the Kählerian setting by using this principle in combination with methods involving plurisubharmonic potential functions. This approach is implicit in considerations of [Heinzner et al. 1994], and is carried out in general in [Heinzner and Loose 1994]. Kählerian quotients are also constructed in [Sjamaar 1995], in the presence of an appropriate, proper Morse function.

Although we carry out proofs in the singular case, for introductory purposes it is sufficient to consider a Kähler manifold (X, ω) equipped with a holomorphic action $G \times X \to X$ of a reductive group. For a choice of a maximal compact subgroup K of G it is assumed that ω is K-invariant and that there exists a K-moment map $\mu : X \to \mathrm{Lie}^*(K)$.

The 0-fiber $X_0 := \mu^{-1}(0)$, referred to as the Kempf–Ness set, provides a foundation for the entire study.

First, the set for which a quotient can be constructed, the set of semi-stable points, is defined by

$$X(\mu) := \{x \in X : \overline{Gx} \cap X_0 \neq \varnothing\}.$$

Using exhaustion properties of plurisubharmonic potential functions ρ with $\omega = dd^c\rho$ it is shown that $X(\mu)$ is open, that the Hausdorff quotient $\pi : X(\mu) \to Q$ exists and that the inclusion $X_0 \hookrightarrow X(\mu)$ induces a homeomorphism $X_0/K = Q$.

The complex structure on the quotient, which is given by the invariant part $U \mapsto \pi_*\mathcal{O}(U)^G$ of the direct image sheaf, is understood via an analytic version of Luna's slice theorem. In this way, using plurisubharmonic potential functions as controlling devices, we return to invariant theory. Hence the classical notation $X(\mu)//G := Q$ is used. Since $\pi : X(\mu) \to X(\mu)//G$ is a Stein map [Heinzner et al. 1998] and is locally an invariant theoretic quotient, this is an example of an analytic Hilbert quotient.

Although it is almost always the case that $X \neq X(\mu)$, there are certain important situations where $X = X(\mu)$. For example, if X is Stein, $\rho : X \to \mathbb{R}$ is a K-invariant, proper exhaustion and $\omega := dd^c\rho$, then $X = X(\mu)$ and $X(\mu) \to X(\mu)//G$ is the invariant theory quotient.

One of the main goals of this paper is to make the basic methods for actions of reductive groups accessible to non-specialists. In particular, discussing from the point of view of complex geometry, in Sections 2 and 3 we attempt to systematically build a foundation for the developments in the last 10 years in complex analytic quotient theory. The existence of the quotient and its essential properties are proved in Section 4.

2. Stein Homogeneous Spaces of Reductive Groups

In the invariant theoretic as well as the Hamiltonian approach, Stein homogeneous spaces play a central role in the construction and understanding of analytic Hilbert quotients. Here we describe the basic properties of these spaces via an analysis of invariant plurisubharmonic functions and Kählerian reduction at the group level.

2.1. Decomposition Theorems. Let G be a reductive complex Lie group. A maximal compact subgroup K determines a decomposition $\mathfrak{g} = \mathfrak{k} \oplus \mathfrak{p}$, with $\mathfrak{p} := i\mathfrak{k}$. Define $P := \exp \mathfrak{p}$.

THE KP-DECOMPOSITION. *The map* $\exp : \mathfrak{p} \to P$ *is a diffeomorphism onto a closed submanifold P of G. Group multiplication, $(k.p) \mapsto k \cdot p$, then defines a diffeomorphism $K \times P \to G$.*

EXAMPLE (THE KP-DECOMPOSITION OF $\mathrm{GL}_n(\mathbb{C})$). Let $G := \mathrm{GL}_n(\mathbb{C})$ and $\sigma : G \to G$ be the anti-holomorphic involution defined by $\sigma(A) = {}^t\bar{A}^{-1}$. The real form $K := U_n = \mathrm{Fix}(\sigma)$ is a maximal compact subgroup of G.

Recall that $\mathfrak{g} = \mathrm{Mat}(n \times n, \mathbb{C})$ and note that at the Lie algebra level $\sigma_* : \mathfrak{g} \to \mathfrak{g}$ is given by $A \mapsto -{}^t\bar{A}$. Thus $\mathfrak{k} = \{A \in \mathrm{Mat}(n \times n, \mathbb{C}) : A + {}^t\bar{A} = 0\}$ and $\mathfrak{p} = i\mathfrak{k}$ is the set of Hermitian matrices.

The exponential map $\exp : \mathfrak{g} \to G$, $A \mapsto e^A = \sum_{n=0}^{\infty} A^n/n!$, maps $\mathfrak{p}$ into the closed submanifold $H^{>0}$ of Hermitian positive-definite matrices. For $h \in H^{>0}$ there exists $k \in K$ such that khk^{-1} is a diagonal matrix.

Since $\mathfrak{p}$ is invariant under the $\mathrm{Int}(K)$-action, i.e., $p \mapsto kpk^{-1}$, and the diagonal elements of $H^{>0}$ are clearly in $\exp(\mathfrak{p})$, it follows that $P = H^{>0}$.

That $\exp : \mathfrak{p} \to P$ and $KP \to G$ are diffeomorphisms follows from concrete calculations with matrices; see [Chevalley 1946].

The theorem on the KP-decomposition for an arbitrary reductive group is proved via this example. For this, first embed K in U_n by a faithful unitary representation $\tau : K \to U_n$. It can be shown that τ can be uniquely extended to a holomorphic representation $\tau^{\mathbb{C}} : G \to \mathrm{GL}_n(\mathbb{C})$ which is in fact biholomorphic onto its image; see [Hochschild 1965]. The $U_n \cdot H^{>0}$-decomposition of $\mathrm{GL}_n(\mathbb{C})$ restricts to G to give its KP-decomposition, as desired.

For notational convenience assume that G is connected and let T be a maximal torus in K. Let K act on $\mathfrak{k}$ by the adjoint representation. It follows that $K \cdot \mathfrak{t} = \mathfrak{k}$. Since $\mathfrak{p} = i\mathfrak{k}$, this can be interpreted in the context of the KP-decomposition. Before doing so, we recall several basic facts concerning the Weyl group. As a basic reference, see [Wallach 1973], for instance.

If for $k \in K$ there exists $\xi \in \mathfrak{t}$ with $k(\xi) \in \mathfrak{t}$, then it in fact follows that $k(\mathfrak{t}) = \mathfrak{t}$ and k is in the normalizer $N_K(T)$. Of course T acts trivially on $\mathfrak{t}$. So the $N_K(T)$ action factors through the action of the Weyl group $W = W_K(T) = N_K(T)/T$.

Since the tori do not possess continuous families of Lie group automorphisms and T is maximal, it follows that $W = W_K(T)$ is finite.

In fact W is generated in a natural way by reflections and therefore has a closed fundamental region $\mathfrak{t}^+$ which is an intersection of finitely many closed half-spaces. Consequently the Weyl-chamber $\mathfrak{t}^+$ serves as a fundamental region for the $\mathrm{Ad}(K)$-representation: The map $K \times \mathfrak{t}^+ \to \mathfrak{t}$, $(k, \xi) \mapsto \mathrm{Ad}(k)(\xi)$, is bijective.

THE KAK-DECOMPOSITON. *Let G be a connected reductive group, K a maximal compact subgroup, $T < K$ a maximal torus and let $A := \exp(i\mathfrak{t}) \subset P$. Let $K \times K$ act on G by multiplication, $g \mapsto k_1 g k_2^{-1}$. Then:*

(i) *$(K \times K)A = G$.*
(ii) *For $A^+ := \exp(i\mathfrak{t}^+)$ it follows that the map $K \times K \times A^+ \to G$, $(k_1, k_2, a) \mapsto k_1 a k_2^{-1}$, is bijective.*
(iii) *Restriction defines an isomorphism $\mathcal{E}(G)^{K \times K} \cong \mathcal{E}(A)^W$.*

PROOF. Since $P = \exp(i\mathfrak{t})$, K acts on $P \cong \mathfrak{t}$ by its adjoint representation. Thus A^+ is a fundamental region for its action and $(K \times K) \cdot A^+ = K \cdot \mathrm{Ad}(K) \cdot A^+ = K \cdot P = G$. It follows from the KP-decomposition and the fact that A^+ is a fundamental region that the map $K \times A^+ \times K \to G$ is injective. This proves (i) and (ii).

If $f \in \mathcal{E}(G)^{K \times K}$, then its restriction is clearly W-invariant. Since $G = KAK$, the restriction map is injective. For the surjectivity, given $f \in \mathcal{E}(A)^W$, let h be any smooth extension to G and define $F \in \mathcal{E}(G)^{K \times K}$ by averaging:

$$F(x) = \int h(k_1 x k_2^{-1}) \, dV,$$

where dV is an invariant probability measure on $K \times K$. It follows that $F|A = f$. $\qquad\square$

REMARK. Since $G = K \cdot G^\circ$, the assumption of connectivity in the KAK-decomposition is of no essential relevance for applications.

2.2. Invariant Plurisubharmonic Functions.

a. Critical points. Throughout this section G is a reductive group, H a closed complex subgroup and $X = G/H$ the associated complex homogeneous manifold. Fix a maximal compact subgroup $K < G$. Note that, since $\mathfrak{g} = \mathfrak{k} + i\mathfrak{k}$, it follows that the real dimension $\dim_\mathbb{R} Kx$ of an arbitrary K-orbit in X is at least the complex dimension $\dim_\mathbb{C} X$.

LEMMA 2.2.1. *Suppose that H has finitely many components. Then it is reductive if and only if there exists $x \in X$ with $\dim_\mathbb{R} Kx = \dim_\mathbb{C} X$.*

PROOF. If H is reductive, then it has a maximal compact subgroup L so that $H = L^\mathbb{C}$. After replacing H by a conjugate, we may assume that $L < K$. Thus the associated orbit of the neutral point $x \in X$ is $Kx = K/L$ and

$$\dim_{\mathbb{R}} Kx = \dim_{\mathbb{R}} K - \dim_{\mathbb{R}} L = \dim_{\mathbb{C}} G - \dim_{\mathbb{C}} H = \dim_{\mathbb{C}} X.$$

Conversely, if $\dim_{\mathbb{R}} Kx = \dim_{\mathbb{C}} X$, then, defining $L := H \cap K$, the preceding string of equalities shows that $\dim_{\mathbb{C}} L^{\mathbb{C}} = \dim_{\mathbb{C}} H$. (For this we also use the fact that $\mathfrak{l}$ is totally real.) Thus H^0 is reductive and, since H has at most finitely many components, it follows that H is reductive. □

Numerous arguments of the present work involve critical points of K-invariant strictly plurisubharmonic functions $\rho : X \to \mathbb{R}$. It is important in this regard to underline the role of the moment map.

As above let $X = G/H$ be a complex homogeneous space of a reductive group $G = K^{\mathbb{C}}$. Let D be a K-invariant open set in X equipped with a smooth, K-invariant Kähler form ω. Assume that there exits an equivariant moment map $\mu : D \to (\text{Lie } K)^*$ with $\mu^{-1}(0) \neq \varnothing$.

Given $x_0 \in \mu^{-1}(0)$ let N be a convex neighborhood of $0 \in \mathfrak{p}$ so that the open set $U = K \exp(iN)x_0$ is contained in D.

LEMMA 2.2.2. $\mu^{-1}(0) \cap U = Kx_0$.

PROOF. If for $k \in K$ and $\xi \in N$ the point $x = k \exp(i\xi)x_0$ is in $\mu^{-1}(0) \cap U$, then the same is true of $x_1 := \exp(i\xi)x_0$.

Now let $x_t := \exp(i\xi t)x_0$. From the defining property $d\mu_\xi = i_{\xi_D}\omega$ of the moment function μ_ξ it follows that $J\xi_D$ is the gradient field of μ_ξ with respect to the Riemannian metric induced by ω and the complex structure J. Thus, either $t \mapsto \mu_\xi(x_t)$ is strictly increasing or $x_t = x_0$ is the constant curve. Since $\mu_\xi(x_0) = \mu_\xi(x_1) = 0$, we are in the latter situation and therefore $x_1 = k(x_0) \in Kx_0$ □

By definition, the K-orbits in $\mu^{-1}(0)$ are isotropic, i.e., the pull-back of ω to such an orbit vanishes identically.

Isotropic submanifolds of maximal dimension, i.e., of half the dimension of the ambient symplectic manifold, are called Lagrangian. For Kählerian symplectic structures, $\omega(v, Jv) > 0$ for all tangent vectors v. Thus Lagrangian submanifolds are totally real.

COROLLARY 2.2.3. The orbit $Kx_0 = \mu^{-1}(0) \cap U$ is Lagrangian and therefore totally real. Furthermore, $G_{x_0}^0 = (K_{x_0}^0)^{\mathbb{C}}$.

PROOF. It remains to prove the last statement. However, it follows immediately from the fact that $\dim_{\mathbb{R}} Kx_0 = \dim_{\mathbb{C}} D$. □

If ρ is a K-invariant, strictly plurisubharmonic function on D, then we may apply the preceding observations to the Kähler form $\omega := dd^c\rho$ and the equivariant moment map defined by $\mu_\xi := -J\xi_D(\rho)$. In this case $x_0 \in \mu^{-1}(0)$ if and only if $d\rho(x_0) = 0$.

A direct translation then yields the following result.

COROLLARY 2.2.4. *If D is a K-invariant domain in the homogeneous space $X = G/H$ of a complex reductive group G and ρ is a K-invariant, strictly plurisubharmonic function on D with $d\rho(x_0) = 0$, then, in a K-invariant neighborhood U of x_0 in D, the critical set $\{x \in U : d\rho(x) = 0\}$ consists of exactly the totally real orbit Kx_0 where ρ has its minimum, i.e., $\min\{\rho(x) : x \in U\} = \rho(x_0)$. Furthermore, if H has finitely many components, then it is reductive.*

PROOF. The statement concerning the minimum of ρ follows immediately from the positive definiteness of the complex Hessian. The other statements are translations of results above. $\square$

REMARK. If D is a K-invariant Stein domain, then it possesses a K-invariant exhaustion $\rho : D \to \mathbb{R}^{\geq 0}$. It follows that at least H^0 is reductive. Of course H itself may not be reductive. For example, if H is an infinite discrete group, every K-orbit has a K-invariant Stein neighborhood.

At the group level the convexity argument above is in fact global.

PROPOSITION 2.2.5 (EXHAUSTION THEOREM FOR REDUCTIVE GROUPS). *Let G be a reductive group, K a maximal compact subgroup and $\rho : G \to \mathbb{R}$ a K-invariant strictly plurisubharmonic function. Then ρ is a proper exhaustion function if and only if $\{d\rho = 0\} \neq \varnothing$. In this case, if $d\rho(x_0) = 0$ and $\rho(x_0) =: c_0$, then $\rho : X \to [c_0, \infty)$ and $\{d\rho = 0\} = Kx_0$.*

PROOF. In this case we have $U = X$. Thus, either $\{d\rho = 0\} = \varnothing$, in which case ρ is clearly not a proper exhaustion, or $\{d\rho = 0\} = Kx_0$ is a K-orbit where ρ takes on its minimum.

Using the KP-decomposition we may then regard $\rho : \mathfrak{p} \to \mathbb{R}$ as a function on the vector space $\mathfrak{p}$ which has a minimum at $0 \in \mathfrak{p}$.

Since for all $\xi \in \mathfrak{k}$ the function $z \mapsto \rho(\exp(\xi z)x_0)$ is a strictly subharmonic $\mathbb{R}$-invariant function on the complex plane, it follows that, regarded as a function on $\mathfrak{p}$, ρ is strictly convex along lines through the origin. Consequently, if m denotes the minimum of the normal derivatives of ρ along the unit sphere in $\mathfrak{p}$, it follows that $\rho(v) \geq m \|v\|$. $\square$

REMARK. The above very useful convexity argument was brought to our attention by Azad and Loeb; see [Azad and Loeb 1993]. As in that reference, we will also apply it to the case of homogeneous spaces.

b. The Theorem of Matsushima–Onitshick. For homogeneous spaces $X = G/H$ of reductive groups there is a close connection between holomorphic and algebraic phenomena. For this the key ingredient is the density of G-finite holomorphic functions.

DEFINITION. Let G be a group acting linearly on a $\mathbb{C}$-vector space V. A vector $v \in V$ is called G-finite if the linear span $\langle g(v) : g \in G\rangle_{\mathbb{C}}$ is finite-dimensional.

The next result is a consequence of the theorem of Peter and Weyl; see [Akhiezer 1995], for example.

PROPOSITION 2.2.6. *Let G be a reductive group acting holomorphically on a complex space X. Then the G-finite holomorphic functions are dense in $\mathcal{O}(X)$.*

Another fundamental property of complex reductive groups is the algebraic nature of their representations; see [Chevalley 1946], for example.

PROPOSITION 2.2.7. *If G is a reductive group realized as a locally closed complex subgroup of some $\mathrm{GL}_n(\mathbb{C})$, then it is an affine subvariety. This is the unique affine structure on G which is compatible with the group and complex manifold structure. A holomorphic representation $\tau : G \to \mathrm{GL}(V)$ is automatically algebraic.*

As a consequence, the existence of sufficiently many independent holomorphic functions implies the algebraicity of X.

PROPOSITION 2.2.8. *Let $X = G/H$ be a complex homogeneous space of the reductive group G. Let $n := \dim_{\mathbb{C}} X$ and suppose that there exist $f_1, \ldots, f_n \in \mathcal{O}(X)$ such that $df_1 \wedge \cdots \wedge df_n \not\equiv 0$. Then H is an algebraic subgroup of G.*

PROOF. By Proposition 2.2.6 we may assume that the f_j are G-finite. Let $V = \langle Gf_1, \ldots, Gf_n \rangle_{\mathbb{C}}$ be the vector space spanned by the G-orbits and define $F : X \to V^*$ by $F(x)(f) = f(x)$. It follows that F is G-equivariant and generically of maximal rank. The image $F(x)$ is therefore a G-orbit $Gv = G/H_1$ with H_1/H discrete. But the G-representation on V is algebraic and consequently H_1 is an algebraic subgroup. Thus, being a subgroup of finite index in a $\mathbb{C}$-algebraic group, H is likewise $\mathbb{C}$-algebraic. $\qquad\square$

REMARK. Using a theorem of Grauert and Remmert or a version of Zariski's main theorem [1984], the finite cover $X = G/H_1 = Gv$ can be extended to a finite G-equivariant ramified covering $Z \to \overline{Gv} \subset V$ of an affine closure Z of X. In particular X is quasi-affine.

Here is another important tool for the analysis of holomorphic functions in the presence of reductive actions:

IDENTITY PRINCIPLE. *Let G be a reductive group, K a maximal compact subgroup and $G \times X \to X$ a holomorphic action. It follows that $\mathcal{O}(X)^G = \mathcal{O}(X)^K$.*

PROOF. For $f \in \mathcal{O}(X)^K$ regard $B_x(g) := f(g(x))$ as a holomorphic K-invariant function on G. Since the submanifold K is totally real in G with $\dim_{\mathbb{R}} K = \dim_{\mathbb{C}} G$ and K has non-empty intersection with every component of G, it follows that $B_x(g) \equiv B_x(e)$, i.e., $f(g(x)) = f(x)$ for all $x \in X$. $\qquad\square$

The *averaging process* is a useful way of constructing invariant holomorphic functions. For this, let G, K and $G \times X \to X$ be as above. For dk an invariant probability measure on K, define $A : \mathcal{O}(X) \to \mathcal{O}(X)$ by $f \mapsto \int_K k^*(f) dk$. It follows from the identity principle that $A : \mathcal{O}(X) \to \mathcal{O}(X)^G$ is a projection.

EXTENSION PRINCIPLE. *Let X be Stein and Y be a closed G-invariant complex subspace. Then the restriction map $r : \mathcal{O}(X)^G \to \mathcal{O}(Y)^G$ is surjective.*

PROOF. Given $f \in \mathcal{O}(Y)$ define $\tilde{f} := A(F)$, where $F \in \mathcal{O}(X)$ is an arbitrary holomorphic extension of f. It follows that $r(\tilde{f}) = f$.   □

THEOREM 2.2.9 (MATSUSHIMA–ONITSHICK). *A complex homogeneous* $X = G/H$ *of a reductive group* G *is Stein if and only if* H *is reductive.*

PROOF. If X is Stein, then by Proposition 2.2.8 it follows that H is an algebraic subgroup of G and in particular has only finitely many components. An application of Corollary 2.2.4 for the case $D = X$ shows that H is reductive.

Conversely, suppose that $H = L^{\mathbb{C}}$ is reductive. For a discrete sequence of points $\{x_n\}$ in X, let $Y := \dot{\bigcup} Y_n$ be the preimage $\pi^{-1}\{x_n\}$ in G via the canonical quotient $\pi : G \to G/H$. Define $\tilde{f} \in \mathcal{O}(Y)^H$ by $f|Y_n \equiv n$. Regard $\tilde{f} \in \mathcal{O}(X)$ and observe that $\tilde{f}(x_n) = n$. This proves both the holomorphic convexity of X and the fact that $\mathcal{O}(X)$ separates points. Thus X is Stein.   □

c. Exhaustions associated to Ad**-invariant inner products.** Let K be a connected compact Lie group. If it is semi-simple, then the Killing form $b : \mathfrak{k} \times \mathfrak{k} \to \mathbb{R}$, $(\xi, \eta) \mapsto \mathrm{Tr}(\mathrm{ad}(\xi) \cdot \mathrm{ad}(\eta))$, is an $\mathrm{Ad}(K)$-invariant, negative-definite inner product; see [Helgason 1978]. Of course, for an arbitrary compact group, the degeneracy of b is exactly the center $\mathfrak{z}$.

Recall that $\mathfrak{k} = \mathfrak{z} \oplus \mathfrak{k}_{\mathrm{ss}}$, where $\mathfrak{k}_{\mathrm{ss}}$ is the Lie algebra of a maximal (compact) semi-simple subgroup K_{ss}. Furthermore, if Z is the connected component of the center $Z(K)$ at the identity, then $K = Z \cdot K_{\mathrm{ss}}$ and $Z \cap K_{\mathrm{ss}}$ is finite.

Given a positive definite Ad-invariant bilinear form on k_{ss}, any extension to $\mathfrak{k} = \mathfrak{z} \oplus \mathfrak{k}_{\mathrm{ss}}$ is Ad-invariant. Thus there exist $\mathrm{Ad}(K)$-invariant inner products $B : \mathfrak{k} \times \mathfrak{k} \to \mathbb{R}$. We now show how to associate a $(K \times K)$-invariant strictly plurisubharmonic exhaustion $\rho : G \to \mathbb{R}$ to such an inner product.

The natural map $(K \times K) \times \mathfrak{k} \to G$, $(k_1, k_2, \xi) \mapsto k_1 \exp(i\xi) k_2^{-1}$, factors through the quotient $(K \times K) \times_K \mathfrak{k}$ by the free diagonal K-action $k(k_1, k_2, \xi) := (k_1 k^{-1}, k_2 k^{-1}, \mathrm{Ad}(k)(\xi))$. In fact, using $KP = G$, a direct computation shows that in this way G is $(K \times K)$-equivariantly identified with the total space of the $(K \times K)$-vector bundle $(K \times K) \times_K \mathfrak{k} \to (K \times K)/K \cong K$, where the base is diffeomorphic to the group K equipped with the standard $(K \times K)$-action, $k \mapsto k_1 k k_2^{-1}$.

Now the representation of the isotropy $K \hookrightarrow K \times K$, $k \mapsto (k, k^{-1})$, on the fiber $\mathfrak{k}$ over the identity e is simply the adjoint representation. Since the definition of $\mathrm{Ad} : K \to \mathrm{GL}(\mathfrak{k})$, $\mathrm{Ad}(k) = \mathrm{int}_*(k)$, is given by the natural induced action on the tangent space $T_e K \cong \mathfrak{k}$, it follows that $(K \times K) \times_K \mathfrak{k} \to (K \times K)/K$ is just the tangent bundle TK.

Summarizing, we have the following result.

PROPOSITION 2.2.10. *Via the KP-decomposition the reductive group* G *is* $(K \times K)$-*equivariantly identifiable with the tangent bundle* TK.

If $B : \mathfrak{k} \times \mathfrak{k} \to \mathbb{R}$ is $\mathrm{Ad}(K)$-invariant, then it defines a $(K \times K)$-invariant metric on $TK = (K \times K) \times_K \mathfrak{k}$; on the neutral fiber this is given by $(\xi, \eta) \mapsto B(\xi, \eta)$

and it is extended to the full space by the $(K \times K)$-action. Let $\rho : TK \to \mathbb{R}$, $\xi \mapsto B(\xi, \xi)$, denote the associated norm function.

The main goal of this section is to prove the following observation.

PROPOSITION 2.2.11. *Under the canonical identification $TK \cong G$, the function $\rho : G \to \mathbb{R}$ is a $(K \times K)$-invariant, strictly plurisubharmonic exhaustion.*

In fact we prove a slightly more general statement which is a special case of Loeb's variation on a theme of Lassalle. For this recall the KAK-decomposition and the fact that a $(K \times K)$-invariant function on G is uniquely determined by a Weyl-group invariant function on $\mathfrak{a} = \text{Lie } A$.

PROPOSITION 2.2.12. *A $(K \times K)$-invariant function $\rho : G \to \mathbb{R}$ determined by a W-invariant strictly convex function $\rho_{\mathfrak{a}} : \mathfrak{a} \to \mathbb{R}$ is strictly plurisubharmonic.*

The proof follows immediately from an elementary lemma in the 3-dimensional case. This requires some notational preparation.

Let S be a 3-dimensional complex semi-simple Lie group, K_S a maximal compact subgroup and let $L := K_S \times K_S$ act on S by left- and right multiplication. With one exception, an orbit $\Sigma = Ls$ is a strictly pseudoconvex hypersurface in S. The exception Σ_0 is totally real with $\dim_{\mathbb{R}} \Sigma_0 = \dim_{\mathbb{C}} S$.

Let T be a maximal torus in K. If we regard it in L via the diagonal embedding, we write T_Δ. The connected component Y of the 1-dimensional complex submanifold $\text{Fix}(T_\Delta)$ which contains the identity $e \in S$, is the subgroup $T^{\mathbb{C}} < S$.

Let $\rho : S \to \mathbb{R}$ be a smooth, L-invariant function such that $\rho|Y$ is a strictly plurisubharmonic exhaustion. Since ρ is in particular invariant by left multiplication by elements of T, it follows from the strict convexity of the pull-back $t \mapsto \rho(\exp(it\xi)y)$, $\langle \xi \rangle = \mathfrak{t}$, that $\rho|Y$ has an absolute minimum along exactly one orbit $Y_0 = Ty_0$ and otherwise $d\rho|Y \neq 0$.

Suppose $\Sigma_0 = Ly_0$ is a hypersurface. If $y_0 \in Y_0$, then, moving along the curve $y_t := \rho(\exp(it\xi))y_0$, the Levi form of Σ_t changes signs at Σ_0. This is contrary to the strong pseudoconvexity of Σ_t for all t. Thus $\{d\rho = 0\} = \Sigma_0$ is the totally real L-orbit.

LEMMA 2.2.13. *The function $\rho : S \to \mathbb{R}$ is strictly plurisubharmonic exhaustion.*

PROOF. Let $y \in Y$ and suppose $\Sigma = Ly$ is a hypersurface. Then $d\rho \neq 0$ in some neighborhood of Σ. Since $\rho(\exp(i\xi t)y)$ is increasing, ρ defines Σ as a strictly pseudoconvex hypersurface. Thus $\omega = dd^c\rho$ is positive-definite on the complex tangent space V of Σ at y.

Of course V is T_Δ-invariant and thus its complement $V^{\perp\alpha}$ in T_yS is likewise T_Δ-invariant. But the only possibility for this is the tangent space T_yY, where α is known to be positive-definite.

If $y = y_0 \in Y_0$, then $\Sigma = \Sigma_0$; in particular, L_{y_0} is 3-dimensional and acts irreducibly on $T_{y_0}G$. Since ω is non-degenerate on $T_{y_0}Y$, it follows that it is non-degenerate on the full space $T_{y_0}G$. Since mixed signature is an open condition, it follows from the preceding discussion along generic L-orbits that ω is positive-definite.

The exhaustion property follows from the fact that from the $(K \times K)$-invariance and the fact that $\rho|Y$ is an exhaustion.   □

PROOF OF PROPOSITION 2.2.12. We begin by showing that $\omega = dd^c\rho$ is non-degenerate. It is enough to do so at points $t_0 \in T^{\mathbb{C}}$, the maximal torus of G associated to the Lie algebra $\mathfrak{t}^{\mathbb{C}} = \langle \mathfrak{a} \rangle_{\mathbb{C}}$.

Since ρ is strictly plurisubharmonic on $T^{\mathbb{C}}$, it follows that $E := (T_{t_0}G)^{\perp\omega}$ is a direct sum $l_*(t_0)\left(\sum_{\alpha \in I} \mathfrak{g}_\alpha\right)$ of certain root spaces when regarded as an $\mathrm{Int}(T)$-module.

Arguing by contradiction, suppose that $E \neq \{0\}$ and let $\mathfrak{s}_\alpha \cong \mathfrak{sl}_2$ be a standard 3-dimensional subalgebra associated to any one of the $\mathfrak{g}_\alpha$'s. Let S be the associated (algebraic) subgroup of G.

Let $\rho : S \to \mathbb{R}$ denote the restriction of the given function to the S-orbit St_0. By assumption $\omega = dd^c\rho$ is degenerate along $l_*(t_0)(\mathfrak{g}_\alpha)$ which is tangent to this S-orbit. But this is contrary to Lemma 2.2.13, i.e., ω is indeed non-degenerate.

Since ω is non-degenerate, it suffices to prove the positive definiteness at the identity $e \in G$. For this observe that the decomposition $T_eG = \mathfrak{t}^{\mathbb{C}} \oplus \sum_\alpha \mathfrak{g}_\alpha$ is α-orthogonal. By assumption $\omega > 0$ on $\mathfrak{t}^{\mathbb{C}}$ and by Lemma 2.2.13 we have the same conclusion on each root space $\mathfrak{g}_\alpha$. Thus ω is positive-definite.   □

2.3. Moment Maps at the Group Level.

In the previous sections we observed that an $\mathrm{Ad}(K)$-invariant bilinear form B leads in a canonical way to a $(K \times K)$-invariant strictly plurisubharmonic exhaustion $\rho : G \to \mathbb{R}^{\geq 0}$ of the reductive group G. Here we give a precise description of the moment map $\mu : X \to (\mathrm{Lie}\, K)^*$ of the group manifold $X := G$ equipped with the Kählerian structure $\omega = dd^c\rho$ which is defined by right-multiplication by elements of K.

a. Generalities on moment maps. For the moment let (M, ω) be an arbitrary connected symplectic manifold equipped with a Hamiltonian action of a connected Lie group K, i.e., there exists an equivariant moment map $\mu : M \to (\mathrm{Lie}\, K)^* = \mathfrak{k}^*$.

The basic formula. For $\xi \in \mathfrak{k}$ let $\xi_M \in \mathrm{Vect}_\omega(M)$ be the associated vector field and μ_ξ the associated momentum function. Since $d\mu_\xi(x) = \omega(x)(\xi_M(x), \cdot)$, it follows that the differential μ_* can be calculated by

$$\mu_*(x)(v_x)(\xi) = \omega(x)(\xi_M(x), v_x),$$

where $v_x \in T_xM$ and $\mu_*(x)(v_x) \in T_{\mu(x)}\mathfrak{k}^* \cong \mathfrak{k}^*$.

In other words, this basic formula shows how $\mu_*(x)(v_x)$ acts as a functional on $\mathfrak{k}$.

There are several direct consequences:

(i): $\mathrm{Ker}(\mu_*(x)) = (T_xKx)^{\perp\omega}$.

(ii): If $\dim_{\mathbb{R}} Kx =: k$ is constant, then $\text{Rank}(\mu) = k$ is likewise constant. In particular, if the action $K \times M \to M$ is locally free, then μ is an open immersion.

(iii): $\text{Im}(\mu_*(x)) = \mathfrak{k}^0_{\mu(x)}$, i.e., the annihilator of the algebra of the isotropy group $K_{\mu(x)}$.

The moment maps under consideration in this section are defined by a strictly plurisubharmonic potential function ρ of a Kähler-form α, i.e., $\alpha = dd^c\rho$ and $\mu_\xi := -\frac{1}{2}d^c\rho(\xi_M) = -\frac{1}{2}J\xi_M(\rho)$. The choice of the coefficient $-\frac{1}{2}$ only puts us in tune with classical mechanics. Moment maps of this type are denoted by $\mu : X \to \mathfrak{k}^*$. We leave it as an exercise to check that such moment maps are equivariant.

b. The moment map associated to an Ad-invariant bilinear form. Let G be a connected reductive group, K a maximal compact subgroup and $B : \mathfrak{k} \times \mathfrak{k} \to \mathbb{R}$ an $\text{Ad}(K)$-invariant symmetric, positive-definite bilinear form. Using the canonical identification of G and the tangent bundle $TK = (K \times K) \times_K \mathfrak{k}$, the norm-function $\xi \mapsto B(\xi, \xi)$ defines a $(K \times K)$-invariant function $\rho : G = KP \to \mathbb{R}^{\geq 0}$ by $\rho(k \exp(i\xi)) := B(\xi, \xi)$.

It follows from Proposition 2.2.12 that ρ is a strictly plurisubharmonic exhaustion of the group manifold $X = G$ with minimum set $\{\rho = 0\} = K$.

Our goal here is to compute the moment map μ^ρ associated to the action of K which is defined by right multiplication.

PROPOSITION 2.3.1. *Let K act on the group manifold by right multiplication and ρ be the strictly plurisubharmonic function associated to an $\text{Ad}(K)$-invariant, symmetric positive-definite bilinear form B. Then*

$$\mu^\rho_\xi(\exp(i\eta)) = B(\xi, \eta).$$

PROOF.

$$\mu^\rho_\xi(\exp(i\eta)) = -\frac{1}{2}\frac{d}{dt}\Big|_{t=0}\rho(\exp(i\eta)\exp(-i\xi t)) = -\frac{1}{2}\frac{d}{dt}\Big|_{t=0}\rho(\exp(i(\eta - \xi t)))$$

$$= -\frac{1}{2}\frac{d}{dt}\Big|_{t=0}B(\eta - \xi t, \eta - \xi t) = B(\xi, \eta).$$

For the second equality we use the Campbell–Hausdorff formula:

$$\exp(i\eta)\exp(-i\xi t) = g(t)\exp(i(\eta - \xi t))),$$

where, due to K-invariance,

$$\rho(g(t)\exp(-i(\eta - \xi t))) = \rho(\tilde{g}(t)\exp(i(\eta - \xi t)))$$

for

$$\tilde{g}(t) = \exp(-[\xi, \eta]t)g(t) = \exp(O(t^2))\exp(O(t^2)).$$

The result, i.e., the second equation, follows from the fact that the curves $\gamma_{\mathfrak{p}}(t) := \exp(i(\eta - \xi t))$ and $\gamma(t) := \tilde{g}(t)\gamma_{\mathfrak{p}}(t)$ are tangent at $\exp(i\eta)$. $\qquad\square$

2.4. Symplectic Reduction and Consequences for Stein Homogeneous Spaces

Generalities on reductions. We consider the following situation: X is Kählerian with respect to $\omega = dd^c \rho$, where ρ is a strictly plurisubharmonic potential function which is invariant by the action of a connected compact group of holomorphic automorphisms, and $\mu^\xi : X \to \mathfrak{k}^*$ is the associated moment map.

Note that if $L < K$ is a subgroup, connected or not, we have the L-equivariant moment map

$$\mu_L^\rho : X \to \mathfrak{l}^*$$

defined by the inclusion $\mathfrak{l} \hookrightarrow \mathfrak{k}$. Unless it might lead to confusion we drop the dependency on ρ and L in the notation and simply write $\mu : X \to \mathfrak{l}^*$.

As is the case for the group manifold $X = G$, we assume in addition that the K-action extends to a holomorphic action $G \times X \to X$ of its complexification $G = K^{\mathbb{C}}$. Assume that $X_0 := \mu^{-1}\{0\} \neq \varnothing$ and define the associated set of semi-stable points by

$$X(\mu) := \{x \in X : \overline{Gx} \cap X_0 \neq \varnothing\}.$$

In the setting of analytic Hilbert quotients (Section 4), the condition $\mu = \mu^\rho$ for some Kählerian potential can only be locally achieved on a covering of the Kempf–Ness set X_0. However, for all practical purposes this is adequate.

The goal is to show that $X(\mu)$ is an open subset of X (in many situations it is in fact Zariski open), and that an equivalence relation is defined on $X(\mu)$ by $x \sim y$ if and only if $\overline{Gx} \cap \overline{Gy} \neq \varnothing$. We refer to the resulting quotient $X(\mu) \to X(\mu)/\!/G := X(\mu)/\!\sim$ as the analytic Hilbert quotient, because its structure sheaf as a reduced complex space is constructed locally on G-invariant Stein neighborhoods of X_0 by invariant theoretic means.

While the complex analytic structure of $X(\mu)/\!/G$ is described in terms of holomorphic invariant theory, the Kählerian structure arises via the symplectic reduction $X_0 \to X_0/K$. Here we are dealing with simple orbit space quotient by the compact group K, but of course the singularities of X_0 present difficulties. One of the main points is to show that the embedding $X_0 \hookrightarrow X(\mu)$ induces a homeomorphism $X_0/K \cong X(\mu)/\!/G$.

In the case where $\omega = dd^c \rho$ for a K-invariant strictly plurisubharmonic function $\rho : X \to \mathbb{R}$, one is handed a quotient structure on a silver platter: Define ρ_{red} by pushing down the K-invariant restriction $\rho|X_0$.

It can be shown that ρ_{red} is a continuous strictly plurisubharmonic function which is smooth on a natural stratification of $X(\mu)/\!/G$ [Heinzner et al. 1994]. The induced singular form $\omega_{\mathrm{red}} = dd^c \rho_{\mathrm{red}}$ is a Kählerian version of the singular symplectic reduced structure [Sjamaar and Lerman 1991].

As we indicated above, these matters are discussed in substantial detail in Section 4. In the present section we consider the case of the right K-action on the group manifold $X = G$, where ρ is a $(K \times K)$-invariant, strictly plurisubharmonic exhaustion associated to an $\mathrm{Ad}(K)$-invariant form. For brevity we refer to this as the group manifold setting.

LEMMA 2.4.1. *In the group manifold setting let $L < K$ be a compact subgroup and $\mu : X \to \mathfrak{l}^*$ the associated moment map. Then:*

(i) *The moment map μ is an open immersion of rank equal to $l := \dim_{\mathbb{R}} L$.*

(ii) *The Kempf–Ness set X_0 is a smooth, generic Cauchy–Riemann submanifold of X.*

(iii) *For $x_0 \in X_0$ there exists the canonical splitting $T_{x_0} X_0 = T_{x_0}(Lx_0) \oplus T_{x_0}^{\mathbb{C}} X_0$, where $T_{x_0}^{\mathbb{C}} X_0 = T_{x_0} X_0 \cap J(T_{x_0} X_0)$.*

(iv) *The orbits of L in the Kempf–Ness set are isotropic and the complex tangent space $T_{x_0}^{\mathbb{C}} X_0$ is the Riemannian orthogonal complement $(T_{x_0}(Lx_0))^{\perp_g}$.*

PROOF. The point (i) has been discussed previously as an immediate consequence of the basic formula. It is therefore clear that X_0 is a smooth submanifold.

Since L-orbits in X_0 are mapped by μ to $0 \in \mathfrak{l}$, it follows that they are isotropic and are therefore totally real. The splitting in (iii) and the genericity statement in (ii) therefore follow by a dimension count.

The Riemannian complement to $T_{x_0}(Lx_0)$ is the Hermitian complement to $T_{x_0}(L^{\mathbb{C}} x_0)$ and is therefore a complex subspace. Again using the basic formula, one observes that this is also in $\mathrm{Ker}(\mu_*(x_0))$. Consequently, $T_{x_0}(Lx_0)^{\perp_g}$ is a complex subspace of $T_{x_0}^{\mathbb{C}} X_0$. Equality follows from a dimension count.   □

Let $H := L^{\mathbb{C}}$ be the smallest complex Lie subgroup of G which contains L; in fact, H is an affine algebraic subgroup. We regard H as acting by right multiplication, $h(x) = xh^{-1}$, i.e., the natural extension of the L-action.

LEMMA 2.4.2. *For an arbitrary point $x \in X$ it follows that the H-orbit Hx intersects X_0 in exactly one L-orbit: $C_x := Hx \cap X_0 \neq \varnothing$ and for $x_0 \in C_x$ it follows that $C_x = Lx_0$. Furthermore, $C_x = \{p \in Hx : d(\rho|Hx)(p) = 0\} = \{p \in Hx : \rho(p) = \min(\rho|Hx)\}$.*

PROOF. It follows from the invariance of ρ that $\xi_M(\rho) \equiv 0$ and, by the definition of μ^ρ, $J\xi_M(x_0)(\rho) = 0$ for all $\xi \in \mathfrak{l}$.

Thus $x_0 \in C_x$ if and only if $d(\rho|Hx)(x_0) = 0$. From the exhaustion theorem at the group level, Proposition 2.2.12, it follows that C_x is exactly the set where $\rho|Hx$ takes on its minimum. (In this case we know already that ρ is an exhaustion. However, if we did not — and this is a key point for general potential functions — it would nevertheless follow at this point.) Furthermore, the exhaustion theorem also guarantees us that this minimum set consists of exactly one L-orbit.   □

Let $\pi : G \to G/H =: X /\!/ L$ be the natural projection. It follows from the preceding lemma that $\pi|X_0$ induces a bijective continuous map $i_x : X_0/L \to X /\!/ L$.

LEMMA 2.4.3. *The mapping $\pi|X_0$ is proper, i.e., $i_x : X_0/L \to X//L$ is a homeomorphism.*

PROOF. Since $\rho : G \to \mathbb{R}^{\geq 0}$ is a proper exhaustion, this is clear: Suppose $\{x_n\} \subset X_0$ is a divergent sequence. It follows that $\{\rho(x_n)\}$ is divergent. If $y_n := \pi(x_n) \to y_0 \in X//L$, let $X_0 \in C_x$ be a critical point in $\pi^{-1}(y_0) = Hx$. For $n \gg 0$ it follows that $\rho(C_{x_n}) \leq \rho(x_0) + \varepsilon$ contrary to $\{\rho(x_n)\}$ being divergent. □

Exhaustions and associated Kählerian structures on reductions. In the setting above the map $\pi|X_0 : X_0 \to X_0/L = X(\mu)//H = X_{\text{red}}$ induces a surjective algebra morphism

$$\pi_* : \mathcal{E}(X)^L \to \mathcal{E}(X_{\text{red}}), \quad f \mapsto f_{\text{red}} := f|X_0.$$

The L-invariant function $f|X_0$ is interpreted as a smooth function on the base X_{red}.

For the analogous statement for differential forms the following result is of use.

LEMMA 2.4.4. *Let $\pi : M \to N$ be a surjective immersion of smooth manifolds with connected fibers. Then*

$$\pi^*(\mathcal{E}^k(N)) = \{\eta \in \mathcal{E}^k(M) : i_V \eta = 0 \text{ and } \mathcal{L}_V \eta = 0 \text{ for all vertical } V \in \text{Vect}(M)\}.$$

PROOF. It is only a matter of pushing down forms which satisfy $i_V \eta = 0$ and $\mathcal{L}_V \eta = 0$ for all vertical fields V. For $q \in N$ and $p \in M$ with $\pi(p) = q$, define $\pi_*(\eta)(q)(v_1, \ldots, v_n) := \eta(p)(\tilde{v}_1, \ldots, \tilde{v}_k)$. The first condition shows that this is a well-defined independent of the choice of $\tilde{v}_j \in T_pM$ with $\pi_*(\tilde{v}_j) = v_j$, for $j = 1, \ldots, k$. Since the π-fibers are connected and any two points p_1, p_2 can be connected by a curve which is piecewise the integral curve of a vertical field, the second condition guarantees that the definition does not depend on the choice of p with $\pi(p) = q$.

Finally, the smoothness of $\pi_*(\eta)$ is proved by identifying it with $\sigma^*(\eta)$, where σ is a local section.   □

If in the context above a Lie group L acts smoothly and transitively on the π-fibers, then the second condition can be replaced by invariance.

COROLLARY 2.4.5. *Let $L \times M \to M$ be a smooth Lie group action, $\pi : M \to N$ a surjective immersion with not necessarily connected fibers which are L-orbits. Then*

$$\pi^*(\mathcal{E}^k(N)) = \{\eta \in \mathcal{E}^k(M)^L : i_V \eta = 0 \text{ for all vertical fields } V\}.$$

PROOF. The independence of definition of $\pi_*(\eta)(q)$ on the choice of $p \in \pi^{-1}\{q\}$ follows from the L-invariance.   □

Now we return to our concrete context: $X = G$ is the group manifold of the complex reductive group $G, \rho : X \to \mathbb{R}$ is a strictly plurisubharmonic function which is invariant by the L-action defined by right multiplication and $X_{\text{red}} = X_0/L = X(\mu)//H = G/H$, where $\mu = \mu^\rho$.

PROPOSITION 2.4.6. *The function $\pi_*(\rho) = \rho_{\text{red}} : G/H \to \mathbb{R}$ is strictly plurisubharmonic and $\omega_{\text{red}} := dd^c\rho_{\text{red}}$ is the reduced symplectic structure of Marsden–Weinstein.*

PROOF. Let $i : X_0 \hookrightarrow X$ be the canonical injection. Define $\tilde{\omega} = i^*(\omega)$, where $\omega = dd^c\rho$ is the associated Kählerian structure on X. Since the L-orbits in X_0 are isotropic and $\tilde{\omega}$ is L-invariant, it follows that $\tilde{\omega} = \pi^*(\omega_{\text{red}})$, where ω_{red} is a smooth 2-form on the base.

Note that the distribution of complex tangent spaces $T_{x_0}^{\mathbb{C}} X_0$ of the Cauchy–Riemann manifold X_0 serves as an invariant connection for $\pi : X_0 \to X_0/L = X_{\text{red}}$. Since $\omega = dd^c\rho$ is positive definite on these horizontal complex vector spaces, it follows that ω is non-degenerate. In fact, except that they do not have a canonical choice for the connection, this is exactly the reduction of Marsden and Weinstein [1974]. To complete the proof we show that $\omega_{\text{red}} = dd^c\rho_{\text{red}}$.

For this, for $y \in X_{\text{red}}$ and $x_0 \in X_0$ with $\pi(x_0) = y$, let $\sigma : \Delta \to G$ be a local holomorphic section of $G = X(\mu) \to X(\mu)//H = X_0/L$ defined near y with $\sigma(y) = x_0$ and with $\sigma_* : T_y\Delta \xrightarrow{\sim} T_{x_0}^{\mathbb{C}} X_0$.

For $v \in T_y\Delta$ compute

$$d^c\rho_{\text{red}}(v) = J_{\text{red}}v(\rho_{\text{red}}) = d\rho_{\text{red}}(J_{\text{red}}v) = d\rho(\sigma_*(J_{\text{red}}v))$$
$$= d\rho(J\sigma_*(v)) = d^c\rho(\sigma_*(v)) = \pi_*(d^c\rho(v)).$$

Thus

$$dd^c\rho_{\text{red}} = d\pi_*(d^c\rho) = \pi_*(dd^c\rho) = \pi_*(\omega) = \omega_{\text{red}}. \qquad \square$$

If ρ is an exhaustion of X, as is the case for those functions which are associated to $\text{Ad}(K)$-invariant bilinear forms, then ρ_{red} is likewise an exhaustion. Note furthermore that, since the action of K by right multiplication commutes with the L-action, it follows that X_0 is K-invariant, $\pi : X_0 \to X_0/L = G/H$ is K-equivariant and ρ_{red} is K-invariant.

COROLLARY 2.4.7. *To every $\text{Ad}(K)$-invariant, positive-definite, symmetric bilinear form $B : \mathfrak{k} \times \mathfrak{k} \to \mathbb{R}$ is canonically associated to a K-invariant, strictly plurisubharmonic exhaustion $\rho_{\text{red}} : G/H \to \mathbb{R}^{\geq 0}$.*

c. The Mostow fibration. An explicit computation of $X_0 := \mu^{-1}\{0\}$ yields the description of the G-homogeneous space G/H as a K-vector bundle over the orbit $Kx_0 = K/L$ of the neutral point in X.

LEMMA 2.4.8. *Let $\mu = \mu^\rho : G \to \mathfrak{l}^*$ be the moment map which is defined by the strictly plurisubharmonic exhaustion ρ associated to an $\text{Ad}(K)$-invariant bilinear form $B : \mathfrak{k} \times \mathfrak{k} \to \mathbb{R}$. Let $\mathfrak{m} := \mathfrak{l}^{\perp B}$. Then $X_0 = K\exp(i\mathfrak{m})$.*

PROOF. It is enough to compute $\mu^{-1}\{0\} \cap P$. But, by Proposition 2.3.1, $\mu_\xi(\exp(i\eta)) = B(\xi, \eta)$. Thus

$$\mu^{-1}\{0\} \cap P = \{\exp(i\eta) : B(\xi, \eta) = 0 \quad \text{for all } \xi \in \mathfrak{l}\} = \exp(i\mathfrak{m}). \qquad \square$$

Of course the decomposition $X_0 = K\exp(i\mathfrak{m})$ is lined up with the KP-decomposition and the action of L by right multiplication in the identification $X_0 = K \times \mathfrak{m}$ is given by $l(k, \xi) = (kl^{-1}, \mathrm{Ad}(l)(\xi))$. Thus we have described a realization of G/H as a K-vector bundle.

THEOREM 2.4.9 (MOSTOW FIBRATION). *Let G be a connected complex reductive Lie group, K a maximal compact subgroup, L a closed subgroup of K and $H = L^{\mathbb{C}}$. Let $B : \mathfrak{k} \times \mathfrak{k} \to \mathbb{R}$ be an $\mathrm{Ad}(K)$-invariant, positive-definite, symmetric bilinear form and $\mathfrak{m} := \mathfrak{l}^{\perp_B}$, then, via the mapping $K \times \mathfrak{m} \to X_0 \hookrightarrow G$, $(k, \xi) \mapsto k\exp(i\xi)$, Kählerian reduction of G realizes the Stein homogeneous G/H as the K-vector bundle $K \times_L \mathfrak{m} \to K/L$ over the K-orbit $Kx_0 = K/L$ of the neutral point.*

The argument for the exhaustion theorem at the level of groups can now be carried out for Stein homogeneous spaces G/H by replacing $\mathfrak{p}$ by the Mostow fiber $\mathfrak{m}$; see [Azad and Loeb 1993].

PROPOSITION 2.4.10 (EXHAUSTION THEOREM). *Let G be a (not necessarily connected) reductive complex Lie group, K a maximal compact subgroup, H a reductive subgroup of G and $X = G/H$ be associated Stein homogeneous space. A K-invariant, strictly plurisubharmonic function $\rho : X \to \mathbb{R}$ is a proper exhaustion $\rho : X \to [m, \infty)$, $m := \min\{x \in X : \rho(x)\}$, if and only if $\{d\rho = 0\} \neq \varnothing$. In this case $\{d\rho = 0\} = Kx_0$, where $\rho(x_0) = m$. Furthermore, $G_{x_0} = (K_{x_0})^{\mathbb{C}}$.*

PROOF. It is enough to prove this for the case where G is connected. Furthermore, if $\{d\rho = 0\} = \varnothing$, then ρ is clearly not a proper exhaustion. Thus we may let $x_0 \in \{d\rho = 0\}$ and recall that near x_0 the function ρ takes on its absolute minimum $m = \rho(x_0)$ exactly on the orbit Kx_0 (see Proposition 2.2.4). Without loss of generality we may assume that $H = G_{x_0}$. Let $L_1 = K_{x_0}$ and note that $H_1 := L_1^{\mathbb{C}}$ is of finite index in H.

Let $\rho_1 : X_1 := G/H_1 \to \mathbb{R}$ be the induced function on the finite covering space X_1 of X and $x_1 \in X_1$ be a neutral point over x_0. Since $L_1^{\mathbb{C}} = H_1$, we may apply the Mostow-fibration to X_1 with base point x_1. In complete analogy to the case of groups, since $\tilde{\rho}_1 : \mathfrak{m} \to \mathbb{R}$ is strictly convex along the lines through $0 \in \mathfrak{m}$ and has a local minimum at 0, it follows that $\tilde{\rho}_1$ is a proper exhaustion with absolute minimum at its only critical point $0 \in \mathfrak{m}$. Consequently, the same can be said of ρ_1: It is a proper exhaustion with $\{d\rho_1 = 0\} = Kx_1$ the set where it takes on its minimum.

Since ρ_1 is the lift of $\rho : X \to \mathbb{R}$ via $\pi : X_1 \to X$ and its critical set is the lift of the critical set $\{d\rho = 0\}$, it follows that

$$\pi^{-1}(Kx_0) = Kx_1.$$

In other words, the entire fiber $\pi^{-1}(x_0)$ is contained in Kx_1. Consequently Kx_0 is a minimal K-orbit in X and $L_1^{\mathbb{C}} = H$. ☐

3. Local Models and Exhaustions by Kählerian Potentials

Here we carry out the basic preparatory work for the construction of analytic Hilbert quotients. The goal is to prove the existence of an étale Stein local model where the potential function of the given Kählerian structure is a proper exhaustion along the fibers of the invariant theoretic quotient. This is a group action version with parameters of the exhaustion theorem, Proposition 2.2.12, for Stein homogeneous spaces.

We begin with a discussion of actions on Stein spaces, where, for holomorphic actions of reductive groups, due to the density of the G-finite functions, the situation is quite close to that of algebraic invariant theory.

3.1. Actions on Stein Spaces

Local algebraicity. If X is a complex space equipped with a holomorphic action $G \times X \to X$ of a complex Lie group, then an orbit Gx is said to be Zariski open in its closure $\overline{Gx} = W$ whenever W is a closed complex subspace and the difference $W \setminus Gx$ is the (locally finite) union of nowhere dense analytic subsets. For G connected this is equivalent to G having an open orbit in W.

PROPOSITION 3.1.1. *If X is Stein, G is reductive and $G \times X \to X$ is holomorphic, then every G-orbit in X is Zariski open in its closure.*

PROOF. Since G has only finitely many components, it is enough to prove this in the case where it is connected. Let x be given, and, just as in the proof of Proposition 2.2.8, construct a holomorphic equivariant map $F : X \to V^*$ to a representation space which is biholomorphic in a neighborhood of x. Of course F may not be injective on Gx, but it is finite-to-one.

Let h be a G-finite, holomorphic function which separates some fiber of $F|Gx$, $V_1 = \langle g(f) \rangle_{g \in G}$ and $F_1 : X \to V_1^*$ the associated map. It follows that $F \oplus F_1$ is injective on Gx and locally biholomorphic at each $z \in Gx$. By changing notation we may assume that F already had this property.

Since G is reductive, its representation on V^* is algebraic and therefore $G(F(x))$ is Zariski open in its closure Z. Let W be the irreducible of $F^{-1}(Z)$ which contains x. By construction it follows that Gx is open in W. ☐

b. Invariant theoretic quotients. If G is reductive, X is affine and $G \times X \to X$ is an algebraic action, then the ring $\mathcal{O}_{\mathrm{alg}}(X)^G$ of invariant regular functions is finitely generated; see [Kraft 1984], for example. Thus $X/\!/G := \mathrm{Spec}(\mathcal{O}_{\mathrm{alg}}(X)^G)$ is affine and there is a canonical surjective, invariant, regular morphism $\pi : X \to X/\!/G$. We view this as a complex analytic quotient.

Since the G-representation on $\mathcal{O}_{\mathrm{alg}}(X)$ is locally finite, it follows that X can be algebraically and equivariantly realized as a G-invariant subvariety of a representation space V. Now, by restriction, the ring of polynomials $\mathbb{C}[V]$ is dense in the ring of holomorphic functions $\mathcal{O}(X)$. Thus, by averaging, $\mathcal{O}_{\mathrm{alg}}(X)^G$ is dense in $\mathcal{O}(X)^G$. Consequently, the Hilbert quotient $\pi : X \to X//G$ is also defined (at least at the set-theoretic level) by the equivalence relation $x \sim y$ if and only if $f(x) = f(y)$ for all invariant holomorphic function $f \in \mathcal{O}(X)^G$ and the points of $X//G$ can be regarded as maximal ideals $\mathfrak{m} < \mathcal{O}(X)^G$.

The above equivalence relation is defined for any action of a group of holomorphic transformations on a complex space. If X is Stein, G is reductive and $G \times X \to X$ is holomorphic, then, if there is no confusion, we also denote this by $\pi : X \to X//G$. In fact, even in the non-reduced case, equipped with the direct image sheaf $U \mapsto \mathcal{O}_X(\pi^{-1}(U))^G$, $X//G$ is a Stein space [Snow 1982; Heinzner 1988; 1989; Hausen and Heinzner 1999]. In the reduced case this is the analytic Hilbert quotient associated to the moment map μ^ρ of any K-invariant strictly plurisubharmonic exhaustion $\rho : X \to \mathbb{R}$ (see Section 4).

Of course at this point $X//G$ is only a Hausdorff topological space. The complex structure will be constructed in the sequel. In situations where the complex structure is known to exist with the preceding properties, we refer to $X \to X//G$ as the holomorphic invariant theoretic quotient.

Since at this point the quotient $X//G$ only carries the structure of a topological space, we temporarily refer to $\pi : X \to X//G =: Q$ as the formal invariant theory quotient. The π-fibers are of course closed analytic subspaces of X.

PROPOSITION 3.1.2. *Let G be reductive, X Stein, $G \times X \to X$ a holomorphic action and $\pi : X \to Q$ the formal invariant theoretic quotient. Then every π-fiber contains a unique closed G-orbit.*

PROOF. Let $Z = \pi^{-1}(\pi\{x\})$ be a π-fiber. It follows from Proposition 3.1.1 that orbits of minimal dimension are closed. Thus Z contains at least one closed orbit $Y = Gy$.

If $\tilde{Y} = G\tilde{x}$ is an additional closed orbit in X, then, by using the extension principle (see 2.2), one can construct $f \in \mathcal{O}(X)^G$ with $f|Y \equiv 0$ and $f|\tilde{Y} \equiv 1$. In particular $\tilde{Y} \cap Z = \varnothing$, i.e., Z contains exactly one closed orbit. $\quad\square$

From the point of view of simply constructing a Hausdorff quotient, it is natural to attempt to define an equivalence relation by $x \sim y$ if and only if $\bar{G}x \cap \bar{G}y \neq \varnothing$. This is in fact the equivalence relation of the analytic Hilbert quotient. The invariant theoretic quotient of Stein spaces is also of this type.

COROLLARY 3.1.3. *Let G be reductive, X Stein and $G \times X \to X$ be a holomorphic G-action. Then $f(x) = f(y)$ for all $f \in \mathcal{O}(X)^G$ if and only if $\bar{G}x \cap \bar{G}y \neq \varnothing$.*

PROOF. In every equivalence class Z there is exactly one closed orbit Y, i.e., $x \sim y$ if and only if $\bar{G}x \cap \bar{G}y$ contains such a Y. $\quad\square$

REMARK. In fact the π-fibers have canonical affine algebraic structure with G acting algebraically; see 3.3.7.

Even if it is connected and smooth, a Stein space X equipped with a holomorphic action of a reductive group G may not be holomorphically, equivariantly embeddable in a G-representation space; see [Heinzner 1988]. Locally, however, there always exist G-equivariant, closed holomorphic embeddings (3.3.14). As a result, the following is of particular use.

THEOREM 3.1.4. *If X is a G-invariant closed complex subspace of a G-representation space V with Hilbert quotient $\pi : V \to V/\!/G$, then the restriction $\pi|X$ has closed complex analytic image $\pi(X) =: X/\!/G$ and $\pi : X \to X/\!/G$ is the holomorphic invariant theoretic quotient.*

PROOF. Let $S := \{v \in V : \overline{Gv} \cap X \neq \varnothing\}$. It follows that S is a closed complex subspace defined by the ideal $I(X)^G := \{f \in \mathcal{O}(V)^G : f|X = 0\}$ and the image $\pi(X)$ is the zero set of that ideal regarded as a space of functions on $X/\!/G$. In particular, $\pi(X)$ is a closed complex subspace of $V/\!/G$. Using coherence arguments [Heinzner 1991], one shows that $\mathcal{O}(X)^G = \mathcal{O}(Z)^G/I(X)^G$ and thus the image is equipped with the right set of functions. $\square$

c. The Kempf–Ness set. Let X be a complex space, G a reductive group $K < G$ a maximal compact subgroup and $\rho : X \to \mathbb{R}$ a K-invariant, smooth strictly plurisubharmonic function. An equivariant moment map $\mu^\rho : X \to \mathfrak{k}$ is defined by $\mu^\rho_\xi = -J\xi_X(\rho)$, where, for $\xi \in \mathfrak{k}$, ξ_X denotes the associated vector field on X. We refer to $X_0 := \mu^{-1}\{0\}$ as the Kempf–Ness set associated to ρ and the action.

The next result is essential.

PROPOSITION 3.1.5. *Let X be Stein and $\rho : X \to \mathbb{R}$ a K-invariant, strictly plurisubharmonic function. For $x_0 \in X_0$ it follows that Gx_0 is closed and $Gx_0 \cap X_0 = Kx_0$. If ρ is an exhaustion, then every closed orbit intersects X_0 and the inclusion $X_0 \hookrightarrow X$ induces a bijective continuous map $X_0/K \to Q$ to the invariant theoretic quotient.*

PROOF. If $x_0 \in X_0$, then x_0 is a critical point of $\rho|Gx_0$. Consequently $\rho|Gx_0$ is an exhaustion of Gx_0 with its only critical points being along the K-orbit Kx_0 (Proposition 2.4.10). Thus Gx_0 is closed and $Gx_0 \cap X_0 = Kx_0$.

If ρ is an exhaustion, then $\rho|Gx$ clearly has critical points along any closed orbit Gx. $\square$

It is of basic importance to show, e.g., in the Stein case for an exhaustion ρ, that the induced map $X_0/K \to Q$ is in fact a homeomorphism. A key notion for the discussion of such questions is that of orbit convexity.

DEFINITION. *Let G be a reductive group acting on a set X and K be a maximal compact subgroup. A K-invariant subset Y in X is said to be orbit convex if for every $y \in Y$ and $\xi \in \mathfrak{k}$ with $\exp(i\xi)y \in Y$ it follows that $\exp(i\xi t)y \in Y$ for $t \in [0, 1]$.*

We now will prove a useful technical result on the existence of orbit convex neighborhood bases at points of the Kempf–Ness set. For later applications we carry this out in a more general setting than that of the formal invariant theoretic quotient of Stein spaces.

Let X be a complex space equipped with a holomorphic action of a reductive group G. A surjective, continuous, G-invariant map $\pi : X \to Q$ to a locally compact Hausdorff topological space is called a Hausdorff quotient if it defines the equivalence relation $x \sim y$ if and only if $\bar{G}x \cap \bar{G}y \neq \varnothing$.

In the Kählerian setting (see Section 4) we deal with strictly plurisubharmonic potentials $\rho : X \to \mathbb{R}$ which are exhaustions along π-fibers. The appropriate properness is defined in terms of the join: ρ is said to be a relative exhaustion if its restriction to every fiber is bounded from below and $\pi \times \rho$ is proper.

If $\pi : X \to Q$ is a Hausdorff quotient and $\rho : X \to \mathbb{R}$ is a K-invariant, strictly plurisubharmonic function whose restriction to each fiber is bounded from below and an exhaustion, then, as was shown above, $\pi|X_0$ is surjective and induces a continuous bijective map $X_0/K \to Q$. If ρ is a relative exhaustion, then it is in fact a homeomorphism. For this we prove the result mentioned above on the existence of orbit convex neighborhoods.

Let $q \in Q$ and $x_q \in X_0$ with $\pi(x_q) = q$. Define $r_q := \rho(x_q)$. For $r \in \mathbb{R}$ let $D_\rho(r) = \{x \in X : \rho(x) < r\}$.

PROPOSITION 3.1.6. *The set $D_\rho(r)$ is orbit convex. If the restriction of ρ to every π-fiber is bounded from below and an exhaustion, then $GD_\rho(r)$ is π-saturated. If ρ is a relative exhaustion, then sets of the form $\pi^{-1}(V) \cap \{x : r_q - \varepsilon < \rho(x) < r_q + \varepsilon\}$, where $V = V(q)$ is an open neighborhood of q and $\varepsilon > 0$, form a neighborhood basis of $\pi^{-1}\{q\} \cap X_0 = Kx_q$.*

PROOF. If $x \in D_\rho(r)$, $\xi \in \mathfrak{k}$, then we consider the $\mathbb{R}$-invariant plurisubharmonic function $\rho_\xi(z) := \rho(\exp(\xi z)x)$. It follows that $\rho_\xi|i\mathbb{R}$ is convex and thus, if $\rho_\xi(i) < r$, then $\rho_\xi(it) < r$ for $t \in [0, 1]$.

To prove that $GD_\rho(r)$ is π-saturated, observe that if a π-fiber Z has non-empty intersection with $D_\rho(r)$, then since $\rho|Z$ is a proper exhaustion, it achieves its minimum in $D_\rho(r)$, i.e., $X_0 \cap Z \subset D_\rho(r)$ and in particular the unique closed orbit $Y \subset Z$ satisfies $Y \cap D_\rho(r) \neq \varnothing$. Since every G-orbit in Z has Y in its closure, it follows that

$$G(Z \cap D_\rho(r)) = Z.$$

Now suppose that ρ is a relative exhaustion and let U be an open neighborhood of Kx_q in X. For convenience, replace Q by the compact closure of an open neighborhood of q; in particular we may assume that ρ is a relative exhaustion and therefore $(\pi \times \rho)^{-1}(q, r_q) = Kx_q$.

Thus there exists open neighborhoods $V = V(q) \subset Q$ (open also in the original Hausdorff quotient Q) and a number $\varepsilon > 0$ so that

$$\pi^{-1}(V) \cap \{x \in X : r_q - \varepsilon < \rho(x) < r_q + \varepsilon\} \subset U.$$

If $V \subset Q \setminus \overline{\pi(D_\rho(r_q - \frac{\varepsilon}{2}))}$, then in fact

$$\pi^{-1}(V) \cap D_\rho(r_q + \varepsilon) = \pi^{-1}(V) \cap \rho^{-1}((r_q - \varepsilon, r_q + \varepsilon)),$$

which would complete the proof.

We prove the existence of such a V by contradiction. Thus, suppose that there exists a sequence $\{y_n\} \subset X$ such that $q = \lim \pi(y_n)$ and $\rho(y_n) < r_q - \frac{\varepsilon}{2}$. By the properness of $\pi \times \rho$ we may assume that $y_n \to y_q \in \pi^{-1}(q)$. But

$$r_q - \frac{\varepsilon}{2} \geq \rho(y_q) \geq \rho(x_q) = r_q$$

which is a contradiction. $\qquad\qquad\qquad\qquad\qquad\qquad\qquad\qquad\qquad\qquad\square$

COROLLARY 3.1.7. *If $\pi : X \to Q$ is a Hausdorff quotient, $\rho : X \to \mathbb{R}$ is a K-invariant strictly plurisubharmonic function and ρ is a relative exhaustion, then the induced map $X_0/K \to Q$ is a homeomorphism.*

REMARK. The main goal of this chapter is to show that, in a certain local setting related to the analytic Hilbert quotient, ρ is a relative exhaustion. In this case ρ is a certain Kählerian potential. The local model for these considerations is the étale Stein covering which is constructed in section 3.2, and the main result of section 2.3 is the relative exhaustion property for ρ.

3.2. Étale Stein Coverings. If $\mu : X \to \mathfrak{k}$ is a moment map of a Kählerian G-space with respect to a K-invariant Kähler form ω, then there exists a covering $\mathcal{U} = \{U_\alpha\}$ of X_0 by G-invariant neighborhoods so that $\omega|U_\alpha$ has a strictly plurisubharmonic K-invariant, potential function ρ_α with $\mu = \mu^{\rho_\alpha}$ (see Section 4) Thus, for $x_0 \in X_0 \cap U_\alpha$, Gx_0 is a closed affine homogeneous space, and in particular $G_{x_0} = K_{x_0}^{\mathbb{C}}$ is reductive.

The goal of the present section is, in the context of actions of reductive group actions, i.e., independent of a Kählerian setting, to construct equivariant étale Stein neighborhoods of points with reductive isotropy.

a. Local product structure. Throughout this paragraph X is a complex space, G is a complex Lie group acting holomorphically on X and L is a compact subgroup. The analysis takes place at a point $x_0 \in X$, where $G_{x_0} =: H = L^{\mathbb{C}}$

It follows that the natural representation of H on the Zariski tangent space $T_{x_0}X$ is holomorphic and there exists an L-invariant neighborhood $U = U(x_0)$ which can be biholomorphically, equivariantly identified with a closed analytic subset A of an L-invariant ball $B \subset T_{x_0}X$ [Kaup 1967]. Let $i : U \to A$ be this identification; of course $i(x_0) = 0$.

PROPOSITION 3.2.1 (LOCAL PRODUCT DECOMPOSITION). *Let*

$$T_{x_0}X = T_{x_0}Gx_0 \oplus V$$

be an H-invariant decomposition, $S_{\mathrm{loc}} := i^{-1}(A \cap V)$ and N an $\mathrm{Int}(L)$-invariant local submanifold at $e \in G$ so that $T_eG = T_eN \oplus T_eH$. Then, after shrinking *all sets appropriately, it follows that $N \times S_{\mathrm{loc}} \to X$, $(g, s) \mapsto g(s)$, is an L-equivariant biholomorphic map onto an open neighborhood of x_0 in X.*

PROOF. Let $A_V := A \cap V$. After shrinking N and A to sets N_1 and A_1, we have the local holomorphic action $N_1 \times A_1 \to A$. Replacing N and A by N_1 and A_1 respectively, we have the induced holomorphic map $\varphi : N \times A_V \to U$. Since the local N-orbit of $0 \in T_{x_0} X$ is transversal to V, by shrinking even further we may assume that φ is biholomorphic onto its image.

Now the desired result is local, and N is connected. Thus we may argue one component at a time to show that locally $\mathrm{Im}(\varphi) = A$.

So assume A is irreducible and let A_V^0 be a component of A_V at 0. Since A_V^0 is a component of an analytic set in A which is defined by the linear functions which define V in $T_{x_0} X$, it follows that

$$\dim A \leq \dim A_V^0 + \mathrm{codim}\, V = \dim \varphi(N \times A_V^0).$$

Since $\mathrm{Im}\,\varphi \subset A$, it follows that $\mathrm{Im}\,\varphi$ is a neighborhood of $0 \in A$.   □

b. Local complexification. In the previous section we constructed an L-invariant local slice S_{loc} transversal to an orbit Gx_0 where $G_{x_0} = L^{\mathbb{C}} = H$. Of course S_{loc} is in general not H-invariant and $H \cdot S_{\mathrm{loc}} \subset X$ is possibly quite wild. Thus, in order to globalize the H-action on S_{loc}, we regard S_{loc} as an L-invariant subvariety of a ball in T_{x_0} where the H is much easier to control. As usual, the controlling device is a strictly plurisubharmonic function.

We formulate these general results on complexification in the original notation, i.e., for a reductive group G and a fixed maximal compact subgroup K. These are very special cases of the results in [Heinzner 1991; Heinzner and Iannuzzi 1997].

The basic question here is, given a real form K of a complex Lie group G and an action of K as a group of holomorphic transformations on a complex space X, does there exist a complex space $X^{\mathbb{C}}$, a holomorphic G-action $G \times X^{\mathbb{C}} \to X^{\mathbb{C}}$ and an open, K-equivariant embedding $i : X \hookrightarrow X^{\mathbb{C}}$. Optimally, this should have the obvious universality property: Holomorphic K-equivariant maps $X \to Y$ to holomorphic G-spaces should factor through $i : X \hookrightarrow X^{\mathbb{C}}$. If this universality property is fulfilled, then $X^{\mathbb{C}}$ is referred to as a G-complexification of X. If X is Stein [Heinzner 1991] or holomorphically convex [Heinzner and Iannuzzi 1997], such a complexification indeed exists.

In this paragraph we develop enough of this theory to construct $S := S_{\mathrm{loc}}^{\mathbb{C}}$.

A convenient notion for these considerations is that of *orbit connectedness*: Let $G \times X \to X$ be a holomorphic action of a reductive group and K a maximal compact subgroup. A set K-invariant subset $U \subset X$ is said to be orbit connected if for every $x \in U$ and $B_x : G \to X$, $g \mapsto g(x)$, the preimage $B_x^{-1}(U)$ is K-connected, i.e., $B_x^{-1}(U)/K$ is connected.

LEMMA 3.2.2. *If $G \times X \to X$ is a holomorphic action of the reductive group G and U is a K-invariant, orbit connected, open subset, then GU is a G-complexification.*

PROOF. Let $\varphi : U \to Y$ be a K-equivariant holomorphic map to a complex space Y equipped with a holomorphic G-action. We must show that it extends to a G-equivariant map $\varphi^{\mathbb{C}} : GU \to Y$. Since the map $G \times U \to Y$, $(g, u) \mapsto g\varphi(u)$, is holomorphic, it is only a question if it factors through $G \times U \to GU$.

For $x \in U$ and $U_x := B_x^{-1}(U)$ it follows from the orbit connectedness and the identity principle that $\psi_1 : G \to Y$, $g \mapsto g(\varphi(x))$, gives a well-defined holomorphic extension of $\psi_2 : U_x \to Y$, $g \mapsto \varphi(g(x))$. So for $y = g_1(x_1) = g_2(x_2)$, with $x_1, x_2 \in U$, it follows that $g_2^{-1} g_1 \in U_{x_1}$ and therefore

$$g_1(\varphi(x_1)) = (g_2(g_2^{-1} g_1))(\varphi(x_1)) = g_2(\varphi(g_2^{-1} g_1(x_1))) = g_2(\varphi(x_2)). \qquad \square$$

If U is orbit convex (a fortiori orbit connected), then the preceding result holds for K-invariant analytic subsets.

LEMMA 3.2.3. *If U is orbit convex and $A \subset U$ is a K-invariant analytic subset, then GA is an analytic subset of $U^{\mathbb{C}} = GU$ and is a G-complexification $A^{\mathbb{C}}$.*

PROOF. The orbit convexity of U is clearly inherited by A. Thus it is enough to show that GA is an analytic subset of GU. In that case, since A is orbit connected in GA, the result follows from the previous lemma.

Note that $GA \subset \bigcup_{g \in G} g(U)$. Thus, to prove that GA is an analytic subset of GU, it is enough to show that $(GA) \cap g(U) = g(A)$ or equivalently that $(GA) \cap U = A$. However, this is just the orbit connectedness of A. $\qquad \square$

We now come to the result which will allow us to complexify an appropriately chosen local slice S_{loc}.

PROPOSITION 3.2.4. *Let $G \times X \to X$ a holomorphic action of a reductive group G on a Stein complex space which has a holomorphic Hausdorff quotient $\pi : X \to Q$, i.e., Q is itself a complex space and π is holomorphic. Assume in addition that $\rho : X \to \mathbb{R}$ is a K-invariant strictly plurisubharmonic relative exhaustion function. If x_q is a point in the Kempf–Ness set X_0 and A is a K-invariant analytic subset of some open subset of X_0 with $x_q \in A$, then there is a basis of K-invariant Stein neighborhoods U of Kx_q such that $G(A \cap U)$ is a Stein G-complexification of A.*

PROOF. Let V run through a Stein neighborhood basis of $q = \pi(x_q)$. In the notation of Proposition 3.1.6 let $U = \pi^{-1}(V) \cap \rho^{-1}((r_q - \varepsilon, r_q + \varepsilon))$ be orbit convex with GU being π-saturated. We may choose V sufficiently small so that the π-saturation GU is just $\pi^{-1}(V)$.

If $\{x_n\}$ is a divergent sequence in $\pi^{-1}(V)$ which is not divergent in X, then $\{\pi(x_n)\}$ is divergent in V. In the former case there exists $f \in \mathcal{O}(X)$ with

$\lim |f(x_n)| = \infty$ and in the latter $f \in \mathcal{O}(\pi^{-1}(V))^G$ with the same property. Thus $\pi^{-1}(V) = GU$ is Stein.

Now, by the previous lemma the orbit convexity of U implies that $G(U \cap A)$ is a complex subspace of GU which is a G-complexification of $U \cap A$. Since $GU = \pi^{-1}(V)$ is Stein, it is likewise Stein.   □

Our main application is the complexification of local slice of Proposition 3.2.1.

PROPOSITION 3.2.5. *Let $G \times X \to X$ be the holomorphic action of a complex Lie group, $x_0 \in X$ and assume that $G_{x_0} = L^{\mathbb{C}} =: H$ is a reductive group. Then there exists a Stein local slice S_{loc} with a Stein H-complexification S.*

PROOF. In fact S_{loc} is constructed as a K-invariant analytic subset of a ball B about $0 \in S_{\mathrm{loc}}$ in an H- representation space $V \hookrightarrow T_{x_0} X$. The algebraic Hilbert quotient $\pi : V \to V//H = Q$ satisfies the conditions of the previous proposition.   □

c. The local model as an étale Stein covering. The étale Stein model is constructed as a quotient $G \times_H S$ of the Stein space $G \times S$ by the diagonal H-action defined by

$$h(g, s) := (gh^{-1}, h(s)).$$

Since this action is free and proper, the following is of use.

LEMMA 3.2.6. *Let $G \times X \to X$ be a free, proper holomorphic action of a complex Lie group G on a complex space X. Then, equipped with the quotient topology, the orbit space X/G has a unique structure of a complex space so that $\pi : X \to X/G$ is holomorphic.*

PROOF. Let $x \in X$ and $Y = Gx$. Since the action is holomorphic and proper, Y is a closed complex submanifold of X. Let S_{loc} be the local slice of Proposition 3.2.1. (In this case the isotropy is trivial.)

The map $\alpha : G \times S_{\mathrm{loc}} \to X$ is biholomorphic in some neighborhood $N \times S_{\mathrm{loc}}$. Since it is equivariant, α is therefore everywhere locally biholomorphic with open G-invariant image $U \subset X$.

If $\alpha(g_1, s_1) = \alpha(g_2, s_2)$ with $g := g_1^{-1} g_2$, then $g(s_2) = s_1$. If $g \neq e$, then, since $\alpha | N \times S_{\mathrm{loc}}$ is biholomorphic, it follows that $g \in G \setminus N$. Now, if we could construct such pairs of points for S_{loc} arbitrarily small, then we would have a sequence $\{s_n\} \subset S_{\mathrm{loc}}$ with $s_n \to x_0$ and $g_n \in G \setminus N$ with $g_n(s_n) \to x_0$ as well. By the properness of the action this would imply that, after going to a subsequence, $g_n \to g \in G \setminus N$ with $g(x_0) = x_0$. This is contrary to G acting freely.

Thus, $\alpha : G \times S_{\mathrm{loc}} \to U$ is biholomorphic and S_{loc} can be identified with a neighborhood of Gx_0 in the orbit space X/G. Since $\pi : X \to X/G$ is required to be holomorphic, this realization of S_{loc} in X/G must be a holomorphic chart.

Finally, given two local sections S_α and S_β over the same chart in X/G, the change of coordinates $\varphi_{\alpha\beta} : S_\alpha \to S_\beta$ is given by a G-valued holomorphic map $S_\alpha \to G$.   □

REMARK. Of course this is just another description of a G-principal bundle. There would be little difference in the discussion if the action were only required to be locally free: the local models in that case are the finite group quotients S_{loc}/Γ, where $\Gamma = G_{x_0}$ is the possibly non-trivial isotropy group.

COROLLARY 3.2.7. *Let* $G \times X \to X$ *be a free, proper, holomorphic action of a reductive group on a Stein complex space* X. *Then* X/G *is Stein and* $\pi : X \to X/G$ *is an invariant theoretic quotient.*

PROOF. The proof that X/G is Stein goes exactly as the proof that G/H is Stein in the Theorem of Matsushima–Onitshick on reductive pairs (see 2.2).

Universality means that every invariant holomorphic map $F : X \to Y$ factors through $\pi : X \to X/G$. By localizing to coordinate neighborhoods in Y this is reduced to the same question for invariant functions and this follows immediately from the fact that $\pi^* : \mathcal{O}(X/G) \to \mathcal{O}(X)^G$ is an isomorphism. □

As an application, let G be a complex Lie group, H a closed subgroup and suppose that $H \times S \to S$ is a holomorphic action on a complex space S. Let $G \times_H S$ denote the quotient by the free, proper H-action. Leaning on the language of representation theory, we might refer to the holomorphic fiber bundle $G \times_H S \to G/H$ with fiber S, base G/H and structure group H as a sort of geometric induction of going from an H-action on S to a G-action on $G \times_H S$.

We now turn to the description of the étale Stein local model. In the following S_{loc} refers to the local slice of Proposition 3.2.1.

PROPOSITION 3.2.8. *Let* $G \times X \to X$ *be a holomorphic action of a Stein complex Lie group on a complex space. For* $x_0 \in X$ *suppose that* $G_{x_0} = L^{\mathbb{C}} = H$ *is a reductive group. Let* S *be a Stein* H-*complexification of the local slice* S_{loc} *and let* $i : S \to X$ *be the* H-*equivariant holomorphic mapping guaranteed by the universality property. Then the canonical holomorphic map* $\tilde{\alpha} : G \times S \to X$, $(g, s) \mapsto g(i(s))$ *factors through the Stein space* $G \times_H S$ *and the induced map* $\alpha : G \times_H S \to X$ *is everywhere locally biholomorphic.*

REMARKS. (1) We refer to $\alpha : G \times_H S \to X$ as an étale Stein model of X at x_0. Although locally biholomorphic, the map α could be very complicated.
(2) The assumption that G is Stein is only needed to insure that $G \times_H S$ is Stein.
(3) Whether or not a complex Lie group is Stein is well understood. For example, if G is connected, then there is a uniquely defined closed, connected, central subgroup M with $\mathcal{O}(M) \cong \mathbb{C}$ so that G/M is Stein.

PROOF OF THE PROPOSITION. Since

$$\tilde{\alpha}(gh^{-1}, i(h(s))) = (gh^{-1})(h(i(s))) = g(i(s)) = \tilde{\alpha}(g, s),$$

it is clear that $\tilde{\alpha}$ is H-invariant and therefore factors through the Stein quotient $G \times_H S$.

Now, when restricted to $N \times S_{\text{loc}} \hookrightarrow G \times_H S$, this map is in fact biholomorphic. By its G-equivariance and the fact that $G(N \times S_{\text{loc}}) = G \times_H S$, it follows that α is everywhere locally biholomorphic. $\qquad\qquad\qquad\qquad\qquad\qquad\qquad\qquad\Box$

3.3. The Relative Exhaustion Property. The main goal of this section is to prove that if $\pi : X \to Q$ is a Hausdorff quotient for the action of a reductive group G with maximal compact subgroup K and $\rho : X \to \mathbb{R}$ is a K-invariant, strictly plurisubharmonic function which, when restricted to a fiber $\pi^{-1}\{q_0\}$ has a local minimum at x_0, then, after replacing Q by an appropriately chosen neighborhood of q_0 and X by its saturation, it follows that ρ is a relative exhaustion. This then shows that, after localizing, the restriction $\pi|X_0$ is proper. In other words, $\pi|X_0$ is an open mapping. In Section 4 this will be applied to a potential of a Kähler form lifted to the étale Stein local model.

a. The Hilbert Lemma for algebraic actions. We must to analyze the fibers of a Hausdorff quotient using a strictly plurisubharmonic function to control the action. In the end (item c on page 342) we will show that such a fiber possesses a natural structure of an affine algebraic variety where the reductive group at hand is acting algebraically. We begin by analyzing the algebraic case.

Let X be an affine algebraic variety equipped with an algebraic action $G \times X \to X$ of a reductive group. Assume that $\mathcal{O}(X)^G \cong \mathbb{C}$ and let $Y \hookrightarrow X$ be the unique closed G-orbit. It follows that $Y \subset \overline{Gx}$ for every $x \in X$. The Hilbert Lemma provides an organized way for finding limit points in Y.

For this let T be a fixed maximal torus in a fixed maximal compact subgroup K of G. In this context a 1-parameter subgroup $\lambda \in \Lambda(T)$ is an algebraic morphism[1] $\lambda : \mathbb{C}^* \to G$ with $\lambda(S^1) \subset T$, i.e., after lifting to the Lie algebra level, $z \mapsto \exp(\xi z)$ for $\xi \in \mathfrak{t}$ with integral periods with respect to $\exp : \mathfrak{t} \to T$.

HILBERT LEMMA. *Let X be affine, $G \times X \to X$ an algebraic action of a reductive group with $\mathcal{O}(X)^G \cong \mathbb{C}$ and K a maximal compact subgroup with a fixed maximal torus T. Let Y be the unique closed G-orbit in X. Then there exist finitely many 1-parameter subgroups $\lambda_1, \ldots, \lambda_m \in \Lambda(T)$ so that, for any $x \in X$ there exists $\lambda \in \{\lambda_1, \ldots, \lambda_m\}$ and $k \in K$ such that $\lim_{t \to 0} \lambda(t)(k(x)) = y \in Y$.*

The proof requires a bit of preparation. First, notice that $\Lambda(T)$ is countable and, for any given $\lambda \in \Lambda(T)$, the saturation

$$S_\lambda(Y) := \{x : \overline{\lambda(\mathbb{C}^*)}(x) \cap Y \neq \varnothing\}$$

is the preimage $\pi^{-1}(\pi(Y))$ defined via the algebraic Hilbert quotient $X \to X/\!/\mathbb{C}^*$, where $\mathbb{C}^*$ acts via the morphism $\lambda : \mathbb{C}^* \to G$. In particular, $S_\lambda(Y)$ is a closed, algebraic subvariety.

---

[1] Holomorphic morphisms with values in an affine algebraic group are automatically algebraic.

It is convenient to formulate the conclusion of the Hilbert Lemma as follows:

$$X = K \bigcup_{j=1}^{m} \mathcal{S}_{\lambda_j}(Y).$$

The essential step in the proof of the Hilbert Lemma is a reduction to closures of $T^{\mathbb{C}}$-orbits.

PROPOSITION 3.3.1. *Under the assumptions of the Hilbert Lemma it follows that*

$$\overline{T^{\mathbb{C}} Kx} \cap Y \neq \varnothing.$$

PROOF. Suppose to the contrary that $\overline{T^{\mathbb{C}} k(x)} \cap Y = \varnothing$ for all $k \in K$. Therefore, for every $k \in K$, there exists a $T^{\mathbb{C}}$-invariant regular function $f_k \in \mathcal{O}(X)^{T^{\mathbb{C}}}$ such that $f_k | \overline{T^{\mathbb{C}} k(x)} \equiv 1$ and $f_k | Y \equiv 0$. Consequently there are finitely many such functions $f_{k_1}, \ldots, f_{k_m}$ such that

$$f := \sum_{j=1}^{m} |f_{k_j}|^2$$

satisfies $f \geq m > 0$ on Kx and $f | Y \equiv 0$. Thus $f \geq m$ on $\overline{T^{\mathbb{C}} Kx}$ as well.

Hence it follows that $\overline{T^{\mathbb{C}} Kx} \cap Y = \varnothing$ which implies that $K.\overline{T^{\mathbb{C}}.Kx} \cap Y = \varnothing$ as well. But $K.\overline{T^{\mathbb{C}}.Kx} = \overline{Gx}$ and every G-orbit in X has Y in its closure. $\square$

REMARK. For this and related information on algebraic actions, see [Kraft 1984]. The proof above is due to Richardson.

The next step is to go from closures of $T^{\mathbb{C}}$-orbits to closures of orbits in $\Lambda(T)$. We only state this result. The proof (see [Kraft 1984]) amounts to proving it for toral groups of matrices with $Y = \{0\}$.

PROPOSITION 3.3.2.

$$\mathcal{S}_{T^{\mathbb{C}}}(Y) = \bigcup_{\lambda \in \Lambda(T)} \mathcal{S}_{\lambda}(Y).$$

PROOF OF THE HILBERT LEMMA. By Proposition 3.3.1 and Lemma 3.3.2 it follows that

$$X = K\mathcal{S}_{T^{\mathbb{C}}}(Y) = K \bigcup_{\lambda \in \Lambda(T)} \mathcal{S}_{\lambda}(Y).$$

From the countability of $\Lambda(T)$ and the fact that $\mathcal{S}_{\lambda}(Y)$ is a closed analytic subset, it follows that there exist $\lambda_j \in \Lambda(T)$, $j = 1, \ldots, m$, so that

$$\bigcup_{\lambda \in \Lambda(T)} \mathcal{S}_{\lambda}(Y) = \bigcup_{j=1}^{m} \mathcal{S}_{\lambda_j}(Y).$$

$\square$

b. The exhaustion property in the case $\mathcal{O}(X)^G \cong \mathbb{C}$. Throughout this paragraph $G \times X \to X$ denotes the holomorphic action of a reductive group on a Stein space. It will always be assumed that $\mathcal{O}(X)^G \cong \mathbb{C}$ and therefore that X possesses a unique closed G-orbit Y.

DEFINITION. Under the preceding assumptions, a G-space X is said to have the exhaustion property if, for every maximal compact subgroup K in G, every K-invariant, strictly plurisubharmonic function $\rho : X \to \mathbb{R}$ which possesses a local minimum $x_0 \in X$ is a proper exhaustion.

REMARK. It follows that if X has the exhaustion property, then any K-invariant, strictly plurisubharmonic function $\rho : X \to \mathbb{R}$ with $\rho(x_0) = m_0$ as local minimum is a proper exhaustion $\rho : X \to [m_0, \infty)$ and the differential $d(\rho|Gx)$ vanishes only on Gx_0 and there it vanishes exactly along Kx_0. In the language of momentum geometry, $X_0 = Kx_0$ is the Kempf–Ness set.

The main goal of this section is to prove the following result:

PROPOSITION 3.3.3. *A holomorphic action* $G \times X \to X$ *of a reductive group on a Stein space with* $\mathcal{O}(X)^G \cong \mathbb{C}$ *has the exhaustion property.*

For this it is important to understand the connection to the Hilbert Lemma in the analytic setting.

PROPOSITION 3.3.4. *A Stein space* X *equipped with the action* $G \times X \to X$ *of a reductive group with* $\mathcal{O}(X)^G \cong \mathbb{C}$ *has the exhaustion property if and only if the Hilbert Lemma is valid.*

REMARKS. (1) The validity of the Hilbert Lemma in this context has the obvious meaning: For K and T fixed there exist finitely many $\lambda_j \in \Lambda(T)$, $j = 1, \ldots, m$, so that

$$X = K \bigcup_{j=1}^{n} \mathcal{S}_{\lambda_j}(Y).$$

(2) In [Heinzner and Huckleberry 1996] we showed that, if X and the action are algebraic, then X has the exhaustion property; in fact, only the Hilbert Lemma was used.

Suppose the Hilbert Lemma is valid and let $\{x_n\}$ be a divergent sequence. Since the Hilbert Lemma only requires finitely many 1-parameter subgroups, we may assume without loss of generality that there exists $\lambda \in \Lambda(T)$ such that $\lim_{t \to 0} \lambda(t)(x_n) \in Y$ for all n.

Of course we wish to control the region where these limit points land.

MONOTONICITY LEMMA. *Let* $\mathbb{C}^* \times X \to X$ *be a holomorphic* $\mathbb{C}^*$-*action, and let* $x \in X$ *with* $\mathbb{C}^*x$ *Zariski open in its closure* Z. *If* $K = S^1 < \mathbb{C}^*$ *is the maximal compact subgroup,* $\rho : X \to \mathbb{R}$ *is a* K-*invariant, plurisubharmonic function and* $z_0 := \lim_{t \to 0} \mathbb{C}^*x$, *then* $\rho(z_0) \leq \rho(x)$.

PROOF. The normalization of Z yields a $\mathbb{C}^*$-equivariant holomorphic map $\varphi :$ $\mathbb{C} \to Z$ from the standard $\mathbb{C}^*$-representation with $\varphi(0) = z_0$. The function $\varphi^*(\rho)$ is S^1-invariant and plurisubharmonic. Thus the result follows from the mean-value theorem.   □

Now we return to the setting Proposition 3.3.4. From the Monotonicity Lemma it follows that if $\{\rho(x_n)\}$ is bounded, then so is the collection of limit points $\{y_n\}$, $y_n := \lim_{t \to 0} \lambda(t)(x_n)$. It should not be forgotten that these are $\mathbb{C}^*$-fixed points.

Consider for a moment a linear $\mathbb{C}^*$-action on a vector space V. Let W be the set of $v \in V$ so that $\lim_{t \to 0} t(v) = w_0$ exists and let $W_0 \subset W$ be the fixed points.

By using the idea of approximation by G-finite functions, given an action $G \times X \to X$ of a reductive group on a Stein space and a relatively compact domain $D \subset X$, there is a G-equivariant holomorphic map $F : X \to V$ with $F|D$ biholomorphic onto its image (see Section 2.2b).

For our purposes choose D to contain the closure of the set of fixed points $\{y_n\}$.

Let $\mathbb{C}^*$ act on V by transporting the λ-induced action by the equivariant map F and $\Sigma(r)$ be an S^1-invariant normal sphere bundle of radius $r > 0$ in W around the fixed point set W_0, i.e., $\Sigma(r)$ is the boundary of a tubular neighborhood of W_0 in W.

Choose $r > 0$ sufficiently small so that every $\mathbb{C}^*$-orbit which has a point in the closure of the set $F(\{y_n\})$ in its closure has non-empty intersection with $\Sigma(r) \cap D$. In fact we may choose a compact set $C \subset \Sigma(r) \cap D$ with this property and regard it in X via the mapping F; in particular, there exists $t_n \in \mathbb{C}^*$ such that $\lambda(t_n)(x_n) = c_n \in C$ for all n.

PROOF OF PROPOSITION 3.3.4. First, suppose that the Hilbert Lemma is valid and set things up as above. Consider the plurisubharmonic functions $\rho_n : \mathbb{C}^* \to \mathbb{R}$, $t \mapsto \rho(\lambda(t)c_n)$. These functions are S^1-invariant and yield strictly convex functions $\tilde{\rho}_n : \mathbb{R} \to \mathbb{R}$, $s \mapsto \rho(\lambda(e^s)c_n)$, with $\tilde{\rho}'_n(0) =: m_n > 0$.

By the construction of C, i.e., its compactness in $D \setminus Fix(\mathbb{C}^*)$, it follows that there exists $m > 0$ with $m_n \geq m$. Consequently, $\rho_n(t_n) \geq M + m t_n$, where $M := \min_C \rho$, and therefore, contrary to assumption, $\rho(x_n) = \rho_n(t_n) \to \infty$.

Conversely, suppose that X has the exhaustion property. The usual techniques with G-finite functions yield the existence equivariant holomorphic map $F : X \to V$ to a representation space which is locally biholomorphic along the closed open Y and such that $F|Y$ is injective (see Section 2.2b).

By averaging we may assume that V is a unitary representation for the compact group K. Let η be an invariant norm function and $\rho := \eta \circ F$. Since every G-orbit has Y in its closure and F is locally biholomorphic along Y, it follows that ρ is strictly plurisubharmonic. By explicit construction, it is possible to insure that the restriction $\rho|Y$ has critical points along a minimal K-orbit, e.g., Kx_0, where x_0 is a base point of the Mostow fibration.

Now X has the exhaustion property and therefore $\rho : X \to \mathbb{R}$ is a proper exhaustion. Using this, the Hilbert Lemma on the algebraic closure $\overline{F(X)}$ can be transported back to X.

Note that $F^{-1}(F(Y)) = Y$, because, if not, there would be an additional K-orbit of minima, i.e., another closed orbit. Thus if, $\lim_{t \to 0} \lambda(t)F(x) = z \in F(Y)$, we must only show that $\{\lambda(t)(x)\}$ is not divergent. However, if it were divergent, then, by the exhaustion property, $\lim_{t \to 0} \rho(\lambda(t)(x)) = \infty$. But $\rho = \eta \circ F$ and consequently this is not the case. $\qquad\square$

In this proof we made use of a holomorphic G-equivariant map $F : X \to Z$ to an affine G-space with the property that $F|Y$ is a closed embedding and F is locally biholomorphic along Y. In that situation, and under the assumption that $\mathcal{O}(X)^G \cong \mathbb{C}$, so that in particular $F^{-1}(F(Y)) = Y$, we say that X has the *embedding property* if and only if every such map is in fact an embedding $F : X \hookrightarrow Z$ onto a closed subvariety.

THEOREM 3.3.5. *Let $G \times X \to X$ be a holomorphic action of a reductive group on a Stein complex space with $\mathcal{O}(X)^G \cong \mathbb{C}$. Then the following are equivalent*:

(i) *X has the embedding property.*
(ii) *The Hilbert Lemma is valid.*
(iii) *X has the exhaustion property.*

PROOF. In the previous proposition we proved the equivalence $(ii) \Leftrightarrow (iii)$. Since there is always an equivariant holomorphic map $F : X \to V$ to a representation space with the desired conditions along Y, if X has the embedding property, then F is an embedding onto an algebraic subvariety of V and the Hilbert Lemma is obviously valid.

Conversely, suppose that $F : X \to Z$ satisfies the conditions along Y and (ii) and (iii) are fulfilled. Now Z can be equivariantly embedded in a representation space. Hence, by pulling back a K-invariant norm from that space and the further pulling this back to X, we obtain a K-invariant, strictly plurisubharmonic function $\rho = \eta \circ F$ which attains its minimum at some point $x_0 \in Y$.

Consequently, $\rho : X \to \mathbb{R}$ is a proper exhaustion and it follows that the image $F(X)$ is closed.

It therefore remains to prove the injectivity of F. For this, suppose that $F(x_1) = F(x_2) = z \in Z$ and choose $\lambda \in \Lambda(T)$ so that $\lim_{t \to 0} \lambda(t)(z) =: z_0 \in F(Y)$. From the exhaustion property, applied to the restriction $\rho|\lambda(t)(x_j)$, it follows that $\lim_{t \to 0} \lambda(t)(x_j) =: y_j \in Y$, $j = 1, 2$.

But $F(y_1) = F(y_2)$. Thus $y_1 = y_2 =: y$. Since F is biholomorphic near y, it follows that, for t sufficiently small, $\lambda(t)(x_1) = \lambda(t)(x_2)$. Hence, this holds for all $t \in \mathbb{C}^*$ and as a result $x_1 = x_2$. $\qquad\square$

In fact, as we shall now show, the Hilbert Lemma is valid.

THEOREM 3.3.6. *Let $G \times X \to X$ be a holomorphic action of a reductive group on a Stein complex space with $\mathcal{O}(X)^G \cong \mathbb{C}$. Then the Hilbert Lemma is valid.*

PROOF. (By induction on $\dim X$) The case of $\dim X = 0$ is clear. Thus we assume that the result is valid for all complex spaces with dimension at most $n - 1$.

Let X be given with $\dim X = n$. We must only show that given $x \in X$ there exists $\lambda \in \Lambda(T)$ so that $\lim_{t \to 0} \lambda(t)(x) = y \in Y$, where Y is the unique closed G-orbit.

Since G-orbits are Zariski open in their closures and $Y \subset \overline{Gx}$, it is enough to consider the case where $X = \overline{Gx}$, i.e., if Gx were not locally open, then the desired result would follow from the induction assumption.

Furthermore, by the induction assumption it follows that the Hilbert Lemma is valid on $E := \overline{Gx} \setminus Gx$. Now let $F : X \to V$ be a holomorphic equivariant map to a representation space which is every locally biholomorphic and embeds Y. By adding a map F_1 generated by G-finite functions which separate the finite fibers of the map $F|Gx$, we may assume that $F|Gx$ is injective.

Now let $\lambda \in \Lambda(T)$ be such that $\lim_{t \to 0} \lambda(t)(F(x)) =: z_0 = F(y_0)$ for $y_0 \in Y$. Since F is biholomorphic near y_0, there is a unique local complex curve C_{y_0} through y_0 with $F|C_{y_0}$ biholomorphic onto a piece of the closure of $F(\mathbb{C}^*)(x)$ through z_0. Thus by equivariance and identity principle, it follows that there exists $x_1 \in \mathbb{C}^* C_{y_0}$ such that $F(x_1) = F(x)$. By the injectivity of F, it follows that $x_1 = x$ and therefore $\lim_{t \to 0} \lambda(t)(x) = y_0 \in Y$ as desired. $\square$

It of course follows that all three properties of Theorem 3.3.5 are fulfilled. As a consequence we have the following basic result of Snow [1982].

COROLLARY 3.3.7. *Let $G \times X \to X$ be a holomorphic action of a reductive group on a Stein complex space with $\mathcal{O}(X)^G \cong \mathbb{C}$. Then X possesses the structure of an affine algebraic variety with algebraic G-action.*

PROOF. It follows from the preceding discussion that X can be holomorphically, equivariantly embedded as a closed complex analytic subvariety of a G-representation space V. Let Z be the Zariski closure of X in such an embedding

For essentially the same reason as that for the algebraicity of a holomorphic G-representation, since $\mathcal{O}(Z)^G \cong \mathbb{C}$, it follows that the G-finite holomorphic functions on Z are algebraic. Since the G-finite functions are dense in $I(X) := \{f \in \mathcal{O}(X) : f|X = 0\}$, it follows that X is defined as the zero set of algebraic functions. $\square$

The following result takes its name from its usefulness in applications to the study of fibers of analytic Hilbert quotients.

COROLLARY 3.3.8 (FIBER EXHAUSTION). *Let $G \times X \to X$ be a holomorphic action of a reductive group on a Stein space with $\mathcal{O}(X)^G \cong \mathbb{C}$. If K is a maximal compact subgroup of G and $\rho : X \to \mathbb{R}$ is a K-invariant, strictly subharmonic function which has a local minimum value $\rho(x_0) = m_0$, then $\rho : X \to [m_0, \infty)$ is a proper exhaustion with $\{x : \rho(x) = m_0\} = Kx_0$. Furthermore, the differential*

$d(\rho|Gx)$ of the restriction to an orbit vanishes only on Gx_0 and there only along Kx_0.

c. The relative exhaustion property of ρ. The exhaustion property for Stein G-spaces with $\mathcal{O}(X)^G \cong \mathbb{C}$ should be regarded as a properness statement for $\pi \times \rho$, where $\pi : X \to (*)$ is a map to a point. Here we extend this to a local theorem for invariant maps with values in a complex space.

Let $\pi : X \to Q$ be a Hausdorff quotient and $\rho_0 : X \to \mathbb{R}$ a K-invariant, strictly plurisubharmonic (background) function which is a relative exhaustion. In this situation we call $\pi : X \to Q$ a *gauged* quotient.

In applications we are interested in the behavior of a K-invariant, strictly plurisubharmonic function ρ which is known to be an exhaustion of the neutral fiber $F_{q_0} := \pi^{-1}(q_0)$.

THEOREM 3.3.9 (RELATIVE EXHAUSTION PROPERTY). *Let $G \times X \to X$ be a holomorphic action of a reductive action with a gauged Hausdorff quotient $\pi : X \to Q$. Let $\rho : X \to \mathbb{R}$ be a K-invariant, strictly plurisubharmonic function such that $\rho|F_{q_0}$ has a local minimum for a π-fiber F_{q_0}. Then, after shrinking Q to an appropriate neighborhood of q_0, it follows that ρ is a relative exhaustion.*

After the preparation in the previous sections the proof is almost immediate. We begin by making several reductions.

First, note that by replacing ρ_0 by $\rho_0 + c$, where $c > 0$ is an appropriate constant, we may assume that $\rho_0 > 0$.

Secondly, if e^ρ is a relative exhaustion, then so is ρ. Of course the function e^ρ also satisfies the assumptions of the theorem. Thus we may assume in addition that $\rho : X \to \mathbb{R}^{\geq 0}$.

In the sequel $\chi : \mathbb{R} \to \mathbb{R}^{\geq 0}$ denotes a smooth function with $\mathrm{supp}(\chi) = [M, \infty)$ and which is strictly convex on (M, ∞). If D is a relatively compact domain in X and $M = \max_D \rho_0$, then the function $\rho + \chi \circ \rho_0$ is still a gauge for $\pi : X \to Q$.

LEMMA 3.3.10. *After replacing ρ_0 by a function of the type $\rho + \chi \circ \rho_0$, it may be assumed that the Kempf-Ness sets agree in a neighborhood of x_0. In particular, after replacing Q by an appropriate neighborhood of $q_0 \in Q$, it follows that $\pi|X_0(\rho)$ is proper and induces a homeomorphism $X_0/K \cong Q$.*

PROOF. Choose D to be a relatively compact neighborhood of Kx_0. The result then follows from Corollary 3.1.7.   □

This result already shows that the restriction of ρ to the π-fibers near F_{q_0} is bounded from below and an exhaustion. A similar argument yields the properness. For later applications we state the relevant technical results.

LEMMA 3.3.11. *If $G \times X \to X$ is an action of a reductive group on a complex space with Hausdorff quotient $\pi : X \to Q$ and $\rho : X \to \mathbb{R}$ is a K-invariant strictly plurisubharmonic function which is a relative exhaustion, then for every*

$R > 0$ and open connected set $V \subset Q$ the set $T_R := \pi^{-1}(V) \cap \{x : \rho(x) < R\}$ is K-connected.

PROOF. Let X_0 be the Kempf–Ness set over V. Since X_0/K is homeomorphic to V, we must only prove that a point in $x \in T_R$ can be connected to X_0. It is therefore sufficient to discuss the case where Q is just a point.

Let Y be the closed G-orbit in X and note that, by the Hilbert Lemma, for a maximal torus T in K there exits a 1-parameter group $\lambda \in \Lambda_T$ so that $\lim_{t \to 0} \lambda(t)(x) = y \in Y$. By the Monotonicity Lemma $\rho(\lambda(t)(x)) < \rho(x) < R$ for $|t| < 1$. In particular we can connect x to $T_R \cap X$ by a curve in T_R.

Now $\rho|(T_R \cap Y)$ has a minimum in each of its components, i.e., X_0 has non-empty intersections with every such component, and therefore x can be connected to X_0 by a curve in T_R. $\square$

Given a radius $R > 0$ we now define a comparison domain D to be used in proving Theorem 3.3.9. As usual let ρ_0 denote the background function for which it is known that $\rho_0 \times \pi$ is proper.

To clarify the notation, let q_0 be the neutral point in Q and, for $q \in Q$, let $F_q := \pi^{-1}(q)$. Let R_1 be the maximum of ρ_0 on the set $\{x \in F_{q_0} : \rho(x) \leq 2R\}$. Let $D := \pi^{-1}(V) \cap \rho_0^{-1}[0, R_1)$ for V a connected neighborhood of q_0 which we will now choose.

Note that $\rho|\partial(D \cap F_{q_0}) \geq 2R$. Thus this condition essential holds for q sufficiency close to q_0: For some small $\varepsilon > 0$ we choose V so that $\rho|\partial(D \cap F_q) \geq 2R - \varepsilon$ for all $q \in V$. For W relatively compact in V it follows that $\pi^{-1}(W) \cap \rho^{-1}[0, R) \cap D = T_R \cap D$ is relatively compact in D.

If D is constructed in this way, in particular containing $T_R \cap D$ as a relatively compact subset, we refer to it as being adapted to ρ and R.

LEMMA 3.3.12. If D is adapted to ρ and R, then $T_R \subset D$.

PROOF. If not, then for some $q \in W$ the set $\{x \in F_q : \rho(x) < R\}$ would be disconnected, contrary to the previous lemma. $\square$

Theorem 3.3.9 is now a consequence of the following properness result.

LEMMA 3.3.13. If for every $q \in Q$ the restriction $\rho|Gx$ of ρ to some orbit in F_q has a critical point, then ρ is a relative exhaustion.

PROOF. The existence of a divergent sequence $x_n \in X$ so that $\pi(x_n) \to q_0$ with $\rho(x_n) < R$ would be in violation to the previous lemma. $\square$

Using the relative exhaustion property it is possible to prove the local embedding theorem for Stein spaces.

THEOREM 3.3.14. Let $G \times X \to X$ be a holomorphic action of a reductive group on a Stein space. Then every point $x \in X$ possesses a G-invariant Stein neighborhood U which is saturated with respect to the formal invariant theoretic quotient and which can be holomorphically and G-equivariantly embedded as a closed complex subspace of a G-representation space.

PROOF. Let $\pi : X \to Q$ be the formal invariant theory quotient and $F := \pi^{-1}(\pi(x))$ its fiber through x. By Corollary 3.3.7 there exits a equivariant, holomorphic embedding $\varphi : F \to V$ onto to a closed (algebraic) subvariety of a G-representation space.

Extend φ to a holomorphic map $\varphi : X \to V$ by the same name which is locally biholomorphic along F. Taking the average

$$A(\varphi)(x) := \int_{k \in K} k\varphi(k^{-1}(x)) \, dk,$$

we obtain a G-equivariant extension with the same properties. Let φ denote this extension.

Now let η be a K-invariant norm function on V and $\rho := \eta \circ \varphi$. It follows from Theorem 3.3.9 that, after shrinking Q to a perhaps smaller neighborhood W of $\pi(x)$, the function ρ is a relative exhaustion.

Furthermore, by adding additional G-finite functions and further shrinking if necessary, we may assume that φ is biholomorphic on the (compact) Kempf–Ness set X_0. Since it is therefore a diffeomorphism on the K-orbits $Kx_0 = K/L$ in X_0, and since $H = L^{\mathbb{C}}$, it is injective on the closed G-orbits $Y = Gx_0$ for $x_0 \in X_0$.

Consequently, the restriction $\varphi|Y$ to every closed G-orbit is a closed embedding (Theorem 3.3.5).

Thus, since we may assume that φ is biholomorphic in a neighborhood of the Kempf–Ness set X_0, we have organized a situation where φ is an injective immersion whose restriction to every π-fiber is a closed embedding. Furthermore, since the Kempf–Ness set X_0 is embedded by φ, if W is a sufficiently small neighborhood of $\varphi(F)$ which is $\pi : V \to V//G$ saturated and $U := \pi^{-1}(W)$, then $\varphi : U \to W$ is a closed embedding.

Now we may choose W is the preimage of a Stein open set W_0 in the Hilbert quotient $V//G$ which we embed as a closed complex subspace of the trivial representation V_0 by a holomorphic map F_0. Letting $\varphi_0 := F_0 \circ \pi \circ \varphi$, it follows that $\varphi \times \varphi_0 : U \to V \times V_0$ is a closed holomorphic embedding. $\qquad \square$

COROLLARY 3.3.15. *If $G \times X \to X$ is a holomorphic action of a reductive group on a Stein space, then every $x \in X$ possesses a G-invariant, neighborhood U which is saturated with respect to the formal invariant theoretic quotient $\pi : X \to Q$ and which possesses a Stein holomorphic invariant theoretic quotient $U \to U//G$.*

PROOF. This is now an immediate consequence of 3.1.4. $\qquad \square$

4. Kähler spaces

4.1. The Slice Theorem.
Let X be a holomorphic $G = K^{\mathbb{C}}$ space and assume that K acts on X in a Hamiltonian fashion. The corresponding equivariant

moment map $\mu : X \to \mathfrak{k}^*$ defines the set $X(\mu)$ of semistable points. We show in this section that at every point $x_0 \in \mu^{-1}(0)$ there is a slice S, i.e., a local Stein G_{x_0}-stable subvariety of X through x_0 such that the natural map $G \times_{G_{x_0}} S \to X$ is biholomorphic onto its open image.

We already know that the isotropy group $G_{x_0} = (K_{z_0})^{\mathbb{C}}$ is reductive at every point $x_0 \in \mu^{-1}(0)$ and therefore there is a Slice $Z := G \times_{G_{x_0}} S$ at x_0 up to local biholomorphy. After pulling back the Kähler form ω on X to Z we obtain a Kähler form $\tilde{\omega}$ on Z with a moment map $\tilde{\mu}$ which is given by pulling back μ to Z. Let $z_0 = [e, 0]$ be the point in Z which corresponds to x_0. Since S can be chosen such that Gz_0 is a strong deformation retract of Z and Kz_0 is a strong deformation retract of Gz_0, it follows that the cohomology class of $\tilde{\omega}$ is determined by the pull back of $\tilde{\omega}$ to the orbit Kz_0. The moment map condition $d\tilde{\mu}_\xi = \imath_{\xi_X} \tilde{\omega}$ implies that $\tilde{\omega}$ is exact. Moreover, since Z can be chosen to be a Stein space, it follows that $\tilde{\omega} = 2i\partial\bar{\partial}\tilde{\rho}$ for some smooth function $\tilde{\rho} : Z \to \mathbb{R}$. Further, after averaging $\tilde{\rho}$ over the compact group K we may assume $\tilde{\rho}$ to be K-invariant. In this setting,there is the moment map $\mu^{\tilde{\rho}}$ which differs only by a K-invariant constant $a \in \mathfrak{k}^*$ from the original moment map $\tilde{\mu}$. Of course, if the group K is semisimple, then a is zero and we have $\tilde{\mu} = \mu^{\tilde{\rho}}$ (see [Heinzner et al. 1994] for more details). In general, after adding to $\tilde{\rho}$ a pluriharmonic K-invariant function h that we can always arrange that $\tilde{\mu} = \mu^{\tilde{\rho}}$; see [Heinzner et al. 1994]. Now by the previous results in the Stein case we are in the following setup.

- There is a locally biholomorphic G-equivariant map $\phi : Z \to X$ such that $\phi(z_0) = x_0$ and whose restriction to Gz_0 is biholomorphic onto its image Gx_0.
- $\tilde{\mu} = \mu \circ \phi = \mu^{\tilde{\rho}}$ for some K-invariant strictly plurisubharmonic positive function $\tilde{\rho} : Z \to \mathbb{R}$.
- The analytic Hilbert quotient $\pi_Z : Z \to Z/\!/G$ exists and $\pi \times \tilde{\rho} : Z \to Z/\!/G \times \mathbb{R}$ is proper.

In order to show that ϕ is biholomorphic we introduce the following terminology.

Let A be a K-invariant subset in X. Consider for any $a \in A$ and $\xi \in \mathfrak{k}$ the set $I(a, \xi) = \{t \in \mathbb{R} : \exp(it\xi)a \in A\}$. We call A μ-adapted if for the closure of any bounded connected component of $I(a, \xi)$, say $[t_-, t_+]$, we have $\mu_\xi(\exp it_-\xi a) < 0$ and $\mu_\xi(\exp it_+\xi a) > 0$. If the closure of the connected component is $[t_-, \infty)$, we require just $\mu_\xi(\exp it_-\xi a) < 0$ and if it is $(-\infty, t_+]$, then we require $\mu_\xi(\exp it_+\xi a) > 0$.

Note that, since $t \to \mu_\xi(\exp it\xi a)$ is increasing, for a μ-adapted subset A of X the set $\{t \in \mathbb{R} : \exp it\xi a \in A\}$ is connected for every $a \in A$. This proves the following

LEMMA 4.1.1. *A μ-adapted subset A of X is orbit convex in X.* $\qquad\square$

The next Lemma is the crucial step in the construction of a slice at x_0.

LEMMA 4.1.2. *After replacing Z with a G-stable open neighborhood of z_0 the map ϕ is biholomorphic onto its open image.*

PROOF. Since ϕ is locally biholomorphic and injective on Kz_0 there is an K-stable neighborhood of $\tilde{U}$ of z_0 which is mapped by ϕ biholomorphically onto its image U. If we can show that U can be chosen to be orbit convex, then the map $(\phi|\tilde{U})^{-1} : U \to Z$ extends to GU and gives an inverse to $\phi|G\tilde{U}$. Thus it is sufficient to show that Kx_0 has arbitrary small K-stable μ-adapted open neighborhoods.

Now we may choose $\tilde{U}$ such that $\tilde{\omega} = 2i\partial\bar{\partial}\tilde{\rho}$ and $\tilde{\mu} = \mu^{\tilde{\rho}}$ on $G\tilde{U}$. In particular we have $\omega = 2i\partial\bar{\partial}\rho$ and $\mu = \mu^\rho$ on U where $\rho \circ \phi = \tilde{\rho}|\tilde{U}$. Moreover, since $\pi \times \tilde{\rho}$ is a relative exhaustion, we may choose an open neighborhood $Q \subset Z//G$ and a $r \in \mathbb{R}$ so that $\bar{V} := \pi_Z^{-1}(Q) \cap \bar{D}_{\tilde{\rho}}(r)$ is contained in $\tilde{U}$. Here $\bar{D}_{\tilde{\rho}}(r) := \{z \in Z : \tilde{\rho}(z) \le r\}$. Let $V := D_{\tilde{\rho}}(r) := \{z \in Z : \tilde{\rho}(z) < r\}$. It is sufficient to show this:

CLAIM. $\phi(V)$ is μ-adapted.

This is a consequence of the definition of the moment map, as follows:

Consider the smallest $t_+ > 0$ such that $\exp it\xi a \in \phi(V)$ for $t \in [0, t_+)$. If $t_+ \ne +\infty$, then by construction of V we have $\rho(\exp it_+\xi a) = r$. Since

$$\mu_\xi(\exp it_+\xi a) = d\rho(J\xi_X(\exp it\xi a)),$$

this implies that $\mu_\xi(\exp it_+\xi a) > 0$. The function $t \to \mu_\xi(\exp it\xi a)$ is strictly increasing. Thus it follows by using again the equality just displayed that $\{t \in \mathbb{R} : t \ge 0$ and $\exp it\xi a \in \phi(V)\}$ is connected. By a similar argument applied to the smallest t_- such that for all negative t bigger then t_- the curve $\exp it\xi a$ is contained in $\phi(V)$ it follows that $\{t \in \mathbb{R} : \exp it\xi a \in \phi(V)\}$ is connected. $\qquad\square$

COROLLARY 4.1.3. *Every point* $x_0 \in X_0 = \mu^{-1}(0)$ *has a neighborhood basis of open μ-adapted neighborhoods.* $\qquad\square$

Now let V be a K-invariant open neighborhood of z_0 such that $\phi_V := \phi|V$ is biholomorphic and maps V biholomorphically onto an μ-adapted open subset $U := \phi(V)$. It follows that the inverse ϕ_V^{-1} extends to a G-equivariant holomorphic map $\psi : U^c \to Z$ where $U^c := GU$. Of course ψ is the inverse of $\phi|GV$. This shows that x_0 has an open G-stable Stein neighborhood U^c which is biholomorphically isomorphic to $G \times_{G_{x_0}} S$, where S is a slice at x_0. Hence we have the following

THEOREM 4.1.4. *At every point* $x_0 \in X$ *such that* $\mu(x_0)$ *is a K-fixed point there exists a slice.*

PROOF. This follows from the preceding discussion, since by adding a constant to μ if necessary, we may assume that $\mu(x) = 0$. $\qquad\square$

4.2. The Quotient Theorem. Given a moment map $\mu : X \to \mathfrak{k}^*$, then, under the assumption that $G = K^{\mathbb{C}}$ acts holomorphically on X, there is an associated set $X(\mu)$ of semistable points. The goal here is to show that the analytic Hilbert quotient $X(\mu)$ exists. We first show that the relation $x \sim y$ if and only if the intersection of the closures in $X(\mu)$ of the corresponding G orbits is non trivial is in fact an equivalence relation.

LEMMA 4.2.1. *Let $V_1, V_2 \subset X$ be μ-adapted subsets of X. Then*

$$G(V_1 \cap V_2) = GV_1 \cap GV_2.$$

PROOF. We have to show that $GV_1 \cap GV_2 \subset G(V_1 \cap V_2)$. Thus let $z = g_1 v_1 = g_2 v_2$ with $g_j \in G$ and $v_j \in V_j$ be given. There exist $k \in K$ and $\xi \in \mathfrak{k}$ so that $g_2^{-1} g_1 = k \exp i\xi$. Consider the path $\alpha \colon [0,1] \to X$, $t \to \exp it\xi v_1$. It is sufficient to show that $\alpha(t_0) \in V_1 \cap V_2$ for some $t_0 \in [0,1]$. Since this is obvious if $v_1 \in V_2$ or $v_2 \in V_1$, we may assume that there exists t_1 and t_2 in $[0,1]$ where α leaves V_1 and enters V_2. But μ_ξ is increasing on α. Thus, from $\mu_\xi(\alpha(t_1)) > 0$ and $\mu_\xi(\alpha(t_2)) < 0$ we conclude that $t_1 > t_2$, i.e., $\alpha(t) \in V_1 \cap V_2$ for $t \in [t_2, t_1]$. □

Since every point $x_0 \in X_0 = \mu^{-1}(0)$ has a μ-adapted neighborhood and the union of μ adapted sets remains μ-adapted, the lemma implies the following

COROLLARY 4.2.2. *If $\overline{Gx} \cap X_0 \neq \varnothing$, then $\overline{Gx} \cap X_0 = Kx_0$ for some $x_0 \in X_0$.* □

Note that $x \sim y$ if and only if $\overline{Gx} \cap \overline{Gy} = Ky_0$ where one has to take the closure in $X(\mu)$.

COROLLARY 4.2.3. *The Hausdorff quotient $X(\mu)/\sim$ exists.* □

Let $X(\mu)//G$ be topologically defined as $X/\sim$ and let $\pi \colon X(\mu) \to X(\mu)//G$ be the quotient map.

THEOREM 4.2.4. *The quotient $X(\mu)//G$ is an analytic Hilbert quotient.*

PROOF. We already know that $X(\mu)//G$ is well defined as a topological space. In order to endow $X(\mu)//G$ with a complex structure we use the slice theorem.

If we fix a point $x_0 \in X_0$, then, by the slice theorem, we find a K-invariant μ-convex Stein neighborhood U of x_0 such that $U^c = GU$ is a complexification of the K-action on U. Moreover, by construction of U we have $(\mu|U)^{-1}(0) = \mu^{-1}(0) \cap U^c$. In particular, U^c is contained in $X(\mu)$ and is π-saturated, i.e., $\pi^{-1}(\pi(U^c)) = U^c$. By Corollary 3.3.15 the analytic Hilbert quotient $U^c//G$ exists and, since U^c is saturated, is naturally identified with an open subset of $X(\mu)//G$. □

Acknowledgement

We are extremely grateful for the opportunity to publish our text in this series and are indebted to the scientific committee for organizing a stimulating year devoted to several complex variables at MSRI. It is our sad duty to dedicate this work to the memory of one member of that committee, our colleague and friend, Michael Schneider.

References

[Akhiezer 1995] D. N. Akhiezer, *Lie group actions in complex analysis*, Aspects of Mathematics **E27**, Vieweg, Braunschweig, 1995.

[Azad and Loeb 1993] H. Azad and J.-J. Loeb, "Plurisubharmonic functions and the Kempf–Ness theorem", *Bull. London Math. Soc.* **25**:2 (1993), 162–168.

[Chevalley 1946] C. Chevalley, *Theory of Lie Groups, I*, Princeton Mathematical Series **8**, Princeton University Press, Princeton, NJ, 1946.

[Guillemin and Sternberg 1984] V. Guillemin and S. Sternberg, "Convexity properties of the moment mapping, II", *Invent. Math.* **77**:3 (1984), 533–546.

[Hausen and Heinzner 1999] J. Hausen and P. Heinzner, "Actions of compact groups on coherent sheaves", *Transform. Groups* **4**:1 (1999), 25–34.

[Heinzner 1988] P. Heinzner, "Linear äquivariante Einbettungen Steinscher Räume", *Math. Ann.* **280**:1 (1988), 147–160.

[Heinzner 1989] P. Heinzner, "Kompakte Transformationsgruppen Steinscher Räume", *Math. Ann.* **285**:1 (1989), 13–28.

[Heinzner 1991] P. Heinzner, "Geometric invariant theory on Stein spaces", *Math. Ann.* **289**:4 (1991), 631–662.

[Heinzner and Huckleberry 1996] P. Heinzner and A. Huckleberry, "Kählerian potentials and convexity properties of the moment map", *Invent. Math.* **126**:1 (1996), 65–84.

[Heinzner and Iannuzzi 1997] P. Heinzner and A. Iannuzzi, "Integration of local actions on holomorphic fiber spaces", *Nagoya Math. J.* **146** (1997), 31–53.

[Heinzner and Loose 1994] P. Heinzner and F. Loose, "Reduction of complex Hamiltonian *G*-spaces", *Geom. Funct. Anal.* **4**:3 (1994), 288–297.

[Heinzner et al. 1994] P. Heinzner, A. T. Huckleberry, and F. Loose, "Kählerian extensions of the symplectic reduction", *J. Reine Angew. Math.* **455** (1994), 123–140.

[Heinzner et al. 1998] P. Heinzner, L. Migliorini, and M. Polito, "Semistable quotients", *Ann. Scuola Norm. Sup. Pisa Cl. Sci.* (4) **26**:2 (1998), 233–248.

[Helgason 1978] S. Helgason, *Differential geometry, Lie groups, and symmetric spaces*, Pure and Applied Mathematics **80**, Academic Press, New York, 1978.

[Hochschild 1965] G. Hochschild, *The structure of Lie groups*, Holden-Day Inc., San Francisco, 1965.

[Kaup 1967] W. Kaup, "Reelle Transformationsgruppen und invariante Metriken auf komplexen Räumen", *Invent. Math.* **3** (1967), 43–70.

[Kirwan 1984] F. C. Kirwan, *Cohomology of quotients in symplectic and algebraic geometry*, Math. Notes **31**, Princeton University Press, Princeton, NJ, 1984.

[Kraft 1984] H. Kraft, *Geometrische Methoden in der Invariantentheorie*, Aspects of Mathematics **D1**, Vieweg, Braunschweig, 1984.

[Marsden and Weinstein 1974] J. Marsden and A. Weinstein, "Reduction of symplectic manifolds with symmetry", *Rep. Mathematical Phys.* **5**:1 (1974), 121–130.

[Mumford et al. 1994] D. Mumford, J. Fogarty, and F. Kirwan, *Geometric invariant theory*, third ed., Ergebnisse der Mathematik und ihrer Grenzgebiete (2) **34**, Springer, Berlin, 1994.

[Neeman 1985] A. Neeman, "The topology of quotient varieties", *Ann. of Math.* (2) **122**:3 (1985), 419–459.

[Sjamaar 1995] R. Sjamaar, "Holomorphic slices, symplectic reduction and multiplicities of representations", *Ann. of Math.* (2) **141**:1 (1995), 87–129.

[Sjamaar and Lerman 1991] R. Sjamaar and E. Lerman, "Stratified symplectic spaces and reduction", *Ann. of Math.* (2) **134**:2 (1991), 375–422.

[Snow 1982] D. M. Snow, "Reductive group actions on Stein spaces", *Math. Ann.* **259**:1 (1982), 79–97.

[Wallach 1973] N. R. Wallach, *Harmonic analysis on homogeneous spaces*, Pure and Applied Mathematics **19**, Marcel Dekker, New York, 1973.

PETER HEINZNER
FAKULTÄT FÜR MATHEMATIK
RUHR-UNIVERSITÄT BOCHUM
UNIVERSITÄTSSTRASSE 150
D - 44780 BOCHUM
GERMANY
heinzner@cplx.ruhr-uni-bochum.de

ALAN HUCKLEBERRY
FAKULTÄT FÜR MATHEMATIK
RUHR-UNIVERSITÄT BOCHUM
UNIVERSITÄTSSTRASSE 150
D-44780 BOCHUM
GERMANY
ahuck@cplx.ruhr-uni-bochum.de

Several Complex Variables
MSRI Publications
Volume **37**, 1999

Varieties of Minimal Rational Tangents on Uniruled Projective Manifolds

JUN-MUK HWANG AND NGAIMING MOK

ABSTRACT. On a polarized uniruled projective manifold we pick an irreducible component $\mathcal{K}$ of the Chow space whose generic members are free rational curves of minimal degree. The normalized Chow space of minimal rational curves marked at a generic point is nonsingular, and its strict transform under the tangent map gives a variety of minimal rational tangents, or VMRT. In this survey we present a systematic study of VMRT by means of techniques from differential geometry (distributions, G-structures), projective geometry (the Gauss map, tangency theorems), the deformation theory of (rational) curves, and complex analysis (Hartogs phenomenon, analytic continuation). We give applications to a variety of problems on uniruled projective manifolds, especially on irreducible Hermitian symmetric manifolds S of the compact type and more generally on rational homogeneous manifolds G/P of Picard number 1, including the deformation rigidity of S and the same for homogeneous contact manifolds of Picard number 1, the characterization of S of rank at least 2 among projective uniruled manifolds in terms of G-structures, solution of Lazarsfeld's Problem for finite holomorphic maps from G/P of Picard number 1 onto projective manifolds, local rigidity of finite holomorphic maps from a fixed projective manifold onto G/P of Picard number 1 other than $\mathbb{P}^n$, and a proof of the stability of tangent bundles of certain Fano manifolds.

CONTENTS

Hwang was supported by BSRI and by the KOSEF through the GARC at Seoul National University. Mok was supported by a grant of the Research Grants Council, Hong Kong.

Rational curves play a crucial role in the study of Fano manifolds. By Mori's theory, Fano manifolds are uniruled. We consider more generally uniruled projective manifolds. Fixing an ample line bundle and considering only components of the Chow space whose generic members are free rational curves, we introduce the notion of minimal rational curves by minimizing the degree of a generic member. The normalized Chow space of minimal rational curves marked at a generic point is nonsingular, and its strict transform under the tangent map gives the variety of minimal rational tangents. In [Hwang and Mok 1997; 1998a; 1998b; 1999; Hwang 1997; 1998] we have put forth the idea of recapturing complex-analytic properties of Fano manifolds from varieties of minimal rational tangents and holomorphic distributions spanned at generic points by them. In this survey we present a systematic treatment of the fundamental notions, and examine a number of applications primarily in the context of rational homogeneous manifolds of Picard number 1.

The scope of problems we consider covers deformation rigidity, algebro-geometric characterizations (of Grassmannians, etc.), stability of the tangent bundle, and holomorphic mappings. We also give complex-analytic and geometric proofs of results from the theory of geometric structures of Tanaka, as given by Ochiai [1970] as well as Tanaka and Yamaguchi [Yamaguchi 1993], which we needed for various problems, making our presentation essentially self-contained. As to the techniques we employ, an important role is played by holomorphic distributions and the Frobenius condition. Distributions spanned by minimal rational tangents are first of all studied using the deformation theory of rational curves. Then, projective geometry enters the picture in various ways, in the problem of integrability of such distributions, in vanishing theorems related to flatness of G-structures and in the study of stability of tangent bundles. Complex-analytic techniques enter, in the form of analytic continuation and Hartogs extension, in conjunction with the use of the Gauss map, when we study varieties of minimal rational tangents $\mathcal{C}_x$ as x varies. Further study of deformations of curves, in the context of finite holomorphic maps to Fano manifolds, leads to the notion of varieties of distinguished tangents. For the study of rational homogeneous manifolds, we will need basics for graded Lie algebras associated to simple Lie groups, a summary of which will be given.

For the general theory, varieties of minimal rational tangents are first studied as projective subvarieties. The first motivation for studying their projective-geometric properties stemmed from [Hwang and Mok 1998b], where we proved the rigidity of irreducible Hermitian symmetric manifolds of the compact type S under Kähler deformation. There we reduced the problem to the study of the distribution W spanned by varieties of minimal rational tangents $\mathcal{C}_x$ at the central fiber X, and derived a sufficient condition for the integrability in terms of the projective geometry of $\mathcal{C}_x$, namely, W is integrable whenever the variety $\mathcal{T}_x$ of lines tangent to $\mathcal{C}_x$ is linearly nondegenerate in $\mathbb{P}(\bigwedge^2 W_x)$ for a generic point x.

In Section 1 we study $\mathcal{C}_x$ as projective subvarieties, and give first applications of such results and their methods of proof. We prove the algebro-geometric characterization of irreducible Hermitian symmetric manifolds of the compact type and of rank at least 2 as the only uniruled projective manifolds admitting G-structures for some reductive G. After identifying varieties of highest weight tangents $\mathcal{W}_x$ with varieties $\mathcal{C}_x$ of minimal rational tangents $\mathcal{C}_x$, the proof is obtained by vanishing theorems for the obstruction to flatness of G-structures, which reduce to projective-geometric properties of $\mathcal{C}_x$. We further discuss the question of stability of the tangent bundle of Fano manifolds by applying variants of Zak's theorem on tangencies on $\mathcal{C}_x$. In Section 2 we return to deformation rigidity. As we may restrict to the case where S is of rank at least 2, the problem reduces to recovering an S-structure at the central fiber X. The latter is possible, whenever $\mathcal{C}_x$ is linearly nondegenerate at a generic point of X. Otherwise we have a proper distribution $W \subsetneq T(X)$ spanned at generic points by $\mathcal{C}_x$. We prove the nonexistence of W by studying its integrability in terms of $\mathcal{C}_x$, as mentioned. We give further a generalization [Hwang 1997] of deformation rigidity to the case of homogeneous contact manifolds, where there is the new element of deformations of contact distributions.

In Section 3 we study varieties of minimal rational tangents $\mathcal{C}_x$ as the base points vary, by considering the tautological 1-dimensional multi-foliation $\mathcal{F}$ defined at generic points of $\mathcal{C}$ by the tautological lifting of minimal rational curves. Assuming that the Gauss map on $\mathcal{C}_x$ to be generically finite for generic x, we prove the univalence of the multi-foliations, resulting in the birationality of the tangent map. This uniqueness result implies that a local biholomorphism f preserving the varieties $\mathcal{C}_x$ must also be $\mathcal{F}$-preserving. The latter constitutes the first step towards a complex-analytic and geometric proof of Ochiai's characterization of S (as above) in terms of flat S-structures, which says that f is the restriction of a biholomorphic automorphism F of S. To prove Ochiai's result, we introduce the method of analytic continuation of $\mathcal{F}$-preserving meromorphic maps along minimal rational curves, and exploit the rational connectedness of S.

In Section 4 we move to rational homogeneous manifolds S of Picard number 1. For the nonsymmetric case a new element arises, namely, there exist nontrivial homogeneous holomorphic distributions. In analogy to Ochiai's result we have the results of Tanaka and Yamaguchi in terms of varieties of highest weight tangents $\mathcal{W}_x$. In the nonsymmetric and noncontact case their results go further, stating that a local biholomorphism f must extend to a biholomorphic automorphism, provided that f preserves the minimal homogeneous distribution D. We give a proof of the result of Tanaka and Yamaguchi, by showing that a D-preserving local biholomorphism already preserves $\mathcal{W}$ and by resorting to methods of Section 3.

In the last two sections we have primarily the study of finite holomorphic maps onto Fano manifolds in mind. In Section 5 we introduce the notion of varieties of distinguished tangents. They generalize varieties of minimal rational tangents, and are of particular relevance in the context of finite holomorphic maps into Fano manifolds, since preimages of varieties of minimal rational tangents give varieties of distinguished tangents. We give an application for rational homogeneous target manifolds S of Picard number 1 distinct from the projective space, proving that any finite holomorphic map into S is locally rigid. In Section 6 we consider the case where the domain manifold is S, and prove that any surjective holomorphic map of S onto a projective manifold X distinct from the projective space is necessarily a biholomorphism, resolving Lazarsfeld's problem.

While in the applications we concentrate on rational homogeneous manifolds, the general theory has been developed to be applicable in much wider contexts. Such applications, especially to the case of Fano complete intersections, will constitute one further step towards developing a theory of "variable geometric structures" modeled on varieties of minimal rational tangents.

1. Minimal Rational Curves, Varieties of Minimal Rational Tangents and Associated Distributions

1.1. For the study of Fano manifolds and more generally uniruled manifolds a basic tool is the deformation theory of rational curves. We will only sketch the basic ideas and refer the reader to [Kollár 1996] for a systematic and rigorous treatment of the general theory. Let X be a projective manifold. By a parametrized rational curve we mean a nonconstant holomorphic map $f : \mathbb{P}^1 \to X$. The image of f is called a rational curve. Given a holomorphic family $f_t : \mathbb{P}^1 \to X$ of rational curves, parametrized by $t \in \triangle := \{t \in \mathbb{C} : |t| < 1\}$, the derivative $\frac{d}{dt}\big|_0 f_t$ defines a holomorphic section of $f_0^* T(X)$. However, given a member f_0 of the space $\mathrm{Hol}(\mathbb{P}^1, X)$ of parametrized rational curves in X, and $\sigma \in \Gamma(\mathbb{P}^1, f_0^* T(X))$, it is not always possible to fit f_0 into a holomorphic family of $f_t \in \mathrm{Hol}(\mathbb{P}^1, X)$, such that $\frac{d}{dt}\big|_0 f_t = \sigma$. Setting in power series $f_t = f + \sigma t + g_2 t^2 + \cdots$ locally, the obstruction of lifting to higher coefficients lies in $H^1(\mathbb{P}^1, f_0^* T(X))$. In case the latter vanishes, $\mathrm{Hol}(\mathbb{P}^1, X)$ is smooth in a neighborhood U of $[f_0]$, and the tangent space at $[f] \in U$ can be identified with $\Gamma(\mathbb{P}^1, f^* T(X))$.

By the Grothendieck splitting theorem any holomorphic vector bundle on $\mathbb{P}^1$ splits into a direct sum $\mathcal{O}(a_1) \oplus \cdots \oplus \mathcal{O}(a_n)$. When $f^* T(X)$ is semipositive, that is, $a_i \geq 0$, then $f^* T(X)$ is spanned by global sections, $H^1(\mathbb{P}^1, f^* T(X)) = 0$, and deformations of f sweeps out some open neighborhood of $C = f(\mathbb{P}^1)$. We call f a free rational curve. A projective manifold is said to be uniruled if it possesses a free rational curve.

Each irreducible component of $\mathrm{Hol}(\mathbb{P}^1, X)$ can be endowed the structure of a quasi-projective variety. It covers some Zariski-open subset of X if and only if some member is a free rational curve. Consequently, there is an at most countable union Z of proper subvarieties of X such that any rational curve passing through $x \notin Z$ is necessarily free. A point x lying outside Z is called a very general point.

Fix an ample line bundle L on X and consider all irreducible components $\mathcal{H}$ of $\mathrm{Hol}(\mathbb{P}^1, X)$ whose generic member is a free rational curve. As degrees of members of a fixed $\mathcal{H}$ with respect to L are the same we may speak of the degree of the component $\mathcal{H}$. A member of a component $\mathcal{H}$ of minimal degree will be called a minimal rational curve. By Mori's break-up trick [1979] a generic member of $\mathcal{H}$ is an immersed rational curve $f : \mathbb{P}^1 \to X$ such that $f^*T(X) \cong \mathcal{O}(2) \oplus [\mathcal{O}(1)]^p \oplus \mathcal{O}^q$ (compare [Mok 1988; Hwang and Mok 1998b]); otherwise one can obtain an algebraic one-parameter family of curves in $\mathcal{H}$ fixing a pair of very general points, which must break up in the limit, contradicting minimality. A minimal rational curve with the splitting type as described is called a standard minimal rational curve.

We fix an irreducible component $\mathcal{H}$ of minimal rational curves. At a generic point $x \in X$ all such curves passing through x are free. Consider the subvariety $\mathcal{H}_x \subset \mathcal{H}$ of all $[f] \in \mathcal{H}$ such that $f(o) = x$, where $o \in \mathbb{P}^1$ is a base point. The isotropy subgroup of $\mathbb{P}^1$ at $f(o) \in x\, o$ acts on $\mathcal{H}_x$, making its normalization into a principal bundle over a nonsingular quasi-projective variety $\mathcal{M}_x$. We called $\mathcal{M}_x$ the normalized Chow space of minimal rational curves marked at x. By minimality $\mathcal{M}_x$ must be compact, that is, a projective manifold which may have several connected components. By Mori's break-up trick a generic member $[f]$ of $\mathcal{H}_x$ is unramified at o. We have therefore a rational map $\Phi_x : \mathcal{M}_x \dashrightarrow \mathbb{P}T_x(X)$ defined by

$$\Phi_x\big([f(\mathbb{P}^1)]\big) = \big[df(T_o(\mathbb{P}^1))\big]$$

at generic points of $\mathcal{M}_x$. We call Φ_x the tangent map at x.

Fix a base point $x \in X$ and consider now the space $\mathrm{Hol}((\mathbb{P}^1, o); (X, x))$ consisting of all parametrized rational curves f sending o to x. For a holomorphic family f_t, $t \in \triangle$, of such curves

$$\frac{d}{dt}\Big|_0 f_t \in \Gamma(\mathbb{P}^1, f_0^*T(X) \otimes \mathcal{O}(-1)),$$

where $\mathcal{O}(-1)$ corresponds to the maximal ideal sheaf $\mathfrak{m}_o$ of o on $\mathbb{P}^1$. Given σ in the latter space of sections, the obstruction to extending f_o to a holomorphic family f_t, $f_t(o) = x$, lies in $H^1(\mathbb{P}^1, f_o^*T(X) \otimes \mathcal{O}(-1))$, which vanishes whenever f_o is a free rational curve, since $H^1(\mathbb{P}^1, \mathcal{O}(a)) = 0$ whenever $a \geq -1$. In this case $\mathrm{Hol}((\mathbb{P}^1, o), (X, x))$ is smooth in a neighborhood U of $[f_o]$, and the tangent space at $[f] \in U$ can be identified with $\Gamma(\mathbb{P}^1, f^*T(X) \otimes \mathcal{O}(-1))$.

Since $[f] \in \mathcal{H}_x$ is standard for a generic $[f]$, Φ_x is generically finite. Let $\mathcal{C}_x \subset \mathbb{P}T_x(X)$ be the closure of the image of the tangent map. We call $\mathcal{C}_x$ the variety of minimal rational tangents at x. It may have several components.

For C smooth we can identify $T_{[C]}(\mathcal{M}_x)$ with $\Gamma(C, N_{C|X} \otimes \mathfrak{m}_x)$ for the normal bundle $N_{C|X}$ of C in X and for $\mathfrak{m}_x$ denoting the maximal ideal sheaf of x on C. For C standard and smooth, let $T_x(C) = \mathbb{C}\alpha \subset T_x(X)$. From the description of $T_{[C]}(\mathcal{M}_x)$ we see that the tangent space $T_{[\alpha]}(\mathcal{C}_x) = P_\alpha/\mathbb{C}\alpha$, where P_α is the positive part $(\mathcal{O}(2) \oplus [\mathcal{O}(1)]^p)_x$ of a Grothendieck decomposition of $T(X)$ over

C. With obvious modifications the preceding discussion applies to all standard minimal rational curves, which are necessarily immersed.

1.2. From now on we assume that for our choice of $\mathcal{H}$ and for a generic point $x \in X$, $\mathcal{C}_x$ is irreducible. For our study of uniruled projective manifolds via the varieties of minimal rational tangents, an important element is the distribution W spanned at generic points by the homogenization $\tilde{\mathcal{C}}_x \subset T_x(X)$ of $\mathcal{C}_x$. For the problem of deformation rigidity, a key question is the question of integrability of such distributions. By Frobenius, W is integrable if the Frobenius form

$$[\,,] : \bigwedge^2 W \to T(X)/W$$

vanishes, where $\varphi_x(u, v) = [\tilde{u}, \tilde{v}] \mod W_x$ for local holomorphic sections $\tilde{u}$, $\tilde{v}$ such that $\tilde{u}(x) = u$, $\tilde{v}(x) = v$.

A line tangent to $\mathcal{C}_x$ at a generic point $[\alpha] \in \mathcal{C}_x$ defines a point in $\mathbb{P} \bigwedge^2 W_x$. The closure of such points will be denoted by $\mathcal{T}_x$ and will be called the variety of tangent lines. The linear span E_x of the homogenization $\tilde{\mathcal{T}}_x \subset \bigwedge^2 W_x$ is then given by $E_x = \mathrm{Span}\{\alpha \wedge \xi : [\alpha] \in \mathcal{C}_x \text{ smooth point}, \xi \in P_\alpha\}$. One of the main results of [Hwang and Mok 1998b] is this:

PROPOSITION 1.2.1. *The distribution W is integrable if $\mathcal{T}_x$ is linearly nondegenerate in $\mathbb{P} \bigwedge^2 W_x$ for a generic point $x \in X$, that is, if $E_x = \bigwedge^2 W_x$ for x generic.*

PROOF. From the nondegeneracy condition, it suffices to check the vanishing of $[\alpha, \xi]$ for $\alpha \wedge \xi \in \bigwedge^2 W_x$, where $[\alpha] \in \mathcal{C}_x$ is a generic point and $\xi \mod \mathbb{C}\alpha$ is tangent to $\mathcal{C}_x$ at $[\alpha]$. By Frobenius' condition, $[\alpha, \xi] = 0$ if we can find a local surface in X passing through x tangent to the distribution W such that the tangent space of the surface at x is generated by α and ξ. By the definition of $\mathcal{C}_x$ and the description of its tangent spaces, we can find a standard minimal rational curve C which is tangent to α at x such that ξ lies in $(\mathcal{O}(2) \oplus [\mathcal{O}(1)]^p)_x = P_\alpha$ in the splitting of $T(X)$ over C. Then, we can choose a point $y \neq x$ on C and deform C with y fixed so that the derivative of this deformation is parallel to ξ at x. The locus Σ of this deformation will give an integral surface Σ of W at x so that α and ξ generate $T_x(\Sigma)$. It follows that we can find W-valued vector fields $\tilde{\alpha}, \tilde{\xi}$ in a neighborhood of x which are tangent to Σ, $\tilde{\alpha}(x) = \alpha$, $\tilde{\xi}(x) = \xi$. This implies the desired vanishing $[\alpha, \xi] = 0$ at x. $\square$

If W is integrable, it defines a foliation on X outside a proper subvariety $\mathrm{Sing}(W) \subset X$ of codimension at least 2. Any minimal rational curve which is not contained in $\mathrm{Sing}(W)$ is contained in a leaf of W. Pick a generic point x, then the leaf of W containing x can be compactified to a subvariety of X in the following way. Consider the subvariety covered by all minimal rational curves through x. Enlarge this subvariety by adjoining all minimal rational curves through generic points on it. Repeat this adjoining process. This process must stop after a finite number of steps and the resulting enlarged subvariety gives the

compactification of the leaf through x [Hwang and Mok 1998b, Proposition 11].
Using this fact, we have the following topological obstruction to the integrability
of W.

PROPOSITION 1.2.2. *Let X be a uniruled projective manifold such that $b_2(X) = 1$.
Suppose a choice of $\mathcal{H}$ can be made so that the generic variety of minimal rational
tangents $\mathcal{C}_x$ is linearly degenerate. Then, the distribution W spanned at generic
point by $\tilde{\mathcal{C}}_x$ cannot be integrable.*

PROOF. Assuming integrability, compactified leaves of W define a rational fibration of X over a projective variety X' of smaller dimension. The exceptional
locus of this fibration is contained in $\mathrm{Sing}(W)$ and of codimension at least 2.
But a generic minimal rational curve is disjoint from $\mathrm{Sing}(W)$ [Hwang and Mok
1998b, Proposition 12]. and is contained in a leaf of the fibration. Taking a
very ample divisor on X' and pulling it back to X, we find a hypersurface in X
disjoint from a generic minimal rational curve. This is a contradiction, since any
effective divisor on X is ample as X is of Picard number 1. $\qquad\square$

1.3. In this section we consider meromorphic distributions W spanned by varieties of minimal rational tangents, as in Section 1.2, and give sufficient conditions
for the integrability of W, in terms of properties of the generic variety of minimal
rational tangents $\mathcal{C}_x$.

PROPOSITION 1.3.1. *Suppose the generic variety of minimal rational tangents
$\mathcal{C}_x \subset \mathbb{P}W \subset \mathbb{P}T_x$ is irreducible and the second fundamental form $\sigma_{[\alpha]} : T_{[\alpha]}(\mathcal{C}_x) \times
T_{[\alpha]}(\mathcal{C}_x) \longrightarrow N_{\mathcal{C}_x|\mathbb{P}W_x,[\alpha]}$ in the sense of projective geometry is surjective at a
generic smooth point $[\alpha]$ of $\mathcal{C}_x$. Then, W is integrable.*

PROOF. Let $\alpha \in \tilde{\mathcal{C}}_x$ be a generic point and let $\{\alpha(t) : t \in \mathbb{C}, |t| < 1\}$ be a local
holomorphic curve. We write

$$\alpha(t) = \alpha + t\xi + t^2\zeta + O(t^3).$$

Denote by σ_α the second fundamental form $\sigma_\alpha : T_\alpha(\tilde{\mathcal{C}}_x) \times T_\alpha(\tilde{\mathcal{C}}_x) \longrightarrow N_{\tilde{\mathcal{C}}_x|W_x,\alpha}$
in the sense of Euclidean geometry. Then $\sigma_{[\alpha]}$ is surjective if and only if σ_α is
surjective. We have $\xi \in P_\alpha$ and $\sigma_\alpha(\xi,\xi) = \zeta \mod P_\alpha$. From now on we will
fix a choice of Euclidean metric on W_x and identify the normal space $N_{\tilde{\mathcal{C}}_x|W_x,\alpha}$
with the orthogonal complement $P_\alpha^\perp$ of P_α in W_x. With this convention we may
now choose the expansion for $\alpha(t)$ such that $\zeta \in P_\alpha^\perp$, so that $\sigma_\alpha(\xi,\xi) = \zeta$. We
fix α and write σ for σ_α. Now

$$\alpha(t) \wedge \alpha'(t) = \left(\alpha + t\xi + t^2\zeta + O(t^3)\right) \wedge \left(\xi + 2t\zeta + O(t^2)\right) = \alpha \wedge \xi + 2t\alpha \wedge \zeta + O(t^2).$$

It follows that $\alpha \wedge \zeta = \alpha \wedge \sigma(\xi,\xi)$ lies on $\mathrm{Span}\{\beta \wedge P_\beta : \beta \in \tilde{\mathcal{C}}_x\} = E_x \subset \bigwedge^2 W_x$.
Since σ is symmetric, by polarization we have $\alpha \wedge \sigma(\xi,\eta) \in E$ for any $\xi, \eta \in P_\alpha$.
The hypothesis of Proposition 1.3.1 then implies that $\alpha \wedge W_x \in E$ for any
$\alpha \in \tilde{\mathcal{C}}_x$. Varying α we conclude that $E_x = \bigwedge^2 W_x$. Since x is a generic point, W
is integrable, by Proposition 1.2.1. $\qquad\square$

PROPOSITION 1.3.2. *Suppose the generic cone* $\mathcal{C}_x \subset \mathbb{P}W_x \subset \mathbb{P}T_x$ *is irreducible and smooth and* $\dim(\mathcal{C}_x) > \frac{1}{2}\operatorname{rank}(W) - 1$. *Then* W *is integrable.*

For the proof of Proposition 1.3.2, we will need Zak's theorem on tangencies in Projective Geometry, for the case of projective submanifolds.

THEOREM 1.3.3 (SPECIAL CASE OF ZAK'S THEOREM ON TANGENCIES [ZAK 1993]). *Let* $Z \subset \mathbb{P}^N$ *be a* k-*dimensional complex submanifold and* $\mathbb{P}E \subset \mathbb{P}^N$ *be a* p-*dimensional projective subspace,* $p \geq k$. *Then, the set of points on* Z *at which* $\mathbb{P}E$ *is tangent to* Z *is at most of complex dimension* $p - k$.

PROOF OF PROPOSITION 1.3.2. In the proof of Proposition 1.3.1 we use the expansion of a 1-parameter family of minimal rational tangents $\alpha(t)$. In the notations there consider now a 2-dimensional local complex submanifold of minimal rational tangent vectors $\{\alpha(t, s) : t, s \in \mathbb{C}, |t|, |s| < 1\}$. Write $r^2 = |t|^2 + |s|^2$. We have

$$\alpha(t, s) = \alpha + t\xi + s\eta + t^2\sigma(\xi, \xi) + 2ts\sigma(\xi, \eta) + s^2\sigma(\eta, \eta) + O(r^3).$$

Taking partial derivatives we have

$$\alpha(t, s) \wedge \partial_t \alpha(t, s) = \big(\alpha + t\xi + s\eta + O(r^2)\big) \wedge \big(\xi + 2t\sigma(\xi, \xi) + 2s\sigma(\xi, \eta) + O(r^2)\big)$$
$$= \alpha \wedge \xi + s\big(\eta \wedge \xi + 2\alpha \wedge \sigma(\xi, \eta)\big) + 2t\big(\alpha \wedge \sigma(\xi, \xi)\big) + O(r^2)$$

and

$$\partial_s\big(\alpha(t, s) \wedge \partial_t \alpha(t, s)\big)(o) = \eta \wedge \xi + 2\alpha \wedge \sigma(\xi, \eta) \in E_x.$$

From the proof of Proposition 1.3.1 we know that $\alpha \wedge \sigma(\xi, \eta) \in E_x$, from which we conclude that $\xi \wedge \eta \in E_x$, that is, $\bigwedge^2 P_\alpha \subset E_x$ for each $\alpha \in \tilde{\mathcal{C}}_x$. Suppose now $E_x \subsetneq \bigwedge^2 W_x$. Then, there exists $\mu \in \bigwedge^2 W_x^*$ such that $\mu(e) = 0$ for any $e \in E_x$. It follows that for each α, $P_\alpha \subset W_x$ is an isotropic subspace with respect to μ.

If $\dim(\mathcal{C}_x) > \frac{1}{2}\operatorname{rank}(W) - 1$, then $\dim(\tilde{\mathcal{C}}_x) > \frac{1}{2}\operatorname{rank}(W)$, and any such μ must be degenerate. We claim that this leads to a contradiction. Let Q be the kernel of μ, so that $\mu(\lambda, \xi) = 0$ for any $\lambda \in Q, \xi \in W_x$. Let $\pi : \mathbb{P}W \dashrightarrow \mathbb{P}(W/Q)$ be the linear projection. For x generic $\mathcal{C}_x$ is linearly nondegenerate in W and $\pi\big|_{\mathcal{C}_x}$ is well-defined as a rational map. Consider a point $[\alpha] \in \mathcal{C}_x, \alpha \notin Q$, where $\pi\big|_{\mathcal{C}_x}$ is of maximal rank and $A = \pi([\alpha])$ is a smooth point of the strict transform $\pi(\mathcal{C}_x)$. Write $T_o \subset \mathbb{P}(W/Q)$ for the projective tangent subspace at A, $T_o = \mathbb{P}S_o$. Let $S \subset W_x$ be the linear subspace such that $S \supset Q$ and $S/Q = S_o$. Then, T is tangent to $\mathcal{C}_x$ along the fiber F of $\pi^{-1}(A)$. μ induces a (nondegenerate) symplectic form $\bar{\mu}$ on W_x/Q, with respect to which S_o is isotropic. Hence

$$\dim S_o \leq \frac{1}{2}\dim(W_x/Q),$$

so that

$$\dim F = \dim \mathcal{C}_x - \dim \mathbb{P}S_o > \frac{1}{2}\dim W_x - \frac{1}{2}\dim(W_x/Q) = \frac{1}{2}\dim Q.$$

On the other hand, for $T = \mathbb{P}S$, we have

$$\dim T = \dim T_o + \dim Q = (\dim \mathcal{C}_x - \dim F) + \dim Q.$$

Since the projective space T is tangent to $\mathcal{C}_x$ along F, by Zak's Theorem on Tangencies we have

$$\dim F \le \dim T - \dim \mathcal{C}_x = \dim Q - \dim F,$$

that is, $\dim F \le \frac{1}{2} \dim Q$, a contradiction.   □

1.4. To illustrate the results of Section 1.3, we look at homogeneous contact manifolds of Picard number 1. Recall that a contact structure on a complex manifold X of odd dimension $n = 2m + 1$, is a holomorphic subbundle $D \subset T(X)$ of rank $2m$ such that the Frobenius bracket tensor $\omega : \bigwedge^2 D \to L := T(X)/D$ defines a symplectic form on D_x for each x. A homogeneous contact manifold is a rational homogeneous manifold with a contact structure. According to Boothby's classification [1961], any homogeneous contact manifold is associated to a complex simple Lie algebra in the following way. For a simple Lie algebra $\mathfrak{g}$, the highest weight orbit in $\mathfrak{g}$ under the adjoint representation has a symplectic structure induced by the Lie bracket of $\mathfrak{g}$, so-called Kostant–Kirillov symplectic structure. This induces a contact structure on the projectivization $X \subset \mathbb{P}\mathfrak{g}$ of the highest weight orbit, making it into a homogeneous contact manifold. When $\mathfrak{g}$ is of type A, X is the projectivized cotangent bundle of a projective space and has Picard number 2. When $\mathfrak{g}$ is of type C, X is an odd dimensional projective space regarded as a homogeneous space of the symplectic group. These two cases are not interesting in our study.

We look at a homogeneous contact manifold X associated to an orthogonal or an exceptional simple Lie algebra. In this case, X has Picard number 1 and the line bundle $L = T(X)/D$ is an ample generator of $\text{Pic}(X)$. In fact, L is the $\mathcal{O}(1)$-bundle of the embedding $X \subset \mathbb{P}\mathfrak{g}$. There are lines of $\mathbb{P}\mathfrak{g}$ lying on X and they are minimal rational curves on X. When we represent X as G/P, where G is the adjoint group of $\mathfrak{g}$ and P is the isotropy subgroup at one point $x \in X$, the set of all lines on X is homogeneous under G and the set of all lines through x is homogeneous under P. Since a minimal rational curve is actually a line under the embedding $X \subset \mathbb{P}g$, we see that $\mathcal{C}_x$ is smooth and homogeneous under the isotropy group P.

Let $\theta : T(X) \to L = T(X)/D$ be the quotient map and $\omega : \bigwedge^2 D \to L$ be the Frobenius bracket. Then $\theta \wedge \omega^m, n = 2m+1$, defines a nowhere vanishing section of $K_X \otimes L^{m+1}$. This shows that for a line $C \subset X$, $T(X)|_C = \mathcal{O}(2) \oplus [\mathcal{O}(1)]^{m-1} \oplus \mathcal{O}^{m+1}$. The symplectic form θ on D induces an isomorphism $D \cong D^* \otimes L$, which gives $D|_C = \mathcal{O}(2) \oplus [\mathcal{O}(1)]^{m-1} \oplus \mathcal{O}^{m-1} \oplus \mathcal{O}(-1)$ using $T(X)/D|_C = L|_C = \mathcal{O}(1)$. This shows that the $\mathcal{O}(2)$-component of $T(X)|_C$ is contained in D. Thus $\mathcal{C}_x \subset \mathbb{P}T_x(X)$ is linearly degenerate and is contained in $\mathbb{P}D_x$. In fact, the isotropy representation of P on $T_x(X)$ is irreducible on D_x and $\mathcal{C}_x$ must be the projectivization of the orbit of a highest weight vector because $\mathcal{C}_x$ is compact and homogeneous under P.

Since X has Picard number 1 and $\mathcal{C}_x$ is degenerate $W_x = D_x$, we have $E_x \neq \bigwedge^2 D_x$ from Section 1.2. The dimension of $\mathcal{C}_x$ is $m-1$, which is smaller than half the rank of D, as expected from Section 1.3. In fact, here we have the symplectic form ω and we can see that $\tilde{\mathcal{C}}_x$ is Lagrangian with respect to ω, as follows. Given two tangent vectors u, v to $\tilde{\mathcal{C}}_x$ at a point on $T(C)$ for a line C through x, we can extend them to sections $\tilde{u}, \tilde{v}$ of $(\mathcal{O}(2) \oplus [\mathcal{O}(1)]^{m-1})$-part of $T(X)|_C$ vanishing at some points of C by Section 1.1. Then $\tilde{u}, \tilde{v}$ are sections of $D|_C$ and $\omega(\tilde{u} \wedge \tilde{v})$ is a section of L having two zeros. From $L|_C = \mathcal{O}(1)$, we see $\omega(\tilde{u} \wedge \tilde{v}) = 0$. This explains $E_x \neq \bigwedge^2 D_x$, because $E_x \subset \mathrm{Ker}(\omega) \subset \bigwedge^2 D_x$.

1.5. The splitting type of the tangent bundle restricted to a minimal rational curve can be used to get information about principal bundles associated to the tangent bundle. For this purpose, we need the full statement of Grothendieck's splitting theorem [1957]. Let $\mathcal{O}(1)^*$ be the principal $\mathbb{C}^*$-bundle on $\mathbb{P}^1$, which is just the complement of the zero section of $\mathcal{O}(1)$. Given a connected reductive complex Lie group G, choose a maximal algebraic torus $H \subset G$.

THEOREM 1.5.1 [Grothendieck 1957]. *Let $\mathcal{P}$ be a principal G-bundle on $\mathbb{P}^1$. Then there exists an algebraic one-parameter subgroup $\rho : \mathbb{C}^* \to H$ such that $\mathcal{P}$ is equivalent to the G-bundle associated to $\mathcal{O}(1)^*$ via the action ρ. Furthermore, let $\mathcal{V}$ be a vector bundle associated to $\mathcal{G}$ via a representation $\mu : G \to \mathrm{GL}(V)$ on a finite dimensional vector space V. Then $\mathcal{V}$ splits as the direct sum of line bundles $\mathcal{O}(\langle \mu_i, \rho \rangle)$, where $\mu_i : H \to \mathbb{C}^*$ are the weights of μ and $\langle \mu_i, \rho \rangle$ denotes the integral exponent of the homomorphism $\mu_i \circ \rho : \mathbb{C}^* \to \mathbb{C}^*$.*

This theorem can be used in the following situation. We are given a principal G-bundle $\mathcal{P}$ on a uniruled manifold X and an associated vector bundle $\mathcal{V}$ via a representation $\mu : G \to \mathrm{GL}(V)$ on a vector space V. Usually the vector bundle $\mathcal{V}$ is related to the tangent bundle $T(X)$ so that from the splitting $T(X)|_C = \mathcal{O}(2) \oplus [\mathcal{O}(1)]^p \oplus \mathcal{O}^q$ on a standard minimal rational curve C, we have information about the splitting type of $\mathcal{V}|_C$. This information helps us understand $\mathcal{P}$ and μ by Theorem 1.5.1.

Consider the case when $\mathcal{V}$ is the tangent bundle $T(X)$ itself. The natural $\mathrm{GL}(V)$-principal bundle associated to $T(X)$ is the frame bundle $\mathcal{F}$. Here V is an n-dimensional complex vector space. Theorem 1.5.1 applied to $\mathcal{F}$ does not say much. A more interesting case is when there is a reduction of the structure group of the frame bundle, namely when there exists a subgroup $G \subset \mathrm{GL}(V)$ and a G-subbundle $\mathcal{G}$ of $\mathcal{F}$. In this case, we say that X has a G-structure. By Theorem 1.5.1, the splitting type of $T(X)$ on a minimal rational curve gives a nontrivial restriction on the possibility of $G \subset \mathrm{GL}(V)$. One particularly simple case is when G is a connected reductive proper subgroup and the representation $\mu : G \subset \mathrm{GL}(V)$ is irreducible. In this case, we can get a complete classification of uniruled manifolds admitting a G-structure in the following manner [Hwang and Mok 1997].

The key point here is the coincidence of two subvarieties in $\mathbb{P}T_x(X)$ for a generic $x \in X$, which are *a priori* of different nature. On the one hand we have the variety of minimal rational tangents $\mathcal{C}_x \subset \mathbb{P}T_x(X)$. On the other hand, the G-structure defines the variety of highest weight tangents $\mathcal{W}_x \subset \mathbb{P}T_x$. Indeed, since G is reductive and μ is irreducible, there is a unique highest weight among μ_i with multiplicity one and the orbit of the highest vector in $\mathbb{P}V$ defines a subvariety $\mathcal{W}_x \subset \mathbb{P}T_x(X)$. These two subvarieties $\mathcal{C}_x$ and $\mathcal{W}_x$ coincide. The proof is as follows.

By Theorem 1.5.1, the existence of a unique highest weight vector and the existence of a unique line subbundle of highest degree $\mathcal{O}(2)$ in the splitting of $V = T(X)$ on a minimal rational curve C imply that the tangent direction of the curve C belongs to the orbit of a highest weight vector. Thus, $\mathcal{C}_x \subset \mathcal{W}_x$. When G is a proper subgroup of $\mathrm{GL}(V)$, we can easily show that $\mathcal{W}_x$ is a proper nondegenerate subvariety of $\mathbb{P}T_x(X)$. To prove that $\mathcal{C}_x = \mathcal{W}_x$, we need to show that $q \leq \mathrm{codim}(\mathcal{W}_x \subset \mathbb{P}T_x(X))$ where $T(X)|_C = \mathcal{O}(2) \oplus [\mathcal{O}(1)]^p \oplus \mathcal{O}^q$. For this, we look at the splitting type of $\mathrm{End}(T(X))$ over a minimal rational curve C:

$$\mathrm{End}(T(X))|_C = [\mathcal{O}(2)]^q \oplus [\mathcal{O}(1)]^{p(q+1)} \oplus \mathcal{O}^{p^2+q^2+1} \oplus [\mathcal{O}(-1)]^{p(q+1)} \oplus [\mathcal{O}(-2)]^q.$$

From the reductivity of G, we have a direct sum decomposition of the Lie algebra $\mathfrak{gl}(V) = \mathfrak{g} \oplus \mathfrak{g}^\perp$ with respect to the trace form. This induces a decomposition $\mathrm{End}(T(X)) = U \oplus U^\perp$. At a generic point $x \in C$, the $\mathcal{O}(2)_x$-factor in $\mathrm{End}(T(X))_x$ corresponds to endomorphisms with image in $\mathbb{C}\alpha = T_x(C)$. Suppose the bundle $U|_C$ corresponding to $\mathfrak{g}$ contains an $\mathcal{O}(2)$-factor. Then for any $\gamma \in \mathcal{W}_x$, the tangent space $\mathfrak{g}\gamma$ to $\mathcal{W}_x$ contains the point $\alpha \in \mathcal{W}_x$. This is a contradiction to the nondegeneracy of $\mathcal{W}_x$. It follows that all $\mathcal{O}(2)$-factors are in $U^\perp$. From the orthogonality of $\mathfrak{g}$ and $\mathfrak{g}^\perp$, endomorphisms in $\mathfrak{g}^\perp$ which have images in $\mathbb{C}\alpha$ must annihilate $\mathfrak{g}\alpha$, the tangent space to $\mathcal{W}_x$ at α. This means that $[\mathcal{O}(2)]_x^q$ annihilates the tangent space to $\mathcal{W}_x$ at α, implying $q \leq \mathrm{codim}(\mathcal{W}_x \subset \mathbb{P}T_x(X))$.

Now we identify $\mathcal{C}_x = \mathcal{W}_x$ for a generic $x \in X$. $\rho : \mathbb{C}^* \to G$ in Theorem 1.5.1 tells us that for each $\alpha \in \mathcal{C}_x$, there exists a $\mathbb{C}^*$-action on $T_x(X)$ under which $T_x(X)$ decomposes as $\mathbb{C}\alpha \oplus \mathcal{H}_\alpha \oplus \mathcal{N}_\alpha$ where $t \in \mathbb{C}^*$ acts as t^2 on $\mathbb{C}\alpha$, as t on $\mathcal{H}_\alpha$, and as 1 on $\mathcal{N}_\alpha$. By rescaling, we get a $\mathbb{C}^*$-action on V_x preserving $\mathcal{C}_x$ which fixes $\mathbb{C}\alpha$, acts as t on $\mathcal{H}_\alpha$, and as t^2 on $\mathcal{N}_\alpha$. Moreover, $\mathbb{C}\alpha \oplus \mathcal{H}_\alpha$ corresponds to the tangent space of $\mathcal{C}_x$ at α. This fact has an interesting implication on $\mathcal{T}_x$, that the linear span E_x of $\mathcal{T}_x$ contains $\alpha \wedge \mathcal{H}_\alpha$ and $\alpha \wedge \mathcal{N}_\alpha$ for all $\alpha \in \mathcal{C}_x$. In fact, Choose a generic point $\alpha + \xi + \zeta$ on $\mathcal{C}_x$. The orbit of the $\mathbb{C}^*$-action is $\alpha + t\xi + t^2\zeta$. At $t = t_0$, we further consider the curve $\alpha + e^s t_0 \xi + e^{2s} t_0^2 \zeta$. Taking derivative with respect to s, we get the tangent vector $t_0 \xi + 2t_0^2 \zeta$ to $\mathcal{C}_x$ at the point $\alpha + t_0 \xi + t_0^2 \zeta$. The corresponding element of $\mathcal{T}_x$ is

$$(\alpha + t_0 \xi + t_0^2 \zeta) \wedge (t_0 \xi + 2t_0^2 \zeta) = t_0 \alpha \wedge \xi + 2t_0^2 \alpha \wedge \zeta + t_0^3 \xi \wedge \zeta.$$

It follows that the linear span of $\mathcal{T}_x$ contains vectors of the form $\alpha \wedge \xi, \alpha \wedge \zeta$. As ξ takes values in the tangent space of $\mathcal{C}_x$ at α, ζ takes independent values in $\mathcal{N}_\alpha$. Thus $\mathcal{T}_x$ contains $\alpha \wedge \mathcal{H}_\alpha$ and $\alpha \wedge \mathcal{N}_\alpha$. This implies that $E_x = \bigwedge^2 T_x(X)$.

Moreover, putting $t = -1$ in the $\mathbb{C}^*$-action on $\mathcal{C}_x$ considered above, we see that $\mathcal{C}_x$ is a Hermitian symmetric space of rank at least 2. It is not hard to see from this that on a uniruled manifold, if an irreducible reductive G-structure with $G \neq \mathrm{GL}(n, \mathbb{C})$ is given, then $G = K^{\mathbb{C}}$ where K is the isotropy subgroup of the isometry group of an irreducible Hermitian symmetric space S of the compact type of rank at least 2. Such a G-structure will be called an S-structure.

A G-structure $\mathcal{G}$ is flat if there exists a local coordinate system whose coordinate frames belong to $\mathcal{G}$ regarded as a subbundle of the frame bundle. The following result of Ochiai will be proved in Section 3.2 below.

PROPOSITION 1.5.2 [Ochiai 1970]. *Let S be an irreducible Hermitian symmetric space of the compact type and of rank at least 2. Let M be a compact simply-connected complex manifold with a flat S-structure. Then, M is biholomorphic to S.*

The flatness of an S-structure is equivalent to the vanishing of certain holomorphic tensors, just as the flatness of a Riemannian metric is equivalent to the vanishing of the Riemannian curvature tensor in Riemannian geometry. By restricting these tensors to minimal rational curves and considering the splitting type, it is easy to show the vanishing of these tensors from $E_x = \bigwedge^2 T_x(X)$; see [Hwang and Mok 1997] for details. As a result:

THEOREM 1.5.3 [Hwang and Mok 1997]. *A uniruled projective manifold with an irreducible reductive G-structure, $G \neq \mathrm{GL}(V)$, is an irreducible Hermitian symmetric space of the compact type.*

This theorem gives an algebro-geometric characterization of irreducible Hermitian symmetric spaces of the compact type without the assumption of homogeneity. For example, Grassmannians of rank at least 2 can be characterized as the only uniruled manifolds whose tangent bundle can be written as the tensor product of two vector bundles of rank at least 2.

1.6. To construct a reasonable moduli space of vector bundles on projective manifolds, we have to restrict ourselves to a special class of vector bundles, called semistable bundles. On a Fano manifold X of Picard number 1, they can be defined as follows. Fix a component $\mathcal{K}$ of the Chow space of rational curves on X, so that a generic member is a free rational curve. Given a torsion-free sheaf $\mathcal{F}$ on X choose a generic member C of $\mathcal{K}$ so that $\mathcal{F}|_C$ is locally free. We can always make such a choice because the singular loci of a torsion-free sheaf has codimension at least 2 (see [Hwang and Mok 1998b, Proposition 12], for example). Let $\mathcal{F}|_C = \mathcal{O}(a_1) \oplus \cdots \oplus \mathcal{O}(a_k), a_1 \geq \cdots \geq a_k$, be the splitting. We define the slope of $\mathcal{F}$ as the rational number $\mu(\mathcal{F}) := \sum a_i / k$. A vector bundle

V of rank r on X is if stable $\mu(\mathcal{F}) < \mu(V)$ for every subsheaf $\mathcal{F}$ of rank k, with $0 < k < r$. It is stable if $\mu(\mathcal{F}) \leq \mu(V)$ for every such $\mathcal{F}$.

A well-known problem in Kähler geometry of Fano manifolds is the Calabi problem on the existence of Kähler–Einstein metrics. The semistability of the tangent bundle is a necessary condition for the existence of Kähler–Einstein metric. As a result, people have been interested in the stability or the semistability of $T(X)$ for a Fano manifold X with Picard number 1. By taking wedge product, the instability of $T(X)$, or equivalently the instability of $T^*(X)$, implies the nonvanishing of $H^0(X, \Omega^r(k))$ for certain r, k. Thus one sufficient condition for the stability of $T(X)$ is the vanishing of these cohomology groups. Peternell and Wiśniewski [1995] proved the stability of $T(X)$ for many examples of X, including all X of dimension at most 4 by using this idea. It looks hard to generalize this method to higher dimensions.

This problem can be studied by relating it to the geometry of $\mathcal{C}_x$ for a generic $x \in X$ [Hwang 1998]. Choose $\mathcal{H}$ as in Section 1.1 and the corresponding Chow space $\mathcal{K}$. Suppose $T(X)$ is not stable. Choose a subsheaf $F \subset T(X)$ with maximal value of $\mu(F) \geq \mu(T(X)) = \frac{p+2}{n}$. The maximality of $\mu(F)$ implies the minimality of $\mu(T(X)/F)$, and the vanishing of the Frobenius bracket tensor $\bigwedge^2 F \to T(X)/F$. Thus, F defines a meromorphic foliation on X. If $\mathcal{C}_x$ is contained in F_x for a generic $x \in X$, we can get a contradiction to the Picard number of X as in Proposition 1.2.2. Thus $\mathbb{P}F_x \subset \mathbb{P}T_x(X)$ is a linear subspace which does not contain $\mathcal{C}_x$. On the other hand, the condition that $\mu(F) \geq \frac{p+2}{n}$ reads $\sum a_i \geq \frac{r(p+2)}{n}$, where $F|_C = \mathcal{O}(a_1) \oplus \cdots \oplus \mathcal{O}(a_r)$ for a generic minimal rational curve C. This can be rephrased as "the intersection of $F|_C$ with the positive part of $T(X)|_C$ has larger dimension than the one expected from the rank of F". Since the positive part of $T(X)|_C$ corresponds to the tangent space to $\mathcal{C}_x$, the assumption that $T(X)$ is not stable implies that the tangent spaces of $\mathcal{C}_x$ have an excessive intersection with a linear subspace $\mathbb{P}F_x$. More precisely:

PROPOSITION 1.6.1. *Suppose that $T(X)$ is not stable. Then, there exists an integrable meromorphic distribution $F \subset T(X)$ of rank r so that for a generic $x \in X$, the intersection of the projective tangent space at a generic point of $\mathcal{C}_x$ with $\mathbb{P}F_x$ has dimension greater than $\frac{r}{n}(p+2) - 2$.*

Thus by studying the projective geometry of $\mathcal{C}_x$, we can prove the stability of $T(X)$. Although very little is known about the geometry of $\mathcal{C}_x$ in general, there are many cases where Proposition 1.6.1 can be used to show the stability of $T(X)$. For example, in low dimension, the excessive intersection property gives a heavy restriction on $\mathcal{C}_x$ which gives a contradiction easily. The main results of [Hwang 1998] that Fano 5-folds with Picard number 1 have stable tangent bundles and Fano 6-folds with Picard number 1 have semistable tangent bundles were obtained this way.

Another interesting case is when we know that $\mathcal{C}_x$ is smooth and of small codimension in $\mathbb{P}T_x(X)$.

THEOREM 1.6.2. *Assume that $\mathcal{C}_x$ is an irreducible submanifold of $\mathbb{P}T_x(X)$ for a generic $x \in X$. If $2p \geq n - 2$, then $T(X)$ is stable.*

From the discussion in Section 1.3, our assumption shows that $\mathcal{C}_x$ is nondegenerate in $\mathbb{P}T_x(X)$. Theorem 1.6.2 follows directly from Proposition 1.6.1 using the following lemma, which is a variation of Zak's theorem on tangencies just as in the proof of Proposition 1.3.2.

LEMMA 1.6.3. *Let $Y \subset \mathbb{P}^{n-1}$ be a nondegenerate irreducible subvariety of dimension $p \geq \frac{n-2}{2}$. If there exists a linear subspace $E \subset \mathbb{P}^{n-1}$ of dimension $r - 1$ such that for a generic point $y \in Y$, the projective tangent space to Y at y intersects E on a subspace of dimension $q - 1$ for $q \geq \frac{r}{n}(p + 2)$, then Y is not smooth.*

As a corollary of Theorem 1.6.2, we get the stability of $T(X)$ in the following cases, most of which have not been proved previously:

- smooth linear sections of codimension 1 or 2 of Grassmannians of rank 2;
- smooth hyperplane sections of Grassmannian of rank 3 of dimension 9;
- smooth linear sections of dimension at least 10 of the 16-dimensional E_6 symmetric space;
- smooth linear sections of dimension at least 20 of the 27-dimensional E_7 symmetric space.

2. Deformation Rigidity of Irreducible Hermitian Symmetric Spaces and Homogeneous Contact Manifolds

2.1. We propose to study deformation of certain Fano manifolds of Picard number 1 by considering deformations of their bundles of varieties of minimal rational tangents and distributions spanned by them. As a first step we deal with irreducible Hermitian symmetric spaces of the compact type, and proved in [Hwang and Mok 1998b] their rigidity under Kähler deformation, as follows.

THEOREM 2.1.1. *Let S be an irreducible Hermitian symmetric space of the compact type. Let $\pi : \mathcal{X} \to \triangle$ be a regular family of compact complex manifolds over the unit disk $\triangle$. Suppose $X_t := \pi^{-1}(t)$ is biholomorphic to S for $t \neq 0$ and the central fiber X_0 is Kähler. Then, X_0 is also biholomorphic to S.*

The case of $S \cong \mathbb{P}^n$ being a consequence of the classical result of Hirzebruch and Kodaira [1957], we restrict ourselves to S of rank at least 2. (The case of the hyperquadric also follows from [Brieskorn 1964].) S is associated to holomorphic G-structures, as explained in Section 1.5. On S the varieties of minimal rational tangents $\mathcal{C}_x \subset \mathbb{P}T_x(S)$ are highest weight orbits of isotropy representations as discussed in Section 1.5. They turn out to be themselves Hermitian symmetric manifolds of the compact type of rank 1 or 2, and irreducible except in the case of Grassmannians $G(p, q)$ of rank at least 2, where $\mathcal{C}_x$ is isomorphic to $\mathbb{P}^{p-1} \times \mathbb{P}^{q-1}$, embedded in $\mathbb{P}^{pq-1}$ by the Segre embedding.

Under the hypothesis of Theorem 2.1.1 we proceed to prove that the central fiber $X_0 \cong S$. Denote by T the relative tangent bundle of $\pi : \mathfrak{X} \to \triangle$. Pick some $t_o \neq 0$ and a minimal rational curve C on X_{t_o}. By considering the deformation of C as a curve in $\mathfrak{X}$, we obtain a subvariety $\mathcal{C} \subset \mathbb{P}T$ such that for a point y on X_t; $t \neq 0$; the fiber $\mathcal{C}_y \subset \mathbb{P}T_y$ is isomorphic to the embedded standard variety of minimal rational tangents $\mathcal{C}_o \subset \mathbb{P}T_o(S)$. At a generic point x of the central fiber every minimal rational curve is free and the normalized Chow space $\mathfrak{M}_x$ of such curves marked at x is projective and nonsingular. Over such a point $\mathcal{C}_x$ is the same as the variety of minimal rational tangents. Here we make use of the hypothesis that X_0 is Kähler, which implies that the deformations of C situated on X_0 are irreducible and of degree 1 with respect to the positive generator of $\mathrm{Pic}(X_0) \cong \mathbb{Z}$.

Suppose we are able to prove that, for a generic point x on X_0,

(A) $\mathfrak{M}_x \cong \mathcal{C}_o$ and

(B) $\mathcal{C}_x \subset \mathbb{P}T_x$ is linearly nondegenerate.

Then, it follows readily that the latter is isomorphic to the model $\mathcal{C}_o \subset \mathbb{P}T_o(S)$ as projective submanifolds. From this one can readily recover a holomorphic S-structure on the complement of some subvariety E of X_0. As a subvariety of $\mathfrak{X}$, E is of codimension at least 2, and a Hartogs-type extension theorem resulting from [Matsushima and Morimoto 1960] allows us to obtain a holomorphic S-structure on X_0. Since flatness of holomorphic G-structures is a closed condition, the S-structure on X_0 is flat, leading to a biholomorphism $X_0 \cong S$, as a consequence of Ochiai's theorem, Proposition 1.5.2.

(A) can be established by induction except for the case of $G(p,q)$; $p, q > 1$ where $\mathcal{C}_x \cong \mathbb{P}^{p-1} \times \mathbb{P}^{q-1}$. In the abstract case it is possible to have nontrivial deformation of $\mathbb{P}^{p-1} \times \mathbb{P}^{q-1}$, as exemplified by the deformation of $\mathbb{P}^1 \times \mathbb{P}^1$ to Hirzebrach surfaces. However, in our situation the individual factors $\mathbb{P}^{p-1}$ and $\mathbb{P}^{q-1}$ correspond to projective spaces $\cong \mathbb{P}^p$ and $\mathbb{P}^q$ (respectively) of degree 1 in S. By cohomological considerations we prove that limits of such projective spaces cannot decompose in the central fiber, thus establishing (A) even in the special case of $S = G(p,q)$, with $p, q > 1$. We refer the reader to [Hwang and Mok 1998b, §3] for details.

2.2. It remains now to prove (B) that on the central fiber X_0, generic varieties of minimal rational tangents $\mathcal{C}_x \subset \mathbb{P}T_x$ are linearly nondegenerate. From (A) we can identify $\mathfrak{M}_x$ with $\mathcal{C}_o \subset \mathbb{P}T_o(S)$, and realize the tangent map $\Phi_x : \mathfrak{M}_x \to \mathcal{C}_x$ as the restriction of a linear projection $\mathbb{P}T_o(S) \to \mathbb{P}W_x$, where $\mathbb{P}W_x$ is the projective-linear span of $\mathcal{C}_x$. We proceed to prove (B) by contradiction. The assignment of W_x to each generic x on X_0 defines a (meromorphic) distribution on X_0. Since X_0 is of Picard number 1 by Proposition 1.2.2 if $W_x \neq T_x$ at generic points the distribution W is not integrable. On the other hand, by Proposition 1.2.1, W is integrable whenever the variety of tangential lines $\mathcal{T}_x \subset \mathbb{P} \bigwedge^2 W_x$

is linearly nondegenerate at generic points x. To prove (B) by contradiction it remains therefore only to check the condition of linear nondegeneracy on $\mathcal{T}_x \subset \mathbb{P} \bigwedge^2 T_x$. As explained in Section 1.5, $\mathcal{T}_o \subset \mathbb{P} \bigwedge^2 T_o(S)$ is nondegenerate for the model $\mathcal{C}_o$. Since the tangent map $\Phi_x : \mathcal{M}_x \to \mathcal{C}_x$; $\mathcal{M}_x \cong \mathcal{C}_o$; is the restriction of a linear projection $\mathbb{P} T_o(S) \to \mathbb{P} W_x$, it follows that $\mathcal{T}_x \subset \mathbb{P} \bigwedge^2 W_x$ is linearly nondegenerate, as desired.

2.3. Deformation rigidity can be asked for other rational homogeneous spaces, too.

CONJECTURE 2.3.1. *Let S be a rational homogeneous space with Picard number 1. Let $\pi : \mathcal{X} \to \Delta$ be a regular family of compact complex manifolds with $X_t \cong S$ for all $t \neq 0$. If X_0 is Kähler, $X_0 \cong S$.*

We expect that the method of Sections 2.1 and 2.2 can be generalized to a general S, although the details will not be straightforward. When S is a homogeneous contact manifold, this was done in [Hwang 1997]. We will sketch the main ideas here.

Let S be a homogeneous contact manifold of dimension $n = 2m + 1$ associated to an orthogonal or an exceptional simple Lie algebra as in Section 1.4. We consider $\pi : \mathcal{X} \to \Delta$ as in Conjecture 2.3.1. There exists a distribution D of rank $2m$ on $\mathcal{X}$ which may have singularity on X_0 and gives the contact distribution on $X_t, t \neq 0$. Just as the deformation rigidity of Hermitian symmetric spaces was obtained by the recovery of the S-structure at a generic point of X_0, the deformation rigidity of homogeneous contact manifolds can be obtained by showing that D defines a contact structure at a generic point of X_0. Here, in place of Ochiai's result, we can just look at the Kodaira–Spencer class [LeBrun 1988].

By the Lagrangian property of $\mathcal{C}_s$, for $s \in S$, discussed in Section 1.4, $\mathcal{T}_s$ is contained in the kernel of

$$\omega : \bigwedge^2 D_s \to L_s.$$

Now the key point of the proof of the deformation rigidity of homogeneous contact manifold is the fact that $\mathcal{T}_s$ is nondegenerate in $\mathbb{P} \operatorname{Ker}(\omega) \subset \mathbb{P} \bigwedge^2 D_s$. This can be checked case by case as in [Hwang 1997, Section 2]. In other words, the symplectic structure on D_s is completely determined by $\mathcal{C}_s$.

To prove the deformation rigidity, we argue as in Sections 2.1 and 2.2. Let $x \in X_0$ be a generic point and choose a section $\sigma : \Delta \to \mathcal{X}$ with $\sigma(0) = x$. If the family $\mathcal{C}_{\sigma(t)} \subset \mathbb{P} T_{\sigma(t)}(X_t)$ remains unchanged as projective subvarieties as $t \to 0$, then the linear span of $\mathcal{T}_{\sigma(t)}$ is isomorphic to $\operatorname{Ker}(\omega)$ of S. This implies that D defines a contact structure at x and we are done. As in Sections 2.1 and 2.2, an induction argument reduces the proof of the rigidity of $\mathcal{C}_{\sigma(t)}$ to showing that $\mathcal{C}_x$ is linearly nondegenerate in D_x. But linear degeneracy would imply the integrability of the distribution D using the linear nondegeneracy of $\mathcal{T}_s$ in $\operatorname{Ker}(\omega)$, just as in Section 2.2. Integrability of D gives a contradiction to Section 1.2, thus $\mathcal{C}_{\sigma(t)}$ is rigid.

3. Tautological Foliations on Varieties of Minimal Rational Tangents and the Method of Analytic Continuation

3.1. In this section, we consider the following problem. For two Fano manifolds X_1, X_2 with bundles of varieties of minimal rational tangents $\mathcal{C}_1 \subset \mathbb{P}T(X_1), \mathcal{C}_2 \subset \mathbb{P}T(X_2)$ and a biholomorphism f from an open subset $U_1 \subset X_1$ to an open subset $U_2 \subset X_2$ satisfying $df(\mathcal{C}_1) = \mathcal{C}_2$, when can we say that f sends the intersection of a minimal rational curve with U_1 to the intersection of a minimal rational curve with U_2. In other words, when does $\mathcal{C} \subset \mathbb{P}T(X)$ determine the minimal rational curves locally? One sufficient condition is that $\mathcal{C}_x$ has generically finite Gauss map as a projective subvariety of $\mathbb{P}T_x(X)$ for generic x and U_1, U_2 are sufficiently generic. This is contained in Corollary 3.1.4 below.

We start with some notations. In this section, we will view a distribution D on a manifold as a subsheaf of the tangent sheaf. We will be always looking at generic points of the manifold where all distributions concerned are locally free, and we regard the distribution as a subbundle of the tangent bundle at such points. We will not make notational distinction between sheaves and bundles in this case. Given a distribution D on a manifold, the derived system of D is the distribution $\partial D := D + [D, D]$, and its Cauchy characteristic $\mathrm{Ch}(D)$ is the distribution defined by $\mathrm{Ch}(D)(U) := \{f \in D(U), [f, g] \in D(U), \forall g \in D(U)\}$, for any open subset U of the manifold. $\mathrm{Ch}(D)$ is always integrable.

Let X be any complex manifold and $U \subset X$ be a sufficiently general small open set. Given any subvariety $\mathcal{C} \subset \mathbb{P}T(U)$, we consider two distributions $\mathcal{J}$ and $\mathcal{P}$ on the smooth part of $\mathcal{C}$ defined by

$$\mathcal{J}_\alpha := (d\pi)^{-1}(\mathbb{C}\alpha),$$
$$\mathcal{P}_\alpha := (d\pi)^{-1}(P_\alpha),$$

where $d\pi : T_\alpha(\mathcal{C}) \to T_x(U)$ is the differential of the natural projection $\pi : \mathcal{C} \to U$ at $\alpha \in \mathcal{C}$, $x = \pi(\alpha)$, and $P_\alpha \subset T_x(X)$ is the linear tangent space of $\mathcal{C}_x := \pi^{-1}(x) \subset \mathbb{P}T_x(X)$ at α. Both $\mathcal{J}$ and $\mathcal{P}$ are canonically determined by $\mathcal{C}$. $\mathcal{J}$ has rank $p + 1$ and $\mathcal{P}$ has rank $2p + 1$, where p is the fiber dimension of $\pi : \mathcal{C} \to U$. Also we have the trivial vertical distribution $\mathcal{V}$ of rank p on $\mathcal{C}$ defining the fibers of π. Clearly, $\mathcal{V} \subset \mathcal{J} \subset \mathcal{P}$.

Now assume that X is a uniruled projective manifold and $\mathcal{C}$ is part of the variety of minimal rational tangents on X. Then we have a meromorphic multi-valued foliation $\mathcal{F}$ on $\mathcal{C}$ defined by lifting minimal rational curves. When we work on a small open set of $\mathcal{C}$, we may assume $\mathcal{F}$ is a foliation by curves on that open set, by choosing a specific branch of the multi-valued foliation. We call $\mathcal{F}$ the tautological foliation. Since $\mathcal{F}$ is defined by lifting curves, $\mathcal{J} = \mathcal{V} + \mathcal{F}$ at generic points of $\mathcal{C}$.

PROPOSITION 3.1.1. $\mathcal{P} = \partial \mathcal{J}$.

PROOF. For notational simplicity, we will work over $\Gamma = T(X) \setminus (0\text{-section})$. Let $\gamma : \Gamma \to \mathbb{P}T(X)$ be the natural $\mathbb{C}^*$-bundle. Let $\mathcal{C}' := d\gamma^{-1}(\mathcal{C}), \mathcal{J}' := d\gamma^{-1}\mathcal{J}, \mathcal{P}' := d\gamma^{-1}\mathcal{P}, \mathcal{V}' := d\gamma^{-1}\mathcal{V}$, and $\mathcal{F}' := d\gamma^{-1}\mathcal{F}$. It suffices to check that $\mathcal{P}' = \partial\mathcal{J}'$.

We start with $\partial\mathcal{J}' \subset \mathcal{P}'$. It suffices to show $[\mathcal{V}', \mathcal{F}'] \subset \mathcal{P}'$. Let $x_1, \ldots, x_n$ be a local coordinate system on U. Let $v_1 = dx_1, \ldots, v_n = dx_n$ be linear coordinates in the vertical direction of Γ. Let $v = \sum_i e^i \frac{\partial}{\partial v_i}$ be a local section of $\mathcal{V}'$ and $f = \sum_i f^i \frac{\partial}{\partial v_i} + \zeta \sum_j v_j \frac{\partial}{\partial x_j}$ be a local section of $\mathcal{F}'$ over a small open set in $\mathcal{C}'$. Here e^i, f^i, ζ are suitable local holomorphic functions. Dividing by ζ and looking at generic points outside the zero set of ζ, we may assume that $\zeta \equiv 1$. Then $[v, f] = \sum_i e^i \frac{\partial}{\partial x_i}$ modulo $\mathcal{V}'$. But this is precisely the vector v viewed as the tangent vector to X. Hence $[v, f] \in \mathcal{P}'$.

From the above expression of $[v, f]$ modulo $\mathcal{V}'$, we see that the rank of $\partial\mathcal{J}'$ is higher than the rank of $\mathcal{J}'$ by at least p, which shows $\partial\mathcal{J}' = \mathcal{P}'$.   □

PROPOSITION 3.1.2. $\mathcal{F} \subset \mathrm{Ch}(\mathcal{P})$.

PROOF. Let $\mathcal{D}$ be the Chow space parametrizing minimal rational curves. Let $U \subset \mathcal{C}$ be a sufficiently generic small open set. Locally, we have a morphism $\rho : U \to \mathcal{D}$ whose fibers are leaves of $\mathcal{F}$. In a neighborhood of a generic point $[C_0] \in \mathcal{D}$ corresponding to a minimal rational curve C_0, we have a distribution $\hat{\mathcal{P}}$ on $\mathcal{D}$ defined as follows. Note that the tangent space to $\mathcal{D}$ at $[C]$ near $[C_0]$ is naturally isomorphic to $H^0(C, N_C)$ where N_C is the normal bundle of C in X. We know that $N_C \cong [\mathcal{O}(1)]^p \oplus [\mathcal{O}]^q$. Let $\hat{\mathcal{P}}_{[C]}$ be the subspace of $H^0(C, N_C)$ consisting of sections of $[\mathcal{O}(1)]^p$-part. This gives a distribution $\hat{\mathcal{P}}$ in a neighborhood of $[C_0]$. From Section 1.1, $\mathcal{P} = d\gamma^{-1}\hat{\mathcal{P}}$. So the result follows from the following easy lemma.   □

LEMMA 3.1.3. *Let $f : W \to Y$ be a submersion between two complex manifolds and $\mathbb{N}$ be a distribution on Y. Let $\mathcal{K}$ be the distribution defined by the fibers of f. Then $\mathcal{K}$ is contained in the Cauchy characteristic of the distribution $df^{-1}\mathbb{N}$.*

THEOREM 3.1.4. *Suppose that a component of $\mathcal{C}_x$ has generically finite Gauss map regarded as a projective subvariety in $\mathbb{P}T_x(X)$. Then $\mathcal{F} = \mathrm{Ch}(\mathcal{P})$.*

PROOF. Suppose that for some $v \in \mathcal{V}(U), h \in \mathcal{J}(U)$ and $f \in \mathcal{F}(U)$, we have $[v, f] + h \in \mathrm{Ch}(\mathcal{P})(U)$. Then it is easy to see that $v \in \mathrm{Ch}(\mathcal{P})(U)$. In fact, to show $[v, p] \in \mathcal{P}(U)$ for any $p \in \mathcal{P}(U)$, it suffices to show $[v, [w, f]] \in \mathcal{P}(U)$ for any $w \in \mathcal{V}(U)$, by using Proposition 3.1.1. From $[v, [w, f]] = [w, [v, f]] + [f, [w, v]]$ and $[h, w] \in \mathcal{P}(U)$, we see that $[v, [w, f]] \in \mathcal{P}(U)$.

So it suffices to show that there is no nonzero $v \in \mathcal{V}(U) \cap \mathrm{Ch}(\mathcal{P})(U)$. We will work on Γ as in the proof of Proposition 3.1.1. We need to show that any $v \in \mathcal{V}'(\rho^{-1}(U)) \cap \mathrm{Ch}(\mathcal{P}')(\rho^{-1}(U))$ is tangent to a fiber of γ. Suppose there exists such a $v = \sum_i a_i \frac{\partial}{\partial v_i}$. Then $[v, f] \in \mathrm{Ch}(\mathcal{P}')$ where we used the same letter f to denote the section of $\mathcal{F}'$ lifting a section f of $\mathcal{F}$. For any section k of $\mathcal{V}'$, $k = \sum_i e_i \frac{\partial}{\partial v_i}$, we have $[[v, f], k] \in \mathcal{P}'$. Modulo vertical part, $[[v, f], k] = \sum_i k(a_i) \frac{\partial}{\partial x_i}$.

For this to be a section of $\mathcal{P}'$, $\sum_i k(a_i)\frac{\partial}{\partial v_i}$ must be a section of $\mathcal{V}'$. Since $[v, k]$ is a section of $\mathcal{V}'$, we conclude that $\sum_i v(e_i)\frac{\partial}{\partial v_i}$ is a section of $\mathcal{V}'$. In other words, for any vector field $k = \sum_i e_i \frac{\partial}{\partial v_i}$ tangent to $\mathcal{C}'_x$, $\sum_i v(e_i)\frac{\partial}{\partial v_i}$ remains tangent to $\mathcal{C}'_x$. This implies that v is in the kernel of the differential of the Gauss map of $\mathcal{C}'_x$. By assumption on the Gauss map of $\mathcal{C}_x$, such v must be tangent to the fibers of γ on $\mathcal{C}'$. $\qquad\square$

COROLLARY 3.1.5. *Under the assumption of Theorem 3.1.4 on $\mathcal{C}$, $\mathcal{F}$ is a single-valued meromorphic foliation on $\mathcal{C}$ uniquely determined by $\mathcal{C}$. In particular, the tangent map $\Phi_x : \mathcal{M}_x \to \mathcal{C}_x$ is birational.*

We note that the Gauss map on $\mathcal{C}_x$ is generically finite, whenever the latter is irreducible, nonsingular and distinct from the projective space, by Zak's Theorem on tangencies [Zak 1993]. This is in particular the case at generic points for Fano complete intersections $X \subset \mathbb{P}^n$ of dimension at least 3 provided that $c_1(X) \geq 3$ (X being of Picard number 1); see [Kollár 1996, Section V.4].

3.2. In Section 1.5 we have stated in Proposition 1.5.2 the result of Ochiai's characterizing irreducible Hermitian symmetric spaces S of the compact type and of rank at least 2 in terms of flat S-structures. We will prove the result here, which follows from this proposition:

PROPOSITION 3.2.1. *Let S be an irreducible Hermitian symmetric manifold of the compact type and of rank at least 2. Denote by $\mathcal{C} \to S$ the bundle of varieties of minimal rational tangents. Let U_1, $U_2 \subset S$ be two connected open sets and $f_{12} : U_1 \to U_2$ be a biholomorphism such that $(f_{12})_* \mathcal{C}|_{U_1} = \mathcal{C}|_{U_2}$. Then, f_{12} extends to a biholomorphic automorphism of S.*

Proposition 3.2.1 implies Proposition 1.5.2, as follows. Since the S-structure on M is flat, given any $x \in M$ there exists a neighborhood U_x of x and a biholomorphism $f : U_x \to S$ of U_x onto some open subset U of S such that $f_* \mathcal{W}|_{U_x} = \mathcal{C}|_U$, where $\mathcal{W} \to M$ is the bundle of varieties of highest weight tangents defined by the S-structures. Starting with one choice of x and f, Proposition 3.2.1 allows us to continue f holomorphically along any continuous curve, by matching different f_y on U_y on intersecting regions using global automorphisms. This leads to a developing map, which is well-defined on M since M is simply connected. The resulting unramified holomorphic map $F : M \to S$ is necessarily a biholomorphism, since S is simply connected.

Ochiai's original proof of 3.2.1 [1970] used harmonic theory of Lie algebra cohomologies. We will give an alternate proof of Proposition 3.2.1, by making use of analytic continuation. The bundle $\pi : \mathcal{C} \to S$ of varieties of minimal rational tangents is equipped with a one-dimensional foliation $\mathcal{F}$, as in Section 3.1. Recall that for $[\alpha] \in \mathcal{C}_x$, $T_{[\alpha]}\mathcal{C}_x = P_\alpha \mod \mathbb{C}\alpha$, by definition. Since f_{12} preserves $\mathcal{C}$, it also preserves $\mathcal{P}_\alpha = (d\pi)^{-1}(P_\alpha)$. By Theorem 3.1.4, $(f_{12})_* \mathcal{F} = \mathrm{Ch}((f_{12})_* \mathcal{P}) = \mathrm{Ch}(\mathcal{P}) = \mathcal{F}$. In other words, f_{12} preserves the holomorphic foliation $\mathcal{F}$, that is, where defined f_{12} sends open sets on lines to open sets on lines.

To explain our approach note that the problem of analytic continuation of $\mathcal{F}$-preserving germs of holomorphic maps also makes sense for the case of $\mathbb{P}^n$. For $n = 2$ and denoting by $B^2 \subset \mathbb{C}^2 \subset \mathbb{P}^2$ the unit ball, any such map $f : B^2 \to \mathbb{P}^2$ defines a holomorphic mapping $f^\#$ on some open subset $D \subset (\mathbb{P}^2)^*$ of the dual projective space, where D is the open set of all lines having nonempty (and automatically connected) intersection with B^2. In this case D is the complement of a closed Euclidean ball in $(\mathbb{P}^2)^*$, and $f^\#$ extends holomorphically to $(\mathbb{P}^2)^*$ by Hartogs extension, from which we can recover an extension of f from B^2 to $\mathbb{P}^2$, by regarding a point $x \in \mathbb{P}^2$ as the intersection of all lines passing through x. In place of implementing this argument in the case of irreducible Hermitian symmetric spaces, we will adopt here a related but more direct argument, by analytically continuing along lines. This approach was adopted in [Mok and Tsai 1992] in a similar context, but the argument there was incomplete due to the possibility of multivalence of analytically continued functions. We will complete the argument by making use of $\mathbb{C}^*$-actions on S.

We start with a lemma concerning analytic continuation along chains of lines. By a chain of lines K on S we mean the union of a finite number of distinct lines $C_1, \ldots, C_m$ such that $C_j \cap C_{j+1}$ is a single point for $1 \leq j < m - 1$. We write $K = C_1 + C_2 + \cdots + C_m$. We will say that K is nonoverlapping to mean $C_j \cap C_k = \varnothing$ whenever $|j - k| \geq 2$. We will more generally be dealing with $\mathcal{F}$-preserving meromorphic maps $f : \Omega \dashrightarrow S$ on a domain $\Omega \subset S$. By this we mean that at a generic point, f is a local biholomorphism and $\mathcal{F}$-preserving.

LEMMA 3.2.2. *Let $K = C_1 + C_2 + \cdots + C_m$ be a nonoverlapping chain of lines on S, $o \in C_1$, and f be a germ of $\mathcal{F}$-preserving meromorphic map at o. Then, there exists a tubular neighborhood U of K and an $\mathcal{F}$-preserving meromorphic map $\hat{f} : U \to S$ such that $\hat{f}$ extends the germ f.*

The assumption that K is nonoverlapping is not essential. In general, one can replace K by a chain $\tilde{K}$ of $\mathbb{P}^1$ and a holomorphic immersion $\pi_o : \tilde{K} \to S$, $\pi_o(\tilde{K}) = K$. The analogue of Lemma 3.2.2 says that there is a Riemann domain $\pi : U \to S$ including $\pi_o : \tilde{K} \to S$ such that f extends to $\hat{f} : U \to S$.

PROOF. Let $\Omega_o \Subset \Omega \subset S$ be open subsets and $f : \Omega \dashrightarrow S$ be an $\mathcal{F}$-preserving meromorphic map. Suppose $C_o \subset S$ is a line such that $C_o \cap \Omega_o$ is nonempty and irreducible. Denote by $F(S)$ the Fano variety of lines on S. Since C_o is reduced and irreducible, for $[C] \in F(S)$ sufficiently close to $[C_o]$, $C \cap \Omega_o$ is nonempty and irreducible. The meromorphic map $f : \Omega \to S$ gives rise to a meromorphic map $f^\# : D \to F(S)$ on some open neighborhood D of $[C]$ in $F(S)$. Denote by $\rho : \mathcal{C} \to F(S)$ the universal family of lines on S. Then, $f^\# \circ \rho$ is defined on $\rho^{-1}(D)$.

Over Ω, f can be recovered from $f^\# \circ \rho$, as follows. For $x \in \Omega$, let σ_1 and σ_2 be two germs of holomorphic sections of $\mathcal{C}$ at x, $\sigma_1(x) \neq \sigma_2(x)$. If f is locally biholomorphic at x, then $f(y) = (f^\# \circ \rho)(\sigma_1(y)) \cap (f^\# \circ \rho)(\sigma_2(y))$ for y sufficiently near x, where a point $[\alpha] \in \mathcal{C}$ is identified as a line $\mathbb{C}\alpha \subset T_x(X)$. In other words,

$f(x)$ is simply the point of intersection of image lines of two distinct lines passing through x. For $x \in \Omega$ in general, let Σ_o be $\mathrm{Graph}(f^{\#} \circ \rho \circ \sigma_1) \cap \mathrm{Graph}(f^{\#} \circ \rho \circ \sigma_2)$, $\sigma_1(x) \neq \sigma_2(x)$, and Σ be the unique germ of irreducible component of Σ_o which dominates the germ of Ω at x. Then, Σ is the germ of graph of the meromorphic map f at x. But the same procedure can be used to define a meromorphic map $\hat{f}$ for x lying in a neighborhood U of C, provided that $f^{\#} \circ \rho$ is defined in a neighborhood of $\sigma_1(x)$ and $\sigma_2(x)$. This observation, together with the following obvious Lemma, implies readily that f admits an extension to a meromorphic map $\hat{f} : U \to S$ on some tubular neighborhood U of C, which is necessarily $\mathcal{F}$-preserving, since it is $\mathcal{F}$-preserving on $\Omega \cap U$.

LEMMA 3.2.3. *Let $V_o \subset V \subset S$ be nonempty connected open subsets of S. Let $g : V_o \to S$ be an $\mathcal{F}$-preserving meromorphic map. Suppose $g^{\#} \circ \rho$ is defined on the graph of two nonintersecting holomorphic sections $\sigma_1, \sigma_2 : V \to \mathcal{C}$ over V. Define now $\Sigma \subset V \times S$ to be the unique irreducible component of $\mathrm{Graph}(g^{\#} \circ \rho \circ \sigma_1) \cap \mathrm{Graph}(g^{\#} \circ \rho \circ \sigma_2)$ which projects onto V. Then, Σ is the graph of an $\mathcal{F}$-preserving meromorphic map $\hat{g} : V \to S$ such that $\hat{g}|_{V_o} \equiv g$.*

We continue with the proof of Lemma 3.2.2. In the application of Lemma 3.2.3, the important thing is to have some holomorphic section of $\mathcal{C}$ over V. In the application to prove Proposition 3.2.2, there is no difficulty with finding such local sections on tubular neighborhoods of pieces of rational curves C (taking $m = 1$ and $C_1 = C$) since the lift $\hat{C}$ of C to $\mathcal{C}$ already lies in the domain of definition of $f^{\#} \circ \rho$. $\square$

We remark that in place of Lemma 3.2.3 one can also take intersections of algebraic families of lines, by first extending the domain of definition of $f^{\#} \circ \rho$ to $\mathcal{C}|_U$ for some tubular neighborhood U of C, using Oka's Theorem on Hartogs radii (see [Mok and Tsai 1992] and the references there).

By Lemma 3.2.2, any $f \in \Omega$ can be analytically continued along tubular neighborhoods of chains of lines. Since S is rationally connected by (nonoverlapping) chains of lines, f can be extended to any point on S. However, it is not obvious that given $y \in S$, the germ $\hat{f}_y$ of an extension $\hat{f}$ at y obtained along a nonoverlapping chain of lines K, $y \in K$, emanating from $x \in \Omega$ (not necessarily o) will be independent of x and independent of the chain of lines. We will show that this is indeed the case, by making use of $\mathbb{C}^*$-actions on S. This will yield the following result:

LEMMA 3.2.4. *In the notation of Proposition 3.2.1, $f_{12} : U_1 \to U_2$ extends to a birational map $F : S \to S$.*

PROOF. Let $y \in S$ and K, with $K' \subset S$, be two (nonoverlapping) chains of lines joining x and x' on Ω to y. We may choose a Harish-Chandra chart $\mathbb{C}^n \subset S$ such that $\Omega \Subset \mathbb{C}^n$, $y \in \mathbb{C}^n$, no irreducible component of K or K' lies on $S - \mathbb{C}^n$ and all points $C_i \cap C_{i+1}$ and $C_j' \cap C_{j+1}'$ lie on $\mathbb{C}^n$. We can now join x to y by a continuous path on $\mathbb{C}^n$ consisting of paths on C_i; similarly x' can be joined to y

by a continuous path consisting of paths on C'_j. Joining x' to x by a continuous path on Ω we obtain a closed continuous path $\gamma(t)$, $\gamma : [0, 1] \to \mathbb{C}^n \subset S$. f can then be analytically continued along λ to obtain $\hat{f}$, such that the germ $\hat{f}_x$ (by abuse of notations) at $t = 1$ may *a priori* be distinct from the germ f_x at $t = 0$. We are going to exclude the latter possibility by making use of $\mathbb{C}^*$-actions on S. For $\lambda \in \mathbb{C}^*$, f can be analytically continued along the path γ_λ given by $\gamma_\lambda(t) = \lambda(\gamma(t))$. Denote by $\hat{f}^\lambda$, with $\hat{f}^1 = \hat{f}$, the analytic continuation of f as a meromorphic map on a tubular neighborhood of γ_λ. For λ small enough, γ_λ lies on Ω. As f is defined on Ω for the germs of the extended maps at $t = 1$ we have $\hat{f}^\lambda_x = f_x$ for λ small, hence for $\lambda = 1$, by the identity theorem on holomorphic functions. With this we have proven that f can be analytically continued from Ω to S. Applying this to the $\mathcal{F}$-preserving biholomorphism $f_{12} : U_1 \to U_2$ in Proposition 3.2.1 and to its inverse $f_{21} : U_2 \to U_1$, we conclude that f_{12} can be extended to a birational map $F : S \dashrightarrow S$. The proof of Lemma 3.2.4 is complete. $\square$

For the proof of Proposition 3.2.1 it remains to establish one more fact:

LEMMA 3.2.5. *Let S be an irreducible Hermitian symmetric manifold and $F :$ $S \dashrightarrow S$ be a birational self-map. Suppose for a generic line C on S, $F|_C$ maps C onto a line C'. Then, F is a biholomorphism.*

PROOF. We denote by $B \subset S$ the subvariety on which F fails to be a local biholomorphism and call B the bad locus of F. Suppose B is of codimension at least 2 (and the same applies to F^{-1}), then F induces a linear isomorphism θ on $\Gamma(S, K_S^{-1})$ by pulling back. Identifying S with its image under the projective embedding by K_S^{-1}, F is nothing other than the restriction of the projectivization

$$[\theta^*] : \mathbb{P}\Gamma(S, K_S^{-1})^* \to \mathbb{P}\Gamma(S, K_S^{-1})^*$$

to S, thus a biholomorphism.

It remains to show that the bad locus B of F is of codimension at least 2. Otherwise let $R \subset B$ be an irreducible component of codimension 1. Choose a connected open subset U on which F is an open embedding. Let $x_o \in U$ and C be a line passing through x_o. C is standard and small deformations of C fill up a tubular neighborhood G of C. Write $Z \subset B$ for the set of indeterminacies of F. Since X is of Picard number 1, C must intersect R. Deforming $x_o \in U$ and hence C slightly without loss of generality we may assume that C intersects R at a point $x_1 \in R - Z$. Since F is holomorphic and ramified at x_1 there exists a nonzero tangent vector $\eta \in T_{x_1}(S)$ such that $dF(\eta) = 0$. For any $x \in C$ we denote by $\alpha(x)$ some nonzero vector tangent to C at x.

Either $\eta \notin P_{\alpha(x_1)}$ or $\eta \in P_{\alpha(x_1)}$. In both cases we are going to obtain a contradiction. Since $T(S)|_C$ is semipositive there exists $s \in \Gamma(C, T(S)|_C)$ such that $s(x_1) = \eta$. Suppose $\eta \notin P_{\alpha(x_1)}$, then $s(x) \notin P_{\alpha(x)}$ for a generic $x \in C$. On the other hand, since $F|_C : C \to S$ is a biholomorphism onto a line

$C' = F(C) \subset S$, F_*s is a well-defined holomorphic section in $\Gamma(C', T(S))$ such that for $y_1 = F(x_1)$, $F_*s(y_1) = dF(s(x_1)) = dF(\eta) = 0$. From $F_*s(y_1) = 0$ it follows that $F_*s(y) \in P_{\beta(y)}$ for any $y \in C', y = F(x), \beta(y)$ being a nonzero vector tangent to C' at y. Choosing x generic on C we get a contradiction from $s(x) \notin P_{\alpha(x)}, F_*s(y) \in P_{\beta(y)}$ and from $dF(P_{\alpha(x)}) = P_{\beta(y)}$, which follows from $dF(\tilde{\mathcal{C}}_x) = \tilde{\mathcal{C}}'_y$.

Suppose now $\eta \in P_{\alpha(x_1)}$. Write $T(S)|_C \cong \mathcal{O}(2) \oplus [\mathcal{O}(1)]^p \oplus \mathcal{O}^q$. Let $s \in \Gamma(C, \mathcal{O}(2) \oplus [\mathcal{O}(1)]^p)$ be such that $s(x_1) = \eta$. We may choose s so that s vanishes at some point $x_0 \in C$. Then, for the corresponding decomposition $T(S)|_{C'} \cong \mathcal{O}(2) \oplus [\mathcal{O}(1)]^p \oplus \mathcal{O}^q$, since F preserves $\mathcal{C}$ along C, we have $F_*s \in \Gamma(C', \mathcal{O}(2) \oplus [\mathcal{O}(1)]^p)$ such that F_*s is not tangent to C' and such that $F_*s(y_0) = F_*s(y_1) = 0$, a plain contradiction. Since $\mathrm{Ker}(dF(x_1)) \neq 0$ leads in any event to a contradiction, we have proven that the bad set of F is of codimension at least 2 in S. The proof of Lemma 3.2.5 is complete. $\qquad\square$

With Lemma 3.2.5 we have completed the proof of Proposition 3.2.1.

4. Minimal Rational Tangents and Holomorphic Distributions on Rational Homogeneous Manifolds of Picard Number 1

4.1. In the study of Fano manifold of Picard number 1 through their varieties of minimal rational tangents, next to irreducible Hermitian symmetric manifolds we have the rational homogeneous manifolds S of Picard number 1. The non-symmetric ones are distinguished by the existence of nontrivial homogeneous holomorphic distributions.

In Sections 1.4 and 2.3 we studied the case of homogeneous contact manifolds. As will be seen, the contact case is special among the nonsymmetric ones. By the Tanaka–Yamaguchi theory of differential systems on S, Ochiai-type theorems hold for any $S \neq \mathbb{P}^n$. On nonsymmetric S, there is a natural distribution D^1, whose definition will be recalled shortly. In case S is neither of symmetric nor of contact type, it follows from [Yamaguchi 1993] that any D^1-preserving local holomorphic map extends to an automorphism of S. Yamaguchi's proof uses harmonic theory of Lie algebra cohomologies, just as Ochiai's proof of Proposition 1.5.2. We will give an alternate proof of relevant results from Tanaka–Yamaguchi, by the method of analytic continuation as in Section 3. We start by recalling some basic facts concerning S.

Fixing a base point $o \in S$, we may write $S = G/P$, where G is a connected and simply-connected simple complex Lie group and $P \subset G$ is the maximal parabolic subgroup fixing o. Let $\mathfrak{g}$ be the Lie algebra of G and $\mathfrak{p}$ be the parabolic subalgebra corresponding to P. Write $\mathfrak{u} \subset \mathfrak{p}$ for the nilpotent radical and let $\mathfrak{p} = \mathfrak{u} + \mathfrak{l}$ be a choice of Levi decomposition. The center $\mathfrak{z}$ of $\mathfrak{l}$ is one-dimensional. We fix a Cartan subalgebra $\mathfrak{h} \subset \mathfrak{l}$, which is also a Cartan subalgebra of $\mathfrak{g}$. We have the root system $\triangle \subset \mathfrak{h}^*$ of $\mathfrak{g}$ with respect to $\mathfrak{h}$. We can uniquely determine a set $\triangle^+$

of positive roots by requiring that $\mathfrak{u}$ is contained in the span of negative root spaces. (Here our sign convention is opposite to the choice in some references, e.g. [Yamaguchi 1993]. As many geometers do, we prefer this choice for the reason that positive roots correspond to positive line bundles.) Fix a system of simple roots $\Sigma = \{\alpha_1, \ldots, \alpha_r\}$. The maximality of $\mathfrak{p}$ implies that there is a unique simple root α_i satisfying $\alpha_i(\mathfrak{z}) \neq 0$. We say that S is of type $(\mathfrak{g}, \alpha_i)$.

Conversely, given a Cartan subalgebra $\mathfrak{h}$, a simple root system of $(\mathfrak{g}, \mathfrak{h})$ and a distinguished simple root α_i, we can recover $\mathfrak{p} \subset \mathfrak{g}$ and hence $S = G/P$, as follows. For an integer k, $-m \leq k \leq m$, we define $\triangle_k$ to be the set of all roots $\sum_{q=1}^{r} m_q \alpha_q$ with $m_i = k$. Here m is the largest integer such that $\triangle_m \neq 0$. For $\alpha \in \triangle$ we denote by $\mathfrak{g}_\alpha$ the corresponding root space. Write

$$\mathfrak{g}_0 = \mathfrak{h} \oplus \bigoplus_{\alpha \in \triangle_0} \mathfrak{g}_\alpha,$$

$$\mathfrak{g}_k = \bigoplus_{\alpha \in \triangle_k} \mathfrak{g}_\alpha, \quad \text{for } k \neq 0$$

for the eigenspace decomposition with respect to $ad(\mathfrak{z})$. More precisely, there exists an element $\theta \in \mathfrak{z}$ such that $[\theta, v] = kv$ for $v \in \mathfrak{g}_k$, so that the eigenspace decomposition $\mathfrak{g} = \bigoplus_{k=-m}^{m} \mathfrak{g}_k$ endows $\mathfrak{g}$ with the structure of a graded Lie algebra. We denote by $(\mathfrak{g}, \alpha_i)$ the Lie algebra $\mathfrak{g}$ with this graded structure and say that $(\mathfrak{g}, \alpha_i)$ (and S) is of depth m. We have

$$\mathfrak{p} = \mathfrak{g}_0 \oplus \mathfrak{g}_{-1} \oplus \cdots \oplus \mathfrak{g}_{-m};$$

$$\mathfrak{l} = \mathfrak{g}_0;$$

$$\mathfrak{u} = \mathfrak{g}_{-1} \oplus \cdots \oplus \mathfrak{g}_{-m}.$$

Identify $T_o(S)$ with $\mathfrak{g}/\mathfrak{p} \cong \mathfrak{g}_1 \oplus \cdots \oplus \mathfrak{g}_m$. For $1 \leq k \leq m$, the translates of $\mathfrak{g}_1 + \cdots + \mathfrak{g}_k$ under G defines a homogeneous holomorphic distribution D^k on S, so that $D^1 \subsetneq D^2 \subsetneq \cdots \subsetneq D^m = T(S)$ defines a filtration of the holomorphic tangent bundle.

For $x \in S$ we denote by $P_x \subset G$ the maximal parabolic subgroup fixing x (so that $P_o = P$). Denote by $U_x \subset P_x$ the unipotent radical, $L_x = P_x/U_x$, and regard D_x^1 as an L_x-representation space. Consider the set of all highest weight vectors ξ of D_x^1 as an L_x-representation space and denote by $\mathcal{W}_x \subset \mathbb{P}D_x^1$ the collection of projectivizations $[\xi]$. L_x acts transitively on $\mathcal{W}_x$, so that $\mathcal{W}_x \subset \mathbb{P}D_x^1$ is a rational homogeneous projective submanifold. We call $\mathcal{W}_x$ the variety of highest weight tangents. The collection of $\mathcal{W}_x$ as x ranges over S defines a homogeneous holomorphic fiber bundle $\mathcal{W} \to S$. We denote by $\mathcal{L}_x^1$ the image of L_x in the bundle of automorphisms $\mathcal{GL}(D_x^1)$ and denote by $\mathcal{L}^1 \to S$, $\mathcal{L}^1 \subset \mathcal{GL}(D^1)$, the fiber bundle thus obtained.

We proceed to relate varieties of highest weight tangents $\mathcal{W}_x \subset \mathbb{P}D_x^1$ with minimal rational curves. More generally, we discuss the construction of rational curves associated to roots. For $\rho \in \triangle^+$, let $H_\rho \in \mathfrak{h}_\rho = [\mathfrak{g}_\rho, \mathfrak{g}_{-\rho}]$ be such that

$\rho(H_\rho) = 2$. We call H_ρ the coroot of ρ. Basis vectors $E_\rho \in \mathfrak{g}_\rho$, $E_{-\rho} \in \mathfrak{g}_{-\rho}$ can be chosen such that $[E_\rho, E_{-\rho}] = H_\rho$, $[H_\rho, E_\rho] = 2E_\rho$, $[H, E_{-\rho}] = -2E_\rho$ so that the triple $(H_\rho, E_\rho, E_{-\rho})$ defines an isomorphism of $\mathfrak{s}_\rho = \mathfrak{h}_\rho \oplus \mathfrak{g}_\rho \oplus \mathfrak{g}_{-\rho}$ with $\mathfrak{sl}_2(\mathbb{C})$; see [Serre 1966, VI, Theorem 2, p. 43 ff.]. Let now $C_\rho \subset X$ be the $\mathbb{P}\,\mathrm{SL}(2,\mathbb{C})$ orbit of $o = eP$ under the Lie group $S_\rho \cong \mathbb{P}\,\mathrm{SL}(2,\mathbb{C})$, $S_\rho \subset G$ with Lie algebra $\mathfrak{s}_\rho$. Consider S of type $(\mathfrak{g}, \alpha_i)$. Write H_j for H_{α_j} and let ω_i be the i-th fundamental weight with $\omega_i(H_j) = \delta_{ij}$, and E be the underlying vector space of the representation of G, with lowest weight $-\omega_i$, defining the first canonical embedding $\tau : S \hookrightarrow \mathbb{P}E$. We have $Hv = -\omega_i(H)v$ for any $H \in \mathfrak{h}$ and a lowest weight vector $v \in E$. For the rational curve C_ρ with $\omega_i(H_\rho) = s$ we have $H_\rho v = -sv$. Since H_ρ is a generator of the weight lattice of $\mathfrak{s}_\rho$, the pull-back of $\mathcal{O}(1)$ on $\mathbb{P}E$ to C_ρ, which is the dual of the tautological line bundle, gives a holomorphic line bundle $\cong \mathcal{O}(s)$. In particular, for $\rho = \alpha_i$ we have $\omega_i(H_i) = 1$ so that $\tau(C_{\alpha_i})$ is a line, and $C_{\alpha_i} \subset S$ represents a generator of $H_2(S, \mathbb{Z}) \cong \mathbb{Z}$. We have

$$\tau : H_2(S, \mathbb{Z}) \xrightarrow{\cong} H_2(\mathbb{P}E, \mathbb{Z}) \cong \mathbb{Z}.$$

In general, $C_\rho \subset S$ is a rational curve of degree $s = \omega_i(H_\rho)$.

A minimal rational curve $C \subset S$ is of degree 1, and will also be called a line. C is called a highest weight line if and only if $[T_x(C)] \in \mathcal{W}_x$ at every $x \in C$. Since the lowest weight orbit in $\mathbb{P}\mathfrak{g}_1$ agrees with the highest weight orbit, $C_{\alpha_i} \subset S$ is a highest weight line. We will see that in case all roots of $\mathfrak{g}$ are of equal length, any line is a highest weight line. This is not the case in general.

From now on we will assume $S \neq \mathbb{P}^n$. We have the following result of Tanaka [1979] and Yamaguchi [1993] and its immediate corollary (see [Hwang and Mok 1999]).

PROPOSITION 4.1.1. *Let $U \subset S$ be a connected open set. Then a holomorphic vector field on U can be extended to a global holomorphic vector field on S if it preserves $\mathcal{W}|_U$. Furthermore, if S is neither of symmetric type nor of contact type, then a holomorphic vector field on U can be extended to a global holomorphic vector field on S if it preserves $D^1|_U$.*

COROLLARY 4.1.2. *Let U_1, $U_2 \subset S$ be connected open sets and $f_{12} : U_1 \to U_2$ a biholomorphic map preserving the distribution $D^1 \subset T(S)$. If S is neither of symmetric nor of contact type, then f_{12} can be extended to a biholomorphic automorphism of S. When S is of symmetric type or of contact type, f_{12} can be extended to a biholomorphic automorphism of S, if f_{12} preserves the fiber subbundle $\mathcal{W} \subset \mathbb{P}D^1$.*

The proof of Proposition 4.1.1 as given in [Tanaka 1979; Yamaguchi 1993] requires algebraic machinery that are quite distinct from techniques explained in this survey. For S of symmetric type, this is just Ochiai's theorem which we proved in Section 3.2. The same proof works for S of contact type. In Section 4.2 we will give directly a proof of Corollary 4.1.2 for the case when all roots of $\mathfrak{g}$

are of equal length, by showing that a local D^1-preserving holomorphic map necessarily preserves the bundle $\mathcal{W} \subset \mathbb{P}D^1$ of varieties of highest weight tangents and by applying the method of analytic continuation in Sections 3.1 and 3.2. An adaptation of the argument will apply even in the case with roots of unequal lengths. We will need the following obvious interpretation of the Frobenius form.

LEMMA 4.1.3. *Let S be a rational homogeneous manifold of Picard number 1 and of depth $m \geq 2$. Let k be a positive integer $1 \leq k < m$ and write F^k : $D_o^k \otimes D_o^k \to T_o(S)/D_o^k$ for the Frobenius form for the distribution D^k at $o \in S$. Under an identification of $T_o(S)$ with $\mathfrak{g}_1 \oplus \cdots \oplus \mathfrak{g}_m$, we have $F^k(\xi, \zeta) = [\xi, \zeta]$ mod $\mathfrak{g}_1 \oplus \cdots \oplus \mathfrak{g}_k$.*

Under the hypothesis of Corollary 4.1.2, it follows readily that $f_{12} : U_1 \to U_2$ preserves the set of $\xi \neq 0$ for which the rank of (F_ξ^1) is minimal. For $[\xi] \in \mathcal{W}_o$ it is easy to see that $\mathrm{rank}(F_\xi^1) \leq \mathrm{rank}(F_\eta^1)$ for any nonzero $\eta \in D_o^1$. In the contact case $\mathrm{rank}(F_\xi^1) = 1$ for any $\xi \neq 0$, and f_{12} does not necessarily preserve $\mathcal{W}$. For the noncontact case it is however not straightforward in the case of exceptional Lie algebras $\mathfrak{g} = E_6, E_7, E_8$ to determine $\mathrm{rank}(F_\xi^1)$.

4.2. We consider in what follows the case of simple Lie algebras $\mathfrak{g}$ for which all roots are of equal length, including $\mathfrak{g} = D_n$ $(n \geq 4)$, E_6, E_7, E_8, (for which there are associated $(\mathfrak{g}, \alpha_i)$ neither of symmetric nor of contact type). We start with a discussion of the root space decomposition for $\mathfrak{g}_1$. Consider a highest weight line C, $o \in C$, $T_x(C) = \mathbb{C}E_\mu$ for a root vector E_μ corresponding to a highest weight $\mu \in \mathfrak{h}^*$ of $\mathfrak{g}_1$. Define

$$\triangle_1'(\mu) = \{\rho \in \triangle_1 : \mu - \rho \in \triangle\},$$
$$\triangle_1''(\mu) = \{\rho \in \triangle_1 : \mu + \rho \in \triangle\},$$
$$\triangle_1^\perp(\mu) = \{\rho \in \triangle_1 : \mu - \rho, \mu + \rho \notin \triangle\}.$$

When all roots of $\mathfrak{g}$ are of equal length, any ρ-chain attached to μ is of length at most 2, so that $\triangle_1 = \{\mu\} \cup \triangle_1'(\mu) \cup \triangle_1^\perp(\mu) \cup \triangle_1''(\mu)$ is a disjoint union. We have the following corresponding lemma on the Grothendieck splitting of D^1 over C.

LEMMA 4.2.1. *Let S be a rational homogeneous manifold of the above type, and $C \subset S$ be a rational curve tangent to the distribution D^1. Then, $D^1|_C$ is of the form $\mathcal{O}(2) \oplus [\mathcal{O}(1)]^u \oplus \mathcal{O}^v \oplus [\mathcal{O}(-1)]^r$ for some nonnegative integers u, v and r.*

PROOF. Since H_μ is a generator for the weight lattice of $\mathfrak{s}_\mu = \mathfrak{h}_\mu \oplus \mathfrak{g}_\mu \oplus \mathfrak{g}_{-\mu}$, the root space decomposition of D_o^1 gives rise to a Grothendieck splitting of $D^1|_C$, with $\mathfrak{g}_\rho$ corresponding to the direct summand $\mathcal{O}(d_\rho)$, where $[H_\mu, E_\rho] = d_\rho E_\rho$, that is, $d_\rho = \rho(H_\mu)$. For $\mathfrak{g}$ with roots of equal length, $d_\rho = 2, 1, 0, -1$, corresponding to the decomposition $\triangle_1 = \{\mu\} \cup \triangle_1'(\mu) \cup \triangle_1^\perp(\mu) \cup \triangle_1''(\mu)$. □

We write P_α for the positive part at o, Z_α for $\mathcal{O}_o^v$ and N_α for $[\mathcal{O}(-1)]_o^r$. The proof of Lemma 4.2.1 also shows that $D^k/D^{k-1}|_C$ can have only summands of degree 1, 0 and -1 for $k > 1$. We know that $T(S)|_C$ must be of the form

$\mathcal{O}(2) \oplus [\mathcal{O}(1)]^p \oplus \mathcal{O}^q$. The quotient bundle $T(S)/D^1|_C$ is semipositive. From the knowledge of splitting types of $D^k/D^{k-1}|_C$ and using composition series, we see that $T(S)/D^1|_C$ has at most summands of degree 1 and 0, and that the number of $\mathcal{O}(1)$'s in the Grothendieck splitting is exactly r, the number of roots in $\triangle_1''(\mu)$. This implies $u = p$, namely, that every deformation of a highest weight line is a highest weight line. We may thus take $\mathcal{W} \to S$ to be a bundle of varieties of minimal rational tangents.

The Grothendieck decomposition of $D^1|_C$ as in Lemma 4.2.1 implies that $\mathcal{W}_o$ is the closure of the graph of a vector-valued cubic polynomial in p variables. More precisely, let $\Theta \subset \triangle_0$ be the set of positive roots such that $\mu - \theta \in \triangle_1'$. Then $|\Theta| = p$. Write $\Theta = \{\theta_1, \ldots, \theta_p\}$ and E_{-a} for $E_{-\theta_a}$. In a neighborhood of $[\alpha]$, $\alpha = E_\mu$, we have the cubic expansion of $\mathcal{W}_o$ as the closure of the image of $[\Phi] : \mathbb{C}^p \longrightarrow \mathbb{P}T_o(S)$ for the vector-valued cubic polynomial $\Phi : \mathbb{C}^p \longrightarrow T_o(S)$ defined by

$$\Phi(z) = E_\mu + \sum_a [E_\mu, E_{-a}] z^a + \frac{1}{2!} \sum_{a,b} [[E_\mu, E_{-a}], E_{-b}] z^a z^b$$
$$+ \frac{1}{3!} \sum_{a,b,c} [[[E_\mu, E_{-a}], E_{-b}], E_{-c}] z^a z^b z^c.$$

We are now ready to state the following result which reduces distribution-preserving local maps to those preserving varieties of highest weight tangents.

PROPOSITION 4.2.2. *Let S be a rational homogeneous manifold of Picard number 1. Assume that S is neither of the symmetric nor of the contact type, and that it is of type $(\mathfrak{g}, \alpha_i)$ for some simple Lie algebra $\mathfrak{g}$ for which all roots are of equal length. Denote by $D^1 \subset T(S)$ the homogeneous holomorphic distribution corresponding to $\mathfrak{g}_1$, and by $\mathcal{W} \subset \mathbb{P}D^1$ the homogeneous holomorphic fiber bundle of varieties of highest weight tangents. Then, any D^1-preserving germ of holomorphic maps must preserve $\mathcal{W}$.*

For the proof of Proposition 4.2.2 we will need a number of lemmas.

LEMMA 4.2.3. *Let S be a rational homogeneous manifold as in Proposition 4.2.2, $o \in S$ be a fixed base point, and $F : D_o^1 \times D_o^1 \to D_o^2$ be the Frobenius form. For $\xi \in D_o^1$ denote by $F_\xi : D_o^1 \to D_o^2$ the linear map defined by $F_\xi(\zeta) = F(\xi, \zeta)$. Let α be a highest weight vector of D_o^1 as an L_o-representation space. Then, there exists some nonzero vector $\eta \in D_o^1$ such that $\mathrm{rank}(F_\eta) > \mathrm{rank}(F_\alpha)$.*

PROOF. For $k \geq 1$ let μ_k and $\lambda_k \in \triangle_k$ denote, respectively, the highest and lowest weight of $\mathfrak{g}_k$. For any $(\mathfrak{g}, \alpha_i)$ of the contact type, $\mathfrak{g}_2$ is 1-dimensional and $\lambda_2 = \lambda$ is the only weight in $\triangle_2$, while $\lambda - \rho$ is a positive root for any $\rho \in \triangle_1$. For S as in the Lemma, in particular not of the contact type, by a straightforward checking, $\lambda_2 - \mu_1 \notin \triangle_1$. In fact, λ_2 does not even dominate μ_1. Write $\alpha = E_{\mu_1}$ and $\triangle_1''(\mu_1) = \{\rho(1), \ldots, \rho(r)\}$. Then, $\mathrm{rank}(F_\alpha) = r$. Since $[\mathfrak{g}_1, \mathfrak{g}_1] = \mathfrak{g}_2$ we have $\lambda_2 = \varphi_1 + \psi_1$ for some $\varphi_1, \psi_1 \in \triangle_1$. There are two possibilities. Either both $\varphi_1, \psi_1 \in \Delta_1^+(\mu_1)$, or we may take $\varphi_1 \in \Delta_1''(\mu_1)$, $\psi_1 \in \Delta_1'(\mu_1)$. Writing $\beta = E_{\varphi_1}$ and $\gamma = E_{\psi_1}$ we have in both cases $[\alpha, \gamma] = 0$. For $t \in \mathbb{C}$ consider the vector $\eta_t \in \mathfrak{g}_1$ given by $\eta_t = \alpha + t\beta$. We may choose basis vectors E_ρ of $\mathfrak{g}_\rho$ such that $[\alpha, E_{\rho(i)}] = \pm E_{\mu_1 + \rho(i)}$; $[\eta_t, \gamma] = [\alpha, \gamma] + t[\beta, \gamma] = \pm t E_{\lambda_2}$. Since $\lambda_2 - \mu_1 \notin \Delta_1$, E_{λ_2} is not proportional to any $E_{\mu_1 + \rho(i)}$ and is linearly independent of $[\alpha, \mathfrak{g}_1] = Im(F_\alpha)$. As $E_{\lambda_2} \in Im(F_{\eta_t})$ for $t \neq 0$ it follows that $[\eta_t, \mathfrak{g}_1]$ contains at least $p + 1$ linearly independent elements for $t \neq 0$ sufficiently small, so that $\mathrm{rank}(F_{\eta_t}) > \mathrm{rank}(F_\alpha)$, as desired.   □

LEMMA 4.2.4. *Let S be a rational homogeneous manifold as in Proposition 4.2.2. Suppose there exists some D^1-preserving germ of holomorphic map which does not preserve $\mathcal{W}$. Then, there exists a holomorphic bundle of connected reductive Lie groups $\mathcal{H} \to S$, $\mathcal{L}^1 \subsetneq \mathcal{H} \subsetneq \mathcal{GL}(D^1)$ such that, denoting by $\mathcal{V} \subset \mathbb{P}D^1$ the orbit $\mathcal{H} \cdot \mathcal{W}$ under the natural action of $\mathcal{H}$ on $\mathbb{P}D^1$, the canonical map $\mathcal{V} \to S$ realizes $\mathcal{V}$ as a holomorphic fiber bundle for which the fibers $\mathcal{V}_x \subset \mathbb{P}D_x^1$ are rational homogeneous submanifolds conjugate to each other under projective linear transformations. Moreover, $\mathcal{W} \subsetneq \mathcal{V} \subsetneq \mathbb{P}D^1$.*

PROOF. Consider the group Q of germs of holomorphic maps $f : (S, o) \to (S, o)$ such that f preserves D^1. We can identify the maximal parabolic P as a subgroup of Q. Let $A \subset \mathrm{GL}(D_o^1)$ be the algebraic subgroup consisting of all $d\varphi(o) \in \mathrm{GL}(D_o^1)$, $\varphi \in Q$. Any A-invariant subvariety contains $\mathcal{W}_o$ and hence $A \cdot \mathcal{W}_o$, the orbit of $\mathcal{W}_o$ under A. $A \cdot \mathcal{W}_o$ is a constructible set and its Zariski closure $\overline{A \cdot \mathcal{W}_o}$ is again A-invariant. The complement B of $A \cdot \mathcal{W}_o$ in $\overline{A \cdot \mathcal{W}_o}$ is A-invariant. B is constructible and its Zariski closure $\overline{B} \subset \overline{A \cdot \mathcal{W}_o}$ is a proper A-invariant subvariety. It follows that B must be empty, otherwise $\overline{B}$ would contain $A \cdot \mathcal{W}_o$, so that $\overline{B} = \overline{A \cdot \mathcal{W}_o}$, a plain contradiction. Thus, $\mathcal{V}_o := A \cdot \mathcal{W}_o$ is a closed subvariety in $\mathbb{P}D_o^1$. By assumption there exists $\nu \in A$ such that $\nu(\mathcal{W}_o) \neq \mathcal{W}_o$, so that $\mathcal{W}_o \subsetneq \mathcal{V}_o$. Since $\mathcal{V}_o$ is homogeneous under A, it must be smooth. As each component of $\mathcal{V}_o$ is P-invariant and contains $\mathcal{W}_o$, we conclude that $\mathcal{V}_o$ is an irreducible homogeneous submanifold of $\mathbb{P}D_o^1$. Let $H_o \subset \mathrm{GL}(D_o^1)$ be the identity component of A. Then, $\mathcal{V}_o = H_o \cdot \mathcal{V}_o \subset \mathbb{P}D_o^1$ is a rational homogeneous manifold equivariantly embedded in $\mathbb{P}D_o^1$. Passing to projectivizations it follows readily from Borel's fixed point theorem that $H_o \subset \mathrm{GL}(D_o^1)$ is reductive.

To prove Lemma 4.2.4 it remains to show that $\mathcal{V}_o \subsetneq \mathbb{P}D_o^1$. Denote by $Z_o \subset \mathbb{P}D_o^1$ the subset consisting of all $[\eta]$ such that $\mathrm{rank}(F_\eta) = \mathrm{rank}(F_\alpha)$, with $[\alpha] \in \mathcal{W}_o$. By Lemma 4.2.3, $Z_o \subsetneq \mathbb{P}D_o^1$. On the other hand, for any D^1-preserving $\varphi \in Q$, $\mathrm{rank}(F_{d\varphi(\eta)}) = \mathrm{rank}(F_\eta)$, so that $\mathcal{V}_o = A \cdot \mathcal{W}_o \subset Z_o \subsetneq \mathbb{P}D_o^1$, as desired. □

We will prove Proposition 4.2.2 by getting a contradiciton to $\mathcal{W} \subsetneq \mathcal{V} \subsetneq \mathbb{P}D^1$. The idea is a variation of the proof of $\mathcal{C}_x = \mathcal{W}_x$ in Section 1.5.

LEMMA 4.2.5. *Let $\mathcal{W} \subsetneq \mathcal{V} \subsetneq \mathbb{P}D^1$ be as given in Lemma 4.2.4. Then, for each $[\alpha] \in \mathcal{W}_o, T_{[\alpha]}(\mathcal{V}_o) \subset (P_\alpha \oplus Z_\alpha)/\mathbb{C}\alpha$.*

PROOF. Let $C \subset S$ be a minimal rational curve passing through o. In what follows we write D for D^1. By Lemma 4.2.1 we have

$$(D^* \otimes D)|_C \cong ([\mathcal{O}(1)]^r \oplus \mathcal{O}^v \oplus [\mathcal{O}(-1)]^p \oplus \mathcal{O}(-2)]) \otimes (\mathcal{O}(2) \oplus [\mathcal{O}(1)]^p \oplus \mathcal{O}^v \oplus [\mathcal{O}(-1)]^r).$$

Write $T_o(C) = \mathbb{C}\alpha$ and take $\omega^* \in D_o^*$ to be a covector annihilating $P_\alpha \oplus Z_\alpha$. Then, $\omega^* \in D_o^*$ lies in the well-defined direct summand $[\mathcal{O}(1)]^r$ of $D^*|_C$. To prove the Lemma it suffices to prove that $\omega^*(\eta) = 0$ whenever $\eta \mod \mathbb{C}\alpha \in T_{[\alpha]}(\mathcal{V}_o)$. As in Section 1.5 let $U \subset D^* \otimes D$ be the holomorphic subbundle where $U_x \subset D_x^* \otimes D_x = \mathfrak{gl}(D_x)$ is the Lie algebra of H_x for any $x \in S$. Consider the direct sum decomposition $D^* \otimes D = U \oplus U^\perp$. The decomposable tensor $\alpha \otimes \omega^*$ lies in $F := \mathcal{O}(2) \otimes [\mathcal{O}(1)]^r \subset (D^* \otimes D)|_C$. Since $F \cong [\mathcal{O}(3)]^r$ and every direct summand of $(D^* \otimes D)|_C$ is of degree at most 3, we have $F = F' \oplus F''$ with $F' \subset U|_C$ and $F'' \subset U^\perp|_C$ from the uniqueness of Grothendieck decompositions. Since every element of F is of the form $\alpha \otimes \tau^*$ for some $\tau^* \in [\mathcal{O}(1)]_x^r$, we must have correspondingly a decomposition $[\mathcal{O}(1)]_x^r = Q' \oplus Q''$ such that $F' = \mathbb{C}\alpha \otimes Q'$ and $F'' = \mathbb{C}\alpha \otimes Q''$. The arguments of Section 1.5 show that U_x contains no nonzero decomposable tensor element, implying therefore that $Q' = 0$ and hence $F = F'' \subset U^\perp|_C$, which means that $\omega^*(\eta) = 0$ whenever $\eta \mod \mathbb{C}\alpha$ is tangent to $\mathcal{V}_o$ at $[\alpha]$, as desired. $\qquad \square$

LEMMA 4.2.6. *Let $\zeta \in Z_\alpha$ be such that $[\zeta, P_\alpha] = 0$ for the Lie bracket $[\cdot, \cdot]$: $\mathfrak{g}_1 \times \mathfrak{g}_1 \to \mathfrak{g}_2$. Then $\zeta = 0$.*

PROOF. Write μ for the highest weight in $\mathfrak{g}_1$ and choose $\alpha = E_\mu$. Recall that Θ is the set of positive roots θ in $\triangle_o$ such that $\mu - \theta = \rho \in \triangle_1'(\mu)$. Then, for any $\theta \in \Theta$ we have $[\zeta, E_{-\theta}] = \pm[\zeta, [\bar{\alpha}, E_\rho]]$. Since $[\bar{\alpha}, \zeta] = 0$ and by hypothesis $[\zeta, P_\alpha] = 0$ we conclude from the Jacobi identity that $[\zeta, E_{-\theta}] = 0$. We proceed to deduce $\zeta = 0$ from $[\zeta, P_\alpha] = 0$ by showing that the latter implies $[\zeta, Z_\alpha] = [\zeta, N_\alpha] = 0$, so that $[\zeta, \mathfrak{g}_1] = 0$, implying $\zeta = 0$. To see this, by the cubic expansion of $\mathcal{W}_o$ and writing $\xi_a = [\alpha, E_{-a}]$, Z_α is spanned by $\zeta_{ab} = [[\alpha, E_{-a}], E_{-b}]] = [\xi_a, E_{-b}]$, $a, b \in \Theta$, so that $[\zeta, \zeta_{ab}] = [[\zeta, \xi_a], E_{-b}] - [[\zeta, E_{-b}], \xi_a] = 0$. Similarly, N_α is spanned by $\omega_{abc} = [[[\alpha, E_{-a}], E_{-b}], E_{-c}] = [\zeta_{ab}, E_{-c}]$ so that $[\zeta, \omega_{abc}] = [[\zeta, \zeta_{ab}], E_{-c}] - [[\zeta, E_{-c}], \zeta_{ab}] = 0$, as desired. $\qquad \square$

PROOF OF PROPOSITION 4.2.2. It suffices now to prove that $\mathcal{V}$ as constructed in Lemma 4.2.4, $\mathcal{W} \subsetneqq \mathcal{V} \subsetneqq \mathbb{P}(D)$ cannot possibly exist. Suppose otherwise. Then for $[\alpha] \in \mathcal{W}_o \subsetneqq \mathcal{V}_o$ we have by Lemma 4.2.5, $T_{[\alpha]}(\mathcal{V}_o) = E \mod \mathbb{C}\alpha$ for some vector subspace E of D_o such that

$$P_\alpha \subsetneqq E \subset P_\alpha \oplus Z_\alpha.$$

By the polarization argument of Section 1.3 E is isotropic with respect to the vector-valued symplectic form $[\cdot, \cdot]$. It follows that there exists some nonzero vector $\zeta \in Z_\alpha$ such that $[\zeta, P_\alpha] = 0$, contradicting Lemma 4.2.6. The proof of Proposition 4.2.2 is complete. $\qquad \square$

For S of type $(\mathfrak{g}, \alpha_i)$ as in Proposition 4.2.2, the arguments of analytic continuation of Sections 3.1 and 3.2 apply to show that any D^1-preserving germ of holomorphic map extends to an automorphism of S, as in Corollary 4.1.2. The proof of Proposition 4.2.2 is also valid for $\mathfrak{g} = B_n$ and for (F_4, α_4). The remaining cases of (C_n, α_i), (F_4, α_2), (F_4, α_3), (G_2, α_2) are characterized by the fact that $\mathcal{W} \subsetneq \mathcal{C}$, that is, by the existence of minimal rational curves other than highest weight lines. We will say for short that $(\mathfrak{g}, \alpha_i)$ is of excessive type. We will exclude (G_2, α_2) from our discussion, since the underlying complex manifold is biholomorphic to the 5-dimensional hyperquadric. For the rest, $\mathcal{C}_s$ is the closure of the isotropy orbit of the highest weight vector in $\mathfrak{g}_2$, from which it can be shown that $\mathcal{W} = \mathcal{C} \cap \mathbb{P}D^1$.

For $(\mathfrak{g}, \alpha_i)$ of excessive type with $\mathfrak{g} = C_n$, F_4; we can still apply the method of analytic continuation to prove Corollary 4.1.2, provided that we prove (1) the analogue of Proposition 4.2.2 and (2) that any $\mathcal{W}$-preserving germ of holomorphic map is $\mathcal{E}$-preserving for the foliation $\mathcal{E}$ on $\mathcal{W}$ defined by highest weight lines, $\mathcal{E} = \mathcal{F}|_{\mathcal{W}}$. (1) can be done by a straightforward verification that highest weight vectors ξ are characterized by the minimality of $\mathrm{rank}(F_\xi^1)$, which we omit. (2) can be done by an adaptation of Section 3.1, as follows. We consider the distribution $\mathcal{R}$ on $\mathcal{W}$ defined by $\mathcal{R} = (d\sigma)^{-1}\mathcal{P}$ for $\sigma : \mathcal{W} \to S$ the restriction of $\pi : \mathcal{C} \to S$ to $\mathcal{W}$. Then, $\mathcal{E} \subset \mathrm{Ch}(\mathcal{R})$. If φ is a germ of $\mathcal{W}$-preserving holomorphic map, then $\varphi^*\mathcal{E}$ is a foliation such that $\varphi^*\mathcal{E} \subset \mathrm{Ch}(\mathcal{R})$. If $\mathcal{E} \neq \varphi^*\mathcal{E}$ then at a generic point $[\alpha]$ of $\mathcal{W}$ we have some vertical vector $\eta \neq 0$ tangent to $\mathcal{W}_x$ at $[\alpha]$, $x = \sigma([\alpha])$, such that $\eta \in \mathrm{Ch}(\mathcal{R})$, by comparing leaves of $\mathcal{E}$ and $\varphi^*\mathcal{E}$ through the same point. Writing $\eta = v \mod \mathbb{C}\alpha$ the arguments of Theorem 3.1.3 then shows that for $k = \sum_i e_i \frac{\partial}{\partial v_i}$ tangent to $\mathcal{W}'_x$, $\sum_i v(e_i)\frac{\partial}{\partial v_i}$ remains tangent to $\mathcal{C}'_x$ (not $\mathcal{W}'_x$). However, since $\mathcal{W}'_x \subset \gamma^{-1}D^1_x$, we conclude that

$$\sum_i v(e_i)\frac{\partial}{\partial v_i} \in \gamma^{-1}D^1_x \cap \mathcal{C}'_x = \mathcal{W}'_x,$$

so that η lies in the kernel of the Gauss map of $\mathcal{W}_x$ in $\mathbb{P}D^1_x$. As $\mathcal{W}_x \subsetneq \mathbb{P}D^1_x$ is linearly nondegenerate this leads to a contradiction.

5. Varieties of Distinguished Tangents and an Application to Finite Holomorphic Maps

5.1. We hope that readers who have followed this note so far would agree, at least partially, that it is quite rewarding to study $\mathcal{C}_x$. It will be very nice to construct something like $\mathcal{C}_x$ using nonrational curves, because general projective manifolds do not have rational curves at all. We can proceed as follows. For a given projective manifold Y, fix a component $\mathcal{M}$ of the Chow space of curves. Let $\mathcal{M}_y$ be the subscheme corresponding to curves through $y \in Y$. We have the tangent map $\Phi_y : \mathcal{M}_y \to \mathbb{P}T_y(Y)$ defined on those points corresponding to curves smooth at y. Then the closure of the image of Φ_y would play the role of $\mathcal{C}_x$. The problem is that quite often this would give the whole $\mathbb{P}T_y(Y)$ and we cannot get anything interesting out of it. However, by taking a special piece of the image of Φ_y, we get an interesting object, which plays an important role when we study generically finite holomorphic maps to uniruled manifolds. Here we will recall some basic definitions and main results of [Hwang and Mok 1999, Section 1].

Let $g : M \to Z$ be a regular map between two quasi-projective complex algebraic varieties. We can stratify M and Z into finitely many nonsingular quasi-projective subvarieties. On the other hand, given $g : M \to Z$ with both M and Z smooth, we can stratify M into finitely many quasi-projective subvarieties on each of which g has constant rank. Applying these two stratifications repeatedly, we can stratify M naturally into finitely many irreducible quasi-projective nonsingular subvarieties $M = M_1 \cup \cdots \cup M_k$, such that for each i, the reduced image $g(M_i)$ is nonsingular and the holomorphic map $g|_{M_i} : M_i \to g(M_i)$ is of constant rank. It will be called the g-stratification of M. The following two properties of this stratification are immediate:

(i) Any tangent vector to $g(M_i)$ can be realized as the image of the tangent vector to a local holomorphic arc in M_i.

(ii) When a connected Lie group acts on M and Z, and g is equivariant, M_i and $g(M_i)$ are invariant under this group action.

For a given projective manifold Y, fix $\mathcal{M}$ as above and let $\Phi_y : \mathcal{M}_y \to \mathbb{P}T_y(Y)$ be the tangent map, which is well-defined on a subset $\mathcal{M}_y^o \subset \mathcal{M}_y$ corresponding to curves smooth at y. Let $\{M_i\}$ be the Φ_y-stratification of $\mathcal{M}_y^o$. A subvariety of $\mathbb{P}T_y(Y)$ will be called a variety of distinguished tangents in $\mathbb{P}T_y(Y)$, if it is the closure of the image $\Phi_y(M_i)$ for some choice of $\mathcal{M}_y$ and M_i. Note that there exist only countably many subvarieties in $\mathbb{P}T_y(Y)$ which can serve as varieties of distinguished tangents, because the Chow space has only countably many components.

Given an irreducible reduced curve l in Y and a smooth point $y \in l$, consider $\mathcal{M}_y$ which parametrizes deformations of l fixing y. $[l]$ is contained in $\mathcal{M}_y^o$, where the tangent map is well-defined. Let M_1 be the component of the stratification of $\mathcal{M}_y^o$ associated to the tangent map, so that $[l] \in M_1$. The variety of distinguished tangents corresponding to M_1 is called the variety of distinguished tangents associated to l at y and is denoted by $\mathcal{D}_y(l)$. It is an irreducible subvariety and $\mathbb{P}T_y(l)$ is a smooth point on it. $\mathcal{D}_y(l)$ is a generalization of $\mathcal{C}_x$ for a general curve l.

Although we do not have Grothendieck splitting for general l, we can get partial information as follows. In the splitting for a standard minimal rational curve C,

$$T(X)|_C = \mathcal{O}(2) \oplus [\mathcal{O}(1)]^p \oplus [\mathcal{O}]^q,$$

the sum of the $\mathcal{O}(1)$-part and the $\mathcal{O}$-part can be replaced by the normal bundle of the general curve l. The $\mathcal{O}$-part alone can be studied as the part generated by sections of the conormal bundle of l. In general we have to be careful about the singularity of the curve. Let $N_l^* = \mathfrak{J}/\mathfrak{J}^2$ be the conormal sheaf of l, where $\mathfrak{J}$ denotes the ideal sheaf of l. We have a natural map $j : N_l^* \to \Omega(Y)|_l$, where $\Omega(Y) = \mathcal{O}(T^*(Y))$. j is injective if l is an immersed curve. In general, $\mathrm{Ker}(j)$ is a sheaf supported on finitely many points. Let N_l' be the image of j in $\Omega(Y)$. If l is a standard minimal rational curve, sections of N_l' correspond to sections of $\mathcal{O}^q$. So the dimension of $\mathcal{C}_x$ is $n - 1 - h^0(l, N_l')$. Using the property (i) of the stratification and general deformation theory, we can get a partial result for general l (see [Hwang and Mok 1999] for details):

PROPOSITION 5.1.1. *Let $y \in Y$ be a sufficiently general point and l be a curve smooth at y. Then the tangent space of $\mathcal{D}_y(l)$ at the point $\mathbb{P}T_y(l)$ has dimension at most $n - 1 - h^0(l, N_l')$, where $n = \dim(Y)$.*

In general, the variety of distinguished tangents for a nonrational curve is not as useful as $\mathcal{C}_x$, because we do not have a good choice of "minimal" $\mathcal{M}$ as in the case of uniruled manifolds. So far, their main interest is in connection with the study of finite morphisms to uniruled manifolds by the following theorem.

THEOREM 5.1.2. *Let $f : Y \to X$ be a generically finite morphism from a projective manifold Y to a uniruled manifold X. Choose $x \in X$ and $y \in f^{-1}(x)$ so that y is sufficiently general and $df : T_y(Y) \to T_x(X)$ is an isomorphism. Then each irreducible component of $df^{-1}(\mathcal{C}_x) \subset \mathbb{P}T_y(Y)$ is a variety of distinguished tangents $\mathcal{D}_y(l)$ for a suitable choice of a curve l through y.*

SKETCH OF PROOF. Choose a generic point $x \in X$ and a component $\mathcal{C}_1$ of $\mathcal{C}_x$. Choose a minimal rational curve C through x so that $\mathbb{P}T_x(C)$ is a generic point of $\mathcal{C}_1$. Let l be an irreducible component of $f^{-1}(C)$ through $y \in f^{-1}(x)$. For simplicity, assume that l is smooth so that $N_l' = N_l^*$. A nonzero section of the conormal bundle of C can be lifted to a nonzero section of the conormal bundle of l. Thus $h^0(l, N_l') \geq h^0(C, N_C^*)$.

Obviously $\mathbb{P}df_y^{-1}(T_x(C)) \in \mathcal{D}_y(l)$. Thus each generic point of $df_y^{-1}(\mathcal{C}_1)$ is contained in some $\mathcal{D}_y(l)$ for a suitable choice of a curve l, depending on C, satisfying $h^0(l, N_l') \geq h^0(C, N_C^*)$. Since there are only countably many subvarieties in $\mathbb{P}T_y(Y)$ which can serve as a variety of distinguished tangents, we can assume that $df_y^{-1}(\mathcal{C}_1) \subset \mathcal{D}_y(l)$, by choosing l generically. We have $\dim(\mathcal{C}_1) = n - 1 - h^0(C, N_C^*)$. Applying the previous proposition,

$$n - 1 - h^0(C, N_C^*) = \dim(df_y^{-1}(\mathcal{C}_1)) \leq \dim(\mathcal{D}_y(l))$$
$$\leq n - 1 - h^0(l, N_l') \leq n - 1 - h^0(C, N_C^*),$$

which implies $df_y^{-1}(\mathcal{C}_1) = \mathcal{D}_y(l)$. $\square$

5.2. As an application of the results of Section 5.1, we will prove the following rigidity theorem for generically finite holomorphic maps over rational homogeneous spaces of Picard number 1.

THEOREM 5.2.1. *Let S be a rational homogeneous space of Picard number 1 different from $\mathbb{P}^n$, and Y be any n-dimensional compact complex manifold. Given a family of surjective holomorphic maps $f_t : Y \to S$ parametrized by $\Delta = \{t \in \mathbb{C}, |t| < \varepsilon\}$, we have a holomorphic map $g : \Delta \to \mathrm{Aut}_o(S)$ with $g_0 = id_S$ so that $f_t = g_t \circ f_0$.*

PROOF. Choose a sufficiently small open set $U \subset Y$ so that $f_t|_U$ is biholomorphic for any $t \in \Delta$. Let $\mathcal{C} \subset \mathbb{P}T(S)$ be the variety of minimal rational tangents. By Theorem 5.1.2, $df_t^{-1}(\mathcal{C}_{f(y)})$ is a family of varieties of distinguished tangents for each $y \in U$. From the discreteness of varieties of distinguished tangents, we see that $df_t^{-1}(\mathcal{C}_{f_t(y)}) = df_0^{-1}(\mathcal{C}_{f_0(y)})$ for all $t \in \Delta$. Thus the biholomorphic map $g_t := f_t \circ f_0^{-1}$ from $f_0(U)$ to $f_t(U)$ preserves $\mathcal{C}$. By Corollary 4.1.2, g_t can be extended to an automorphism of S. Since $f_t = g_t \circ f_0$ on U, it must hold on the whole Y.   □

6. Lazarsfeld's Problem on Rational Homogeneous Manifolds of Picard Number 1

6.1.   When we have a surjective holomorphic map $f : Y \to Z$ between two projective manifolds, it is a general principle of complex geometry that the target Z is more positively curved than the source Y in a suitable sense. Among all projective manifolds, the projective space is most positively curved in the sense that projective manifolds with ample tangent bundles are projective spaces, a result of Mori ([Mori 1979]). Combining these two, one may ask: if a projective manifold Z is the image of a projective space under a holomorphic map, is Z itself a projective space? This was a conjecture of Remmert and Van de Ven [1960], proved by Lazarsfeld [1984]. Not surprisingly, Lazarsfeld used this result of Mori:

THEOREM 6.1.1 [Mori 1979]. *Let X be a Fano manifold and $P \in X$ be a point. If the restrictions of $T(X)$ to all minimal rational curves through P are ample, then X is a projective space.*

The idea of Lazarsfeld's proof is as follows. Given $f : \mathbb{P}^n \to Z$, it is immediate that Z is Fano. Choose a generic $P \in Z$ and consider any minimal rational curve C through P. Then $f^{-1}(C)$ must have ample normal sheaf, because it is a curve in $\mathbb{P}^n$. This forces C to have ample normal sheaf, and Z is a projective space by Theorem 6.1.1.

It is expected that rational homogeneous manifolds of Picard number 1 are the next most positively curved manifolds after projective spaces. So the following question of Lazarsfeld is a natural generalization of Remmert and Van de Ven's conjecture:

CONJECTURE 6.1.2 [Lazarsfeld 1984]. *Let S be a rational homogeneous manifold of Picard number 1. For any surjective holomorphic map $f : S \to X$ to a* *projective manifold* X, *either* X *is a projective space, or* $X \cong S$ *and* f *is a biholomorphism.*

Applying Theorem 6.1.1, as in the case of $S = \mathbb{P}^n$, we see that the problem is to understand the curves on S on which the restrictions of $T(S)$ are not ample. Of course, minimal rational curves are such examples. But in general, there are a lot of other curves with this property. When S is a hyperquadric, Paranjape and Srinivas [1989] showed that minimal rational curves are the only curves on S with this property, and using this, they settled the conjecture. When S is a Hermitian symmetric space, Tsai [1993] had classified certain classes of curves on S where the restrictions of $T(S)$ are not ample, and settled the conjecture. For this, he needed a very detailed study of the global geometry of curves on S, using fine structure theory of Hermitian symmetric spaces [Wolf 1972]. Generalizing his methods to other S looks hopelessly complicated. To start with, very little is known about global structure of curves on S. Furthermore, even the local picture, say the structure of isotropy representation of the parabolic group, has completely different features from the symmetric case. In [Hwang and Mok 1999], we have settled the conjecture in full generality by a different approach. We will survey this work in this section.

6.2. First, we reduce Conjecture 6.1.2 to the following extension problem of holomorphic maps.

THEOREM 6.2.1. *Let* S *be a rational homogeneous space of Picard number 1 different from* $\mathbb{P}^n$ *and* $f : S \to X$ *be a finite morphism to a projective manifold* X *different from* $\mathbb{P}^n$. *Let* $s, t \in S$ *be an arbitrary pair of distinct points such that* $f(s) = f(t)$ *and* f *is unramified at* s *and* t. *Write* φ *for the unique germ of holomorphic map at* s, *with target space* S, *such that* $\varphi(s) = t$ *and* $f \circ \varphi = f$. *Then* φ *extends to a biholomorphic automorphism of* S.

In fact, once Theorem 6.2.1 is proved, we can use automorphisms of S arising from various choices of φ to conclude that $f : S \to X$ is a quotient map by a finite group action on S. Then Lazarsfeld's conjecture follows from the following.

PROPOSITION 6.2.2. *Let* S *be a rational homogeneous space of Picard number 1 of dimension* $n \geq 3$, *different from* $\mathbb{P}^n$. *Suppose there exists a nontrivial finite cyclic group* $F \subset \mathrm{Aut}(S)$ *which fixes a hypersurface* $E \subset S$ *pointwise. Then* S *is the hyperquadric,* E *is equal to an* $\mathcal{O}(1)$-*hypersurface, and the quotient of* S *by* F, *endowed with the standard normal complex structure, is a projective space.*

In principle, Proposition 6.2.2 can be checked case by case. It can be proved also using induction on the dimension by showing that a suitable deformation of E is itself homogeneous and preserved by the F-action.

To prove Theorem 6.2.1, it suffices to show that φ preserves $\mathcal{C} \subset \mathbb{P}T(S)$ for S of symmetric type or contact type, and the distribution D^1 for the other S, by Corollary 4.1.2. For simplicity, we will assume that the Fano manifold X has the property that $\mathcal{C}_x \subset \mathbb{P}T_x(X)$ is a proper subvariety, namely $q > 0$. In the case $q = 0$ and X different from $\mathbb{P}^n$, essentially the same argument works when combined with the result in [Mok 1988, 2.4].

6.3. We will consider S of symmetric type or of contact type first. We need to show that φ sends $\mathcal{C}_{s_1}$ to $\mathcal{C}_{t_1}$ for s_1 sufficiently close to s. From Theorem 5.1.2, φ sends a variety of distinguished tangents $\mathcal{D}_{s_1}$ in $\mathbb{P}T_{s_1}(S)$ to a variety of distinguished tangents $\mathcal{D}_{t_1}$ in $\mathbb{P}T_{t_1}(S)$. From the property (ii) of g-stratification mentioned in Section 5.1, $\mathcal{D}_{s_1}$ and $\mathcal{D}_{t_1}$ are invariant under the action of isotropy groups at s_1 and t_1 respectively. Moreover, from the countability of varieties of distinguished tangents, we can assume that $\mathcal{D}_{s_1}$ and $\mathcal{D}_{t_1}$ are conjugate under the G-action. After G-conjugation, φ induces an automorphism of $\mathbb{P}T_{s_1}(S)$ preserving $\mathcal{D}_{s_1}$, and we need to show that it preserves $\mathcal{C}_{s_1}$.

When S is of symmetric type, an automorphism of $\mathbb{P}T_s(S)$ preserving an isotropy-invariant proper subvariety must preserve the highest weight orbit $\mathcal{C}_s$ from the fine structure theory of Hermitian symmetric spaces ([Wolf 1972]). In fact, one can show that the highest weight orbit is a singularity stratum of any other isotropy-invariant subvariety. Thus φ must preserve $\mathcal{C}$, and we are done. Alternatively, the argument in Section 6.4 case (1) gives a different proof without using the fine structure theory.

When S is of contact type, we can show that $\mathcal{D}_{s_1}$ must be equal to $\mathcal{C}_{s_1}$ directly. It is easy to see that any proper isotropy-invariant subvariety of $\mathbb{P}T_s(S)$ is contained in $\mathbb{P}D_s$, the contact hyperplane. From the basic structure theory of isotropy orbits, $\mathcal{C}_s \subset \mathcal{D}_s \subset \mathbb{P}D_s$. The Lagrangian property of $\mathcal{C}_s$ in Section 1.4 can be used to show that the variety of tangential lines to $\mathcal{D}_s$ is nondegenerate in $\mathbb{P} \bigwedge^2 D_s$ unless $\mathcal{D}_s = \mathcal{C}_s$. But if the variety of tangential lines to $\mathcal{D}_s$ is nondegenerate in $\mathbb{P} \bigwedge^2 D_s$, then the distribution D must be integrable by arguing as in Section 1.2, since $\mathcal{D}_s$ is the pull back of the variety of minimal rational tangents in X. This is contradictory to the definition of D. This proves Conjecture 6.1.2 in the contact case.

6.4. For the proof of Theorem 6.2.1, it remains to consider the case of S of depth $m \geq 2$ and of noncontact type. We will also again exclude the unnecessary case of (G_2, α_2). By Corollary 4.1.2 it suffices to show that φ preserves D^1, or equivalently that φ preserves the bundle of varieties of highest weight tangents $\mathcal{W}$, by Proposition 4.2.2. Suppose otherwise. By the argument of Lemma 4.2.4 there exists a holomorphic fiber bundle $\mathcal{V} \to S$, $\mathcal{V} \subset \mathbb{P}T(S)$ preserved by φ, such that the fibers $\mathcal{V}_x \subset \mathbb{P}T_x(S)$ are rational homogeneous submanifolds conjugate to each other under projective transformations. Either

(1) $\mathcal{V}_x \subset \mathbb{P}T_x(S)$ is linearly nondegenerate, or
(2) $\mathcal{V} \subset \mathbb{P}D^k$ for some k for $2 \leq k < m$, and $\mathcal{V} \not\subset \mathbb{P}D^1$.

In the linear nondegenerate case (1), since φ preserves some isotropy-invariant proper subvariety as in Section 6.3, $\mathcal{V}_x \subsetneqq \mathbb{P}T_x(S)$ and $\mathcal{V} \to S$ defines a G-

structure over S, with $\mathrm{G} \subset \mathbb{P}\,\mathrm{GL}(T_o(S))$ being the identity component of the group of projective linear transformations on $\mathbb{P}T_o(S)$ leaving $\mathcal{V}_o$ invariant. As $\mathcal{V}_o$ is linearly nondegenerate in $\mathbb{P}T_o(S)$, G is reductive by Borel's fixed point theorem. It follows from Theorem 1.5.3 that S must be biholomorphic to an irreducible Hermitian symmetric manifold of the compact type and of rank at least 2, contradicting the assumption on S.

We are going to rule out the linearly degenerate case (2), $\mathcal{V} \subset \mathbb{P}D^k$ for some k with $2 \leq k < m$, $\mathcal{V} \not\subset \mathbb{P}D^1$. Only the cases of depth at least 3 matter, with $\mathfrak{g} = F_4, E_6, E_7$ or E_8. By Corollary 4.1.2 it suffices to show that if φ is D^k-preserving for some $k \geq 2$, then it must already be D^1-preserving. Consider the Frobenius form $F^k : D_o^k \times D_o^k \to T_o(S)/D_o^k$. From Lemma 4.1.3, φ must preserve the subvariety of $[\xi] \in \mathbb{P}D_o^k$ for which $\mathrm{rank}(F_\xi^k)$ is minimum. To deduce that φ is necessarily D^1-preserving it remains therefore to establish this result:

PROPOSITION 6.4.1. *Let η_1 be a highest weight vector of $\mathfrak{g}_1 = D_o^1$ as an L_o-representation space and $\xi \in D_o^k - \mathfrak{g}_1$. Then, $\mathrm{rank}\, F_\xi^k > \mathrm{rank}\, F_{\eta_1}^k$.*

The rank of F_ξ^k is constant along the P-orbit of ξ and is lower semicontinuous in $\xi \in D_o^k$. Consider the $\mathbb{C}^*$-action on D_o^k defined by the centre of L_o, given by

$$t \cdot \xi = t\xi_1 + t^2\xi_2 + \cdots + t^k\xi_k$$

according to the decomposition $\xi = \xi_1 + \xi_2 + \cdots + \xi_k$, $\xi_j \in \mathfrak{g}_j$. From lower semicontinuity we conclude that $\mathrm{rank}(F_\xi^k) \geq \mathrm{rank}(F_{\xi_i}^k)$ whenever i is the largest index for which $\xi_i \neq 0$. Consider $\mathfrak{g}_j$ as an L_o-representation space. Noting that the highest weight orbit in $\mathbb{P}\mathfrak{g}_j$ lies in the Zariski closure of any orbit in $\mathbb{P}\mathfrak{g}_j$ to prove the Proposition it suffices to show that $\mathrm{rank}(F_{\eta_j}^k) > \mathrm{rank}(F_{\eta_1}^k)$ for highest weight vectors η_j of $\mathfrak{g}_j$, $j = 2, \ldots, k$. Furthermore, as η_j lies on the same P-orbit of some $\eta_j + \theta_{j-1}$, $0 \neq \theta_{j-1} \in \mathfrak{g}_{j-1}$, using the $\mathbb{C}^*$-action as described it follows readily that $\mathrm{rank}(F_{\eta_k}^k) \geq \cdots \geq \mathrm{rank}(F_{\eta_2}^k)$, thus reducing the proof of Proposition 6.4.1 to the special case of $\xi = \eta_2$. For the case of $\mathfrak{g} = F_4$ a straight-forward checking shows that indeed

$$\mathrm{rank}(F_{\eta_2}^k) > \mathrm{rank}(F_{\eta_1}^k)$$

is always valid. For the exceptional cases $\mathfrak{g} = E_6, E_7, E_8$, for which roots are of equal length, in place of tedious checking we have the following statement with a uniform proof.

PROPOSITION 6.4.2. *Let $\mathfrak{g} = E_6, E_7, E_8$ and $(\mathfrak{g}, \alpha_i)$ be of depth at least 3. For $k \geq 2$ and for η_j highest weight vectors of $\mathfrak{g}_j$ as an L_o-representation space, we have $\mathrm{rank}(F_{\eta_2}^k) = 2\,\mathrm{rank}(F_{\eta_1}^k)$.*

PROOF. We will interpret $\mathrm{rank}(F_{\eta_s}^k)$, for $s = 1, 2$, as Chern numbers. In the notations of Section 4.1, for S of type $(\mathfrak{g}, \alpha_i)$ and for $\rho \in \triangle^+$, the rational curve $C_\rho \subset S$ is of degree $\omega_i(H_\rho)$. When all roots of $\mathfrak{g}$ are of equal length, the root system is self-dual, and for $\rho = \sum_{j=1}^{m} m_j \alpha_j$, we have $H_\rho = \sum_{j=1}^{m} m_j H_{\alpha_j}$, so that for $\rho \in \triangle_s$, C_ρ is of degree s. For $s = 1, 2$ write C_s for C_{μ_s}. We have

$$\operatorname{rank}(F_{\eta_s}^k) = \dim\left(\left[\eta_s, \bigoplus_{j=1}^{k} \mathfrak{g}_j\right] \mod \bigoplus_{j=1}^{k} \mathfrak{g}_j\right). \tag{$*$}$$

Denote by $\mathcal{E}^k$ the holomorphic vector bundle D^m/D^k. We claim that

$$\operatorname{rank}(F_{\eta_s}^k) = c_1(\mathcal{E}^k) \cdot C_s,$$

from which $\operatorname{rank}(F_{\eta_2}^k) = 2\operatorname{rank}(F_{\eta_1}^k)$ follows, since C_s is of degree s. $\mathcal{E}^k$ admits a composition series with factors D^l/D^{l-1}, $k < l \le m$. To prove the claim by the argument on splitting types of D^l/D^{l-1} as in Section 4.2, we have

$$c_1(D^l/D^{l-1}) \cdot C_s = \left|\{\rho_l \in \triangle_l : \rho_l - \mu_s \in \triangle_{l-s}\}\right| - \left|\{\rho_l \in \triangle_l : \rho_l + \mu_s \in \triangle_{l+s}\}\right|.$$

Observe that for $l > k$, whenever $\rho_l + \mu_s = \rho_{l+s} \in \triangle_{l+s}$ we also have $\rho_{l+s} - \mu_s = \rho_l \in \triangle_l$. From this and adding up Chern numbers we have

$$c_1(\mathcal{E}^k) \cdot C_s = \left|\{\rho \in \triangle_1 \cup \cdots \cup \triangle_{k-1} : \rho + \mu_s \in \triangle_k \cup \cdots \cup \triangle_m\}\right| = \operatorname{rank}(F_{\eta_s}^k),$$

by $(*)$, as claimed. The proof of Proposition 6.4.2 is complete. $\qquad\square$

We remark that, using the composition series, splitting types of D^l/D^{l-1} over C_s, and the fact that D^m/D^l is nonnegative, one can easily verify that $\mathcal{E}^k|_{C_s} \cong [\mathcal{O}(1)]^{r_s} \oplus \mathcal{O}^{q_s}$, $r_s = \operatorname{rank}(F_{\eta_s}^k)$, and $q_s = \operatorname{rank}(\mathcal{E}^k) - r_s$.

Proposition 6.4.1 follows from Proposition 6.4.2. From this we also rule out alternative (2) (that $\mathcal{V} \subset \mathbb{P}D^k$ for $2 \le k < m$ but $\mathcal{V} \not\subset \mathbb{P}D^1$). We have thus proven by contradiction that in the nonsymmetric and noncontact case, the local biholomorphism φ on S must preserve varieties of highest weight tangents. By Corollary 4.1.2 we conclude that φ extends to a biholomorphic automorphism on S. By Proposition 6.2.2 the finite map $f : S \to X$ must be a biholomorphism unless $X \cong \mathbb{P}^n$. With this we have resolved Conjecture 6.1.2 of Lazarsfeld's.

Acknowledgements

In 1995–96 the first author stayed at MSRI on the occasion of the Special Year in Several Complex Variables. The second author stayed there during part of February, 1996. On various occasions both authors presented results covered by the present survey. We would like to thank MSRI for its invitation and hospitality, and the organizers for asking us to write up the survey article. We learnt with deep regret that Professor Michael Schneider, one of the organizers of the Special Year, passed away in August 1997, and wish to dedicate this article to the memory of his work and his friendliness.

References

[Boothby 1961] W. M. Boothby, "Homogeneous complex contact manifolds", pp. 144–154 in *Differential geometry*, edited by C. B. Allendoerfer, Proc. Sympos. Pure Math. **3**, Amer. Math. Soc., Providence, RI, 1961.

[Brieskorn 1964] E. Brieskorn, "Ein Satz über die komplexen Quadriken", *Math. Ann.* **155** (1964), 184–193.

[Grothendieck 1957] A. Grothendieck, "Sur la classification des fibrés holomorphes sur la sphère de Riemann", *Amer. J. Math.* **79** (1957), 121–138.

[Hirzebruch and Kodaira 1957] F. Hirzebruch and K. Kodaira, "On the complex projective spaces", *J. Math. Pures Appl.* (9) **36** (1957), 201–216.

[Hwang 1997] J.-M. Hwang, "Rigidity of homogeneous contact manifolds under Fano deformation", *J. Reine Angew. Math.* **486** (1997), 153–163.

[Hwang 1998] J.-M. Hwang, "Stability of tangent bundles of low-dimensional Fano manifolds with Picard number 1", *Math. Ann.* **312**:4 (1998), 599–606.

[Hwang and Mok 1997] J.-M. Hwang and N. Mok, "Uniruled projective manifolds with irreducible reductive G-structures", *J. Reine Angew. Math.* **490** (1997), 55–64.

[Hwang and Mok 1998a] J.-M. Hwang and N. Mok, "Characterization and deformation-rigidity of compact irreducible Hermitian symmetric spaces of rank ≥ 2 among Fano manifolds", in *Algebra and geometry* (Taiwan, 1995), edited by M.-C. Kang, International Press, Cambridge, MA, 1998.

[Hwang and Mok 1998b] J.-M. Hwang and N. Mok, "Rigidity of irreducible Hermitian symmetric spaces of the compact type under Kähler deformation", *Invent. Math.* **131**:2 (1998), 393–418.

[Hwang and Mok 1999] J.-M. Hwang and N. Mok, "Holomorphic maps from rational homogeneous spaces of Picard number 1 onto projective manifolds", *Invent. Math.* **136**:1 (1999), 209–231.

[Kollár 1996] J. Kollár, *Rational curves on algebraic varieties*, Ergebnisse der Mathematik und ihrer Grenzgebiete (3. Folge) **32**, Springer, Berlin, 1996.

[Lazarsfeld 1984] R. Lazarsfeld, "Some applications of the theory of positive vector bundles", pp. 29–61 in *Complete intersections* (Acireale, 1983), edited by S. Greco and R. Strano, Lecture Notes in Math. **1092**, Springer, Berlin, 1984.

[LeBrun 1988] C. LeBrun, "A rigidity theorem for quaternionic-Kähler manifolds", *Proc. Amer. Math. Soc.* **103**:4 (1988), 1205–1208.

[Matsushima and Morimoto 1960] Y. Matsushima and A. Morimoto, "Sur certains espaces fibrés holomorphes sur une variété de Stein", *Bull. Soc. Math. France* **88** (1960), 137–155.

[Mok 1988] N. Mok, "The uniformization theorem for compact Kähler manifolds of nonnegative holomorphic bisectional curvature", *J. Differential Geom.* **27**:2 (1988), 179–214.

[Mok and Tsai 1992] N. Mok and I.-H. Tsai, "Rigidity of convex realizations of irreducible bounded symmetric domains of rank ≥ 2", *J. Reine Angew. Math.* **431** (1992), 91–122.

[Mori 1979] S. Mori, "Projective manifolds with ample tangent bundles", *Ann. of Math.* (2) **110**:3 (1979), 593–606.

[Ochiai 1970]   T. Ochiai, "Geometry associated with semisimple flat homogeneous spaces", *Trans. Amer. Math. Soc.* **152** (1970), 159–193.

[Paranjape and Srinivas 1989]   K. H. Paranjape and V. Srinivas, "Self-maps of homogeneous spaces", *Invent. Math.* **98**:2 (1989), 425–444.

[Peternell and Wiśniewski 1995]   T. Peternell and J. A. Wiśniewski, "On stability of tangent bundles of Fano manifolds with $b_2 = 1$", *J. Algebraic Geom.* **4**:2 (1995), 363–384.

[Remmert and van de Ven 1960]   R. Remmert and T. van de Ven, "Über holomorphe Abbildungen projektiv-algebraischer Mannigfaltigkeiten auf komplexe Räume", *Math. Ann.* **142** (1960), 453–486.

[Serre 1966]   J.-P. Serre, *Algèbres de Lie semi-simples complexes*, W. A. Benjamin, New York, 1966. Translated as *Complex semisimple Lie algebras*, Springer, New York, 1987.

[Tanaka 1979]   N. Tanaka, "On the equivalence problems associated with simple graded Lie algebras", *Hokkaido Math. J.* **8**:1 (1979), 23–84.

[Tsai 1993]   I. H. Tsai, "Rigidity of holomorphic maps from compact Hermitian symmetric spaces to smooth projective varieties", *J. Algebraic Geom.* **2**:4 (1993), 603–633.

[Wolf 1972]   J. A. Wolf, "Fine structure of Hermitian symmetric spaces", pp. 271–357 in *Symmetric spaces* (St. Louis, MO, 1969–1970), edited by W. M. Boothby and G. L. Weiss, Pure and App. Math. **8**, Dekker, New York, 1972.

[Yamaguchi 1993]   K. Yamaguchi, "Differential systems associated with simple graded Lie algebras", pp. 413–494 in *Progress in differential geometry*, edited by K. Shiohama, Adv. Stud. Pure Math. **22**, Math. Soc. Japan and Kinokuniya, Tokyo, 1993.

[Zak 1993]   F. L. Zak, *Tangents and secants of algebraic varieties*, Transl. Math. Monographs **127**, Amer. Math. Soc., Providence, RI, 1993. Translated from the Russian manuscript by the author.

JUN-MUK HWANG
SEOUL NATIONAL UNIVERSITY
SEOUL 151-742
KOREA
jmhwang@math.snu.ac.kr

NGAIMING MOK
THE UNIVERSITY OF HONG KONG
POKFULAM ROAD
HONG KONG
nmok@hkucc.hku.hk

Several Complex Variables
MSRI Publications
Volume **37**, 1999

Recent Developments in Seiberg–Witten Theory and Complex Geometry

CHRISTIAN OKONEK AND ANDREI TELEMAN

We dedicate this paper to our wives Christiane and Roxana
for their invaluable help and support during the past two years.

ABSTRACT. In this article, written at the end of 1996, we survey some
of the most important results in Seiberg–Witten Theory which are directly
related to Algebraic or Kählerian Geometry. We begin with an introduction
to abelian Seiberg–Witten Theory, with special emphasis on the generalized
Seiberg–Witten invariants, which take also into account 1-homology classes
of the base manifold. The more delicate case of manifolds with $b_+ = 1$ is
discussed in detail; we present our universal wall-crossing formula which
shows that, crossing a wall in the parameter space, produces jumps of the
invariants which are of a purely topological nature.

Next we introduce nonabelian Seiberg–Witten equations associated with
very general compact Lie groups, and we describe in detail some of the properties of the moduli spaces of $PU(2)$-monopoles. The latter play an important role in our approach to prove Witten's conjecture. Then we specialize
to the case where the base manifold is a Kähler surface, and we present
the complex geometric interpretation of the corresponding moduli spaces of
monopoles. This interpretation is another instance of a Kobayashi–Hitchin
correspondence, which is based on the analysis of various types of vortex
equations. Finally we explain our strategy for a proof of Witten's conjecture
in an abstract setting, using the algebraic geometric "coupling principle"
and "master spaces" to relate the relevant correlation functions.

CONTENTS

The authors were partially supported by AGE (Algebraic Geometry in Europe) contract
no. ERBCHRXCT940557 (BBW 93.0187), and by SNF grant no. 21-36111.92.

Introduction

In October 1994, E. Witten revolutionized the theory of 4-manifolds by introducing the now famous Seiberg–Witten invariants [Witten 1994]. These invariants are defined by counting gauge equivalence classes of solutions of the Seiberg–Witten monopole equations, a system of nonlinear PDE's which describe the absolute minima of a Yang–Mills–Higgs type functional with an abelian gauge group.

Within a few weeks after Witten's seminal paper became available, several long-standing conjectures were solved, many new and totally unexpected results were found, and much simpler and more conceptional proofs of already established theorems were given.

Among the most spectacular applications in this early period are the solution of the Thom conjecture [Kronheimer and Mrowka 1994], new results about Einstein metrics and Riemannian metrics of positive scalar curvature [LeBrun 1995a; 1995b], a proof of a $\frac{10}{8}$ bound for intersection forms of Spin manifolds [Furuta 1995], and results about the $\mathcal{C}^\infty$-classification of algebraic surfaces [Okonek and Teleman 1995a; 1995b; 1997; Friedman and Morgan 1997; Brussee 1996]. The latter include Witten's proof of the $\mathcal{C}^\infty$-invariance of the canonical class of a minimal surface of general type with $b_+ \neq 1$ up to sign, and a simple proof of the Van de Ven conjecture by the authors. Moreover, combining results in [LeBrun 1995b; 1995a] with ideas from [Okonek and Teleman 1995b], P. Lupaşcu recently obtained [1997] the optimal characterization of complex surfaces of Kähler type admitting Riemannian metrics of nonnegative scalar curvature.

In two of the earliest papers on the subject, C. Taubes found a deep connection between Seiberg–Witten theory and symplectic geometry in dimension four: He first showed that many aspects of the new theory extend from the case of Kähler surfaces to the more general symplectic case [Taubes 1994], and then he went on to establish a beautiful relation between Seiberg–Witten invariants and Gromov–Witten invariants of symplectic 4-manifolds [Taubes 1995; 1996].

A report on some papers of this first period can be found in [Donaldson 1996].

Since the time this report was written, several new developments have taken place:

The original Seiberg–Witten theory, as introduced in [Witten 1994], has been refined and extended to the case of manifolds with $b_+ = 1$. The structure of the Seiberg–Witten invariants is more complicated in this situation, since the invariants for manifolds with $b_+ = 1$ depend on a chamber structure. The general theory, including the complex-geometric interpretation in the case of Kähler surfaces, is now completely understood [Okonek and Teleman 1996b].

At present, three major directions of research have emerged:

– Seiberg–Witten theory and symplectic geometry
– Nonabelian Seiberg–Witten theory and complex geometry
– Seiberg–Witten–Floer theory and contact structures

In this article, which had its origin in the notes for several lectures which we gave in Berkeley, Bucharest, Paris, Rome and Zürich during the past two years, we concentrate mainly on the second of these directions.

The reader will probably notice that the nonabelian theory is a subject of much higher complexity than the original (abelian) Seiberg–Witten theory; the difference is roughly comparable to the difference between Yang–Mills theory and Hodge theory. This complexity accounts for the length of the article. In rewriting our notes, we have tried to describe the essential constructions as simply as possible but without oversimplifying, and we have made an effort to explain the most important ideas and results carefully in a nontechnical way; for proofs and technical details precise references are given.

We hope that this presentation of the material will motivate the reader, and we believe that our notes can serve as a comprehensive introduction to an interesting new field of research.

We have divided the article in three chapters. In Chapter 1 we give a concise but complete exposition of the basics of abelian Seiberg–Witten theory in its most general form. This includes the definition of refined invariants for manifolds with $b_1 \neq 0$, the construction of invariants for manifolds with $b_+ = 1$, and the universal wall crossing formula in this situation.

Using this formula in connection with vanishing and transversality results, we calculate the Seiberg–Witten invariant for the simplest nontrivial example, the projective plane.

In Chapter 2 we introduce nonabelian Seiberg–Witten theories for rather general structure groups G. After a careful exposition of Spin^G-structures and G-monopoles, and a short description of some important properties of their moduli spaces, we explain one of the main results of the Habiliationsschrift of the second author [Teleman 1996; 1997]: the fundamental Uhlenbeck type compactification of the moduli spaces of $\mathrm{PU}(2)$-monopoles.

Chapter 3 deals with complex-geometric aspects of Seiberg–Witten theory: We show that on Kähler surfaces moduli spaces of G-monopoles, for unitary structure groups G, admit an interpretation as moduli spaces of purely holomorphic objects. This result is a Kobayashi–Hitchin type correspondence whose proof depends on a careful analysis of the relevant vortex equations. In the abelian case it identifies the moduli spaces of twisted Seiberg–Witten monopoles with certain Douady spaces of curves on the surface [Okonek and Teleman 1995a]. In the nonabelian case we obtain an identification between moduli spaces of $\mathrm{PU}(2)$-monopoles and moduli spaces of stable oriented pairs; see [Okonek and Teleman 1996a; Teleman 1997].

The relevant stability concept is new and makes sense on Kähler manifolds of arbitrary dimensions; it is induced by a natural moment map which is closely related to the projective vortex equation. We clarify the connection between this new equation and the parameter dependent vortex equations which had been studied in the literature [Bradlow 1991]. In the final section we construct moduli spaces of stable oriented pairs on projective varieties of any dimension with GIT methods [Okonek et al. 1999]. Our moduli spaces are projective varieties which come with a natural $\mathbb{C}^*$-action, and they play the role of master spaces for stable pairs. We end our article with the description of a very general construction principle which we call "coupling and reduction". This fundamental principle allows to reduce the calculation of correlation functions associated with vector bundles to a computation on the space of reductions, which is essentially a moduli space of lower rank objects.

Applied to suitable master spaces on curves, our principle yields a conceptional new proof of the Verlinde formulas, and very likely also a proof of the Vafa–Intriligator conjecture. The gauge theoretic version of the same principle can be used to prove Witten's conjecture, and more generally, it will probably also lead to formulas expressing the Donaldson invariants of arbitrary 4-manifolds in terms of Seiberg–Witten invariants.

1. Seiberg–Witten Invariants

1.1. The Monopole Equations. Let (X, g) be a closed oriented Riemannian 4-manifold. We denote by Λ^p the bundle of p-forms on X and by $A^p := A^0(\Lambda^p)$ the corresponding space of sections. Recall that the Riemannian metric g defines a Hodge operator $* : \Lambda^p \longrightarrow \Lambda^{4-p}$ with $*^2 = (-1)^p$. Let $\Lambda^2 = \Lambda^2_+ \oplus \Lambda^2_-$ be the corresponding eigenspace decomposition.

A Spin^c-structure on (X, g) is a triple $\tau = (\Sigma^\pm, \iota, \gamma)$ consisting of a pair of $U(2)$-vector bundles $\Sigma^\pm$, a unitary isomorphism $\iota : \det \Sigma^+ \longrightarrow \det \Sigma^-$ and an orientation-preserving linear isometry $\gamma : \Lambda^1 \longrightarrow \mathbb{R} \mathrm{SU}(\Sigma^+, \Sigma^-)$. Here

$$\mathbb{R} \mathrm{SU}(\Sigma^+, \Sigma^-) \subset \mathrm{Hom}_{\mathbb{C}}(\Sigma^+, \Sigma^-)$$

is the subbundle of real multiples of (fibrewise) isometries of determinant 1. The spinor bundles $\Sigma^\pm$ of τ are — up to isomorphism — uniquely determined by their first Chern class $c := c_1(\det \Sigma^\pm)$, the Chern class of the Spin^c(4)-structure τ. This class can be any integral lift of the second Stiefel–Whitney class $w_2(X)$ of X, and, given c, we have

$$c_2(\Sigma^\pm) = \frac{1}{4}(c^2 - 3\sigma(X) \mp 2e(X)).$$

Here $\sigma(X)$ and $e(X)$ denote the signature and the Euler characteristic of X.

The map γ is called the Clifford map of the Spin^c-structure τ. We denote by Σ the total spinor bundle $\Sigma := \Sigma^+ \oplus \Sigma^-$, and we use the same symbol γ also for the induced the map $\Lambda^1 \longrightarrow \mathrm{su}(\Sigma)$ given by

$$u \longmapsto \begin{pmatrix} 0 & -\gamma(u)^* \\ \gamma(u) & 0 \end{pmatrix}.$$

Note that the Clifford identity

$$\gamma(u)\gamma(v) + \gamma(v)\gamma(u) = -2g(u,v)$$

holds, and that the formula

$$\Gamma(u \wedge v) := \tfrac{1}{2}[\gamma(u), \gamma(v)]$$

defines an embedding $\Gamma : \Lambda^2 \longrightarrow \mathrm{su}(\Sigma)$ which maps $\Lambda^2_{\pm}$ isometrically onto $\mathrm{su}(\Sigma^{\pm}) \subset \mathrm{su}(\Sigma)$.

The second cohomology group $H^2(X, \mathbb{Z})$ acts on the set of equivalence classes $\mathfrak{c}$ of $\mathrm{Spin}^c(4)$-structures on (X, g) in a natural way: Given a representative $\tau = (\Sigma^{\pm}, \iota, \gamma)$ of $\mathfrak{c}$ and a Hermitian line bundle M representing a class $m \in H^2(X, \mathbb{Z})$, the tensor product $(\Sigma^{\pm} \otimes M, \iota \otimes \mathrm{id}_{M^{\otimes 2}}, \gamma \otimes \mathrm{id}_M)$ defines a Spin^c-structure τ_m. Endowed with the $H^2(X, \mathbb{Z})$-action given by $(m, [\tau]) \longmapsto [\tau_m]$, the set of equivalence classes of Spin^c-structures on (X, g) becomes a $H^2(X, \mathbb{Z})$-torsor, which is independent of the metric g up to canonical isomorphism [Okonek and Teleman 1996b]. We denote this $H^2(X, \mathbb{Z})$-torsor by $\mathrm{Spin}^c(X)$.

Recall that the choice of a $\mathrm{Spin}^c(4)$-structure $(\Sigma^{\pm}, \iota, \gamma)$ defines an isomorphism between the affine space $\mathcal{A}(\det \Sigma^+)$ of unitary connections in $\det \Sigma^+$ and the affine space of connections in $\Sigma^{\pm}$ which lift the Levi-Civita connection in the bundle $\Lambda^2_{\pm} \simeq \mathrm{su}(\Sigma^{\pm})$. We denote by $\hat{a} \in \mathcal{A}(\Sigma)$ the connection corresponding to $a \in \mathcal{A}(\det \Sigma^+)$.

The <u>Dirac operator</u> associated with the connection $a \in \mathcal{A}(\det \Sigma^+)$ is the composition

$$\not{D}_a : A^0(\Sigma^{\pm}) \xrightarrow{\nabla_{\hat{a}}} A^1(\Sigma^{\pm}) \xrightarrow{\gamma} A^0(\Sigma^{\mp})$$

of the covariant derivative $\nabla_{\hat{a}}$ in the bundles $\Sigma^{\pm}$ and the Clifford multiplication $\gamma : \Lambda^1 \otimes \Sigma^{\pm} \longrightarrow \Sigma^{\mp}$.

Note that, in order to define the Dirac operator, one needs a Clifford map, not only a Riemannian metric and a pair of spinor bundles; this will later become important in connection with transversality arguments. The Dirac operator $\not{D}_a : A^0(\Sigma^{\pm}) \longrightarrow A^0(\Sigma^{\mp})$ is an elliptic first order operator with symbol $\gamma : \Lambda^1 \longrightarrow \mathbb{R}\,\mathrm{SU}(\Sigma^{\pm}, \Sigma^{\mp})$. The direct sum-operator $\not{D}_a : A^0(\Sigma) \longrightarrow A^0(\Sigma)$ on the total spinor bundle is selfadjoint and its square has the same symbol as the rough Laplacian $\nabla_{\hat{a}}^* \nabla_{\hat{a}}$ on $A^0(\Sigma)$.

The corresponding <u>Weitzenböck formula</u> is

$$\not{D}_a^2 = \nabla_{\hat{a}}^* \nabla_{\hat{a}} + \tfrac{1}{2}\Gamma(F_a) + \tfrac{1}{4}s\,\mathrm{id}_{\Sigma},$$

where $F_a \in iA^2$ is the curvature of the connection a, and s denotes the scalar curvature of (X, g) [Lawson and Michelsohn 1989].

To write down the Seiberg–Witten equations, we need the following notations: For a connection $a \in \mathcal{A}(\det \Sigma^+)$ we let $F_a^{\pm} \in iA^2_{\pm}$ be the (anti) self-dual components of its curvature. Given a spinor $\Psi \in A^0(\Sigma^+)$, we denote by $(\Psi\bar{\Psi})_0 \in A^0(\mathrm{End}_0(\Sigma^{\pm}))$ the trace free part of the Hermitian endomorphism

$\Psi \otimes \bar\Psi$. Now fix a $\mathrm{Spin}^c(4)$-structure $\tau = (\Sigma^\pm, \iota, \gamma)$ for (X, g) and a closed 2-form $\beta \in A^2$. The β-twisted monopole equations for a pair $(a, \Psi) \in A(\det\Sigma^+) \times A^0(\Sigma^+)$ are

$$\displaystyle{\not{D}}_a \Psi = 0,$$
$$\Gamma(F_a^+ + 2\pi i \beta^+) = (\Psi\bar\Psi)_0. \qquad (\mathrm{SW}_\beta^\tau)$$

These β-twisted Seiberg–Witten equations should not be regarded as perturbations of the equations (SW_0^τ) since later the cohomology class of β will be fixed. The twisted equations arise naturally in connection with nonabelian monopoles (see Section 2.2). Using the Weitzenböck formula one easily gets the following fact:

LEMMA 1.1.1. *Let β be a closed 2-form and $(a, \Psi) \in A(\det\Sigma^+) \times A^0(\Sigma^+)$. Then*

$$\|{\not{D}}_a\Psi\|^2 + \tfrac{1}{4}\|(F_a^+ + 2\pi i\beta^+) - (\Psi\bar\Psi)_0\|^2$$
$$= \|\nabla_{A_a}\Psi\|^2 + \tfrac{1}{4}\|F_a^+ + 2\pi i\beta^+\|^2 + \tfrac{1}{8}\|\Psi\|_{L^4}^4 + \int_X ((\tfrac{1}{4}s\,\mathrm{id}_{\Sigma^+} - \Gamma(\pi i\beta^+))\Psi, \Psi).$$

COROLLARY 1.1.2 [Witten 1994]. *On manifolds (X, g) with nonnegative scalar curvature s the only solutions of (SW_0^τ) are pairs $(a, 0)$ with $F_a^+ = 0$.*

1.2. Seiberg–Witten Invariants for 4-Manifolds with $b_+ > 1$.

Let (X, g) be a closed oriented Riemannian 4-manifold, and let $\mathfrak{c} \in \mathrm{Spin}^c(X)$ be an equivalence class of Spin^c-structures of Chern class c, represented by the triple $\tau = (\Sigma^\pm, \iota, \gamma)$. The configuration space for Seiberg–Witten theory is the product $A(\det\Sigma^+) \times A^0(\Sigma^+)$ on which the gauge group $\mathcal{G} := \mathcal{C}^\infty(X, S^1)$ acts by

$$f \cdot (a, \Psi) := (a - 2f^{-1}df, \; f\Psi).$$

Let $\mathcal{B}(c) := \big(A(\det\Sigma^+) \times A^0(\Sigma^+)\big)/\mathcal{G}$ be the orbit space; up to homotopy equivalence, it depends only on the Chern class c. Since the gauge group acts freely in all points (a, Ψ) with $\Psi \neq 0$, the open subspace

$$\mathcal{B}(c)^* := \big(A(\det\Sigma^+) \times \big(A^0(\Sigma^+) \setminus \{0\}\big)\big)/\mathcal{G}$$

is a classifying space for $\mathcal{G}$. It has the weak homotopy type of a product $K(\mathbb{Z}, 2) \times K(H^1(X, \mathbb{Z}), 1)$ of Eilenberg–Mac Lane spaces and there is a natural isomorphism

$$\nu : \mathbb{Z}[u] \otimes \Lambda^*(H_1(X, \mathbb{Z})/\mathrm{Tors}) \longrightarrow H^*(\mathcal{B}(c)^*, \mathbb{Z}),$$

where the generator u is of degree 2. The $\mathcal{G}$-action on $A(\det\Sigma^+) \times A^0(\Sigma^+)$ leaves the subset $[A(\det\Sigma^+) \times A^0(\Sigma^+)]^{\mathrm{SW}_\beta^\tau}$ of solutions of (SW_β^τ) invariant; the orbit space

$$\mathcal{W}_\beta^\tau := [A(\det\Sigma^+) \times A^0(\Sigma^+)]^{\mathrm{SW}_\beta^\tau}/\mathcal{G}$$

is the moduli space of β-twisted monopoles. It depends, up to canonical isomorphism, only on the metric g, on the closed 2-form β, and on the class $\mathfrak{c} \in \mathrm{Spin}^c(X)$ [Okonek and Teleman 1996b].

Let $\mathcal{W}_\beta^{\tau\,*} \subset \mathcal{W}_\beta^\tau$ be the open subspace of monopoles with nonvanishing spinor-component; it can be described as the zero-locus of a section in a vector-bundle over $\mathcal{B}(c)^*$. The total space of this bundle is

$$\left[\mathcal{A}(\det \Sigma^+) \times (A^0(\Sigma^+) \setminus \{0\})\right] \times_{\mathcal{G}} \left[iA_+^2 \oplus A^0(\Sigma^-)\right],$$

and the section is induced by the $\mathcal{G}$-equivariant map

$$\mathrm{SW}_\beta^\tau : \mathcal{A}(\det \Sigma^+) \times \left(A^0(\Sigma^+) \setminus \{0\}\right) \longrightarrow iA_+^2 \oplus A^0(\Sigma^-)$$

given by the equations (SW_β^τ).

Completing the configuration space and the gauge group with respect to suitable Sobolev norms, we can identify $\mathcal{W}_\beta^{\tau\,*}$ with the zero set of a real analytic Fredholm section in the corresponding Hilbert vector bundle on the Sobolev completion of $\mathcal{B}(c)^*$, hence we can endow this moduli space with the structure of a finite dimensional real analytic space. As in the instanton case, one has a Kuranishi description for local models of the moduli space around a given point $[a, \Psi] \in \mathcal{W}_\beta^\tau$ in terms of the first two cohomology groups of the elliptic complex

$$0 \longrightarrow iA^0 \overset{D_p^0}{\longrightarrow} iA^1 \oplus A^0(\Sigma^+) \overset{D_p^1}{\longrightarrow} iA_+^2 \oplus A^0(\Sigma^-) \longrightarrow 0 \qquad (\mathcal{C}_p)$$

obtained by linearizing in $p = (a, \Psi)$ the action of the gauge group and the equivariant map SW_β^τ. The differentials of this complex are

$$D_p^0(f) = (-2df, f\Psi),$$
$$D_p^1(\alpha, \psi) = \left(d^+\alpha - \Gamma^{-1}[(\Psi\bar\psi)_0 + (\psi\bar\Psi)_0], \slashed{D}_a(\psi) + \gamma(\alpha)(\Psi)\right),$$

and its <u>index</u> w_c depends only on the Chern class c of the Spin^c-structure τ and on the characteristic classes of the base manifold X:

$$w_c = \tfrac{1}{4}(c^2 - 3\sigma(X) - 2e(X)).$$

The moduli space $\mathcal{W}_\beta^\tau$ is compact. This follows, as in [Kronheimer and Mrowka 1994], from the following consequence of the Weitzenböck formula and the maximum principle.

PROPOSITION 1.2.1 (A PRIORI $\mathcal{C}^0$-BOUND OF THE SPINOR COMPONENT). *If* (a, Ψ) *is a solution of* (SW_β^τ), *then*

$$\sup |\Psi|^2 \leq \max\left(0, \sup_X(-s + |4\pi\beta^+|)\right).$$

Moreover, let $\tilde{\mathcal{W}}^\tau$ be the moduli space of triples

$$(a, \Psi, \beta) \in \mathcal{A}(\det \Sigma^+) \times A^0(\Sigma^+) \times Z_{\mathrm{DR}}^2(X)$$

solving the Seiberg–Witten equations above now regarded as equations for the triple (a, Ψ, β). Two such triples define the same point in $\tilde{\mathcal{W}}^\tau$ if they are congruent modulo the gauge group $\mathcal{G}$ acting trivially on the third component. Using the proposition above and arguments of [Kronheimer and Mrowka 1994], one can easily see that the natural projection $\tilde{W}^\tau \xrightarrow{p} Z^2_{DR}(X)$ is proper. Moreover, one has the following transversality results:

LEMMA 1.2.2. *After suitable Sobolev completions the following results hold:*

1 [Kronheimer and Mrowka 1994]. *The open subspace $[\tilde{W}^\tau]^* \subset \tilde{W}^\tau$ of points with nonvanishing spinor component is smooth.*
2 [Okonek and Teleman 1996b]. *For any de Rham cohomology class $b \in H^2_{DR}(X)$ the moduli space $[\tilde{W}^\tau_b]^* := [\tilde{W}^\tau]^* \cap p^{-1}(b)$ is also smooth.*

Now let $c \in H^2(X, \mathbb{Z})$ be a characteristic element, that is, an integral lift of $w_2(X)$. A pair $(g, b) \in \text{Met}(X) \times H^2_{DR}(X)$ consisting of a Riemannian metric g on X and a de Rham cohomology class b is called c-good when the g-harmonic representant of $c - b$ is not g-antiselfdual. This condition guarantees that $W^\tau_\beta = W^{\tau*}_\beta$ for every Spin^c-structure τ of Chern class c and every 2-form β in b. Indeed, if $(a, 0)$ would solve (SW$^\tau_\beta$), then the g-antiselfdual 2-form $\frac{i}{2\pi} F_a - \beta$ would be the g-harmonic representant of $c - b$.

In particular, using the transversality results above, one gets:

THEOREM 1.2.3 [Okonek and Teleman 1996b]. *Let $c \in H^2(X, \mathbb{Z})$ be a characteristic element and suppose $(g, b) \in \text{Met}(X) \times H^2_{DR}(X)$ is c-good. Let τ be a Spin^c-structure of Chern class c on (X, g), and $\beta \in b$ a general representant of the cohomology class b. Then the moduli space $W^\tau_\beta = W^{\tau*}_\beta$ is a closed manifold of dimension $w_c = \frac{1}{4}(c^2 - 3\sigma(X) - 2e(X))$.*

Fix a maximal subspace $H^2_+(X, \mathbb{R})$ of $H^2(X, \mathbb{R})$ on which the intersection form is positive definite. The dimension $b_+(X)$ of such a subspace is the number of positive eigenvalues of the intersection form. The moduli space W^τ_β can be oriented by the choice of an orientation of the line $\det H^1(X, \mathbb{R}) \otimes \det H^2_+(X, \mathbb{R})^\vee$.

Let $[W^\tau_\beta]_\mathfrak{o} \in H_{w_c}(\mathcal{B}(c)^*, \mathbb{Z})$ be the fundamental class associated with the choice of an orientation $\mathfrak{o}$ of the line $\det H^1(X, \mathbb{R}) \otimes \det H^2_+(X)^\vee$.

The Seiberg-Witten form associated with the data $(g, b, \mathfrak{c}, \mathfrak{o})$ is the element $\text{SW}^{(g,b)}_{X,\mathfrak{o}}(\mathfrak{c}) \in \Lambda^* H^1(X, \mathbb{Z})$ defined by

$$\text{SW}^{(g,b)}_{X,\mathfrak{o}}(\mathfrak{c})(l_1 \wedge \cdots \wedge l_r) := \left\langle \nu(l_1) \cup \cdots \cup \nu(l_r) \cup u^{(w_c - r)/2}, [W^\tau_\beta]_\mathfrak{o} \right\rangle$$

for decomposable elements $l_1 \wedge \cdots \wedge l_r$ with $r \equiv w_c \pmod 2$. Here τ is a Spin^c-structure on (X, g) representing the class $\mathfrak{c} \in \text{Spin}^c(X)$, and β is a general form in the class b.

One shows, using again transversality arguments, that the Seiberg–Witten form $\text{SW}^{(g,b)}_{X,\mathfrak{o}}(\mathfrak{c})$ is well defined, independent of the choices of τ and β. Moreover, if any two c-good pairs (g_0, b_0), (g_1, b_1) can be joined by a smooth path of c-good pairs, then $\text{SW}^{(g,b)}_{X,\mathfrak{o}}(\mathfrak{c})$ is also independent of (g, b) [Okonek and Teleman 1996b].

Note that the condition "(g, b) is not c-good" is of codimension $b_+(X)$ for a fixed class c. This means that for manifolds with $b_+(X) > 1$ we have a well defined map

$$\mathrm{SW}_{X,\mathfrak{o}} : \mathrm{Spin}^c(X) \longrightarrow \Lambda^* H^1(X, \mathbb{Z})$$

which associates to a class of Spin^c-structures $\mathfrak{c}$ the form $\mathrm{SW}_{X,\mathfrak{o}}^{(g,b)}(\mathfrak{c})$ for any $b \in H^2_{\mathrm{DR}}(X)$ such that (g, b) is c-good. This map, which is functorial with respect to orientation preserving diffeomorphisms, is the $\underline{\text{Seiberg–Witten}}$ $\underline{\text{invariant}}$.

Using the identity in Lemma 1.1.1, one can easily prove the next result:

REMARK 1.2.4 [Witten 1994]. *Let X be an oriented closed 4-manifold with $b_+(X) > 1$. Then the set of classes $\mathfrak{c} \in \mathrm{Spin}^c(X)$ with nontrivial Seiberg–Witten invariant is finite.*

In the special case $b_+(X) > 1$, $b_1(X) = 0$, $\mathrm{SW}_{X,\mathfrak{o}}$ is simply a function

$$\mathrm{SW}_{X,\mathfrak{o}} : \mathrm{Spin}^c(X) \longrightarrow \mathbb{Z}.$$

The values $\mathrm{SW}_{X,\mathfrak{o}}(\mathfrak{c}) \in \mathbb{Z}$ are refinements of the numbers $n_c^{\mathfrak{o}}$ defined by Witten [1994]. More precisely:

$$n_c^{\mathfrak{o}} = \sum_{\mathfrak{c}} \mathrm{SW}_{X,\mathfrak{o}}(\mathfrak{c}),$$

the summation being over all classes of Spin^c-structures $\mathfrak{c}$ of Chern class c. It is easy to see that the indexing set is a torsor for the subgroup $\mathrm{Tors}_2 H^2(X, \mathbb{Z})$ of 2-torsion classes in $H^2(X, \mathbb{Z})$.

The structure of the Seiberg–Witten invariants for manifolds with $b_+(X) = 1$ is more complicated and will be described in the next section.

1.3. The Case $b_+ = 1$ and the Wall Crossing Formula. Let X be a closed oriented differentiable 4-manifold with $b_+(X) = 1$. In this situation the Seiberg–Witten forms depend on a $\underline{\text{chamber}}$ $\underline{\text{structure}}$: Recall first that in the case $b_+(X) = 1$ there is a natural map $\mathrm{Met}(X) \longrightarrow \mathbb{P}(H^2_{\mathrm{DR}}(X))$ which sends a metric g to the line $\mathbb{R}[\omega_+] \subset H^2_{\mathrm{DR}}(X)$, where ω_+ is any nontrivial g-selfdual harmonic form. Let

$$\mathbf{H} := \{ h \in H^2_{\mathrm{DR}}(X) : h^2 = 1 \}$$

be the hyperbolic space. This space has two connected components, and the choice of one of them orients the lines $\mathbb{H}^2_{+,g}(X)$ of selfdual g-harmonic forms, for all metrics g. Furthermore, once we fix a component $\mathbf{H}_0$ of $\mathbf{H}$, every metric defines a unique g-selfdual form ω_g of length 1 with $[\omega_g] \in \mathbf{H}_0$; see Figure 1.

Let $c \in H^2(X, \mathbb{Z})$ be characteristic. The $\underline{\text{wall}}$ associated with c is the hypersurface

$$c^\perp := \{ (h, b) \in \mathbf{H} \times H^2_{\mathrm{DR}}(X) : (c - b) \cdot h = 0 \},$$

and the connected components of $[\mathbf{H} \times H^2_{\mathrm{DR}}(X)] \setminus c^\perp$ are called $\underline{\text{chambers}}$ of type c.

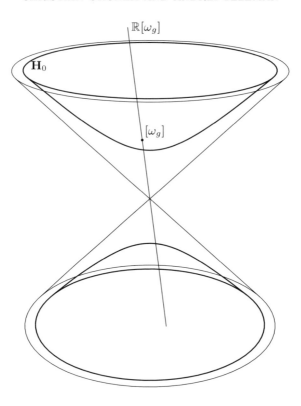

Figure 1.

Notice that the walls are nonlinear. Each characteristic element c defines precisely four chambers of type c, namely

$$C_{\mathbf{H}_0,\pm} := \{(h,b) \in \mathbf{H} \times H^2_{\mathrm{DR}}(X) : \pm(c-b)\cdot h < 0\}, \ \mathbf{H}_0 \in \pi_0(\mathbf{H}),$$

and each of these four chambers contains elements of the form $([\omega_g], b)$ with $g \in \mathrm{Met}(X)$.

Let $\mathfrak{o}_1$ be an orientation of $H^1(X, \mathbb{R})$. The choice of $\mathfrak{o}_1$ together with the choice of a component $\mathbf{H}_0 \in \pi_0(\mathbf{H})$ defines an orientation $\mathfrak{o} = (\mathfrak{o}_1, \mathbf{H}_0)$ of $\det(H^1(X,\mathbb{R})) \otimes \det(H^2_+(X,\mathbb{R})^\vee)$. Set

$$\mathrm{SW}^\pm_{X,(\mathfrak{o}_1,\mathbf{H}_0)}(\mathfrak{c}) := \mathrm{SW}^{(g,b)}_{X,\mathfrak{o}}(\mathfrak{c}),$$

where (g,b) is a pair such that $([\omega_g], b)$ belongs to the chamber $C_{\mathbf{H}_0,\pm}$. The map

$$\mathrm{SW}_{X,(\mathfrak{o}_1,\mathbf{H}_0)} : \mathrm{Spin}^c(X) \longrightarrow \Lambda^* H^1(X,\mathbb{Z}) \times \Lambda^* H^1(X,\mathbb{Z})$$

which associates to a class $\mathfrak{c}$ of Spin^c-structures on the oriented manifold X the pair of forms $(\mathrm{SW}^+_{X,(\mathfrak{o}_1,\mathbf{H}_0)}(\mathfrak{c}), \mathrm{SW}^-_{X,(\mathfrak{o}_1,\mathbf{H}_0)}(\mathfrak{c}))$ is the <u>Seiberg-Witten invariant</u> of X with respect to the orientation data $(\mathfrak{o}_1, \mathbf{H}_0)$. This invariant is functorial with respect to orientation-preserving diffeomorphisms and behaves as follows with respect to changes of the orientation data:

$$\mathrm{SW}_{X,(-\mathfrak{o}_1,\mathbf{H}_0)}(\mathfrak{c}) = -\,\mathrm{SW}_{X,(\mathfrak{o}_1,\mathbf{H}_0)}(\mathfrak{c}), \quad \mathrm{SW}^{\pm}_{X,(\mathfrak{o}_1,-\mathbf{H}_0)}(\mathfrak{c}) = -\,\mathrm{SW}^{\mp}_{X,(\mathfrak{o}_1,\mathbf{H}_0)}(\mathfrak{c}).$$

More important, however, is the fact that the difference

$$\mathrm{SW}^{+}_{X,(\mathfrak{o}_1,\mathbf{H}_0)}(\mathfrak{c}) - \mathrm{SW}^{-}_{X,(\mathfrak{o}_1,\mathbf{H}_0)}(\mathfrak{c})$$

is a topological invariant of the pair (X, c). To be precise, consider the element $u_c \in \Lambda^2\big(H_1(X,\mathbb{Z})/\,\mathrm{Tors}\big)$ defined by

$$u_c(a \wedge b) := \tfrac{1}{2}\langle a \cup b \cup c,\, [X]\rangle$$

for elements $a, b \in H^1(X,\mathbb{Z})$. The following <u>universal</u> <u>wall</u>-crossing formula generalizes results of [Witten 1994; Kronheimer and Mrowka 1994; Li and Liu 1995].

THEOREM 1.3.1 (WALL-CROSSING FORMULA [Okonek and Teleman 1996b]). *Let* $l_{\mathfrak{o}_1} \in \Lambda^{b_1} H^1(X,\mathbb{Z})$ *be the generator defined by the orientation* $\mathfrak{o}_1$, *and let* $r \geq 0$ *with* $r \equiv w_c \pmod 2$. *For every* $\lambda \in \Lambda^r\big(H_1(X,\mathbb{Z})/\,\mathrm{Tors}\big)$ *we have*

$$\big[\mathrm{SW}^{+}_{X,(\mathfrak{o}_1,\mathbf{H}_0)}(\mathfrak{c}) - \mathrm{SW}^{-}_{X,(\mathfrak{o}_1,\mathbf{H}_0)}(\mathfrak{c})\big](\lambda) = \frac{(-1)^{(b_1-r)/2}}{(\tfrac{1}{2}(b_1-r)/2)!}\langle \lambda \wedge u_c^{(b_1-r)/2},\, l_{\mathfrak{o}_1}\rangle$$

when $r \leq \min(b_1, w_c)$, *and the difference vanishes otherwise.*

We illustrate these results with the simplest possible example, the projective plane.

EXAMPLE. Let $\mathbb{P}^2$ be the complex projective plane, oriented as a complex manifold, and denote by h the first Chern class of $\mathcal{O}_{\mathbb{P}^2}(1)$. Since $h^2 = 1$, the hyperbolic space $\mathbf{H}$ consists of two points $\mathbf{H} = \{\pm h\}$. We choose the component $\mathbf{H}_0 := \{h\}$ to define orientations.

An element $c \in H^2(\mathbb{P}^2,\mathbb{Z})$ is characteristic if and only if $c \equiv h \pmod 2$. In Figure 2 we have drawn (as vertical intervals) the two chambers

$$C_{\mathbf{H}_0,\pm} = \{(h,b) \in \mathbf{H}_0 \times H^2_{\mathrm{DR}}(\mathbb{P}^2) : \pm(c-b)\cdot h < 0\}$$

of type c, for every $c \equiv h \pmod 2$.

The set $\mathrm{Spin}^c(\mathbb{P}^2)$ can be identified with the set $(2\mathbb{Z}+1)h$ of characteristic elements under the map which sends a Spin^c-structure $\mathfrak{c}$ to its Chern class c. The corresponding virtual dimension is $w_c = \tfrac{1}{4}(c^2 - 9)$. Note that, for any metric g, the pair $(g,0)$ is c-good for all characteristic elements c. Also recall that the Fubini–Study metric g is a metric of positive scalar curvature which can be normalized such that $[\omega_g] = h$. We can now completely determine the Seiberg–Witten invariant $\mathrm{SW}_{\mathbb{P}^2,\mathbf{H}_0}$ using three simple arguments:

(i) For $c = \pm h$ we have $w_c < 0$, hence $\mathrm{SW}^{\pm}_{\mathbb{P}^2,\mathbf{H}_0}(c) = 0$, by the transversality results of Section 1.2.

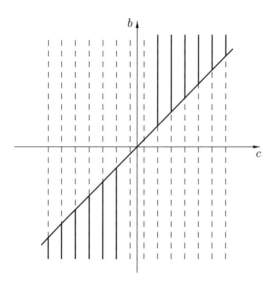

Figure 2.

(ii) Let c be a characteristic element with $w_c \geq 0$. Since the Fubini–Study metric g has positive scalar curvature and $(g, 0)$ is c-good, we have $\mathcal{W}_0^\tau = \mathcal{W}_0^{\tau^*} = \varnothing$ by Corollary 1.1.2. But this moduli space can be used to compute $\mathrm{SW}_{\mathbb{P}^2, \mathbf{H}_0}^{\pm}(c)$ for characteristic elements c with $\pm c \cdot h < 0$. Thus we find $\mathrm{SW}_{\mathbb{P}^2, \mathbf{H}_0}^{\pm}(c) = 0$ when $w_c \geq 0$ and $\pm c \cdot h < 0$.

(iii) The remaining values, $\mathrm{SW}_{\mathbb{P}^2, \mathbf{H}_0}^{\mp}(c) = 0$ for classes satisfying $w_c \geq 0$ and $\pm c \cdot h < 0$, are determined by the wall-crossing formula. Altogether we get

$$\mathrm{SW}_{\mathbb{P}^2, \mathbf{H}_0}^{+}(c) = \begin{cases} 1 & \text{if } c \cdot h \geq 3, \\ 0 & \text{if } c \cdot h < 3, \end{cases} \qquad \mathrm{SW}_{\mathbb{P}^2, \mathbf{H}_0}^{-}(c) = \begin{cases} -1 & \text{if } c \cdot h \leq -3, \\ 0 & \text{if } c \cdot h > -3. \end{cases}$$

2. Nonabelian Seiberg–Witten Theory

2.1. G-Monopoles. Let V be a Hermitian vector space, and let $U(V)$ be its group of unitary automorphisms. For any closed subgroup $G \subset U(V)$ which contains the central involution $- \mathrm{id}_V$, we define a new Lie group by

$$\mathrm{Spin}^G(n) := \mathrm{Spin}(n) \times_{\mathbb{Z}_2} G.$$

By construction one has the exact sequences

$$1 \longrightarrow \mathrm{Spin} \longrightarrow \mathrm{Spin}^G \xrightarrow{\ \delta\ } G/\mathbb{Z}_2 \longrightarrow 1,$$

$$1 \longrightarrow G \longrightarrow \mathrm{Spin}^G \xrightarrow{\ \pi\ } \mathrm{SO} \longrightarrow 1,$$

$$1 \longrightarrow \mathbb{Z}_2 \longrightarrow \mathrm{Spin}^G \xrightarrow{(\pi, \delta)} \mathrm{SO} \times G/\mathbb{Z}_2 \longrightarrow 1,$$

where Spin, Spin^G, SO denote one of the groups $\mathrm{Spin}(n)$, $\mathrm{Spin}^G(n)$, $\mathrm{SO}(n)$, respectively.

Given a Spin^G-principal bundle P^G over a topological space, we form the following associated bundles:

$$\delta(P^G) := P^G \times_\delta (G/\mathbb{Z}_2), \quad \mathbb{G}(P^G) := P^G \times_{\mathrm{Ad}} G, \quad \mathfrak{g}(P^G) := P^G \times_{\mathrm{ad}} \mathfrak{g},$$

where $\mathfrak{g}$ stands for the Lie algebra of G. The group $\mathcal{G}$ of sections of the bundle $\mathbb{G}(P^G)$ can be identified with the group of automorphism of P^G over the associated SO-bundle $P^G \times_\pi$ SO.

Consider now an oriented manifold (X, g), and let P_g be the SO-bundle of oriented g-orthonormal coframes. A $\underline{\mathrm{Spin}^G\text{-structure}}$ in P_g is a principal bundle morphism $\sigma : P^G \longrightarrow P_g$ of type π [Kobayashi and Nomizu 1963]. An isomorphism of Spin^G-structures σ, σ' in P_g is a bundle isomorphism $f : P^G \longrightarrow P'^G$ with $\sigma' \circ f = \sigma$. One shows that the data of a Spin^G-structure in (X, g) is equivalent to the data of a linear, orientation-preserving isometry $\gamma : \Lambda^1 \longrightarrow P^G \times_\pi \mathbb{R}^n$, which we call the $\underline{\mathrm{Clifford}}$ $\underline{\mathrm{map}}$ of the Spin^G-structure [Teleman 1997].

In dimension 4, the spinor group $\mathrm{Spin}(4)$ splits as

$$\mathrm{Spin}(4) = \mathrm{SU}(2)_+ \times \mathrm{SU}(2)_- = \mathrm{Sp}(1)_+ \times \mathrm{Sp}(1)_- .$$

Using the projections

$$p_\pm : \mathrm{Spin}(4) \longrightarrow \mathrm{SU}(2)_\pm$$

one defines the adjoint bundles

$$\mathrm{ad}_\pm(P^G) := P^G \times_{\mathrm{ad}_\pm} \mathrm{su}(2).$$

Coupling $p_\pm$ with the natural representation of G in V, we obtain representations $\lambda_\pm : \mathrm{Spin}^G(4) \longrightarrow U(\mathbb{H}_\pm \otimes_\mathbb{C} V)$ and associated $\underline{\mathrm{spinor}}$ $\underline{\mathrm{bundles}}$

$$\Sigma^\pm(P^G) := P^G \times_{\lambda_\pm} (\mathbb{H}_\pm \otimes_\mathbb{C} V).$$

The Clifford map $\gamma : \Lambda^1 \longrightarrow P^G \times_\pi \mathbb{R}^4$ of the Spin^G-structure yields identifications

$$\Gamma : \Lambda_\pm^2 \longrightarrow \mathrm{ad}_\pm(P^G).$$

An interesting special case occurs when V is a Hermitian vector space over the quaternions and G is a subgroup of $\mathrm{Sp}(V) \subset U(V)$. Then one can define $\underline{\mathrm{real}}$ $\underline{\mathrm{spinor}}$ $\underline{\mathrm{bundles}}$

$$\Sigma_\mathbb{R}^\pm(P^G) := P^G \times_{\rho_\pm} (\mathbb{H}_\pm \otimes_\mathbb{H} V),$$

associated with the representations

$$\rho_\pm : \mathrm{Spin}^G(4) \longrightarrow \mathrm{SO}(\mathbb{H}_\pm \otimes_\mathbb{H} V).$$

EXAMPLES. Let (X, g) be a closed oriented Riemannian 4-manifold with coframe bundle P_g.

$G = S^1$: A Spin^{S^1}-structure is just a Spin^c-structure as described in Chapter 1 [Teleman 1997].

$G = \mathrm{Sp}(1)$: $\mathrm{Spin}^{\mathrm{Sp}(1)}$-structures have been introduced in [Okonek and Teleman 1996a], where they were called Spin^h-structures. The map which associates to a $\mathrm{Spin}^{\mathrm{Sp}(1)}$-structure $\sigma : P^h \longrightarrow P_g$ the first Pontrjagin class $p_1(\delta(P^h))$ of the associated $\mathrm{SO}(3)$-bundle $\delta(P^h)$, induces a bijection between the set of isomorphism classes of $\mathrm{Spin}^{\mathrm{Sp}(1)}$-structures in (X, g) and the set

$$\{p \in H^4(X, \mathbb{Z}) : p \equiv w_2(X)^2 \ (\mathrm{mod}\ 4)\}.$$

There is a 1-to-1 correspondence between isomorphism classes of $\mathrm{Spin}^{\mathrm{Sp}(1)}$-structures in (X, g) and equivalence classes of triples $(\tau : P^{S^1} \longrightarrow P_g, E, \iota)$ consisting of a Spin^{S^1}-structure τ, a unitary vector bundle E of rank 2, and an unitary isomorphism $\iota : \det \Sigma_\tau^+ \longrightarrow \det E$. The equivalence relation is generated by tensorizing with Hermitian line bundles [Okonek and Teleman 1996a; Teleman 1997]. The associated bundles are — in terms of these data — given by

$$\delta(P^h) = P_E/S^1, \quad \mathbb{G}(P^h) = \mathrm{SU}(E), \quad \mathfrak{g}(P^h) = su(E),$$
$$\Sigma^\pm(P^h) = (\Sigma_\tau^\pm)^\vee \otimes E, \quad \Sigma_\mathbb{R}^\pm(P^h) = \mathbb{R}\mathrm{SU}(\Sigma_\tau^\pm, E),$$

where P_E denotes the principal $U(2)$-frame bundle of E.

$G = U(2)$: In this case $G/\mathbb{Z}_2$ splits as $G/\mathbb{Z}_2 = \mathrm{PU}(2) \times S^1$, and we write δ in the form $(\bar\delta, \det)$. The map which associates to a $\mathrm{Spin}^{U(2)}$-structure $\sigma : P^u \longrightarrow P_g$ the characteristic classes $p_1(\bar\delta(P^u))$, $c_1(\det P^u)$ identifies the set of isomorphism classes of $\mathrm{Spin}^{U(2)}$-structures in (X, g) with the set

$$\{(p, c) \in H^4(X, \mathbb{Z}) \times H^2(X, \mathbb{Z}) : p \equiv (w_2(X) + \bar c)^2 \ (\mathrm{mod}\ 4)\}.$$

There is a one-to-one correspondence between isomorphism classes of $\mathrm{Spin}^{U(2)}$-structures in (X, g) and equivalence classes of pairs $(\tau : P^{S^1} \longrightarrow P_g, E)$ consisting of a Spin^{S^1}-structure τ and a unitary vector bundle of rank 2. Again the equivalence relation is given by tensorizing with Hermitian line bundles [Teleman 1997]. If $\sigma : P^u \longrightarrow P_g$ corresponds to the pair $(\tau : P^{S^1} \longrightarrow P_g, E)$, the associated bundles are now

$$\bar\delta(P^u) = P_E/S^1, \quad \det P^u = \det[\Sigma_\tau^+]^\vee \otimes \det E,$$
$$\mathbb{G}(P^u) = U(E), \quad \mathfrak{g}(P^u) = u(E), \quad \Sigma^\pm(P^u) = (\Sigma_\tau^\pm)^\vee \otimes E.$$

We will later also need the subbundles $\mathbb{G}_0(P^u) := P^u \times_{\mathrm{Ad}} \mathrm{SU}(2) \simeq \mathrm{SU}(E)$ and $\mathfrak{g}_0 := P^u \times_{\mathrm{ad}} su(2) \simeq su(E)$. The group of sections $\Gamma(X, \mathbb{G}_0)$ can be identified with the group of automorphisms of P^u over $P_g \times_X \det(P^u)$.

Now consider again a general Spin^G-structure $\sigma : P^G \longrightarrow P_g$ in the 4-manifold (X, g). The spinor bundle $\Sigma^\pm(P^G)$ has $\mathbb{H}_\pm \otimes_\mathbb{C} V$ as standard fiber, so that the standard fiber $su(2)_\pm \otimes \mathfrak{g}$ of the bundle $\mathrm{ad}_\pm(P^G) \otimes \mathfrak{g}(P^G)$ can be viewed as real subspace of $\mathrm{End}(\mathbb{H}_\pm \otimes_\mathbb{C} V)$. We define a quadratic map

$$\mu_{0G} : \mathbb{H}_\pm \otimes_\mathbb{C} V \longrightarrow su(2)_\pm \otimes \mathfrak{g}$$

by sending $\psi \in \mathbb{H}_{\pm} \otimes_{\mathbb{C}} V$ to the orthogonal projection $pr_{su(2)_{\pm} \otimes \mathfrak{g}}(\psi \otimes \bar{\psi})$ of the Hermitian endomorphism $(\psi \otimes \bar{\psi}) \in \operatorname{End}(\mathbb{H}_{\pm} \otimes_{\mathbb{C}} V)$. One can show that $-\mu_{0G}$ is the total (hyperkähler) moment map for the G-action on the space $\mathbb{H}_{\pm} \otimes_{\mathbb{C}} V$ endowed with the natural hyperkähler structure given by left multiplication with quaternionic units [Teleman 1997].

These maps give rise to quadratic bundle maps

$$\mu_{0G} : \Sigma^{\pm}(P^G) \longrightarrow \operatorname{ad}_{\pm}(P^G) \otimes \mathfrak{g}(P^G).$$

In the case $G = U(2)$ one can project $\mu_{0U(2)}$ on $\operatorname{ad}_{\pm}(P^G) \otimes \mathfrak{g}_0(P^G)$ and gets a map

$$\mu_{00} : \Sigma^{\pm}(P^u) \longrightarrow \operatorname{ad}_{\pm}(P^u) \otimes \mathfrak{g}_0(P^u).$$

Note that a fixed Spin^G-structure $\sigma : P^G \longrightarrow P_g$ defines a bijection between connections $A \in \mathcal{A}(\delta(P^G))$ in $\delta(P^G)$ and connections $\hat{A} \in \mathcal{A}(P^G)$ in the Spin^G-bundle P^G which lift the Levi-Civita connection in P_g via σ. This follows immediately from the third exact sequence above. Let

$$\not{D}_A : A^0(\Sigma^{\pm}(P^G)) \longrightarrow A^0(\Sigma^{\mp}(P^G))$$

be the associated <u>Dirac operator</u>, defined by

$$\not{D}_A : A^0(\Sigma^{\pm}(P^G)) \xrightarrow{\nabla_{\hat{A}}} A^1(\Sigma^{\pm}(P^G)) \xrightarrow{\gamma} A^0(\Sigma^{\mp}(P^G)).$$

Here $\gamma : \Lambda^1 \otimes \Sigma^{\pm}(P^G) \longrightarrow \Sigma^{\mp}(P^G)$ is the Clifford multiplication corresponding to the embeddings $\gamma : \Lambda^1 \longrightarrow P^G \times_{\pi} \mathbb{R}^4 \subset \operatorname{Hom}_{\mathbb{C}}(\Sigma^{\pm}(P^G), \Sigma^{\mp}(P^G))$.

DEFINITION 2.1.1. Let $\sigma : P^G \longrightarrow P_g$ be a Spin^G-structure in the Riemannian manifold (X, g). The <u>G-monopole equations</u> for a pair (A, Ψ), with $A \in \mathcal{A}(\delta(P^G))$ and $\Psi \in A^0(\Sigma^+(P^G))$, are

$$\not{D}_A \Psi = 0,$$
$$\Gamma(F_A^+) = \mu_{0G}(\Psi). \tag{SW^σ}$$

The solutions of these equations will be called <u>G-monopoles</u>. The symmetry group of the G-monopole equations is the gauge group $\mathcal{G} := \Gamma(X, \mathbb{G}(P^G))$. If the Lie algebra of G has a nontrivial center $z(\mathfrak{g})$, then one has a family of $\mathcal{G}$-equivariant "twisted" G-monopole equations $(\mathrm{SW}_\beta^\sigma)$ parameterized by $iz(\mathfrak{g})$-valued 2-forms $\beta \in A^2(iz(\mathfrak{g}))$:

$$\not{D}_A \Psi = 0,$$
$$\Gamma((F_A + 2\pi i\beta)^+) = \mu_{0G}(\Psi). \tag{SW_β^σ}$$

We denote by $\mathcal{M}^\sigma$ and $\mathcal{M}_\beta^\sigma$, respectively, the corresponding moduli spaces of solutions modulo the gauge group $\mathcal{G}$.

Since in the case $G = U(2)$ there exists the splitting

$$U(2)/\mathbb{Z}_2 = \operatorname{PU}(2) \times S^1,$$

the data of a connection in $\delta(P^u) = \bar{\delta}(P^u) \times_X \det P^u$ is equivalent to the data of a pair of connections $(A, a) \in \mathcal{A}(\bar{\delta}(P^u)) \times \mathcal{A}(\det P^u)$. This can be used to introduce new important equations, obtained by fixing the abelian connection $a \in \mathcal{A}(\det P^u)$ in the $U(2)$-monopole equations, and regarding it as a parameter. One gets in this way the equations

$$\begin{aligned} \not{D}_{A,a}\Psi &= 0, \\ \Gamma(F_A^+) &= \mu_{00}(\Psi) \end{aligned} \qquad (\text{SW}_a^\sigma)$$

for a pair

$$(A, \Psi) \in \mathcal{A}(\bar{\delta}(P^u)) \times A^0(\Sigma^+(P^u)),$$

which will be called the <u>PU(2)-monopole equations</u>. These equations should be regarded as a twisted version of the quaternionic monopole equations introduced in [Okonek and Teleman 1996a], which coincide in our present framework with the SU(2)-monopole equations. Indeed, a $\text{Spin}^{U(2)}$-structure $\sigma : P^u \longrightarrow P_g$ with trivialized determinant line bundle can be regarded as $\text{Spin}^{SU(2)}$-structure, and the corresponding quaternionic monopole equations are $(\text{SW}_\theta^\sigma)$, where θ is the trivial connection in $\det P^u$.

The PU(2)-monopole equations are only invariant under the group $\mathcal{G}_0 := \Gamma(X, \mathbb{G}_0)$ of automorphisms of P^u over $P_g \times_X \det P^u$. We denote by $\mathcal{M}_a^\sigma$ the <u>moduli space</u> of PU(2)-monopoles modulo this gauge group. Note that $\mathcal{M}_a^\sigma$ comes with a natural <u>S^1-action</u> given by the formula $\zeta \cdot [A, \Psi] := [A, \zeta^{\frac{1}{2}}\Psi]$.

<u>Comparing with other formalisms:</u>

1. For $G = S^1$, $V = \mathbb{C}$ one recovers the original abelian Seiberg–Witten equations and the twisted abelian Seiberg–Witten equations of [LeBrun 1995b; Brussee 1996; Okonek and Teleman 1996b].

2. For $G = S^1$, $V = \mathbb{C}^{\oplus k}$ one gets the so called "multimonopole equations" studied by J. Bryan and R. Wentworth [1996].

3. In the case $G = U(2)$, $V = \mathbb{C}^2$ one obtains the $U(2)$-monopole equations which were studied in [Okonek and Teleman 1995a] (see also Chapter 3).

4. In the case of a Spin-manifold X and $G = \text{SU}(2)$ the corresponding monopole equations were introduced in [Okonek and Teleman 1995c]; they have been studied from a physical point of view in [Labastida and Mariño 1995].

5. If X is simply connected, the S^1-quotient $\mathcal{M}_a^\sigma/S^1$ of a moduli space of PU(2)-monopoles can be identified with a moduli space of "nonabelian monopoles" as defined in [Pidstrigach and Tyurin 1995]. Note that in the general non-simply connected case, one has to use our formalism.

REMARK 2.1.2. Let $G = \text{Sp}(n) \cdot S^1 \subset U(\mathbb{C}^{2n})$ be the Lie group of transformations of $\mathbb{H}^{\oplus n}$ generated by left multiplication with quaternionic matrices in $\text{Sp}(n)$ and by right multiplication with complex numbers of modulus 1. Then $G/\mathbb{Z}_2$ splits as $\text{PSp}(n) \times S^1$. In the same way as in the PU(2)-case one defines the

PSp(n)-monopole equations (SW$_a^\sigma$) associated with a Spin$^{\mathrm{Sp}(n)\cdot S^1}$(4)-structure $\sigma : P^G \longrightarrow P_g$ in (X, g) and an abelian connection a in the associated S^1-bundle.

The solutions of the (twisted) G- and PU(2)-monopole equations are the absolute minima of certain gauge invariant functionals on the corresponding configuration spaces $\mathcal{A}(\delta(P^G)) \times A^0(\Sigma^+(P^G))$ and $\mathcal{A}(\bar{\delta}(P^u)) \times A^0(\Sigma^+(P^G))$.

For simplicity we describe here only the case of nontwisted G-monopoles. The Seiberg-Witten functional SW$^\sigma : \mathcal{A}(\delta(P^G)) \times A^0(\Sigma^+(P^G)) \longrightarrow \mathbb{R}$ associated to a Spin^G-structure is defined by

$$\mathrm{SW}^\sigma(A, \Psi) := \|\nabla_{\hat{A}}\Psi\|^2 + \tfrac{1}{4}\|F_A\|^2 + \tfrac{1}{2}\|\mu_{0G}(\Psi)\|^2 + \tfrac{1}{4}\int_X s|\Psi|^2.$$

The Euler-Lagrange equations describing general critical points are

$$d_A^* F_A + J(A, \Psi) = 0,$$
$$\Delta_{\hat{A}}\Psi + \mu_{0G}(\Psi)(\Psi) + \tfrac{1}{4}s\Psi = 0,$$

where the current $J(A, \Psi) \in A^1(\mathfrak{g}(P^G))$ is given by $\sqrt{32}$ times the orthogonal projection of the End$(\Sigma^+(P^G))$-valued 1-form $\nabla_{\hat{A}}\Psi \otimes \bar{\Psi} \in A^1(\mathrm{End}(\Sigma^+(P^G)))$ onto $A^1(\mathfrak{g}(P^G))$.

In the abelian case $G = S^1$, $V = \mathbb{C}$, a closely related functional and the corresponding Euler–Lagrange equations have been investigated in [Jost et al. 1996].

2.2. Moduli Spaces of PU(2)-Monopoles. We retain the notations of the previous section. Let $\sigma : P^u \longrightarrow P_g$ be a Spin$^{U(2)}$-structure in a closed oriented Riemannian 4-manifold (X, g), and let $a \in \mathcal{A}(\det P^u)$ be a fixed connection. The PU(2)-monopole equations

$$\begin{aligned}\displaystyle{\not{D}}_{A,a}\Psi &= 0,\\ \Gamma(F_A^+) &= \mu_{00}(\Psi)\end{aligned} \qquad (\mathrm{SW}_a^\sigma)$$

associated with these data are invariant under the action of the gauge group $\mathcal{G}_0$, and hence give rise to a closed subspace $\mathcal{M}_a^\sigma \subset \mathcal{B}(P^u)$ of the orbit space $\mathcal{B}(P^u) := \left(\mathcal{A}(\bar{\delta}(P^u)) \times A^0(\Sigma^+(P^u))\right)/\mathcal{G}_0$.

The moduli space $\mathcal{M}_a^\sigma$ can be endowed with the structure of a ringed space with local models constructed by the well-known Kuranishi method [Okonek and Teleman 1995a; 1996a; Donaldson and Kronheimer 1990; Lübke and Teleman 1995]. More precisely: The linearization of the PU(2)-monopole equations in a solution $p = (A, \Psi)$ defines an elliptic deformation complex

$$0 \to A^0(\mathfrak{g}_0(P^u)) \xrightarrow{D_p^0} A^1(\mathfrak{g}_0(P^u)) \oplus A^0(\Sigma^+(P^u)) \xrightarrow{D_p^1} A_+^2(\mathfrak{g}_0(P^u)) \oplus A^0(\Sigma^-(P^u)) \to 0$$

whose differentials are given by $D_p^0(f) = (-d_A f, f\Psi)$ and

$$D_p^1(\alpha, \psi) = (d_A^+\alpha - \Gamma^{-1}[m(\psi, \Psi) + m(\Psi, \psi)], {\not{D}}_{A,a}\psi + \gamma(\alpha)\Psi).$$

Here m denotes the sesquilinear map associated with the quadratic map μ_{00}. Let $\mathbb{H}_p^i$, for $i = 0, 1, 2$, denote the harmonic spaces of the elliptic complex above. The stabilizer $\mathcal{G}_{0p}$ of the point $p \in \mathcal{A}(\bar{\delta}(P^u)) \times A^0(\Sigma^+(P^u))$ is a finite dimensional Lie group, isomorphic to a closed subgroup of SU(2), which acts in a natural way on the spaces $\mathbb{H}_p^i$.

PROPOSITION 2.2.1 [Okonek and Teleman 1996a; Teleman 1997]. *For every point $p \in \mathcal{M}_a^\sigma$ there exists a neighborhood $V_p \subset \mathcal{M}_a^\sigma$, a $\mathcal{G}_{0p}$-invariant neighborhood U_p of $0 \in \mathbb{H}_p^1$, an $\mathcal{G}_{0p}$-equivariant map $K_p : U_p \longrightarrow \mathbb{H}_p^2$ with $K_p(0) = 0$ and $dK_p(0) = 0$, and an isomorphism of ringed spaces*

$$V_p \simeq Z(K_p)/\mathcal{G}_{0p}$$

sending p to $[0]$.

The local isomorphisms $V_p \simeq Z(K_p)/\mathcal{G}_{0p}$ define the structure of a smooth manifold on the open subset

$$\mathcal{M}_{a,\text{reg}}^\sigma := \{[A, \Psi] \in \mathcal{M}_a^\sigma : \mathcal{G}_{0p} = \{1\}, \ \mathbb{H}_p^2 = \{0\}\},$$

and a real analytic orbifold structure in the open set of points $p \in \mathcal{M}_a^\sigma$ with $\mathcal{G}_{0p}$ finite. The dimension of $\mathcal{M}_{a,\text{reg}}^\sigma$ coincides with the <u>expected dimension</u> of the PU(2)-monopole moduli space, which is given by the index $\chi(\text{SW}_a^\sigma)$ of the elliptic deformation complex:

$$\chi(\text{SW}_a^\sigma) = \tfrac{1}{2}(-3p_1(\bar{\delta}(P^u)) + c_1(\det P^u)^2) - \tfrac{1}{2}(3e(X) + 4\sigma(X)).$$

Our next goal is to describe the fixed point set of the S^1-action on $\mathcal{M}_a^\sigma$ introduced above.

First consider the closed subspace $\mathcal{D}(\bar{\delta}(P^u)) \subset \mathcal{M}_a^\sigma$ of points of the form $[A, 0]$. It can be identified with the Donaldson moduli space of anti-selfdual connections in the PU(2)-bundle $\bar{\delta}(P^u)$ modulo the gauge group $\mathcal{G}_0$. Note however, that if $H^1(X, \mathbb{Z}_2) \neq \{0\}$, $\mathcal{D}(\bar{\delta}(P^u))$ does not coincide with the usual moduli space of PU(2)-instantons in $\bar{\delta}(P^u)$ but is a finite cover of it.

The stabilizer $\mathcal{G}_{0p}$ of a Donaldson point $(A, 0)$ contains always $\{\pm \text{id}\}$, hence $\mathcal{M}_a^\sigma$ has at least $\mathbb{Z}_2$-orbifold singularities in the points of $\mathcal{D}(\bar{\delta}(P^u))$.

Secondly, consider S^1 as a subgroup of PU(2) via the standard embedding $S^1 \ni \zeta \longmapsto [(\begin{smallmatrix} \zeta & 0 \\ 0 & 1 \end{smallmatrix})] \in \text{PU}(2)$. Note that any S^1-reduction $\rho : P \longrightarrow \bar{\delta}(P^u)$ of $\bar{\delta}(P^u)$ defines a reduction $\tau_\rho : P^\rho := P^u \times_{\bar{\delta}(P^u)} P \longrightarrow P^u \xrightarrow{\sigma} P_g$ of the $\text{Spin}^{U(2)}$-structure σ to a $\text{Spin}^{S^1 \times S^1}$-structure, hence a pair of Spin^c-structures $\tau_\rho^i : P^{\rho_i} \longrightarrow P_g$. One has natural isomorphisms

$$\det P^{\rho_1} \otimes \det P^{\rho_2} = (\det P^u)^{\otimes 2}, \quad \det P^{\rho_1} \otimes (\det P^{\rho_2})^{-1} = P^{\otimes 2},$$

and natural embeddings $\Sigma^\pm(P^{\rho_i}) \longrightarrow \Sigma^\pm(P^u)$ induced by the bundle morphism $P^{\rho_i} \longrightarrow P^u$. A pair (A, Ψ) will be called <u>abelian</u> if it lies in the image of $\mathcal{A}(P) \times A^0(\Sigma^+(P^{\rho_1}))$ for a suitable S^1-reduction ρ of $\bar{\delta}(P^u)$.

PROPOSITION 2.2.2. *The fixed point set of the S^1-action on $\mathcal{M}_a^\sigma$ is the union of the Donaldson locus $\mathcal{D}(\bar{\delta}(P^u))$ and the locus of abelian solutions. The latter can be identified with the disjoint union*

$$\coprod_\rho \mathcal{W}^{\tau^1_\rho}_{\frac{i}{2\pi} F_a},$$

where the union is over all S^1-reductions of the $\mathrm{PU}(2)$-bundle $\bar{\delta}(P^u)$.

This result suggests to use the S^1-quotient of $\mathcal{M}_a^\sigma \setminus (\mathcal{M}_a^\sigma)^{S^1}$ for the comparison of Donaldson invariants and (twisted) Seiberg–Witten invariants, as explained in [Okonek and Teleman 1996a].

Note that only using moduli spaces $\mathcal{M}_\theta^\sigma$ of quaternionic monopoles one gets, by the proposition above, moduli spaces of <u>non</u>-twisted abelian monopoles in the fixed point locus of the S^1-action. This was one of the motivations for studying the quaternionic monopole equations in [Okonek and Teleman 1996a]. There it has been shown that one can use the moduli spaces of quaternionic monopoles to relate certain Spin^c-polynomials to the original nontwisted Seiberg–Witten invariants.

The remainder of this section is devoted to the description of the <u>Uhlenbeck compactification</u> of the moduli spaces of PU(2)-monopoles [Teleman 1996].

First of all, the Weitzenböck formula and the maximum principle yield a bound on the spinor component, as in the abelian case. More precisely, one has the a priori estimate

$$\sup_X |\Psi|^2 \le C_{g,a} := \max\left(0,\ C\sup(-\frac{1}{2}s + |F_a^+|)\right)$$

on the space of solutions of (SW_a^σ), where C is a universal positive constant.

The construction of the Uhlenbeck compactification of $\mathcal{M}_a^\sigma$ is based, as in the instanton case, on the following three essential results.

1. A <u>compactness</u> theorem for the subspace of solutions with suitable bounds on the curvature of the connection component.
2. A <u>removable singularities</u> theorem.
3. Controlling <u>bubbling</u> phenomena for an arbitrary sequence of points in the moduli space $\mathcal{M}_a^\sigma$.

1. A compactness result.

THEOREM 2.2.3. *There exists a positive number $\delta > 0$ such that for every oriented Riemannian manifold (Ω, g) endowed with a $\mathrm{Spin}^{U(2)}(4)$-structure $\sigma : P^u \longrightarrow P_g$ and a fixed connection $a \in \mathcal{A}(\det P^u)$, the following holds:*

If (A_n, Ψ_n) is a sequence of solutions of (SW_a^σ), such that any point $x \in \Omega$ has a geodesic ball neighborhood D_x with

$$\int_{D_x} |F_{A_n}|^2 < \delta^2$$

for all large enough n, then there is a subsequence $(n_m) \subset \mathbb{N}$ and gauge transformations $f_m \in \mathcal{G}_0$ such that $f_m^*(A_{n_m}, \Psi_{n_m})$ converges in the $\mathcal{C}^\infty$-topology on Ω.

2. Removable singularities.
Let g be a metric on the 4-ball B, and let

$$\sigma : P^u = B \times \mathrm{Spin}^{U(2)}(4) \longrightarrow P_g \simeq B \times \mathrm{SO}(4)$$

be a $\mathrm{Spin}^{U(2)}$-structure in (B, g). Fix $a \in iA_B^1$ and put $B^\bullet := B \backslash \{0\}$, $\sigma^\bullet := \sigma|_{B^\bullet}$.

THEOREM 2.2.4. *Let (A_0, Ψ_0) be a solution of the equations $(\mathrm{SW}_a^{\sigma^\bullet})$ on the punctured ball such that*

$$\|F_{A_0}\|_{L^2}^2 < \infty.$$

Then there exists a solution (A, Ψ) of (SW_a^σ) on B and a gauge transformation $f \in \mathcal{C}^\infty(B^\bullet, \mathrm{SU}(2))$ such that $f^*(A|_{B^\bullet}, \Psi|_{B^\bullet}) = (A_0, \Psi_0)$.

3. Controlling bubbling phenomena.
The main point is that the selfdual components $F_{A_n}^+$ of the curvatures of a sequence of solutions $([A_n, \Psi_n])_{n \in \mathbb{N}}$ in $\mathcal{M}_a^\sigma$ cannot bubble.

DEFINITION 2.2.5. *Let $\sigma : P^u \longrightarrow P_g$ be a $\mathrm{Spin}^{U(2)}$-structure in (X, g) and fix $a \in \mathcal{A}(\det P^u)$. An* ideal monopole *of type (σ, a) is a pair $([A', \Psi'], \{x_1, \ldots, x_l\})$ consisting of a point $[A', \Psi'] \in \mathcal{M}_a^{\sigma_l'}$, where $\sigma_l' : P'^u \longrightarrow P_g$ is a $\mathrm{Spin}^{U(2)}$-structure satisfying*

$$\det P'^u = \det P^u, \quad p_1(\bar{\delta}(P'^u)) = p_1(\bar{\delta}(P^u)) + 4l,$$

and $\{x_1, \ldots, x_l\} \in S^l X$. The set of ideal monopoles of type (σ, a) is

$$I\mathcal{M}_a^\sigma := \coprod_{l \geq 0} \mathcal{M}_a^{\sigma_l} \times S^l X.$$

THEOREM 2.2.6. *There exists a metric topology on $I\mathcal{M}_a^\sigma$ such that the moduli space $\mathcal{M}_a^\sigma$ becomes an open subspace with compact closure $\overline{\mathcal{M}_a^\sigma}$.*

SKETCH OF PROOF. Given a sequence $([A_n, \Psi_n])_{n \in \mathbb{N}}$ of points in $\mathcal{M}_a^\sigma$, one finds a subsequence $([A_{n_m}, \Psi_{n_m}])_{m \in \mathbb{N}}$, a finite set of points $S \subset X$, and gauge transformations f_m such that $(B_m, \Phi_m) := f_m^*(A_{n_m}, \Psi_{n_m})$ converges on $X \setminus S$ in the $\mathcal{C}^\infty$-topology to a solution (A_0, Ψ_0). This follows from the compactness theorem above, using the fact that the total volume of the sequence of measures $|F_{A_n}|^2$ is bounded. The set S consists of points in which the measure $|F_{A_{n_m}}|^2$ becomes concentrated as m tends to infinity.

By the Removable Singularities theorem, the solution (A_0, Ψ_0) extends after gauge transformation to a solution (A, Ψ) of $(\mathrm{SW}_a^{\sigma'})$ on X, for a possibly different $\mathrm{Spin}^{U(2)}$-structure σ' with the same determinant line bundle. The curvature of A satisfies

$$|F_A|^2 = \lim_{m \to \infty} |F_{A_{n_m}}|^2 - 8\pi^2 \sum_{x \in S} \lambda_x \delta_x,$$

where δ_x is the Dirac measure of the point x. Now it remains to show that the λ_x's are natural numbers and

$$\sum_{x \in S} \lambda_x = \tfrac{1}{4}(p_1(\bar{\delta}(P'^u)) - p_1(\bar{\delta}(P^u))).$$

This follows as in the instanton case, if one uses the fact that the measures $|F_{A_{n_m}}^+|^2$ cannot bubble in the points $x \in S$ as $m \to \infty$ and that the integral of $|F_{A_n}^-|^2 - |F_{A_n}^+|^2$ is a topological invariant of $\bar{\delta}(P^u)$. In this way one gets an ideal monopole $\mathfrak{m} := ([A, \Psi], \{\lambda_1 x_1, \ldots, \lambda_k x_k\})$ of type (σ, a). With respect to a suitable topology on the space of ideal monopoles, one has $\lim_{m \to \infty} [A_{n_m}, \Psi_{n_m}] = \mathfrak{m}$.

$\square$

3. Seiberg–Witten Theory and Kähler Geometry

3.1. Monopoles on Kähler Surfaces. Let (X, J, g) be an almost Hermitian surface with associated Kähler form ω_g. We denote by Λ^{pq} the bundle of (p, q)-forms on X and by A^{pq} its space of sections. The Hermitian structure defines an orthogonal decomposition

$$\Lambda_+^2 \otimes \mathbb{C} = \Lambda^{20} \oplus \Lambda^{02} \oplus \Lambda^{00} \omega_g$$

and a canonical Spin^c-structure τ. The spinor bundles of τ are

$$\Sigma^+ = \Lambda^{00} \oplus \Lambda^{02}, \quad \Sigma^- = \Lambda^{01},$$

and the Chern class of τ is the first Chern class $c_1(T_J^{10}) = c_1(K_X^\vee)$ of the complex tangent bundle. The complexification of the canonical Clifford map γ is the standard isomorphism

$$\gamma : \Lambda^1 \otimes \mathbb{C} \longrightarrow \mathrm{Hom}(\Lambda^{00} \oplus \Lambda^{02}, \Lambda^{01}), \quad \gamma(u)(\varphi + \alpha) = \sqrt{2}(\varphi u^{01} - i\Lambda_g u^{10} \wedge \alpha),$$

and the induced isomorphism $\Gamma : \Lambda^{20} \oplus \Lambda^{02} \oplus \Lambda^{00} \omega_g \longrightarrow \mathrm{End}_0(\Lambda^{00} \oplus \Lambda^{02})$ acts by

$$(\lambda^{20}, \lambda^{02}, f\omega_g) \overset{\Gamma}{\longmapsto} 2 \begin{bmatrix} -if & -*(\lambda^{20} \wedge \cdot) \\ \lambda^{02} \wedge \cdot & if \end{bmatrix} \in \mathrm{End}_0(\Lambda^{00} \oplus \Lambda^{02}).$$

Recall from Section 1.1 that the set Spin$^c(X)$ of equivalence classes of Spin^c-structures in (X, g) is a $H^2(X, \mathbb{Z})$-torsor. Using the class of the canonical Spin^c-structure $\mathfrak{c} := [\tau]$ as base point, Spin$^c(X)$ can be identified with the set of isomorphism classes of S^1-bundles: When M is an S^1-bundle with $c_1(M) = m$, the Spin^c-structure τ_m has spinor bundles $\Sigma^\pm \otimes M$ and Chern class $2c_1(M) - c_1(K_X)$. Let $\mathfrak{c}_m$ be the class of τ_m.

Suppose now that (X, J, g) is Kähler, and let $k \in \mathcal{A}(K_X)$ be the Chern connection in the canonical line bundle. In order to write the (abelian) Seiberg–Witten equations associated with the Spin^c-structure τ_m in a convenient form, we make the variable substitution $a = k \otimes e^{\otimes 2}$ for a connection $e \in \mathcal{A}(M)$ in the S^1-bundle M, and we write the spinor Ψ as a sum $\Psi = \varphi + \alpha \in A^0(M) \oplus A^{02}(M)$.

LEMMA 3.1.1 [Witten 1994; Okonek and Teleman 1996b]. *Let (X, g) be a Kähler surface, $\beta \in A_{\mathbb{R}}^{11}$ a closed real $(1,1)$-form in the de Rham cohomology class b, and let M be a S^1-bundle with $(2c_1(M) - c_1(K_X) - b) \cdot [\omega_g] < 0$. The pair $(k \otimes e^{\otimes 2}, \varphi + \alpha) \in \mathcal{A}(\det(\Sigma^+ \otimes M)) \times A^0(\Sigma^+ \otimes M)$ solves the equations $(\mathrm{SW}_\beta^{\tau_m})$ if and only if $\alpha = 0$, $F_e^{20} = F_e^{02} = 0$, $\bar{\partial}_e \varphi = 0$, and*

$$i\Lambda_g F_e + \tfrac{1}{4}\varphi\bar{\varphi} + (\tfrac{1}{2}s - \pi\Lambda_g\beta) = 0. \tag{*}$$

Note that the conditions $F_e^{20} = F_e^{02} = 0$, $\bar{\partial}_e \varphi = 0$ mean that e is the Chern connection of a holomorphic structure in the Hermitian line bundle M and that φ is a holomorphic section with respect to this holomorphic structure. Integrating the relation $(*)$ and using the inequality in the hypothesis, one sees that φ cannot vanish identically.

To interpret the condition $(*)$ consider an arbitrary real valued function function $t : X \longrightarrow \mathbb{R}$, and let

$$m_t : \mathcal{A}(M) \times A^0(M) \longrightarrow iA^0$$

be the map defined by

$$m_t(e, \varphi) := \Lambda_g F_e - \tfrac{1}{4}i\varphi\bar{\varphi} + it.$$

It is easy to see that (after suitable Sobolev completions) $\mathcal{A}(M) \times A^0(M)$ has a natural symplectic structure, and that m_t is a <u>moment map</u> for the action of the gauge group $\mathcal{G} = \mathcal{C}^\infty(X, S^1)$. Let $\mathcal{G}^{\mathbb{C}} = \mathcal{C}^\infty(X, \mathbb{C}^*)$ be the complexification of $\mathcal{G}$, and let $\mathcal{H} \subset \mathcal{A}(M) \times A^0(M)$ be the closed set

$$\mathcal{H} := \{(e, \varphi) \in \mathcal{A}(M) \times A^0(M) : F_e^{02} = 0, \ \bar{\partial}_e \varphi = 0\}$$

of integrable pairs. For any function t put

$$\mathcal{H}_t := \{(e, \varphi) \in \mathcal{H} : \mathcal{G}^{\mathbb{C}}(e, \varphi) \cap m_t^{-1}(0) \neq \varnothing\}.$$

Using a general principle in the theory of symplectic quotients, which also holds in our infinite dimensional framework, one can prove that the $\mathcal{G}^{\mathbb{C}}$-orbit of a point $(e, \varphi) \in \mathcal{H}_t$ intersects the zero set $m_t^{-1}(0)$ of the moment map m_t precisely along a $\mathcal{G}$-orbit (see Figure 3). In other words, there is a natural bijection of quotients

$$[m_t^{-1}(0) \cap \mathcal{H}]/\mathcal{G} \simeq \mathcal{H}_t/\mathcal{G}^{\mathbb{C}}. \tag{1}$$

Now take $t := -(\tfrac{1}{2}s - \pi\Lambda_g\beta)$ and suppose again that the assumptions in Lemma 3.1.1 hold. We have seen that $m_t^{-1}(0) \cap \mathcal{H}$ cannot contain pairs of the form $(e, 0)$, hence $\mathcal{G}$ ($\mathcal{G}^{\mathbb{C}}$) acts freely on $m_t^{-1}(0) \cap \mathcal{H}$ ($\mathcal{H}_t$). Using this fact one can show that $\mathcal{H}_t$ is open in the space $\mathcal{H}$ of integrable pairs, and endowing the two quotients in (1) with the natural real analytic structures, one proves that (1) is a real analytic isomorphism. By the lemma, the first quotient is precisely the moduli space $\mathcal{W}_\beta^{\tau_m}$. The second quotient is a complex-geometric object, namely an open subspace in the moduli space of simple holomorphic pairs $\mathcal{H} \cap \{\varphi \neq 0\}/\mathcal{G}^{\mathbb{C}}$. A point in this moduli space can be regarded as an

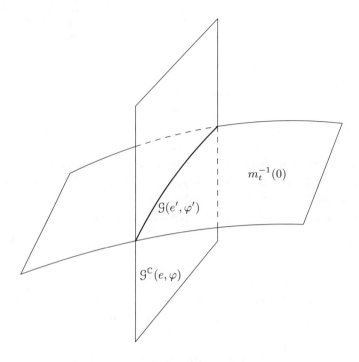

Figure 3.

isomorphism class of pairs $(\mathcal{M}, \varphi)$ consisting of a holomorphic line bundle $\mathcal{M}$ of topological type M, and a holomorphic section in $\mathcal{M}$. Such a pair defines a point in $\mathcal{H}_t/\mathcal{G}^{\mathbb{C}}$ if and only if $\mathcal{M}$ admits a Hermitian metric h satisfying the equation

$$i\Lambda F_h + \tfrac{1}{4}\varphi\bar{\varphi}^h = t.$$

This equation for the unknown metric h is the <u>vortex</u> <u>equation</u> associated with the function t: it is solvable if and only if the <u>stability</u> <u>condition</u>

$$(2c_1(M) - c_1(K_X) - b)\cdot[\omega_g] < 0$$

is fulfilled. Let $\mathcal{D}ou(m)$ be the Douady space of effective divisors $D \subset X$ with $c_1(\mathcal{O}_X(D)) = m$. The map $Z : \mathcal{H}_t/\mathcal{G}^{\mathbb{C}} \longrightarrow \mathcal{D}ou(m)$ which associates to an orbit $[e, \varphi]$ the zero-locus $Z(\varphi) \subset X$ of the holomorphic section φ is an isomorphism of complex spaces.

Putting everything together, we have the following interpretation for the monopole moduli spaces $\mathcal{W}_{\beta}^{\tau_m}$ on Kähler surfaces.

THEOREM 3.1.2 [Okonek and Teleman 1995a; 1996b]. *Let (X, g) be a compact Kähler surface, and let τ_m be the Spin^c-structure defined by the S^1-bundle M. Let $\beta \in A_{\mathbb{R}}^{11}$ be a closed 2-form representing the de Rham cohomology class b such that*

$$(2c_1(M) - c_1(K_X) - b)\cdot[\omega_g] < 0 \quad (alternatively, \ > 0).$$

If $c_1(M) \notin \mathrm{NS}(X)$, were $\mathrm{NS}(X)$ is the Néron–Severi group of X, then $\mathcal{W}_\beta^{\tau m} = \varnothing$. When $c_1(M) \in \mathrm{NS}(X)$, there is a natural real analytic isomorphism

$$\mathcal{W}_\beta^{\tau m} \simeq \mathrm{Dou}(m) \quad (\text{respectively, } \mathrm{Dou}(c_1(K_X) - m)).$$

A moduli space $\mathcal{W}_\beta^{\tau m} \neq \varnothing$ is smooth at the point corresponding to $D \in \mathrm{Dou}(m)$ if and only if $h^0(\mathcal{O}_D(D)) = \dim_D \mathrm{Dou}(m)$. This condition is always satisfied when $b_1(X) = 0$. If $\mathcal{W}_\beta^{\tau m}$ is smooth at a point corresponding to $D \in \mathrm{Dou}(m)$, then it has the expected dimension in this point if and only if $h^1(\mathcal{O}_D(D)) = 0$.

The natural isomorphisms $\mathcal{W}_\beta^{\tau m} \simeq \mathrm{Dou}(m)$ respects the orientations induced by the complex structure of X when $(2c_1(M) - c_1(K_X) - b) \cdot [\omega_g] < 0$. If $(2c_1(M) - c_1(K_X) - b) \cdot [\omega_g] > 0$, then the isomorphism $\mathcal{W}_\beta^{\tau m} \simeq \mathrm{Dou}(c_1(K_X) - m)$ multiplies the complex orientations by $(-1)^{\chi(M)}$ [Okonek and Teleman 1996b].

EXAMPLE. Consider again the complex projective plane $\mathbb{P}^2$, polarized by

$$h = c_1(\mathcal{O}_{\mathbb{P}^2}(1)).$$

The expected dimension of $\mathcal{W}_\beta^{\tau m}$ is $m(m + 3h)$. The theorem above yields the following explicit description of the corresponding moduli spaces:

$$\mathcal{W}_\beta^{\tau m} \simeq \begin{cases} |\mathcal{O}_{\mathbb{P}^2}(m)| & \text{if } (2m + 3h - [\beta]) \cdot h < 0, \\ |\mathcal{O}_{\mathbb{P}^2}(-(m+3))| & \text{if } (2m + 3h - [\beta]) \cdot h > 0. \end{cases}$$

E. Witten [1994] has shown that on Kählerian surfaces X with geometric genus $p_g > 0$ all nontrivial Seiberg–Witten invariants $\mathrm{SW}_{X,\mathfrak{o}}(\mathfrak{c})$ satisfy $w_\mathfrak{c} = 0$.

In the case of Kählerian surfaces with $p_g = 0$ one has a different situation. Suppose for instance that $b_1(X) = 0$. Choose the standard orientation $\mathfrak{o}_1$ of $H^1(X, \mathbb{R}) = 0$ and the component $\mathbf{H}_0$ containing Kähler classes to orient the moduli spaces of monopoles. Then, using the previous theorem and the wall-crossing formula, we get:

PROPOSITION 3.1.3. Let X be a Kähler surface with $p_g = 0$ and $b_1 = 0$. If $m \in H^2(X, \mathbb{Z})$ satisfies $m(m - c_1(K_X)) \geq 0$, that is, the expected dimension $w_{2m - c_1(K_X)}$ is nonnegative, then

$$\mathrm{SW}_{X, \mathbf{H}_0}^+(\mathfrak{c}_m) = \begin{cases} 1 & \text{if } \mathrm{Dou}(m) \neq \varnothing, \\ 0 & \text{if } \mathrm{Dou}(m) = \varnothing, \end{cases}$$

$$\mathrm{SW}_{X, \mathbf{H}_0}^-(\mathfrak{c}_m) = \begin{cases} 0 & \text{if } \mathrm{Dou}(m) \neq \varnothing, \\ -1 & \text{if } \mathrm{Dou}(m) = \varnothing. \end{cases}$$

Our next goal is to show that the PU(2)-monopole equations on a Kähler surface can be analyzed in a similar way. This analysis yields a complex geometric description of the moduli spaces whose S^1-quotients give formulas relating the Donaldson invariants to the Seiberg–Witten invariants. If the base is projective, one also has an algebro-geometric interpretation [Okonek et al. 1999], which leads to explicitly computable examples of moduli spaces of PU(2)-monopoles

[Teleman 1996]. Such examples are important, because they illustrate the general mechanism for proving the relation between the two theories, and help to understand the geometry of the ends of the moduli spaces in the more difficult $\mathcal{C}^\infty$-category.

Recall that, since (X, g) comes with a canonical Spin^c-structure τ, the data of of a $\mathrm{Spin}^{U(2)}$-structure in (X, g) is equivalent to the data of a Hermitian bundle E of rank 2. The bundles of the corresponding $\mathrm{Spin}^{U(2)}$-structure $\sigma : P^u \longrightarrow P_g$ are given by $\bar{\delta}(P^u) = P_E/S^1$, $\det P^u = \det E \otimes K_X$, and $\Sigma^\pm(P^u) = \Sigma^\pm \otimes E \otimes K_X$.

Suppose that $\det P^u$ admits an <u>integrable</u> connection $a \in \mathcal{A}(\det P^u)$. Let $k \in \mathcal{A}(K_X)$ be the Chern connection of the canonical bundle, and let $\lambda := a \otimes k^\vee$ be the induced connection in $L := \det E$. We denote by $\mathcal{L} := (L, \bar{\partial}_\lambda)$ the holomorphic structure defined by λ. Now identify the affine space $\mathcal{A}(\bar{\delta}(P^u))$ with the space $\mathcal{A}_{\lambda \otimes k^{\otimes 2}}(E \otimes K_X)$ of connections in $E \otimes K_X$ which induce $\lambda \otimes k^{\otimes 2} = a \otimes k$ in $\det(E \otimes K_X)$, and identify $A^0(\Sigma^+(P^u))$ with

$$A^0(E \otimes K_X) \oplus A^0(E) = A^0(E \otimes K_X) \oplus A^{02}(E \otimes K_X).$$

PROPOSITION 3.1.4. *Fix an integrable connection $a \in \mathcal{A}(\det E \otimes K_X)$. A pair $(A, \varphi + \alpha) \in \mathcal{A}_{\lambda \otimes k^{\otimes 2}}(E \otimes K_X) \times [A^0(E \otimes K_X) \oplus A^{02}(E \otimes K_X)]$ solves the* PU(2)-*monopole equations* (SW_a^σ) *if and only if A is integrable and one of the following conditions is satisfied*:

(I) $\alpha = 0$, *and* $\bar{\partial}_A \varphi = 0$, *and* $i\Lambda_g F_A^0 + \frac{1}{2}(\varphi\bar{\varphi})_0 = 0$.
(II) $\varphi = 0$, $\partial_A \alpha = 0$, *and* $i\Lambda_g F_A^0 - \frac{1}{2} * (\alpha \wedge \bar{\alpha})_0 = 0$.

Note that solutions (A, φ) of type I give rise to holomorphic pairs $(\mathcal{F}_A, \varphi)$, consisting of a holomorphic structure in $F := E \otimes K$ and a holomorphic section φ in $\mathcal{F}_A$. The remaining equation $i\Lambda_g F_A^0 + \frac{1}{2}(\varphi\bar{\varphi})_0 = 0$ can again be interpreted as the vanishing condition for a moment map for the $\mathcal{G}_0$-action in the space of pairs $(A, \varphi) \in \mathcal{A}_{\lambda \otimes k^{\otimes 2}}(F) \times A^0(F)$. We shall study the corresponding stability condition in the next section.

The analysis of the solutions of type II can be reduced to the investigation of the type I solutions: Indeed, if $\varphi = 0$ and $\alpha \in A^{02}(E \otimes K_X)$ satisfies $\partial_A \alpha = 0$, we see that the section $\psi := \bar{\alpha} \in A^0(\bar{E})$ must be holomorphic, that is, it satisfies $\bar{\partial}_{A \otimes [a^\vee]}\psi = 0$. On the other hand one has $-*(\alpha \wedge \bar{\alpha})_0 = *(\bar{\alpha} \wedge \bar{\bar{\alpha}})_0 = (\psi\bar{\psi})_0$.

3.2. Vortex Equations and Stable Oriented Pairs.
Let (X, g) be a compact Kähler manifold of arbitrary dimension n, and let E be a differentiable vector bundle of rank r, endowed with a fixed holomorphic structure $\mathcal{L} := (L, \bar{\partial}_\mathcal{L})$ in $L := \det E$.

An <u>oriented</u> <u>pair</u> of type $(E, \mathcal{L})$ is a pair $(\mathcal{E}, \varphi)$, consisting of a holomorphic structure $\mathcal{E} = (E, \bar{\partial}_\mathcal{E})$ in E with $\bar{\partial}_{\det \mathcal{E}} = \bar{\partial}_\mathcal{L}$, and a holomorphic section $\varphi \in H^0(\mathcal{E})$. Two oriented pairs are isomorphic if they are equivalent under the natural action of the group $\mathrm{SL}(E)$ of differentiable automorphisms of E with determinant 1.

An oriented pair $(\mathcal{E}, \varphi)$ is <u>simple</u> if its stabilizer in $\mathrm{SL}(E)$ is contained in the center $\mathbb{Z}_r \cdot \mathrm{id}_E$ of $\mathrm{SL}(E)$; it is <u>strongly simple</u> if this stabilizer is trivial.

PROPOSITION 3.2.1 [Okonek and Teleman 1996a]. *There exists a (possibly non-Hausdorff) complex analytic orbifold $\mathcal{M}^{\mathrm{si}}(E, \mathcal{L})$ parameterizing isomorphism classes of simple oriented pairs of type $(E, \mathcal{L})$. The open subset $\mathcal{M}^{\mathrm{ssi}}(E, \mathcal{L}) \subset \mathcal{M}^{\mathrm{si}}(E, \mathcal{L})$ of classes of strongly simple pairs is a complex analytic space, and the points $\mathcal{M}^{\mathrm{si}}(E, \mathcal{L}) \setminus \mathcal{M}^{\mathrm{ssi}}(E, \mathcal{L})$ have neighborhoods modeled on $\mathbb{Z}_r$-quotients.*

Now fix a Hermitian background metric H in E. In this section we use the symbol $(\mathrm{SU}(E))\, U(E)$ for the groups of (special) unitary automorphisms of (E, H), and not for the bundles of (special) unitary automorphisms.

Let λ be the Chern connection associated with the Hermitian holomorphic bundle $(\mathcal{L}, \det H)$. We denote by $\bar{\mathcal{A}}_{\bar{\partial}_\lambda}(E)$ the affine space of semiconnections in E which induce the semiconnection $\bar{\partial}_\lambda = \bar{\partial}_{\mathcal{L}}$ in $L = \det E$, and we write $\mathcal{A}_\lambda(E)$ for the space of unitary connections in (E, H) which induce λ in L. The map $A \longmapsto \bar{\partial}_A$ yields an identification $\mathcal{A}_\lambda(E) \longrightarrow \bar{\mathcal{A}}_{\bar{\partial}_\lambda}(E)$, which endows the affine space $\mathcal{A}_\lambda(E)$ with a complex structure. Using this identification and the Hermitian metric H, the product $\mathcal{A}_\lambda(E) \times A^0(E)$ becomes — after suitable Sobolev completions — an infinite dimensional Kähler manifold. The map

$$m : \mathcal{A}_\lambda(E) \times A^0(E) \longrightarrow A^0(su(E))$$

defined by $m(A, \varphi) := \Lambda_g F_A^0 - \frac{i}{2}(\varphi \bar{\varphi})_0$ is a <u>moment map</u> for the $\mathrm{SU}(E)$-action on the Kähler manifold $\mathcal{A}_\lambda(E) \times A^0(E)$.

We denote by $\mathcal{H}_\lambda(E) := \{(A, \varphi) \in \mathcal{A}_\lambda(E) \times A^0(E) |\ F_A^{02} = 0,\ \bar{\partial}_A \varphi = 0\}$ the space of integrable pairs, and by $\mathcal{H}_\lambda(E)^{\mathrm{si}}$ the open subspace of pairs

$$(A, \varphi) \in \mathcal{H}_\lambda(E)$$

with $(\bar{\partial}_A, \varphi)$ simple. The quotient

$$\mathcal{V}_\lambda(E) := \big(\mathcal{H}_\lambda(E) \cap m^{-1}(0)\big) / \mathrm{SU}(E)$$

is called the moduli space of <u>projective vortices</u>, and

$$\mathcal{V}_\lambda^*(E) := \big(\mathcal{H}_\lambda^{\mathrm{si}}(E) \cap m^{-1}(0)\big) / \mathrm{SU}(E)$$

is called the moduli space of <u>irreducible projective vortices</u>. Note that a vortex (A, φ) is irreducible if and only if $\mathrm{SL}(E)_{(A,\varphi)} \subset \mathbb{Z}_r\, \mathrm{id}_E$. Using again an infinite dimensional version of the theory of symplectic quotients (as in the abelian case), one gets a homeomorphism

$$j : \mathcal{V}_\lambda(E) \xrightarrow{\simeq} \mathcal{H}_\lambda^{\mathrm{ps}}(E) / \mathrm{SL}(E)$$

where $\mathcal{H}_\lambda^{\mathrm{ps}}(E)$ is the subspace of $\mathcal{H}_\lambda(E)$ consisting of pairs whose $\mathrm{SL}(E)$-orbit meets the vanishing locus of the moment map. $\mathcal{H}_\lambda^{\mathrm{ps}}(E)$ is in general not open, but $\mathcal{H}_\lambda^s(E) := \mathcal{H}_\lambda^{ps}(E) \cap \mathcal{H}_\lambda^{si}(E)$ is open, and restricting j to $\mathcal{V}_\lambda^*(E)$ yields an isomorphism of real analytic orbifolds

$$\mathcal{V}_\lambda^*(E) \xrightarrow{\simeq} \mathcal{H}_\lambda^s(E)\big/\operatorname{SL}(E) \subset \mathcal{M}^{si}(E,\mathcal{L}).$$

The image

$$\mathcal{M}^s(E,\mathcal{L}) := \mathcal{H}_\lambda^s(E)\big/\operatorname{SL}(E)$$

of this isomorphism can be identified with the set of isomorphism classes of simple oriented holomorphic pairs $(\mathcal{E},\varphi)$ of type $(E,\mathcal{L})$, with the property that $\mathcal{E}$ admits a Hermitian metric with $\det h = \det H$ which solves the <u>projective vortex equation</u>

$$i\Lambda_g F_h^0 + \tfrac{1}{2}(\varphi\bar{\varphi}^h)_0 = 0.$$

Here F_h is the curvature of the Chern connection of $(\mathcal{E},h)$.

The set $\mathcal{M}^s(E,\mathcal{L})$ has a purely holomorphic description as the subspace of elements $[\mathcal{E},\varphi] \in \mathcal{M}^{si}(E,\mathcal{L})$ which satisfy a suitable <u>stability</u> condition.

This condition is rather complicated for bundles E of rank $r > 2$, but it becomes very simple when $r = 2$.

Recall that, for any torsion free coherent sheaf $\mathcal{F} \neq 0$ over a n-dimensional Kähler manifold (X,g), one defines the g-<u>slope</u> of $\mathcal{F}$ by

$$\mu_g(\mathcal{F}) := \frac{c_1(\det \mathcal{F}) \cup [\omega_g]^{n-1}}{\operatorname{rk}(\mathcal{F})}.$$

A holomorphic bundle $\mathcal{E}$ over (X,g) is called <u>slope-stable</u> if $\mu_g(\mathcal{F}) < \mu_g(\mathcal{E})$ for all proper coherent subsheaves $\mathcal{F} \subset \mathcal{E}$. The bundle $\mathcal{E}$ is <u>slope-polystable</u> if it decomposes as a direct sum $\mathcal{E} = \oplus\mathcal{E}_i$ of slope-stable bundles with $\mu_g(\mathcal{E}_i) = \mu_g(\mathcal{E})$.

DEFINITION 3.2.2. Let $(\mathcal{E},\varphi)$ be an oriented pair of type $(E,\mathcal{L})$ with $\operatorname{rk} E = 2$ over a Kähler manifold (X,g). The pair $(\mathcal{E},\varphi)$ is <u>stable</u> if $\varphi = 0$ and $\mathcal{E}$ is slope-stable, or $\varphi \neq 0$ and the divisorial component D_φ of the zero-locus $Z(\varphi) \subset X$ satisfies $\mu_g(\mathcal{O}_X(D_\varphi)) < \mu_g(E)$. The pair $(\mathcal{E},\varphi)$ is <u>polystable</u> if it is stable or $\varphi = 0$ and $\mathcal{E}$ is slope-polystable.

EXAMPLE. Let $D \subset X$ be an effective divisor defined by a section

$$\varphi \in H^0(\mathcal{O}_X(D)) \setminus \{0\},$$

and put $\mathcal{E} := \mathcal{O}_X(D) \oplus [\mathcal{L} \otimes \mathcal{O}_X(-D)]$. The pair $(\mathcal{E},\varphi)$ is stable if and only if $\mu_g(\mathcal{O}_X(2D)) < \mu_g(\mathcal{L})$.

The following result gives a metric characterization of polystable oriented pairs.

THEOREM 3.2.3 [Okonek and Teleman 1996a]. *Let E be a differentiable vector bundle of rank 2 over (X,g) endowed with a Hermitian holomorphic structure $(\mathcal{L},l)$ in $\det E$. An oriented pair of type $(E,\mathcal{L})$ is polystable if and only if $\mathcal{E}$ admits a Hermitian metric h with $\det h = l$ which solves the projective vortex equation*

$$i\Lambda_g F_h^0 + \tfrac{1}{2}(\varphi\bar{\varphi}^h)_0 = 0.$$

If $(\mathcal{E}, \varphi)$ is stable, then the metric h is unique.

This result identifies the subspace $\mathcal{M}^s(E, \mathcal{L}) \subset \mathcal{M}^{si}(E, \mathcal{L})$ as the subspace of isomorphism classes of <u>stable</u> <u>oriented</u> <u>pairs</u>.

Theorem 3.2.3 can be used to show that the moduli spaces $\mathcal{M}_a^\sigma$ of PU(2)-monopoles on a Kähler surface have a natural complex geometric description when the connection a is integrable. Recall from Section 3.1 that in this case $\mathcal{M}_a^\sigma$ decomposes as the union of two Zariski-closed subspaces

$$\mathcal{M}_a^\sigma = (\mathcal{M}_a^\sigma)_I \cup (\mathcal{M}_a^\sigma)_{II}$$

according to the two conditions I, II in Proposition 3.1.4. By this proposition, both terms of this union can be identified with moduli spaces of projective vortices. Using again the symbol * to denote subsets of points with central stabilizers, one gets the following Kobayashi–Hitchin type description of $(\mathcal{M}_a^\sigma)^*$ in terms of stable oriented pairs.

THEOREM 3.2.4 [Okonek and Teleman 1996a; 1997]. *If $a \in \mathcal{A}(\det P^u)$ is integrable, the moduli space $\mathcal{M}_a^\sigma$ decomposes as a union $\mathcal{M}_a^\sigma = (\mathcal{M}_a^\sigma)_I \cup (\mathcal{M}_a^\sigma)_{II}$ of two Zariski closed subspaces isomorphic with moduli spaces of projective vortices, which intersect along the Donaldson moduli space $\mathcal{D}(\bar{\delta}(P^u))$. There are natural real analytic isomorphisms*

$$(\mathcal{M}_a^\sigma)_I^* \xrightarrow{\simeq} \mathcal{M}^s(E \otimes K_X, \mathcal{L} \otimes \mathcal{K}_X^{\otimes 2}), \quad (\mathcal{M}_a^\sigma)_{II}^* \xrightarrow{\simeq} \mathcal{M}^s(E^\vee, \mathcal{L}^\vee),$$

where $\mathcal{L}$ denotes the holomorphic structure in $\det E = \det P^u \otimes K_X^\vee$ defined by $\bar{\partial}_a$ and the canonical holomorphic structure in K_X.

EXAMPLE (R. Plantiko). On $\mathbb{P}^2$, endowed with the standard Fubini–Study metric g, we consider the $\mathrm{Spin}^{U(2)}(4)$-structure $\sigma : P^u \longrightarrow P_g$ defined by the standard $\mathrm{Spin}^c(4)$-structure $\tau : P^c \longrightarrow P_g$ and the $U(2)$-bundle E with $c_1(E) = 7$, $c_2(E) = 13$, and we fix an integrable connection $a \in \mathcal{A}(\det P^u)$. This $\mathrm{Spin}^{U(2)}(4)$-structure is characterized by $c_1(\det(P^u)) = 4$, $p_1(\bar{\delta}(P^u)) = -3$, and the bundle $F := E \otimes K_{\mathbb{P}^2}$ has Chern classes $c_1(F) = 1$, $c_2(F) = 1$. It is easy to see that every stable oriented pair $(\mathcal{F}, \varphi)$ of type $(F, \mathcal{O}_{\mathbb{P}^2}(1))$ with $\varphi \neq 0$ fits into an exact sequence of the form

$$0 \longrightarrow \mathcal{O} \xrightarrow{\varphi} \mathcal{F} \longrightarrow J_{Z(\varphi)} \otimes \mathcal{O}_{\mathbb{P}^2}(1) \longrightarrow 0,$$

where $\mathcal{F} = \mathcal{T}_{\mathbb{P}^2}(-1)$ and the zero locus $Z(\varphi)$ of φ consists of a simple point $z_\varphi \in \mathbb{P}^2$. Two such pairs $(\mathcal{F}, \varphi)$, $(\mathcal{F}, \varphi')$ define the same point in the moduli space $\mathcal{M}^s(F, \mathcal{O}_{\mathbb{P}^2}(1))$ if and only if $\varphi' = \pm\varphi$. The resulting identification

$$\mathcal{M}^s(F, \mathcal{O}_{\mathbb{P}^2}(1)) = H^0(\mathcal{T}_{\mathbb{P}^2}(-1))/\{\pm \mathrm{id}\}$$

is a complex analytic isomorphism.

Since every polystable pair of type $(F, \mathcal{O}_{\mathbb{P}^2}(1))$ is actually stable, and since there are no polystable oriented pairs of type $(E^\vee, \mathcal{O}_{\mathbb{P}^2}(-7))$, Theorem 3.2.4 yields a real analytic isomorphism

$$\mathcal{M}_a^\sigma = H^0(\mathcal{T}_{\mathbb{P}^2}(-1))/\{\pm \mathrm{id}\},$$

where the origin corresponds to the unique stable oriented pair of the form $(\mathcal{T}_{\mathbb{P}^2}(-1), 0)$. The quotient $H^0(\mathcal{T}_{\mathbb{P}^2}(-1))/\{\pm \mathrm{id}\}$ has a natural algebraic compactification $\mathcal{C}$, given by the cone over the image of $\mathbb{P}(H^0(\mathcal{T}_{\mathbb{P}^2}(-1)))$ under the Veronese map to $\mathbb{P}(S^2 H^0(\mathcal{T}_{\mathbb{P}^2}(-1)))$. This compactification coincides with the Uhlenbeck compactification $\overline{\mathcal{M}_a^\sigma}$ (see Section 2.2 and [Teleman 1996]). More precisely, let $\sigma' : P'^u \longrightarrow P_g$ be the $\mathrm{Spin}^{U(2)}(4)$-structure with $\det P'^u = \det P^u$ and $p_1(P'^u) = 1$. This structure is associated with τ and the $U(2)$-bundle E' with Chern classes $c_1(E') = 7$, $c_2(E') = 12$. The moduli space $\mathcal{M}_a^{\sigma'}$ consists of one (abelian) point, the class of the abelian solution corresponding to the <u>stable</u> oriented pair $(\mathcal{O}_{\mathbb{P}^2} \oplus \mathcal{O}_{\mathbb{P}^2}(1), \mathrm{id}_{\mathcal{O}_{\mathbb{P}^2}})$ of type $(E' \otimes K_{\mathbb{P}^2}, \mathcal{O}_{\mathbb{P}^2}(1))$. $\mathcal{M}_a^{\sigma'}$ can be identified with the moduli space

$$\mathcal{W}_{\frac{i}{2\pi} F_a}^\tau$$

of $[\frac{i}{2\pi} F_a]$-twisted abelian Seiberg–Witten monopoles. Under the identification $\mathcal{C} = \overline{\mathcal{M}_a^\sigma}$, the vertex of the cone corresponds to the unique Donaldson point which is given by the stable oriented pair $(\mathcal{T}_{\mathbb{P}^2}(-1), 0)$. The base of the cone corresponds to the space $\mathcal{M}_a^{\sigma'} \times \mathbb{P}^2$ of ideal monopoles concentrated in one point.

We close this section by explaining the stability concept which describes the subset $\mathcal{M}_X^s(E, \mathcal{L}) \subset \mathcal{M}_X^{si}(E, \mathcal{L})$ in the general case $r \geq 2$. This stability concept does <u>not</u> depend on the choice of parameter and the corresponding moduli spaces can be interpreted as "master spaces" for holomorphic pairs (see next section); in the projective framework they admit Gieseker type compactifications [Okonek et al. 1999].

We shall find this stability concept by relating the $SU(E)$-moment map

$$m : \mathcal{A}_\lambda(E) \times A^0(E) \longrightarrow A^0(su(E))$$

to the universal family of $U(E)$-moment maps $m_t : \mathcal{A}(E) \times A^0(E) \longrightarrow A^0(u(E))$ defined by

$$m_t(A, \varphi) := \Lambda_g F_A - \tfrac{1}{2} i(\varphi \bar{\varphi}) + \tfrac{1}{2} it \, \mathrm{id}_E,$$

where $t \in A^0$ is an arbitrary real valued function. Given t, we consider the system of equations

$$F_A^{02} = 0,$$
$$\bar{\partial}_A \varphi = 0, \qquad\qquad (V_t)$$
$$i\Lambda_g F_A + \tfrac{1}{2}(\varphi \bar{\varphi}) = \tfrac{1}{2} t \, \mathrm{id}_E$$

for pairs $(A, \varphi) \in A(E) \times A^0(E)$. Put

$$\rho_t := \frac{1}{4\pi n} \int_X t\omega_g^n.$$

To explain our first result, we have to recall some classical stability concepts for holomorphic pairs.

For any holomorphic bundle $\mathcal{E}$ over (X, g) denote by $\mathcal{S}(\mathcal{E})$ the set of reflexive subsheaves $\mathcal{F} \subset \mathcal{E}$ with $0 < \mathrm{rk}(\mathcal{F}) < \mathrm{rk}(\mathcal{E})$, and for a fixed section $\varphi \in H^0(\mathcal{E})$ put

$$\mathcal{S}_\varphi(\mathcal{E}) := \{\mathcal{F} \in \mathcal{S}(\mathcal{E}) : \varphi \in H^0(\mathcal{F})\}.$$

Define real numbers $\underline{m}_g(\mathcal{E})$ and $\overline{m}_g(\mathcal{E}, \varphi)$ by

$$\underline{m}_g(\mathcal{E}) := \max(\mu_g(\mathcal{E}), \sup_{\mathcal{F}' \in \mathcal{S}(\mathcal{E})} \mu_g(\mathcal{F}')), \quad \overline{m}_g(\mathcal{E}, \varphi) := \inf_{\mathcal{F} \in \mathcal{S}_\varphi(\mathcal{E})} \mu_g(\mathcal{E}/\mathcal{F}).$$

A bundle $\mathcal{E}$ is φ-stable in the sense of S. Bradlow when $\underline{m}_g(\mathcal{E}) < \overline{m}_g(\mathcal{E}, \varphi)$. Let $\rho \in \mathbb{R}$ be any real parameter. A holomorphic pair $(\mathcal{E}, \varphi)$ is called ρ-stable if ρ satisfies the inequality

$$\underline{m}_g(\mathcal{E}) < \rho < \overline{m}_g(\mathcal{E}, \varphi).$$

The pair $(\mathcal{E}, \varphi)$ is ρ-polystable if it is ρ-stable or $\mathcal{E}$-splits holomorphically as $\mathcal{E} = \mathcal{E}' \oplus \mathcal{E}''$ such that $\varphi \in H^0(\mathcal{E}')$, $(\mathcal{E}', \varphi)$ is ρ-stable and $\mathcal{E}''$ is a slope-polystable vector bundle with $\mu_g(\mathcal{E}'') = \rho$ [Bradlow 1991]. Let $\mathrm{GL}(E)$ be the group of bundle automorphisms of E. With these definitions one proves [Okonek and Teleman 1995a] (see [Bradlow 1991] for the case of a constant function t):

PROPOSITION 3.2.5.   *The complex orbit* $\mathrm{GL}(E) \cdot (A, \varphi)$ *of an integrable pair* $(A, \varphi) \in A(E) \times A^0(E)$ *contains a solution of* (V_t) *if and only if the pair* $(\mathcal{E}_A, \varphi)$ *is* ρ_t-*polystable.*

Now fix again a Hermitian metric H in E and an integrable connection λ in the Hermitian line bundle $L := (\det E, \det H)$. Consider the system of equations

$$F_A^{02} = 0,$$
$$\bar{\partial}_A \varphi = 0, \qquad\qquad (V^0)$$
$$i\Lambda_g F_A^0 + \tfrac{1}{2}(\varphi\bar{\varphi})_0 = 0$$

for pairs $(A, \varphi) \in A_\lambda(E) \times A^0(E)$. Then one can prove

PROPOSITION 3.2.6.   *Let* $(A, \varphi) \in A_\lambda(E) \times A^0(E)$ *be an integrable pair. The following assertions are equivalent:*

(i)   *The complex orbit* $\mathrm{SL}(E) \cdot (A, \varphi)$ *contains a solution of* (V^0).
(ii) *There exists a function* $t \in A^0$ *such that the* $\mathrm{GL}(E)$-*orbit* $\mathrm{GL}(E) \cdot (A, \varphi)$ *contains a solution of* (V_t).
(iii) *There exists a real number* ρ *such that the pair* $(\mathcal{E}_A, \varphi)$ *is* ρ-*polystable.*

COROLLARY 3.2.7. *The open subspace* $\mathcal{M}^s(E, \mathcal{L}) \subset \mathcal{M}^{\mathrm{si}}(E, \mathcal{L})$ *is the set of isomorphism classes of simple oriented pairs which are ρ-polystable for some* $\rho \in \mathbb{R}$.

REMARK 3.2.8. There exist stable oriented pairs $(\mathcal{E}, \varphi)$ whose stabilizer with respect to the $\mathrm{GL}(E)$-action is of positive dimension. Such pairs cannot be ρ-stable for any $\rho \in \mathbb{R}$.

Note that the moduli spaces $\mathcal{M}^s(E, \mathcal{L})$ have a natural $\mathbb{C}^*$-action defined by $z \cdot [\mathcal{E}, \varphi] := [\mathcal{E}, z^{1/r}\varphi]$. This action is well defined since r-th roots of unity are contained in the complex gauge group $\mathrm{SL}(E)$.

There exists an equivalent definition for stability of oriented pairs, which does not use the parameter dependent stability concepts of [Bradlow 1991]. The fact that it is expressible in terms of ρ-stability is related to the fact that the moduli spaces $\mathcal{M}^s(E, \mathcal{L})$ are master spaces for moduli spaces of ρ-stable pairs.

3.3. Master Spaces and the Coupling Principle. Let $X \subset \mathbb{P}^N_{\mathbb{C}}$ be a smooth complex projective variety with hyperplane bundle $\mathcal{O}_X(1)$. All degrees and Hilbert polynomials of coherent sheaves will be computed corresponding to these data.

We fix a torsion-free sheaf $\mathcal{E}_0$ and a holomorphic line bundle $\mathcal{L}_0$ over X, and we choose a Hilbert polynomial P_0. By $P_{\mathcal{F}}$ we denote the Hilbert polynomial of a coherent sheaf $\mathcal{F}$. Recall that any nontrivial torsion free coherent sheaf $\mathcal{F}$ admits a unique subsheaf $\mathcal{F}_{\max}$ for which $P_{\mathcal{F}'}/\mathrm{rk}\,\mathcal{F}'$ is maximal and whose rank is maximal among all subsheaves $\mathcal{F}'$ with $P_{\mathcal{F}'}/\mathrm{rk}\,\mathcal{F}'$ maximal.

An $\mathcal{L}_0$-oriented pair of type $(P_0, \mathcal{E}_0)$ is a triple $(\mathcal{E}, \varepsilon, \varphi)$ consisting of a torsion free coherent sheaf $\mathcal{E}$ with determinant isomorphic to $\mathcal{L}_0$ and Hilbert polynomial $P_{\mathcal{E}} = P_0$, a homomorphism $\varepsilon : \det \mathcal{E} \longrightarrow \mathcal{L}_0$, and a morphism $\varphi : \mathcal{E} \longrightarrow \mathcal{E}_0$. The homomorphisms ε and φ will be called the orientation and the framing of the oriented pair. There is an obvious equivalence relation for such pairs. When $\ker \varphi \neq 0$, we set

$$\delta_{\mathcal{E}, \varphi} := P_{\mathcal{E}} - \frac{\mathrm{rk}\,\mathcal{E}}{\mathrm{rk}\left[\ker(\varphi)_{\max}\right]} P_{\ker(\varphi)_{\max}}.$$

An oriented pair $(\mathcal{E}, \varepsilon, \varphi)$ is semistable if either

1. φ is injective, or
2. ε is an isomorphism, $\ker \varphi \neq 0$, $\delta_{\mathcal{E}, \varphi} \geq 0$, and for all nontrivial subsheaves $\mathcal{F} \subset \mathcal{E}$ the inequality

$$\frac{P_{\mathcal{F}}}{\mathrm{rk}\,\mathcal{F}} - \frac{\delta_{\mathcal{E}, \varphi}}{\mathrm{rk}\,\mathcal{F}} \leq \frac{P_{\mathcal{E}}}{\mathrm{rk}\,\mathcal{E}} - \frac{\delta_{\mathcal{E}, \varphi}}{\mathrm{rk}\,\mathcal{E}}.$$

holds.

The corresponding stability concept is slightly more complicated; see [Okonek et al. 1999]. Note that the (semi)stability definition above does not depend on a parameter. It is, however, possible to express (semi)stability in terms of the parameter dependent Gieseker-type stability concepts of [Huybrechts and Lehn 1995]. For example, $(\mathcal{E}, \varepsilon, \varphi)$ is semistable if and only if φ is injective, or $\mathcal{E}$ is Gieseker semistable, or there exists a rational polynomial δ of degree smaller than $\dim X$ with positive leading coefficient, such that $(\mathcal{E}, \varphi)$ is δ-semistable in the sense of [Huybrechts and Lehn 1995].

For all stability concepts introduced so far there exist analogous notions of slope-(semi)stability. In the special case when the reference sheaf $\mathcal{E}_0$ is the trivial sheaf $\mathcal{O}_X$, slope stability is the algebro-geometric analog of the stability concept associated with the projective vortex equation.

THEOREM 3.3.1 [Okonek et al. 1999]. *There exists a <u>projective</u> scheme*

$$\mathcal{M}^{\mathrm{ss}}(P_0, \mathcal{E}_0, \mathcal{L}_0)$$

whose closed points correspond to gr-*equivalence classes of Gieseker semistable* $\mathcal{L}_0$-*oriented pairs of type* $(P_0, \mathcal{E}_0)$. *This scheme contains an open subscheme* $\mathcal{M}^s(P_0, \mathcal{E}_0, \mathcal{L}_0)$ *which is a coarse moduli space for stable* $\mathcal{L}_0$-*oriented pairs.*

It is also possible to construct moduli spaces for stable oriented pairs where the orienting line bundle is allowed to vary [Okonek et al. 1999]. This generalization is important in connection with Gromov–Witten invariants for Grassmannians [Bertram et al. 1996].

Note that $\mathcal{M}^{\mathrm{ss}}(P_0, \mathcal{E}_0, \mathcal{L}_0)$ possesses a natural $\mathbb{C}^*$-action, given by

$$z \cdot [\mathcal{E}, \varepsilon, \varphi] := [\mathcal{E}, \varepsilon, z\varphi],$$

whose fixed point set can be explicitly described. The fixed point locus

$$[\mathcal{M}^{\mathrm{ss}}(P_0, \mathcal{E}_0, \mathcal{L}_0)]^{\mathbb{C}^*}$$

contains two distinguished subspaces, $\mathcal{M}_0$ defined by the equation $\varphi = 0$, and $\mathcal{M}_\infty$ defined by $\varepsilon = 0$. $\mathcal{M}_0$ can be identified with the Gieseker scheme $\mathcal{M}^{\mathrm{ss}}(P, \mathcal{L}_0)$ of equivalence classes of semistable $\mathcal{L}_0$-oriented torsion free coherent sheaves with Hilbert polynomial P_0. The subspace $\mathcal{M}_\infty$ is the Grothendieck Quot-scheme $\mathrm{Quot}^{\mathcal{E}_0, \mathcal{L}_0}_{P_{\mathcal{E}_0} - P_0}$ of quotients of $\mathcal{E}_0$ with fixed determinant isomorphic with $(\det \mathcal{E}_0) \otimes \mathcal{L}_0^\vee$ and Hilbert polynomial $P_{\mathcal{E}_0} - P_0$.

In the terminology of [Białynicki-Birula and Sommese 1983], $\mathcal{M}_0$ is the source $\mathcal{M}_{\mathrm{source}}$ of the $\mathbb{C}^*$-space $\mathcal{M}^{\mathrm{ss}}(P_0, \mathcal{E}_0, \mathcal{L}_0)$, and $\mathcal{M}_\infty$ is its sink when nonempty.

The remaining subspace of the fixed point locus

$$\mathcal{M}_R := [\mathcal{M}^{\mathrm{ss}}(P_0, \mathcal{E}_0, \mathcal{L}_0)]^{\mathbb{C}^*} \setminus [\mathcal{M}_0 \cup \mathcal{M}_\infty],$$

the so-called space of <u>reductions</u>, consists of objects which are of the same type but essentially of lower rank.

Note that the Quot scheme $\mathcal{M}_\infty$ is empty if $\mathrm{rk}(\mathcal{E}_0)$ is smaller than the rank r of the sheaves $\mathcal{E}$ under consideration, in which case the sink of the moduli space is a closed subset of the space of reductions.

Recall from [Białynicki-Birula and Sommese 1983] that the closure of a general $\mathbb{C}^*$-orbit connects a point in $\mathcal{M}_{\text{source}}$ with a point in $\mathcal{M}_{\text{sink}}$, whereas closures of special orbits connect points of other parts of the fixed point set.

The flow generated by the $\mathbb{C}^*$-action can therefore be used to relate data associated with $\mathcal{M}_0$ to data associated with $\mathcal{M}_\infty$ and $\mathcal{M}_R$.

The technique of computing data on $\mathcal{M}_0$ in terms of $\mathcal{M}_R$ and $\mathcal{M}_\infty$ is a very general principle which we call <u>coupling</u> <u>and</u> <u>reduction</u>. This principle has already been described in a gauge theoretic framework in Section 2.2 for relating monopoles and instantons. However, the essential ideas may probably be best understood in an abstract Geometric Invariant Theory setting, where one has a very simple and clear picture.

Let G be a complex reductive group, and consider a linear representation $\rho_A : G \longrightarrow \mathrm{GL}(A)$ in a finite dimensional vector space A. The induced action $\bar{\rho}_A : G \longrightarrow \mathrm{Aut}(\mathbb{P}(A))$ comes with a natural linearization in $\mathcal{O}_{\mathbb{P}(A)}(1)$, hence we have a stability concept, and thus we can form the GIT quotient

$$\mathcal{M}_0 := \mathbb{P}(A)^{\text{ss}}/\!\!/G.$$

Suppose we want to compute "correlation functions"

$$\Phi_I := \langle \mu_I, [\mathcal{M}_0] \rangle ;$$

that is, we want to evaluate suitable products of canonically defined cohomology classes μ_i on the fundamental class $[\mathcal{M}_0]$ of $\mathcal{M}_0$. Usually the μ_i's are slant products of characteristic classes of a "universal bundle" $\mathcal{E}_0$ on $\mathcal{M}_0 \times X$ with homology classes of X. Here X is a compact manifold, and $\mathcal{E}_0$ comes from a tautological bundle $\tilde{\mathcal{E}}_0$ on $A \times X$ by applying Kempf's Descend Lemma.

The main idea is now to couple the original problem with a simpler one, and to use the $\mathbb{C}^*$-action which occurs naturally in the resulting GIT quotients to express the original correlation functions in terms of simpler data. More precisely, consider another representation $\rho_B : G \longrightarrow \mathrm{GL}(B)$ with GIT quotient $\mathcal{M}_\infty := \mathbb{P}(B)^{\text{ss}}/\!\!/G$. The direct sum $\rho := \rho_A \oplus \rho_B$ defines a naturally linearized G-action on the projective space $\mathbb{P}(A \oplus B)$. We call the corresponding quotient

$$\mathcal{M} := \mathbb{P}(A \oplus B)^{\text{ss}}/\!\!/G$$

the <u>master</u> <u>space</u> associated with the coupling of ρ_A to ρ_B.

The space $\mathcal{M}$ comes with a natural $\mathbb{C}^*$-action, given by

$$z \cdot [a, b] := [a, z \cdot b],$$

and the union $\mathcal{M}_0 \cup \mathcal{M}_\infty$ is a closed subspace of the fixed point locus $\mathcal{M}^{\mathbb{C}^*}$.

Now make the simplifying assumptions that $\mathcal{M}$ is smooth and connected, the $\mathbb{C}^*$-action is free outside $\mathcal{M}^{\mathbb{C}^*}$, and suppose that the cohomology classes μ_i extend to $\mathcal{M}$. This condition is always satisfied if the μ_i's were obtained by the procedure described above, and if Kempf's lemma applies to the pull-back bundle $p_A^*(\mathcal{E}_0)$ and provides a bundle on $\mathcal{M} \times X$ extending $\mathcal{E}_0$.

Under these assumptions, the complement

$$\mathcal{M}_R := \mathcal{M}^{\mathbb{C}^*} \setminus (\mathcal{M}_0 \cup \mathcal{M}_\infty)$$

is a closed submanifold of $\mathcal{M}$, disjoint from $\mathcal{M}_0$, and $\mathcal{M}_\infty$. We call $\mathcal{M}_R$ the manifold of <u>reductions</u> of the master space. Now remove a sufficiently small S^1-invariant tubular neighborhood U of $\mathcal{M}^{\mathbb{C}^*} \subset \mathcal{M}$, and consider the S^1-quotient $W := [\mathcal{M} \setminus U]/S^1$. This is a compact manifold whose boundary is the union of the projectivized normal bundles $\mathbb{P}(N_{\mathcal{M}_0})$ and $\mathbb{P}(N_{\mathcal{M}_\infty})$, and a differentiable projective fiber space P_R over $\mathcal{M}_R$. Note that in general P_R has no natural holomorphic structure. Let n_0, n_∞ be the complex dimensions of the fibers of $\mathbb{P}(N_{\mathcal{M}_0})$, $\mathbb{P}(N_{\mathcal{M}_\infty})$, and let $u \in H^2(W, \mathbb{Z})$ be the first Chern class of the S^1-bundle dual to $\mathcal{M} \setminus U \longrightarrow W$. Let μ_I be a class as above. Then, taking into account orientations, we compute:

$$\Phi_I := \langle \mu_I, [\mathcal{M}_0] \rangle = \langle \mu_I \cup u^{n_0}, [\mathbb{P}(N_{\mathcal{M}_0})] \rangle$$
$$= \langle \mu_I \cup u^{n_0}, [\mathbb{P}(N_{\mathcal{M}_\infty})] \rangle - \langle \mu_I \cup u^{n_0}, [P_R] \rangle.$$

In this way the coupling principle reduces the calculation of the original correlation functions on $\mathcal{M}_0$ to computations on $\mathcal{M}_\infty$ and on the manifold of reductions $\mathcal{M}_R$. A particular important case occurs when the GIT problem given by ρ_B is trivial, that is, when $\mathbb{P}(B)^{ss} = \varnothing$. Under these circumstances the functions Φ_I are completely determined by data associated with the manifold of reductions $\mathcal{M}_R$.

Of course, in realistic situations, our simplifying assumptions are seldom satisfied, so that one has to modify the basic idea in a suitable way.

One of the realistic situations which we have in mind is the coupling of coherent sheaves with morphisms into a fixed reference sheaf $\mathcal{E}_0$. In this case, the original problem is the classification of stable torsion-free sheaves, and the corresponding Gieseker scheme $\mathcal{M}^{ss}(P_0, \mathcal{L}_0)$ of $\mathcal{L}_0$-oriented semistable sheaves of Hilbert polynomial P_0 plays the role of the quotient $\mathcal{M}_0$. The corresponding master spaces are the moduli spaces $\mathcal{M}^{ss}(P_0, \mathcal{E}_0, \mathcal{L}_0)$ of semistable $\mathcal{L}_0$-oriented pairs of type $(P_0, \mathcal{E}_0)$.

Coupling with $\mathcal{E}_0$-valued homomorphisms $\varphi : \mathcal{E} \longrightarrow \mathcal{E}_0$ leads to two essentially different situations, depending on the rank r of the sheaves $\mathcal{E}$ under consideration:

1. When $\mathrm{rk}(\mathcal{E}_0) < r$, the framings $\varphi : \mathcal{E} \longrightarrow \mathcal{E}_0$ can never be injective, i.e. there are no semistable homomorphisms. This case correspond to the GIT situation $\mathcal{M}_\infty = \varnothing$.

2. As soon as $\mathrm{rk}(\mathcal{E}_0) \geq r$, the framings φ can become injective, and the Grothendieck schemes $\mathrm{Quot}_{P_{\mathcal{E}_0} - P_0}^{\mathcal{E}_0, \mathcal{L}_0}$ appear in the master space $\mathcal{M}^{ss}(P_0, \mathcal{E}_0, \mathcal{L}_0)$. These Quot schemes are the analoga of the quotients $\mathcal{M}_\infty$ in the GIT situation.

In both cases the spaces of reductions are moduli spaces of objects which are of the same type but essentially of lower rank.

Everything can be made very explicit when the base manifold is a curve X with a trivial reference sheaf $\mathcal{E}_0 = \mathcal{O}_X^{\oplus k}$. In the case $k < r$, the master spaces relate correlation functions of moduli spaces of semistable bundles with fixed determinant to data associated with reductions. When $r = 2$, $k = 1$, the manifold of reductions are symmetric powers of the base curve, and the coupling principle can be used to prove the <u>Verlinde</u> <u>formula</u>, or to compute the volume and the characteristic numbers (in the smooth case) of the moduli spaces of semistable bundles.

The general case $k \geq r$ leads to a method for the computation of <u>Gromov-Witten</u> invariants for Grassmannians. These invariants can be regarded as correlation functions of suitable Quot schemes [Bertram et al. 1996], and the coupling principle relates them to data associated with reductions and moduli spaces of semistable bundles. In this case one needs a master space $\mathfrak{M}^{\mathrm{ss}}(P_0, \mathcal{E}_0, \mathcal{L})$ associated with a Poincaré line bundle $\mathcal{L}$ on $\mathrm{Pic}(X) \times X$ which set theoretically is the union over $\mathcal{L}_0 \in \mathrm{Pic}(X)$ of the master spaces $\mathfrak{M}^{\mathrm{ss}}(P_0, \mathcal{E}_0, \mathcal{L}_0)$ [Okonek et al. 1999]. One could try to prove the <u>Vafa</u>–<u>Intriligator</u> formula along these lines.

Note that the use of master spaces allows us to avoid the sometimes messy investigation of <u>chains</u> of <u>flips</u>, which occur whenever one considers the family of all possible $\mathbb{C}^*$-quotients of the master space [Thaddeus 1994; Bradlow et al. 1996].

The coupling principle has been applied in two further situations.

Using the coupling of vector bundles with twisted endomorphisms, A. Schmitt has recently constructed projective moduli spaces of <u>Hitchin pairs</u> [Schmitt 1998]. In the case of curves and twisting with the canonical bundle, his master spaces are natural compactifications of the moduli spaces introduced in [Hitchin 1987].

Last but not least, the coupling principle can also be used in certain gauge theoretic situations:

The coupling of instantons on 4-manifolds with Dirac-harmonic spinors has been described in detail in Chapter 2. In this case the instanton moduli spaces are the original moduli spaces $\mathfrak{M}_0$, the Donaldson polynomials are the original correlation functions to compute, and the moduli spaces of $\mathrm{PU}(2)$-monopoles are master spaces for the coupling with spinors. One is again in the special situation where $\mathfrak{M}_\infty = \varnothing$, and the manifold of reductions is a union of moduli spaces of twisted abelian monopoles. In order to compute the contributions of the abelian moduli spaces to the correlation functions, one has to give explicit descriptions of the master space in an S^1-invariant neighborhood of the abelian locus.

Finally consider again the Lie group $G = \mathrm{Sp}(n) \cdot S^1$ and the $\mathrm{PSp}(n)$-monopole equations (SW_a^σ) for a $\mathrm{Spin}^{\mathrm{Sp}(n) \cdot S^1}(4)$-structure $\sigma : P^G \longrightarrow P_g$ in (X, g) and an abelian connection a in the associated S^1-bundle (see Remark 2.1.2). Regarding the compactification of the moduli space $\mathfrak{M}_a^\sigma$ as master space associated with the coupling of $\mathrm{PSp}(n)$-instantons to harmonic spinors, one should get a relation between Donaldson $\mathrm{PSp}(n)$-theory and Seiberg–Witten type theories.

References

[Bertram et al. 1996] A. Bertram, G. Daskalopoulos, and R. Wentworth, "Gromov invariants for holomorphic maps from Riemann surfaces to Grassmannians", *J. Amer. Math. Soc.* **9**:2 (1996), 529–571.

[Białynicki-Birula and Sommese 1983] A. Białynicki-Birula and A. J. Sommese, "Quotients by $\mathbb{C}^*$ and $SL(2, \mathbb{C})$ actions", *Trans. Amer. Math. Soc.* **279**:2 (1983), 773–800.

[Bradlow 1991] S. B. Bradlow, "Special metrics and stability for holomorphic bundles with global sections", *J. Differential Geom.* **33**:1 (1991), 169–213.

[Bradlow et al. 1996] S. B. Bradlow, G. D. Daskalopoulos, and R. A. Wentworth, "Birational equivalences of vortex moduli", *Topology* **35**:3 (1996), 731–748.

[Brussee 1996] R. Brussee, "The canonical class and the C^∞ properties of Kähler surfaces", *New York J. Math.* **2** (1996), 103–146.

[Bryan and Wentworth 1996] J. A. Bryan and R. Wentworth, "The multi-monopole equations for Kähler surfaces", *Turkish J. Math.* **20**:1 (1996), 119–128.

[Donaldson 1996] S. K. Donaldson, "The Seiberg-Witten equations and 4-manifold topology", *Bull. Amer. Math. Soc. (N.S.)* **33**:1 (1996), 45–70.

[Donaldson and Kronheimer 1990] S. K. Donaldson and P. B. Kronheimer, *The geometry of four-manifolds*, The Clarendon Press Oxford University Press, New York, 1990. Oxford Science Publications.

[Friedman and Morgan 1997] R. Friedman and J. W. Morgan, "Algebraic surfaces and Seiberg-Witten invariants", *J. Algebraic Geom.* **6**:3 (1997), 445–479.

[Furuta 1995] M. Furuta, "Monopole equation and the $\frac{11}{8}$-conjecture", preprint, Res. Inst. Math. Sci., Kyoto, 1995.

[Hitchin 1987] N. J. Hitchin, "The self-duality equations on a Riemann surface", *Proc. London Math. Soc. (3)* **55**:1 (1987), 59–126.

[Huybrechts and Lehn 1995] D. Huybrechts and M. Lehn, "Framed modules and their moduli", *Internat. J. Math.* **6**:2 (1995), 297–324.

[Jost et al. 1996] J. Jost, X. Peng, and G. Wang, "Variational aspects of the Seiberg-Witten functional", *Calc. Var. Partial Differential Equations* **4**:3 (1996), 205–218.

[Kobayashi and Nomizu 1963] S. Kobayashi and K. Nomizu, *Foundations of differential geometry*, vol. I, Interscience, New York, 1963.

[Kronheimer and Mrowka 1994] P. B. Kronheimer and T. S. Mrowka, "The genus of embedded surfaces in the projective plane", *Math. Res. Lett.* **1**:6 (1994), 797–808.

[Labastida and Mariño 1995] J. M. F. Labastida and M. Mariño, "Non-abelian monopoles on four-manifolds", *Nuclear Phys. B* **448**:1-2 (1995), 373–395.

[Lawson and Michelsohn 1989] J. Lawson, H. Blaine and M.-L. Michelsohn, *Spin geometry*, Princeton Math. Series **38**, Princeton Univ. Press, Princeton, NJ, 1989.

[LeBrun 1995a] C. LeBrun, "Einstein metrics and Mostow rigidity", *Math. Res. Lett.* **2**:1 (1995), 1–8.

[LeBrun 1995b] C. LeBrun, "On the scalar curvature of complex surfaces", *Geom. Funct. Anal.* **5**:3 (1995), 619–628.

[Li and Liu 1995] T. J. Li and A. Liu, "General wall crossing formula", *Math. Res. Lett.* **2**:6 (1995), 797–810.

[Lübke and Teleman 1995] M. Lübke and A. Teleman, *The Kobayashi–Hitchin correspondence*, World Scientific, River Edge, NJ, 1995.

[Lupaşcu 1997] P. Lupaşcu, "Metrics of nonnegative scalar curvature on surfaces of Kähler type", *Abh. Math. Sem. Univ. Hamburg* **67** (1997), 215–220.

[Okonek and Teleman 1995a] C. Okonek and A. Teleman, "The coupled Seiberg-Witten equations, vortices, and moduli spaces of stable pairs", *Internat. J. Math.* **6**:6 (1995), 893–910.

[Okonek and Teleman 1995b] C. Okonek and A. Teleman, "Les invariants de Seiberg-Witten et la conjecture de van de Ven", *C. R. Acad. Sci. Paris Sér. I Math.* **321**:4 (1995), 457–461.

[Okonek and Teleman 1995c] C. Okonek and A. Teleman, "Quaternionic monopoles", *C. R. Acad. Sci. Paris Sér. I Math.* **321**:5 (1995), 601–606.

[Okonek and Teleman 1996a] C. Okonek and A. Teleman, "Quaternionic monopoles", *Comm. Math. Phys.* **180**:2 (1996), 363–388.

[Okonek and Teleman 1996b] C. Okonek and A. Teleman, "Seiberg-Witten invariants for manifolds with $b_+ = 1$ and the universal wall crossing formula", *Internat. J. Math.* **7**:6 (1996), 811–832.

[Okonek and Teleman 1997] C. Okonek and A. Teleman, "Seiberg-Witten invariants and rationality of complex surfaces", *Math. Z.* **225**:1 (1997), 139–149.

[Okonek et al. 1999] C. Okonek, A. Schmitt, and A. Teleman, "Master spaces for stable pairs", *Topology* **38**:1 (1999), 117–139.

[Pidstrigach and Tyurin 1995] V. Y. Pidstrigach and A. N. Tyurin, "Localisation of the Donaldson invariants along the Seiberg–Witten classes", Technical report, 1995. Available at http://xxx.lanl.gov/abs/dg-ga/9507004.

[Schmitt 1998] A. Schmitt, "Projective moduli for Hitchin pairs", *Internat. J. Math.* **9**:1 (1998), 107–118.

[Taubes 1994] C. H. Taubes, "The Seiberg-Witten invariants and symplectic forms", *Math. Res. Lett.* **1**:6 (1994), 809–822.

[Taubes 1995] C. H. Taubes, "The Seiberg-Witten and Gromov invariants", *Math. Res. Lett.* **2**:2 (1995), 221–238.

[Taubes 1996] C. H. Taubes, "SW $\implies$ Gr: from the Seiberg-Witten equations to pseudo-holomorphic curves", *J. Amer. Math. Soc.* **9**:3 (1996), 845–918.

[Teleman 1996] A. Teleman, "Moduli spaces of PU(2)-monopoles", preprint, Universität Zürich, 1996. To appear in *Asian J. Math.*

[Teleman 1997] A. Teleman, "Non-abelian Seiberg–Witten theory and stable oriented pairs", *Internat. J. Math.* **8**:4 (1997), 507–535.

[Thaddeus 1994] M. Thaddeus, "Stable pairs, linear systems and the Verlinde formula", *Invent. Math.* **117**:2 (1994), 317–353.

[Witten 1994] E. Witten, "Monopoles and four-manifolds", *Math. Res. Lett.* **1**:6 (1994), 769–796.

CHRISTIAN OKONEK
INSTITUT FÜR MATHEMATIK
UNIVERSITÄT ZÜRICH
WINTERTHURERSTRASSE 190
CH-8057 ZÜRICH
SWITZERLAND
  okonek@math.unizh.ch

ANDREI TELEMAN
FACULTATEA DE MATEMATICĂ
UNIVERSITATEA DEN BUCUREŞTI
STRADA ACADEMICI 14
70109 BUCAREST
RUMANIA
  teleman@math.unizh.ch

Several Complex Variables
MSRI Publications
Volume **37**, 1999

Recent Techniques in Hyperbolicity Problems

YUM-TONG SIU

ABSTRACT. We explain the motivations and main ideas regarding the new techniques in hyperbolicity problems recently introduced by the author and Sai-Kee Yeung and by Michael McQuillan. Streamlined proofs and alternative approaches are given for previously known results.

We say that a complex manifold is *hyperbolic* if there is no nonconstant holomorphic map from $\mathbb{C}$ to it. This paper discusses the new techniques in hyperbolicity problems introduced in recent years in a series of joint papers which I wrote with Sai-Kee Yeung [Siu and Yeung 1996b; 1996a; 1997] and in a series of papers by Michael McQuillan [McQuillan 1996; 1997]. The goal is to explain the motivations and the main ideas of these techniques. In the process we examine known results using new approaches, providing streamlined proofs for them.

The paper consists of three parts: an Introduction, Chapter 1, and Chapter 2. The Introduction provides the necessary background, states the main problems, and discusses the motivations and the main ideas of the recent new techniques. Chapter 1 presents a proof of the following theorem, using techniques from diophantine approximation.

THEOREM 0.0.1. *Let $\hat{m}$ be a positive integer. Let V_λ ($1 \leq \lambda \leq \Lambda$) be regular complex hypersurfaces in $\mathbb{P}_n$ of degree δ in normal crossing. Let $\varphi : \mathbb{C}^{\hat{m}} \to \mathbb{P}_n$ be a holomorphic map whose image is not contained in any hypersurface of $\mathbb{P}_n$. Then the sum of the defects $\sum_{\lambda=1}^\Lambda \mathrm{Defect}(\varphi, V_\lambda)$ is no more than ne for any $\delta \geq 1$ and is no more than $n + 1$ for $\delta = 1$.*

Chapter 2 presents a streamlined proof of the following result:

THEOREM 0.0.2 [Siu and Yeung 1996a]. *The complement in $\mathbb{P}_2$ of a generic curve of sufficiently high degree is hyperbolic.*

An overview of the proof of these two theorems can be found in Section 0.10 (page 446).

Partially supported by a grant from the National Science Foundation.

Introduction

0.1. Statement of Hyperbolicity Problems.

Hyperbolicity problems have two aspects, the qualitative aspect and the quantitative aspect. The easier qualitative aspect of the hyperbolicity problems is to prove that certain classes of complex manifolds are hyperbolic in the following sense. A complex manifold is *hyperbolic* if there is no nonconstant holomorphic map from $\mathbb{C}$ to it. There are two classes of manifolds which are usually used to test techniques introduced to prove hyperbolicity. One class is the complement of an ample divisor in an abelian variety, or a submanifold of an abelian variety containing no translates of abelian subvarieties. The second class is the complement of a generic hypersurface of high degree (at least $2n + 1$) in the n-dimensional projective space $\mathbb{P}_n$ or a generic hypersurface of high degree (at least $2n - 1$ for $n \geq 3$) in $\mathbb{P}_n$. The general conjecture is that any holomorphic map from $\mathbb{C}$ to a compact complex manifold with ample canonical line bundle (or even of general type) must be algebraically degenerate in the sense that its image is contained in a complex hypersurface of the manifold.

The harder quantitative aspect of the hyperbolicity problems is to get a defect relation. The precise definition of defect will be given below. Again there are two situations which are usually used to test new techniques to get defect relations. The first situation is to show that the defect for an ample divisor in an abelian variety is zero. The second situation is to show that for any algebraically nondegenerate holomorphic map from $\mathbb{C}$ to $\mathbb{P}_n$ the sum of the defects for a collection of hypersurfaces of degree δ in normal crossing is no more than $(n + 1)/\delta$. The general conjecture is that, for any algebraically nondegenerate holomorphic map from $\mathbb{C}$ to a compact complex manifold M and for a positive line bundle L on M, the sum of the defects for a collection of hypersurfaces in normal crossing is no more than γ if each hypersurface is the divisor of a holomorphic section of L and if $(\gamma + \varepsilon)L + K_M$ is positive for any positive rational number ε. Here K_M means the canonical line bundle of M and the positivity of the $\mathbb{Q}$-bundle $(\gamma + \varepsilon)L + K_M$ means that some high integral multiple of $(\gamma + \varepsilon)L + K_M$ is a positive line bundle.

So far as hyperbolicity problems are concerned, whatever can be done for abelian varieties can also usually be done, with straightforward modifications, for semi-abelian varieties. So we will confine ourselves in this paper only to abelian varieties and not worry about the seemingly more general situation of semi-abelian varieties.

We now state more precisely what has been recently proved and what conjectures remain unsolved. We do not include here a number of very recent results available in preprint form whose proofs are still in the process of being studied and verified.

Since at this point the major difficulties of the hyperbolicity problems already occur in the case of abelian varieties and the complex projective space, we will confine ourselves to abelian varieties and the complex projective space and will not elaborate further on the case of a general compact projective algebraic manifold.

THEOREM 0.1.1 [McQuillan 1996; Siu and Yeung 1996b; 1997]. *The defect of an ample divisor in an abelian variety is zero. In particular, the complement of an ample divisor in an abelian variety is hyperbolic.*

CONJECTURE 0.1.2. *The complement in $\mathbb{P}_n$ of a generic hypersurface of degree at least $2n + 1$ is hyperbolic.*

CONJECTURE 0.1.3. *A generic hypersurface of degree at least $2n - 1$ in $\mathbb{P}_n$ is hyperbolic for $n \geq 3$.*

For dimensions higher than 1, one known case for Conjecture 0.1.2 is the following.

THEOREM 0.1.4 [Siu and Yeung 1996a]. *The complement in $\mathbb{P}_2$ of a generic curve of sufficiently high degree is hyperbolic.*

There are many partial results in cases when the hypersurface in Conjecture 0.1.2 or Conjecture 0.1.3 is not generic and either has many components or is of a special form such as defined by a polynomial of high degree and few nonzero terms. Since there are already quite a number of survey papers about such partial results for non generic hypersurfaces (for example [Siu 1995]), we will not discuss them here.

In the formal analogy between Nevanlinna theory and diophantine approximation [Vojta 1987], Conjecture 0.1.2 corresponds to the theorem of Roth [Roth 1955; Schmidt 1980] and Conjecture 0.1.3 corresponds to the Mordell Conjecture [Faltings 1983; 1991; Vojta 1992]. For that reason very likely a proof of Conjecture 0.1.3 may require some techniques different from those used in a proof of Conjecture 0.1.2. For example, the analog of Theorem 0.1.4 for the setting of Conjecture 0.1.3 is still open. The most difficult step in the proof of Theorem 0.1.4, which involves the argument of log-pole jet differentials and touching order, uses in an essential way the disjointness of the entire holomorphic curve from the generic curve of sufficiently high degree (see Remarks 0.3.1 and 0.3.2 and also Section 2.8).

For quantitative results involving defects the basic conjecture in the complex projective space is the following.

CONJECTURE 0.1.5. *Let V_λ $(1 \leq \lambda \leq \Lambda)$ be regular complex hypersurfaces in $\mathbb{P}_n$ of degree δ in normal crossing. Let $\varphi : \mathbb{C} \to \mathbb{P}_n$ be a holomorphic map whose image is not contained in any hypersurface of $\mathbb{P}_n$. Then the sum of defects $\sum_{\lambda=1}^{\Lambda} \text{Defect}(\varphi, V_\lambda)$ is no more than $(n+1)/\delta$.*

The main difficulty of the conjecture occurs already for a single hypersurface. If there is a method to handle the case of a single hypersurface for Conjecture

0.1.5, very likely the same method works for the general case of a collection of hypersurfaces in normal crossing. Though the conjecture for a single hypersurface does not imply immediately Conjecture 0.1.2, it is very likely that its proof can be modified to give Conjecture 0.1.2. An example by Biancofiore [1982] shows that the algebraic nondegeneracy condition in Conjecture 0.1.5 cannot be replaced by the weaker condition that the image of φ is not contained in any hypersurface of degree δ.

0.2. Characteristic Functions, Counting Functions, Proximity Functions, and Defects.

We now give certain definitions needed for precise discussion. Let M be a compact complex manifold with a positive holomorphic line bundle L whose positive definite curvature form is θ. Let s be a holomorphic section of L over M whose zero-divisor is W. Let $\varphi : \mathbb{C} \to M$ be a holomorphic map. We multiply the metric of L by a sufficiently large positive constant so that the pointwise norm $\|s\|$ of s with respect to the metric of L is less than 1 at every point of M. The characteristic function is defined by

$$T(r, \varphi, \theta) = \int_{\rho=0}^{r} \frac{d\rho}{\rho} \int_{|\zeta|<\rho} \varphi^* \theta$$

which changes by a bounded term as $r \to \infty$ when another positive definite curvature form of L is used. Let $n(\rho, \varphi^* W)$ denote the number of zeroes (with multiplicities) of the divisor $\varphi^* W$ in $\{|\zeta| < \rho\}$. The counting function is defined as

$$N(r, \varphi, W) = \int_{\rho=0}^{r} n(\rho, \varphi^* W) \frac{d\rho}{\rho}$$

which we also denote by $N(r, \varphi, s)$. When Z is a divisor in $\mathbb{C}$, we also denote by $n(\rho, Z)$ the number of zeroes (with multiplicities) of the divisor Z in $\{|\zeta| < \rho\}$ and define

$$N(r, Z) = \int_{\rho=0}^{r} n(\rho, Z) \frac{d\rho}{\rho}.$$

Let $\oint_{|\zeta|=r}$ denote the average over the circle $\{|\zeta| = r\}$. The proximity function is defined by

$$m(r, \varphi, s) = \oint_{|\zeta|=r} \log \frac{1}{\|\varphi^* s\|}$$

which changes by a bounded term as $r \to \infty$ when another metric of L is used. We will denote $m(r, \varphi, s)$ also by $m(r, \varphi, W)$. The defect is defined as

$$\text{Defect}(\varphi, s) = \liminf_{r \to \infty} \frac{m(r, \varphi, s)}{T(r, \varphi, \theta)}$$

which we also denote by $\text{Defect}(\varphi, W)$. Let σ be a positive number and let $\tilde{\varphi}_\sigma(\zeta) = \varphi(\sigma\zeta)$. Then from the definitions we have

$$T(r, \varphi, \theta) = T\left(\frac{r}{\sigma}, \tilde{\varphi}_\sigma, \theta\right), \quad N(r, \varphi, s) = N\left(\frac{r}{\sigma}, \tilde{\varphi}_\sigma, s\right), \quad m(r, \varphi, s) = m\left(\frac{r}{\sigma}, \tilde{\varphi}_\sigma, s\right).$$

When $M = \mathbb{P}_n$ and L is the hyperplane section line bundle of $\mathbb{P}_n$ and θ is the Fubini–Study form, we simply denote $T(r, \varphi, \theta)$ by $T(r, \varphi)$. In the case a holomorphic map from $\mathbb{C}^{\hat{m}}$ to M, its characteristic function, counting function and proximity function is defined by computing those of the restriction of the map to a complex line in the complex vector space $\mathbb{C}^{\hat{m}}$ and then averaging over all such complex lines. Its defect is defined in the same way from its proximity function and its characteristic function as in the case $\hat{m} = 1$.

There is an alternative description of the characteristic function in the case of the complex projective space and we need this alternative description for the dimension one case later. For a holomorphic map φ from $\mathbb{C}$ to $\mathbb{P}_n$ we can use the homogeneous coordinates of $\mathbb{P}_n$ and represent φ in the form $[\varphi_0, \ldots, \varphi_n]$ by $n + 1$ holomorphic functions φ_j $(0 \leq j \leq n)$ without common zeroes on $\mathbb{C}$. Let θ be the Fubini–Study form on $\mathbb{P}_n$. Then

$$\varphi^* \theta = \frac{\sqrt{-1}}{2\pi} \partial \bar{\partial} \log \left(\sum_{j=0}^{n} |\varphi_j|^2 \right)$$

and two integrations give

$$T(r, \varphi, \theta) = \oint_{|\zeta|=r} \tfrac{1}{2} \log \left(\sum_{j=0}^{n} |\varphi_j|^2 \right) - \tfrac{1}{2} \log \left(\sum_{j=0}^{n} |\varphi_j(0)|^2 \right).$$

Since

$$\max_{0 \leq j \leq n} \log |\varphi_j| \leq \tfrac{1}{2} \log \left(\sum_{j=0}^{n} |\varphi_j|^2 \right) \leq \tfrac{1}{2} \log \left((n+1) \max_{0 \leq j \leq n} \log |\varphi_j|^2 \right)$$

$$\leq \max_{0 \leq j \leq n} \log |\varphi_j| + \tfrac{1}{2} \log(n+1),$$

it follows that up to a bounded term the characteristic function $T(r, \varphi, \theta)$ can be described by $\oint_{|\zeta|=r} \max_{0 \leq j \leq n} \log |\varphi_j|$.

Consider the special case $n = 1$. The characteristic function $T(r, \varphi)$ up to a bounded term is equal to

$$\oint_{|\zeta|=r} \max \left(|\varphi_0|, |\varphi_1| \right) = \oint_{|\zeta|=r} \log |\varphi_0| + \oint_{|\zeta|=r} \max \left(1, \log \left| \frac{\varphi_1}{\varphi_0} \right| \right)$$

$$= \log |\varphi(0)| + N(r, \varphi_0, 0) + \oint_{|\zeta|=r} \log^{+} \left| \frac{\varphi_1}{\varphi_0} \right|.$$

Here $\log^{+}$ means the maximum of $\log$ and 0. Thus for a single meromorphic function F the characteristic function for the map $\mathbb{C} \to \mathbb{P}_1$ defined by F is equal to

$$\oint_{|\zeta|=r} \log^{+} |F| + N(r, F, \infty)$$

up to a bounded term.

0.3. The Approach of Jet Differentials. There are two different approaches to proving hyperbolicity. One originated with Bloch [1926], who introduced the use of holomorphic jet differentials vanishing on some ample divisor. Another has its origin from the theory of diophantine approximation. From our present understanding of the so-called Ahlfors–Schwarz lemma for jet differentials, the technique of jet differentials and the technique of diophantine approximation share the same origin of using meromorphic functions of low pole order with high vanishing order, as explained later in this section by means of the logarithmic derivative lemma.

A holomorphic (respectively meromorphic) k-jet differential ω of total weight m on a complex manifold M with local coordinates $z_1, \ldots, z_n$ is locally a polynomial, with holomorphic (respectively meromorphic) functions as coefficients, in the variables $d^l z_j$ ($1 \leq l \leq k$, $1 \leq j \leq n$) and of homogeneous weight m when $d^l z_j$ is given the weight l. A meromorphic k-jet differential M is said to be a log-pole k-jet differential M if it is locally a polynomial, with holomorphic functions as coefficients, in the variables $d^l z_j, d^\nu \log g_\lambda$ ($1 \leq l \leq k$, $1 \leq j \leq n$, $1 \leq \nu \leq k$, $1 \leq \lambda \leq \Lambda$), where the g_λ ($1 \leq \lambda \leq \Lambda$) are local holomorphic functions whose zero-divisors are contained in a finite number of global nonnegative divisors of M.

The key step in the approach using holomorphic jet differentials is what is usually referred to as the Ahlfors–Schwarz lemma or simply as the Schwarz lemma which says the following. If φ is a holomorphic map from $\mathbb{C}$ to a complex manifold M and if ω is a holomorphic (or log-pole) k-jet differential on M which vanishes on an ample divisor of M (and the image of φ is disjoint from the log-pole of ω), then $\varphi^*\omega$ is identically zero on $\mathbb{C}$.

REMARK 0.3.1. In the Schwarz lemma for log-pole jet differentials, the image of the map has to be disjoint from the log-pole of the jet differential. This is one of the main reasons why Conjecture 0.1.3 may require some techniques different from those used in a proof of Conjecture 0.1.2. It is the same reason why the proof of Theorem 0.1.4 cannot be readily modified to yield its analog in the setting of Conjecture 0.1.3.

REMARK 0.3.2. Nevanlinna's original theory already makes use of the log-pole differential

$$\left(\prod_{j=1}^m 1/(z - a_j) \right) dz$$

on $\mathbb{P}_1$ with affine coordinate z for $m \geq 3$. Note that, in the Schwarz lemma, the vanishing of the pullback of a meromorphic jet differential vanishing on some ample divisor requires the following two key ingredients. The first one is that only log-pole singularities are allowed. Other kinds of pole orders are not allowed. The second one is that the image of the map has to be disjoint from the log-pole. Since the two key ingredients are already essential in the case $M = \mathbb{P}_1$, one cannot weaken the two requirements by simply assuming that the poles of the meromorphic jet differential are in some normal form.

We denote by $J_k(M)$ the bundle of all k-jets of M so that $J_1(M)$ is simply the tangent bundle of M. An element of $J_k(M)$ at a point P of M is defined by a holomorphic map $\gamma : U \to M$ for some open neighborhood U of 0 in $\mathbb{C}$ with $\gamma(0) = P$ and another $\tilde{\gamma}$ defines the same element of $J_k(M)$ if γ and $\tilde{\gamma}$ agree up to order k at 0.

Define the map $d^k\varphi : \mathbb{C} \to J_k(M)$ so that its value at $\zeta \in \mathbb{C}$ is the k-jet at $\varphi(\zeta) \in M$ defined by the curve $\varphi : \mathbb{C} \to M$. The Schwarz lemma means that the image of $\mathbb{C}$ under φ satisfies the differential equation $\omega = 0$. For this, it suffices to have the k-jet differential ω defined as a function on $(d^k\varphi)(\mathbb{C})$ instead of on all of $J_k(M)$. When we have enough independent differential equations of such a kind, we can eliminate the derivatives of φ from the differential equations to get the constancy of the map φ and conclude hyperbolicity. An equivalent way of looking at it is to get hyperbolicity by constructing a holomorphic (or log-pole) k-jet differential on the Zariski closure in $J_k(M)$ of $\varphi(\mathbb{C})$ which vanishes on an ample divisor. It suffices also to construct a collection of local holomorphic (or log-pole) k-jet differential on M vanishing on an ample divisor so that they can be pieced together to give a well defined function on the Zariski closure of $(d^k\varphi)(\mathbb{C})$ in $J_k(M)$. Here the Zariski closure of $(d^k\varphi)(\mathbb{C})$ in $J_k(M)$ means the intersection with $J_k(M)$ of the Zariski closure of $(d^k\varphi)(\mathbb{C})$ in the compactification of $J_k(M)$.

The geometric reason for the Schwarz lemma can be heuristically explained as follows. The existence of a holomorphic section ω of the k-jet bundle $J_k(M)$ which vanishes on an ample divisor D means that $J_k(M)$ carries certain positivity. The pullback $\varphi^*\omega$ is a holomorphic section of $J_k(\mathbb{C})$ and vanishes on the pullback of the zero divisor of ω. On the other hand, since the bundle $J_k(\mathbb{C})$ over $\mathbb{C}$ is globally trivial, there is no positivity of $J_k(\mathbb{C})$ to support the zero divisor of the holomorphic section $\varphi^*\omega$ which contains φ^*D if $\varphi^*\omega$ is not identically zero.

A so-called pointwise version of the Schwarz lemma could be formulated and proved by using arguments involving curvature or some generalized notion of it (see for example [Siu and Yeung 1997]). Such a pointwise version implies the Schwarz lemma just stated. However, the most natural proof of the Schwarz lemma is from the use of the logarithmic derivative lemma in Nevanlinna theory. Let $F(\zeta)$ denote the value of ω at $(d^k\varphi)(\zeta) \in J_k(M)$. Assume that $\varphi^*\omega$ is not identically zero and we will get a contradiction. For some suitable coordinate ζ of $\mathbb{C}$, the holomorphic function $F(\zeta)$ is not identically zero. The characteristic function $T(r, F)$ of F is computed by

$$T(r, F) = \oint_{|\zeta|=r} \log^+ |F(\zeta)|.$$

The key point is that ω is dominated in absolute value by a polynomial with constant coefficients of a finite number of variables of the form $d^l \log g$ with

$1 \leq l \leq k$ for some meromorphic functions g on M. The logarithmic derivative lemma says that

$$\oint_{|\zeta|=r} \log^+ |d^l \log g(\varphi(\zeta))| = O(\log T(r, \varphi))$$

for $l \geq 1$. (Note that later on, when we have inequalities derived from the logarithmic derivative lemma, they will hold only outside a set of finite measure with respect to dr/r. This is not made explicit in the notation, but it should not cause confusion.) Hence $T(r, F) = O(\log T(r, \varphi))$. On the other hand, since ω vanishes on an ample divisor of M, we must have $T(r, F) \geq N(r, F, 0) \geq cT(r, \varphi)$ for some positive c, giving $T(r, \varphi) = O(\log T(r, \varphi))$ which contradicts φ being a nonconstant map. This proof works also when ω is a k-jet differential with at most log-pole singularities vanishing on an ample divisor if the image of φ is disjoint from the log-pole. The idea of this proof in the case of an abelian variety was already in [Bloch 1926] and for the case of a general complex manifold was already in [Ru and Wong 1995]. The proof can be interpreted by the pole-order and the vanishing order in the spirit of the method of diophantine approximation as follows. The pullback of the holomorphic 1-jet differential when regarded as a holomorphic function must vanish because the logarithmic derivative lemma takes care of the differentials so that the characteristic function is less than the case of the pole order of any ample divisor but the counting function is like the case of the vanishing order of an ample divisor.

When it comes to the quantitative aspect involving defects, the approach of jet differentials uses jet differentials with low pole-order but high vanishing order along the hypersurfaces whose defects are under consideration. There are two difficulties, the first difficulty is to construct a jet differential with low pole order but high vanishing order along the hypersurfaces. The second difficulty is to make sure that the pullback, to the entire holomorphic curve, of the constructed jet differential is not identically zero.

To handle the first difficulty, when we construct jet differentials we can adjoin many variables of the form $d^l \log g$, with $l \geq 1$ and g holomorphic, to increase the available degrees of freedom to get more vanishing order along the hypersurfaces, without essentially increasing the growth order of the pullback of the constructed jet differential. What makes this possible is the logarithmic derivative lemma. The troublesome point is that we have to make sure that, after adjoining variables of the form $d^l \log g$, the counting function for the pole order is somehow still under control. The situation is much easier in the case of an abelian variety, because we can use the differentials

$$d^l z_j = d^l \exp z_j$$

of coordinates of $\mathbb{C}^n$ as $d^l \log g$ and the nowhere vanishing of the exponential function $\exp z_j$ makes it unnecessary for us to worry about the difficulty of the increased growth of the counting function for the pole order.

When the difficulty of constructing a jet differential with low pole order and high vanishing order along the hypersurfaces and the difficulty of making sure that its pullback to the entire holomorphic curve is not identically zero are both overcome, the above proof of the Schwarz lemma by Nevanlinna theory is easily adapted to give a defect relation.

The second difficulty of making sure the non identical vanishing of the pullback of the jet differential to the entire holomorphic curve corresponds to the step in the proof of Roth's theorem [Roth 1955; Schmidt 1980] of making sure that the constructed polynomial of low degree and high vanishing order has low vanishing order at a point whose components are all equal to the given algebraic number. In the proof of Roth's theorem it was originally done by using Roth's lemma [Roth 1955; Schmidt 1980] and could also be handled by methods introduced later such as the product theorem of Faltings [1991].

For function theory, so far there are two ways of handling the difficulty. One is the use of the translational invariance of the Zariski closure of the differential of a Zariski dense entire curve [Siu and Yeung 1996a; 1997]. Another is the independent slight rescaling of the parameters of the component functions of an entire curve in a product of copies of an abelian variety [McQuillan 1997] which we will discuss more in Section 0.4. Both were introduced to prove Theorem 0.1.1.

Probably the correct way of handling the situation is to use the product theorem of Faltings [1991], but so far there is no way to overcome the following difficulty of adapting Faltings's product theorem to the function theory case. For the application of Faltings's product formula, the ratio of the degrees of the constructed polynomial in consecutive sets of variables has to be greater than some appropriate constant. For diophantine approximation the sequence of approximating rational numbers are chosen to have heights and proximities corresponding to the degrees. An analogous situation for function theory is that, for the component functions of an entire curve in a product of copies of the target manifold, one chooses a rescaling of the parameters to make the characteristic functions and *at the same time* the proximity functions correspond to the degrees of the constructed polynomial in various sets of variables. However, unlike the case of diophantine approximation where a finite sum is used for the corresponding situation, in function theory the proximity function is defined by an integral, which gives rise to a more complicated technical difficulty, so far not overcome.

0.4. The Approach Motivated by Diophantine Approximation. Now

we discuss the second approach of using techniques motivated from those of diophantine approximation. The key feature of this second approach is that the k-jet bundle $J_k(M)$ of the target manifold M in the jet differential approach is replaced by a product $M^{\times(k+1)}$ of $k + 1$ copies of M. A jet differential in the first approach is replaced by a section of a certain positive line bundle L over

$M^{\times (k+1)}$ in the second approach. For example, in the case where M is an abelian variety A, one can use as L the pullback under

$$A^{\times (k+1)} \to A^{\times (k+1)},$$

$$(x_0, \ldots, x_k) \mapsto (x_0, x_1 - x_0, \ldots, x_k - x_{k-1})$$

of the tensor product of appropriate ample line bundles on the factors of $A^{\times (k+1)}$. For the defect of a hypersurface D in M or the hyperbolicity of $M - D$, this approach involves constructing holomorphic sections s of L over $M^{\times (k+1)}$ so that the sections vanish to high order along $D^{\times (k+1)}$ and yet the characteristic function, with respect to the positive curvature form of L, of the diagonal map $\tilde{\varphi} : \mathbb{C} \to M^{\times (k+1)}$ of the holomorphic map $\varphi : \mathbb{C} \to M$ has slow growth.

For the abelian variety A, the use of $x_j - x_k$ in the approach of diophantine approximation corresponds to the use of dx_j in the approach of jet differentials. It gives us more available degrees of freedom to get more vanishing order, without essentially increasing the growth order of the pullback of the constructed section by the diagonal map, because $x_j - x_k$ vanishes on the diagonal map.

As in the approach of jet differentials, there are in the approach of diophantine approximation the same two major difficulties. The first difficulty is to construct a holomorphic section of a line bundle on the product space with high vanishing order along certain subvarieties so that its pullback to the entire holomorphic curve has low pole order (*i.e.* small characteristic function). The second difficulty is to make sure that the pullback $\tilde{\varphi}^* s$ of the section s to the entire holomorphic curve is not identically zero.

One advantage of the approach of diophantine approximation is that it is easier to use the assumption of algebraic nondegeneracy of the map φ to handle the difficulty of the identical vanishing of $\tilde{\varphi}^* s$. When M is an abelian variety A, for this step McQuillan [1996; 1997] introduced the technique of considering the map $\mathbb{C}^{k+1} \to A^{\times (k+1)}$ induced by φ and rescaling separately the variable of each factor of $\mathbb{C}^{k+1}$. He chose the difference between the rescaling factors and 1 to be of the order of the reciprocal of some high power of the characteristic function at r when integration over the circle $\{|\zeta| = r\}$ is considered.

On the other hand, for the approach of diophantine approximation it can be very hard to construct a holomorphic section of a line bundle on the product space with high vanishing order along certain subvarieties whose pullback to the entire holomorphic curve has low pole order. How hard it is depends on which subvarieties the section is required to vanish along to high order. For example, in the case of the complex projective space it is not possible to require vanishing to high order along the product $D^{\times (k+1)}$ of one single hypersurface D, but it is easy to require vanishing to high order only along the diagonal of $D^{\times (k+1)}$. In order to use rescaling techniques to rule out identical vanishing of the pullback to the entire holomorphic curve, the vanishing along $D^{\times (k+1)}$, instead of merely its diagonal, is needed. That is the reason why for Theorem 0.0.1 only the case of many hypersurfaces gives nontrivial results. For the case of many hypersurfaces $D = \bigcup_\lambda V_\lambda$, the argument goes through also when vanishing to high order along $\bigcup_\lambda V_\lambda^{\times(k+1)}$ is used instead of $D^{\times(k+1)}$.

The abelian case is special in that there is an addition so that for a holomorphic map φ from $\mathbb{C}$ to an abelian variety, the rescaled map $\varphi_\lambda(\zeta) := \varphi(\lambda\zeta)$ gives the following inequality concerning the characteristic function of the difference of two rescaled maps:

$$T(\varphi_\lambda - \varphi_\mu, r) \leq \frac{|\lambda - \mu| r}{(R - |\lambda| r)(R - |\mu| r)} T(\varphi, R) + O(1)$$

when $\max(|\lambda|, |\mu|)r < R$, which enables one to control the characteristic function after separate rescaling. Note that, when one has a holomorphic map $\varphi : \mathbb{C} \to \mathbb{C}^n$, this inequality for the characteristic functions of the difference of two rescaled maps does not hold for the difference operation in $\mathbb{C}^n$. In the case of the abelian variety A we can use the difference operation in A to construct a holomorphic section of a line bundle on $A^{\times(k+1)}$ with high vanishing order along $D^{\times(k+1)}$ whose pullback to the entire holomorphic curve has low pole order. The above inequality makes sure that after the perturbation by rescaling, there is no essential increase in the pole order of the pullback.

One also has to control the effect of the separate rescaling on the counting function which was worked out in [McQuillan 1997]. That particular control works in the case of the projective variety as well as for the abelian variety and it is explained in Section 1.3.

For the first approach of jet differentials, Pit-Mann Wong with his collaborators Min Run and Julie Wang also started introducing the perturbation of $(d^k\varphi)(\mathbb{C})$ to handle the difficulty that $\varphi^*\omega$ is identically zero. The difficulties with such perturbation methods for the approach of jet differentials are the same as those occurring in the approach of diophantine approximation when one requires a constructed section to vanish to high order only along the diagonal of $D^{\times(k+1)}$. So far such difficulties are essential and cannot yet be overcome. We will explain more about them later in Section 0.8.

To see how the techniques mentioned above are applied to hyperbolicity problems and to understand the major obstacles for further progress, we discuss the hyperbolicity problems of the abelian variety which by now have been completely proved and understood. The starting point is the following theorem of Bloch.

THEOREM 0.4.1 [Bloch 1926; Green and Griffiths 1980; Ochiai 1977; Wong 1980; Kawamata 1980; Noguchi and Ochiai 1990]. *Let A be an abelian variety and $\varphi : \mathbb{C} \to A$ be a holomorphic map. Let X be the Zariski closure of the image of φ. Then X is the translate of an abelian subvariety of A.*

0.5. Proof of Bloch's Theorem. Denote by $\mathcal{X}$ the Zariski closure of $(d^k\varphi)(\mathbb{C})$ in $J_k(A)$. Here and in the rest of this discussion the Zariski closure in $J_k(A)$

means the intersection with $J_k(A)$ of the Zariski closure of $(d^k\varphi)(\mathbb{C})$ in the
compactification $A \times \mathbb{P}_{nk}$ of $J_k(A) = A \times \mathbb{C}^{nk}$. Consider the diagram

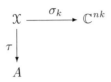

where σ_k is induced by the natural projection map $J_k(A) = A \times \mathbb{C}^{nk} \to \mathbb{C}^{nk}$
and τ comes from the composite of the map $J_k(X) \to X$ and $X \hookrightarrow A$.

The proof of Bloch's theorem depends on two observations of Bloch.

OBSERVATION 0.5.1 (BLOCH).   *For $k \geq n$ if the map $\sigma_k : \mathcal{X} \to \mathbb{C}^{nk}$ is not
generically finite onto its image, then X is invariant under the translation by
some nonzero element of A.*

PROOF. Take a point $\zeta_0 \in \mathbb{C}$ so that $\varphi(\zeta_0)$ is a regular point of X. Let N be the
complex codimension of X in A. Let $\omega_1, \ldots, \omega_N$ be local holomorphic 1-forms
on A whose common zero-set is the tangent bundle of X near $\varphi(\zeta_0)$. There is
a tangent vector ξ to $J_k(X)$ at the point $(d^k\varphi)(\zeta_0)$ which is mapped to zero by
σ_k. The tangent vector ξ is given by a one-parameter local perturbation $\Phi(\zeta, t)$
of the curve φ inside X defined near the point $(\zeta, t) = (\zeta_0, 0)$. The vanishing
of $\sigma_k(\xi)$ means that the tangent vector field $\frac{\partial \Phi}{\partial t}(\zeta, 0)$ has zero derivative up to
order k along $\varphi(\mathbb{C})$ at $\varphi(\zeta_0)$. Here the differentiation of a tangent vector field
of A is with respect to the flat connection for A. Then the fact that $\xi \in J_k(X)$
implies that the value of the derivatives of ω_j up to order k along $\varphi(\mathbb{C})$ vanishes
at the tangent vector $\frac{\partial \Phi}{\partial t}(\zeta_0, 0)$. Thus the $((k+1)N) \times n$ matrix formed by the
derivatives up to order k, of $\omega_j(\frac{\partial}{\partial z_\nu})$ $(1 \leq \nu \leq n, 1 \leq j \leq N)$ along $\varphi(\mathbb{C})$ at
ζ_0 has rank less than n. Since this holds when ζ_0 is replaced by an arbitrary ζ
near ζ_0, it follows from the standard Wronskian argument that there is a nonzero
constant tangent vector η on A such that $\omega_j(\eta)$ is identically zero along $\varphi(\zeta)$ near
$\zeta = \zeta_0$. The Zariski density of the image of φ in X implies that X is invariant
under the translation in the direction of the tangent vector η.   □

OBSERVATION 0.5.2 (BLOCH).   *If $\sigma_k : \mathcal{X} \to \mathbb{C}^{nk}$ is generically finite onto its
image, then for any ample divisor D of A there exists some polynomial of $d^l z_j$
$(1 \leq l \leq k, 1 \leq j \leq n)$ with constant coefficients which vanishes on $\tau^{-1}(D)$ but
does not vanish identically on $\mathcal{X}$.*

PROOF. The existence of P is verified as follows. For q sufficiently large, there
exists a meromorphic function F on A whose divisor is $E - qD$ so that $E \cap X$
and $D \cap X$ do not have any common branch. Since τ is surjective and σ_k is
generically finite onto its image, $F \circ \tau$ belongs to a finite extension of the field of
all rational functions of $\mathbb{C}^{nk}$. Thus there exist polynomials P_j $(0 \leq j \leq p)$ with

constant coefficients in the variables $d^l z_\nu$ $(1 \le l \le k, 1 \le \nu \le n)$ such that

$$\sum_{j=0}^{p} (\sigma_k^* P_j)(\tau^* F)^j = 0$$

on X and $\sigma_k^* P_p$ is not identically zero on X. Then P_p must vanish on $\tau^{-1}(D)$ and the holomorphic jet differential P_p on X must vanish on $\tau^{-1}(D)$. We need only set $P = P_p$. $\qquad \square$

Bloch's theorem now follows easily from the two observations in the following way. Assume that X is not a translate of an abelian subvariety of A. Let A' be the quotient of A by the subgroup of all elements whose translates leave X invariant. By replacing φ by its composite with the quotient map $A \to A'$, we can assume without loss generality that X is not invariant by the translation of any element of A. From Bloch's first observation σ_k is generically finite onto its image. From Bloch's second observation and the Schwarz's lemma $\varphi^* P$ is identically zero, which contradicts the non identical vanishing of P on X.

In Observation 0.5.1 the significance of the number n in the inequality $k \ge n$ is that there are n coefficients in each $\omega_1, \ldots, \omega_N$, which means that $k \ge$ the dimension of X. The zero-dimensionality of the generic fiber of σ_k corresponds to the following statement used in diophantine approximation [Vojta 1996, Lemma 5.1].

PROPOSITION 0.5.3. *Suppose A is an abelian variety and X is a subvariety of A which is not invariant under the translation of any nonzero element of A. Then for any $m > \dim X$ the map $X^{\times m} \to A^{\times (m(m-1)/2)}$ defined by $(x_j)_{1 \le j \le m} \mapsto (x_j - x_k)_{1 \le j < k \le m}$ is generically finite onto its image.*

0.6. Proof of Hyperbolicity of Complement of an Ample Divisor in an Abelian Variety.

Bloch's argument is modified in [Siu and Yeung 1996a] with the introduction of a log-pole jet differential to give the hyperbolicity of the complement of $A - D$ for any ample divisor D of the abelian variety A. Suppose there is a nonconstant holomorphic map $\varphi : \mathbb{C} \to A - D$ and we will derive a contradiction. By Bloch's theorem we can assume that the image of φ is Zariski dense in A. Let E be the largest subspace of $\mathbb{C}^n$ such that the lifting of φ to $\mathbb{C} \to \mathbb{C}^n$ is contained in a translate of E. A basis of E is given by $\partial/\partial z_{\nu_1}, \ldots,$ $\partial/\partial z_{\nu_q}$. Let $k = q + 1$. Let θ be a theta function defining the ample divisor D. The locally defined k-jet differential

$$\det \begin{pmatrix} d \log \theta & dz_{\nu_1} & dz_{\nu_2} & \cdots & dz_{\nu_q} \\ d^2 \log \theta & d^2 z_{\nu_1} & d^2 z_{\nu_2} & \cdots & d^2 z_{\nu_q} \\ \vdots & \vdots & \vdots & \ddots & \vdots \\ d^{q+1} \log \theta & d^{q+1} z_{\nu_1} & d^{q+1} z_{\nu_2} & \cdots & d^{q+1} z_{\nu_q} \end{pmatrix}$$

gives a well-defined function Θ on the Zariski closure X of $(d^k \varphi)(\mathbb{C})$ in $J_k(A)$. Now add the function Θ to the nk coordinates of the map $\sigma_k : X \to \mathbb{C}^{nk}$ to form $\tilde{\sigma}_k : X \to \mathbb{C}^{nk+1}$. We now use $\tilde{\sigma}_k$ instead of σ_k in Bloch's two observations. Bloch's second observation shows that the map σ_k cannot be generically finite onto its image. Bloch's first observation shows that there exists some nonzero constant direction $\sum_{\alpha=1}^{n} c_\alpha \frac{\partial}{\partial z_\alpha}$ such that $\varphi^*\left(\sum_{\alpha=1}^{n} c_\alpha \frac{\partial}{\partial z_\alpha}\right)\Theta$ is identically zero. The standard Wronskian argument then shows that $\varphi^*\left(\sum_{\alpha=1}^{n} c_\alpha \frac{\partial}{\partial z_\alpha}\right)^2 \log\theta$ is identically zero on $\mathbb{C}$. Because of the Zariski density of $\varphi(\mathbb{C})$ in A, this implies that $\left(\sum_{\alpha=1}^{n} c_\alpha \frac{\partial}{\partial z_\alpha}\right)^2 \log\theta$ is identically zero on A, which is a contradiction.

0.7. Proof of the Defect Relation for Ample Divisors of Abelian Varieties.

The defect relation in Theorem 0.1.1 for an ample divisor D in an abelian variety A was proved in [Siu and Yeung 1997] by using the following generalization of Bloch's theorem. If the image of a holomorphic map $\varphi : \mathbb{C} \to A$ is Zariski dense in an abelian variety A, then the Zariski closure $\overline{(d^k\varphi)(\mathbb{C})}$ of $(d^k\varphi)(\mathbb{C})$ in $\overline{J_k(A)} = A \times \mathbb{P}_{nk}$ is invariant under the translation by any element of A.

The translational invariance of $\overline{(d^k\varphi)(\mathbb{C})}$ by elements of A means that $\overline{(d^k\varphi)(\mathbb{C})}$ is of the form $A \times W$ for some complex subvariety $W \subset \mathbb{P}_{nk}$ of complex dimension d. When $k \geq n$, since the dimension of $J_k(D) \cap (A \times W)$ is at most $(n + d) - (k + 1) \leq d - 1$ which is less than the complex dimension of W, by the theorem of Riemann–Roch, for any $\varepsilon > 0$ we obtain the following. There exist positive integers p, q with $p/q < \varepsilon$ and there exist pD-valued holomorphic k-jet differentials on A vanishing to order at least q on $J_k(D)$ so that they give a non identically zero well-defined function on $\overline{(d^k\varphi)(\mathbb{C})}$. Then the following standard application of the First Main Theorem technique and the logarithmic derivative lemma yields the upper bound ε for the defect $\mathrm{Defect}(\varphi, D)$ of the map $\varphi : \mathbb{C} \to A$ and the ample divisor D.

Let $\mathcal{A}_r(\cdot)$ denote the operator which averages over the circle in $\mathbb{C}$ of radius r centered at the origin. Let $A = \mathbb{C}^n/\Lambda$ for some lattice Λ and let the divisor D be defined by the theta function θ on $\mathbb{C}^n$ which satisfies the transformation equation

$$\theta(z + u) = \theta(z) \exp\left(\pi H(z, u) + \frac{\pi}{2} H(u, u) + 2\pi\sqrt{-1}\, K(u)\right)$$

for some positive definite Hermitian form $H(z, w)$ and some real-valued function $K(u)$ for $u \in \Lambda$ so that $\exp\left(2\pi\sqrt{-1}K(u)\right)$ is a character on the lattice Λ. Let L_θ be the line bundle on A associated to the divisor D. We choose the global trivialization of the pullback of L_θ to $\mathbb{C}^n$ so that the theta function θ on $\mathbb{C}^n$ corresponds to a holomorphic section of L_θ whose divisor is D. We give L_θ a Hermitian metric so that with respect to our global trivialization of the pullback of L_θ to $\mathbb{C}^n$, it is given by $\exp(-\pi H(z, z))$. The connection from the Hermitian metric is given by $\mathcal{D}g = \partial g - \pi H(dz, z)g$ on $\mathbb{C}^n$. In particular,

$$\mathcal{D}\theta = d\theta + \sum_{\mu,\nu=1}^{n} h_{\mu,\bar{\nu}} \bar{z}_\nu dz_\mu \theta,$$

where $H(z,z) = \sum_{\mu,\nu=1}^{n} h_{\mu,\bar{\nu}} z_\mu \bar{z}_\nu$, and

$$\frac{D\theta}{\theta} = d\log\theta + \sum_{\mu,\nu=1}^{n} h_{\mu,\bar{\nu}} \bar{z}_\nu d\log\exp z_\mu.$$

Let

$$\vec{\nu} = (\nu_{\alpha,\beta})_{1\le\alpha\le k, 1\le\beta\le n}, \qquad \text{weight}(\vec{\nu}) = \sum_{\substack{1\le\alpha\le k \\ 1\le\beta\le n}} \alpha\nu_{\alpha,\beta},$$

and

$$d^{\vec{\nu}} z = \prod_{\substack{1\le\alpha\le k \\ 1\le\beta\le n}} \left(d^\alpha z_\beta\right)^{\nu_{\alpha,\beta}}.$$

An pD-valued holomorphic k-jet differential on A vanishing to order at least q on $J_k(D)$ means

$$P = \sum_{\text{weight}(\vec{\nu})=p} \tau_{\vec{\nu}} \left(d^{\vec{\nu}} z\right),$$

where $\tau_{\vec{\nu}}$ is an entire function on $\mathbb{C}^n$ so that $\frac{\tau_{\vec{\nu}}}{\theta^p}$ defines a meromorphic function on the abelian variety A. In other words, $\tau_{\vec{\nu}}$ defines a holomorphic section of $p L_\theta$ over A. Moreover, P vanishes to order at least q along

$$\{\theta = d\theta = \cdots = d^k\theta = 0\},$$

which means that we can write

$$P = \sum_{\nu_0+\nu_1+\cdots+\nu_k=q} a_{\nu_0,\nu_1,\dots,\nu_k} \theta^{\nu_0} \left(D\theta\right)^{\nu_1} \cdots \left(D^k\theta\right)^{\nu_k}$$

with smooth functions $a_{\nu_0,\nu_1,\dots,\nu_k}$ on $\mathbb{C}^n$ so that $\frac{a_{\nu_0,\nu_1,\dots,\nu_k}}{\theta^{p-q}}$ defines a function on A. In other words, $a_{\nu_0,\nu_1,\dots,\nu_k}$ is a smooth section of $(p-q)L_\theta$ over A. Then

$$\frac{P}{\theta^q} = \sum_{\nu_0+\nu_1+\cdots+\nu_k=q} a_{\nu_0,\nu_1,\dots,\nu_k} \left(\frac{D\theta}{\theta}\right)^{\nu_1} \cdots \left(\frac{D^k\theta}{\theta}\right)^{\nu_k}$$

Let $\tilde{\varphi}$ be the lifting of φ to $\mathbb{C} \to \mathbb{C}^n$. Now we compute the characteristic function of $\tilde{\varphi}^* P$ which is regarded as a meromorphic function on $\mathbb{C}$ (by identifying it with the coefficient of $(d\zeta)^m$ with $\zeta \in \mathbb{C}$. By the logarithmic derivative lemma

$$\mathcal{A}_r \left(\log^+ |\tilde{\varphi}^*(dz_\nu)|\right) = O\left(\log r + \log T(r,\varphi)\right).$$

Since

$$|\tau_{\vec{\nu}}|^2 \exp\left(-p\pi H(z,z)\right)$$

is smooth bounded function on $\mathbb{C}^n$, it follows that

$$\mathcal{A}_r \left(\log^+ |\tilde{\varphi}^* \tau_{\vec{\nu}}|\right) = \mathcal{A}_r \left(\frac{p\pi}{2} H(z,z)\right) \le p T(r,\varphi)$$

and

$$(0.7.1) \qquad T(r,\tilde{\varphi}^* P) = p\, T(r,\varphi) + O\left(\log r + \log T(r,\varphi)\right).$$

We also need the estimate for $A_r\left(\log^+\left|\tilde{\varphi}^*\left(\frac{P}{\theta^q}\right)\right|\right)$. From

$$(0.7.2) \qquad \frac{P}{\theta^q} = \sum_{\nu_0+\nu_1+\cdots+\nu_k=q} a_{\nu_0,\nu_1,\ldots,\nu_k}\left(\frac{\mathcal{D}\theta}{\theta}\right)^{\nu_1}\cdots\left(\frac{\mathcal{D}^k\theta}{\theta}\right)^{\nu_k}$$

and

$$(0.7.3) \qquad \frac{\mathcal{D}\theta}{\theta} = d\log\theta + \sum_{\mu,\nu=1}^n h_{\mu,\bar{\nu}}\bar{z}_\nu d\log\exp z_\mu$$

it follows that

$$A_r\left(\log^+\left|\tilde{\varphi}^*\left(\frac{\mathcal{D}\theta}{\theta}\right)\right|\right) = O\left(\log r + \log T(r,\varphi)\right).$$

Since

$$|a_{\nu_0,\nu_1,\ldots,\nu_k}|^2\exp\left(-p\pi H(z,z)\right)$$

is smooth bounded function on $\mathbb{C}^n$, it follows that

$$A_r\left(\log^+|\tilde{\varphi}^*a_{\nu_0,\nu_1,\ldots,\nu_k}|\right) = A_r\left(\frac{p\pi}{2}H(z,z)\right) \le pT(r,\varphi).$$

Thus

$$(0.7.4) \qquad A_r\left(\log^+\left|\tilde{\varphi}^*\left(\frac{P}{\theta^q}\right)\right|\right) \le pT(r,\varphi) + O\left(\log r + \log T(r,\varphi)\right).$$

The vanishing of the defect $\mathrm{Defect}(\varphi,D)$ now follows from $\frac{p}{q} < \varepsilon$ and from

$$qm(r,\theta,0) = A_r\left(\log^+\left|\tilde{\varphi}^*\left(\frac{1}{\theta^q}\right)\right|\right) \le A_r\left(\log^+\left|\tilde{\varphi}^*\left(\frac{P}{\theta^q}\right)\right|\right) + T\left(r,\tilde{\varphi}^*\left(\frac{1}{P}\right)\right)$$

$$\le A_r\left(\log^+\left|\tilde{\varphi}^*\left(\frac{P}{\theta^q}\right)\right|\right) + T\left(r,\tilde{\varphi}^*P\right)$$

which by Equations (0.7.1) and (0.7.4) is no more than

$$2pT(r,\varphi) + O\left(\log r + \log T(r,\varphi)\right).$$

S.-K. Yeung observed that the proof in [Siu and Yeung 1997] could be slightly refined as follows to give the following stronger Second Main Theorem for an ample divisor D in an abelian variety A and for any positive number ε.

$$m(r,\varphi,D) + (N(r,\varphi,D) - N_n(r,\varphi,D)) \le \varepsilon T(r,\varphi) + O\left(\log r + \log T(r,\varphi)\right),$$

where $N_n(r,\varphi,D)$ is defined in the same as the counting function $N(r,\varphi,D)$ except that the counting is truncated at multiplicity n so that multiplicity greater than n is counted only as n. The refinement is as follows. From Equations (0.7.2) and (0.7.3) it follows that

$$N\left(r,\tilde{\varphi}^*\left(\frac{P}{\theta^q}\right),\infty\right) \le qN_n(r,D,0).$$

and

$$T\left(r, \tilde{\varphi}^*\left(\frac{P}{\theta^q}\right)\right) = A_r\left(\log^+\left|\varphi^*\left(\frac{P}{\theta^q}\right)\right|\right) + N\left(r, \varphi^*\left(\frac{P}{\theta^q}\right), \infty\right).$$

Moreover, it follows from (0.7.4) that

$$T\left(r, \tilde{\varphi}^*\left(\frac{P}{\theta^q}\right)\right) \leq pT(r, \varphi) + qN(r, D, 0) + O\left(\log r + \log T(r, \varphi)\right).$$

and

$$
\begin{aligned}
q\, m\,(r, \varphi, D) + qN\,(r, \varphi, D) &= T\left(r, \tilde{\varphi}^*\left(\frac{1}{\theta^q}\right)\right) + O(1) \\
&= T\left(r, \tilde{\varphi}^*\left(\frac{P}{\theta^q}\frac{1}{P}\right)\right) + O(1) \\
&\leq T\left(r, \tilde{\varphi}^*\left(\frac{P}{\theta^q}\right)\right) + T\left(r, \tilde{\varphi}^*\left(\frac{1}{P}\right)\right) + O(1) \\
&\leq T\left(r, \tilde{\varphi}^*\left(\frac{P}{\theta^q}\right)\right) + T\left(r, \tilde{\varphi}^* P\right) + O(1) \\
&\leq 2pT(r, \varphi) + qN_n\,(r, \varphi, D) + O\left(\log r + \log T(r, \varphi)\right).
\end{aligned}
$$

Dividing both sides by q and using p/q yields the stronger Second Main Theorem.

0.8. Perturbation of Holomorphic Maps. By the second approach of using techniques motivated by diophantine approximation, McQuillan [1996] gives an alternative proof of Bloch's theorem and obtains [1997] the zero defect of an ample divisor D of A. He uses different rescalings of variables of $\mathbb{C}$ to handle the problem of the identical vanishing of the pullback of a section constructed for an appropriate line bundle. It comes as a great surprise that his method of perturbation by rescaling of variables works, but in fact it does. Since in Chapter 1 of this paper we will apply the rescaling method to the complex projective space to get a proof of Theorem 0.0.1, we will not elaborate further on that method here.

We make a remark about the difficulty of using perturbation for the approach by jet differentials. For hyperbolicity problems Pit-Mann Wong with his collaborators introduces the method of perturbing the map $d^k\varphi : \mathbb{C} \to J_k(M)$ into another map $\Phi_k : \mathbb{C} \to J_k(M)$ so that the composite of Φ_k and the natural projection $J_k(M) \to M$ is φ. The main difficulty with such a perturbation is that, unlike the case of using the product of a number of copies of the target manifold, there is yet no known good way of perturbation which could control the change of the proximity term, even when the perturbation is done by rescaling. The problem can be illustrated by the simple case of $k = 1$ and M being an abelian variety A whose universal cover has coordinates $z_1, \ldots, z_n$. Suppose

$$\varphi(\zeta) = (\varphi_1(\zeta), \ldots, \varphi_n(\zeta))$$

in terms of $z_1, \ldots, z_n$ and we perturb $d\varphi$ to

$$(d\varphi)(\zeta) = \left(\varphi(\zeta), \left(\frac{\partial \varphi_1}{\partial \zeta} \right)(\xi_1 \zeta), \ldots, \left(\frac{\partial \varphi_n}{\partial \zeta} \right)(\xi_n \zeta) \right) \in A \times \mathbb{C}^n$$

with some rescaling factors $\xi_1, \ldots, \xi_n$. When we estimate the effect of the perturbation on the proximity function for some theta function s_D defining an ample divisor D, even with the possible use of another rescaling factor ξ' there is no way to handle the difficulty coming from the discrepancy between

$$\left(\frac{\partial s_D}{\partial \zeta} \right)(\xi' \zeta) \quad \text{and} \quad \sum_{\nu=1}^{n} \left(\frac{\partial s_D}{\partial z_\nu} \right)(\varphi(\zeta)) \left(\frac{\partial \varphi_\nu}{\partial \zeta} \right)(\xi_\nu \zeta).$$

0.9. Since the main ideas of the streamlined version of the proof of Theorem 0.1.4 will be discussed in the overview in Chapter 2, here in the Introduction we will confine ourselves to only a couple of comments on the relation between number theory and the easier first step of finding meromorphic 1-jet differentials whose pullback on the entire holomorphic curve vanishes.

The construction of 2-jet differentials of certain explicit forms given in Chapter 2 is accomplished by using polynomials whose terms contain the factors f, df, d^2f to a certain order, where f is the polynomial defining the plane curve C of degree δ (see 2.1.2). This means that the constructed jet differential vanishes to that order along $J_2(C)$. This requirement is related to the techniques discussed above.

On the branched cover X over $\mathbb{P}_2$ with branching along C, the construction of holomorphic 2-jet differentials is possible because there are more divisors on $J_k(X)$ and some factors from the additional ways of factorization become holomorphic jet differentials; see Section 2.3. This is analogous to the following observation due to Vojta in number theory. The finiteness of rational points for a subvariety of abelian varieties not containing the translate of an abelian variety is the consequence of the fact that in the product space of many copies of the subvariety there are more line bundles or divisors than constructed from the factors which are copies of the subvariety [Faltings 1991; Vojta 1992].

On the other hand, the existence of more divisors in $J_k(X)$ and more ways of factorization mean that it is easier for two jet differentials to share a common factor and as a result it is more difficult to conclude that the zero-sets of two jet differentials do not have a branch in common.

0.10. Overview of the Proofs. We conclude this introduction with a brief discussion of the proofs of the main results. The proof of Theorem 0.0.1 is parallel to that of Roth's Theorem [Roth 1955; Schmidt 1980]. It provides more tangible evidence to support the formal analogy between Nevanlinna theory and diophantine approximation pointed out by Osgood [1985] and Vojta [1987]. It also introduces a new approach to the hyperbolicity problem of the complement of a generic hypersurface of high degree in a complex projective space, which might hold a better promise than other approaches for an eventual solution to the full conjecture with optimal bounds involving such complements of hypersurfaces. There is no attempt to get the optimal bound from the proof of Theorem 0.0.1. Some small improvements in the bounds may be possible from that argument.

Theorem 0.0.1 is not a new result. The case of $\hat{m} = 1$ of Theorem 0.0.1 is contained in the defect relation of Cartan [1933] and Ahlfors [1941] and the following result of Eremenko and Sodin. The case of general $\hat{m}$ of Theorem 0.0.1 follows from the standard process of averaging over the complex lines in the complex vector space $\mathbb{C}^{\hat{m}}$.

THEOREM 0.10.1 [Erëmenko and Sodin 1991, p. 111, Theorem 1]. *If Q_ν ($1 \leq \nu \leq q$) are homogeneous polynomials of degree d_ν in $n+1$ variables so that no more than n of them have a common zero in $\mathbb{C}^{n+1} - 0$ and if $\varphi : \mathbb{C} \to \mathbb{P}_n$ so that $\varphi^* Q_\nu$ is not identically zero for $1 \leq \nu \leq q$, then*

$$(q - 2n)T(r, \varphi) \leq \sum_{\nu=1}^{q} \frac{1}{d_\nu} N(r, Q_k, 0) + o(T(r, \varphi))$$

where $T(r, \varphi)$ is the characteristic function, $N(r, Q_k, 0)$ is the counting function, and the inequality holds outside a subset of the real line with finite measure with respect to dr/r.

Chapter 2 is devoted to the proof of Theorem 0.0.2, which contains two main steps. The first is to produce a meromorphic 1-jet differential h whose pullback to the entire holomorphic curve is zero; see Sections 2.2 to 2.5. When the degree of h in the affine variables is at least 4 times its degree in the differentials of those variables, the proof is rather easily finished by using arguments of Riemann–Roch to construct some holomorphic 1-jet differential defined only on a branched cover of the zero-set of h which vanishes on an ample divisor of $\mathbb{P}_2$; see Section 2.6. The second step is to deal with the most difficult remaining case. When the curve C is defined by a polynomial f of the affine coordinates, the main idea is to use an appropriate meromorphic 1 form η of low degree and consider the restriction of $\frac{\eta}{f}$ to the zero-set of h. When there is a good upper bound for the touching order of the "integral curves" of h and C, the argument for the Ahlfors–Schwarz lemma for log-pole jet differentials finishes the proof; see Section 2.8. The main streamlining is some new ingredients in the touching order argument in the difficult last step; see Section 2.7. A less important streamlining is that we employ more the cleaner language of cohomology theory, instead of the direct arguments of using polynomials, in the first step of constructing the meromorphic 1-jet differential h whose pullback to the entire holomorphic curve is zero. The method of proof is chosen and presented in a way which facilitates possible generalizations to the higher dimensional case.

1. Sum of Defects of Hypersurfaces in the Projective Space

We prove in this chapter the following theorem, which is the case of $\hat{m} = 1$ in Theorem 0.0.1. All the principal difficulties of the proof of Theorem 0.0.1 already occur in the special case of $\hat{m} = 1$. So for notational simplicity we give only the details for the case of $\hat{m} = 1$ and then present the minor modifications needed for the case of a general $\hat{m}$ after the proof of Theorem 1.0.1.

THEOREM 1.0.1. *Let* V_λ $(1 \leq \lambda \leq \Lambda)$ *be regular complex hypersurface in* $\mathbb{P}_n$ *of degree* δ *in normal crossing. Let* $\varphi : \mathbb{C} \to \mathbb{P}_n$ *be a holomorphic map whose image is not contained in any hypersurface of* $\mathbb{P}_n$. *Then the sum of the defects* $\sum_{\lambda=1}^{\Lambda} \text{Defect}(\varphi, V_\lambda)$ *is no more than* ne *for any* $\delta \geq 1$ *and is no more than* $n+1$ *for* $\delta = 1$.

The method of proof uses techniques motivated by diophantine approximation. We construct holomorphic section s of low degree on the product $\mathbb{P}_n^{\times m}$ of m copies of $\mathbb{P}_n$ which vanishes to high order at points of $\bigcup_{\lambda \in \Lambda} V_\lambda^{\times m}$. Then we use McQuillan's estimate [1997] for the proximity function with a rescaling of the variable of $\mathbb{C}$. The m different rescalings on $\mathbb{C}$ for the map from $\mathbb{C}$ to $\mathbb{P}_n^{\times m}$ induced by φ guarantee the non identical vanishing of the pullback to $\mathbb{C}$ of s by the perturbed map. The defect relation then follows from the standard argument of the Poisson–Jensen formula or the First Main Theorem. The normal crossing condition is required to make sure that the product of the multi-order ideal sheaves for $V_\lambda^{\times m}$ is equal to their intersection.

1.1. Preliminaries on Combinatorics and Integrals

LEMMA 1.1.1. *Let* n *be a positive integer. For any positive number* $\tau > 1$ *let* $\Theta_n(\tau)$ *be*

$$\overline{\lim}_{m \to \infty} \left(\int_{\left\{ \substack{x_1 + \cdots + x_m < \frac{m}{\tau(n+1)} \\ 0 < x_1 < 1, \ldots, 0 < x_m < 1} \right\}} (1-x_1)^{n-1} \cdots (1-x_m)^{n-1} \, dx_1 \cdots dx_m \right)^{1/m}.$$

Then

$$\Theta_n(\tau) \leq \min \left(\frac{e}{\tau(n+1)}, \frac{1}{n} e^{-\frac{1}{4(n+1)^2} \left(1 - \frac{1}{\tau}\right)^2} \right).$$

PROOF. First we show that

$$\Theta_n(\tau) \leq \frac{1}{n} e^{-\frac{1}{4(n+1)^2} \left(1 - \frac{1}{\tau}\right)^2}.$$

We need the following combinatorial lemma, which follows from [Schmidt 1980, p. 122, Lemma 4C] and the fact that the number of n-tuples of nonnegative integers $i_1, \ldots, i_n$ with $i_1 + \cdots + i_n = r$ is equal to $\binom{r+n-1}{r}$.

LEMMA 1.1.2. *Let $d_1, \ldots, d_m$ be positive integers, $0 < \varepsilon < 1$, and n be a positive integer. Then*

$$\sum_{\left|\left(\frac{j_1}{d_1} + \cdots + \frac{j_m}{d_m}\right) - \frac{m}{n+1}\right| \geq \varepsilon m} \binom{d_1 - j_1 + n - 1}{n-1} \cdots \binom{d_m - j_m + n - 1}{n-1}$$

$$\leq \binom{d_1 + n}{n} \cdots \binom{d_m + n}{n} \cdot 2e^{-\frac{\varepsilon^2 m}{4}}.$$

Setting $\varepsilon = \frac{1}{n+1}\left(1 - \frac{1}{\tau}\right)$ and $d_1 = \cdots = d_m = d$, we get

$$\sum_{j_1 + \cdots + j_m < \frac{md}{\tau(n+1)}} \binom{d - j_1 + n - 1}{n-1} \cdots \binom{d - j_m + n - 1}{n-1} \leq \binom{d+n}{n}^m 2e^{-\frac{m}{4(n+1)^2}\left(1 - \frac{1}{\tau}\right)^2}.$$

Forming the Riemann sum by choosing $1/d$ as the size of an increment for each variable and choosing the points $x_\nu = j_\nu/d$ for $1 \leq j_\nu \leq d$ from each rectangular parallelpiped of size $1/d$ and passing to limit as $d \to \infty$, we get

$$\lim_{d \to \infty} \frac{1}{d^{nm}} \sum_{j_1 + \cdots + j_m < \frac{md}{\tau(n+1)}} \binom{d - j_1 + n - 1}{n-1} \cdots \binom{d - j_m + n - 1}{n-1}$$

$$= \frac{1}{((n-1)!)^m} \int_{\left\{ \substack{x_1 + \cdots + x_m < \frac{m}{\tau(n+1)} \\ 0 < x_1 < 1, \ldots, 0 < x_m < 1} \right\}} (1 - x_1)^{n-1} \cdots (1 - x_m)^{n-1} \, dx_1 \cdots dx_m.$$

On the other hand,

$$\lim_{d \to \infty} \frac{1}{d^{nm}} \binom{d+n}{n}^m 2e^{-\frac{m}{4(n+1)^2}\left(1 - \frac{1}{\tau}\right)^2} = \frac{1}{(n!)^m} 2e^{-\frac{m}{4(n+1)^2}\left(1 - \frac{1}{\tau}\right)^2}.$$

After taking the m-th root in the above two limits and using 1.1.2 and letting $m \to \infty$, we get

$$\Theta_n(\tau) \leq \frac{1}{n} e^{-\frac{1}{4(n+1)^2}\left(1 - \frac{1}{\tau}\right)^2}.$$

For the other inequality, $\Theta_n(\tau) \leq \frac{e}{\tau(n+1)}$, we make the substitution $x_\nu = \frac{y_\nu}{\tau}$ and get

$$\int_{\left\{ \substack{\frac{y_1 + \cdots + y_m}{\tau(n+1)} < m \\ 0 < x_1 < 1, \ldots, 0 < x_m < 1} \right\}} (1 - x_1)^{n-1} \cdots (1 - x_m)^{n-1} \, dx_1 \cdots dx_m$$

$$= \frac{1}{\tau^m} \int_{\left\{ \substack{y_1 + \cdots + y_m < \frac{m}{n+1} \\ 0 < y_1 < \tau, \ldots, 0 < y_m < \tau} \right\}} \left(1 - \frac{y_1}{\tau}\right)^{n-1} \cdots \left(1 - \frac{y_m}{\tau}\right)^{n-1} \, dy_1 \cdots dy_m$$

$$\leq \frac{1}{\tau^m} \text{ Volume of } \left\{ y_1 + \cdots + y_m < \frac{m}{n+1} : y_1 > 0, \ldots, y_m > 0 \right\}$$

$$\leq \frac{m^m}{m!} \left(\frac{1}{\tau(n+1)}\right)^m.$$

Taking the m-th root and letting $m \to \infty$ and using

$$\lim_{m \to \infty} \frac{m!\, e^m}{m^m \sqrt{2\pi m}} = 1,$$

from Stirling's formula, we get $\Theta_n(\tau) \le \frac{e}{\tau(n+1)}$. □

LEMMA 1.1.3. *Let δ, Λ be positive integers and τ be a number > 1 such that $\delta n \Theta_n(\tau) < 1$. Then there exists m_0 such that for $m \ge m_0$ there exists d_0 depending on m with the property that for $d \ge d_0$ one has*

$$\Lambda \delta^m \sum_{j_1 + \cdots + j_m < \frac{md}{\tau(n+1)}} \binom{d - j_1 + n - 1}{n-1} \cdots \binom{d - j_m + n - 1}{n-1} < \binom{d+n}{n}^m.$$

PROOF. Let $0 < \eta < 1$ such that $\delta n \Theta_n(\tau) < 1 - \eta$. There exists m_0 such that $\Lambda (1-\eta)^m < 1$ for $m \ge m_0$ and such that for $m \ge m_0$ we have

$$(\delta n)^m \int_{\left\{ \substack{x_1 + \cdots + x_m < \frac{m}{\tau(n+1)} \\ 0 < x_1 < 1, \ldots, 0 < x_m < 1} \right\}} (1 - x_1)^{n-1} \cdots (1 - x_m)^{n-1}\, dx_1 \cdots dx_m < (1-\eta)^m.$$

Choose any $m \ge m_0$. Forming the Riemann sum by choosing $1/d$ as the size of an increment for each variable and choosing the points $x_\nu = j_\nu/d$ for $1 \le j_\nu \le d$ from each rectangular parallelpiped of size $1/d$ and passing to limit as $d \to \infty$, we get

$$\lim_{d \to \infty} \frac{1}{d^{nm}} \sum_{j_1 + \cdots + j_m < \frac{md}{\tau(n+1)}} \binom{d - j_1 + n - 1}{n-1} \cdots \binom{d - j_m + n - 1}{n-1}$$

$$= \frac{1}{((n-1)!)^m} \int_{\left\{ \substack{x_1 + \cdots + x_m < \frac{m}{\tau(n+1)} \\ 0 < x_1 < 1, \ldots, 0 < x_m < 1} \right\}} (1 - x_1)^{n-1} \cdots (1 - x_m)^{n-1}\, dx_1 \cdots dx_m.$$

Since

$$\lim_{d \to \infty} \frac{1}{d^{nm}} \binom{d+n}{n}^m = \frac{1}{(n!)^m},$$

it follows that there exists d_0 depends on m such that for $d \ge d_0$ one has

$$\delta^m \sum_{j_1 + \cdots + j_m < \frac{md}{\tau(n+1)}} \binom{d - j_1 + n - 1}{n-1} \cdots \binom{d - j_m + n - 1}{n-1} < \binom{d+n}{n}^m (1-\eta)^m$$

and

$$\Lambda \delta^m \sum_{j_1 + \cdots + j_m < \frac{md}{\tau(n+1)}} \binom{d - j_1 + n - 1}{n-1} \cdots \binom{d - j_m + n - 1}{n-1} < \binom{d+n}{n}^m. \qquad \square$$

1.2. Construction of Sections of Low Degree and High Vanishing Order

PROPOSITION 1.2.1. *Let V_λ $(1 \leq \lambda \leq \Lambda)$ be nonsingular hypersurfaces of degree δ in $\mathbb{P}_n$ in normal crossing. Let $\tau > 1$ satisfy $\delta n \Theta_n(\tau) < 1$. There exists m_0 and for $m \geq m_0$ there exists d_0 depending on m such that for $d \geq d_0$ there exists an element*

$$F \in H^0\big(\mathbb{P}_n^{\times m}, \mathcal{O}_{\mathbb{P}_n^{\times m}}(d, \ldots, d)\big)$$

which vanishes at each $V_\lambda^{\times m}$ to every multi-order $(j_1, \ldots, j_m)$ which satisfies

$$j_1 + \cdots + j_m < \frac{dm}{\tau(n+1)}.$$

PROOF. The space of all homogeneous polynomials of degree r on V_λ is equal to the space of all polynomials of degree r on $\mathbb{P}_n$ quotiented by the ideal generated by the defining polynomial for V_λ. Thus

$$\dim_{\mathbb{C}} H^0(V_\lambda, \mathcal{O}_{V_\lambda}(r)) = \binom{r+n}{n} - \binom{r-\delta+n}{n}.$$

It follows from the following identity for binomial coefficients

$$\binom{b+1}{c+1} - \binom{b}{c+1} = \binom{b}{c}$$

that

$$\dim_{\mathbb{C}} H^0(V_\lambda, \mathcal{O}_{V_\lambda}(r)) = \sum_{\nu=1}^{\delta} \binom{r-\delta+\nu+n}{n} - \binom{r-\delta+\nu+n-1}{n}$$

$$= \sum_{\nu=1}^{\delta} \binom{r-\delta+\nu+n-1}{n-1} \leq \delta \binom{r+n-1}{n-1},$$

where we use the definition

$$\binom{a}{b} = \frac{\prod_{\nu=1}^{b}(a-b+\nu)}{b!}$$

so that $\binom{a}{b} = 0$ for $a < b$ and we use the inequality

$$\binom{a}{b} < \binom{c}{b}$$

for integers $b \leq a < c$. By Künneth's formula we have

$$\dim_{\mathbb{C}} H^0(V_\lambda^{\times m}, \mathcal{O}_{V_\lambda^{\times m}}(d_1, \ldots, d_m)) \leq \delta^m \prod_{\nu=1}^{m} \binom{d_\nu + n - 1}{n - 1}.$$

Let $z_{\nu,0}, \ldots, z_{\nu,n}$ be the homogeneous coordinates for the ν-th factor of $\mathbb{P}_n^{\times m}$. An element

$$F \in H^0\big(\mathbb{P}_n^{\times m}, \mathcal{O}_{\mathbb{P}_n^{\times m}}(d, \ldots, d)\big)$$

is represented by a polynomial in the $m(n+1)$ variables

$$z_{1,0}, \ldots, z_{1,n}, \ldots, z_{m,0}, \ldots, z_{m,n},$$

which is homogeneous of degree d_ν in the variables $z_{\nu,0}, \ldots, z_{\nu,n}$ for $1 \le \nu \le m$. Assume that the complex line $z_{\nu,1} = \cdots = z_{\nu,n} = 0$ is not contained in any V_λ. Then the vanishing of F on V_λ to every multi-order $(j_1, \ldots, j_m)$ with

$$j_1 + \cdots + j_m < \frac{md}{\tau(n+1)}$$

means that

$$\frac{\partial^{j_1 + \cdots + j_m}}{\partial z_{1,0}^{j_1} \cdots \partial z_{m,0}^{j_m}} F$$

as an element of

$$H^0\big(\mathbb{P}_n^{\times m}, \mathcal{O}_{\mathbb{P}_n^{\times m}}(d - j_1, \ldots, d - j_m)\big)$$

vanishes identically on V_λ for every multi-order $(j_1, \ldots, j_m)$ satisfying

$$j_1 + \cdots + j_m \le \frac{md}{\tau(n+1)}.$$

There exists

$$F \in H^0\big(\mathbb{P}_n^{\times m}, \mathcal{O}_{\mathbb{P}_n^{\times m}}(d, \ldots, d)\big)$$

which vanishes at each $V_\lambda^{\times m}$ to every multi-order $(j_1, \ldots, j_m)$ which satisfies

$$j_1 + \cdots + j_m < \frac{md}{\tau(n+1)}$$

if

$$\Lambda \delta^m \sum_{j_1 + \cdots + j_m < \frac{md}{\tau(n+1)}} \binom{d - j_1 + n - 1}{n-1} \cdots \binom{d - j_m + n - 1}{n-1} < \binom{d+n}{n}^m,$$

which is the case by Lemma 1.1.3 and the assumption $\delta n \Theta_n(\tau) < 1$. $\qquad \square$

1.3. Effect of Rescaling on Proximity Term.

For the estimate of the effect of rescaling on the proximity term we follow the method of [McQuillan 1997]. Let

$$G_{R,a}(\zeta) = \frac{R^2 - \bar{a}\zeta}{R(\zeta - a)},$$

so that

$$\frac{1}{G_{R,a}(\rho\zeta)} = \frac{R(\rho\zeta - a)}{R^2 - \bar{a}\rho\zeta}.$$

We have

$$\frac{1}{G_{R,a}(\rho_1\zeta)} - \frac{1}{G_{R,a}(\rho_2\zeta)} = \frac{R(\rho_1\zeta - a)}{R^2 - \bar{a}\rho_1\zeta} - \frac{R(\rho_2\zeta - a)}{R^2 - \bar{a}\rho_2\zeta}$$

$$= R\left\{\frac{(\rho_1 - \rho_2)R^2\zeta + (\rho_2 - \rho_1)a\bar{a}\zeta}{(R^2 - \bar{a}\rho_1\zeta)(R^2 - \bar{a}\rho_2\zeta)}\right\}$$

$$= R(\rho_1 - \rho_2)\zeta\frac{R^2 - \bar{a}a}{(R^2 - \bar{a}\rho_1\zeta)(R^2 - \bar{a}\rho_2\zeta)}.$$

Now we impose the conditions

$$|\rho_1| < R, \quad |\rho_2| < R, \quad |a| \le R.$$

Let

$$\gamma_{R,\rho_1,\rho_2} = \frac{R|\rho_1 - \rho_2|}{(R - |\rho_1|)(R - |\rho_2|)}.$$

For $|\zeta| = 1$ we have

$$\left|\frac{1}{G_{R,a}(\rho_1\zeta)} - \frac{1}{G_{R,a}(\rho_2\zeta)}\right| \le R|\rho_1 - \rho_2| \cdot \frac{R^2}{(R^2 - |\rho_1 a|)(R^2 - |\rho_2 a|)}$$

$$\le \frac{R|\rho_1 - \rho_2|}{(R - |\rho_1|)(R - |\rho_2|)} = \gamma_{R,\rho_1,\rho_2}$$

and

$$\left|\frac{G_{R,a}(\rho_2\zeta)}{G_{R,a}(\rho_1\zeta)} - 1\right| \le \gamma_{R,\rho_1,\rho_2}|G_{R,a}(\rho_2\zeta)|, \quad \left|\frac{G_{R,a}(\rho_2\zeta)}{G_{R,a}(\rho_1\zeta)}\right| \le 1 + \gamma_{R,\rho_1,\rho_2}|G_{R,a}(\rho_2\zeta)|.$$

Poisson's integral formula states that for $h(\zeta)$ meromorphic on $\{|\zeta| \le R\}$ we have

$$\log|h(\zeta)| = \int_{\theta=0}^{2\pi}\log|h(R\,e^{i\theta})|\,\mathrm{Re}\left(\frac{R\,e^{i\theta} + \zeta}{R\,e^{i\theta} - \zeta}\right)\frac{d\theta}{2\pi} - \log\prod_{|a|\le R}\left|\frac{R^2 - \bar{a}\zeta}{R(\zeta - a)}\right|^{\mathrm{ord}_a h}.$$

In particular, when $\zeta = 0$ we have

$$\log|h(0)| = \int_{\theta=0}^{2\pi}\log|h(R\,e^{i\theta})|\frac{d\theta}{2\pi} - \log\prod_{|a|\le R}\left|\frac{R}{a}\right|^{\mathrm{ord}_a h}.$$

Apply the last equation to the special case

$$h(\zeta) = \prod_{|a|\le R}\left(\frac{R^2 - \bar{a}\zeta}{R(\zeta - a)}\right)^{\mathrm{ord}_a h}$$

with R replaced by $r < R$ in the formula. Then

$$\log \prod_{|a| \leq R} \left| \frac{R}{a} \right|^{\operatorname{ord}_a h} = \frac{1}{2\pi} \int_{|\zeta|=r} \log \prod_{|a| \leq R} \left| \frac{R^2 - \bar{a}\zeta}{R(\zeta - a)} \right|^{\operatorname{ord}_a h} - \log \prod_{|a| \leq r} \left| \frac{r}{a} \right|^{\operatorname{ord}_a h}.$$

If Z is a divisor on $\mathbb{C}$ and $Z \cap \{|\zeta| < t\} = \{a_1, \ldots, a_N\}$ with multiplicity, then

$$N(R, Z) = \int_{t=0}^{R} n(t, Z) \frac{dt}{t} \sum_{\nu=1}^{N} \log \left| \frac{R}{a_\nu} \right|.$$

Hence

$$\frac{1}{2\pi} \int_{|\zeta|=r} \log \prod_{|a| \leq R} \left| \frac{R^2 - \bar{a}\zeta}{R(\zeta - a)} \right|^{\operatorname{ord}_a h}$$
$$= (N(R, \{h = 0\}) - N(r, \{h = 0\})) - (N(R, \{h = \infty\}) - N(r, \{h = \infty\})).$$

Now for $|\rho_1| < R, |\rho_2| < R$ we have

$$\log \left| \frac{h(\rho_1 \zeta)}{h(\rho_2 \zeta)} \right| = \int_{\theta=0}^{2\pi} \log |h(R e^{i\theta})| \operatorname{Re} \left(\frac{R e^{i\theta} + \rho_1 \zeta}{R e^{i\theta} - \rho_1 \zeta} - \frac{R e^{i\theta} + \rho_2 \zeta}{R e^{i\theta} - \rho_2 \zeta} \right) \frac{d\theta}{2\pi}$$
$$- \log \prod_{|a| \leq R} \left| \frac{G_{R,a,\rho_1 \zeta}}{G_{R,a}(\rho_2 \zeta)} \right|^{\operatorname{ord}_a h}.$$

To estimate the right-hand side, we observe that

$$\frac{R e^{i\theta} + \rho_1 \zeta}{R e^{i\theta} - \rho_1 \zeta} - \frac{R e^{i\theta} + \rho_2 \zeta}{R e^{i\theta} - \rho_2 \zeta} = \frac{2(\rho_1 - \rho_2)\zeta R e^{i\theta}}{(R e^{i\theta} - \rho_1 \zeta)(R e^{i\theta} - \rho_2 \zeta)}.$$

Hence

$$\left| \operatorname{Re} \left(\frac{R e^{i\theta} + \rho_1 \zeta}{R e^{i\theta} - \rho_1 \zeta} - \frac{R e^{i\theta} + \rho_2 \zeta}{R e^{i\theta} - \rho_2 \zeta} \right) \right| \leq \frac{2|\rho_1 - \rho_2|R}{(R - |\rho_1|)(R - |\rho_2|)}.$$

So

$$\log^+ \left| \frac{h(\rho_1 \zeta)}{h(\rho_2 \zeta)} \right| \leq \int_{\theta=0}^{2\pi} \log^+ |h(R e^{i\theta})| \frac{2|\rho_1 - \rho_2|R}{(R - |\rho_1|)(R - |\rho_2|)} \frac{d\theta}{2\pi}$$
$$+ \log \prod_{|a| \leq R, \operatorname{ord}_a h > 0} (1 + \gamma_{R,\rho_1,\rho_2}) |G_{R,a}(\rho_1 \zeta)|^{\operatorname{ord}_a h}$$
$$+ \log \prod_{|a| \leq R, -\operatorname{ord}_a h > 0} (1 + \gamma_{R,\rho_1,\rho_2}) |G_{R,a}(\rho_2 \zeta)|^{-\operatorname{ord}_a h}.$$

Now averaging over $\{|\zeta| = 1\}$ gives us

$$\oint_{|\zeta|=1} \log^+ \left| \frac{h(\rho_1\zeta)}{h(\rho_2\zeta)} \right|$$

$$\leq \frac{2|\rho_1 - \rho_2|R}{(R - |\rho_1|)(R - |\rho_2|)} \int_{\theta=0}^{2\pi} \log^+ |h(Re^{i\theta})| \frac{d\theta}{2\pi}$$

$$+ \sum_{\substack{|a|\leq R \\ \mathrm{ord}_a h > 0}} \log(1 + \gamma_{R,\rho_1,\rho_2}) + \sum_{\substack{|a|\leq R \\ \mathrm{ord}_a h > 0}} \oint_{|\zeta|=1} (\mathrm{ord}_a h) \log|G_{R,a}(\rho_1\zeta)|$$

$$+ \sum_{\substack{|a|\leq R \\ -\mathrm{ord}_a h > 0}} \log(1 + \gamma_{R,\rho_1,\rho_2}) + \sum_{\substack{|a|\leq R \\ -\mathrm{ord}_a h > 0}} \oint_{|\zeta|=1} (-\mathrm{ord}_a h) \log|G_{R,a}(\rho_2\zeta)|$$

$$= \frac{2|\rho_1 - \rho_2|R}{(R - |\rho_1|)(R - |\rho_2|)} \int_{\theta=0}^{2\pi} \log^+ |h(Re^{i\theta})| \frac{d\theta}{2\pi}$$

$$+ \sum_{\substack{|a|\leq R \\ \mathrm{ord}_a h \neq 0}} \log(1 + \gamma_{R,\rho_1,\rho_2}) + (N(R, \{h = 0\}) - N(|\rho_1|, \{h = 0\}))$$

$$+ (N(R, \{h = \infty\}) - N(|\rho_2|, \{h = \infty\})).$$

Observe that if Z is a divisor in $\mathbb{C}$ whose support does not contain the origin, then

$$n(R, Z) = \sum_{\substack{a \in Z \\ 0 < |a| < R}} \mathrm{ord}_a Z \leq \frac{1}{\log \frac{\tilde{R}}{R}} \sum_{\substack{a \in Z \\ 0 < |a| < R}} (\mathrm{ord}_a Z) \log \frac{\tilde{R}}{|a|}$$

$$\leq \frac{1}{\log \frac{\tilde{R}}{R}} N(\tilde{R}, Z).$$

Moreover, for $0 < \rho < R$ we have

$$N(R, Z) - N(\rho, Z) = \sum_{0 < |a| < \rho} \mathrm{ord}_a Z \log \frac{R}{\rho} + \sum_{\rho \leq |a| < R} \mathrm{ord}_a Z \log \frac{R}{|a|}$$

$$\leq \sum_{0 < |a| < R} \mathrm{ord}_a Z \log \frac{R}{\rho}$$

$$= \log \frac{R}{\rho} n(R, Z) \leq \frac{\log \frac{R}{\rho}}{\log \frac{\tilde{R}}{R}} N(\tilde{R}, Z).$$

Using $\log(1 + x) \leq x$ for $x \geq 0$, we now summarize our result in the following proposition.

PROPOSITION 1.3.1. *Let $h(\zeta)$ be a holomorphic function on $\{\zeta \in \mathbb{C} : |\zeta| \le R\}$ and let ρ_1, ρ_2 be complex numbers such that $|\rho_1| < R, |\rho_2| < R$. Let $\tilde{R} > R$. Then*

$$\oint_{|\zeta|=1} \log^+ \left| \frac{h(\rho_1 \zeta)}{h(\rho_2 \zeta)} \right|$$

$$\le \frac{|\rho_1 - \rho_2| R}{(R - |\rho_1|)(R - |\rho_2|)}$$

$$\times \left(2 \int_{\theta=0}^{2\pi} \log^+ |h(R\, e^{i\theta})| \frac{d\theta}{2\pi} + n(\tilde{R}, \{h = 0\}) + n(\tilde{R}, \{h = \infty\}) \right)$$

$$+ (N(R, \{h = 0\}) - N(|\rho_1|, \{h = 0\}))$$

$$+ (N(R, \{h = \infty\}) - N(|\rho_2|, \{h = \infty\}))$$

$$\le \frac{|\rho_1 - \rho_2| R}{(R - |\rho_1|)(R - |\rho_2|)}$$

$$\times \left(2 \int_{\theta=0}^{2\pi} \log^+ |h(R\, e^{i\theta})| \frac{d\theta}{2\pi} + \frac{N(\tilde{R}, \{h = 0\})}{\log \frac{\tilde{R}}{R}} + \frac{N(\tilde{R}, \{h = \infty\})}{\log \frac{\tilde{R}}{R}} \right)$$

$$+ (N(R, \{h = 0\}) - N(|\rho_1|, \{h = 0\}))$$

$$+ (N(R, \{h = \infty\}) - N(|\rho_2|, \{h = \infty\})) .$$

1.4. Lower Bound of Some Derivative at One Point. Now we make precise what rescaling is required for the perturbation of the holomorphic map to make sure that the pullback of the constructed section to $\mathbb{C}$ is not identically zero. Let

$$\tilde{\varphi}_m : \mathbb{C}^{\times m} \to \mathbb{P}_n^{\times m}$$

be defined by

$$\tilde{\varphi}_m(\zeta_1, \ldots, \zeta_m) = (\varphi(\zeta_1), \ldots, \varphi(\zeta_m)) .$$

We expand $\tilde{\varphi}_m^* F$ into homogeneous components $\tilde{\varphi}_m^* F = \sum_{\mu=0}^{\infty} G_\mu$ in the m variables $(\zeta_1, \ldots, \zeta_n)$. Since the image of φ is not contained in any hypersurface of $\mathbb{P}_n$, it follows that there exists the smallest l such that G_l is not identically zero. We now consider the worst case where $F(\varphi(\zeta), \ldots, \varphi(\zeta))$ is identically zero. In particular, $G_l(1, \ldots, 1) = 0$. Choose positive numbers $\tau_1, \ldots, \tau_m$ less than $\frac{1}{2}$ such that $G_l(1 + \tau_1, \ldots, 1 + \tau_m)$ is nonzero. Since $G_l(\zeta_1, \ldots, \zeta_m)$ is homogeneous in the m variables $\zeta_1, \ldots, \zeta_m$, we can write

$$G_l(1 + \tau_1 \zeta, \ldots, 1 + \tau_m \zeta) = \chi_p \zeta^p + \chi_{p+1} \zeta^{p+1} + \cdots + \chi_l \zeta^l$$

with $0 \ne \chi_p \in \mathbb{C}$. Let η_0 be a positive number such that

$$\left| \chi_{p+1} \zeta + \cdots + \chi_l \zeta^{l-p} \right| \le \frac{1}{2} |\chi_p|$$

for $|\zeta| \leq \eta_0$. Suppose $A > 1/\eta_0$. Let r be a positive number and let $\rho_\nu = r(1 + \tau_\nu/A)$. Let

$$\varphi_{\rho_1,\ldots,\rho_m}(\zeta) = (\varphi(\rho_1\zeta), \ldots, \varphi(\rho_m\zeta)).$$

Then

$$\left| \lim_{\zeta \to 0} \frac{1}{\zeta^l} \left(\varphi^*_{\rho_1,\ldots,\rho_m} F \right)(\zeta) \right| \geq \frac{r^l}{2} |\chi_p| \frac{1}{A^p},$$

because

$$G_l(\rho_1\zeta, \ldots, \rho_m\zeta) = r^l \zeta^l F_l \left(1 + \frac{\tau_1}{A}, \ldots, 1 + \frac{\tau_m}{A} \right)$$

$$= r^l \zeta^l \frac{1}{A^p} \left(\chi_p + \chi_{p+1}\left(\frac{1}{A}\right) + \cdots + \chi_l \left(\frac{1}{A}\right)^{l-p} \right).$$

In our application we will use $A = \frac{1}{r^2 T(r,\varphi)^\kappa}$ with $\kappa > 4$.

1.5. Computation of Defect and the Proof of Theorem 0.0.1. Let $s_{V_\lambda} \in H^0(\mathbb{P}_n, \mathcal{O}_{\mathbb{P}_n}(\delta))$ $(1 \leq \lambda \leq \Lambda)$ define the smooth hypersurface V_λ in $\mathbb{P}_n$. By Lemma 1.1.3 we can choose $\tau > 1$ such that $\delta n \Theta_n(\tau) < 1$. Then we can choose m sufficiently large and then choose d sufficiently large such that there exists

$$F \in H^0\left(\mathbb{P}_n^{\times m}, \mathcal{O}_{\mathbb{P}_n^{\times m}}(d, \ldots, d)\right)$$

so that F vanishes to any multi-order $(j_1, \ldots, j_m)$ at $V_\lambda^{\times m}$ $(1 \leq \lambda \leq \Lambda)$ which satisfies

$$j_1 + \cdots + j_m < \frac{md}{\tau(n+1)}.$$

Let x be an $(n+1)$-tuple of functions which form the coordinate system of the affine part $\mathbb{C}^n$ of $\mathbb{P}_n$. When $\mathbb{P}_n$ is the j-th factor of $\mathbb{P}_n^{\times m}$ we relabel x as x_j so that $(x_1, \ldots, x_m)$ form the affine coordinate system of the affine part of $\mathbb{P}_n^{\times m}$. We rescale the coordinate ζ of $\mathbb{C}$ to $\rho_\nu \zeta$ to get from φ another map from $\mathbb{C}$ to $\mathbb{P}_n$ for $1 \leq \nu \leq m$, where $\rho_1, \ldots, \rho_m$ are from Section 1.4. We let $\tilde{x}_\nu = x_\nu(\varphi_\nu(\rho_\nu\zeta))$ and $\hat{x} = x(\varphi(r\zeta))$. Let q be the largest integer less than $\frac{md}{\tau(n+1)}$.

We now make the following trivial observation. Let Λ, m, N be positive integers such that $\Lambda n \leq N$. Let $z_1, \ldots, z_N$ be the coordinates of $\mathbb{C}^N$. For $1 \leq \lambda \leq \Lambda$ let $I(\lambda, j_{\lambda,1}, \ldots, j_{\lambda,n})$ be the principal ideal generated by $\prod_{\nu=1}^{n} z_{\lambda n+\nu}^{j_{\lambda,\nu}}$ over the local ring $\mathcal{O}_{\mathbb{C}^N,0}$ of $\mathbb{C}^N$ at the origin. Then

$$\bigcap_{\lambda=1}^{\Lambda} I(\lambda, j_{\lambda,1}, \ldots, j_{\lambda,n}) = \prod_{\lambda=1}^{\Lambda} I(\lambda, j_{\lambda,1}, \ldots, j_{\lambda,n})$$

for any nonnegative integers $j_{\lambda,\nu}$ $(1 \leq \lambda \leq \Lambda, 1 \leq \nu \leq n)$, because both are equal to the principal ideal generated by the single element

$$\prod_{\substack{1 \leq \lambda \leq \Lambda \\ 1 \leq \nu \leq m}} z_{\lambda n+\nu}^{j_{\lambda,\nu}}.$$

Since the hypersurfaces V_λ $(1 \leq \lambda \leq \Lambda)$ of $\mathbb{P}_n$ are in normal crossing, the trivial observation implies that the ideal sheaf of germs of holomorphic functions on $\mathbb{P}_n^{\times m}$ which vanish to multi-order $(j_1, \ldots, j_m)$ on each $V_\lambda^{\times m}$ is generated by

$$\prod_{\substack{1 \leq \lambda \leq \Lambda \\ 1 \leq \nu \leq m}} (\pi_\nu^* s_{V_\lambda})^{j_\nu},$$

where $\pi_\nu : \mathbb{P}_n^{\times m} \to \mathbb{P}_n$ is the projection onto the ν-th factor.

Let $l_\mu(x_1, \ldots, x_m)$ $(1 \leq \mu \leq k)$ be a product of m generic polynomials respectively of degree 1 in the affine coordinates $x_1, \ldots, x_m$ of $\mathbb{P}_n^{\times m}$. For N sufficiently large we can write

$$F(x_1, \ldots, x_m) l_\mu(x_1, \ldots, x_m)^N$$

$$= \sum_{\substack{j_{1,1} + \cdots + j_{1,m} = q \\ \cdots\cdots\cdots \\ j_{\Lambda,1} + \cdots + j_{\Lambda,m} = q}} \left(\prod_{\substack{1 \leq \lambda \leq \Lambda \\ 1 \leq \nu \leq m}} s_{V_\lambda}(x_\nu)^{j_{\lambda,\nu}} \right) G_{\mu, \{j_{\lambda,\nu}\}_{\substack{1 \leq \lambda \leq \Lambda \\ 1 \leq \nu \leq m}}}(x_1, \ldots, x_m).$$

We have

$$\frac{\prod_{\nu=1}^m \left(1 + |\tilde{x}_\nu|^2\right)^{\frac{d+N}{2}}}{|F(\tilde{x}_1, \ldots, \tilde{x}_m)| \sum_{\mu=1}^k |l_\mu(\tilde{x}_1, \ldots, \tilde{x}_m)|^N}$$

$$\geq \frac{\left(1 + |\hat{x}|^2\right)^{\frac{q\delta\Lambda}{2}}}{\prod_{\lambda=1}^\Lambda |s_{V_\lambda}(\hat{x})|^q} \frac{\prod_{\nu=1}^m \left(1 + |\tilde{x}_\nu|^2\right)^{\frac{d+N}{2}} / \left(1 + |\hat{x}|^2\right)^{\frac{q\delta\Lambda}{2}}}{\sum_{\substack{j_{1,1} + \cdots + j_{1,m} = q \\ \cdots\cdots\cdots \\ j_{\Lambda,1} + \cdots + j_{\Lambda,m} = q \\ 1 \leq \mu \leq k}} \left|\prod_{\substack{1 \leq \lambda \leq \Lambda \\ 1 \leq \nu \leq m}} s_{V_\lambda}(\tilde{x}_\nu)^{j_{\lambda,\nu}}\right| \left|G_{\mu, \{j_{\lambda,\nu}\}_{\substack{1 \leq \lambda \leq \Lambda \\ 1 \leq \nu \leq m}}}\right|}.$$

Note that instead of using $l_\mu(x_1, \ldots, x_m)$ $(1 \leq \mu \leq k)$, one could also write $F(x_1, \ldots, x_m)$ as a linear combination of

$$\prod_{\substack{1 \leq \lambda \leq \Lambda \\ 1 \leq \nu \leq m}} s_{V_\lambda}(x_\nu)^{j_{\lambda,\nu}}$$

with smooth sections of

$$\mathcal{O}_{\mathbb{P}_n^{\times m}}(d - q, \ldots, d - q)$$

over $\mathbb{P}_n^{\times m}$ as coefficients as in Section 0.7. We consider the following long string of inequalities:

(1.5.1)

$$\log \frac{\left(1 + |\hat{x}|^2\right)^{\frac{q\delta\Lambda}{2}}}{\prod_{\lambda=1}^\Lambda |s_{V_\lambda}(\hat{x})|^q}$$

$$\leq \log \frac{\prod_{\nu=1}^{m}\left(1+|\tilde{x}_\nu|^2\right)^{\frac{d+N}{2}}}{|F(\tilde{x}_1,\ldots,\tilde{x}_m)|\sum_{\mu=1}^{k}|l_\mu(\tilde{x}_1,\ldots,\tilde{x}_m)|^N}$$

$$+\log \frac{\sum_{\substack{j_{1,1}+\cdots+j_{1,m}=q\\ \cdots\cdots\cdots\\ j_{\Lambda,1}+\cdots+j_{\Lambda,m}=q\\ 1\leq\mu\leq k}} \prod_{\substack{1\leq\lambda\leq\Lambda\\ 1\leq\nu\leq m}}\left|\frac{s_{V_\lambda}(\tilde{x}_\nu)}{s_{V_\lambda}(\hat{x})}\right|^{j_{\lambda,\nu}}\left|G_{\mu,\{j_{\lambda,\nu}\}_{\substack{1\leq\lambda\Lambda\\ 1\leq\nu\leq m}}}(\tilde{x}_1,\ldots,\tilde{x}_m)\right|}{\prod_{\nu=1}^{m}\left(1+|\tilde{x}_\nu|^2\right)^{\frac{d+N}{2}}\Big/\left(1+|\hat{x}|^2\right)^{\frac{q\delta\Lambda}{2}}}$$

$$\leq \log \frac{\prod_{\nu=1}^{m}\left(1+|\tilde{x}_\nu|^2\right)^{\frac{d+N}{2}}}{|F(\tilde{x}_1,\ldots,\tilde{x}_m)|\sum_{\mu=1}^{k}|l_\mu(\tilde{x}_1,\ldots,\tilde{x}_m)|^N}$$

$$+\log \sum_{\substack{j_{1,1}+\cdots+j_{1,m}=q\\ \cdots\cdots\cdots\\ j_{\Lambda,1}+\cdots+j_{\Lambda,m}=q}} \prod_{\substack{1\leq\lambda\leq\Lambda\\ 1\leq\nu\leq m}}\left|\frac{s_{V_\lambda}(\tilde{x}_\nu)}{s_{V_\lambda}(\hat{x})}\right|^{j_{\lambda,\nu}}$$

$$\times\sum_{\mu=1}^{k}\left(\frac{\left|G_{\mu,\{j_{\lambda,\nu}\}_{\substack{1\leq\lambda\leq\Lambda\\ 1\leq\nu\leq m}}}\right|}{\prod_{\nu=1}^{m}\left(1+|\tilde{x}_\nu|^2\right)^{\frac{d+N-(j_{1,\nu}+\cdots+j_{\Lambda,\nu})\delta}{2}}}\prod_{\nu=1}^{m}\left(\frac{1+|\hat{x}|^2}{1+|\tilde{x}_\nu|^2}\right)^{\frac{(j_{1,\nu}+\cdots+j_{\Lambda,\nu})\delta}{2}}\right)$$

$$\leq \log \frac{\prod_{\nu=1}^{m}\left(1+|\tilde{x}_\nu|^2\right)^{\frac{d+N}{2}}}{|F(\tilde{x}_1,\ldots,\tilde{x}_m)|\sum_{\mu=1}^{k}|l_\mu(\tilde{x}_1,\ldots,\tilde{x}_m)|^N}+\sum_{\substack{j_{1,1}+\cdots+j_{1,m}=q\\ \cdots\cdots\cdots\\ j_{\Lambda,1}+\cdots+j_{\Lambda,m}=q}}\log^{+}\prod_{\substack{1\leq\lambda\leq\Lambda\\ 1\leq\nu\leq m}}\left|\frac{s_{V_\lambda}(\tilde{x}_\nu)}{s_{V_\lambda}(\hat{x})}\right|^{j_{\lambda,\nu}}$$

$$+\sum_{\substack{j_{1,1}+\cdots+j_{1,m}=q\\ \cdots\cdots\cdots\\ j_{\Lambda,1}+\cdots+j_{\Lambda,m}=q\\ 1\leq\mu\leq k}}\log^{+}\frac{\left|G_{\mu,\{j_{\lambda,\nu}\}_{\substack{1\leq\lambda\leq\Lambda\\ 1\leq\nu\leq m}}}\right|}{\prod_{\nu=1}^{m}\left(1+|\tilde{x}_\nu|^2\right)^{\frac{d+N-(j_{1,\nu}+\cdots+j_{\Lambda,\nu})\delta}{2}}}$$

$$+\sum_{\substack{j_{1,1}+\cdots+j_{1,m}=q\\ \cdots\cdots\cdots\\ j_{\Lambda,1}+\cdots+j_{\Lambda,m}=q\\ 1\leq\mu\leq k,1\leq\nu\leq m}}\log^{+}\left(\frac{1+|\hat{x}|^2}{1+|\tilde{x}_\nu|^2}\right)^{\frac{(j_{1,\nu}+\cdots+j_{\Lambda,\nu})\delta}{2}}+C_{m,q}$$

$$= \log \frac{A\prod_{\nu=1}^{m}\left(1+|\tilde{x}_\nu|^2\right)^{\frac{d}{2}}}{|F(\tilde{x}_1,\ldots,\tilde{x}_m)|}+\log \frac{\prod_{\nu=1}^{m}\left(1+|\tilde{x}_\nu|^2\right)^{\frac{N}{2}}}{\sum_{\mu=1}^{k}|l_\mu(\tilde{x}_1,\ldots,\tilde{x}_m)|^N}$$

$$+\sum_{\substack{j_{1,1}+\cdots+j_{1,m}=q\\ \cdots\cdots\cdots\\ j_{\Lambda,1}+\cdots+j_{\Lambda,m}=q}}(j_{1,\nu}+\cdots+j_{\Lambda,\nu})\log^{+}\left|\frac{s_{V_\lambda}(\tilde{x}_\nu)}{s_{V_\lambda}(\hat{x})}\right|$$

$$+\sum_{\substack{j_{1,1}+\cdots+j_{1,m}=q\\ \cdots\cdots\cdots\\ j_{\Lambda,1}+\cdots+j_{\Lambda,m}=q\\ 1\leq\mu\leq k}}\log^{+}\frac{\left|G_{\mu,\{j_{\lambda,\nu}\}_{\substack{1\leq\lambda\leq\Lambda\\ 1\leq\nu\leq m}}}\right|}{\prod_{\nu=1}^{m}\left(1+|\tilde{x}_\nu|^2\right)^{\frac{d+N-(j_{1,\nu}+\cdots+j_{\Lambda,\nu})}{2}}}$$

$$+\sum_{\substack{j_{1,1}+\cdots+j_{1,m}=q\\ \cdots\cdots\cdots\\ j_{\Lambda,1}+\cdots+j_{\Lambda,m}=q\\ 1\leq\mu\leq k,1\leq\nu\leq m}}\tfrac{1}{2}(j_{1,\nu}+\cdots+j_{\Lambda,\nu})\delta\log^{+}\left(\frac{1+|\hat{x}|^2}{1+|\tilde{x}_\nu|^2}\right)+C_{m,q}-\log A,$$

where $C_{m,q}$ is a constant depending only on m and q and A is a positive constant chosen so large that

$$\log \frac{A \prod_{\nu=1}^m \left(1 + |\tilde{x}_\nu|^2\right)^{\frac{d}{2}}}{|F(\tilde{x}_1, \ldots, \tilde{x}_m)|} > 0$$

at every point of $\mathbb{P}_n^{\times m}$. We will average the left-hand side and the right-hand side of 1.5.1 over the unit circle $\{|\zeta| = 1\}$. We will consider a lower bound for the averaged left-hand side of 1.5.1 and also conisder the an upper bound for each of the averaged term on the right-hand side of 1.5.1, in order to get the defect relation stated in Theorem 1.0.1.

First we look at an upper bound for each of the averaged term on the right-hand side of 1.5.1. Both terms

$$\log \frac{\prod_{\nu=1}^m \left(1 + |\tilde{x}_\nu|^2\right)^{\frac{N}{2}}}{\sum_{\mu=1}^k |l_\mu(\tilde{x}_1, \ldots, \tilde{x}_m)|^N}$$

and

$$\log^+ \frac{\left|G_{\mu, \{j_{\lambda,\nu}\}_{\substack{1 \le \lambda \le \Lambda \\ 1 \le \nu \le m}}}(\tilde{x}_1, \ldots, \tilde{x}_m)\right|}{\prod_{\nu=1}^m \left(1 + |\tilde{x}_\nu|^2\right)^{\frac{md + mN - (j_{1,\nu} + \cdots + j_{\Lambda,\nu})\delta}{2}}}$$

are uniformly bounded on $\mathbb{P}_n^{\times m}$.

To get an upper bound of the average of

$$\log \frac{A \prod_{\nu=1}^m \left(1 + |\tilde{x}_\nu|^2\right)^{\frac{d}{2}}}{|F(\tilde{x}_1, \ldots, \tilde{x}_m)|}$$

over the circle $\{|\zeta| = 1\}$, we apply the standard First Main Theorem argument of two integrations to

$$\partial\bar{\partial} \log \left|\frac{F(\tilde{x}_1, \ldots, \tilde{x}_m)}{\zeta^l}\right|$$

which is nonzero at $\zeta = 0$, we get

$$\oint_{|\zeta|=1} \log \frac{A \prod_{\nu=1}^m \left(1 + |\tilde{x}_\nu|^2\right)^{\frac{d}{2}}}{|F(\tilde{x}_1, \ldots, \tilde{x}_m)|} \le d \sum_{\nu=1}^m T(\rho_\nu, \varphi) + \lim_{\zeta \to 0} \log \frac{|\zeta^l|}{|F(\tilde{x}_1, \ldots, \tilde{x}_m)|} + O(1)$$

$$\le d \sum_{\nu=1}^m T(\rho_\nu, \varphi) + \log \left(\frac{r^l}{2} |\chi_p| r^2 T(r, \varphi)^\kappa\right) + O(1)$$

$$\le d \sum_{\nu=1}^m T(\rho_\nu, \varphi) + O\left(\log r + \log T(r, \varphi)\right),$$

where l, p, χ_p come from Section 1.4.

To get an upper bound for

$$\log^+ \left(\frac{1 + |\hat{x}|^2}{1 + |\tilde{x}_\nu|^2}\right)$$

we use the following trivial inequality

$$\frac{1 + a_1 + \cdots + a_n}{1 + b_1 + \cdots + b_n} \leq 1 + \frac{a_1}{b_1} + \cdots + \frac{a_n}{b_n}.$$

for positive numbers $a_1, \ldots, a_n, b_1, \ldots, b_n$. Let $x = (z_1, \ldots, z_n)$. Then $\hat{x}(\zeta) = (z_1(\varphi(r\zeta)), \ldots, z_n(\varphi(r\zeta)))$ and $\tilde{x}_\nu(\zeta) = (z_1(\varphi(\rho_\nu\zeta)), \ldots, z_n(\varphi(\rho_\nu\zeta)))$. We have

$$\begin{aligned}
\log^+\left(\frac{1 + |\hat{x}|^2}{1 + |\tilde{x}_\nu|^2}\right) &= \log^+\left(\frac{1 + \sum_{\lambda=1}^n |z_\lambda(\varphi(r\zeta))|^2}{1 + \sum_{\lambda=1}^n |z_\lambda(\varphi(\rho_\nu\zeta))|^2}\right) \\
&\leq \log^+\left(1 + \sum_{\lambda=1}^n \left|\frac{z_\lambda(\varphi(r\zeta))}{z_\lambda(\varphi(\rho_\nu\zeta))}\right|^2\right) \\
&\leq \sum_{\lambda=1}^n \log^+\left(\left|\frac{z_\lambda(\varphi(r\zeta))}{z_\lambda(\varphi(\rho_\nu\zeta))}\right|^2\right) + \log(n + 1).
\end{aligned}$$

To estimate the discrepancy from rescaling of the coordinate of $\mathbb{C}$, we need to compare at the same time both the characteristic function and the counting function at a pair of points whose distance is of the order of the reciprocal of the characteristic function. For that we need the following simple lemma on real functions, which is modified from [Hayman 1964, p. 14] so that the conclusion is valid at the same time for several functions.

LEMMA 1.5.2 (REAL FUNCTIONS [Hayman 1964, p. 14]). *Suppose that $S_1(r)$, $\ldots, S_k(r)$ are positive nondecreasing functions for $r_0 \leq r < \infty$ which are bounded in every interval $[r_0, r_1]$ for $r_0 \leq r_1 < \infty$. Then given $K > 1$, $B_1 > 1$, and $B_2 > 1$ with $B_2 \sum_{\nu=1}^k S_\nu(r_0) > 1$ there exists a sequence $r_\mu \to \infty$ such that*

$$S_\nu(r) < K\, S_\nu(r_\mu) \qquad for\ r_\mu < r < r_\mu + \frac{B_1}{\left(\log\left(B_2 \sum_{\nu=1}^k S_\nu(r_\mu)\right)\right)^K}.$$

PROOF. Assume that our conclusion is false. Then for all sufficiently large r we can find ρ such that

$$r < \rho < r + \frac{B_1}{\left(\log\left(B_2 \sum_{\nu=1}^k S_\nu(r)\right)\right)^K}$$

and $S_\nu(\rho) \geq K\, S_\nu(r)$ for some ν with $1 \leq \nu \leq k$.

Choose r_1 so that this holds for $r \geq r_1$. Then if r_μ has already been defined we define $r_{\mu+1}$ so that

$$r_\mu < r_{\mu+1} < r_\mu + \frac{B_1}{\left(\log\left(B_2 \sum_{\nu=1}^k S_\nu(r_\mu)\right)\right)^K}$$

and $S_{\nu_\mu}(r_{\mu+1}) \geq K\, S_{\nu_\mu}(r_\mu)$ for some ν_μ with $1 \leq \nu_\mu \leq k$.

Let $p_{\nu,\mu} = 1$ if $\nu = \nu_\mu$ and $p_{\nu,\mu} = 0$ for $\nu \neq \nu_\mu$ and $1 \leq \nu \leq k$. Then

$$S_\nu(r_{\mu+1}) \geq K^{p_{\nu,\mu}} S_\nu(r_\mu) \quad for\ \ 1 \leq \nu \leq k.$$

We have

$$\sum_{\nu=1}^{k} S_\nu(r_{\mu+1}) \geq \sum_{\nu=1}^{k} K^{p_{\nu,1}+\cdots+p_{\nu,\mu}} S_\nu(r_0) \geq K^\mu \min\left(S_1(r_0), \ldots, S_k(r_0)\right).$$

Thus

$$r_{\mu+1} - r_\mu \leq \frac{B_1}{\left(\mu \log K + \log\left(B_2 \min\left(S_1(r_0), \ldots, S_k(r_0)\right)\right)\right)^K}$$

and $\sum_{\mu=1}^{\infty}(r_{\mu+1}-r_\mu)$ converges so that $\sup_\mu r_\mu$ is finite. On the other hand, there exists some ν_0 such that there are infinitely many $p_{\nu_0,m_l} = 1$ with $1 \leq m_l < \infty$ and from the nondecreasing property of $S_{\nu_0}(r)$ we have

$$S_{\nu_0}(r_{\mu+1}) \geq K^{q_\mu} S_{\nu_0}(r_1),$$

where q_μ is the number of m_l less than μ. Since $q_\mu \to \infty$ as $n \to \infty$, we conclude that $S_{\nu_0}(r)$ is unbounded on the finite interval $[r_0, \sup_\mu r_\mu]$, which is a contradiction. $\qquad \square$

COROLLARY 1.5.3. *Given any $K > 1$ and $B > 1$ there exists a sequence $r_\mu \to \infty$ such that*

$$T\left(r_\mu + \frac{B}{T(r_\mu, \varphi)}, \varphi\right) \leq KT(r_\mu), \qquad N\left(r_\mu + \frac{B}{T(r_\mu, \varphi)}, \varphi\right) \leq KN(r_\mu).$$

PROOF. If $T(r, \varphi)$ is bounded, the statement is trivial. If $T(r, \varphi)$ is unbounded, we have

$$\frac{B}{T(r, \varphi)} < \frac{B}{\log\left(2T(r, \varphi)\right)} < \frac{B+1}{\log\left(T(r, \varphi) + N(r, \varphi)\right)}$$

for r sufficiently large. $\qquad \square$

Let η be an arbitrary positive number and we choose $\kappa > 4$. Now choose a sequence $\{r_\mu\}_{1 \leq \mu < \infty}$ going to infinity such that

$$r_\mu \geq 2, \qquad T(r_\mu, \varphi) \geq 2, \qquad T\left(r_\mu + \frac{1}{T(r_\mu, \varphi)}, \varphi\right) \leq (1+\eta)T(r_\mu, \varphi),$$

$$N\left(r_\mu + \frac{1}{T(r_\mu, \varphi)}, \varphi\right) \leq (1+\eta)N(r_\mu, \varphi), \qquad R = r_\mu + \frac{1}{2T(r_\mu, \varphi)},$$

$$\tilde{R} = r_\mu + \frac{1}{T(r_\mu, \varphi)}, \qquad \rho_\nu = r_\mu + \frac{\tau_\nu}{r_\mu\, T(r_\mu, \varphi)^\kappa},$$

where $\tau_1, \ldots, \tau_m$ are from Section 1.4. From here to the end of the section r will be a member of the sequence $\{r_\mu\}_{1 \leq \mu < \infty}$ though for notational simplicity we suppress the subscript μ of r_μ. Since $\log(1+\eta) \geq \eta - \frac{\eta^2}{2}$ for $\eta < 1$, it follows that both $\log \frac{\tilde{R}}{R}$ and $\log \frac{R}{\rho_\nu}$ are at most

$$\frac{1}{rT(r, \varphi)} - \frac{1}{2(rT(r, \varphi))^2} - \frac{1}{2rT(r, \varphi)} \geq \frac{1}{4rT(r, \varphi)}.$$

Moreover, both

$$\frac{(\rho_\nu - \rho_\mu)R}{(R - \rho_\nu)(R - \rho_\mu)} \quad \text{and} \quad \frac{(\rho_\nu - r)R}{(R - \rho_\nu)(R - \rho_\mu)}$$

are no less than

$$\frac{\frac{\tau_\nu - \tau_\mu}{r^2 T(r,\varphi)^\kappa}(r + \frac{1}{4})}{\left(\frac{1}{4}T(r,\varphi)\right)^2} \leq \frac{32}{r\,T(r,\varphi)^{\kappa-2}}.$$

By Proposition 1.3.1,

$$\oint_{|\zeta|=1} \log^+ \left| \frac{z_\lambda \circ \varphi(r\zeta)}{z_\lambda \circ \varphi(\rho_\nu\zeta)} \right|^2$$

$$\leq \frac{32}{r\,T(r,\varphi)^{\kappa-2}}\left(2(1+\eta)T(r,\varphi) + 8r\,T(r,\varphi)^2\right) + 2\eta T(r,\varphi) + O(1)$$

$$\leq 2\eta T(r,\varphi) + O(1),$$

because $\kappa > 4$. Hence

$$\oint_{|\zeta|=1} \log^+ \left(\frac{1 + |\hat{x}|^2}{1 + |\tilde{x}_\nu|^2} \right) \leq 4n\eta T(r,\varphi) + O(1)$$

and

$$\oint_{|\zeta|=1} \log^+ \left| \frac{s_{V_\lambda}(\tilde{x}_\nu)}{s_{V_\lambda}(\hat{x})} \right| \leq 2\delta\eta T(r,\varphi) + O(1).$$

Thus we have the upper bounds

$$\oint_{|\zeta|=1} \sum_{\substack{j_{1,1}+\cdots+j_{1,m}=q \\ \cdots\cdots\cdots \\ j_{\Lambda,1}+\cdots+j_{\Lambda,m}=q \\ 1\leq\mu\leq k, 1\leq\nu\leq m}} \tfrac{1}{2}\left(j_{1,\nu} + \cdots + j_{\Lambda,\nu}\right)\delta \log^+\left(\frac{1+|\hat{x}|^2}{1+|\tilde{x}_\nu|^2}\right)$$

$$\leq 2n\delta km\Lambda q\binom{q+m-1}{m-1}^\Lambda \eta T(r,\varphi) + O(1).$$

and

$$\oint_{|\zeta|=1} \sum_{\substack{j_{1,1}+\cdots+j_{1,m}=q \\ \cdots\cdots\cdots \\ j_{\Lambda,1}+\cdots+j_{\Lambda,m}=q}} \left(j_{1,\nu} + \cdots + j_{\Lambda,\nu}\right)\log^+\left|\frac{s_{V_\lambda}(\tilde{x}_\nu)}{s_{V_\lambda}(\hat{x})}\right|$$

$$\leq 2\delta\Lambda q\binom{q+m-1}{m-1}^\Lambda \eta T(r,\varphi) + O(1).$$

To get a lower bound for the left-hand side of 1.5.1, we use the definition of defect and get

$$\oint_{|\zeta|=1} \log \frac{(1 + |\hat{x}|^2)^{\frac{q\delta\Lambda}{2}}}{\prod_{\lambda=1}^\Lambda |s_{V_\lambda}(\hat{x})|^q} \geq q\delta\left(\sum_{\lambda=1}^\Lambda \text{Defect}(\varphi, s_{V_\lambda}) - \eta\right)T(r,\varphi)$$

for r sufficiently large.

We now put together the lower bound for the averaged left-hand side of 1.5.1 and the upper bounds for the averaged terms on the right-hand side of 1.5.1. We get

$$q\delta\left(\sum_{\lambda=1}^{\Lambda}\text{Defect}(\varphi,s_{V_\lambda})-\eta\right)T(r,\varphi)$$

$$\leq d(1+\eta)\sum_{\nu=1}^{m}T(r,\varphi_\nu)+2\delta(nmk+1)\Lambda q\binom{q+m-1}{m-1}^{\Lambda}\eta T(r,\varphi)+O(1).$$

Since η is an arbitrary positive number, it follows that

$$\sum_{\lambda=1}^{\Lambda}\text{Defect}(\varphi,s_{V_\lambda})\leq\frac{md}{q\delta}.$$

The number q is chosen so that q is the largest integer less than $\frac{md}{\tau(n+1)}$ with $\delta n\Theta_n(\tau)<1$. Hence

$$\sum_{\lambda\in\Lambda}\text{Defect}(\varphi,V_\lambda)\leq\frac{n+1}{\delta}\Theta_n^{-1}\left(\frac{1}{n\delta}\right).$$

This gives Theorem 1.5.4 below for the case $q=1$. It now follows from

$$\Theta_n(\tau)\leq\min\left(\frac{e}{\tau(n+1)},\frac{1}{n}e^{-\frac{1}{4(n+1)^2}\left(1-\frac{1}{\tau}\right)^2}\right)$$

that $\sum_{\lambda\in\Lambda}\text{Defect}(\varphi,V_\lambda)$ is no more than ne for any $\delta\geq 1$ and is no more than $n+1$ for $\delta=1$. This proves the Theorem 1.0.1. The modification needed to prove Theorem 0.0.1 and Theorem 1.5.4 is standard. The modification is to restrict φ to a complex line in the complex vector space $\mathbb{C}^{\hat{m}}$ and then compute the proximity term by restricting and average over the complex line with respect to the Fubini–Study volume form of $\mathbb{P}_{\hat{m}-1}$.

THEOREM 1.5.4. *For $\tau>0$ let $\Theta_n(\tau)$ be*

$$\overline{\lim}_{m\to\infty}\left(\int_{\left\{\substack{x_1+\cdots+x_m<\frac{m}{\tau(n+1)}\\0<x_1<1,\ldots,0<x_m<1}\right\}}(1-x_1)^{n-1}\cdots(1-x_m)^{n-1}\,dx_1\cdots dx_m\right)^{1/m},$$

which is bounded by the minimum of $\frac{e}{\tau(n+1)}$ and $\frac{1}{n}e^{-\frac{1}{4(n+1)^2}\left(1-\frac{1}{\tau}\right)^2}$. Let V_λ ($1\leq\lambda\leq\Lambda$) be regular complex hypersurfaces in $\mathbb{P}_n$ of degree δ in normal crossing. Let $\varphi:\mathbb{C}\to\mathbb{P}_n$ is a holomorphic map whose image is not contained in any hypersurface of $\mathbb{P}_n$. Then

$$\sum_{\lambda=1}^{\Lambda}\text{Defect}(\varphi,V_\lambda)\leq\frac{n+1}{\delta}\Theta_n^{-1}\left(\frac{1}{n\delta}\right).$$

2. Hyperbolicity of the Complement
of a Generic High Degree Plane Curve

2.1. Overview of the Method of Proof. In this Chapter we will give a streamlined version of the proof of Theorem 0.1.4 [Siu and Yeung 1996a]. As explained in the introduction of this paper for the approach of jet differentials, the main difficulty of proving hyperbolicity is how to construct enough holomorphic jet differentials vanishing on an ample divisor which are independent in an appropriate sense.

As discussed in Section 0.10, there are two main steps in the proof. Though the first step is easier, we will spend more time in explaining the techniques in it, because these techniques may be generalizable to the higher dimensional case. In this overview the techniques of the first step are explained from here to the end of 2.1.4 and the techniques of the second step are explained in 2.1.5.

For the first step of constructing a meromorphic 1-jet differential whose pullback to the entire holomorphic curve vanishes, we use the following three ingredients to construct holomorphic 2-jet differentials vanishing on an ample divisor on a branched cover of $\mathbb{P}_2$ (see 2.1.1 and 2.1.2):

(i) meromorphic *nonlinear* connections of low pole order for the tangent bundle,
(ii) the Wronskian, and
(iii) the positivity of the canonical line bundle.

This particular way of constructing holomorphic 2-jet differentials gives us some control over their explicit forms so that by comparing degrees with respect to suitable distinct polarizations we can get the independence of two 2-jet differentials to obtain our desired meromorphic 1-jet differential as their resultant (see 2.1.4). A polarization here means a collection of affine variables and their differentials with respect to which degrees are measured. So far our method works only in the 2-dimensional case. The difficulty of extending it to the case of general dimension is that the algebraic procedure of concluding independence by comparing degrees with respect to suitable distinct polarizations is not yet developed for the case of general dimension. Such an algebraic procedure used in the dimension two case is done in a very *ad hoc* way by brute force.

2.1.1. Use of Linear Connections. To put our construction in the proper context, we first consider the the use of meromorphic *linear* connection of low pole order in some special cases. Let us look at the situation of lifting a connection for the tangent bundle of the base manifold to a branched cover. We assume that the branching is cyclic and the branching locus is smooth. In addition we assume that the second fundamental form of the branching locus with respect to the connection is zero in the sense that with respect to the connection the derivative of a local vector field with another local vector field always vanishes when both vector fields are tangential to the branching. Let $z^1, \ldots, z^n$ be local coordinates for the base manifold and $w^1, \ldots, w^n$ be local coordinates for the branched cover so that $w^n = (z^n)^{\frac{1}{\delta}}$ and $w^\alpha = z^\alpha$ for $1 \le \alpha \le n-1$. Let $\mathcal{D}$ denote a connection for the tangent bundle of the base manifold and let $\Gamma^\gamma_{\alpha\beta}$ be its Christoffel symbol so that

$$\mathcal{D}_{\frac{\partial}{\partial z^\alpha}} \frac{\partial}{\partial z^\beta} = \Gamma^\gamma_{\alpha\beta} \frac{\partial}{\partial z^\gamma}.$$

Here we use the summation convention of summing over an index appearing both in the superscript and subscript positions. Let the lifting of $\mathcal{D}$ to the branched cover be $\tilde{\mathcal{D}}$ with Christoffel symbol $\tilde{\Gamma}^\nu_{\lambda\mu}$ so that

$$\mathcal{D}_{\frac{\partial}{\partial w^\lambda}} \frac{\partial}{\partial w^\mu} = \tilde{\Gamma}^\nu_{\lambda\mu} \frac{\partial}{\partial w^\nu}.$$

From

$$\mathcal{D}_{\frac{\partial}{\partial w^\lambda}} \frac{\partial}{\partial w^\mu} = \mathcal{D}_{\frac{\partial}{\partial w^\lambda}} \left(\frac{\partial z^\beta}{\partial w^\mu} \frac{\partial}{\partial z^\beta} \right) = \frac{\partial^2 z^\beta}{\partial w^\lambda \partial w^\mu} \frac{\partial}{\partial z^\beta} + \frac{\partial z^\alpha}{\partial w^\lambda} \frac{\partial z^\beta}{\partial w^\mu} \Gamma^\gamma_{\alpha\beta} \frac{\partial}{\partial z^\gamma}$$

it follows that

(2.1.1.1) $$\tilde{\Gamma}^\nu_{\lambda\mu} = \frac{\partial^2 z^\beta}{\partial w^\lambda \partial w^\mu} \frac{\partial w^\nu}{\partial z^\beta} + \frac{\partial z^\alpha}{\partial w^\lambda} \frac{\partial w^\nu}{\partial z^\gamma} \frac{\partial z^\beta}{\partial w^\mu} \Gamma^\gamma_{\alpha\beta}.$$

Suppose $\mathcal{D}$ is locally holomorphic. We would like to compute the pole-order of $\tilde{\mathcal{D}}$ by using the condition that the second fundamental form of the branching locus is zero with respect to $\mathcal{D}$. From (2.1.1.1) the only pole contribution comes from

$$\frac{\partial w^n}{\partial z^n} = \frac{1}{\delta} \frac{1}{(w^n)^{\delta-1}}.$$

The pole could occur only in $\tilde{\Gamma}^n_{\lambda\mu}$, which has the two terms

$$T_1 := \frac{\partial^2 z^n}{\partial w^\lambda \partial w^\mu} \frac{\partial w^n}{\partial z^n}, \qquad T_2 := \frac{\partial z^\alpha}{\partial w^\lambda} \frac{\partial w^n}{\partial z^n} \frac{\partial z^\beta}{\partial w^\mu} \Gamma^n_{\alpha\beta}.$$

Since the only term in T_1 that is nonzero is for the case $\lambda = \mu = n$, it follows that

$$T_1 = \frac{\partial^2 z^n}{(\partial w^n)^2} \frac{\partial w^n}{\partial z^n} = (\delta - 1) \frac{1}{w^n}.$$

For the term T_2 the only pole contribution comes from the case $1 \le \lambda, \mu \le n-1$. In that case from the vanishing of the second fundamental form of the branching locus with respect to $\mathcal{D}$ we know that

$$\Gamma^n_{\lambda\mu} = O(z^n) = O((w^n)^\delta)$$

which more than makes up for the pole contribution from $\frac{\partial w^n}{\partial z^n}$. Thus we conclude that the pole of $\tilde{\mathcal{D}}$ is at most order one along the branching locus $\{w^n = 0\}$ of the branched cover.

Let the branching locus be defined locally by a function $f = 0$. We would like to see what the vanishing of the second fundamental form of the branching locus means in terms of the defining function f of the branching locus. Let ξ, η be arbitrary local vector fields tangential to the branching locus. Let $df, \omega_1, \ldots, \omega_{n-1}$ be a local basis of 1-forms. From the vanishing of $\langle df, \xi \rangle$ and $\langle df, \mathcal{D}_\eta \xi \rangle$ the equation

$$d_\eta \langle df, \xi \rangle = \langle \mathcal{D}_\eta df, \xi \rangle + \langle df, \mathcal{D}_\eta \xi \rangle$$

imples that $\langle \mathcal{D} df, \xi \otimes \eta \rangle = 0$. By writing

$$\mathcal{D} df = df \otimes df + df \otimes \sum_{j=1}^{n-1} a_j \omega_j + \left(\sum_{j=1}^{n-1} b_j \omega_j \right) \otimes df + \sum_{j,k=1}^{n-1} c_{j\,k} \omega_j \otimes \omega_k$$

for some local scalar functions $a_j, b_j, c_{j\,k}$, we conclude that the term

$$\sum_{j,k=1}^{n-1} c_{j\,k} \omega_j \otimes \omega_k$$

must vanish on $\{f = 0\}$. Thus on $\{f = 0\}$ we have

$$\mathcal{D} df = df \otimes df + df \otimes \sum_{j=1}^{n-1} a_j \omega_j + \left(\sum_{j=1}^{n-1} b_j \omega_j \right) \otimes df,$$

or in terms of the first-order derivative f_α and the second-order derivatives $f_{\alpha\beta}$ we have scalar functions $A_\beta, B_\alpha, C_{\alpha\beta}$ such that

$$f_{\alpha\beta} - \Gamma^\gamma_{\alpha\beta} f_\gamma = f_\alpha A_\beta + B_\alpha f_\beta + C_{\alpha\beta} f.$$

We now start our construction of holomorphic jet differentials from meromorphic connections. Let $z = \varphi(\zeta)$ represent a local holomorphic curve in an n-dimensional complex manifold X and let $\tilde{\mathcal{D}}$ be the meromorphic connection for the tangent bundle of X. Then

$$\varphi \mapsto d\varphi \wedge \tilde{\mathcal{D}} d\varphi \wedge \cdots \wedge \tilde{\mathcal{D}}^{n-1} d\varphi$$

defines a K_X-valued n-jet differential. Let $\omega \in \Gamma(X, mK_X)$. Then

$$\varphi \mapsto \langle \omega, (d\varphi \wedge \tilde{\mathcal{D}} d\varphi \wedge \cdots \wedge \tilde{\mathcal{D}}^{n-1} d\varphi)^{\otimes m} \rangle$$

defines an n-jet differential. If the pole order of $\tilde{\mathcal{D}}$ is small and the vanishing order of ω is high, then the n-jet differential

$$\varphi \mapsto \langle \omega, (d\varphi \wedge \tilde{\mathcal{D}} d\varphi \wedge \cdots \wedge \tilde{\mathcal{D}}^{n-1} d\varphi)^{\otimes m} \rangle$$

is holomorphic.

Suppose C is a smooth curve in $\mathbb{P}_2$ defined by the polynomial $f(x, y) = 0$ in the affine coordinates x, y of degree δ and X is the branched cover over $\mathbb{P}_2$ with cyclic branching of order δ along C. Suppose $\mathcal{D}$ is a meromorphic connection of low pole order for the tangent bundle of $\mathbb{P}_2$ such that the second fundamental form of C with respect to $\mathcal{D}$ is zero in the sense that the covariant derivatives of tangent vector fields of C in the direction of C with respect to $\mathcal{D}$ are zero. Then the connection $\mathcal{D}$ for the tangent bundle of $\mathbb{P}_2$ can be lifted to a connection $\tilde{\mathcal{D}}$ for the tangent bundle of X. We could define such a connection $\mathcal{D}$ if

$$f_{xx} = a_0 f + a_1 f_x + a_2 f_y,$$
$$f_{xy} = b_0 f + b_1 f_x + b_2 f_y,$$
$$f_{yy} = c_0 f + c_1 f_x + c_2 f_y,$$

by using $z^1 = x$, $z^2 = y$ and defining the Christoffel symbol

$$\Gamma^l_{jk} \otimes \frac{\partial}{\partial z^l} \otimes dz^j \otimes dz^k$$

for the connection $\mathcal{D}$ by

$$\Gamma^1_{1\,1} = a_1, \quad \Gamma^1_{1\,2} = b_1, \quad \Gamma^1_{1\,1} = c_1,$$
$$\Gamma^2_{1\,1} = a_2, \quad \Gamma^2_{1\,2} = b_2, \quad \Gamma^2_{2\,2} = c_2.$$

For a local holomorphic curve $\varphi : U \to X$ parametrized by an open subset U of $\mathbb{C}$, we form

$$\Phi = (\mathcal{D}_\zeta \varphi^\alpha_\zeta)\varphi^\beta_\zeta \left(\frac{\partial}{\partial z^\alpha} \wedge \frac{\partial}{\partial z^\beta} \right) = \tfrac{1}{2}\left((\mathcal{D}_\zeta \varphi^\alpha_\zeta)\varphi^\beta_\zeta - (\mathcal{D}_\zeta \varphi^\beta_\zeta)\varphi^\alpha_\zeta \right) \left(\frac{\partial}{\partial z^\alpha} \wedge \frac{\partial}{\partial z^\beta} \right).$$

Let $s = s_{\alpha\,\beta} dz^\alpha \wedge dz^\beta$ be a 2-form. Then the evaluation of s at Φ gives

$$\langle s, \Phi \rangle = \tfrac{1}{2}\left((\mathcal{D}_\zeta \varphi^\alpha_\zeta)\varphi^\beta_\zeta - (\mathcal{D}_\zeta \varphi^\beta_\zeta)\varphi^\alpha_\zeta \right) s_{\alpha\,\beta}.$$

From

$$\mathcal{D}_\zeta \varphi^\alpha_\zeta = \varphi^\alpha_{\zeta\zeta} + \Gamma^\alpha_{\lambda\,\mu}\varphi^\lambda_\zeta \varphi^\mu_\zeta$$

it follows that

$$(\mathcal{D}_\zeta \varphi^\alpha_\zeta)\varphi^\beta_\zeta - (\mathcal{D}_\zeta \varphi^\beta_\zeta)\varphi^\alpha_\zeta = \left(\varphi^\alpha_{\zeta\zeta}\varphi^\beta_\zeta - \varphi^\beta_{\zeta\zeta}\varphi^\alpha_\zeta \right) + \varphi^\lambda_\zeta \varphi^\mu_\zeta \left(\Gamma^\alpha_{\lambda\,\mu}\varphi^\beta_\zeta - \Gamma^\beta_{\lambda\,\mu}\varphi^\alpha_\zeta \right).$$

For our special case, when we set $s = dx \wedge dy$ and $z^1 = x, z^2 = y$, we get

$$\begin{aligned}
\langle s, \Phi \rangle = \varphi^* \big\{ &(d^2 x\, dy - dx\, d^2 y) \\
&+ dx^2 (a_1\, dy - a_2\, dx) + 2\, dx\, dy(b_1\, dy - b_2\, dx) + dy^2(c_1\, dy - c_2\, dx) \big\} \\
= \varphi^* \big\{ &(d^2 x\, dy - dx\, d^2 y) \\
&+ (a_1\, dx^2 + 2b_1\, dx\, dy + c_1\, dy^2)\, dy - (a_2\, dx^2 + 2b_2\, dx\, dy + c_2\, dy^2)\, dx \big\}
\end{aligned}$$

Let $t^\delta = f(x, y)$. On X the pullback of the 2-form $dx \wedge dy$ yields a holomorphic 2-jet differential after we divide it by an appropriate power of t, because its vanishing order in t along the branching locus more than offsets its pole order along the infinity line of $\mathbb{P}_2$. An analytic way of seeing it is that

$$dx \wedge dy = \frac{dx \wedge df}{f_y} = \frac{\delta t^{\delta-1}}{f_y}\,(dx \wedge dt),$$

which says that

$$\frac{1}{t^{\delta-1}}\,(dx \wedge dy)$$

is a holomorphic 2-jet differential on X. The key point is that $dx \wedge df$ is divisible by f_y as well as by $t^{\delta-1}$.

Now we come back to $\langle s, \Phi \rangle$. Geometrically we know that

$$\frac{g(x,y)}{t^{\delta-2}} \langle s, \Phi \rangle$$

is a holomorphic 2-jet differential if the pole divisor of meromorphic connection $\tilde{\mathcal{D}}$ is contained in the zero divisor of $g(x,y)$ in the affine part, because the pullback of a holomorphic 2-jet differential to the branched cover has at most a simple pole along the branching locus. We would like to see analytically why we the 2-jet differential

$$\frac{g(x,y)}{t^{\delta-2}} \langle s, \Phi \rangle$$

is holomorphic on X. We do it in a way analogous to the analytic proof of the holomorphicity of

$$\frac{1}{t^{\delta-1}} (dx \wedge dy)$$

by the divisibility of $dx \wedge df$ by f_y as well as by $t^{\delta-1}$. Just as in the case of the analytic proof of the holomorphicity of

$$\frac{1}{t^{\delta-1}} (dx \wedge dy),$$

we first convert $d^j y$ to $d^j f$ for $j = 1, 2$. We use

$$d^2 f \, dx - d^2 x \, df = f_y (d^2 y \, dx - d^2 x \, dy) + \mathrm{II} \, dx,$$

where

$$\mathrm{II} = f_{xx} \, dx^2 + 2 f_{xy} \, dx \, dy + f_{yy} \, dy^2.$$

Write

$$\mathrm{II} = (a_0 \, dx^2 + 2b_0 \, dx \, dy + c_0 \, dy^2) f$$
$$+ (a_1 \, dx^2 + 2b_1 \, dx \, dy + c_1 \, dy^2) f_x + (a_2 \, dx^2 + 2b_2 \, dx \, dy + c_2 \, dy^2) f_y$$

and use

$$f_x \, dx = df - f_y \, dy$$

to get

$d^2 f \, dx - d^2 x \, df$

$$
\begin{aligned}
&= f_y(d^2y\,dx - d^2x\,dy) + (a_0\,dx^2 + 2b_0\,dx\,dy + c_0\,dy^2)f\,dx \\
&\quad + (a_1\,dx^2 + 2b_1\,dx\,dy + c_1\,dy^2)f_x\,dx + (a_2\,dx^2 + 2b_2\,dx\,dy + c_2\,dy^2)f_y\,dx \\
&= f_y(d^2y\,dx - d^2x\,dy) + (a_0\,dx^2 + 2b_0\,dx\,dy + c_0\,dy^2)f\,dx \\
&\quad + (a_1\,dx^2 + 2b_1\,dx\,dy + c_1\,dy^2)(df - f_y\,dy) + (a_2\,dx^2 + 2b_2\,dx\,dy + c_2\,dy^2)f_y\,dx \\
&= f_y\big\{(d^2y\,dx - d^2x\,dy) + (a_2\,dx^2 + 2b_2\,dx\,dy + c_2\,dy^2)\,dx \\
&\qquad - (a_1\,dx^2 + 2b_1\,dx\,dy + c_1\,dy^2)\,dy\big\} \\
&\quad + (a_0\,dx^2 + 2b_0\,dx\,dy + c_0\,dy^2)f\,dx + (a_1\,dx^2 + 2b_1\,dx\,dy + c_1\,dy^2)\,df.
\end{aligned}
$$

Thus,

$$
\begin{aligned}
\langle s, \Phi \rangle &= \varphi^*\big\{(d^2x\,dy - dx\,d^2y) \\
&\qquad + (a_1\,dx^2 + 2b_1\,dx\,dy + c_1\,dy^2)\,dy - (a_2\,dx^2 + 2b_2\,dx\,dy + c_2\,dy^2)\,dx\big\} \\
&= \frac{1}{f_y}\big\{d^2f\,dx - d^2x\,df \\
&\qquad -(a_0\,dx^2 + 2b_0\,dx\,dy + c_0\,dy^2)f\,dx - (a_1\,dx^2 + 2b_1\,dx\,dy + c_1\,dy^2)\,df\big\}
\end{aligned}
$$

and

$$
\begin{aligned}
\frac{g(x,y)}{t^{\delta-2}}\,\langle s, \Phi \rangle &= \frac{g(x,y)}{t^{\delta-2}f_y}\big\{d^2f\,dx - d^2x\,df \\
&\qquad -(a_0\,dx^2 + 2b_0\,dx\,dy + c_0\,dy^2)f\,dx - (a_1\,dx^2 + 2b_1\,dx\,dy + c_1\,dy^2)\,df\big\}
\end{aligned}
$$

is holomorphic, because $f = t^\delta$ implies that

$$
df = \delta t^{\delta-1}dt,
$$
$$
d^2f = \delta(\delta-1)t^{\delta-2}dt^2 + \delta t^{\delta-2}d^2t = \delta t^{\delta-2}((\delta-1)(dt)^2 + t\,d^2t).
$$

2.1.2. Use of Nonlinear Connections. In general, we do not have

$$
\begin{aligned}
f_{xx} &= a_0 f + a_1 f_x + a_2 f_y, \\
f_{xy} &= b_0 f + b_1 f_x + b_2 f_y, \\
f_{yy} &= c_0 f + c_1 f_x + c_2 f_y,
\end{aligned}
$$

with low pole order for a_j, b_i, c_j ($j = 0, 1, 2$). On the other hand, we know that the theorem of Riemann–Roch guarantees the existence of holomorphic 2-jet differentials in general. The theorem of Riemann–Roch is just a more refined form of counting the number of unknowns and the number of equations. The disadvantage of the use of the theorem of Riemann–Roch is that we do not have any explicit form of holomorphic 2-jet differentials to obtain any conclusion about independence. For the general case we need to modify our approach of using connections to get holomorphic 2-jet differentials in an explicit form. The connections constructed above for the special cases are linear connections.

When we differentiate a tangent vector field without a connection, we end up with a field of 2-jets. A connection is a way of converting such a field of 2-jets back to a tangent vector field. For the purpose of constructing a holomorphic 2-jet differential we do not have to confine ourselves to a linear connection. The conversion of a field of 2-jets back to a tangent vector field can involve a conversion function which is not linear. For example, the conversion function can be an algebraic function which is a root of a polynomial equation. Geometrically there is no existing interpretation for a connection which is an algebraic function. If we just carry out in a purely analytic way the analog of the argument for a linear connection, we should consider, in the case of a connection which is an algebraic function, a polynomial of the form

$$\Phi = \sum_{k=0}^{m} \omega_{s+3k} f^{2(m-k)} (d^2 f \, dx - d^2 x \, df)^{m-k}$$

which is divisible by f_y, where

$$\omega_\mu = \sum_{\nu_0+\nu_1+\nu_2=\mu} a_{\nu_0\nu_1\nu_2}(x,y)(df)^{\nu_0}(f \, dx)^{\nu_1}(f \, dy)^{\nu_2}$$

and $a_{\nu_0\nu_1\nu_2}(x,y)$ is a polynomial in x and y of degree $\leq p$. The integers s, p, m are chosen so that the counting of the number of coefficients and the number of equations yields the existence of a function Φ which is not identically zero. The powers of f in the above expressions are used so that

$$\frac{1}{f^{s+3m}}\Phi = \sum_{k=0}^{m}\left(\frac{1}{f^{s+3k}}\omega_{s+3k}\right)\left(\frac{d^2 f}{f}\,dx - d^2 x\,\frac{df}{f}\right)^{m-k}$$

is divisible by f_y, where

$$\frac{1}{f^\mu}\omega_\mu = \sum_{\nu_0+\nu_1+\nu_2=\mu} a_{\nu_0\nu_1\nu_2}(x,y)\left(\frac{df}{f}\right)^{\nu_0}(dx)^{\nu_1}(dy)^{\nu_2}.$$

With

$$\frac{df}{f} = \delta\frac{dt}{t},$$

$$\frac{d^2 f}{f} = \delta\left(\frac{d^2 t}{t} + (\delta-1)\left(\frac{dt}{t}\right)^2\right),$$

$$\frac{d^2 f}{f}\,dx - d^2 x\,\frac{df}{f} = \delta\left(\frac{d^2 t}{t}\,dx - \frac{dt}{t}\,d^2 x\right) + \delta(\delta-1)\left(\frac{dt}{t}\right)^2 dx,$$

it means that, when we set

$$\tilde{\Phi} = \frac{1}{f^{s+3m}}\Phi, \qquad \tilde{\omega}_\mu = \frac{1}{f^\mu}\omega_\mu,$$

we are looking for

$$\tilde{\Phi} = \sum_{k=0}^{m} \tilde{\omega}_{s+3k} \left(\delta \left(\frac{d^2 t}{t} \, dx - \frac{dt}{t} \, d^2 x \right) + \delta(\delta - 1) \left(\frac{dt}{t} \right)^2 dx \right)^{m-k}$$

to be divisible by f_y, where

$$\tilde{\omega}_\mu = \sum_{\nu_0 + \nu_1 + \nu_2 = \mu} a_{\nu_0 \nu_1 \nu_2}(x, y) \left(\delta \frac{dt}{t} \right)^{\nu_0} (dx)^{\nu_1} (dy)^{\nu_2}.$$

Thus we can construct a 2-jet differential which has small degrees with respect to

$$dx, \ dy, \ \frac{dt}{t}, \ \frac{d^2 t}{t} \, dx - \frac{dt}{t} \, d^2 x.$$

2.1.3. Independence from Degree Considerations for Different Polarizations.

By interchanging the rôles of x and y, we can also construct a 2-jet differential which has small degrees with respect to

$$dx, \ dy, \ \frac{dt}{t}, \ \frac{d^2 t}{t} \, dy - \frac{dt}{t} \, d^2 y.$$

The expressions $dx, dy, \frac{dt}{t}$ used in the two sets of polarizations above are not completely independent. They are related by

$$\frac{dt}{t} = \frac{1}{\delta} \left(\frac{f_x}{f} \, dx + \frac{f_y}{f} \, dy \right).$$

The difference between the two sets of polarizations

$$dx, \ dy, \ \frac{dt}{t}, \ \frac{d^2 t}{t} \, dx - \frac{dt}{t} \, d^2 x$$

and

$$dx, \ dy, \ \frac{dt}{t}, \ \frac{d^2 t}{t} \, dy - \frac{dt}{t} \, d^2 y$$

is the last component in each, namely

$$\frac{d^2 t}{t} \, dx - \frac{dt}{t} d^2 x \quad \text{and} \quad \frac{d^2 t}{t} \, dy - \frac{dt}{t} d^2 y.$$

They are related by

$$\frac{1}{f_y} \left(\frac{d^2 t}{t} dx - \frac{dt}{t} d^2 x \right) + \frac{1}{f_x} \left(\frac{d^2 t}{t} dy - \frac{dt}{t} d^2 y \right) = \left(\frac{\mathrm{II}}{\delta f} - (\delta - 1) \left(\frac{dt}{t} \right)^2 \right) \left(\frac{1}{f_y} dx - \frac{1}{f_x} dy \right)$$

which has large degree in x, y. From this, for generic affine coordinates x, y and for generic f of sufficiently high degree we get the following statement which will later be given and proved in detail in Section 2.8.

CLAIM 2.1.4. *There exist two affine coordinate systems so that the irreducible branch of the zero-set of one 2-jet differential containing the entire holomorphic curve constructed from one affine coordinate system is different from the one constructed from the other affine coordinate system.*

This statement is actually obtained by using the set of holomorphic 2-jet differentials ω_γ from the action of $\gamma \in \mathrm{SU}(2, \mathbb{C})$ on the affine coordinates x, y and by using the restriction placed on the coefficients of f by the differential equation on f which f is forced to satisfy when the set of holomorphic 2-jet differentials have a common irreducible branch containing the 2-jets of the entire holomorphic curve. We use more than just the high degree in x, y of the relation of the two different sets of polarizations, but we also use the fact that the polarizations involve differentials so that dependence in our sense implies that f satisfies a differential equation which imposes conditions on the coefficients of f, thereby making f not generic.

Since the 2-jet differential is of homogeneous weight in dx, dy, $d^2x\, dy - dx\, d^2y$, its zero-set is of complex dimension 3. The common zero-set of the two irreducible branches is of complex dimension 2.

Because dx, dy, $\frac{dt}{t}$ have the relation

$$\frac{dt}{t} = \frac{1}{\delta}\left(\frac{f_x}{f}\, dx + \frac{f_y}{f}\, dy \right),$$

when we factor any of the two 2-jet differentials we have to worry about losing the property of having small degree with respect to either

$$\left(dx,\ dy,\ \frac{dt}{t},\ \frac{d^2t}{t}\, dx - \frac{dt}{t}\, d^2x \right)$$

and

$$\left(dx,\ dy,\ \frac{dt}{t},\ \frac{d^2t}{t}\, dy - \frac{dt}{t}\, d^2y \right).$$

For that we need the following irreducibility criterion, which is given as Proposition 2.3.2 below.

Suppose $P(x, y, dx, dy, \frac{df}{f}, Z)$ is irreducible as a polynomial of the 6 variables with degree p in x and y and homogeneous degree m in $dx, dy, \frac{df}{f}, Z$. If $p + m + 1 \leq \delta$, then $P\left(x, y, dx, dy, \frac{df}{f}, Z\right)$ is irreducible as a polynomial in dx, dy, Z over the field $\mathbb{C}(x, y)$.

In the application the weight of Z is 3 while the weight of each of $dx, dy, \frac{df}{f}$ is 1. To handle that, we rewrite $P\left(x, y, dx, dy, \frac{df}{f}, Z\right)$ as $P_1\left(x, y, dx, dy, \frac{df}{f}, \frac{Z}{dx^2}\right)$ so that the weight of $\frac{Z}{dx^2}$ is 1 and can be regarded as a new variable $\tilde{Z} = \frac{Z}{dx^2}$.

2.1.5. Touching Order with 1-Jet Differential of Low Pole Order. When we take the resultant of the two irreducible factors of the two 2-jet differentials, we have to use either $\frac{d^2t}{t}\, dx - \frac{dt}{t}\, d^2x$ or $\frac{d^2t}{t}\, dy - \frac{dt}{t}\, d^2y$ at the same for both factors and we end up with a relation among x, y, dx, dy, $\frac{dt}{t}$ which is of small degree with respect to $\left(dx, dy, \frac{dt}{t} \right)$. We use $\delta \frac{dt}{t} = \frac{df}{f}$ to write the relation as a polynomial in x, y, dx, dy which is homogeneous in dx, dy. The pullback of this relation to the entire holomorphic curve is identically zero. So the pullback of one of its factors to the entire holomorphic curve is identically zero. Let $h = h(x, y, dx, dy)$ be that factor. Let q be its degree in x, y and m be its degree as a homogeneous polynomial in dx, dy.

Let $\tilde{h}$ be the pullback of h to the δ-sheeted branched cover X over $\mathbb{P}_2$. Let $V_{\tilde{h}}$ be the zero-set of $\tilde{h}$ as a function on the projectivization of $\mathbb{P}(T_X)$ of the tangent bundle T_X of X. Let L_X be the line bundle over $\mathbb{P}(T_X)$ so that the global sections of rL_X correspond to 1-jet differentials over X of degree r.

We use, for sufficiently large r, the existence of a nontrivial global holomorphic section of $r(F - G)$ over Y if F, G are two ample line bundles over a compact complex variety Y of complex dimension n with $F^n > nF^{n-1}G$. We apply it to the case $rL_X = F - G$ with $F = (r+1)(L_X + 3H_{\mathbb{P}_2})$ and $G = L_X + 3(r+1)H_{\mathbb{P}_2}$ over a branch of $V_{\tilde{h}}$. We use the branch of $V_{\tilde{h}}$ which contains a lifting of the entire holomorphic curve. The cyclic group of order δ which is the Galois group of $X \to \mathbb{P}_2$ acts on the set of all branches of $V_{\tilde{h}}$. When $q > 4m$, by using the Galois group of $X \to \mathbb{P}_2$, for sufficiently large δ we obtain a nontrivial global section s of rL_X over that branch of $V_{\tilde{h}}$ for r sufficiently large. The zero-set of s projects down to an algebraic curve in $\mathbb{P}_2$ which contains the entire holomorphic curve. So the case that remains is $q \leq 4m$.

The number m can be chosen to be independent of δ. There are integers N, δ_0 depending only on q, m such that a generic curve of degree $\delta \geq \delta_0$ cannot be tangential, to order N at any point, to any irreducible 1-jet differential $\theta(x, y, dx, dy)$ of degree q in x, y and of homogeneous degree m in dx, dy. We can choose δ sufficiently large relative to N. We choose a polynomial $S(x, y)$ with degree small relative to δ so that S vanishes to order N at all the points on the zero-set of $f(x, y)$ where the discriminant of $h(x, y, dx, dy)$ as a homogeneous polynomial of dx, dy vanishes. Let η be any meromorphic 1-jet differential of low pole order (for example, a suitable linear combination of dx and dy) whose pullback to the entire holomorphic curve is not identically zero. We then prove an inequality of Schwarz lemma type:

$$\frac{\sqrt{-1}}{2\pi} \partial\bar{\partial} \log \left(\frac{\left\| f^{\frac{N-1}{N}} S(x, y)\eta \right\|^2}{\|f\|^2 \left(\log \|f\|^2 \right)^2} \right) \geq \varepsilon \frac{\left\| f^{\frac{N-1}{N}} S(x, y)\eta \right\|^2}{\|f\|^2 \left(\log \|f\|^2 \right)^2}$$

for some positive number ε when pulled back to $\mathbb{C}$ by the entire holomorphic curve. The Schwarz lemma type inequality implies the nonexistence of the entire holomorphic curve. This concludes the overview of our proof. Now we give the details.

2.2. Construction of Holomorphic 2-Jet Differentials. Let p be a positive integer and s be a nonnegative integer. We are going to construct a 2-jet differential Φ of degree m on X of the form

$$\Phi = \sum_{k=0}^{m} \omega_{s+3k} f^{2(m-k)} (d^2 f\, dx - d^2 x\, df)^{m-k},$$

where

$$\omega_\mu = \sum_{\nu_0+\nu_1+\nu_2=\mu} a_{\nu_0\nu_1\nu_2}(x,y)(df)^{\nu_0}(f\,dx)^{\nu_1}(f\,dy)^{\nu_2}$$

and $a_{\nu_0\nu_1\nu_2}(x,y)$ is a polynomial in x and y of degree $\leq p$. We are going to choose the polynomials $a_{\nu_0\nu_1\nu_2}(x,y)$ so that Φ is divisible by f_y. Then we will conclude that $t^{-N}f_y^{-1}\Phi$ is a holomorphic 2-jet differential on X when certain inequalities involving p, s, δ, m, and N are satisfied. This is done by regarding the coefficients of the polynomials $a_{\nu_0\nu_1\nu_2}(x,y)$ as unknowns and counting the number of linear equations corresponding to divisibility of Φ by f_y and solving the linear equations when the number of unknowns exceeds the number of equations. In order to guarantee that the 2-jet differential Φ obtained by solving the linear equations is not identically zero, we need the following lemma involving the independence of the coefficients of the polynomials $a_{\nu_0\nu_1\nu_2}(x,y)$.

LEMMA 2.2.1. *Let q be a positive integer $< \delta$. Let l be any positive integer. For $\nu_0 + \nu_1 + \nu_2 = l$ let $b_{\nu_0\nu_1\nu_2}(x,y)$ be a polynomial in x and y of degree at most q. If $\sum_{\nu_0+\nu_1+\nu_2=l} b_{\nu_0\nu_1\nu_2}(df)^{\nu_0}(f\,dx)^{\nu_1}(f\,dy)^{\nu_2}$ is identically zero, then $b_{\nu_0\nu_1\nu_2}(x,y)$ is identically zero for $\nu_0 + \nu_1 + \nu_2 = l$.*

PROOF. Regard (x,y) as the affine coordinate for $\mathbb{P}_2$ and introduce the homogeneous coordinates $[\xi, \eta, \zeta]$ for anther $\mathbb{P}_2$. On the product $\mathbb{P}_2 \times \mathbb{P}_2$ consider the hypersurface M of bidegree $(\delta, 1)$ defined by

$$f(x,y)\zeta = f_x(x,y)\xi + f_y(x,y)\eta.$$

Let

$$s \in \Gamma\left(\mathbb{P}_2 \times \mathbb{P}_2, \mathcal{O}_{\mathbb{P}_2 \times \mathbb{P}_2}(q, l)\right)$$

be defined by

$$\sum_{\nu_0+\nu_1+\nu_2=l} b_{\nu_0\nu_1\nu_2}(x,y)\zeta^{\nu_0}\xi^{\nu_1}\eta^{\nu_2}.$$

The assumption of the Lemma means that the restriction of s to M is identically zero. Since $q < \delta$, from the exact sequence

$$0 = H^0\left(\mathbb{P}_2 \times \mathbb{P}_2, \mathcal{O}_{\mathbb{P}_2 \times \mathbb{P}_2}(q - \delta, l - 1)\right) \to$$
$$H^0\left(\mathbb{P}_2 \times \mathbb{P}_2, \mathcal{O}_{\mathbb{P}_2 \times \mathbb{P}_2}(q, l)\right) \to H^0\left(M, \mathcal{O}_{\mathbb{P}_2 \times \mathbb{P}_2}(q, l)|M\right)$$

it follows that s is identically zero. $\qquad\square$

2.2.2. Computation of the numbers of equations and unknowns. On $\mathbb{P}_2 \times \mathbb{P}_2 \times \mathbb{P}_1$ we use the affine coordinate (x,y) for the first factor and use the affine coordinate (dx, dy) for the second factor and then use the affine coordinate $d^2x\,dy - d^2y\,dx$ for the third factor. Then consider

$$\Phi = \sum_{k=0}^{m} \omega_{f,s+3k} f^{2(m-k)}(d^2f\,dx - d^2x\,df)^{m-k},$$

as a holomorphic section of $\mathcal{O}_{\mathbb{P}_2 \times \mathbb{P}_2 \times \mathbb{P}_1}(a, b, c)$ over $\mathbb{P}_2 \times \mathbb{P}_2 \times \mathbb{P}_1$ for suitable integers a, b, c and then restrict to the hypersurface defined by $f_y(x, y) = 0$. We do the counting of the dimensions of the section modules to show that there exists Φ not identically zero whose restriction to $\{f_y(x, y) = 0\}$ is identically zero. Here $\mathbb{P}_2 \times \mathbb{P}_2 \times \mathbb{P}_1$ is regarded as birationally equivalent to the space of special 2-jet differentials over $\mathbb{P}_2$.

We now compute the number of equations involved in setting

$$\Phi = \sum_{k=0}^{m} \omega_{s+3k} f^{2(m-k)} (f_{xx}\, dx^2 + 2f_{xy}\, dx\, dy + f_{yy}\, dy^2)^{m-k} dx^{m-k}$$

equal to zero modulo f_y. Using

$$d^2 f\, dx - d^2 x\, df = (f_{xx}\, dx^2 + 2f_{xy}\, dx\, dy + f_{yy}\, dy^2)\, dx - f_y(d^2 x\, dy - d^2 y\, dx)$$

and expanding Φ, we end up with an expression of the form

$$\sum_{j=0}^{s+3m} b_j(x, y)\, dx^j\, dy^{s+3m-j}$$

modulo f_y, where $b_j = b_j(x, y)$ is a polynomial in x and y of degree at most $p + (s + 3m)\delta$. The number of coefficients in each b_j is at most

$$\tfrac{1}{2}(p + (s + 3m)\delta + 2)(p + (s + 3m)\delta + 1).$$

For each $b_j(x, y)$ we have to rule out expressions of the form $q_j(x, y) f_y(x, y)$ with the degree of $q_j(x, y)$ in x and y no more than $p + (s + 3m)\delta - (\delta - 1)$. So the number of possible constraints for each b_j is at most

$$\tfrac{1}{2}(p + (s+3m)\delta + 2)(p + (s+3m)\delta + 1)$$
$$-\tfrac{1}{2}(p + (s+3m)\delta + 2 - (\delta-1))(p + (s+3m)\delta + 1 - (\delta-1)),$$

which is to say

$$(\delta - 1)(p + (s+3m)\delta) - \tfrac{1}{2}(\delta^2 - 5\delta + 4).$$

There are altogether $s + 3m + 1$ such functions $b_j(x, y)$. Thus the total number of equations is at most

$$(s + 3m + 1)((\delta - 1)(p + (s + 3m)\delta) - \tfrac{1}{2}(\delta^2 - 5\delta + 4)).$$

Now we would like to compute the number of unknowns. The number of unknowns is the sum of the number of unknowns from each ω_μ. For

$$\omega_\mu = \sum_{\nu_0 + \nu_1 + \nu_2 = \mu} a_{\nu_0 \nu_1 \nu_2} (df)^{\nu_0} (f\, dx)^{\nu_1} (f\, dy)^{\nu_2},$$

the number of unknowns from ω_μ is equal to the sum of the number of coefficients in each of the polynomials $a_{\nu_0 \nu_1 \nu_2}$ with $\nu_0 + \nu_1 + \nu_2 = \mu$. There are $\tfrac{1}{2}(\mu+2)(\mu+1)$

such $a_{\nu_0\nu_1\nu}$ and each $a_{\nu_0\nu_1\nu}$ has $\frac{1}{2}(p+2)(p+1)$ coefficients. Hence the number of unknowns in ω_μ is $\frac{1}{4}(\mu+2)(\mu+1)(p+2)(p+1)$. The total number of unknowns is

$$\sum_{k=0}^{m} \tfrac{1}{4}(s+3k+2)(s+3k+1)(p+2)(p+1).$$

When the number of unknowns exceeds the number of equations, for a generic f we can solve the linear equations and the solutions will be rational functions of the coefficients of f. We summarize the result in the following lemma.

LEMMA 2.2.3. *To be able to construct a 2-jet differential Φ which is divisible by f_y and which is of the form*

$$\sum_{k=0}^{m} \omega_{s+3k} f^{2(m-k)}(d^2 f\, dx - d^2 x\, df)^{m-k},$$

where

$$\omega_\mu = \sum_{\nu_0+\nu_1+\nu_2=\mu} a_{\nu_0\nu_1\nu_2}(x,y)(df)^{\nu_0}(f\,dx)^{\nu_1}(f\,dy)^{\nu_2}$$

and $a_{\nu_0\nu_1\nu_2}(x,y)$ is a polynomial in x and y of degree $\leq p$, it suffices to have the following inequalities $p < \delta - 1$ and

$$\sum_{k=0}^{m} \frac{1}{4}(s+3k+2)(s+3k+1)(p+2)(p+1)$$

$$> (s+3m+1)((\delta-1)(p+(s+3m)\delta) - \tfrac{1}{2}(\delta^2 - 5\delta + 4)).$$

Moreover, for a generic f the coefficients of $a_{\nu_0\nu_1\nu_2}(x,y)$ are rational functions of the coefficients of f.

The reason for the last statement of Lemma 2.2.3 is as follows. When we solve the system of homogeneous linear equations for the coefficients of $a_{\nu_0\nu_1\nu_2}(x,y)$, we choose a square submatrix A with nonzero determinant in the matrix of the coefficients of the system of homogeneous linear equations so that A has maximum size among all square submatrices with nonzero determinants and then we apply Cramer's rule to those equations whose coefficients are involved in A to solve for the the coefficients of $a_{\nu_0\nu_1\nu_2}(x,y)$. When we do this process, we can regard the coefficients of the system of homogeneous linear equations as functions of the coefficients of f. The square submatrix A has maximum size among all square submatrices whose determinants are not identically zero as functions of the coefficients of f. A sufficient condition for the genericity of f involved in this process is that the point represented by the coefficients of f is outside the zero set of A when A is regarded as a function of the coefficients of f.

2.2.4. Condition for the holomorphicity of the 2-jet differential. We would like to determine under what condition the constructed 2-jet differential $t^{-N} f_y^{-1} \Phi$ is holomorphic on X and vanishes on some ample curve of X.

First we consider the pole order at infinity of various factors. Recall that $[\zeta^0, \zeta^1, \zeta^2]$ is the homogeneous coordinates of $\mathbb{P}_2$ with $x = \zeta^1/\zeta^0$ and $y = \zeta^2/\zeta^0$. At a point at the infinity line we assume without loss of generality that $\zeta^1 \neq 0$. At that point of the infinity line we use the affine coordinates $u = \zeta^0/\zeta^1 = 1/x$ and $v = \zeta^2/\zeta^1 = y/x$. Thus $x = 1/u$ and $y = xv = v/u$. We have

$$dx = -\frac{du}{u^2}, \quad dy = \frac{dv}{u} - \frac{v\,du}{u^2}, \quad d^2x\,dy - d^2y\,dx = -\frac{1}{u^3}(d^2u\,dv - d^2v\,du).$$

Thus we conclude that the pole order of $d^2x\,dy - d^2y\,dx$ at infinity is 3. From

$$d^2x df - d^2 f\,dx = -f_{xx}\left(-\frac{du}{u^2}\right)^3 - 2f_{xy}\left(-\frac{du}{u^2}\right)^2\left(\frac{dv}{u} - \frac{v\,du}{u^2}\right)$$
$$-f_{yy}\left(-\frac{du}{u^2}\right)\left(\frac{dv}{u} - \frac{v\,du}{u^2}\right)^2 - \frac{f_y}{u^3}(d^2u\,dv - d^2v\,du)$$

we conclude that the pole order of $d^2x df - d^2 f dx$ at infinity is $\delta + 4$. From

$$df = f_x\left(-\frac{du}{u^2}\right) + f_y\left(\frac{dv}{u} - \frac{v\,du}{u^2}\right)$$

we have the pole order $\delta + 1$ for df at infinity.

Since $f = t^\delta$ and $df = \delta t^{\delta-1} dt$ and $d^2 f = \delta t^{\delta-1} d^2 t + \delta(\delta-1)t^{\delta-2} dt^2$, it follows that from ω_μ we can factor out $t^{\mu(\delta-1)}$. From $d^2 f dx - d^2 x df$ we can factor out $t^{\delta-2}$. Hence from the term $\omega_{s+3k} f^{2(m-k)}(d^2 f dx - d^2 x df)^{m-k}$ we can factor out t to the power $(s + 3k)(\delta - 1) + 2(m - k)\delta + (m - k)(\delta - 2)$ which is the same as $(s + 3m)\delta - (s + 2m + k)$ for $0 \le k \le m$. We can only factor out the minimum power of t, namely $(s + 3m)(\delta - 1)$. When we can divide by f_y, we factor out a pole order of $\delta - 1$ which corresponds to the power $\delta - 1$ of t. On the other hand, the pole order at infinity for ω_μ is $p + \mu(\delta + 2)$ and as a result the pole order of the term $\omega_{s+3k} f^{2(m-k)}(d^2 f dx - d^2 x df)^{m-k}$ of Φ at infinity is

$$p + (s + 3k)(\delta + 2) + 2(m - k)\delta + (m - k)(\delta + 4) = p + (s + 3m)\delta + 2s + 4m + 2k$$

for $0 \le k \le m$. We have to take in this case the maximum of the expression for $0 \le k \le m$ and we get $p + (s + 3m)(\delta + 2)$. Take a positive integer q. To end up with a holomorphic jet differential $t^{-(s+3m)(\delta-1)} f_y^{-1} \Phi$ on X with at least q zero order at infinity, we can impose the condition

$$\delta - 1 + (s + 3m)(\delta - 1) \ge q + p + (s + 3m)(\delta + 2)$$

which is the same as $p \le \delta - q - 1 - 3s - 9m$.

2.3. Two Kinds of Irreducibility. In number theory it was first pointed out by Vojta that the finiteness of rational points for a subvariety of abelian varieties not containing the translate of an abelian variety is the consequence of the fact that in the product space of many copies of the subvariety there are more line bundles or divisors than constructed from the factors which are copies of the subvariety [Faltings 1991; Vojta 1992]. In hyperbolicity problems the analog of taking the product of copies of a manifold is to use the space of jets. The analog of the existence of more divisors or line bundles is the existence of more ways of factorization for meromorphic jet differentials. Some factors from the additional ways of factorization become holomorphic jet differentials. In our construction we pullback

$$\Phi = \sum_{k=0}^{m} \omega_{s+3k} f^{2(m-k)} (d^2 f dx - d^2 x df)^{m-k},$$

to the space of 2-jets of the branched cover and obtain a new factor t^N so that one of the other factors becomes a holomorphic 2-jet differntial on the branched cover.

On the other hand, the many more different ways of factorization makes it more difficult to control the factors to get the independence of holomorphic jet differentials. Two meromorphic jet differentials on the complex projective plane constructed in different ways may share a common factor when pulled back to the branched cover, because there are more ways of factorization in the space of jets of the branched cover. We have to strike a balance between having many ways of factorization to get holomorphic jet differentials and having not too many ways of factorization to get the independence of holomorphic jet differentials. The way we handle it is to construct an appropriate intermediate manifold between the space of jets of the complex projective space and the space of jets of the branched cover. On this intermediate manifold we introduce a certain class of meromorphic functions with the following property. Every memomorphic function in that class can be pulled back to the space of jets of the branched cover to give a factor which is a holomorphic jet differential. On the other hand, for that particular class of meromorphic functions the number of ways of factorization is not too numerous that we could construct two meromorphic functions in that class having no common factors before being pulled back to the jet space of the branched cover.

PROPOSITION 2.3.1. *Let $g_j(z_0, z_1, z_2)$ $(0 \leq j \leq 2)$ be homogeneous polynomials of degree δ whose common zero-set consists only of the single point*

$$(z_0, z_1, z_2) = 0.$$

Let $P(x, y, w_0, w_1, w_2, Y)$ be a polynomial of the 6 variables x, y, w_0, w_1, w_2, Y with degree p in x, y and homogeneous degree m in w_1, w_2, w_3 and of degree q in Y. Let Q be obtained from P by replacing w_0 by a function in w_1, w_2, x, y

satisfying $\sum_{j=0}^{2} g_j(1, x, y) w_j = 0$, *in other words,*

$$Q(x, y, w_1, w_2, Y) = P\left(x, y, -\frac{g_1(1, x, y)}{g_0(1, x, y)} w_1 - \frac{g_2(1, x, y)}{g_0(1, x, y)} w_2, w_1, w_2, Y\right).$$

Suppose $P(x, y, w_0, w_1, w_2, Y)$ is irreducible as a polynomial of the 6 variables x, y, w_0, w_1, w_2, Y. If $p < \delta$, then $Q(x, y, w_1, w_2, Y)$ is irreducible as a polynomial of the 3 variables w_1, w_2, Y over the field $\mathbb{C}(x, y)$.

PROOF. Introduce the homogeneous variables z_0, z_1, z_0 of $\mathbb{P}_2$ so that $x = \frac{z_1}{z_0}$ and $y = \frac{z_2}{z_0}$. Introduce the homogeneous variables Z_0, Z_1 of $\mathbb{P}_1$ so that $Y = \frac{Z_1}{Z_0}$. We use the coordinates $\left([z_0, z_1, z_2], [w_0, w_1, w_2], [Z_0, Z_1]\right)$ for the product $\mathbb{P}_2 \times \mathbb{P}_2 \times \mathbb{P}_1$. Let M be the subvariety in $\mathbb{P}_2 \times \mathbb{P}_2 \times \mathbb{P}_1$ defined by

$$\sum_{j=0}^{2} g_j(z_0, z_1, z_2) w_j = 0.$$

Since $g_j(z_0, z_1, z_2)$ $(0 \leq j \leq 2)$ have no common zeroes except the single point $(z_0, z_1, z_2) = 0$, it follows that M is a submanifold of $\mathbb{P}_2 \times \mathbb{P}_2 \times \mathbb{P}_1$. Let $\tilde{\pi}_j$ be the projection of $\mathbb{P}_2 \times \mathbb{P}_2 \times \mathbb{P}_1$ onto its j-th factor $(1 \leq j \leq 3)$. Let π_j be the restriction of $\tilde{\pi}$ to M. Let

$$\tilde{\pi} : \mathbb{P}_2 \times \mathbb{P}_2 \times \mathbb{P}_1 \to \mathbb{P}_2 \times \mathbb{P}_1$$

be the projection

$$\left([z_0, z_1, z_2], [w_0, w_1, w_2], [Z_0, Z_1]\right) \mapsto \left([z_0, z_1, z_2], [Z_0, Z_1]\right);$$

in other words, $\tilde{\pi} = \tilde{\pi}_1 \times \tilde{\pi}_3$. Let $\pi : M \to \mathbb{P}_2 \times \mathbb{P}_1$ be the restriction of $\tilde{\pi}$ to M. Then $\pi : M \to \mathbb{P}_2 \times \mathbb{P}_1$ is a $\mathbb{P}_1$-bundle over $\mathbb{P}_2 \times \mathbb{P}_1$ whose fiber over the point $\left([z_0, z_1, z_2], [Z_0, Z_1]\right)$ is the complex line

$$\sum_{j=0}^{2} g_j(z_0, z_1, z_2) w_j = 0$$

in the projective plane $\mathbb{P}_2$ with homogeneous coordinates $[w_0, w_1, w_2]$.

Clearly the inclusion map $M \subset \mathbb{P}_2 \times \mathbb{P}_2 \times \mathbb{P}_1$ induces the isomorphisms

$$R\tilde{\pi}_*^j \mathbb{Z} \xrightarrow{\approx} R\pi_*^j \mathbb{Z} \qquad (0 \leq j \leq 2),$$

$$R\tilde{\pi}_*^j \mathcal{O}_{\mathbb{P}_2 \times \mathbb{P}_2 \times \mathbb{P}_1} \xrightarrow{\approx} R\pi_*^j \mathcal{O}_M \qquad (0 \leq j \leq 2).$$

From these isomorphisms and the standard spectral sequence arguments the following isomorphisms follow.

$$H^j \left(\mathbb{P}_2 \times \mathbb{P}_2 \times \mathbb{P}_1, \mathbb{Z}\right) \xrightarrow{\approx} H^j \left(M, \mathbb{Z}\right) \qquad (0 \leq j \leq 2),$$

$$H^j \left(\mathbb{P}_2 \times \mathbb{P}_2 \times \mathbb{P}_1, \mathcal{O}_{\mathbb{P}_2 \times \mathbb{P}_2 \times \mathbb{P}_1}\right) \xrightarrow{\approx} H^j \left(M, \mathcal{O}_M\right) \qquad (0 \leq j \leq 2).$$

In particular, we have the isomorphisms between the group of holomorphic line bundles over $\mathbb{P}_2 \times \mathbb{P}_2 \times \mathbb{P}_1$ and the group of holomorphic line bundles over M, namely,

$$(2.3.1.1) \qquad H^1\big(\mathbb{P}_2 \times \mathbb{P}_2 \times \mathbb{P}_1, \mathcal{O}^*_{\mathbb{P}_2 \times \mathbb{P}_2 \times \mathbb{P}_1}\big) \xrightarrow{\approx} H^1\left(M, \mathcal{O}^*_M\right).$$

Then a holomorphic line bundle over M is of the form

$$\mathcal{O}_M\left(k_1, k_2, k_3\right) := (\pi_1)^*\left(\mathcal{O}_{\mathbb{P}_2}(k_1)\right) \otimes (\pi_2)^*\left(\mathcal{O}_{\mathbb{P}_2}(k_2)\right) \otimes (\pi_3)^*\left(\mathcal{O}_{\mathbb{P}_1}(k_3)\right).$$

By Künneth's formula we have

$$(2.3.1.2) \qquad H^1\left(\mathbb{P}_2 \times \mathbb{P}_2 \times \mathbb{P}_1, \mathcal{O}_{\mathbb{P}_2 \times \mathbb{P}_2 \times \mathbb{P}_1}(k_1, k_2, k_3)\right) = 0 \quad \text{for } k_3 \geq 1,$$

because $H^1(\mathbb{P}_2, \mathcal{O}_{\mathbb{P}_2}(k)) = 0$ for every integer k and $H^1(\mathbb{P}_1, \mathcal{O}_{\mathbb{P}_1}(k)) = 0$ for every integer $k \geq -1$. Let

$$\psi \in \Gamma(\mathbb{P}_2 \times \mathbb{P}_2 \times \mathbb{P}_1, \mathcal{O}_{\mathbb{P}_2 \times \mathbb{P}_2 \times \mathbb{P}_1}(\delta, 1, 0))$$

be defined by multiplication by $\sum_{j=0}^{2} g_j(z_0, z_1, z_2) w_j = 0$, From (2.3.1.2) and the short exact sequence

$$0 \to \mathcal{O}_{\mathbb{P}_2 \times \mathbb{P}_2 \times \mathbb{P}_1}(k_1 - \delta, k_2 - 1, k_3) \xrightarrow{\theta} \mathcal{O}_{\mathbb{P}_2 \times \mathbb{P}_2 \times \mathbb{P}_1}(k_1, k_2, k_3)$$
$$\to \mathcal{O}_M(k_1, k_2, k_3) \to 0$$

with θ defined by multiplication by ψ it follows that

$$\Theta_{k_1, k_2, k_3} : \Gamma(\mathbb{P}_2 \times \mathbb{P}_2 \times \mathbb{P}_1, \mathcal{O}_{\mathbb{P}_2 \times \mathbb{P}_2 \times \mathbb{P}_1}(k_1, k_2, k_3)) \to H^0(M, \mathcal{O}_M(k_1, k_2, k_3))$$

is surjective for $k_3 \geq -1$ and that Θ_{k_1, k_2, k_3} is injective for $k_1 < \delta$.

Let s be the meromorphic function on $\mathbb{P}_2 \times \mathbb{P}_2 \times \mathbb{P}_1$ be defined by P/w_0^m. Let $\tilde{H}_1$ (respectively $\tilde{H}_2$, $\tilde{H}_3$) be the hypersurface in $\mathbb{P}_2 \times \mathbb{P}_2 \times \mathbb{P}_1$ defined by $z_0 = 0$ (respectively $w_0 = 0$, $Z_0 = 0$). Let $H_l = M \cap \tilde{H}_l$ for $1 \leq l \leq 3$. The pole divisor of s is $pH_1 + mH_2 + qH_3$.

Suppose $Q(x, y, w_1, w_2, Y)$ is not irreducible as a polynomial of the 3 variables w_1, w_2, Y over the field $\mathbb{C}(x, y)$. Then we can write $Q(x, y, w_1, w_2, Y)$ as a product of two factors $Q_j(x, y, w_1, w_2, Y)$ $(j = 1, 2)$ each of which is a polynomial of positive degree in the 3 variables w_1, w_2, Y over the field $\mathbb{C}(x, y)$. Thus the restriction $s|M$ of s to M can be written as the product of two meromorphic functions $s_1 s_2$ on M with s_j defined by $Q_j(x, y, w_1, w_2, Y)$ $(j = 1, 2)$. Let

$$W_j - V_j - \sum_{l=1}^{3} r'_{j,l} H_l$$

be the divisor of s_j $(j = 1, 2)$, where W_j, V_j are effective divisors with support not contained in $\bigcup_{l=1}^{3} H_l$. We know that $\pi_1(V_j)$ is a proper subvariety of $\mathbb{P}_2$ for $j = 1, 2$, because of the factorization of Q into the product of Q_1 and Q_2 over the field $C(x, y)$. We also know that for $j = 1, 2$ both $r'_{j,2}, r'_{j,3}$ are nonnegative and one of them is positive. The key point is that by (2.3.1.1) there exists a meromorphic function σ_j on M such that the divisor of σ_j is equal to

$$V_j - \sum_{l=1}^{3} r''_{j,l} H_l$$

for some integers $r''_{j,l}$.

The integers $r''_{j,l}$ $(1 \le l \le 3)$ are all nonnegative, because of the following fact.

CLAIM 2.3.1.3. *If u is a non-identically-zero meromorphic function on M whose divisor is $E - \sum_{l=1}^{3} \kappa_l H_l$, where E is an effective divisor of M, then the integers κ_l $(1 \le l \le 3)$ are all nonnegative.*

PROOF. Suppose the contrary. Let $b = \max(-1, \kappa_3)$. Then one of κ_1, κ_2, b is negative. Let

$$\tau \in \Gamma\left(M, \mathcal{O}_M(r_{j,1}, r_{j,2}, b)\right)$$

be defined by $u(z_0)^{\kappa_1}(w_0)^{\kappa_2}(Z_0)^b$. Since $b \ge -1$, it follows from the surjectivity of $\Theta_{\kappa_1, \kappa_2, b}$ that τ can be lifted to an element

$$\tilde{\tau} \in \Gamma\left(\mathbb{P}_2 \times \mathbb{P}_2 \times \mathbb{P}_1, \mathcal{O}_{\mathbb{P}_2 \times \mathbb{P}_2 \times \mathbb{P}_1}(\kappa_1, \kappa_2, b)\right).$$

Since one of κ_1, κ_2, b is negative, it follows that $\tilde{\tau}$ is identically zero, which is a contradiction and concludes the proof of Claim 2.3.1.3.   □

Since $s|M = s_1 s_2$ on M, it follows that the support of the divisor of the meromorphic function $(s|M)(\sigma_1 s_1 \sigma_2 s_2)^{-1}$ on M is contained in $\bigcup_{l=1}^{3} H_l$. By (2.3.1.1) we know that the meromorphic function $(s|M)(\sigma_1 s_1 \sigma_2 s_2)^{-1}$ on M must be a constant.

The divisor of $s_j \sigma_j$ is equal to

$$W_j - \sum_{l=1}^{3}(r'_{j,l} + r''_{j,l}) H_l.$$

At least one of the two integers $r'_{j,2} + r''_{j,2}$ and $r'_{j,3} + r''_{j,3}$ is positive. Both are nonnegative. By Claim (2.3.1.3) the integer $r'_{j,1} + r''_{j,1}$ is nonnegative for $j = 1, 2$. From $s|M = c(s_1 \sigma_1)(s_2 \sigma_2)$ on M for some nonzero constant c it follows that for $j = 1, 2$ we have

$$0 \le r'_{j,1} + r''_{j,1} \le p,$$
$$0 \le r'_{j,2} + r''_{j,2} \le m,$$
$$0 \le r'_{j,3} + r''_{j,3} \le q$$

and one of $r'_{j,2} + r''_{j,2}, r'_{j,3} + r''_{j,3}$ is positive. From $p < \delta$ it follows that

$$\Theta_{r_{j,1}+r'_{j,1}, r_{j,2}+r'_{j,2}, r_{j,3}+r'_{j,3}}$$

is an isomorphism and $s_j \sigma_j$ is induced by a polynomial $R_j(x, y, w_0, w_1, w_2, Y)$ of degree $r'_{j,1} + r''_{j,1} \le p$ in x, y and of degree $r'_{j,2} + r''_{j,2} \le m$ in w_0, w_1, w_2 and of degree $r'_{j,3} + r''_{j,3} \le q$ in Z. From $P = cR_1 R_2$ and one of $r'_{j,2} + r''_{j,2}, r'_{j,3} + r''_{j,3}$

being positive for $j = 1, 2$, we have a contradiction to the irreducibility of P in the six variables x, y, w_0, w_1, w_2, Z. $\square$

PROPOSITION 2.3.2. *Suppose $P\left(x, y, dx, dy, \frac{df}{f}, Z\right)$ is irreducible as a polynomial of the 6 variables with degree p in x and y and homogeneous weight m in dx, dy, $\frac{df}{f}$, Z when each of dx, dy, $\frac{df}{f}$ has weight 1 and Z has weight 3. If $p + m < \delta$, then $P\left(x, y, dx, dy, \frac{df}{f}, Z\right)$ is irreducible as a polynomial in dx, dy, Z over the field $\mathbb{C}(x, y)$ for generic f.*

PROOF. We rewrite $P\left(x, y, dx, dy, \frac{df}{f}\right)$ as

$$P_1\left(x, y, \frac{dx}{x}, \frac{dy}{y}, \frac{df}{f}\right)$$

and introduce the symbols

$$w_0 = \frac{df}{f}, \quad w_1 = \frac{dx}{x}, \quad w_2 = \frac{dy}{y}.$$

The degree p' of $P_1\left(x, y, \frac{dx}{x}, \frac{dy}{y}, \frac{df}{f}\right)$ in x, y can be as high as $p + m$ when $P_1\left(x, y, \frac{dx}{x}, \frac{dy}{y}, \frac{df}{f}\right)$ is regarded as a polynomial of the 5 variables x, y, w_0, w_1, w_2. Let

$$g_0(z_0, z_1, z_2) = -z_0^\delta f\left(\frac{z_1}{z_0}, \frac{z_2}{z_0}\right),$$

$$g_1(z_0, z_1, z_2) = z_0^{\delta-1} z_1 f_x\left(\frac{z_1}{z_0}, \frac{z_2}{z_0}\right),$$

$$g_2(z_0, z_1, z_2) = z_0^{\delta-1} z_2 f_y\left(\frac{z_1}{z_0}, \frac{z_2}{z_0}\right),$$

so that

$$\frac{g_1(z_0, z_1, z_2)}{g_0(z_0, z_1, z_2)} = -\frac{x f_x(x, y)}{f(x, y)}, \qquad \frac{g_2(z_0, z_1, z_2)}{g_0(z_0, z_1, z_2)} = -\frac{y f_y(x, y)}{f(x, y)},$$

with $x = z_1/z_0$ and $y = z_2/z_0$. For a generic f the three polynomials g_0, g_1, g_2 have no common zeroes other than the point $(z_0, z_1, z_2) = 0$, because it is the case for the special $f(x, y) = 1 + x^\delta + y^\delta$, where

$$g_0(z_0, z_1, z_0) = -(z_0^\delta + z_1^\delta + z_2^\delta),$$

$$g_1(z_0, z_1, z_2) = \delta z_1^\delta,$$

$$g_2(z_0, z_1, z_2) = \delta z_2^\delta.$$

The result now follows from Proposition 2.3.1. $\square$

2.4. Degree of Second Order Differential Greater Than One. We factor

$$\Phi = \sum_{k=0}^{m} \omega_{f,s+3k} f^{2(m-k)} (d^2 f\, dx - d^2 x\, df)^{m-k}$$

into irreducible factors

$$\Phi = \Phi_1 \Phi_2 \cdots \Phi_k$$

as polynomials in the independent variables

$$\frac{dx}{x}, \quad \frac{dy}{y}, \quad \frac{df}{f}, \quad \frac{d^2 f\, dx - d^2 x\, df}{f}$$

with coefficients in the field $\mathbb{C}(x, y)$ and then clear the denominators. The polynomial Φ satisfies the following three properties:

(1) Φ has homogeneous total weight $\leq s+3m$ when dx, dy, df are assigned weight 1 and $d^2 f\, dx - df\, d^2 x$ is assigned weight 3.
(2) The degree of Φ as a polynomial in $d^2 f\, dx - df\, d^2 x$ is at most m.
(3) When Φ is written as a polynomial in

$$x, \; y, \; \frac{dx}{x}, \frac{dy}{y}, \frac{df}{f}, \frac{d^2 f\, dx - d^2 x\, df}{f},$$

the degree of Φ in x, y is $\leq p + 3m + s$.

Hence each of the factors Φ_j $(1 \leq j \leq k)$ satisfies the same three properties. The third property means that, when Φ_j is written

$$\Phi_j = \sum_{k=0}^{m_j} \omega_{f, s_j + 3k}^{(j)} f^{2(m_j - k)} (d^2 f\, dx - d^2 x\, df)^{m_j - k},$$

with

$$\omega_{f, \mu}^{(j)} = \sum_{\nu_0 + \nu_1 + \nu_2 = \mu} a_{f, \nu_0 \nu_1 \nu_2}^{(j)}(x, y)(df)^{\nu_0}(f\, dx)^{\nu_1}(f\, dy)^{\nu_2},$$

the degree of the polynomial $a_{f, \nu_0 \nu_1 \nu_2}^{(j)}(x, y)$ in x, y is at most $p + 3m + s$. Since Φ is divisible by f_y, at least one of the factors Φ_j divisible by f_y. We can now replace Φ by that factor Φ_j and assume that Φ is irreducible. One difference is that after this replacement the degree of the polynomial $a_{\nu_0 \nu_1 \nu_2}(x, y)$ in x, y is now at most $p + 3m + s$ instead of at most p.

The degree m of the irreducible new Φ in $f^2(d^2 f\, dx - d^2 x\, df)$ may be equal to 1 or even 0. If m is zero, then we can get a holomorphic 1-jet differential on X which according to Sakai's result [1979] is impossible. We now would like to rule out the case of $m = 1$ for a generic f of sufficiently large degree δ relative to m, p, s. Assume $m = 1$ and we are going to derive a contradiction. The case of $m = 1$ means that we have the divisibility of $\omega_s \mathrm{II} + \omega_{s+3}$ by f_y. We use the following terminology. For a polynomial $g(x, y)$ of degree $\leq k$, by the element of $H^0\left(\mathbb{P}_2, \mathcal{O}_{\mathbb{P}_2}(k)\right)$ defined by g we mean the element defined by the element of $H^0\left(\mathbb{P}_2, \mathcal{O}_{\mathbb{P}_2}(k)\right)$ defined by the homogeneous polynomial $G(z_0, z_1, z_2)$ given by

$$G(z_0, z_1, z_2) = z_0^k g\left(\frac{z_1}{z_0}, \frac{z_2}{z_0}\right).$$

LEMMA 2.4.1.   *Suppose $g(x,y), g_1(x,y), g_2(x,y)$ are polynomials of degree δ in x, y. Let G, G_1, G_2 be elements of $H^0(\mathbb{P}_2, \mathcal{O}_{\mathbb{P}_2}(\delta))$ defined respectively by g, g_1, g_2. Assume that G, G_1, G_2 have no common zeroes on $\mathbb{P}_2$. Let $k \geq \delta$. If $a(x,y), a_1(x,y), a_2(x,y)$ are polynomials of degree $\leq k$ so that $ga = a_1 g_1 + a_2 g_2$, then there exist polynomials b_1, b_2 of degree $\leq k - \delta$ such that $a = b_1 g_1 + b_2 g_2$.*

PROOF.   Let E be the element in $H^0(\mathbb{P}_2, \mathcal{O}_{\mathbb{P}_2}(e))$ (with $0 \leq e \leq \delta$ whose zero-set is the union of all the common branches of the zero-set of G_1 and the zero-set of G_2. Let $\tilde{G}_j = \frac{G_j}{E} \in H^0(\mathbb{P}_2, \mathcal{O}_{\mathbb{P}_2}(\delta - e))$ for $j = 1, 2$. Let $\mathcal{I}$ be the ideal sheaf on $\mathbb{P}_2$ generated by G_1, G_2. Consider the exact sequence

$$0 \to \mathcal{O}_{\mathbb{P}_2}(k - 2\delta + e) \xrightarrow{\sigma} \mathcal{O}_{\mathbb{P}_2}(k - \delta)^{\oplus 2} \xrightarrow{\tau} \mathcal{I}(k) \to 0$$

with σ defined by the 2×1 matrix $\begin{pmatrix} -\tilde{G}_2 \\ \tilde{G}_1 \end{pmatrix}$ and with τ defined by the 1×2 matrix (G_1, G_2). Since $H^1(\mathbb{P}_2, \mathcal{I}(k - 2\delta + e)) = 0$, it follows that the map

$$\tilde{\sigma} : H^0(\mathbb{P}_2, \mathcal{O}_{\mathbb{P}_2}(k - \delta))^{\oplus 2})) \to H^0(\mathbb{P}_2, \mathcal{O}_{\mathbb{P}_2}(k))$$

is surjective. Let A, A_1, A_2 be elements of $H^0(\mathbb{P}_2, \mathcal{O}_{\mathbb{P}_2}(k))$ defined by a, a_1, a_2. It follows from $ga = a_1 g_1 + a_2 g_2$ that $GA = A_1 G_1 + A_2 G_2$. Since G, G_1, G_2 have no common zeroes in $\mathbb{P}_2$, it follows that $A \in H^0(\mathbb{P}_2, \mathcal{I}(k))$. Hence there exist $B_1, B_2 \in H^0(\mathbb{P}_2, \mathcal{O}_{\mathbb{P}_2}(k - \delta))$ such that $A = \tilde{\sigma}(B_1, B_2)$. Let $b_1(x,y), b_2(x,y)$ be polynomials of degree $\leq k - \delta$ corresponding respectively to B_1, B_2. Then $a = b_1 g_1 + b_2 g_2$. $\square$

LEMMA 2.4.2.   *Suppose $g(x,y), g_1(x,y), g_2(x,y)$ are polynomials of degree δ in x, y. Let G, G_1, G_2 be elements of $H^0(\mathbb{P}_2, \mathcal{O}_{\mathbb{P}_2}(\delta))$ defined respectively by g, g_1, g_2. Assume that G, G_1, G_2 have no common zeroes on $\mathbb{P}_2$. Let $a_\mu(x,y)$ $(0 \leq \mu \leq s)$ be polynomials of degree at most p so that $a_\mu(x,y)$ $(0 \leq \mu \leq s)$ are not all identically zero. Let $h(x,y)$ be a polynomial of degree k. Let $b_\mu(x,y)$ $(0 \leq \mu \leq s+1)$ be polynomials of degree at most $p + k - \delta$. Suppose*

$$\left(\sum_{\mu=0}^{q} a_\mu(x,y) g_1(x,y)^{q-\mu} g_2(x,y)^\mu \right) h(x,y) + \left(\sum_{\nu=0}^{q+1} b_\nu(x,y) g_1(x,y)^{q+1-\nu} g_2(x,y)^\nu \right)$$

is divisible by $g(x,y)$. Then there exist non identically zero polynomials $a(x,y)$, $c_1(x,y), c_2(x,y), c(x,y)$ of degree at most p such that

$$a(x,y)h(x,y) = c_1(x,y)g_1(x,y) + c_2(x,y)g_2(x,y) + c(x,y)g(x,y).$$

PROOF.   By replacing $g_1(x,y), g_2(x,y)$ by

$$\tilde{g}_1(x,y) = \alpha_1 g_1(x,y) + \alpha_2 g_2(x,y),$$
$$\tilde{g}_2(x,y) = \beta_1 g_1(x,y) + \beta_2 g_2(x,y),$$

for some suitable constants α_j, β_j $(j = 1, 2)$, we can assume without loss of generality that $a_0(x, y)$ is not identically zero. Then

$$g_1(x, y)^q \left(a_0(x, y)h(x, y) + b_0(x, y)g_1(x, y) \right) = \psi_2(x, y)g_2(x, y) + \psi(x, y)g(x, y),$$

where

$$\psi_2(x, y) = -\sum_{\mu=1}^{q} a_\mu(x, y)g_1^{q-\mu}g_2(x, y)^{\mu-1}h(x, y) - \sum_{\mu=1}^{q+1} b_\mu(x, y)g_1^{q+1-\mu}g_2(x, y)^{\mu-1}$$

and $\psi(x, y)$ are polynomials in x, y of degree at most $q\delta + p + k$.

Applying q times Lemma 2.4.1 gives us polynomials $c_2(x, y), c(x, y)$ of degree at most $p + k - \delta$ such that

$$a_0(x, y)h(x, y) + b_0(x, y)g_1(x, y) = c_2(x, y)g_2(x, y) + c(x, y)g(x, y).$$

It suffices to set $a(x, y) = a_0(x, y)$ and $c_1(x, y) = -b_0(x, y)$.   □

2.4.3.   The case $m = 1$ means that there exist polynomials $a_{\nu_0\nu_1\nu_2}$ of degree at most p such that

$$\sum_{\nu_0+\nu_1+\nu_2=s} a_{\nu_0\nu_1\nu_2}(x, y)(df)^{\nu_0}(f\,dx)^{\nu_1}(f\,dy)^{\nu_2}f^2(d^2f\,dx - d^2x\,df)$$

$$+ \sum_{\nu_0+\nu_1+\nu_2=s+3} a_{\nu_0\nu_1\nu_2}(x, y)(df)^{\nu_0}(f\,dx)^{\nu_1}(f\,dy)^{\nu_2}$$

is divisible by f_y. This means that

$$\sum_{\nu_0+\nu_1+\nu_2=s} a_{\nu_0\nu_1\nu_2}(x, y)(f_x)^{\nu_0}f^{\nu_1+\nu_2+2}(dx)^{\nu_0+\nu_1+1}(dy)^{\nu_2}\mathrm{II}$$

$$+ \sum_{\nu_0+\nu_1+\nu_2=s+3} a_{\nu_0\nu_1\nu_2}(x, y)(f_x)^{\nu_0}f^{\nu_1+\nu_2}(dx)^{\nu_0+\nu_1}(dy)^{\nu_2}$$

is divisible by f_y. Let

$$\xi_l = \sum_{\nu=0}^{l} a_{\nu,l-\nu,s-l}(x, y)(f_x)^\nu f^{s+2-\nu}, \qquad \eta_l = \sum_{\nu=0}^{l} a_{\nu,l-\nu,s+3-l}(x, y)(f_x)^\nu f^{s+3-\nu}.$$

Then

$$\sum_{l=0}^{s} \xi_l(dx)^{l+1}(dy)^{s-l}\left(f_{xx}\,dx^2 + 2f_{xy}\,dx\,dy + f_{yy}\,dy^2 \right) - \sum_{l=0}^{s+3} \eta_l(dx)^l(dy)^{s+3-l}$$

is divisible by f_y.

Let l_0 be the largest l such that the polynomial $\xi_l(x, y)$ is not identically zero. Let l_1 be the smallest l such that the polynomial ξ_l is not identically zero. Then from the coefficient of $(dx)^{l_0+3}(dy)^{s-l_0}$ we conclude that

$$\xi_{l_0}f_{xx} - \eta_{l_0+3}$$

is divisible by f_y. From the coefficient of $(dx)^{l_1+1}(dy)^{s+2-l_1}$ we conclude that

$$\xi_{l_1} f_{yy} - \eta_{l_1+1}$$

is divisible by f_y. From the coefficient of $(dx)^{l_0+2}(dy)^{s-l_0+1}$ we conclude that

$$\xi_{l_0+1} f_{xx} + 2\xi_{l_0} f_{xy} - \eta_{l_0+2}$$

is divisible by f_y. Hence

$$2\xi_{l_0}^2 f_{xy} - \xi_{l_0}\eta_{l_0+2} + \xi_{l_0+1}\eta_{l_0+3}$$

is divisible by f_y.

Choose two polynomials $\lambda_1(x,y), \lambda_2(x,y)$ of degree 1 in x, y such that the elements in $H^0(\mathbb{P}_2, \mathcal{O}_{\mathbb{P}_2}(\delta))$ defined by $\lambda_1(x,y)f_x(x,y)$, $\lambda_2(x,y)f_y(x,y)$, and $f(x,y)$ have no common zeroes on $\mathbb{P}_2$. Let $g_1(x,y) = \lambda_1(x,y)f_x(x,y)$ and $g_2(x,y) = \lambda_2(x,y)f_y(x,y)$, and

$$\tilde{\xi}_l = \lambda_1^l \xi_l = \sum_{\nu=0}^{l} a_{\nu,l-\nu,s-l}(x,y)\lambda_1^{l-\nu}(g_1)^\nu f^{s+2-\nu},$$

$$\tilde{\eta}_l = \lambda_1^l \eta_l = \sum_{\nu=0}^{l} a_{\nu,l-\nu,s+3-l}(x,y)\lambda_1^{l-\nu}(g_1)^\nu f^{s+3-\nu}.$$

Then the three polynomials

$$\lambda_2 \lambda_1^3 \tilde{\xi}_{l_0} f_{xx} - \lambda_2 \tilde{\eta}_{l_0+3},$$
$$\lambda_2 \lambda_1 \tilde{\xi}_{l_1} f_{yy} - \lambda_2 \tilde{\eta}_{l_1+1},$$
$$2\lambda_2 \lambda_1^4 \tilde{\xi}_{l_0}^2 f_{xy} - \lambda_2 \lambda_1^2 \tilde{\xi}_{l_0}\tilde{\eta}_{l_0+2} + \lambda_2 \tilde{\xi}_{l_0+1}\tilde{\eta}_{l_0+3}$$

are all divisible by g_2.

By Lemma 2.4.2 there exist polynomials $c_{i,j}(x,y)$ such that

$$c_{1,0} f_{xx} = c_{1,1}\lambda_1 f_x + c_{1,2}\lambda_2 f_y + c_{1,3}f,$$
$$c_{2,0} f_{xy} = c_{2,1}\lambda_1 f_x + c_{2,2}\lambda_2 f_y + c_{2,3}f,$$
$$c_{3,0} f_{xx} = c_{3,1}\lambda_1 f_x + c_{3,2}\lambda_2 f_y + c_{3,3}f,$$

with

$$\deg c_{1,j} \leq p + l_0 + 4,$$
$$\deg c_{2,j} \leq p + l_1 + 2,$$
$$\deg c_{3,j} \leq p + 2l_0 + 5$$

for $0 \leq j \leq 3$. Consider the above system of linear equations in

$$f_{xx}, \ f_{xy}, \ f_{yy}, \ f_x, \ f_y$$

as a system of linear differential equations for the unknown functions f_x, f_y, f. Counting the degree of freedom for all the polynomials $c_{i,j}$, we conclude from

the uniqueness property of the system of differential equations that the degree
of freedom for f is no more than

$$3 + 4\left(\binom{p+s+4}{2} + \binom{p+s+2}{2} + \binom{p+2s+5}{2}\right).$$

So when

$$\binom{\delta+2}{2} > 3 + 4\left(\binom{p+s+4}{2} + \binom{p+s+2}{2} + \binom{p+2s+5}{2}\right),$$

the case of $m = 1$ cannot occur for a generic f of degree δ.

2.5. Independence of Special 2-Jet Differentials by Invariant Theory.
Let p be a positive integer and s be a nonnegative integer. By solving linear
equations we can generically construct a special 2-jet differential Φ of total weight
$s + 3m$ $(m \geq 1)$ on X of the form

$$\Phi = \sum_{k=0}^{m} \omega_{f,s+3k} f^{2(m-k)} (d^2 f\, dx - d^2 x\, df)^{m-k},$$

where

$$\omega_{f,\mu} = \sum_{\nu_0+\nu_1+\nu_2=\mu} a_{f,\nu_0\nu_1\nu_2}(x,y)(df)^{\nu_0}(f\,dx)^{\nu_1}(f\,dy)^{\nu_2}$$

and $a_{f,\nu_0\nu_1\nu_2}(x,y)$ is a polynomial in x and y of degree $\leq p$ so that Φ is divisible
by f_y and as a consequence $t^{-N} f_y^{-1} \Phi$ is a holomorphic 2-jet differential on X
defined by $t^\delta = f(x,y)$, when certain inequalities involving p, s, δ, m, and N are
satisfied.

We can assume that Φ, as a polynomial in

$$x,\ y,\ dx,\ dy,\ \frac{df}{f},\ \frac{d^2 f\, dx - dx\, d^2 f}{f^2},$$

is irreducible and the coefficients of a_{ν_0,ν_1,ν_2} are rational functions of the coefficients of $f(x,y)$. This assumption is possible because we can replace Φ by the
corresponding irreducible factor which is divisible by f_y. This means that we
can assume without loss of generality that Φ as a polynomial in x, y, dx, dy,
$d^2 x$, $dy - dx$, $d^2 y$ is irreducible.

Consider the space $\mathcal{F}$ of polynomials f. Let $G = SL(2,\mathbb{C})$. Let C be the
curve defined by f. For $\gamma \in G$, the defining function for $\gamma(C)$ is $(\gamma^{-1})^* f$. Let
$(x_\gamma, y_\gamma) = \gamma(x,y)$. We have a procedure which gives us a special 2-jet differential
Ψ_f for $f \in \mathcal{F}$ generically. We can use $\gamma \in SL(2,\mathbb{C})$ to get another $\gamma^* \Psi_{(\gamma^{-1})^* f}$.
Suppose this procedure with the use of $\gamma \in SL(2,\mathbb{C})$ does not give us at least
two independent special 2-jet differentials. By Proposition 3.3.1 each $\gamma^* \Psi_{(\gamma^{-1})^* f}$
is irreducible over $\mathbb{C}(x,y)$ as a polynomial of dx, dy, $d^2 x\, dy - dx\, d^2 y$. Then we
have

$$\gamma^* \Psi_{(\gamma^{-1})^* f} = R_{\gamma,f}(x,y)\Psi_f$$

for some rational function $R_{\gamma,f}(x,y)$ in x,y. To take away $R_{\gamma,f}(x,y)$ we define for every γ the following. Let Z_f be the union of all algebraic complex curves Z_f' in $\mathbb{C}^2$ such that the inverse image of Z_f' in the space of 2-jets is contained in the zero-set of Ψ_f. In other words, Ψ_f is divisible by the polynomial in x,y which defines Z_f'. Let $g_f(x,y)$ be a polynomial in x,y which defines Z_f. In other words, $g_f(x,y)$ is the polynomial (defined up to a nonzero constant) which divides Ψ_f. Then we conclude that

$$\gamma^*\left(\frac{1}{g_{(\gamma^{-1})^*f}}\Psi_{(\gamma^{-1})^*f}\right) = c_{\gamma,f}\frac{1}{g_f}R_{\gamma,f}(x,y)\Psi_f$$

for some nonzero constant $c_{\gamma,f}$. Let $g_{\gamma,f} = \gamma^*g_{(\gamma^{-1})^*f}$.

2.5.1. For $\gamma \in SL(2,\mathbb{C})$ let f_{y_γ} be the partial derivative with respect to y_γ in the coordinate system (x_γ, y_γ). We have

$$\frac{1}{g_{\gamma,f}}f_{y_\gamma}^{-1}\sum_{k=0}^{m}\gamma^*(\omega_{(\gamma^{-1})^*f,s+3k})f^{2(m-k)}(d^2f\,d(x_\gamma) - d^2(x_\gamma)\,df)^{m-k}$$

$$= c_{\gamma,f}\frac{1}{g_{1,f}}f_y^{-1}\sum_{k=0}^{m}\omega_{f,s+3k}f^{2(m-k)}(d^2f\,dx - d^2x\,df)^{m-k}$$

and

$$\gamma^*\omega_{(\gamma^{-1})^*f,\mu} = \sum_{\nu_0+\nu_1+\nu_2=\mu} a_{(\gamma^{-1})^*f,\nu_0\nu_1\nu_2}(x_\gamma,y_\gamma)(df)^{\nu_0}(f\,d(x_\gamma))^{\nu_1}(f\,d(y_\gamma))^{\nu_2}.$$

We use

$$d^2f\,dx - d^2x\,df = f_y(d^2y\,dx - d^2x\,dy) + \mathrm{II}\,dx,$$

where

$$\mathrm{II} = f_{xx}\,dx^2 + 2f_{xy}\,dx\,dy + f_{yy}\,dy^2.$$

Since $d^2y\,dx - d^2x\,dy$ and II are both invariant under $SL(2,\mathbb{C})$, it follows that

$$d^2f\,d(x_\gamma) - d^2(x_\gamma)\,df = \gamma^*\left(d^2((\gamma^{-1})^*f)\,dx - d^2x\,d(\gamma^{-1})^*f)\right)$$

$$= f_{y_\gamma}(d^2y\,dx - d^2x\,dy) + \mathrm{II}\,d(x_\gamma).$$

Thus

(2.5.1.1)

$$\frac{1}{g_{\gamma,f}}f_{y_\gamma}^{-1}\sum_{k=0}^{m}\gamma^*(\omega_{(\gamma^{-1})^*f,s+3k})f^{2(m-k)}\left(f_{y_\gamma}(d^2y\,dx - d^2x\,dy) + \mathrm{II}\,d(x_\gamma)\right)^{m-k}$$

$$= c_{\gamma,f}\frac{1}{g_{1,f}}f_y^{-1}\sum_{k=0}^{m}\omega_{f,s+3k}f^{2(m-k)}(d^2f\,dx - d^2x\,df)^{m-k}.$$

We consider the terms in (2.5.1.1) with the highest power for the factor $d^2y\,dx - d^2x\,dy$ and conclude that

(2.5.1.2) $$\frac{1}{g_{\gamma,f}}\gamma^*(\omega_{(\gamma^{-1})^*f,s})(f_{y_\gamma})^{m-1} = c_{\gamma,f}\frac{1}{g_{1,f}}\omega_{f,s}(f_y)^{m-1}.$$

2.5.2. From Section 2.4 we know that $m > 1$. Let q be the largest integer such that $g_{\gamma,f}$ is divisible by $(f_{y_\gamma})^q$ for a generic γ.

Again we differentiate between two cases. The first case is that $q < m - 1$. Since for any integer $l \geq 2$ the distinct generic elements

$$\gamma_j = \begin{pmatrix} \alpha_j & \beta_j \\ \sigma_j & \tau_j \end{pmatrix} \in SL(2, \mathbb{C}) \qquad (1 \leq j \leq l)$$

the l polynomials $f_{\gamma_j y} = -\beta_j f_x + \alpha_j f_y$ are relatively prime. it follows from (2.5.1.2) that $\omega_{f,s}$ contains the factor $\prod_{j=1}^{l}(-\beta_j f_x + \alpha_j f_y)$ for arbitrarily large l and we have a contradiction.

Now consider the second case of $q \geq m - 1$. Then $(f_{y_\gamma})^{m-1}$ divides $g_{\gamma,f}$ for a generic γ. As a consequence

$$\sum_{k=0}^{m} \gamma^*(\omega_{(\gamma^{-1})^*f,s+3k}) f^{2(m-k)} \left(f_{y_\gamma}(d^2y\,dx - d^2x\,dy) + \mathrm{II}\,d(x_\gamma) \right)^{m-k}$$

is divisible by $(f_{y_\gamma})^m$. Since we consider a generic f, we can assume that γ being equal to the identity element is the generic case. By considering the coefficient of $\left(d^2y\,dx - d^2x\,dy\right)^{m-1}$, we conclude that f_y divides $m\omega_s\mathrm{II} + \omega_{s+3}$ and the 2-jet differential

$$f_y^{-1}\left(m\omega_s f^2(d^2f\,dx - df\,d^2x) + \omega_{s+3}\right)$$

gives rise to a holomorphic 2-jet differential, which means that we have the case of $m = 1$, contradicting the earlier conclusion that the case of $m = 1$ cannot occur.

2.6. Construction of Sections of Multiples of Differences of Ample Line Bundles.
We now take the resultant for the two independent holomorphic 2-jet differentials and get a meromorphic 1-jet differential h whose pullback by the entire holomorphic curve is identically zero. After replacing h by one of its factors, we can also assume without loss of generality that h is and its homogeneous degree q in x, y is m.

LEMMA 2.6.1 (AMPLE LINE BUNDLE DIFFERENCE [Siu 1993]). *Let F and G be ample line bundles over a reduced compact complex space X of complex dimension n. If $F^n > nF^{n-1}G$, then for k sufficiently large there exists a nontrivial holomorphic section of $k(F - G)$ which vanishes on some ample divisor of X.*

PROOF. By replacing F and G by their sufficiently high powers, we can assume without loss of generality that both F and G are very ample. Let k be any positive integer. We select $k+1$ reduced members G_j, $1 \leq j \leq k+1$ in the linear system $|G|$ and consider the exact sequence

$$0 \to H^0(X, kF - \sum_j G_j) \to H^0(X, kF) \to \bigoplus_{j=1}^{k+1} H^0(G_j, kF|G_j).$$

By Kodaira's vanishing theorem and the theorem of Riemann–Roch

$$\dim_{\mathbb{C}} H^0(X, kF - (k+1)G) \geq \frac{k^n}{n!} F^n - \sum_{j=1}^{k+1} \frac{k^{n-1}}{(n-1)!} F^{n-1} G_j - o(k^{n-1})$$

$$\geq \frac{k^n}{n!}(F^n - nF^{n-1}G) - o(k^n).$$

So for k sufficiently large there exists a nontrivial global holomorphic section s of $kF - (k+1)G$ over X. We multiply s by a nontrivial global holomorphic section of G on X to get a nontrivial holomorphic section of $k(F - G)$ over X which vanishes on an ample divisor of X. $\qquad\square$

LEMMA 2.6.2. *Let $h(x, y, dx, dy)$ be an irreducible polynomial in x, y, dx, dy which is of degree q in x, y and is of homogeneous degree $m \geq 1$ in dx, dy. Suppose $q \geq 4m$ and $\delta \geq 1$. Let $f(x, y)$ be a polynomial of degree δ such that the curve C in $\mathbb{P}_2$ defined by f is smooth. Then there exists no holomorphic map $\varphi : \mathbb{C} \to \mathbb{P}_2 - C$ such that the image of φ is Zariski dense in $\mathbb{P}_2$.*

PROOF. Assume that there is a holomorphic map $\varphi : \mathbb{C} \to \mathbb{P}_2 - C$ such that the image of φ is Zariski dense in $\mathbb{P}_2$. We are going to derive a contradiction.

Let X be the surface in $\mathbb{P}_3$ which has affine coordinates x, y, t with $t^\delta = f(x, y)$ so that X is a cyclic branched cover over $\mathbb{P}_2$ with branching along C with projection map $\pi : X \to \mathbb{P}_2$. Let $\tilde{C} = \pi^{-1}(C)$ and $\tilde{\varphi} : \mathbb{C} \to X - \tilde{C}$ be the lifting of φ so that $\pi \circ \tilde{\varphi} = \varphi$.

We use the following notations. For a vector space E over $\mathbb{C}$, we let $\mathbb{P}(E)$ denote the space of all 1-dimensional $\mathbb{C}$-linear subspaces of E. For a vector bundle $\sigma : B \to Y$ we let $\mathbb{P}(B)$ denote the bundle of projective spaces over Y so that the fiber of $\mathbb{P}(B)$ over a point $y \in Y$ is $\mathbb{P}(\sigma^{-1}(y))$. We let L_X denote the line bundle over $\mathbb{P}(T_X)$ whose restriction to the fiber of $\mathbb{P}(T_X) \to X$ over $x \in X$ is the hyperplane section line bundle of $\mathbb{P}(T_{X,x})$, where $T_{X,x}$ is the tangent space of X at x. We regard h as a holomorphic section of $mL_X + qH_{\mathbb{P}_2}$. For the proof we will produce a non identically zero holomorphic section of L_X over the Zariski closure of the image of $\tilde{\varphi}$ which vanishes on ample divisor, which then yields a contradiction by the usual Schwarz lemma argument.

We will compute the Chern classes of L_X and use the following well-known formula of Grothendieck [Fulton 1976; Grothendieck 1958] to do the computation to produce such a holomorphic section of L_X.

FORMULA 2.6.3 (GROTHENDIECK). *Let E be a vector bundle of rank r over X and $p : \mathbb{P}(E^*) \to X$ be the projection from the projectivization of the dual of E. Let L_E be the hyperplane section line bundle over $\mathbb{P}(E^*)$. Then*

$$\sum_{j=0}^{r} (-1)^j p^*(c_j(E^*))(c_1(L_E))^j = 0,$$

where $c_0(E^*)$ means 1. $\qquad\square$

To compute the Chern classes of T_X we use the exact sequence

$$0 \to T_X \to T_{\mathbb{P}_3}|X \to \delta H_{\mathbb{P}_3}|X \to 0.$$

From the Euler sequence

$$0 \to 1 \to H_{\mathbb{P}_3}^{\oplus 4} \to T_{\mathbb{P}_3} \to 0$$

we conclude that the total Chern class of $T_{\mathbb{P}_3}$ is $(1 + H_{\mathbb{P}_3})^4$. Thus the total Chern class of T_X is $(1 + H_{\mathbb{P}_3})^4(1 + \delta H_{\mathbb{P}_3})^{-1}|X$ and the total Chern class of T_X^* is $(1 - H_{\mathbb{P}_3})^4(1 - \delta H_{\mathbb{P}_3})^{-1}|X$. We conclude that $c_1(T_X^*) = (\delta - 4)H_{\mathbb{P}_3}|X$ and $c_2(T_X^*) = (\delta^2 - 4\delta + 6)H_{\mathbb{P}_3}^2|X$. Grothendieck's formula yields $L_X^2 - (\delta - 4)H_{\mathbb{P}_3}L_X + (\delta^2 - 4\delta + 6)H_{\mathbb{P}_3}^2 = 0$ on $\mathbb{P}(T_X)$. Since $H_{\mathbb{P}_3}$ is lifted up from X via the projction map $\mathbb{P}(T_X) \to X$, we have $H_{\mathbb{P}_3}^3|X = 0$. Hence $L_X^2 H_{\mathbb{P}_3} = (\delta - 4)H_{\mathbb{P}_3}^2 L_X$ and

$$\begin{aligned}
L_X^3 &= (\delta - 4)H_{\mathbb{P}_3}L_X^2 - (\delta^2 - 4\delta + 6)H_{\mathbb{P}_3}^2 L_X \\
&= (\delta - 4)^2 H_{\mathbb{P}_3}^2 L_X - (\delta^2 - 4\delta + 6)H_{\mathbb{P}_3}^2 L_X \\
&= (-4\delta + 10)H_{\mathbb{P}_3}^2 L_X.
\end{aligned}$$

Note that $H_{\mathbb{P}_3}|X = \pi^*(H_{\mathbb{P}_2})$ so that we simply write $H_{\mathbb{P}_3}|X = H_{\mathbb{P}_2}$. It follows from $H_{\mathbb{P}_2}^2|\mathbb{P}_2 = 1$ that $H_X^2|X = \delta$ and $L_X H_{\mathbb{P}_2}^2|\mathbb{P}(T_X) = \delta$. Hence $L_X^2 H_{\mathbb{P}_2} = \delta(\delta - 4)$ and $L_X^3 = \delta(-4\delta + 10)$.

We know that $L_X + 3H_{\mathbb{P}_2}$ is positive on $\mathbb{P}(T_X)$ as we can easily see by using dx, dy, dt and considering the order of their poles at infinity. Take a large positive integer r. Now to apply Lemma 2.6.1, we let $F = (r + 1)(L_X + 3H_{\mathbb{P}_2})$ and $G = L_X + 3(r + 1)H_{\mathbb{P}_2}$ so that $rL_X = F - G$. We have to verify $F^2 > 2FG$ on V_h. In other words,

$$\begin{aligned}
((r + 1)(L_X + 3H_{\mathbb{P}_2}))^2 &(mL_X + qH_{\mathbb{P}_2}) \\
&> 2((r + 1)(L_X + 3H_{\mathbb{P}_2}))(L_X + 3(r + 1)H_{\mathbb{P}_2})(mL_X + qH_{\mathbb{P}_2}),
\end{aligned}$$

because V_h as a hypersurface in $\mathbb{P}(T_X)$ is defined by $h = 0$. We rewrite this inequality as

$$\begin{aligned}
(r + 1)^2 &\left(mL_X^3 + (6m + q)L_X^2 H_{\mathbb{P}_2} + (6q + 9m)L_X H_{\mathbb{P}_2}^2\right) \\
&> 2(r + 1)\left(mL_X^3 + (3m(r + 2) + q)L_X^2 H_{\mathbb{P}_2} + (9m(r + 1) + 3q(r + 2))L_X H_{\mathbb{P}_2}^2\right).
\end{aligned}$$

Dividing both sides of the inequality by $(r + 1)\delta$, we get

$$\begin{aligned}
(r + 1)&\left(m(-4\delta + 10) + (6m + q)(\delta - 4) + (6q + 9m)\right) \\
&> 2\left(m(-4\delta + 10) + (3m(r + 2) + q)(\delta - 4) + (9m(r + 1) + 3q(r + 2))\right).
\end{aligned}$$

Since we are free to choose arbitrarily large r, it suffices to consider the coefficients of r on both sides. The coefficient of r on the left-hand side is

$$(2m + q)\delta + 2q - 5m = (2m + q)(\delta - 1) + 3q - 3m$$

and the coefficient of r on the right-hand side is

$$6m\delta + 3q - 15m = 6m(\delta - 1) + 3q - 9m.$$

If $q \geq 4m$, we get the inequality we want for r sufficiently large.

Since the 1-jet differential h on $\mathbb{P}_2$ is irreducible, its zero-set in $\mathbb{P}(T_{\mathbb{P}_2})$ is again irreducible. However, the pullback $\tilde{h}$ of h to the branched cover X over $\mathbb{P}_2$ may not be irreducible. The holomorphic section of rL_X over $V_{\tilde{h}}$ we get may be identically zero on the branch of $V_{\tilde{h}}$ which contains the lifting of the entire holomorphic curve. To deal with this case, we will use the observation that the subvariety $V_{\tilde{h}}$ of $\mathbb{P}(T_X)$ is branched over the subvariety V_h of $\mathbb{P}(T_{\mathbb{P}_2})$ and the branching is cylic. The action of the cyclic group of order δ acting on $V_{\tilde{h}}$ will in the following way help us get a non identically zero section on the branch we want.

We lift $\varphi : \mathbb{C} \to \mathbb{P}_2 - C$ to $\tilde{\varphi} : \mathbb{C} \to X - \tilde{C}$. We consider the projectivization $\mathbb{P}(T_X)$ of the tangent bundle T_X of X and let $p_X : \mathbb{P}(T_X) \to X$ be the projection map. We also consider the projectivization $\mathbb{P}(T_{\mathbb{P}_2})$ of the tangent bundle $T_{\mathbb{P}_2}$ of $\mathbb{P}_2$ and let $p_{\mathbb{P}_2} : \mathbb{P}(T_{\mathbb{P}_2}) \to X$ be the projection map. The projection map $\pi : X \to \mathbb{P}_2$ induces a meromorphic map $\mathbb{P}(\pi) : \mathbb{P}(T_X) \to \mathbb{P}(T_{\mathbb{P}_2})$ whose restriction to $\mathbb{P}(T_{X-\tilde{C}})$ is holomorphic. We have a holomorphic map $\mathbb{P}(d\varphi) : \mathbb{C} \to \mathbb{P}(T_{\mathbb{P}_2})$ which we define first at points of $\mathbb{C}$ where $d\varphi$ is nonzero and then extend by holomorphicity to all of $\mathbb{C}$. Likewise we have a holomorphic map $\mathbb{P}(d\tilde{\varphi}) : \mathbb{C} \to \mathbb{P}(T_X)$ which we define first at points of $\mathbb{C}$ where $d\varphi$ is nonzero and then extend by holomorphicity to all of $\mathbb{C}$. Let W be the Zariski closure in $\mathbb{P}(T_{\mathbb{P}_2})$ of the image of $\mathbb{P}(d\varphi)$. Let $\tilde{W}$ be the Zariski closure in $\mathbb{P}(T_X)$ of the image of $\mathbb{P}(d\tilde{\varphi})$. We let $(\mathbb{P}(\pi))(\tilde{W})$ denote the proper image of $\tilde{W}$ under $\mathbb{P}(\pi)$ in the sense that it is Zariski closure in $\mathbb{P}(T_{\mathbb{P}_2})$ of the image of $\mathbb{P}(\pi)$ of $W \cap \mathbb{P}(T_{X-\tilde{C}})$. Then $W = (\mathbb{P}(\pi))(\tilde{W})$. We know that $W = V_h$. Also we know that $\tilde{W}$ is a branch of $V_{\tilde{h}}$. Let $\hat{W}$ be a branch of $V_{\tilde{h}}$ where the 1-jet differential ω constructed as a section of the difference of ample line bundles is not identically zero. There is a proper subvariety $\tilde{E}$ of X such that the projection under p_X of the intersection of any two distinct branches of $V_{\tilde{h}}$ onto X is contained in $\tilde{Z}$. Let Z be the projection of $\tilde{Z}$ to $\mathbb{P}_2$. Take a point $P_0 \in \mathbb{P}_2 - (C \cap Z)$ such that $V_h \cap \pi_{\mathbb{P}_2}^{-1}(P_0)$ consists of precisely m distinct points $Q_1, \ldots, Q_m$. The inverse image of P_0 under π consists of δ distinct points P_0^ν $(1 \leq \nu \leq \delta)$. The inverse image of Q_j under $\mathbb{P}(\pi)$ consists of δ distinct points $Q_j^{(\nu)}$ $(1 \leq \nu \leq m)$ so that $Q_j^{(\nu)} \in \pi_X^{-1}(P_0^{(\nu)})$. Some $Q_{j_0}^{(\nu_0)} \in \hat{W}$. Then there exists some ν_1 such that $Q_{j_0}^{(\nu_1)} \in \tilde{W}$. There exists an element γ in the Galois group of automorphisms of X over $\mathbb{P}_2$ such that γ maps $P_0^{(\nu_1)}$ to $P_0^{(\nu_0)}$. Then the induced automorphism of $\tilde{\gamma}$ $\mathbb{P}(T_X)$ over $\mathbb{P}(T_{\mathbb{P}_2})$ maps $Q_{j_0}^{(\nu_1)}$ to $Q_{j_0}^{(\nu_0)}$. As a consequence $\tilde{\gamma}^*(\omega)$ is not identically zero on the branch $\tilde{W}$ of $V_{\tilde{h}}$ which is the Zariski closure of the image of $\mathbb{P}(d\tilde{\varphi})$. This forces the pullback by $d\tilde{\varphi}$ of $\tilde{\gamma}^*(\omega)$ to vanish identically on $\mathbb{C}$, which is a contradiction. $\qquad \square$

2.7. An Algebraic Geometric Lemma on Touching Order

LEMMA 2.7.1. *Let $F(x,y) = \sum_{\nu=0}^{m} a_\nu(x)y^\nu$ be an irreducible polynomial in x,y, where the degree of $a_\nu(x)$ in x is no more than q. Let $y_0(x)$ be a polynomial in x such that the vanishing order N of $F(x, y_0(x))$ in x at $x = 0$ is greater than $(2m-1)q$. Let e be the vanishing order of $\frac{\partial F}{\partial y}(x, y_0(x))$ in x at $x = 0$. Then $e \leq (2m-1)q$.*

PROOF. Consider the system of $2m - 1$ linear equations

$$\sum_{\nu=0}^{m} a_\nu(x)y_0^{\nu+j} = x^N g(x)y_0^j \qquad (0 \leq j \leq m - 2),$$

$$\sum_{\nu=0}^{m-1} (\nu+1)a_{\nu+1}(x)y_0^{\nu+j} = x^e h(x)y_0^j \qquad (0 \leq j \leq m - 1).$$

Let $D(x)$ be the resultant of $F(x,y)$ and $\frac{\partial F}{\partial y}$ as polynomials in y. We can solve for the unknowns $1, y(x), \ldots, y(x)^{2m-2}$ in the above system of $2m - 1$ linear equations and get $D(x)y(x)^k \equiv 0 \mod x^{\min(N,e)}$ for $0 \leq k \leq 2m-2$. The degree of the $(2m-1) \times (2m-1)$ determinant $D(x)$ in x is at most $(2m-1)q$. Since $D(x)$ is not identically zero due to the irreducibility of $F(x)$, the vanishing order of $D(x)$ in x is at most $(2m-1)q$. Since $D(x) \equiv 0 \mod x^{\min(N,e)}$, it follows from the case $k = 0$ in $D(x)y(x)^k \equiv 0 \mod x^{\min(N,e)}$ and from $N > (2m-1)q$ that $e \leq (2m-1)q$. ☐

LEMMA 2.7.2. *Let $F(x,y) = \sum_{\nu=0}^{m} a_\nu(x)y^\nu$ be a polyomial in x,y, where the degree of $a_\nu(x)$ in x is no more than q. Let e be the vanishing order of $\Delta(x) = \frac{\partial F}{\partial y}(x, y_0(x))$. Let l be a positive integer $> 2e$. Let $y_0(x)$ be a polynomial in x such that*

$$F(x, y_0(x)) \equiv 0 \mod x^l.$$

Then there exists a convergent power series $\tilde{y}(x)$ in x such that $F(x, \tilde{y}(x)) = 0$ and $\tilde{y}(x) \equiv y_0(x) \mod x^{l-e}$. In particular, if $l > 2(2m-1)q$ and the polynomial $F(x,y)$ is irreducible, then there exists a convergent power series $\tilde{y}(x)$ in x such that $F(x, \tilde{y}(x)) = 0$ and $\tilde{y}(x) \equiv y_0(x) \mod x^{l-(2m-1)q}$.

PROOF (adapted from the proof of [Artin 1968, Lemma 2.8]). Let $\Delta(x) = \frac{\partial F}{\partial y}(x, y_0(x))$. We now apply Taylor's formula and consider the equation

$$0 = F(x, y_0(x) + x^{l-2e}\Delta(x)h(x))$$
$$= F(x, y_0(x)) + \Delta(x)^2 x^{l-2e}h(x) + P(x)\Delta(x)^2 x^{2(l-2e)}h(x)^2.$$

It follows from

$$F(x, y_0(x)) \equiv 0 \mod x^l.$$

that $F(x, y_0(x)) = x^{l-2e}\Delta(x)^2\psi(x)$ for some convergent power series $\psi(x)$. We have

$$0 = x^{l-2e}\Delta(x)^2\psi(x) + x^{l-2e}\Delta(x)^2 h(x) + P(x)\Delta(x)^2 x^{2(l-2e)}h(x)^2$$

for some polynomial $P(x)$. Division by $x^{l-2e}\Delta(x)^2$ yields

$$0 = \psi(x) + h(x) + P(x)x^{l-2e}h(x)^2.$$

From $l > 2e$ it follows that

$$\frac{\partial}{\partial Y}\left(\psi(x) + Y + P(x)x^{l-2e}Y^2\right) = 1 + 2P(x)x^{l-2e}Y^2 = 1$$

at $x = 0$. The implicit function theorem yields a convergent power series $h(x)$ so that

$$0 = F(x, y_0(x) + x^{l-2e}\Delta(x)h(x))$$

It suffices to set $y(x) = y_0(x) + x^{l-2e}\Delta(x)h(x)$. When $F(x, y)$ is irreducible, it follows from $l > (2m-1)q$ and Lemma 2.7.1 that $e \leq (2m-1)q$. $\quad\square$

For the rest of this paper, for any real number u we use $\lfloor u \rfloor$ to denote the round-down of u, which means the largest integer not exceeding u.

LEMMA 2.7.3. *Let $F(x, y) = \sum_{\nu=0}^m a_\nu(x)y^\nu$ be a non identically zero polyomial in x, y, where the degree of $a_\nu(x)$ in x is no more than q. Let l be a positive integer $> 2m(2m-1)q$. Let $y_0(x)$ be a polynomial in x such that*

$$F(x, y_0(x)) \equiv 0 \mod x^l.$$

Then there exists a convergent power series $\tilde{y}(x)$ in x such that $F(x, \tilde{y}(x)) = 0$ and $\tilde{y}(x) \equiv y_0(x) \mod x^{\lfloor l/m \rfloor - (2m-1)q}$.

PROOF. Let

$$F(x, y) = \prod_{\lambda=1}^{\tilde{m}} F_\lambda(x, y)$$

be the decomposition into irreducible factors. Then $1 \leq \tilde{m} \leq m$ and the degree of each $F_\lambda(x, y)$ in x is no more than q and its degree in y is no more than m. It follows from

$$F(x, y_0(x)) \equiv 0 \mod x^l$$

that there exists some $1 \leq \lambda \leq \tilde{m}$ such that

$$F_\lambda(x, y_0(x)) \equiv 0 \mod x^{\lfloor l/m \rfloor}.$$

By Lemma 2.7.2 there exists a convergent power series $\tilde{y}(x)$ in x such that $F_\lambda(x, \tilde{y}(x)) = 0$ and $\tilde{y}(x) \equiv y_0(x) \mod x^{\lfloor l/m \rfloor - (2m-1)q}$. Hence $F(x, \tilde{y}(x)) = 0$ and $\tilde{y}(x) \equiv y_0(x) \mod x^{\lfloor l/m \rfloor - (2m-1)q}$. $\quad\square$

LEMMA 2.7.4. *Let $a_\nu(x)$ be polynomials of degree at most q in x $(0 \leq \nu \leq m)$ not all identically zero. Let N be an integer $> 2m(2m-1)q$. Then in the space of all polynomials $y(x)$ of degree at most N in x the subset defined by*

$$\sum_{\nu=0}^m a_\nu(x)y(x)^\nu \equiv 0 \mod x^N$$

is of codimension at least $\lfloor N/m \rfloor - (2m-1)q$.

PROOF. Let $F(x, y) = \sum_{\nu=0}^{m} a_\nu(x) y^\nu$ and let $y_0(x)$ be an arbitrary polynomial of degree at most N which satisfies

$$\sum_{\nu=0}^{m} a_\nu(x) y_0(x)^\nu \equiv 0 \bmod x^N.$$

By Lemma 2.7.3, there exists a convergent power series $\tilde{y}(x)$ such that

$$\sum_{\nu=0}^{m} a_\nu(x) \tilde{y}(x)^\nu = 0$$

and

$$\tilde{y}(x) \equiv y_0(x) \bmod x^{\lfloor N/m \rfloor - (2m-1)q}.$$

On the other hand, there are only a finite number of convergent power series $\tilde{y}(x)$ which could satisfy the equation

$$\sum_{\nu=0}^{m} a_\nu(x) \tilde{y}(x)^\nu = 0.$$

This means that there are only a finite number of possibilities for the first $\lfloor N/m \rfloor - (2m - 1)q$ terms of $y_0(x)$ if $y_0(x)$ is an arbitrary polynomial of degree at most N in x satisfying

$$\sum_{\nu=0}^{m} a_\nu(x) y(x)^\nu \equiv 0 \bmod x^N. \qquad \square$$

PROPOSITION 2.7.5. *Suppose m, q, N, δ are positive integers such that*

$$\binom{\delta + 2}{2} > N \geq \frac{3}{2}(2m + q)(m + 1)\left((2m - 1)(q + m) + \binom{q + 2}{2}(m + 1) + 2\right).$$

Then a generic polynomial $f(x, y)$ of degree δ in x, y cannot be tangential at any point to order at least N to any 1-jet differential h of the form

$$\sum_{\nu=0}^{m} a_\nu(x, y)(dx)^{m-\nu}(dy)^\nu$$

where $a_\nu(x, y)$ $(0 \leq \nu \leq m)$ is a polynomial in x, y of degree at most q with $a_0(x, y), \ldots, a_m(x, y)$ not all identically zero. Here tangential to order N at a point P means that the restriction, to the zero-set of $f(x, y)$, of

$$\sum_{\nu=0}^{m} a_\nu(x, y)(-f_y)^{m-\nu}(f_x)^\nu$$

vanishes to order at least N at P.

PROOF. Let Ω be the set of all polynomials $f(x,y)$ of degree δ such that the homogeneous polynomial $z_0^{\delta} f\left(\frac{z_1}{z_0}, \frac{z_1}{z_0}\right)$ in the homogeneous coordinates $[z_0, z_1, z_2]$ defines a nonsingular complex curve in $\mathbb{P}_2$. For any nonnegative integer l, any point $P_0 \in \mathbb{C}^2$, and any non identically zero 1-jet differential

$$h := \sum_{\nu=0}^{m} a_{\nu}(x,y)(dx)^{m-\nu}(dy)^{\nu},$$

we let $\mathcal{A}_{h,l,P_0}$ be the set of all $f \in \Omega$ such that $f(P_0) = 0$ and

$$\left(f_y \frac{\partial}{\partial x} - f_x \frac{\partial}{\partial y}\right)^j \left(\sum_{\nu=0}^{m} a_{\nu}(x,y)(-f_y)^{m-\nu}(f_x)^{\nu}\right)$$

vanishes at P_0 for all $0 \leq j < l$. In other words, $\mathcal{A}_{h,l,P_0}$ consists of all $f \in \Omega$ such that h is tangential to the zero-set of f at P_0 to order at least N. The definition of $\mathcal{A}_{h,l,P_0}$ shows how the algebraic set $\mathcal{A}_{h,l,P_0}$ depends algebraically on the coefficients of h and on the coordinates of P_0.

Let $\mathcal{H}_{m,q}$ be the set of all non identically zero polynomials

$$h(x, y, dx, dy) = \sum_{\nu=0}^{m} a_{\nu}(x,y)(dx)^{m-\nu}(dy)^{\nu},$$

in x, y, dx, dy of degree no more than q in x, y and of homogeneous degree no more than m in dx, dy. The complex dimension of $\mathcal{H}_{m,q}$ is $(m+1)\binom{q+2}{2}$. The degree of freedom of the point P_0 is 2 as it varies in $\mathbb{C}^2$. Since the complex dimension of Ω is $\binom{\delta+2}{2}$, to finish the proof of the Proposition it suffices to show that for any fixed $h \in \mathcal{H}_{m,q}$ and $P_0 \in \mathbb{C}^2$, the complex codimension of $\mathcal{A}_{h,N,P_0}$ is greater than $2 + (m+1)\binom{q+2}{2}$, because then

$$\bigcup \{\mathcal{A}_{h,N,P_0} \mid h \in \mathcal{H}_{m,q}, P_0 \in \mathbb{C}^2\}$$

is not Zariski dense in Ω. We will prove

$$\operatorname{codim} \mathcal{A}_{h,N,P_0} > 2 + (m+1)\binom{q+2}{2}$$

at a point $f \in \Omega$ by showing that

$$\operatorname{codim} \mathcal{A}_{h,N,P_0} \cap \mathcal{Z} > k + 2 + (m+1)\binom{q+2}{2}$$

for some subvariety germ $\mathcal{Z}$ of Ω at the point f defined by k local holomorphic functions on Ω at the point f.

Fix $P_0 \in \mathbb{C}^2$. By an affine coordinate change in $\mathbb{C}^2$ we can assume without loss of generality that P_0 is the origin of $\mathbb{C}^2$. For a nonnegative integer l we define $\mathcal{Z}_l$ as the set of all $f \in \Omega$ such that

(1) $f(0,0) = f_x(0,0) = 0$, and

(2) the convergent power series $y_f(x)$ defined by $f(x, y_f(x)) = 0$ satisfies $y_f(x) \equiv 0$ mod x^l.

For the rest of the proof of this proposition we will use $y_f(x)$ to denote such a convergent power series. The subvariety $\mathcal{Z}_l$ of Ω is locally defined by l functions and its codimension in Ω is l when l does not exceed the dimension of Ω. Let $\kappa = \lfloor 2N/(3(2m+q)) \rfloor$. Then

$$\binom{\delta+2}{2} > N \geq (2m+q)\tfrac{3}{2}\kappa$$

and

$$\min\left(\frac{\kappa}{2}, \frac{\kappa}{m} - (2m-1)(q+m)\right) > \binom{q+2}{2}(m+1) + 2.$$

The subvariety germ $\mathcal{Z}$ mentioned above will be $\mathcal{Z}_\kappa$ and the number k mentioned above will be κ.

Fix an element

$$h(x, y, dx, dy) = \sum_{\substack{0 \leq \lambda, \mu \leq q \\ 0 \leq \nu \leq m}} c_{\lambda\mu\nu} x^\lambda y^\mu (dx)^{m-\nu}(dy)^\nu$$

of $\mathcal{H}_{m,q}$. Choose (μ_0, ν_0) so that $\mu_0 + \nu_0$ is the minimum among all $\mu + \nu$ with $c_{\lambda\mu\nu} \neq 0$ for some λ. Let

$$P_\nu(x) = \sum_{\lambda=0}^{q} \sum_{\mu_0+\nu_0=\mu+\nu} c_{\lambda\mu\nu} x^{m-\nu+\lambda}.$$

When $P_\nu(x)$ is not identically zero for some $\nu > 0$, we let $G_1(x), \ldots, G_{\tilde{m}}(x)$ be the set of all convergent power series such that

$$\sum_{\nu=0}^{m} P_\nu(x) G_j(x)^\nu = 0.$$

We know that $\tilde{m} \leq m$. For a given nonnegative integer l we let $\mathcal{W}_l$ be the set of all $f \in \mathcal{Z}_0$ such that

(1) $y_f(x) \equiv 0 \mod x^l$, and
(2) when we write $y_f(x) = x^l \tilde{y}_f(x)$, we have

$$\tilde{y}_f(x) = \tilde{y}_f(0) \exp\left(\int_{\xi=0}^{x} \frac{G_j(\xi) - G_j(0)}{\xi} d\xi\right) \mod x^{\lfloor l/m \rfloor - (2m-1)(q+m)}$$

for some $1 \leq j \leq \tilde{m}$.

The codimension of $\mathcal{W}_l$ in Ω is at least $l + \lfloor l/m \rfloor - (2m-1)(q+m)$ if the dimension of Ω is at least $l + \lfloor l/m \rfloor - (2m-1)(q+m)$, because each choice of the $\tilde{m}$ set of conditions means $\lfloor l/m \rfloor - (2m-1)(q+m)$ independent conditions on the coefficients of $\tilde{y}_f(x)$, which translates to $l + \lfloor l/m \rfloor - (2m-1)(q+m)$ independent conditions on the coefficients of $y_f(x) = x^l \tilde{y}_f(x)$. When $P_\nu(x)$ is identically zero for all $\nu > 0$, we do not define $\mathcal{W}_l$.

CLAIM 2.7.5.1. *If $\mathcal{A}_{h,N,P_0} \cap \mathcal{Z}_\kappa$ is not contained in $\mathcal{Z}_{\lfloor 3\kappa/2 \rfloor}$, then $P_\nu(x)$ is not identically zero for some $\nu > 0$ and*

$$\mathcal{A}_{h,N,P_0} \cap \mathcal{Z}_\kappa \subset \mathcal{Z}_{\lfloor 3\kappa/2 \rfloor} \cup \left(\bigcup_{l=\kappa}^{\lfloor 3\kappa/2 \rfloor} \mathcal{W}_l \right).$$

PROOF. Take $f \in \mathcal{A}_{h,N,P_0} \cap \mathcal{Z}_\kappa$ such that f does not belong to $\mathcal{Z}_{\lfloor 3\kappa/2 \rfloor}$. Let l be the vanishing order at $x = 0$ of the convergent power series $y_f(x)$. Then $l < \frac{3}{2}\kappa$. Write $y_f(x) = x^l \tilde{y}_f(x)$. Then

$$\frac{xy_f'}{y_f} = l + \frac{x\tilde{y}_f'}{\tilde{y}_f}$$

which is equal to l at $x = 0$. We have

$$\sum_{\substack{0 \le \lambda,\mu \le q \\ 0 \le \nu \le m}} c_{\lambda\mu\nu} x^{m-\nu+\lambda} y_f^{\mu+\nu} \left(\frac{xy_f'}{y_f} \right)^\nu = \sum_{\substack{0 \le \lambda,\mu \le q \\ 0 \le \nu \le m}} c_{\lambda\mu\nu} x^{m-\nu+\lambda} y_f^\mu (xy_f')^\nu$$

$$= x^m \sum_{\substack{0 \le \lambda,\mu \le q \\ 0 \le \nu \le m}} c_{\lambda\mu\nu} x^\lambda y_f^\mu (y_f')^\nu \equiv 0 \mod x^{m+N}$$

(which is from the definition of $\mathcal{A}_{h,N,P_0}$). It follows from $l < \frac{3}{2}\kappa$ that $N > (2m + q)l$. Since $\mu_0 \le q$ and $\nu_0 \le m$, we have $N > (\mu_0 + \nu_0 + 1)l$. Hence

$$\sum_{\substack{0 \le \lambda,\mu \le q \\ 0 \le \nu \le m}} c_{\lambda\mu\nu} x^{m-\nu+\lambda} y_f^{\mu+\nu} \left(\frac{xy_f'}{y_f} \right)^\nu \equiv 0 \mod x^{(\mu_0+\nu_0+1)l}.$$

Since $c_{\lambda\mu\nu} = 0$ for $\mu + \nu < \mu_0 + \nu_0$, it follows that we can divide the above congruence relation by $x^{(\mu_0+\nu_0)l}$ and get

$$\sum_{\substack{0 \le \lambda \le q \\ \mu+\nu=\mu_0+\nu_0}} c_{\lambda\mu\nu} x^{m-\nu+\lambda} \left(\frac{xy_f'}{y_f} \right)^\nu \equiv 0 \mod x^l.$$

We cannot have $c_{\lambda\mu\nu} = 0$ zero for all $\mu + \nu = \mu_0 + \nu_0$ and $\nu > 0$, otherwise

$$\sum_{\substack{0 \le \lambda \le q \\ \mu+\nu=\mu_0+\nu_0}} c_{\lambda\mu\nu} x^{m-\nu+\lambda} \equiv 0 \mod x^l,$$

contradicting $l \ge \kappa > m + q$ and $c_{\lambda\mu\nu} \ne 0$ for some $\mu + \nu = \mu_0 + \nu_0$. Thus

$$\sum_{\nu=0}^m P_\nu(x) \left(\frac{xy_f'}{y_f} \right)^\nu \equiv 0 \mod x^l$$

with $P_\nu(x)$ not identically zero for some $\nu > 0$. By Lemma 2.7.3 we know that

$$\frac{xy_f'}{y_f} \equiv G_j(x) \mod x^{\lfloor l/m \rfloor - (2m-1)(m+q)}$$

for some $1 \leq j \leq \tilde{m}$. It follows that

$$\tilde{y}_f(x) = \tilde{y}_f(0) \exp\left(\int_{\xi=0}^x \frac{G_j(\xi) - l}{\xi} d\xi\right) \mod x^{\lfloor l/m \rfloor - (2m-1)(q+m)}.$$

Thus $f \in \mathcal{W}_l$ and Claim (2.7.5.1) is proved. $\qquad\square$

The codimension of $\mathcal{W}_l$ in Ω is at least $\kappa + \lfloor \kappa/m \rfloor - (2m-1)(q+m)$ and the codimension of $\mathcal{Z}_{\lfloor 3\kappa/2 \rfloor}$ in Ω is $\lfloor 3\kappa/2 \rfloor$. Hence the codimension of $\mathcal{A}_{h,N,P_0} \cap \mathcal{Z}_\kappa$ in Ω is at least

$$\min\left(\kappa + \lfloor \kappa/m \rfloor - (2m-1)(q+m), \lfloor 3\kappa/2 \rfloor\right).$$

Since $\mathcal{Z}_\kappa$ is locally defined by κ holomorphic functions, it follows that the codimension of $\mathcal{A}_{h,N,P_0}$ in Ω is at least

$$\min\left(\lfloor \kappa/m \rfloor - (2m-1)(q+m), \lfloor \kappa/2 \rfloor\right) > 2 + (m+1)\binom{q+2}{2}.$$

This concludes the proof of Proposition 2.7.5. $\qquad\square$

2.8. A Schwarz Lemma Using Low Touching Order.
We now resume our argument of the hyperbolicity of the complement of a generic plane curve of sufficiently high degree. We can assume that we have an irreducible meromorphic 1-jet differential $h(x, y, dx, dy)$ whose pullback by the entire holomorphic curve is identically zero. Moreover, the degree of $h(x, y, dx, dy)$ in x, y is q and the homogeneous degree of $h(x, y, dx, dy)$ in dx, dy is m with $q \leq 4m$. We consider the resultant $R(x, y)$ of

$$\frac{h(x, y, dx, dy)}{dx^m} = \sum_{\nu=0}^m h_\nu(x, y) \left(\frac{dy}{dx}\right)^\nu$$

and its derivative with respect to $\frac{dy}{dx}$

$$\sum_{\nu=0}^{m-1} (\nu + 1) h_{\nu+1}(x, y) \left(\frac{dy}{dx}\right)^\nu$$

as polynomials in $\frac{dy}{dx}$. Since $h(x, y)$ is irreducible, the resultant $R(x, y)$ is not identically zero and its degree is no more than $(2m-1)q$. Let Z be the common zero-set of $R(x, y)$ and $f(x, y)$. The number of points in Z is no more than $(2m-1)q\delta$. When a point of $\mathbb{P}_2$ is not a zero of $R(x, y)$ we can have a finite number of families of local integral curves going through that point and the entire holomorphic curve is locally contained in such a local integral curve.

Let N be the smallest integer satisfying

$$N \geq \frac{3}{2}(2m + q)(m + 1)\left((2m - 1)(q + m) + \binom{q+2}{2}(m+1) + 2\right).$$

Assume that $\binom{\delta+2}{2} > N$. Then by Proposition 2.7.5 for our generic f, the touching order of f with $h(x, y, dx, dy)$ is no more than N. Let $S(x, y)$ be a non identically zero polynomial of degree r with

$$\binom{r+2}{2} > (2m-1)q\delta\binom{N+2}{2}$$

such that it vanishes to order at least N at each point of the common zero-set Z of $R(x, y)$ and $f(x, y)$. Let $e^{-\psi_0}$ be a smooth metric for the hyperplane section line bundle $H_{\mathbb{P}_2}$ of $\mathbb{P}_2$ with strictly positive curvature. Let A be a positive number and let locally $\psi = \psi_0 + A$ so that $e^{-\psi}$ is a metric for $H_{\mathbb{P}_2}$. We will later choose A to be sufficiently large for our purpose. Let $\theta_\psi = \frac{\sqrt{-1}}{2\pi}\partial\bar{\partial}\psi$ be the curvature form of the metric $e^{-\psi}$.

For a holomorphic section u of a line bundle with a metric, we use $\|u\|$ to denote its pointwise norm with respect to the metric and we use $|u|$ to denote the absolute value of a function which represents u in a local trivialization of the line bundle. The pointwise norm $\|u\|$ is used to give a globally well defined expression. In proving results involving estimates of the norm, we will use local trivialization of the line bundle and it does not matter which local trivialization of the line bundle is used.

Consider f as a section of the δ-th power of the hyperplane section line bundle so that the pointwise norm of f is given by $\|f\|^2 = |f|^2 e^{-\delta\psi}$. We assume that A is chosen so large that $\|f\| < 1$ on all of $\mathbb{P}_2$. Let (x_j, y_j) $(1 \leq j \leq J)$ be a finite number of affine coordinates of affine open subsets of $\mathbb{P}_2$ so that $dx_1, dy_1, \ldots, dx_J, dy_J$ generate at every point of $\mathbb{P}_2$ the cotangent bundle of $\mathbb{P}_2$ tensored by $2H_{\mathbb{P}_2}$. Let $\{\eta_j\}_j$ denote the set $\{dx_1, dy_1, \ldots, dx_J, dy_J\}$. We use $\|\eta_j\|^2$ to denote $|\eta_j|^2 e^{-2\psi}$, which is a function on the tangent bundle of $\mathbb{P}_2$. Let

$$\left\|f^{\frac{N-1}{N}}S\right\|^2 = \left|f^{\frac{N-1}{N}}S\right|^2 e^{-\left(\frac{(N-1)\delta}{N}+r\right)\psi},$$

which can be geometrically interpreted as the N-th root of the pointwise square norm of the section of

$$N\left(\frac{(N-1)\delta}{N}+r\right)H_{\mathbb{P}_2}$$

over $\mathbb{P}_2$ defined by $\left(f^{\frac{N-1}{N}}S\right)^N$.

PROPOSITION 2.8.1. *Assume* $\delta > (r+2)N$. *Let*

$$\Psi = \frac{\left\|f^{\frac{N-1}{N}}S\right\|^2 \sum_j \|\eta_j\|^2}{\|f\|^2\left(\log\frac{1}{\|f\|^2}\right)^2}.$$

Then, when A is sufficiently large, there exists a positive constant ε such that the pullback of

$$\sqrt{-1}\,\partial\bar{\partial}\log\Psi \geq \varepsilon\Psi$$

to any local holomorphic curve Γ in $\mathbb{P}_2 - \{f = 0\}$ holds if Γ satisfies $h = 0$.

PROOF. From standard direct computation we have the following Poincaré–Lelong formula on $\mathbb{P}_2$ in the sense of currents.

$$\frac{\sqrt{-1}}{2\pi}\partial\bar\partial\log\frac{\left\|f^{\frac{N-1}{N}}S\right\|^2\sum_j\|\eta_j\|^2}{\|f\|^2\left(\log\frac{1}{\|f\|^2}\right)^2}=\left(\frac{\delta}{N}-(r+2)-\frac{2\delta}{\log\frac{1}{\|f\|^2}}\right)\theta_\psi-\frac{1}{N}Z_f$$

$$+Z_S+\frac{\sqrt{-1}}{2\pi}\partial\bar\partial\log\sum_j|\eta_j|^2+\frac{2}{\|f\|^2\left(\log\frac{1}{\|f\|^2}\right)^2}\frac{\sqrt{-1}}{2\pi}Df\wedge\overline{Df}.$$

Here Z_f (respectively Z_S) is the $(1,1)$-current defined by the zero-set of f (respectively S) and Df is the smooth $\delta H_{\mathbb{P}_2}$-valued 1-form on $\mathbb{P}_2$ which is the covariant differentiation of the section of $\delta H_{\mathbb{P}_2}$ defined by f with respect to the metric $e^{-\psi}$ of $H_{\mathbb{P}_2}$.

Since we could change affine coordinates, we need only verify the inequality on any compact subset of the affine plane $\mathbb{C}^2$ with affine coordinates x,y. Fix a point P in the common zero-set Z of $R(x,y)$ and $f(x,y)$ and take a compact neighborhood U_P of P in $\mathbb{C}^2$ disjoint from $Z-\{P\}$. We are going to derive an inequality on U_P (which we may have to shrink to get the inequaltiy). Without loss of generality we can assume that $f_x\neq 0$ on U_P (after shrinking U_P and making an affine coordinate transformation if necessary). We write

$$h=\sum_{\nu=0}^m\hat h_\nu(df)^{m-\nu}(dy)^\nu$$

with

$$\hat h_m=\sum_{\nu=0}^m h_\nu\left(\frac{-f_y}{f_x}\right)^{m-\nu}.$$

We use the following two trivial inequalities for positive numbers a,b and α,β with $\alpha+\beta=1$.

$$a^\alpha b^\beta\leq\alpha a+\beta b,$$

$$a^m+b^m\leq(a+b)^m\leq\big(2\max(a,b)\big)^m\leq 2^m(a^m+b^m).$$

We use C_j to denote positive constants. We consider separately the case of $m>1$ and the case of $m=1$. We first look at the case of $m>1$. For a nonnegative bounded continuous function ρ we have

$$\rho|\hat h_m(dy)^m|^2\leq C_1\left(\rho|h|^2+\sum_{\nu=0}^{m-1}\left(|df|^{\frac{2(m-\nu)}{m}}(\rho^{\frac{1}{\nu}}|dy|^2)^{\frac{\nu}{m}}\right)^m|\hat h_\nu|^2\right)$$

$$\leq C_2\left(\rho|h|^2+|df|^{2m}+\rho^{\frac{m}{m-1}}|dy|^{2m}\right).$$

Hence

$$\rho^{\frac{1}{m}}|\hat h_m|^{\frac{2}{m}}|dy|^2\leq C_2^{\frac{1}{m}}\left(\rho^{\frac{1}{m}}|h|^{\frac{2}{m}}+|df|^2+\rho^{\frac{1}{m-1}}|dy|^2\right),$$

$$\rho^{\frac{1}{m}}\left(|f|^2+|\hat h_m|^{\frac{2}{m}}\right)|dy|^2\leq C_3\left(\rho^{\frac{1}{m}}|h|^{\frac{2}{m}}+|f|^2|dy|^2+|df|^2+\rho^{\frac{1}{m-1}}|dy|^2\right).$$

For $m > 1$ we set $\rho = (2C_3)^{-m(m-1)}(|f|^2 + |\hat{h}_m|^{\frac{2}{m}})^{m(m-1)}$. Then

$$C_3\rho^{\frac{1}{m-1}}|dy|^2 = \tfrac{1}{2}\rho^{\frac{1}{m}}(|f|^2 + |\hat{h}_m|^{\frac{2}{m}})|dy|^2$$

and

$$(|f|^2 + |\hat{h}_m|^2)|dy|^2 \le C_4(|h|^{\frac{2}{m}} + |f|^2|dy|^2 + |df|^2).$$

For $m = 1$ the inequality is obviously true. Since the vanishing order of $\hat{h}_m$ on $\{f = 0\}$ at P is at most N and the vanishing order of $S(x, y)$ at P is at least N, it follows that on U_P (after shrinking U_P if necessary)

$$|S|^2|dy|^2 \le C_5(|f|^2 + |\hat{h}_m|^2)|dy|^2 \le C_6(|h|^{\frac{2}{m}} + |f|^2|dy|^2 + |df|^2).$$

Using the inequalities

$$|df|^2 \le C_7(|f|^2 + |Df|^2)$$

and

$$\frac{|f|^2|dy|^2}{\|f\|^2\left(\log\frac{1}{\|f\|^2}\right)^2} \le \varepsilon_0\theta_\psi$$

for any positive number ε_0 when A is sufficiently large, we conclude from the Poincaré–Lelong formula that

$$\frac{|S|^2|dy|^2}{\|f\|^2\left(\log\frac{1}{\|f\|^2}\right)^2} \le \varepsilon_0\theta_\psi + C_8\frac{|Df|^2}{\|f\|^2\left(\log\frac{1}{\|f\|^2}\right)^2} \le C_9\frac{\sqrt{-1}}{2\pi}\partial\bar\partial\log\frac{\left\|f^{\frac{N-1}{N}}S\right\|^2\sum_j|\eta_j|^2}{\|f\|^2\left(\log\frac{1}{\|f\|^2}\right)^2}$$

when pulled back to any local holomorphic curve in U_P which is disjoint from the zero-set of f and which satisfies $h = 0$. We repeat the same argument for a finite number of other affine coordinates instead of (x, y) and sum up to get the inequality we want to prove on local holomorphic curves in U_P which are disjoint from the zero-set of f and which satisfies $h = 0$.

We can find an open neighborhood W of the zero-set of f so that $W - \bigcup_{P \in Z} U_P$ is disjoint from the zero-set of R. At every point Q of W where R is not zero, we can find an open neighborhood Ω_Q of Q in W so that the equation $h = 0$ gives rise to a finite number of families of integral curves. The vanishing order of f on each such integral curve Γ is at most N. With respect to a local holomorphic coordinate ζ, the function $f(\zeta) = \zeta^l g(\zeta)$ with $g(0) \ne 0$ for some $l \le N$. Since $\frac{\delta}{N} > r + 2$, by choosing A sufficiently large we have

$$\frac{\delta}{N} > r + 2 + \frac{2\delta}{\log\frac{1}{\|f\|^2}}.$$

Hence when pulled back to Γ, at points not on the zero-set of f we have

$$\frac{\sqrt{-1}}{2\pi}\partial\bar\partial\log\frac{\left\|f^{\frac{N-1}{N}}\right\|^2\sum_j\|\eta_j\|^2}{\|f\|^2\left(\log\frac{1}{\|f\|^2}\right)^2} \ge C_{10}\frac{|df|^2}{\|f\|^2\left(\log\frac{1}{\|f\|^2}\right)^2}$$

$$\ge C_{11}\frac{|d\zeta|^2}{\|\zeta\|^2\left(\log\frac{1}{\|f\|^2}\right)^2} \ge C_{12}\frac{\left\|f^{\frac{N-1}{N}}\right\|^2\sum_j\|\eta_j\|^2}{\|f\|^2\left(\log\frac{1}{\|f\|^2}\right)^2}.$$

After shrinking W if necessary, the positive constant C_{12} can be made independent of the integral curve Γ of $h = 0$ as long as it is inside W. This gives us on $W - \bigcup_{P \in Z} U_P$ the inequality stated in the Proposition. On $\mathbb{P}_2 - W$ the inequality stated in the Proposition is clear, because there

$$\frac{\left\| f^{\frac{N-1}{N}} \right\|^2 \sum_j \|\eta_j\|^2}{\|f\|^2 \left(\log \frac{1}{\|f\|^2}\right)^2} \leq C_{13}\theta_\psi$$

and the Poincaré–Lelong formula gives

$$\theta_\psi \leq C_{14} \frac{\sqrt{-1}}{2\pi} \partial\bar{\partial} \log \frac{\left\| f^{\frac{N-1}{N}} \right\|^2 \sum_j \|\eta_j\|^2}{\|f\|^2 \left(\log \frac{1}{\|f\|^2}\right)^2}$$

when pulled back to local holomorphic curves in $\mathbb{P}_2 - \{f = 0\}$ which satisfy $h = 0$. $\square$

COROLLARY 2.8.2. *If $\delta > (r+2)N$, then there is no entire holomorphic curve in $\mathbb{P}_2$ which is disjoint from the curve in $\mathbb{P}_2$ defined by $f = 0$ for a generic f.*

PROOF. The inequality

$$\sqrt{-1}\, \partial\bar{\partial} \log \Psi \geq \varepsilon\Psi$$

from Proposition 2.8.1 implies that the pullback of Ψ to any such entire holomorphic curve must be identically zero. This means that the entire holomorphic curve must be contained in the zero-set of S, which is not possible for a generic f. $\square$

2.9. The Final Step. We now combine all the preceding steps together and formulate our theorem.

THEOREM 2.9.1. *Let δ, p, m, N, r be positive integers and s a nonnegative integer, and set $\tilde{m} = (s + 3m)(2m - 1)$. Assume that the following inequalities are satisfied:*

(a) $\sum_{k=0}^m \frac{1}{4}(s + 3k + 2)(s + 3k + 1)(p + 2)(p + 1) > (s + 3m + 1)$
$$\times ((\delta - 1)(p + (s + 3m)\delta) - \tfrac{1}{2}(\delta^2 - 5\delta + 4)).$$

(b) $p \leq \delta - 2 - 4(s + 3m)$.

(c) $\binom{\delta+2}{2} > 3 + 4\left(\binom{p+s+4}{2} + \binom{p+s+2}{2} + \binom{p+2s+5}{2}\right)$.

(d) $N \geq \frac{3}{2}(6\tilde{m} + 1)(\tilde{m} + 1)\left((2\tilde{m} - 1)(5\tilde{m} + 1) + \binom{4\tilde{m}+3}{2}(\tilde{m} + 1) + 2\right)$.

(e) $\binom{r+2}{2} > (2\tilde{m} - 1)(4\tilde{m} + 1)\delta\binom{N+2}{2}$.

(f) $\delta > (r+2)N$.

Let $f(x, y)$ be any generic polynomial of degree δ in x, y and C be the complex curve in $\mathbb{P}_2$ defined by $f = 0$. Then $\mathbb{P}_2 - C$ is hyperbolic in the sense that there is no nonconstant holomorphic map from $\mathbb{C}$ to $\mathbb{P}_2 - C$.

PROOF. Because of the inequality

$$\sum_{k=0}^{m} \tfrac{1}{4}(s+3k+2)(s+3k+1)(p+2)(p+1)$$
$$> (s+3m+1)((\delta-1)(p+(s+3m)\delta) - \tfrac{1}{2}(\delta^2-5\delta+4)),$$

by Lemma 2.2.1 we can construct a 2-jet differential Φ which is divisible by f_y and which is of the form

$$\sum_{k=0}^{m} \omega_{s+3k} f^{2(m-k)} (d^2 f \, dx - d^2 x \, df)^{m-k},$$

where

$$\omega_\mu = \sum_{\nu_0+\nu_1+\nu_2=\mu} a_{\nu_0\nu_1\nu_2}(x,y)(df)^{\nu_0}(f \, dx)^{\nu_1}(f \, dy)^{\nu_2}$$

and $a_{\nu_0\nu_1\nu_2}(x,y)$ is a polynomial in x and y of degree $\leq p$. By Proposition 2.3.1 and the paragraphs before Lemma 2.4.1, we can factor Φ and get Φ_1 which is divisible by f_y and which is of the form

$$\Phi_1 \sum_{k=0}^{m_1} \omega^{(1)}_{s_1+3k} f^{2(m_1-k)} (d^2 f \, dx - d^2 x \, df)^{m_1-k}$$

which is irreducible as a polynomial in $dx, dy, d^2x \, dy - dx \, d^2y$, where

$$\omega^{(1)}_\mu = \sum_{\nu_0+\nu_1+\nu_2=\mu} a^{(1)}_{\nu_0\nu_1\nu_2}(x,y)(df)^{\nu_0}(f \, dx)^{\nu_1}(f \, dy)^{\nu_2}$$

and $a^{(1)}_{\nu_0\nu_1\nu_2}(x,y)$ is a polynomial in x and y of degree $\leq p+3m_1+s_1$. We know that $s_1 + 3m_1 \leq s + 3m$.

By 2.2.4 it follows from the inequality

$$p + 3m_1 + s_1 \leq \delta - 2 - 3(s_1 + 3m_1)$$

that $t^{-(s_1+3m_1)(\delta-1)} f_h^{-1} \Phi_1$ defines a holomorphic 2-jet differential on X which vanishes on an ample divisor. Thus the pullback of Φ_1 to the entire holomorphic curve in $\mathbb{P}_2 - C$ is identically zero. To emphasize the dependence of Φ_1 on f we denote Φ_1 also by $\Phi_{1,f}$. By Section 2.4 we know that $m_1 > 1$. By Section 2.5 we can choose an element $\gamma \in SL(2,\mathbb{C})$ such that

$$\tilde{\Phi}_1 := \gamma^* \left(\Phi_{1,(\gamma^{-1})^* f} \right)$$

and $\tilde{\Phi}_1$ are independent in the sense that the resultant $h(x,y,dx,dy)$ of Φ_1 and $\tilde{\Phi}_1$ as polynomials in the variable $d^2x \, dy - dx \, d^2y$ is not identically zero. Since $t^{-(s_1+3m_1)(\delta-1)} f_h^{-1} \tilde{\Phi}_1$ also defines a holomorphic 2-jet differential on X which vanishes on an ample divisor, the pullback of $\tilde{\Phi}_1$ to the entire holomorphic curve in $\mathbb{P}_2 - C$ is also identically zero. It follows that the pullback of h to the entire holomorphic curve in $\mathbb{P}_2 - C$ is again identically zero. We factor the polynomial $h(x,y,dx,dy)$ into irreducible factors. Then one of the factors $h_1(x,y,dx,dy)$ satisfies the property that its pullback to the entire holomorphic curve in $\mathbb{P}_2 - C$ is identically zero. Since the homogeneous degree of $h(x, y, dx, dy)$ in the variables dx, dy is at most $(s_1 + 3m_1)(2m_1 - 1)$, the homogeneous degree of $h_1(x, y, dx, dy)$ in the variables dx, dy is at most $(s_1 + 3m_1)(2m_1 - 1)$ which is no more than $(s + 3m)(2m - 1)$ which is $\tilde{m}$. Let q be the degree of $h_1(x, y, dx, dy)$ in x, y. If $q \geq 4\tilde{m}$, then by §6 we know that the entire holomorphic curve in $\mathbb{P}_2 - C$ must be contained in an algebraic curve in $\mathbb{P}_2$. This means that for a generic C there is no entire holomorphic curve in $\mathbb{P}_2 - C$. So we now assume that $q < 4\tilde{m}$. By Proposition 2.7.5 and Corollary 2.8.2 we know that there cannot be any entire holomorphic curve in $\mathbb{P}_2 - C$.   □

2.9.1. Example of the Degree and a Set of Parameters. We could choose $s = 0$ and $m = 145$. Then $\tilde{m} = 3m(2m - 1) = 125715$ and we choose N to be the smallest integer satisfying

$$N \geq \frac{3}{2}(6\tilde{m} + 1)(\tilde{m} + 1)\left((2\tilde{m} - 1)(5\tilde{m} + 1) + \binom{4\tilde{m} + 3}{2}(\tilde{m} + 1) + 2\right).$$

and choose r to be the smallest integer satisfying

$$r \geq (2\tilde{m} - 1)(4\tilde{m} + 1)(N + 2)(N + 1)$$

and finally choose δ as the smallest integer satisfying

$$\delta > ((2\tilde{m} - 1)(4\tilde{m} + 1)(N + 2)(N + 1) + 3) N.$$

The number p is set to be the largest integer not exceeding $\frac{12}{145}\delta$. Such values of $s, p, m, \tilde{m}, N, r, \delta$ satisfy all the inequalities in the statement of Theorem 2.9.1. Note that the dominant term in

$$\sum_{k=0}^{m} \frac{1}{4}(s + 3k + 2)(s + 3k + 1)(p + 2)(p + 1)$$

is $\frac{3}{4}m^3 p^2$ and the dominant term in $(s + 3m + 1)((\delta - 1)(p + (s + 3m)\delta) - \frac{1}{2}(\delta^2 - 5\delta + 4))$ is $9m^2\delta^2$. To make sure that the condition

$$\binom{\delta + 2}{2} > 3 + 4\left(\binom{p + s + 4}{2} + \binom{p + s + 2}{2} + \binom{p + 2s + 5}{2}\right)$$

is satisfied for sufficiently large δ we have to require that $\delta^2 > 12p^2$. Hence the smallest m one should use to get a sufficiently large δ to satisfy the inequality

$\sum_{k=0}^{m} \frac{1}{4}(s + 3k + 2)(s + 3k + 1)(p + 2)(p + 1)$
$$> (s + 3m + 1)((\delta - 1)(p + (s + 3m)\delta) - \frac{1}{2}(\delta^2 - 5\delta + 4)),$$

is $m = 145$.

References

[Ahlfors 1941] L. V. Ahlfors, "The theory of meromorphic curves", *Acta Soc. Sci. Fennicae. Nova Ser. A* **3**:4 (1941), 31 p.

[Artin 1968] M. Artin, "On the solutions of analytic equations", *Invent. Math.* **5** (1968), 277–291.

[Biancofiore 1982] A. Biancofiore, "A hypersurface defect relation for a class of meromorphic maps", *Trans. Amer. Math. Soc.* **270**:1 (1982), 47–60.

[Bloch 1926] A. Bloch, "Sur les systèmes de fonctions uniformes satisfaisant à l'équation d'une variété algébrique dont l'irrégularité dépasse la dimension", *J. de Math.* **5** (1926), 19–66.

[Cartan 1933] H. Cartan, "Sur les zéros des combinaisons linéaires de p fonctions holomorphes données", *Mathematica (Cluj)* **7** (1933), 5–29.

[Erëmenko and Sodin 1991] A. È. Erëmenko and M. L. Sodin, "The value distribution of meromorphic functions and meromorphic curves from the point of view of potential theory", *Algebra i Analiz* **3**:1 (1991), 131–164. In Russian; translated in *St. Petersburg Math. J.* **3** (1992), 109–136.

[Faltings 1983] G. Faltings, "Endlichkeitssätze für abelsche Varietäten über Zahlkörpern", *Invent. Math.* **73**:3 (1983), 349–366.

[Faltings 1991] G. Faltings, "Diophantine approximation on abelian varieties", *Ann. of Math.* (2) **133**:3 (1991), 549–576.

[Fulton 1976] W. Fulton, "Ample vector bundles, Chern classes, and numerical criteria", *Invent. Math.* **32**:2 (1976), 171–178.

[Green and Griffiths 1980] M. Green and P. Griffiths, "Two applications of algebraic geometry to entire holomorphic mappings", pp. 41–74 in *The Chern Symposium* (Berkeley, 1979), edited by W.-Y. Hsiang et al., Springer, New York, 1980.

[Grothendieck 1958] A. Grothendieck, "La théorie des classes de Chern", *Bull. Soc. Math. France* **86** (1958), 137–154.

[Hayman 1964] W. K. Hayman, *Meromorphic functions*, Oxford Mathematical Monographs, Clarendon Press, Oxford, 1964.

[Kawamata 1980] Y. Kawamata, "On Bloch's conjecture", *Invent. Math.* **57**:1 (1980), 97–100.

[McQuillan 1996] M. McQuillan, "A new proof of the Bloch conjecture", *J. Algebraic Geom.* **5**:1 (1996), 107–117.

[McQuillan 1997] M. McQuillan, "A dynamical counterpart to Faltings' 'diophantine approximation on abelian varieties'", preprint, Inst. Hautes Études Sci., Bures-sur-Yvette, 1997.

[Noguchi and Ochiai 1990] J. Noguchi and T. Ochiai, *Geometric function theory in several complex variables*, Amer. Math. Soc., Providence, RI, 1990.

[Ochiai 1977] T. Ochiai, "On holomorphic curves in algebraic varieties with ample irregularity", *Invent. Math.* **43**:1 (1977), 83–96.

[Osgood 1985] C. F. Osgood, "Sometimes effective Thue–Siegel–Roth–Schmidt–Nevanlinna bounds, or better", *J. Number Theory* **21**:3 (1985), 347–389.

[Roth 1955] K. F. Roth, "Rational approximations to algebraic numbers", *Mathematika* **2** (1955), 1–20. Corrigendum, p. 168.

[Ru and Wong 1995] M. Ru and P.-M. Wong, "Holomorphic curves in abelian and semi-abelian varieties", preprint, University of Notre Dame, Notre Dame, IN, 1995.

[Sakai 1979] F. Sakai, "Symmetric powers of the cotangent bundle and classification of algebraic varieties", pp. 545–563 in *Algebraic geometry* (Copenhagen, 1978), edited by K. Lønsted, Lecture Notes in Math. **732**, Springer, Berlin, 1979.

[Schmidt 1980] W. M. Schmidt, *Diophantine approximation*, Lecture Notes in Math. **785**, Springer, Berlin, 1980.

[Siu 1993] Y. T. Siu, "An effective Matsusaka big theorem", *Ann. Inst. Fourier (Grenoble)* **43**:5 (1993), 1387–1405.

[Siu 1995] Y.-T. Siu, "Hyperbolicity problems in function theory", pp. 409–513 in *Five decades as a mathematician and educator: on the 80th birthday of Professor Yung-Chow Wong*, edited by K.-Y. Chan and M.-C. Liu, World Sci. Publishing, Singapore, 1995.

[Siu and Yeung 1996a] Y.-T. Siu and S.-K. Yeung, "A generalized Bloch's theorem and the hyperbolicity of the complement of an ample divisor in an abelian variety", *Math. Ann.* **306**:4 (1996), 743–758.

[Siu and Yeung 1996b] Y.-T. Siu and S.-K. Yeung, "Hyperbolicity of the complement of a generic smooth curve of high degree in the complex projective plane", *Invent. Math.* **124**:1-3 (1996), 573–618.

[Siu and Yeung 1997] Y.-T. Siu and S.-K. Yeung, "Defects for ample divisors of abelian varieties, Schwarz lemma, and hyperbolic hypersurfaces of low degrees", *Amer. J. Math.* **119**:5 (1997), 1139–1172.

[Vojta 1987] P. Vojta, *Diophantine approximations and value distribution theory*, Lecture Notes in Math. **1239**, Springer, Berlin, 1987.

[Vojta 1992] P. Vojta, "A generalization of theorems of Faltings and Thue–Siegel–Roth–Wirsing", *J. Amer. Math. Soc.* **5**:4 (1992), 763–804.

[Vojta 1996] P. Vojta, "Integral points on subvarieties of semiabelian varieties. I", *Invent. Math.* **126**:1 (1996), 133–181.

[Wong 1980] P. M. Wong, "Holomorphic mappings into abelian varieties", *Amer. J. Math.* **102**:3 (1980), 493–502.

YUM-TONG SIU
DEPARTMENT OF MATHEMATICS
HARVARD UNIVERSITY
CAMBRIDGE, MA 02138
UNITED STATES
siu@math.harvard.edu

Several Complex Variables
MSRI Publications
Volume **37**, 1999

Rigidity Theorems in Kähler Geometry and Fundamental Groups of Varieties

DOMINGO TOLEDO

ABSTRACT. We review some developments in rigidity theory of compact
Kähler manifolds and related developments on restrictions on their possible
fundamental groups.

1. Introduction

This article surveys some developments, which started almost twenty years
ago, on the applications of harmonic mappings to the study of topology and
geometry of Kähler manifolds. The starting point of these developments was the
strong rigidity theorem of Siu [1980], which is a generalization of a special case
of the strong rigidity theorem of Mostow [1973] for locally symmetric manifolds.

Siu's theorem introduced for the first time an effective way of using, in a broad
way, the theory of harmonic mappings to study mappings between manifolds.
Many interesting applications of harmonic mappings to the study of mappings of
Kähler manifolds to nonpositively curved spaces have been developed since then
by various authors. More generally the linear representations (and other representations) of their fundamental groups have also been studied. Our purpose
here is to give a general survey of this work.

One interesting by-product of this study is that it has produced new results
on an old an challenging question: what groups can be fundamental groups of
smooth projective varieties (or of compact Kähler manifolds)? These groups
are called *Kähler groups* for short, and have been intensively studied in the
last decade. New restrictions on Kähler groups have been obtained by these
techniques. On the other hand new examples of Kähler groups have also shown
the limitations of some of these methods. We do not discuss these developments
in much detail because we have nothing to add to the recent book [Amorós et al.
1996] on this subject.

The author was partially supported by National Science Foundation Grant DMS-9625463.

Even though the motivation for much of what we cover here came from the general rigidity theory for lattices in Lie groups, we do not attempt to review this important subject. We begin our survey with the statement of Mostow's strong rigidity theorem for hyperbolic space forms, and refer the reader to [Pansu 1995] and the references therein for more information on both the history and the present state of rigidity theory. We also refer the reader to [Amorós et al. 1996; Arapura 1995; Corlette 1995; Katzarkov 1997; Kollár 1995; Simpson 1997] for surveys that have some overlap and give more information on some of the subjects specifically covered here.

2. The Theorems of Mostow and Siu

We begin by recalling the first strong rigidity theorem of all, Mostow's strong rigidity theorem for hyperbolic space forms. We have slightly restated the original formulation found in [Mostow 1968].

THEOREM 2.1. *Let M and N be compact manifolds of constant negative curvature and dimension at least three, and let $f : M \to N$ be a homotopy equivalence. Then f is homotopic to an isometry.*

This theorem says in particular that there are no continuous deformations of metrics of constant negative curvature in dimensions greater than two, in sharp contrast to the situation for Riemann surfaces, where there are deformations. All proofs of this theorem seem to involve the study of an extension to the boundary of hyperbolic n-space of the lift of f to the universal cover of M. Besides the original proof in [Mostow 1968] we mention the proof by Gromov and Thurston (explained in [Thurston 1978] in dimension 3 and now known to be valid in all dimensions). They prove actually more: if in the statement of Theorem 2.1 we assume that f is a map of degree equal to the ratio of the hyperbolic volumes, then f is homotopic to a covering isometry. This stronger statement is also proved in [Besson et al. 1995].

For the purposes of this survey, we note that a natural way to attempt to prove Theorem 2.1 would be the following. First, the basic existence theorem of Eells and Sampson [1964] implies that f is homotopic to a harmonic map (unique in this case because of the strict negativity of the curvature [Hartman 1967]). We can thus assume that the homotopy equivalence f is harmonic, and it is natural to expect that one could prove directly that f is an isometry, thus establishing Mostow's theorem 2.1.

It is very curious to note that this has not been done, and in some sense is one of the outstanding problems in the theory of harmonic maps. All the developments in harmonic map theory that we mention in this article, by the very nature of the methods employed, must leave this case untouched. Of course one knows *a fortiori*, from Mostow's theorem and the uniqueness of harmonic maps that f is an isometry. But it does not seem to be known even how to prove that

f is a diffeomorphism without appealing to Mostow's theorem. In this context it should be noted than in dimension two, where the rigidity theorem 2.1 fails, it is known that a harmonic homotopy equivalence between compact surfaces of constant negative curvature is a diffeomorphism [Sampson 1978; Schoen and Yau 1978].

It is the author's impression that during the 1960's and 1970's several mathematicians attempted to prove Theorem 2.1 by showing that the harmonic map is an isometry. The failure of all these attempts was taken at that time as an indication of the limited applicability of the theory of harmonic maps.

In the early 1970's Mostow proceeded to prove his general rigidity theorem [Mostow 1973], namely the same as Theorem 2.1 with M and N now irreducible compact locally symmetric manifolds, the statement otherwise unchanged. Since what was thought to be the simplest case, namely that of constant curvature manifolds, was not accessible by harmonic maps, no one expected the more general case to be approachable this method. It was thus surprising when Siu [1980] was able to prove, by harmonic maps, the following strengthening of Mostow's rigidity theorem for Hermitian symmetric manifolds:

THEOREM 2.2 (SIU'S RIGIDITY THEOREM). *Let M and N be compact Kähler manifolds. Assume that the universal cover of N is an irreducible bounded symmetric domain other that the unit disc in $\mathbb{C}$. Let $f : M \to N$ be a homotopy equivalence. Then f is homotopic to a holomorphic or anti-holomorphic map.*

This strengthens Mostow's rigidity theorem because only one of the two manifolds is assumed to be locally symmetric. The conclusion may seem weaker (biholomorphic map rather than isometric), but recall that if M is also locally symmetric, that is, its universal cover is a bounded symmetric domain, then f is indeed homotopic to an isometry because biholomorphic maps of bounded domains are isometric for their Bergmann metrics.

Siu proves his theorem by showing that the harmonic map homotopic to f is holomorphic or antiholomorphic. We explain the details, and some extensions, in the next two sections.

We close this section with the remark that this theorem was one of the first two substantial applications of harmonic maps to geometry. The other application, appearing about the same time, was the solution by Siu and Yau [1980] of Frankel's conjecture: a compact Kähler manifold of positive holomorphic bisectional curvature is biholomorphic to complex projective space. This theorem was somewhat overshadowed by Mori's proof [1979], at about the same time, of the more general Hartshorne conjecture in algebraic geometry: a smooth projective variety with ample tangent bundle is biholomorphic to complex projective space. This is another type of rigidity property, in the context of positive curvature rather than negative curvature. It concerns the rigidity properties of Hermitian symmetric spaces of compact, rather than noncompact type. We do not cover this interesting line of development here, but refer the reader to [Hwang

and Mok 1998; 1999; Mok 1988; Siu 1989; Tsai 1993] and the references in these
papers for more information.

It is worth noting that both the theorem of Siu and Yau and the theorem of
Mori are based on producing suitable rational curves. Siu and Yau use harmonic
two-spheres, Mori uses the action of Frobenius in positive characteristic to produce the rational curves. It has been remarked to the author by M. Gromov the
philosophical similarity between elliptic theory and the action of Frobenius, and
the fact that the latter should also be used to study rigidity problems in nonpositively curved situations. This author would not be surprised to find that the
solution to some of the open problems mentioned in this article will eventually
depend on ideas from algebraic geometry in positive characteristic.

3. Harmonic Maps are Pluriharmonic

We explain briefly the proof of Siu's rigidity theorem and some of its extensions, following the exposition in [Amorós et al. 1996; Carlson and Toledo 1989].
Recall that a map $f : M \to N$ between Riemannian manifolds is called *harmonic*
if it is an extremal for the energy functional

$$E(f) = \int_M \|df\|^2 \, dV,$$

where dV is the Riemannian volume element of M. Being an extremal is equivalent to the Euler–Lagrange equation

$$\Delta f := *d_\nabla * df = 0, \tag{3-1}$$

where the symbols have the following meaning. We write $A^k(M, f^*TN)$ to
denote the space of smooth k-forms on M with coefficients in f^*TN, $d_\nabla :$
$A^k(M, f^*TN) \to A^{k+1}(M, f^*TN)$ the exterior differentiation induced by the the the
Levi-Civita connection of N: $d_\nabla(\alpha \otimes s) = d\alpha \otimes s + (-1)^k \alpha \otimes \nabla s$ for $\alpha \in A^k(M)$
and s a smooth section of f^*TN. Then $d_\nabla^2 = -R$, where R is the curvature
tensor of N.

In a Hermitian manifold of complex dimension n one has an identity on one
forms (up to a multiplicative constant) $*\alpha = \omega^{n-1} \wedge J\alpha$, where ω is the fundamental 2-form associated to the metric and J is the complex structure. Thus there
is an identity (up to multiplicative constant) $*df = \omega^{n-1} \wedge Jdf = \omega^{n-1} \wedge d^c f$.
Thus in a Hermitian manifold the harmonic equation (3–1) is equivalent to the
equation

$$d_\nabla(\omega^{n-1} \wedge d^c f) = 0.$$

Thus in a Kähler manifold, since $d\omega = 0$, the harmonic equation is equivalent
to the equation

$$\omega^{n-1} \wedge d_\nabla d^c f = 0. \tag{3-2}$$

Observe that if $n = 1$ then (3–2) is equivalent to $d_\nabla d^c f = 0$, which is independent
of the Hermitian metric on M (depends just on the complex structure of M).

Thus if M is a complex manifold, N is a Riemannian manifold, and $f : M \to N$ is a smooth map, it makes sense to say that f is a *pluriharmonic* map if its restriction to every germ of a complex curve in M is a harmonic map. Clearly f is pluriharmonic if and only if it satisfies the equation

$$d_\nabla d^c f = 0. \tag{3-3}$$

The basic discovery of Siu was that harmonic maps of compact Kähler manifolds to Kähler manifolds with suitable curvature restrictions are pluriharmonic. This was later extended by Sampson to more general targets. The curvature condition on N is called *nonpositive Hermitian curvature* and is defined to be the condition:

$$R(X, Y, \bar{X}, \bar{Y}) \le 0$$

for all $X, Y \in TN \otimes \mathbb{C}$. Here R is the curvature tensor of N, extended by complex multilinearity to complex vectors. The theorem is then the following:

THEOREM 3.1 (SIU–SAMPSON). *Let M be a compact Kähler manifold, let N be a Riemannian manifold of nonpositive Hermitian curvature, and let $f : M \to N$ be a harmonic map. Then f is pluriharmonic.*

We now explain the proof of this theorem. If $n = 1$ there is nothing to prove, since there is no difference between harmonic and pluriharmonic. If $n \ge 2$ the proof proceeds by an integration by parts argument (or Bochner formula) as follows. First, by Stokes's theorem and the compactness of M we have

$$\int_M d(\langle d^c f \wedge d_\nabla d^c f \rangle \wedge \omega^{n-2}) = 0. \tag{3-4}$$

Here, and it what follows, we use the symbol $< \alpha >$ to denote the scalar-valued form obtained from a form α with values in $f^*(TN \otimes TN)$ by composing with the inner product $\langle \, , \rangle : TN \otimes TN \to \mathbb{R}$. Expanding the integrand using the Leibniz rule and $d\omega = 0$, we get a sum of two terms:

$$\langle d_\nabla d^c f \wedge d_\nabla d^c f \rangle \wedge \omega^{n-2} - \langle d^c f \wedge d_\nabla^2 d^c f \rangle \wedge \omega^{n-2}.$$

Now the first term is pointwise negative definite on harmonic maps by the so-called Hodge signature theorem: $\alpha \wedge \alpha \wedge \omega^{n-1} \le 0$ on the space of $(1,1)$-forms α such that $\alpha \wedge \omega^{n-1} = 0$, with equality if and only if $\alpha = 0$. Now the harmonic equation on Kähler manifolds we have just seen is equivalent to (3–2), thus the asserted negativity on harmonic maps.

The second term, when rewritten using the definition of curvature $d_\nabla^2 = -R$, turns out to be the average value of $R(df(X), df(Y), df(\bar{X}), df(\bar{Y}))$ over all unit length decomposable vectors $X \wedge Y \in \bigwedge^2 T^{1,0} M$ (that is, over all two-dimensional subspaces of $T^{1,0} M$). This computation can be found in [Amorós et al. 1996] (in the notation used here), or in equivalent forms in [Siu 1980; Sampson 1986].

Thus if N has nonpositive Hermitian curvature the two terms have the same sign and add to zero, thus each is zero. The vanishing of the first term is the pluriharmonic equation (3–3):

$$d_\nabla d^c f = 0$$

and the vanishing of the second term gives the following equations, which are also directly a consequence (by differentiation) of the pluriharmonic equation:

$$R(df(X), df(Y), df(\bar{X}), df(\bar{Y})) = 0 \ for \ all \ X, Y \in T^{1,0}M. \qquad (3\text{–}5)$$

This concludes the proof of the Siu–Sampson theorem 3.1.

Before proceedings to applications, we point out three generalizations of this theorem that will be needed in the sequel:

GENERALIZATION 1. Theorem 3.1 holds for *twisted harmonic maps*. This means the following. Let X be a Riemannian manifold of nonpositive Hermitian curvature, let G be its group of isometries, and let $\rho : \pi_1(M) \to G$ be a representation. A twisted harmonic map (twisted by ρ) means a ρ-equivariant harmonic map $f : \tilde{M} \to X$, where $\tilde{M}$ denotes the universal cover of M and $\pi_1(M)$ acts on $\tilde{M}$ by covering transformations. Equivariant means as usual that $f(\gamma x) = \rho(\gamma)f(x)$ holds for all $\gamma \in \pi_1(M)$ and all $x \in \tilde{M}$. Equivariant maps are in one to one correspondence with sections of the flat bundle over M with fiber X associated to ρ, and equivariant harmonic maps correspond to harmonic sections of this bundle.

Since the integrand in (3–4) is an invariant form on $\tilde{M}$ (and thus descends to a form on M) for f a ρ-equivariant map, it is clear that the proof of Theorem 3.1 still holds in this context. Thus twisted harmonic maps of compact Kähler manifolds to Riemannian manifolds of nonpositive Hermitian curvature are pluriharmonic (and (3–5) holds).

GENERALIZATION 2. Theorem 3.1 holds under the following variation of its hypotheses: M, rather than a Kähler manifold, is a hermitian manifold whose fundamental form ω satisfies $dd^c(\omega^{n-2}) = 0$, and f, rather than a harmonic map, is a map that satisfies the equation (3–2). This was observed by Jost and Yau [1993b] where they call such manifolds M *astheno-Kählerian* and such maps f *Hermitian harmonic*. The proof of this extension is that the condition $dd^c(\omega^{n-2}) = 0$ is exactly what is needed to carry through the above integration by parts argument, provided of course that f satisfies the equation (3–2) (which differs from the harmonic equation by a lower order term if $d\omega \neq 0$).

GENERALIZATION 3. Theorem 3.1 holds for harmonic maps (or twisted harmonic maps) of compact Kähler manifolds to suitable singular spaces of nonpositive curvature (for example trees, or Bruhat–Tits buildings). This has been proved by Gromov and Schoen [1992]. The main points are, first, to define what is meant by a harmonic map, and then to prove that such a map has sufficient regularity for the integrand in (3–4) to make sense and the argument to go through.

4. Applications of Pluriharmonic Maps

We specialize the considerations of the last section to the case where N is a locally symmetric space of noncompact type. This means that the universal covering manifold of N is a symmetric space G/K, where G is a connected semisimple linear Lie group without compact factors and K is its maximal compact subgroup, and G/K is given the invariant metric determined by the Killing form $\langle\,,\,\rangle$ on $\mathfrak{g}$. All computations can be reduced to Lie algebra computations: We have the Cartan decomposition

$$\mathfrak{g} = \mathfrak{k} \oplus \mathfrak{p},$$

where $\mathfrak{g}$ and $\mathfrak{k}$ are the Lie algebras of G and K respectively and $\mathfrak{p}$ is a K-invariant complement to $\mathfrak{k}$. The Killing form is positive definite on $\mathfrak{p}$ and negative definite on $\mathfrak{k}$. We have the equations

$$[\mathfrak{k}, \mathfrak{p}] \subset \mathfrak{p}, \quad [\mathfrak{p}, \mathfrak{p}] \subset \mathfrak{k}$$

expressing the invariance of $\mathfrak{p}$ and the fact that $\mathfrak{k}$ is the fixed point set of an involution of $\mathfrak{g}$. For our purposes it will be harmless to make the identification $T_x N \cong \mathfrak{p}$ of the tangent space to N at any fixed point $x \in N$ with $\mathfrak{p}$. (Strictly speaking, we should have a varying isotropy subalgebra $\mathfrak{k}$ and thus varying complement $\mathfrak{p}$.)

Under this identification the curvature tensor is given (up to multiplicative constant) by

$$R(X, Y) = [X, Y],$$

and the Hermitian curvature on $TN \otimes \mathbb{C}$ is given by

$$R(X, Y, \bar{X}, \bar{Y}) = \langle [X, Y], [\bar{X}, \bar{Y}] \rangle$$

which is nonpositive, and zero if and only if $[X, Y] = 0$, because the Killing form is negative definite on $\mathfrak{k}$.

Thus if N is a locally symmetric manifold of noncompact type the Siu–Sampson theorem 3.1 applies, the map f is pluriharmonic and satisfies the further equations (also consequence of the pluriharmonic equation):

$$R(df(X), df(Y)) = [df(X), df(Y)] = 0 \ for \ all \ X, Y \in T^{1,0}M. \tag{4–1}$$

This vanishing of curvature has the following interpretation (compare [Amorós et al. 1996; Carlson and Toledo 1989]). Let

$$d''_\nabla : A^{0,k}(M, f^*TN \otimes \mathbb{C}) \to A^{0,k+1}(M, f^*TN \otimes \mathbb{C})$$

denote the Cauchy–Riemann operator induced by the Levi-Civita connection of N. Then $(d''_\nabla)^2 = 0$, thus $f^*TN \otimes \mathbb{C}$ is a holomorphic vector bundle over M. Then, if $d'f$ denotes the restriction of df to $T^{1,0}M$, the pluriharmonic equation (3–3) reads

$$d''_\nabla d'f = 0, \tag{4–2}$$

which means that $d'f$ is a holomorphic section of $Hom(T^{1,0}M, f^*TN \otimes \mathbb{C})$. Moreover, the Lie bracket form of (4–1) means that, if we identify T_xN with $\mathfrak{p}$ as above, then $df(T^{1,0}M)$ *is an abelian subalgebra of* $\mathfrak{p} \otimes \mathbb{C}$.

This last statement, which was observed by Sampson in [Sampson 1986], represents a nontrivial set of equations that must be satisfied by pluriharmonic maps. These equations extend, to targets which are not hermitian symmetric, the equations that Siu used in [Siu 1980] to prove his rigidity theorem. Namely, observe that if G/K is a Hermitian symmetric space, then correspoding to any invariant complex structure on G/K (there are only two if G/K is irreducible) we have the decomposition

$$\mathfrak{p} \otimes \mathbb{C} = \mathfrak{p}^{1,0} \oplus \mathfrak{p}^{0,1},$$

and the integrability condition $[\mathfrak{p}^{1,0}, \mathfrak{p}^{1,0}] \subset \mathfrak{p}^{1,0}$ is equivalent, in view of $[\mathfrak{p}, \mathfrak{p}] \subset \mathfrak{k}$, to $[\mathfrak{p}^{1,0}, \mathfrak{p}^{1,0}] = 0$, thus $\mathfrak{p}^{1,0}$ *is an abelian subalgebra of* $\mathfrak{p} \otimes \mathbb{C}$. The idea of rigidity can thus be explained by saying that the Cauchy–Riemann equations

$$df(T^{1,0}M) \subset \mathfrak{p}^{1,0} = T^{1,0}N$$

can be forced on a pluriharmonic map f if one knows that abelian subalgebras of large dimension are rare. To this end the following algebraic theorem was proved in [Carlson and Toledo 1989].

THEOREM 4.1. *Let* G/K *be a symmetric space of noncompact type that does not contain the hyperbolic plane as a factor. Let* $\mathfrak{a} \subset \mathfrak{p} \otimes \mathbb{C}$ *be an abelian subalgebra. Then* $\dim(\mathfrak{a}) \leq 1/2 \dim(\mathfrak{p} \otimes \mathbb{C})$. *Equality holds in this inequality if and only if* G/K *is hermitian symmetric and* $\mathfrak{a} = \mathfrak{p}^{1,0}$ *for an invariant complex structure on* G/K.

This theorem gives a simple proof of the geometric version of Siu's rigidity theorem, namely the following statement:

THEOREM 4.2. *Let* M *be a compact Kähler manifold, let* N *be a manifold whose universal cover is an irreducible bounded symmetric domain other than the unit disk in* $\mathbb{C}$, *let* $f : M \to N$ *be a harmonic map, and suppose there is a point* $x \in M$ *such that* $df(T_xM) = T_{f(x)}N$. *Then* f *is either holomorphic or antiholomorphic.*

The proof of this theorem is now very simple. By the Siu–Sampson theorem 3.1, f is pluriharmonic. Since, by (4–2), $d'f$ is a holomorphic section of $Hom(T^{1,0}M, f^*TN \otimes \mathbb{C})$, the subset U of M on which df is surjective is the complement of an analytic subvariety. Since, by assumption, U is not empty, it is a dense connected open subset of M. By (4–1) and Theorem 4.1, at each $x \in U$ f satisfies the Cauchy–Riemann equations with respect to one of the two invariant complex structures on N. This complex structure is independent of x by the connectedness of U, hence f is holomorphic on a dense open set, hence holomorphic, with respect to this structure. In other words, f is holomorphic or antiholomorphic with respect to a preassigned complex structure on N and the proof of Theorem 4.2 is complete.

This proof of Theorem 4.2, taken from [Carlson and Toledo 1989] contains two simplifications of Siu's original proof [1980]. The first is the simple way in which the Cauchy–Riemann eauations follow from Theorem 4.1 at the points of maximum rank. The second is the observation (4–2) implies that these points form a dense connected open set, thus obviating one difficult (although interesting) result needed by Siu [1980], namely his unique continuation theorem to the effect that a harmonic map which is holomorphic on a nonempty open set is everywhere holomorphic.

The rigidity theorem 2.2 follows immediately from 4.2 and the existence theorem for harmonic maps of Eells and Sampson [1964]. Namely, since M and N are compact and N has nonpositive curvature, the main theorem of [Eells and Sampson 1964] asserts that any continuous map is homotopic to a harmonic one. Thus one may assume that the homotopy equivalence in Theorem 2.2 is harmonic. Since a smooth homotopy equivalence must have maximal rank at at least one point, Theorem 4.2 implies that it is holomorphic or anti-holomorphic, thus proving Theorem 2.2.

Now it is clear from this proof that knowledge of the abelian subalgebras of $\mathfrak{p}^{\mathbb{C}}$ should place restrictions on the harmonic maps of compact Kähler manifolds to locally symmetric spaces for G/K, and consequently, by the Eells–Sampson theorem, on the possible homotopy classes of maps. This has been done in the following cases:

Large Abelian Subalgebras of Hermitian Symmetric Spaces

THEOREM 4.3 [Siu 1982]. *Let G/K be a Hermitian symmetric space. Then there is an integer $\nu(G/K)$ with the property that if $\mathfrak{a} \subset \mathfrak{p} \otimes \mathbb{C}$ is an abelian subalgebra of dimension larger than $\nu(G/K)$, then $\mathfrak{a} \subset \mathfrak{p}^{1,0}$ for an invariant complex structure in G/K. Thus if M is compact Kähler and $f : M \to \Gamma\backslash\Gamma/K$ is a harmonic map of rank larger than $2\nu(G/K)$, then f is holomorphic with respect to an invariant complex structure on G/K.*

The numbers $\nu(G/K)$ are computed in [Siu 1982] for the irreducible Hermitian symmetric spaces.

These numbers $\nu(G/K)$ turn out to be sharp, because they happen to coincide with the largest (complex) dimension of a totally geodesic complex subspace of G/K that contains the hyperbolic plane as a factor. Thus using the nonrigidity of Riemann surfaces one can readily construct examples of nonholomorphic harmonic maps up to this rank. In this connection the most elementary and interesting case is perhaps that of the unit ball (complex hyperbolic space) where $\nu = 1$ and harmonic maps of real rank larger than two are holomorphic or anti-holomorphic. An immediate topological consequence of 4.3 that any continous map that for some topological reason forces any smooth map in its homotopy class to have rank larger than $2\nu(G/K)$ (for instance, being nontrivial in homology above that dimension) is homotopic to a holomorphic map.

When the target G/K is not Hermitian symmetric the Siu–Sampson theorem still has interesting consequences. For instance an immediate consequence of Theorem 4.1 is the following theorem [Carlson and Toledo 1989]:

THEOREM 4.4. *Let M be a compact Kähler manifold, let N be a locally symmetric space whose universal cover is not Hermitian symmetric, and let $f : M \to N$ be a harmonic map. Then $\operatorname{rank} f < \dim(N)$.*

Now one would like to improve this theorem by giving a sharp upper bound for the rank of harmonic maps. Also one may want to know more about the structure of the harmonic maps of maximum rank:

Maximum-Dimensional Abelian Subalgebras

THEOREM 4.5 [Carlson and Toledo 1993]. *Let G/K be a symmetric space of noncompact type which is not Hermitian symmetric, and let $\mu(G/K)$ be the maximum dimension of an abelian subalgebra of $\mathfrak{p} \otimes \mathbb{C}$. If M is a compact Kähler manifold and $f : M \to \Gamma \backslash G/K$ is a harmonic map, then $\operatorname{rank} f \leq 2\mu(G/K)$.*

The numbers $\mu(G/K)$ are computed in [Carlson and Toledo 1993] for all classical groups G.

The earliest, simplest, and most dramatic computation of μ was for real hyperbolic space by Sampson [1986], where he shows that $\mu = 1$ in this case, thus proving the following theorem:

THEOREM 4.6. *Let M be a compact Kähler manifold, let N be a manifold of constant negative curvature, and let $f : M \to N$ be a harmonic map. Then $\operatorname{rank} f \leq 2$.*

This theorem implies that any continuous map of a compact Kähler manifold to a compact constant curvature manifold has image deformable to a two-dimensional subspace. Thus Kähler geometry and constant negative curvature geometry are incompatible in a very strong sense.

The computations of the numbers $\mu(G/K)$ in [Carlson and Toledo 1993] show that they are typically about $1/4 \dim(G/K)$, thus giving an upper bound of about $\frac{1}{2} \dim(G/K)$ for the rank of harmonic maps. In some cases the bounds coincide with the largest dimension of a totally geodesic Hermitian subspace of G/K, thus they are sharp for suitable choice of discrete group Γ. In other cases the bound is one more than this number. In some of the cases when the two numbers coincide there is a further rigidity phenomenom: any harmonic map of this maximum rank of a compact Kähler manifold must have image contained in a totally geodesic Hermitian symmetric subspace. This is the case, for example, when $G = SO(2p, 2q)$ for $p, q \geq 4$.

There are other numbers, less understood than the numbers in 4.5, which are the analogues of the numbers in 4.3 for the non-Hermitian G/K: any harmonic map of rank larger than twice this number must arise from a variation of Hodge structure (see Section 6). This number is shown to be one for quaternionic hyperbolic space in [Carlson and Toledo 1989], in analogy to the results of Siu and Sampson just discussed: $\nu = 1$ for complex hyperbolic space and $\mu = 1$ for real hyperbolic space. For classical G these numbers are estimated in [Carlson and Toledo 1993]. In contrast with the situation of the numbers in Theorem 4.3 and most of the numbers in Theorem 4.5, these estimates are not always sharp. In some cases they can be improved by more global methods [Jost and Zuo 1996; Zuo 1994] discussed in Section 5. The general picture still has to be worked out.

Finally, we would like to mention the fact that all these results on harmonic maps can be immediately extended to *twisted harmonic maps* as in the last section. Then, thanks to the existence theorem for twisted harmonic maps, all the analogous topological applications hold.

First, the existence theorem asserts that if M is a compact Riemannian manifold, X is a complete manifold of nonpositive curvature with group of isometries G, and if $\rho : \pi_1(M) :\to G$ is a suitable representation, then a ρ-equivariant harmonic map $f : \tilde{M} \to X$ exists. The first theorem of this nature was proved by Diederich and Ohsawa [1985] for X the hyperbolic plane, then in different contexts by other authors [Donaldson 1987; Corlette 1988; Labourie 1991; Jost and Yau 1991]. In more general contexts for X not a manifold it is proved in [Gromov and Schoen 1992; Korevaar and Schoen 1993]. We state here Corlette's theorem because it is the one most relevant to this survey. It was also the first fairly general statement of the existence theorem, and the first that was stated with a broad range of applicability in mind:

THEOREM 4.7. *Let M be a compact Riemannian manifold, let G be a semisimple algebraic group, and let $\rho : \pi_1(M) \to G$ be a representation. Then a ρ-equivariant harmonic map $f : \tilde{M} :\to G/K$ exists if and only if the Zariski closure of the image of ρ is a reductive group.*

This theorem is proved in [Corlette 1988] where an application to rigidity is also given, namely the rigidity in $\mathrm{PSU}(1, n + 1)$ of representations of $\pi_1(M)$ in $\mathrm{PSU}(1, n)$, $n \geq 2$, with nonvanishing volume invariant, thus solving a conjecture of Goldman and Millson. (The corresponding statement for $n = 1$ is stated and proved in [Toledo 1989].) Another application to rigidity in presence of nonvanishing volume invariant is given in [Corlette 1991]. An application in a similar spirit, proving that certain $\mathrm{SO}(2p, 2q)$ representations must factor through $\mathrm{SU}(p, q)$ is given in [Carlson and Toledo 1993, Theorem 9.1].

5. Further Applications of Pluriharmonic Maps

Pluriharmonic maps are very special even when they are not holomorphic. For instance their fibers and their fibration structure are special. This seems to have been first exploited by Jost and Yau [1983], who proved that the fibers of

a pluriharmonic map of constant maximum rank of a compact Kähler manifold
to a Riemann surface are complex manifolds which vary holomorphically. Thus
the harmonic map can be made holomorphic by changing the complex structure
of the target surface. In this way they prove that all deformations of Kodaira
surfaces arise from deformations of the base curve.

More generally, even it the rank of the pluriharmonic map $f : M \to N$ is not
constant, or if N is not a Riemann surface, one can still try to form the quotient
of M by the equivalence relation: two points are equivalent if and only if they lie
in the same connected component of a maximal complex subvariety of a fiber.
In some cases one can show that the quotient of M by this equivalence relation
is a complex space V and that the pluriharmonic map factors as

$$M \to V \to N, \qquad\qquad (5\text{--}1)$$

where the first map is holomorphic and the second is pluriharmonic. This works
very well in case that the generic fiber of f is a divisor, and V is then a Riemann
surface: see [Siu 1987; Carlson and Toledo 1989; Jost and Yau 1991]. Thus one
can prove that harmonic maps $f : M \to N$, where N is a hyperbolic Riemann
surface factor as in (5–1), where V is a Riemann surface of possibly higher genus
than N, the first map in (5–1) is holomorphic and the second is harmonic [Siu
1987]. Similar factorization theorems hold for maps to real hyperbolic space
(consequently strengthening Sampson's theorem 4), for nonholomorphic maps
to complex hyperbolic space, and for maps to quaternionic hyperbolic space
that do not arise from variations of Hodge structure [Carlson and Toledo 1989;
Jost and Yau 1991]. Finally, an analog of this factorization theorem has been
proved by Gromov and Schoen for maps to trees (thus N is a tree rather than a
manifold). See [Gromov and Schoen 1992, § 9].

These factorizations theorems have interesting applications to the study of
Kähler groups. The first is that the property of a compact Kähler manifold of
fibering over a Riemann surface is purely a property of its fundamental group
[Beauville 1991; Catanese 1991; Siu 1987]; compare the general discussion in
[Amorós et al. 1996, Chapter 2]:

THEOREM 5.1. *Let M be a compact Kähler manifold. Then there exists a
surjective holomorphic map $f : M \to N$, where N is a compact Riemann surface
of genus $g \geq 2$ if and only if there exists a surjection $\pi_1(M) \to \Gamma_h$, where Γ_h is
the fundamental group of a compact surface of genus $h \geq 2$ and $h \leq g$.*

The second application is the following restriction on fundamental groups of
compact Kähler manifolds [Carlson and Toledo 1989]:

THEOREM 5.2. *Let Γ be the fundamental group of a compact manifold of constant
negative curvature and dimension at least 3. Then Γ is not a Kähler group.*

The interest of this theorem is that it provided the first application of pluriharmonic maps to the study of Kähler groups. Namely, the results of the last

section can be used to restrict the possible homotopy types of compact Kähler manifolds (see [Carlson and Toledo 1989, Theorem; Amorós et al. 1996, Theorem 6.17] for a concrete restriction of this type), but it is hard to give restrictions on the fundamental group by these methods.

This theorem has been extended in two directions. In [Hernández 1991], Hernández proves the same statement for Γ the fundamental group of a compact pointwise $\frac{1}{4}$-pinched negatively curved manifold. In [Carlson and Toledo 1997] the authors apply an existence theorem of Jost and Yau for Hermitian harmonic maps and Generalization 2 of section 4 to prove that such groups are not fundamental groups of compact complex surfaces.

The third application of these factorization theorems to fundamental groups of compact Kähler manifolds is the Theorem of Gromov and Schoen on amalgamated products [Gromov and Schoen 1992]:

THEOREM 5.3. *Let M be a compact Kähler manifold with $\pi_1(M) = \Gamma_1 *_\Delta \Gamma_2$, where the index of Δ in Γ_1 is at least 2 and its index in Γ_2 is at least 3. Then there exists a representation $\rho : \pi_1(M) \to \mathrm{PSL}(2, \mathbb{R})$ with discrete, cocompact image, and a holomorphic equivariant map $f : \tilde{M} \to D$, where D is the Poincaré disc.*

The interest of this theorem is that it provides restrictions on fundamental groups of compact Kähler manifolds that do not assume (as, for instance, theorem 5.2 does), that the group is linear. We will see in section 7 that there is good reason for doing this. One consequence of this theorem is that it excludes amalgamated products that are not residually finite from being Kähler groups. See [Amorós et al. 1996, § 6.5, 6.6] for further discussion of this point.

So far we have used the existence of factorizations (5–1) in situations where the generic fiber of f is a divisor. For fibers of higher codimension the situation is much more subtle. The (singular) foliation of M by the maximal complex subvarieties of the fibers of f may not have compact leaves. In cases where it can be proved to have compact leaves, the factorization (5–1) need only hold after blowing up M. It is technically much more difficult to obtain factorization theorems. A very careful discussion of such a theorem is given in [Mok 1992], where Mok proves a factorization theorem for discrete $\mathrm{SL}(k, \mathbb{R})$ representations of the fundamental group. This general philosophy makes it plausible that representations of fundamental groups of compact Kähler manifolds should factor through lower dimensional varieties. These ideas are further pursued by several authors, see [Jost and Zuo 1996; Katzarkov and Pantev 1994; Zuo 1994].

A certain picture emerges from these works, and from the work of Simpson [1991] where many of these considerations started: If a representation is not rigid, then it factors through a representation of the fundamental group of a lower-dimensional variety, whose dimension is bounded by the rank of the group. On the other hand there are rigid representations that cannot factor. It is not yet known how to combine these pictures into a picture of the general representation.

We refer to [Simpson 1993] for many examples and for formulation of specific problems that may help in seeing this general picture.

Another subject related to these ideas is the Shafarevich conjecture. This is the name given to the statement that the universal cover of a smooth projective variety is holomorphically convex, and which is posed as a question in the last section of [Shafarevich 1974]. To relate this question to pluriharmonic maps, we first make the following stricly heuristic remark. Suppose that M is compact Kähler and $\rho : \pi_1(M) \to G$ is a discrete, faithful and reductive representation with image Γ, where $G = \mathrm{SL}(k, \mathbb{R})$. Then by Theorem 4.7 there is a harmonic map $f : M \to \Gamma\backslash G/K$. It is easy to see that a pluriharmonic map pulls back convex functions to plurisubharmonic functions. Thus if $\phi : G/K \to \mathbb{R}$ denotes distance from a point, then ϕ is convex and consequently $f^*\phi$ is a plurisubharmonic exhaustion function on $\tilde{M}$. If it were *strictly* plurisubharmonic one would of course prove that $\tilde{M}$ is Stein, hence holomorphically convex. It is however well known that the existence of (weakly) plurisubharmonic exhaustion functions does not imply holomorphic convexity, so this approach does not prove the Shafarevich conjecture for M. But I hope that it makes it plausible that there could be a connection between pluriharmonic maps and the Shafarevich conjecture for (discrete, reductive) linear groups.

In fact there are such connections, of course in more subtle and involved ways. It is now known, thanks to work of Napier, Ramachandran, Lassell, Katzarkov, Pantev that the Shafarevich conjecture holds for *surfaces with linear fundamental group*; see [Katzarkov 1997; Katzarkov and Ramachandran 1998; Lasell and Ramachandran 1996; Napier 1990; Napier and Ramachandran 1995]. Very briefly, pluriharmonic maps to symmetric spaces and to buildings are used to prove the Shafarevich conjecture for linear reductive groups in [Katzarkov and Ramachandran 1998], where a reduction to a criterion of Napier [1990] (no infinite connected chain of compact curves in the universal cover) is used. The nonreductive case case is described in [Katzarkov 1997] by combining the reductive ideas with relative nilpotent completion ideas.

On the other hand, there is a tantalizing idea, due to Bogomolov and further developed in [Bogomolov and Katzarkov 1998; Katzarkov 1997], for possibly giving counterexamples to the Shafarevich conjecture. Part of the idea is to find relations with the free Burnside groups. The groups in question will of course be far from linear. Even though this work has not yet produced the desired counterexamples, this author feels that this type of construction will eventually prove fruitful in producing examples of nontrivial behavior of fundamental groups.

Much of what has been said in this section concerns the factorization of a manifold by a suitable equivalence relation. We point out the paper [Kollár 1993] where Kollár proves that the natural equivalence relation related to the Shafarevich conjecture is generically well-behaved. (See also [Campana 1994].) The book [Kollár 1995] contains many interesting examples and information on this equivalence relation, and in relations of the fundamental group with algebraic geometric properties of varieties.

Finally, we mention another recent application of pluriharmonic maps, namely the solution of Bloch's conjecture by Reznikov [1995]. This is the statement that the higher Chern–Simons classes of a flat vector bundle over a smooth projective variety are torsion classes.

6. Nonabelian Hodge Theory

Closely related to the theory of harmonic maps is the nonabelian Hodge theory of Corlette and Simpson. Since this theory is amply described in [Amorós et al. 1996; Simpson 1992; Simpson 1997], we limit ourselves to a few comments most closely related to this survey.

For simplicity we let $G = GL(m, \mathbb{C})$ and observe that if $\rho : \pi_1(M) :\to G$ is a reductive representation, then by Theorem 4.7 a twisted harmonic map exists, which is pluriharmonic by Theorem 3.1. If we let $\theta = d'f$, then equation (4–2) can be interpreted as saying that the flat $\mathbb{C}^n$-bundle undelying the flat $GL(m, \mathbb{C})$-bundle has a holomorphic structure, so that the induced holomophic structure on $\mathrm{End}(E)$ is the holomorphic structure given by d''_∇ (using the identification $\mathrm{End}(\mathbb{C}^n) = \mathfrak{p} \otimes \mathbb{C}$). Thus

$$\theta \in H^0(M, \Omega^1 \otimes \mathrm{End}(E)), \tag{6–1}$$

and the abelian equations (4–1) are equivalent to

$$[\theta, \theta] = 0 \in H^0(M, \Omega^2 \otimes \mathrm{End}(E)). \tag{6–2}$$

Now the data: a holomorphic vector bundle E over M and a holomorphic one-form θ as in (6–1) satisfying (6–2) is by definition a Higgs bundle over M. This notion was introduced by Hitchin [1987] for M a Riemann surface, where (6–2) is vacuous, and for higher-dimensional M by Simpson [1992].

We have just seen that a reductive representation of $\pi_1(M)$ gives rise to a Higgs bundle, which must satisfy, as a consequence of reductivity, a suitable stability condition (in the sense of geometric invariant theory). Conversely, Simpson proves that a stable Higgs bundle arises from a representation of $\pi_1(M)$. The end result is that the subset $H^1_{\mathrm{red}}(M, G)$ of the first cohomology set $H^1(M, G)$ given by reductive representations is in one to one correspondence with the set of isomorphism classes of stable Higgs bundles.

Simpson uses this correspondence to define a $\mathbb{C}^*$-action on $H^1_{\mathrm{red}}(M, G)$, namely the action such that $t \in \mathbb{C}^*$ sends a Higgs bundle E, θ to the Higgs bundle $E, t\theta$. This action (which is interpreted as the nonabelian analogue of the Hodge filtration on abelian cohomology) has for fixed points the *variations of Hodge structure*. For our purposes these can be defined as the representations of $\pi_1(M)$ to $GL(m, \mathbb{C})$ that have image in a subgroup of type $U(p, q)$ and whose harmonic section, with values in the symmetric space of $U(p,q)$, lifts to a horizontal holomorphic map of a suitable homogeneous complex manifold fibering over this symmetric space.

Simpson proceeds to prove that every reductive representation of $\pi_1(M)$ can be deformed to a variation of Hodge structure. In particular, rigid representations must be variations of Hodge structure. He then uses the fact that the Zarisiki closure of the monodromy of a variation of Hodge structure is what he calls a group of Hodge type (equivalently, a group with a compact Cartan subgroup, equivalent a group where the geodesic symmetry in its symmetric space is in the connected component of the identity) and the infinitesimal rigidity of most lattices to derive the following theorem:

THEOREM 6.1. *Let Γ be a lattice in a simple Lie group G and suppose that Γ is a Kähler group. Then G has a compact Cartan subgroup.*

As an application, we see that lattices in simple complex Lie groups, in $\mathrm{SL}(n, \mathbb{R})$, in $\mathrm{SO}(2p+1, 2q+1)$ are not Kähler groups.

Now if G has a compact Cartan subgroup but is not the group of automorphisms of a bounded symmetric domain (for example, the group $\mathrm{SO}(2p, 2q)$ where $p, q > 1$) it is not known if lattices in G can be Kähler groups. It is conjectured in [Carlson and Toledo 1989] that they are not, but except for the cases $\mathrm{SO}(1, 2n)$, solved in the same paper, and the automorphism group of the Cayley hyperbolic plane, solved in [Carlson and Hernández 1991], this question remains open.

Another open question is the following, Suppose G is the group of automorphisms of an irreducible bounded symmetric domain of dimension at least two, $\Gamma \subset G$ is a lattice, and M is a compact Kähler manifold with fundamental group Γ. Does there exist a *holomorphic* map $f : M \to \Gamma \backslash G / K$ inducing an isomorphism on fundamental group? If G/K is complex hyperbolic space (of dimension at least 2) the answer is affirmative, but in other cases it remains open. One needs to know whether the harmonic map is holomorphic, equivalently whether the variation of Hodge structure given by the proof of Simpson's theorem 6.1 is the standard one.

Finally, we mention that perhaps the first geometric application of nonabelian Hodge theory was the computation by Hitchin of the components of the space of $\mathrm{SL}(2, \mathbb{R})$ (or $\mathrm{PSL}(2, \mathbb{R})$)-representations of a surface group. Recall that the space of representations of the fundamental group of a surface of genus $g > 1$ in $\mathrm{PSL}(2, \mathbb{R})$ has $4g-3$ components, indexed by the value k of the Euler class, which can take any value k such that $|k| \leq 2g-2$ [Goldman 1985]. Let $r = 2g-2-|k|$. It follows from [Hitchin 1987] that the component with Euler class k is the total space of a vector bundle over the rth symmetric power of the base surface. This identification has the draw-back that it requires a fixed complex structure on the surface and does not allow one to draw any conclusions as to the action of the mapping class group. Knowledge of this action on the components of Euler class k, for $|k| < 2g - 2$, is an interesting open problem [Goldman 1985].

7. Nonlinear Kähler Groups

We have given a number of examples of how harmonic map techniques, as well as the nonabelian Hodge theory, can be used to study Kähler groups. These techniques are a natural extension of the classical ones of linear Hodge theory (compare [Amorós et al. 1996, Chapters 1 and 3]). We have seen that the nonlinear harmonic equation and nonabelian Hodge theory can be used effectively to study linear representations of Kähler groups. We have have seen one example where a more general harmonic theory applies to possibly nonlinear groups, namely Theorem 5.3. Other restrictions on Kähler groups that do not assume linearity of the group arise from L_2 harmonic theory. This development started in [Gromov 1989] and we refer to [Amorós et al. 1996, Chapter 4] for discussion of the present state of this particular subject.

Now nonlinear Kähler groups do exist. This means that the nonabelian Hodge theory can only capture part of the fundamental group, and there is indeed good reason for developing methods that apply to nonlinear groups, as the methods just mentioned.

The first example of a non-residually finite, and hence nonlinear, Kähler group was given in [Toledo 1993]. The construction is briefly the following. Let M be a compact locally symmetric variety for the symmetric space of $SO(2,4)$ such that M contains a smooth totally geodesic divisor D corresponding to a standard embedding of $SO(2,3)$ in $SO(2,4)$. It is proved in [Toledo 1993] that there is a smooth projective variety $X \subset M - D$ so that the inclusion induces an isomorphism $\pi_1(X) \cong \pi_1(M - D)$. Now there is an exact sequence

$$1 \to K \to \pi_1(M - D) \to \pi_1(M) \to 1, \qquad (7\text{--}1)$$

where K is a free group of infinite rank, namely $K = \pi_1(\tilde{M} - \pi^{-1}(D))$, where $\tilde{M}$, the universal cover of M, is the symmetric space for $SO(2,4)$ and $\pi^{-1}(D)$ is the disjoint union of countably many copies of the symmetric space of $SO(2,3)$, each totally geodesically embedded in $\tilde{M}$.

Let N denote a tubular neighborhood of D in M, and let ∂N denote its boundary, which is a circle bundle over D. Then there is an exact sequence

$$1 \to \mathbb{Z} \to \pi_1(\partial N) \to \pi_1(D) \to 1, \qquad (7\text{--}2)$$

and it is easily seen that the maps induced by inclusion map each element of (7–2) *injectively* to the corresponding element of (7–1). In particular $\pi_1(\partial N)$ is a subgroup of $\pi_1(M - D) \cong \pi_1(X)$. Now since ∂N is a locally homogeneous circle bundle over D, it is easy to identify this bundle and to show that its fundamental group is a lattice in an infinite cyclic covering group of $SO(2,3)$. Now this covering group is a nonlinear Lie group, and a remarkable theorem of Raghunathan [1984] implies that this lattice is not residually finite. Thus $\pi_1(X)$ contains the non-residually finite subgroup $\pi_1(\partial N)$, thus it is itself not residually finite. From this it is easy to see that the intersection of all subgroups of finite index of $\pi_1(X)$ is a free group of infinite rank.

It is possible to prove, essentially as a consequence of the Margulis superrigidity theorem (see Section 8 for the Margulis theorem), that $\pi_1(\partial N)$ is not a linear group. This is a weaker, but less subtle, result than Raghunathan's theorem. It gives immediately the weaker result that $\pi_1(X)$ is not a linear group. We leave the details of this simpler result to the interested reader.

There have been other constructions of non-residually finite Kähler groups. A construction by Nori and independently by Catanese and Kollár [1992] gives Kähler groups such that the intersection of all subgroups of finite index is a finite cylic group. The present author then constructed examples where this intersection is any finitely generated abelian group. See [Amorós et al. 1996, Chapter 8] for a detailed discussion of all these examples.

It is interesting to note that to date all known examples of non-residually finite Kähler groups are based on Raghunathan's theorem (or similar theorems for lattices in covering groups of automorphism groups of other symmetric domains [Prasad and Rapinchuk 1996]. There is an interesting proposal arising from the work of Bogomolov and Katzarkov and a suggestion of Nori's that may give a different kind of example, where the interesection of all subgroups of finite index is itself not residually finite. However the verification of the proposed examples is still conjectural and depends on the solution of difficult problems in group theory [Bogomolov and Katzarkov 1998].

8. Other rigidity Theorems

Even though this article is concerned mostly with applications of harmonic maps to complex analysis, there are closely related applications of harmonic maps to rigidity theorems that should be mentioned here. We refer to the surveys [Corlette 1995; Pansu 1995], and to the original references [Corlette 1992; Jost and Yau 1993a; Mok et al. 1993] for more details.

In retrospect, one can say that the reason that the Siu–Sampson theorem works is that the holonomy group of a Kähler manifold is contained in the unitary group $U(n)$ which is a proper subgroup of the holonomy group $SO(2n)$ of the general oriented Riemannian manifold of dimension $2n$. On the general Riemannian manifold the only Bochner formula that is available is the original formula of Eells and Sampson [1964] which involves both the curvature of the target *and* the Ricci curvature of the domain. One of the achievements of [Siu 1980] was to find a Bochner identity that did not involve the curvature tensor of the domain. One can now say that the reason Siu was successful was that Kähler manifolds of complex dimension at least two admit a parallel form distinct from the volume form, namely the Kähler form. And this is equivalent to the fact that the holonomy group of a Kähler manifold is contained in $U(n)$.

Once it was realized that the Siu–Sampson theorem was probably related to special holonomy groups, the search began for other Bochner formulas for other holonomy groups. The interest in this search was to complete the superrigidity theorems of Margulis [1975; 1991]. Namely, Margulis had proved his celebrated generalization of the Mostow rigidity theorem for irreducible lattices in a real algebraic groups G *of real rank at least two*, and which says essentially that a homomorphism of such a lattice to a simple algebraic group H over a local field either extends to a homomorphism of algebraic groups or it has relatively compact image. (See [Margulis 1975; 1991; Zimmer 1984] for the precise statement of the theorem and for proofs and applications.) The methods of Margulis used in an essential way the hypothesis of the real rank of G (i.e., the rank of the symmetric space G/K) being at least two. It was known that the theorem failed for the groups $SO(1, n)$ — [Gromov and Piatetski-Shapiro 1988] and the references therein — and (at least for small n) for the groups $SU(1, n)$ of real rank one, but it was possible that Margulis's theorem was still true for lattices in the remaining simple groups of real rank one: $Sp(1, n)$ and the automorphism group of the Cayley hyperbolic plane.

The local field in the statement of Margulis's theorem may be Archimedean or nonarchimedean. In the Archimedean case for the target group, and assuming also that the lattice is cocompact, the existence theorem for equivariant harmonic maps [Corlette 1988], reduces the Margulis theorem to the following statement (where K, K' denote the maximal compact subgroups of G, H respectively):

THEOREM 8.1. *Let $\Gamma \subset G$ be a torsion-free cocompact lattice, let $\rho : \Gamma \to H$ be a representation, let $f : G/K \to H/K'$ be a ρ-equivariant harmonic map. Then f is totally geodesic.*

In [Corlette 1992] Corlette succeeded in this search by proving a Bochner identity for harmonic maps with domain a manifold with a parallel form which implies, in the case that the domain has holonomy $Sp(1)\cdot Sp(n)$, for $n \geq 2$, as in quaternionic hyperbolic space, where there is a parallel 4-form, that harmonic maps are totally geodesic as in 8.1. He also proves 8.1 for G/K the noncompact dual of the Cayley plane, which has a parallel 8-form, thus proving the Archimedean superrigidity for cocompact lattices in these real rank one groups.

If the local field in the statement of Margulis's theorem is nonarchimedean, then then the symmetic space H/K' of the Archimedean case is replaced by the Tits building X, which is a nonpositively curved simplicial complex which plays the analogous role, for p-adic Lie groups, that the symmetric spaces play for real Lie groups. In this case the existence theorem for harmonic maps was developed by Gromov and Schoen [1992] where they reduce the Margulis theorem to the analogous statement to 8.1, where one must note that a totally geodesic map from a symmetric space of noncompact type to a building must be constant. They also prove that Corlette's Bochner formula also applies in this case to give the nonarchimedean version of 8.1 for lattices acting on quaternionic hyperbolic space (of quaternionic dimension at least two) and the hyperbolic Cayley plane. The main interest in the nonarchimedean superrigidity is that it implies the *arithmeticity* of lattices; see [Margulis 1991; Zimmer 1984].

Finally, both these results can be extended to noncocompact lattices. For the existence theorem of equivariant harmonic maps one needs an initial condition of finite energy, and one knows how to do this in the case that the target manifold has negative curvature bounded away from zero, as is the case in finite volume quotients of the rank one symmetric spaces. This requires some understanding of the nature of the cusps, as does the integration by parts argument required for the Bochner formula. All this is understood and explained in [Corlette 1992; Gromov and Schoen 1992], thus completing the Margulis superridity theorem for these rank one groups. I consider the results of these two papers the best applications of harmonic maps to rigidity questions since Siu's original rigidity theorem.

There has been another important development, namely a new proof, by harmonic maps, of most cases of the Margulis theorem. The statement of Theorem 8.1 has now been proved for G any simple noncompact group other than $SO(1, n)$ and $SU(1, n)$ (and H/K' replaced by manifolds with suitable curvature assumptions) in [Jost and Yau 1993a; Mok et al. 1993]. Instead of the Bochner formula these authors use a suitable version of Matsushima's formula, which also exploits the fact that the holonomy of the domain manifold is special.

We mention again (compare Section 2) that these methods cannot prove the original Mostow rigidity theorem for hyperbolic space forms, Theorem 2.1, because the holonomy group of a constant negative curvature manifold is the full orthogonal group, so it does not allow any of the integration by parts formulas that have been used to derive rigidity from harmonic maps. Similarly these methods, even though they easily prove Mostow rigidity for lattices in $SU(1, n)$, $n \geq 2$, by their very nature they cannot shed any light on the open question of the possibility of geometric superrigidity theorem 8.1 for lattices in $SU(1, n)$ for n large. Geometric super-rigidity fails for $n = 2$ and $n = 3$ because of the existence of non-arithmetic lattices; see [Mostow 1980; Deligne and Mostow 1986]. For $n = 2$ there is a further more dramatic failure of super-rigidity due to the existence of "non-standard homomorphisms"; [Mostow 1980, § 22]. What happens for large n seems to be wide open. Since the holonomy of a constant negative holomorphic sectional manifold is the full unitary group, it does not allow any of the additional formulas used to prove superrigidity for other Hermitian symmetric spaces. It seems that the question of geometric superrigidity 8.1 for lattices in $SU(1, n)$, n large, is the main open question in this subject.

From the point of view of the theory of harmonic mappings, two aesthetic problems that one would like to solve are the following: First, the proofs of superrigidity in [Jost and Yau 1993a; Mok et al. 1993] require an intense amount of case by case verification, which would be nice to replace by more conceptual and general arguments. Second, the harmonic map techniques have not yet been successful in proving a known and important part of Margulis's theorem, namely the superrigidity for noncocompact lattices in Lie groups of real rank at least two. The existence theorem for an equivariant harmonic map is not known here because it is not known in all generality how to find an initial condition of finite energy in the heat equation method.

References

[Amorós et al. 1996] J. Amorós, M. Burger, K. Corlette, D. Kotschick, and D. Toledo, *Fundamental groups of compact Kähler manifolds*, Math. Surveys and Monographs **44**, Amer. Math. Soc., Providence, RI, 1996.

[Arapura 1995] D. Arapura, "Fundamental groups of smooth projective varieties", pp. 1–16 in *Current topics in complex algebraic geometry* (Berkeley, CA, 1992/93), edited by H. Clemens and J. Kollár, Cambridge Univ. Press, New York, 1995.

[Beauville 1991] A. Beauville, 1991. Appendix to [Catanese 1991].

[Besson et al. 1995] G. Besson, G. Courtois, and S. Gallot, "Entropies et rigidités des espaces localement symétriques de courbure strictement négative", *Geom. Funct. Anal.* **5**:5 (1995), 731–799.

[Bogomolov and Katzarkov 1998] F. Bogomolov and L. Katzarkov, "Complex projective surfaces and infinite groups", *Geom. Funct. Anal.* **8**:2 (1998), 243–272.

[Campana 1994] F. Campana, "Remarques sur le revêtement universel des variétés kählériennes compactes", *Bull. Soc. Math. France* **122**:2 (1994), 255–284.

[Carlson and Hernández 1991] J. A. Carlson and L. Hernández, "Harmonic maps from compact Kähler manifolds to exceptional hyperbolic spaces", *J. Geom. Anal.* **1**:4 (1991), 339–357.

[Carlson and Toledo 1989] J. A. Carlson and D. Toledo, "Harmonic mappings of Kähler manifolds to locally symmetric spaces", *Inst. Hautes Études Sci. Publ. Math.* **69** (1989), 173–201.

[Carlson and Toledo 1993] J. A. Carlson and D. Toledo, "Rigidity of harmonic maps of maximum rank", *J. Geom. Anal.* **3**:2 (1993), 99–140.

[Carlson and Toledo 1997] J. A. Carlson and D. Toledo, "On fundamental groups of class VII surfaces", *Bull. London Math. Soc.* **29**:1 (1997), 98–102.

[Catanese 1991] F. Catanese, "Moduli and classification of irregular Kaehler manifolds (and algebraic varieties) with Albanese general type fibrations", *Invent. Math.* **104**:2 (1991), 263–289.

[Catanese and Kollár 1992] F. Catanese and J. Kollár, "Trento examples", pp. 134–139 in *Classification of irregular varieties: minimal models and abelian varieties* (Trento, 1990), edited by E. Ballico et al., Lecture Notes in Math. **1515**, Springer, Berlin, 1992.

[Corlette 1988] K. Corlette, "Flat *G*-bundles with canonical metrics", *J. Differential Geom.* **28**:3 (1988), 361–382.

[Corlette 1991] K. Corlette, "Rigid representations of Kählerian fundamental groups", *J. Differential Geom.* **33**:1 (1991), 239–252.

[Corlette 1992] K. Corlette, "Archimedean superrigidity and hyperbolic geometry", *Ann. of Math.* (2) **135**:1 (1992), 165–182.

[Corlette 1995] K. Corlette, "Harmonic maps, rigidity, and Hodge theory", pp. 465–471 in *Proceedings of the International Congress of Mathematicians* (Zürich, 1994), vol. 1, Birkhäuser, Basel, 1995.

[Deligne and Mostow 1986] P. Deligne and G. D. Mostow, "Monodromy of hypergeometric functions and nonlattice integral monodromy", *Inst. Hautes Études Sci. Publ. Math.* **63** (1986), 5–89.

[Diederich and Ohsawa 1985] K. Diederich and T. Ohsawa, "Harmonic mappings and disc bundles over compact Kähler manifolds", *Publ. Res. Inst. Math. Sci.* **21**:4 (1985), 819–833.

[Donaldson 1987] S. K. Donaldson, "Twisted harmonic maps and the self-duality equations", *Proc. London Math. Soc.* (3) **55**:1 (1987), 127–131.

[Eells and Sampson 1964] J. Eells, James and J. H. Sampson, "Harmonic mappings of Riemannian manifolds", *Amer. J. Math.* **86** (1964), 109–160.

[Goldman 1985] W. M. Goldman, "Representations of fundamental groups of surfaces", pp. 95–117 in *Geometry and topology* (College Park, MD, 1983/84), edited by J. Alexander and J. Harer, Lecture Notes in Math. **1167**, Springer, Berlin, 1985.

[Gromov 1989] M. Gromov, "Sur le groupe fondamental d'une variété kählérienne", *C. R. Acad. Sci. Paris Sér. I Math.* **308**:3 (1989), 67–70.

[Gromov and Piatetski-Shapiro 1988] M. Gromov and I. Piatetski-Shapiro, "Nonarithmetic groups in Lobachevsky spaces", *Inst. Hautes Études Sci. Publ. Math.* **66** (1988), 93–103.

[Gromov and Schoen 1992] M. Gromov and R. Schoen, "Harmonic maps into singular spaces and p-adic superrigidity for lattices in groups of rank one", *Inst. Hautes Études Sci. Publ. Math.* **76** (1992), 165–246.

[Hartman 1967] P. Hartman, "On homotopic harmonic maps", *Canad. J. Math.* **19** (1967), 673–687.

[Hernández 1991] L. Hernández, "Kähler manifolds and 1/4-pinching", *Duke Math. J.* **62**:3 (1991), 601–611.

[Hitchin 1987] N. J. Hitchin, "The self-duality equations on a Riemann surface", *Proc. London Math. Soc.* (3) **55**:1 (1987), 59–126.

[Hwang and Mok 1998] J.-M. Hwang and N. Mok, "Rigidity of irreducible Hermitian symmetric spaces of the compact type under Kähler deformation", *Invent. Math.* **131**:2 (1998), 393–418.

[Hwang and Mok 1999] J.-M. Hwang and N. Mok, "Holomorphic maps from rational homogeneous spaces of Picard number 1 onto projective manifolds", *Invent. Math.* **136**:1 (1999), 209–231.

[Jost and Yau 1983] J. Jost and S. T. Yau, "Harmonic mappings and Kähler manifolds", *Math. Ann.* **262**:2 (1983), 145–166.

[Jost and Yau 1991] J. Jost and S.-T. Yau, "Harmonic maps and group representations", pp. 241–259 in *Differential geometry*, edited by B. Lawson and K. Tenenblat, Pitman Monographs and Surveys in Pure and Applied Math. **52**, Longman Sci. Tech., Harlow, 1991.

[Jost and Yau 1993a] J. Jost and S.-T. Yau, "Harmonic maps and superrigidity", pp. 245–280 in *Differential geometry: partial differential equations on manifolds* (Los Angeles, 1990), edited by R. Greene and S. T. Yau, Proc. Symp. Pure Math. **54, pt. 1**, Amer. Math. Soc., Providence, RI, 1993.

[Jost and Yau 1993b] J. Jost and S.-T. Yau, "A nonlinear elliptic system for maps from Hermitian to Riemannian manifolds and rigidity theorems in Hermitian geometry", *Acta Math.* **170**:2 (1993), 221–254. Errata in **173** (1994), 307.

[Jost and Zuo 1996] J. Jost and K. Zuo, "Harmonic maps and Sl(r, $\mathbb{C}$)-representations of fundamental groups of quasiprojective manifolds", *J. Algebraic Geom.* **5**:1 (1996), 77–106.

[Katzarkov 1997] L. Katzarkov, "On the Shafarevich maps", pp. 173–216 in *Algebraic geometry* (Santa Cruz, 1995), vol. 2, edited by J. Kollár et al., Proc. Sympos. Pure Math. **62**, Amer. Math. Soc., Providence, RI, 1997.

[Katzarkov and Pantev 1994] L. Katzarkov and T. Pantev, "Representations of fundamental groups whose Higgs bundles are pullbacks", *J. Differential Geom.* **39**:1 (1994), 103–121.

[Katzarkov and Ramachandran 1998] L. Katzarkov and M. Ramachandran, "On the universal coverings of algebraic surfaces", *Ann. Sci. École Norm. Sup.* (4) **31**:4 (1998), 525–535.

[Kollár 1993] J. Kollár, "Shafarevich maps and plurigenera of algebraic varieties", *Invent. Math.* **113**:1 (1993), 177–215.

[Kollár 1995] J. Kollár, *Shafarevich maps and automorphic forms*, Princeton University Press, Princeton, NJ, 1995.

[Korevaar and Schoen 1993] N. J. Korevaar and R. M. Schoen, "Sobolev spaces and harmonic maps for metric space targets", *Comm. Anal. Geom.* **1**:3-4 (1993), 561–659.

[Labourie 1991] F. Labourie, "Existence d'applications harmoniques tordues à valeurs dans les variétés à courbure négative", *Proc. Amer. Math. Soc.* **111**:3 (1991), 877–882.

[Lasell and Ramachandran 1996] B. Lasell and M. Ramachandran, "Observations on harmonic maps and singular varieties", *Ann. Sci. École Norm. Sup.* (4) **29**:2 (1996), 135–148.

[Margulis 1975] G. A. Margulis, "Discrete groups of motions of manifolds of nonpositive curvature", pp. 21–34 in *Proceedings of the International Congress of Mathematicians* (Vancouver, 1974), vol. 2, Canad. Math. Congress, Montreal, 1975. In Russian.

[Margulis 1991] G. A. Margulis, *Discrete subgroups of semisimple Lie groups*, Ergebnisse der Math. **17**, Springer, Berlin, 1991.

[Mok 1988] N. Mok, "The uniformization theorem for compact Kähler manifolds of nonnegative holomorphic bisectional curvature", *J. Differential Geom.* **27**:2 (1988), 179–214.

[Mok 1992] N. Mok, "Factorization of semisimple discrete representations of Kähler groups", *Invent. Math.* **110**:3 (1992), 557–614.

[Mok et al. 1993] N. Mok, Y. T. Siu, and S.-K. Yeung, "Geometric superrigidity", *Invent. Math.* **113**:1 (1993), 57–83.

[Mori 1979] S. Mori, "Projective manifolds with ample tangent bundles", *Ann. of Math.* (2) **110**:3 (1979), 593–606.

[Mostow 1968] G. D. Mostow, "Quasi-conformal mappings in *n*-space and the rigidity of hyperbolic space forms", *Inst. Hautes Études Sci. Publ. Math.* **34** (1968), 53–104.

[Mostow 1973] G. D. Mostow, *Strong rigidity of locally symmetric spaces*, Annals of Mathematics Studies **78**, Princeton Univ. Press, Princeton, NJ, 1973.

[Mostow 1980] G. D. Mostow, "On a remarkable class of polyhedra in complex hyperbolic space", *Pacific J. Math.* **86**:1 (1980), 171–276.

[Napier 1990] T. Napier, "Convexity properties of coverings of smooth projective varieties", *Math. Ann.* **286**:1-3 (1990), 433–479.

[Napier and Ramachandran 1995] T. Napier and M. Ramachandran, "Structure theorems for complete Kähler manifolds and applications to Lefschetz type theorems", *Geom. Funct. Anal.* **5**:5 (1995), 809–851.

[Pansu 1995] P. Pansu, "Sous-groupes discrets des groupes de Lie: rigidité, arithméticité", pp. Exp. 778, 69–105 in *Séminaire Bourbaki, 1993/94*, Astérisque **227**, Soc. math. France, Paris, 1995.

[Prasad and Rapinchuk 1996] G. Prasad and A. S. Rapinchuk, "Computation of the metaplectic kernel", *Inst. Hautes Études Sci. Publ. Math.* **84** (1996), 91–187.

[Raghunathan 1984] M. S. Raghunathan, "Torsion in cocompact lattices in coverings of Spin(2, *n*)", *Math. Ann.* **266**:4 (1984), 403–419.

[Reznikov 1995] A. Reznikov, "All regulators of flat bundles are torsion", *Ann. of Math.* (2) **141**:2 (1995), 373–386.

[Sampson 1978] J. H. Sampson, "Some properties and applications of harmonic mappings", *Ann. Sci. École Norm. Sup.* (4) **11**:2 (1978), 211–228.

[Sampson 1986] J. H. Sampson, "Applications of harmonic maps to Kähler geometry", pp. 125–134 in *Complex differential geometry and nonlinear differential equations* (Brunswick, ME, 1984), Contemp. Math. **49**, Amer. Math. Soc., Providence, 1986.

[Schoen and Yau 1978] R. Schoen and S. T. Yau, "On univalent harmonic maps between surfaces", *Invent. Math.* **44**:3 (1978), 265–278.

[Shafarevich 1974] I. R. Shafarevich, *Basic algebraic geometry*, Grundlehren der Math. Wissenschaften **213**, Springer, New York, 1974.

[Simpson 1991] C. T. Simpson, "The ubiquity of variations of Hodge structure", pp. 329–348 in *Complex geometry and Lie theory* (Sundance, UT, 1989), edited by J. A. Carlson et al., Proc. Symp. Pure Math. **53**, Amer. Math. Soc., Providence, RI, 1991.

[Simpson 1992] C. T. Simpson, "Higgs bundles and local systems", *Inst. Hautes Études Sci. Publ. Math.* **75** (1992), 5–95.

[Simpson 1993] C. Simpson, "Some families of local systems over smooth projective varieties", *Ann. of Math.* (2) **138**:2 (1993), 337–425.

[Simpson 1997] C. Simpson, "The Hodge filtration on nonabelian cohomology", pp. 217–281 in *Algebraic geometry* (Santa Cruz, 1995), vol. 2, edited by J. Kollár et al., Proc. Sympos. Pure Math. **62**, Amer. Math. Soc., Providence, RI, 1997.

[Siu 1980] Y. T. Siu, "The complex-analyticity of harmonic maps and the strong rigidity of compact Kähler manifolds", *Ann. of Math.* (2) **112**:1 (1980), 73–111.

[Siu 1982] Y. T. Siu, "Complex-analyticity of harmonic maps, vanishing and Lefschetz theorems", *J. Differential Geom.* **17**:1 (1982), 55–138.

[Siu 1987] Y. T. Siu, "Strong rigidity for Kähler manifolds and the construction of bounded holomorphic functions", pp. 124–151 in *Discrete groups in geometry and analysis* (New Haven, 1984)), edited by R. Howe, Birkhäuser, Boston, 1987.

[Siu 1989] Y. T. Siu, "Nondeformability of the complex projective space", *J. Reine Angew. Math.* **399** (1989), 208–219. Errata in **431** (1992), 65–74.

[Siu and Yau 1980] Y. T. Siu and S. T. Yau, "Compact Kähler manifolds of positive bisectional curvature", *Invent. Math.* **59**:2 (1980), 189–204.

[Thurston 1978] W. Thurston, "The geometry and topology of 3-manifolds", lecture notes, Princeton University, 1978. Available at http://www.msri.org/publications/books/gt3m/.

[Toledo 1989] D. Toledo, "Representations of surface groups in complex hyperbolic space", *J. Differential Geom.* **29**:1 (1989), 125–133.

[Toledo 1993] D. Toledo, "Projective varieties with non-residually finite fundamental group", *Inst. Hautes Études Sci. Publ. Math.* **77** (1993), 103–119.

[Tsai 1993] I. H. Tsai, "Rigidity of holomorphic maps from compact Hermitian symmetric spaces to smooth projective varieties", *J. Algebraic Geom.* **2**:4 (1993), 603–633.

[Zimmer 1984] R. J. Zimmer, *Ergodic theory and semisimple groups*, Birkhäuser Verlag, Basel, 1984.

[Zuo 1994] K. Zuo, "Factorizations of nonrigid Zariski dense representations of π_1 of projective algebraic manifolds", *Invent. Math.* **118**:1 (1994), 37–46.

DOMINGO TOLEDO
UNIVERSITY OF UTAH
DEPARTMENT OF MATHEMATICS
155 S 1400 E, ROOM 233
SALT LAKE CITY, UT 84112-0090
UNITED STATES
toledo@math.utah.edu

Several Complex Variables
MSRI Publications
Volume **37**, 1999

Nevanlinna Theory
and Diophantine Approximation

PAUL VOJTA

ABSTRACT. As observed originally by C. Osgood, certain statements in value distribution theory bear a strong resemblance to certain statements in diophantine approximation, and their corollaries for holomorphic curves likewise resemble statements for integral and rational points on algebraic varieties. For example, if X is a compact Riemann surface of genus > 1, then there are no non-constant holomorphic maps $f : \mathbb{C} \to X$; on the other hand, if X is a smooth projective curve of genus > 1 over a number field k, then it does not admit an infinite set of k-rational points. Thus non-constant holomorphic maps correspond to *infinite* sets of k-rational points.

This article describes the above analogy, and describes the various extensions and generalizations that have been carried out (or at least conjectured) in recent years.

When looked at a certain way, certain statements in value distribution theory bear a strong resemblance to certain statements in diophantine approximation, and their corollaries for holomorphic curves likewise resemble statements for integral and rational points on algebraic varieties. The first observation in this direction is due to C. Osgood [1981]; subsequent work has been done by the author, S. Lang, P.-M. Wong, M. Ru, and others.

To begin describing this analogy, we consider two questions. On the analytic side, let X be a connected Riemann surface. Then we ask:

QUESTION 1. Does there exist a non-constant holomorphic map $f : \mathbb{C} \to X$?

The answer, as is well known, depends only on the genus g of the compactification $\bar{X}$ of X, and on the number of points s in $\bar{X} \setminus X$. See Table 1.

On the algebraic side, let k be a number field with ring of integers R, and let X be either an affine or projective curve over k. Let S be a finite set of places of k containing the archimedean places. For such sets S let R_S denote the localization of R away from places in S (that is, the subring of k consisting of elements that can be written in such a way that only primes in S occur in

Supported by NSF grant DMS95-32018 and the Institute for Advanced Study.

the denominator). We assume again that X is nonsingular. If X is affine, then fix an affine embedding; we then define an *S*-**integral point** of X (or just an **integral point**, if it is clear from the context) to be a point whose coordinates are elements of R_S. If X is projective, then we define an integral point on X to be any k-rational point (that is, any point that can be written with homogeneous coordinates in k). In this case, we ask:

QUESTION 2. Do there exist infinitely many integral points on X?

Again, let $\overline{X}$ be a nonsingular projective completion of X, let g be the genus of $\overline{X}$ (which is the same as the genus of the corresponding Riemann surface), and let s be the degree of the divisor $\overline{X} \setminus X$ (the sum of the degrees of the fields of definition of the points; over $\mathbb{C}$ this is just the number of points). The answers to both of the above questions are summarized in the following table:

g	s	Holo. curve?	∞ many integral points?
0	0	Yes	Maybe
	1	Yes	Maybe
	2	Yes	Maybe
	> 2	No	No
1	0	Yes	Maybe
	> 0	No	No
> 1		No	No

Table 1

The entries "Maybe" in the right-hand column require a little explanation. In each case there exists a curve with the given values of g and s with *no* integral points; but, for any curve with the given g and s, over a large enough number field k and with a large enough set S, there are infinitely many integral points. In that spirit, the two columns on the right have exactly the same answers.

This table could be summarized more succinctly by noting that the answer is "No" if and only if $2g - 2 + s > 0$. This condition holds if and only if X is of "logarithmic general type." On the analytic side, there is a single proof of the non-existence of these holomorphic curves, relying on a Second Main Theorem for curves. For integral points, the corresponding finiteness statements were proved separately for $g = 0$, $s > 2$ and $g > 0$, $s > 0$ by Siegel in 1921; and for $g > 1$, $s = 0$ by Faltings in 1983 (the Mordell conjecture). One of the first major applications of the analogy with Nevanlinna theory was to find a finiteness proof that unified these various proofs. This proof consisted of proving an inequality in diophantine approximation that closely parallels the Second Main Theorem.

The analogy goes into more detail on how the statements of Nevanlinna theory and diophantine approximation correspond; this will be described more fully in the first section. It allows the statements of theorems such as the First and Second Main Theorems to be translated into statements of theorems in number theory (and vice versa), but it is not as useful for translating proofs. In particular, the proofs of the first and second main theorems do not translate in this analogy, but some of the derivations of other results from these theorems can be translated. Thus, the analogy is largely formal.

In addition, it is important to note that the analogue of *one* (non-constant) holomorphic map is an *infinite* set of integral points. It is not the same analogy as one would obtain by first looking at diophantine problems over function fields, and then treating the corresponding polynomials or algebraic functions as holomorphic functions.

I thank William Cherry for many thoughtful comments on this paper.

1. The Dictionary

The analogy mentioned above is quite precise, at least as far as the statements of theorems is concerned. Before describing it, though, we briefly review some of the basics of number theory; for more details see any of the standard texts, such as [Lang 1970].

Let k be a number field; that is, a finite field extension of the rational number field $\mathbb{Q}$. Let R be its ring of integers; that is, the integral closure of the rational integers $\mathbb{Z}$ in k. We have a standard set M_k of **places** of k; it consists of **real places**, **complex places**, and **non-archimedean places**. The real places are defined by embeddings $\sigma : k \hookrightarrow \mathbb{R}$; the complex places, by complex conjugate (unordered) pairs $\sigma, \bar{\sigma} : k \hookrightarrow \mathbb{C}$; the non-archimedean places, by non-zero prime ideals $\mathfrak{p} \subseteq R$. The real and complex places are referred to collectively as the **archimedean** places.

Each place has an associated absolute value $\| \cdot \|_v : k \to \mathbb{R}_{\geq 0}$. If v is a real or complex place, corresponding to $\sigma : k \hookrightarrow \mathbb{R}$ or $\sigma : k \hookrightarrow \mathbb{C}$, respectively, then this absolute value is defined by $\|x\|_v = |\sigma(x)|_v$ or $\|x\|_v = |\sigma(x)|^2$, respectively. If v is non-archimedean, corresponding to a prime ideal $\mathfrak{p} \subseteq R$, then we define $\|x\|_v = (R : \mathfrak{p})^{-\operatorname{ord}_{\mathfrak{p}}(x)}$ if $x \neq 0$; here $\operatorname{ord}_{\mathfrak{p}}(x)$ denotes the exponent of $\mathfrak{p}$ occurring in the prime factorization of the fractional ideal (x). We will also write $\operatorname{ord}_v(x) = \operatorname{ord}_{\mathfrak{p}}(x)$. (Of course, we also define $\|0\|_v = 0$.) Here we use a little abuse of terminology when referring to "absolute values," since $\| \cdot \|_v$ does not obey the triangle inequality when v is a complex place.

The simplest example of all of this is $k = \mathbb{Q}$; in that case we have $R = \mathbb{Z}$ and $M_k = \{\infty, 2, 3, 5, 7, \ldots\}$. Here $\|x\|_\infty$ is just the usual absolute value of a rational number, and $\|x\|_p = p^{-m}$ if x can be written as $p^m a/b$ with a and b integers not divisible by p.

We can now describe the most fundamental ingredients of the analogy between Nevanlinna theory and diophantine approximation. This takes the form of a dictionary for translating various concepts between the two fields. This dictionary starts out with just a few ideas, which seem to come from nowhere. These ideas allow one to translate the basic definitions of Nevanlinna theory, and consequently the statements of many of the theorems.

On the complex analytic side of this dictionary, let X be a complex projective variety and let $f : \mathbb{C} \to X$ be a (non-constant) holomorphic curve. On the algebraic side, let k be a number field, let X be a projective variety over k (that is, an irreducible projective scheme over k), and let S be a finite set of places of k, containing all the archimedean places.

We are comparing f to an infinite set of rational points, so it is useful to split f into infinitely many pieces, each of which can be compared to one of the rational points. This is done as follows: for each $r > 0$, let f_r denote the restriction $f_r := f|_{\overline{\mathbb{D}}_r}$. Assume for the moment that $X = \mathbb{P}^1$, so f is a meromorphic map and the rational points are just rational numbers (or ∞).

In this dictionary, the domain $\overline{\mathbb{D}}_r$ of f_r is compared to M_k. Points on the boundary are compared to places $v \in S$, and $|f_r(re^{i\theta})|$ is translated into $\|x\|_v$ (for the rational point x being compared to f_r). Interior points $w \in \mathbb{D}_r$ are compared to places $v \notin S$: we translate $r/|w|$ to $(R : \mathfrak{p})$, where $\mathfrak{p}$ is the prime ideal in R corresponding to v. We also translate $\operatorname{ord}_w(f_r) = \operatorname{ord}_w(f)$ to $\operatorname{ord}_{\mathfrak{p}}(x)$. Then the counterpart of $-\log\|x\|_v$ is $\operatorname{ord}_w(f) \cdot \log(r/|w|)$. Dividing by $|w|$ requires that we rule out $w = 0$ in the above translations; this is an imperfection in the analogy, but a minor one.

Thus, the ring of meromorphic functions on $\overline{\mathbb{D}}_r$ has something close to archimedean absolute values on the boundary, and non-archimedean absolute values on the interior of the domain.

The following table summarizes the dictionary, so far.

Nevanlinna Theory	Number Theory
f meromorphic on $\mathbb{C}$	$\mathscr{S} \subseteq k$
$f|_{\overline{\mathbb{D}}_r}$ $(r > 0)$	$x \in \mathscr{S}$
$\overline{\mathbb{D}}_r$	M_k
θ	$v \in S$
$w \in \mathbb{D}_r^\times$	$v \notin S$
$|f(re^{i\theta})|,\ 0 \le \theta < 2\pi$	$\|x\|_v,\ v \in S$
$\operatorname{ord}_w f$	$\operatorname{ord}_v x$
$\dfrac{r}{|w|}$	$(R : \mathfrak{p})$
$\operatorname{ord}_w f \cdot \log \dfrac{r}{|w|}$	$\operatorname{ord}_v x \cdot \log(R : \mathfrak{p}) = -\log\|x\|_v$

Table 2. Fundamental part of the dictionary.

Let X now be an arbitrary projective variety over $\mathbb{C}$ or k, and let ϕ be a rational function on X whose zero or pole set does not contain the image of f or does not contain infinitely many of the rational points under consideration. Then one can apply the same dictionary above to $\phi \circ f_r$ in the analytic case and to $\phi(x)$ in the number field case.

More generally still, in the analytic case we can let s be a rational section of a metrized line sheaf $\mathscr{L}$ on X. Then we again have $|s(f_r(re^{i\theta}))|$ on the boundary of $\overline{\mathbb{D}}_r$ and $\mathrm{ord}_w(s \circ f_r)$ on the interior. These can be translated into the number field case as follows. Let X be a projective variety over k, let $\mathscr{L}$ be a line sheaf on X, and let s be a section of $\mathscr{L}$. For archimedean places v, let $\|\cdot\|_v$ be a metric on the lifting $\mathscr{L}_{(v)}$ of $\mathscr{L}$ to $X \times_\sigma \mathbb{C}$, where $\sigma : k \hookrightarrow \mathbb{C}$ is an embedding corresponding to v. Then, for a rational point $x \in X(k)$, $\|s(x)\|_v$ is defined via the metric on $\mathscr{L}_{(v)}$. If v is non-archimedean, one can define something similar to a metric, via the absolute values on the completions k_v. These must be done consistently, though, so that infinite sums in these absolute values converge. This can be done either via Weil functions [Lang 1983, Chapter 10] or Arakelov theory. As an example of such a consistent choice of metrics, if $X = \mathbb{P}^n$, if $\mathscr{L} = \mathscr{O}(1)$, if s is the standard section of $\mathscr{O}(1)$ vanishing at infinity, and if x has homogeneous coordinates $[x_0 : \cdots : x_n]$, then

$$\|s(x)\|_v = \frac{\|x_0\|_v}{\max\{\|x_0\|_v, \ldots, \|x_n\|_v\}} \tag{1.1}$$

is one possible choice. By tensoring and pulling back, this example can be used to construct such systems of metrics in general. One can then define $\mathrm{ord}_v(s(x))$ in terms of $\|s(x)\|_v$.

Applying these more general definitions to the dictionary in Table 2 gives the following table translating the proximity, counting, and height (characteristic) functions of Nevanlinna theory into the arithmetic setting. Here s is the canonical section of $\mathscr{O}(D)$, for a divisor D on X.

Nevanlinna Theory	Number Theory
Proximity function	
$m(D, r) = \displaystyle\int_0^{2\pi} -\log\bigl|s(f(re^{i\theta}))\bigr| \, \dfrac{d\theta}{2\pi}$	$m(D, x) = \dfrac{1}{[k : \mathbb{Q}]} \displaystyle\sum_{v \in S} -\log \|s(x)\|_v$
Counting function	
$N(D, r) = \displaystyle\sum_{w \in \mathbb{D}_r^\times} \mathrm{ord}_w \, f^* s \cdot \log \dfrac{r}{|w|}$	$N(D, x) = \dfrac{1}{[k : \mathbb{Q}]} \displaystyle\sum_{v \notin S} -\log \|s(x)\|_v$
Height (characteristic function)	
$T_D(r) = m(D, r) + N(D, r)$	$h_D(x) = \dfrac{1}{[k : \mathbb{Q}]} \displaystyle\sum_{v \in M_k} -\log \|s(x)\|_v$

Table 3. Higher-level entries in the dictionary.

Note that the integrals over the set (of finite measure) of values of θ translate into sums over the finite set S.

Also note that the above functions are additive in D (up to $O(1)$, or assuming compatible choices of metrics). They are also functorial; for example, if $\phi : X \to Y$ is a morphism of varieties and D is a divisor on Y whose support does not contain the image of ϕ, then $m_f(\phi^*D, r) = m_{\phi \circ f}(D, r)$ for holomorphic curves $f : \mathbb{C} \to X$ and $m(\phi^*D, x) = m(D, \phi(x))$ for $x \in X(k)$, again assuming that the metrics on $\mathscr{O}(D)$ and $\phi^*\mathscr{O}(D)$ are compatible.

We conclude this section by considering the case of affine varieties X. Since everything is functorial, we may assume that $X = \mathbb{A}^n$. Regard it as embedded into $\mathbb{P}^n$, and let D be the divisor at infinity. A holomorphic curve in $\mathbb{A}^n$ does not meet D; therefore $N(D, r) = 0$. Likewise, a rational point $x = [1 : x_1 : \cdots : x_n]$ is an S-integral point if and only if $x_1, \ldots, x_n$ lie in R_S. If so, then $\|x_i\|_v \leq 1$ for all i and all $v \notin S$ (corresponding to the coordinates of f_r not having poles); hence by (1.1), we have $N(D, x) = 0$ again.

Thus, it is also true on affine varieties that a non-constant holomorphic curve corresponds to an infinite set of integral points. More generally, let X be any quasi-projective variety, and write $X = \bar{X} \setminus D$, where D is a divisor; then f is a holomorphic curve in X if and only if $N(D, r) = 0$; likewise an infinite collection of rational points $x \in X(k)$ is a set of integral points if $N(D, x) = O(1)$. (The different choices of metrics may lead to $N(D, x)$ varying by a bounded amount; by the same token, different affine embeddings may introduce bounded denominators. Also, the situation is more complicated in function fields, since in that case $N(D, x)$ may be bounded, but the denominators may come from an infinite set of primes.)

The first indication that this dictionary is useful comes from the translation of Jensen's formula

$$\log |f(0)| = \int_0^{2\pi} \log |f(re^{i\theta})| \frac{d\theta}{2\pi} + N(\infty, r) - N(0, r)$$

into the number field case (here we assume that f does not have a zero or pole at the origin). The right-hand side translates (up to a factor $1/[k : \mathbb{Q}]$) into

$$\sum_{v \in S} \log \|x\|_v + \sum_{v \notin S} \log^+ \|x\|_v - \sum_{v \notin S} \log^+ \|1/x\|_v = \sum_{v \in M_k} \log \|x\|_v,$$

which is zero by the product formula [Lang 1970, Chapter V, § 1]. Consequently, the First Main Theorem (which, with the above definitions, asserts that the height $T_D(r)$ depends up to $O(1)$ only on the linear equivalence class of D, and hence we may write $T_{\mathscr{L}}(r)$ for a line sheaf $\mathscr{L}$) translates into the same assertion for the height $h_D(x)$, which is again a standard fact.

Likewise, consider the following weak version of the Second Main Theorem: if X is a smooth complex projective curve, if D is an effective divisor on X with no multiple points, if K is a canonical divisor on X, if A is an ample divisor on

X, and if $\varepsilon > 0$, then any holomorphic curve $f : \mathbb{C} \to X$ satisfies

$$m(D, r) + T_K(r) \leq_{\mathrm{exc}} \varepsilon T_A(r) + O(1). \qquad (1.2)$$

Here the subscript "exc" means that the inequality holds for all $r > 0$ outside a subset of finite Lebesgue measure. This inequality implies all the "No" entries in the middle column of Table 1.

This translates into the number field case as follows [Vojta 1992]: if X is a smooth projective curve over a number field k, if D is an effective divisor on X with no multiple points, if K is a canonical divisor on X, if A is an ample divisor on X, if $\varepsilon > 0$, and if C is a constant, then

$$m(D, x) + h_K(x) \leq \varepsilon h_A(x) + C \qquad (1.3)$$

for all but finitely many $x \in X(k)$. Again, this implies all the "No" entries in the right-hand column of Table 1.

If $\dim X > 1$, note that the above dictionary still refers to holomorphic curves, so that equidimensional results do not play a role here (although they motivate conjectures for holomorphic curves). In this case one may restrict f to be a Zariski-dense holomorphic curve, and correspondingly restrict the set of rational points to be such that every infinite subset is Zariski-dense.

2. Holomorphic Curves in Varieties of Dimension Greater Than 1

In arbitrary dimension, non-existence results for non-constant holomorphic curves mainly concern subvarieties of semiabelian varieties, and quotients of bounded symmetric domains. We first consider the former.

Recall that a **semiabelian variety** over $\mathbb{C}$ is a complex group variety A such that there exists an exact sequence of group varieties,

$$0 \to \mathbb{G}_m^\mu \to A \to A_0 \to 0,$$

where A_0 is an abelian variety. A semiabelian variety over a number field is a group variety over that number field that becomes a semiabelian variety over $\mathbb{C}$ after base change. In the context of this section, it is useful to regard semiabelian varieties as the generalization of Albanese varieties to the case of quasi-projective varieties; hence it is also common to refer to them as **quasi-abelian varieties**.

The main theorem for holomorphic curves in semiabelian varieties is the following.

THEOREM 2.1. *Let A be a semiabelian variety defined over $\mathbb{C}$, let X be a closed subvariety of A, and let D be an effective divisor on X. Then the Zariski closure of the image of any holomorphic curve $f : \mathbb{C} \to X \setminus D$ is the translate of a subgroup of A contained in $X \setminus D$.*

It is useful to think of two special cases of this theorem:

(a) $D = 0$ (Bloch's theorem)

(b) $X = A$

The general case follows from these special cases by an easy argument.

THEOREM 2.2. *Let* k, S, *and* R_S *be as usual. Let* A *be a semiabelian variety defined over* k, *let* X *be a closed subvariety of* A, *and let* D *be an effective divisor on* X. *Let* $\mathscr{Y}$ *be a model for* $X \setminus D$ *over* $\operatorname{Spec} R_S$. *Then the set* $\mathscr{Y}(R_S)$ *of* R_S-*valued points in* $\mathscr{Y}$ *is a finite union*

$$\mathscr{Y}(R_S) = \bigcup_i \mathscr{B}_i(R_S),$$

where each $\mathscr{B}_i$ *is a subscheme of* $\mathscr{Y}$ *whose generic fiber* B_i *is a translated subgroup of* A.

Here the idea of a **model** for a variety comes from Arakelov theory; see [Soulé 1992, §0.2]. This more general notion is necessary because, in general, a semiabelian variety is neither projective nor affine.

Another way to view Theorem 2.2 is that, if Z is the Zariski closure of the set of integral points of $X \setminus D$ (defined relative to some fixed model), then any irreducible component of Z must be the translate of a subgroup of A contained in $X \setminus D$. (In the case of holomorphic curves, the Zariski closure of the image of the curve is already irreducible.)

In the case of holomorphic curves, the special case $D = 0$ was proved by Bloch [1926] if A is an abelian variety; see also [Siu 1995] for a history of the other contributors to this theory, including Green-Griffiths, Kawamata, and Ochiai. The more general case when $D = 0$ and A is semiabelian was proved by Noguchi [1981]. The case $X = A$ was proved by Siu and Yeung [1996a] when A is an abelian variety and by Noguchi [1998] when A is semiabelian. This is one of the few cases in which something was proved in the number field case before the complex analytic case.

In the number field case, Faltings proved the special case in which A is an abelian variety (if $X = A$ is an abelian variety, then D is generally assumed to be ample, and then one obtains finiteness of integral points; this implies the result for general D). The general case was proved by the author. See [Faltings 1991; 1994; Vojta 1996a; 1999].

Bounded symmetric domains. Let D be a bounded symmetric domain. Recall that the underlying real manifold can be realized as a quotient G/K, where G is a semisimple Lie group and K is a maximal compact subgroup. The group G can be identified with the connected component of the group of holomorphic automorphisms of D. A subgroup H of G is called an **arithmetic subgroup** if there exists a map $i : G \to \operatorname{GL}_n(\mathbb{R})$ of Lie groups inducing an isomorphism of G with a closed subvariety of $\operatorname{GL}_n(\mathbb{R})$ defined over $\mathbb{Q}$, such that H is commensurable with $i^{-1}(\operatorname{GL}_n(\mathbb{Z}))$. Here two subgroups H_1 and H_2 of a group G

are **commensurable** if $H_1 \cap H_2$ is of finite index in H_1 and H_2. See [Baily and Borel 1966, 3.3]. Finally, an **arithmetic quotient** of D is a quotient of D by an arithmetic subgroup of G.

The quotient of a bounded symmetric domain $D = G/K$ by an arithmetic subgroup Γ of G does not contain a nontrivial holomorphic curve. Indeed, this follows by lifting the curve to D and applying Liouville's theorem. For a proof in the spirit of Nevanlinna theory, see [Griffiths and King 1973, Corollary 9.22]. Also, for a slightly stronger result, see [Vojta 1987, 5.7.7], using Theorem 5.7.2 instead of Conjecture 5.7.5 of the same reference.

Of course, nothing in the above paragraph made essential use of the fact that Γ is an arithmetic subgroup. The interest in arithmetic subgroups stems from the result of Baily and Borel [1966], showing that an arithmetic quotient of a bounded symmetric domain is a complex quasi-projective variety.

In general there is a wide choice of immersions i, leading to a wide choice of commensurability classes of arithmetic subgroups. Therefore, an arithmetic quotient is not necessarily defined over a number field. When it is, however, the philosophy of Section 1 suggests that the set of integral points on any given model of the quotient would be finite:

CONJECTURE 2.3. *Let X be a quasi-projective variety over a number field k, whose set of complex points is isomorphic to an arithmetic quotient of a bounded symmetric domain. Then, for any S and model $\mathscr{X}$ for X over R_S (where S and R_S are as in the introduction of this chapter), the set $\mathscr{X}(R_S)$ of integral points of $\mathscr{X}$ is finite.*

This is unknown except for one special case. Let $\mathscr{A}_{g,n}$ denote the moduli space of principally polarized abelian varieties of dimension g with level-n structure. For n sufficiently large, $\mathscr{A}_{g,n}$ is a quasi-projective variety defined over a number field, and its set of complex points is isomorphic to an arithmetic quotient of a bounded symmetric domain (in fact, the Siegel upper half plane). By Conjecture 2.3, $\mathscr{A}_{g,n}$ should have only finitely many integral points over R_S, for any number field over which this variety is defined. In fact, this is true, since S-integral points correspond to abelian varieties with good reduction outside S with given level-n structure, and there are only finitely many such varieties for given g, n, k, and S. This was conjectured by Shafarevich and proved by Faltings [1991]. By an extension of the Chevalley–Weil theorem [Vojta 1987, Theorem 5.1.6], this then extends to the quotient by any subgroup commensurable with $\mathrm{Sp}_{2g}(\mathbb{Z})$.

Faltings' proof of the Shafarevich conjecture, however, does not correspond to the proof of Griffiths-King. Of course it is difficult to compare proofs between Nevanlinna theory and number theory, especially for the fundamental results. However, essentially all other results of Second Main Theorem type in the number field case are proved by constructing an auxiliary polynomial. Faltings' proof, on the other hand, uses Hodge theory. Therefore a proof of the Shafarevich conjecture via construction of an auxiliary polynomial would be good to have. Indeed, the proof of this result for holomorphic curves has some moderate differences from other proofs for holomorphic curves, so this may shed more light on the analogy of Section 1.

Conjectures on general varieties. Concerning the qualitative question of existence of non-constant holomorphic curves or infinite sets of integral points, a general conjecture has been formulated by S. Lang [1991, Chapter VIII, Conjecture 1.4]. Let X be a variety defined over a subfield of $\mathbb{C}$. We begin by describing various **special sets** in X.

DEFINITION 2.4. The **algebraic special set** $\mathrm{Sp}_{\mathrm{alg}}(X)$ is the Zariski closure of the union of all images of non-constant rational maps $f : G \to X$ of group varieties into X.

DEFINITION 2.5. If X is defined over $\mathbb{C}$, the **holomorphic special set** $\mathrm{Sp}_{\mathrm{hol}}(X)$ is the Zariski closure of the union of all images of non-constant holomorphic maps $f : \mathbb{C} \to X$.

We have, trivially, $\mathrm{Sp}_{\mathrm{alg}}(X) \subseteq \mathrm{Sp}_{\mathrm{hol}}(X)$.

A general conjecture for rational points on projective varieties is:

CONJECTURE 2.6 [Lang 1991, Chapter VIII, Conjectures 1.3 and 1.4]. *Let X be a projective variety defined over a subfield K of $\mathbb{C}$ finitely generated over $\mathbb{Q}$. Then*
$$\mathrm{Sp}_{\mathrm{alg}}(X) \times_K \mathbb{C} = \mathrm{Sp}_{\mathrm{hol}}(X \times_K \mathbb{C});$$
that is, the algebraic and holomorphic special sets are the same. Moreover, the following are equivalent:

(i) X *is of general type.*

(ii) X *is* **pseudo-Brody hyperbolic**; *that is,* $\mathrm{Sp}_{\mathrm{hol}}(X \times_K \mathbb{C}) \subsetneq X \times_K \mathbb{C}$.

(iii) X *is* **pseudo Mordellic**; *that is,* $\mathrm{Sp}_{\mathrm{alg}}(X) \subsetneq X$ *and for any finitely generated extension field K' of K, $(X \setminus \mathrm{Sp}_{\mathrm{alg}}(X))(K')$ is finite.*

Here we are primarily interested in the case in which K is a number field. However, the above also contains the Green–Griffiths conjecture (implicit in [Green and Griffiths 1980]; see also [Lang 1991, Chapter VIII, § 1]), which says that if X is a complex projective variety of general type, then the image of a holomorphic curve cannot be Zariski dense. Indeed, if X is defined over $\mathbb{C}$, then it can be obtained from a variety X as above; then use the implication (i) $\implies$ (ii).

For integral points, the case is not so clear, since the boundary may affect things in a number of different ways. So fewer implications are conjectured here. See [Lang 1991, Chapter IX, § 5] for explanations.

CONJECTURE 2.7. *Let R be a subring of $\mathbb{C}$, finitely generated over $\mathbb{Z}$, let K be its field of fractions, and let X be a quasi-projective variety over K. Consider the following conditions.*

(i) X *is of logarithmic general type.*

(ii) X *is* **pseudo-Brody hyperbolic**.

(iii) X *is* **pseudo Mordellic**; *that is,* $\mathrm{Sp}_{\mathrm{alg}}(X) \subsetneq X$, *and for every scheme* $\mathscr{X}$ *over* $\mathrm{Spec}\, R$ *with generic fiber isomorphic to* X, *and for any finitely generated extension ring* R' *of* R, *all but finitely many points of* $\mathscr{X}(R')$ *lie in* $\mathrm{Sp}_{\mathrm{alg}}(X)$.

Then (i) $\Longrightarrow$ (ii) *and* (i) $\Longrightarrow$ (iii).

3. Conjectural Second Main Theorems

Motivated by the situation in the equidimensional case, it is generally believed that the Second Main Theorem should hold for holomorphic curves in arbitrary (nonsingular) varieties:

DEFINITION. Let X be a nonsingular complex variety and let D be a divisor on X. We say that D is a **normal crossings divisor** (or that D **has normal crossings**) if each point $P \in X$ has a open neighborhood (in the classical topology) with local coordinates $z_1, \ldots, z_n$ such that D is locally equal to the principal divisor $(z_1 \cdots z_r)$ for some $r \in \{0, \ldots, n\}$. Note that this implies that D is effective and has no multiple components. If X is a variety over a number field k, then a divisor D on X has normal crossings if the corresponding divisor $X \times_k \mathbb{C}$ does, for some embedding $k \hookrightarrow \mathbb{C}$.

CONJECTURE 3.1. *Let X be a nonsingular complex projective variety, let D be a normal crossings divisor on X, let K be a canonical divisor on X, let A be an ample divisor on X, and let $\varepsilon > 0$. Then there exists a proper Zariski-closed subset $Z \subseteq X$, depending only on X, D, A, and ε, such that for any holomorphic curve $f : \mathbb{C} \to X$ whose image is not contained in Z,*

$$m(D, r) + T_K(r) \leq_{\mathrm{exc}} \varepsilon\, T_A(r) + O(1).$$

Here the notation $\leq_{\mathrm{exc}}$ means that the inequality holds for all r outside a set of finite Lebesgue measure.

The corresponding statement in the number field case is also highly conjectural:

CONJECTURE 3.2. *Let X be a nonsingular projective variety over a number field k, let D be a normal crossings divisor on X, let K be a canonical divisor on X, let A be an ample divisor on X, and let $\varepsilon > 0$. Then there exists a proper Zariski-closed subset $Z \subseteq X$, depending only on X, D, A, and ε, such that for all rational points $x \in X(k)$ with $x \notin Z$,*

$$m(D, x) + T_K(x) \leq \varepsilon\, h_A(x) + O(1).$$

Moreover, the set Z should be the same in both these conjectures.

Conjectures 3.1 and 3.2 give the implications (i) $\Longrightarrow$ (ii) and (i) $\Longrightarrow$ (iii) of Conjecture 2.7, respectively.

The set Z must depend on ε: see [Vojta 1989b, Example 8.15].

4. Approximation to Hyperplanes in Projective Space

In dimension > 1, the only case in which Conjectures 3.1 and 3.2 are known in their full strength is when $X = \mathbb{P}^n$ and D is a union of hyperplanes. This was proved for holomorphic curves by Cartan [1933] and for rational points by W. M. Schmidt [1980, Chapter VI, Theorem 1F]. See [Vojta 1987, Chapter 2] for a description of how to formulate Schmidt's theorem in a form similar to Cartan's.

We recall the statement of Cartan's theorem; except for the stronger error term, this is proved in [Cartan 1933]:

THEOREM 4.1. *Let $n > 0$, let $H_1, \ldots, H_q$ be hyperplanes in $\mathbb{P}^n_{\mathbb{C}}$ lying in general position (that is, so that $H_1 + \cdots + H_q$ is a normal crossings divisor), and let $f : \mathbb{C} \to \mathbb{P}^n_{\mathbb{C}}$ be a holomorphic curve not lying in any hyperplane (linearly nondegenerate). Then*

$$\sum_{i=1}^{q} m(H_i, r) \leq_{\mathrm{exc}} (n+1)T(r) + O(\log^+ T(r)) + o(\log r). \qquad (4.1.1)$$

This is a special case of Conjecture 3.1, except for the stronger error term and the weaker condition concerning Z.

Actually, a straightforward translation of Theorem 4.1 into the number field case gives something that is not quite as strong as Schmidt's subspace theorem, due to the fact that Schmidt's theorem allows a different collection of hyperplanes for each $v \in S$, so that the aggregate collection is not necessarily in general position.

To describe this further, let $x_0, \ldots, x_n$ be homogeneous coordinates on $\mathbb{P}^n$. Write $D = H_1 + \cdots + H_q$ and for each i let $a_{i0}x_0 + \cdots + a_{in}x_n$ be a nonzero linear form vanishing on H_i. Let k be a number field and S a finite set of places of k. Then, for $v \in S$ and $x \in \mathbb{P}^n(k) \setminus H_i$, we define the **Weil function** for H_i as

$$\lambda_{H_i,v}(x) = -\log \frac{\|a_{i0}x_0 + \cdots + a_{in}x_n\|_v}{\max\{\|x_0\|_v, \ldots, \|x_n\|_v\}}, \qquad (4.2)$$

These Weil functions depend on the choice of $a_{i0}, \ldots, a_{in}$ only up to $O(1)$; the choice of a linear form $a_{i0}x_0 + \cdots + a_{in}x_n$ amounts to choosing a section s of $\mathcal{O}(1)$; the fraction in (4.2) can then be regarded as a metric of that section. Thus, as in Table 3, we may take

$$m(H_i, x) = \frac{1}{[k : \mathbb{Q}]} \sum_{v \in S} \lambda_{H_i,v}(x)$$

Then Schmidt's Subspace Theorem can be stated as follows.

THEOREM 4.3. *Let k be a number field, let S be a finite set of places of k, let $n > 0$, let $H_1, \ldots, H_q$ be hyperplanes in $\mathbb{P}^n_k$, and let $\varepsilon > 0$. Then*

$$\frac{1}{[k : \mathbb{Q}]} \sum_{v \in S} \max_{L} \sum_{i \in L} \lambda_{H_i,v}(x) \leq (n + 1 + \varepsilon)h(x) + O(1) \qquad (4.3.1)$$

for all $x \in \mathbb{P}^n(k)$ outside of a finite union of proper linear subspaces depending only on k, S, $H_1, \ldots, H_q$, and ε. Here L varies over all subsets of $\{1, \ldots, q\}$ for which the set $\{H_i\}_{i \in L}$ lies in general position.

Its translation into the case of holomorphic curves is also true:

THEOREM 4.4 [Vojta 1997]. *Let $n > 0$, let $H_1, \ldots, H_q$ be hyperplanes in $\mathbb{P}^n_{\mathbb{C}}$, and let $f : \mathbb{C} \to \mathbb{P}^n_{\mathbb{C}}$ be a holomorphic curve not lying in any hyperplane (linearly nondegenerate). Then*

$$\int_0^{2\pi} \max_L \sum_{j \in L} \lambda_{H_j}(f(re^{i\theta})) \frac{d\theta}{2\pi} \leq_{\mathrm{exc}} (n+1)T(r) + O(\log^+ T(r)) + o(\log r). \quad (4.4.1)$$

Note that, if $H_1, \ldots, H_q$ lie in general position (that is, the divisor $H_1 + \cdots + H_q$ has normal crossings), then we may take $L = \{1, \ldots, q\}$, and the maximum occurs up to $O(1)$ at that value of L. Thus, the left-hand side of (4.4.1) is

$$\int_0^{2\pi} \sum_{j=1}^q \lambda_{H_j}(f(re^{i\theta})) \frac{d\theta}{2\pi} + O(1) = \sum_{j=1}^q m(H_j, r) + O(1),$$

which coincides with the left-hand side of (4.1.1).

Similar comments apply to Theorem 4.3.

Exceptional sets. In [Vojta 1989b; 1997], the conditions on exceptional sets or nondegeneracy were sharpened as follows.

THEOREM 4.5. *In the notation of Theorem 4.3, there exists a proper Zariski-closed subset $Z \subseteq \mathbb{P}^n_k$, depending only on $H_1, \ldots, H_q$, such that (4.3.1) holds for all $x \in \mathbb{P}^n(k) \setminus Z$.*

THEOREM 4.6. *In the notation of Theorem 4.4, there exists a proper Zariski-closed subset $Z \subseteq \mathbb{P}^n_{\mathbb{C}}$, depending only on $H_1, \ldots, H_q$, such that (4.4.1) holds for all holomorphic curves f whose image does not lie in Z.*

Thus, for hyperplanes in projective space, the condition regarding Z is sharper than in Conjectures 3.1 and 3.2, because it does not depend on ε.

Cartan's conjecture. A conjecture of Cartan concerns the question of what happens in Theorem 4.1 if the holomorphic curve is linearly degenerate. If it is degenerate, say if the linear span of its image has codimension t, then Cartan conjectured that the $n + 1$ factor in front of $T(r)$ should increase to $n + t + 1$. This was finally proved by Nochka [1982; 1983]; it was subsequently improved by Chen, and converted to the number field case by Ru and Wong [1991]. See also [Vojta 1997].

THEOREM 4.7. *Let $n > 0$, let $H_1, \ldots, H_q$ be hyperplanes in $\mathbb{P}^n_{\mathbb{C}}$ in general position, let $f : \mathbb{C} \to \mathbb{P}^n_{\mathbb{C}}$ be a holomorphic curve, and let t be the codimension* *of the linear span of f. Then*

$$\sum_{i=1}^{q} m(H_i, r) \leq_{\mathrm{exc}} (n + t + 1)T(r) + O(\log^+ T(r)) + o(\log r).$$

This was proved by assigning a **Nochka weight** ω_i to each hyperplane H_i. Shiffman observed that one of the conditions on these weights can be interpreted as the $\mathbb{Q}$-divisor $\omega_1 H_1 + \cdots + \omega_q H_q$ being **log canonical**; see [Shiffman 1996].

Additional refinements. The fact that one is working with hyperplanes in projective space means that one can work largely in linear algebra. The proof of Cartan's conjecture, for example, is largely a question of some (very difficult) linear algebra. Some other, simpler results of this nature have also been known.

THEOREM 4.8 [Dufresnoy 1944, Theorem XVI]. *Let $n > 0$ and $k > 0$, let $H_1, \ldots, H_{n+k}$ be hyperplanes in $\mathbb{P}^n_{\mathbb{C}}$ in general position, and let $f : \mathbb{C} \to \mathbb{P}^n_{\mathbb{C}}$ be a holomorphic curve that does not meet $H_1, \ldots, H_{n+k}$. Then the image of f is contained in a linear subspace of dimension $\leq [n/k]$, where the brackets denote greatest integer.*

This result has also been independently rediscovered by other authors.

COROLLARY 4.9. *A holomorphic curve that misses $2n+1$ hyperplanes in general position must be constant.*

Consequently, the complement of those hyperplanes is **Brody hyperbolic**.

These results hold also in the number field case; see [Ru and Wong 1991].

A diophantine inequality and semistability. Faltings and Wüstholz prove a finiteness result involving the notion of semistability of a filtration on a vector space. This requires a few definitions to state.

Let K be a number field, let L be a finite extension of K, and let S be a finite set of places of K. For each $w \in S$ let I_w be a finite index set. For each $\alpha \in I_w$ let $s_{w,\alpha}$ be a nonzero section of $\Gamma(\mathbb{P}^n_K, \mathscr{O}(1))$; that is, a nonzero linear form in $X_0, \ldots, X_n$. Also choose a real number $c_{w,\alpha} \geq 0$ for each w and α.

Let $V = \Gamma(\mathbb{P}^n_K, \mathscr{O}(1))$ and $V_L = V \otimes_K L$. For each $w \in S$ the choices of $s_{w,\alpha}$ and $c_{w,\alpha}$ define a filtration

$$V_L = W_w^0 \supseteq W_w^1 \supseteq \cdots \supseteq W_w^{e+1} = 0$$

Indeed, for $p \in \mathbb{R}$ let $W_{w,p}$ be the subspace spanned by $\{s_{w,\alpha} : c_{w,\alpha} \geq p\}$. For $j = 0, \ldots, e$ let $p_{w,e}$ be the smallest value of p for which $W_w^j = W_{w,p}$, and let $p_{w,e+1} = p_{w,e} + 1$.

DEFINITION 4.10. With the above notation, and for all nonzero linear subspaces W of V_L, let

$$\mu_w(W) = \frac{1}{\dim W} \sum_{j=1}^{e} p_j \dim((W \cap W_w^j)/(W \cap W_w^{j+1})).$$

DEFINITION 4.11. With the above notation, we say that the data $s_{w,\alpha}$ and $c_{w,\alpha}$ define a **jointly semistable** fibration on V_L if, for all nonzero linear subspaces $W \subseteq V_L$, we have

$$\sum_{w \in S} \mu_w(W) \leq \sum_{w \in S} \mu_w(V_L).$$

The main theorem of Faltings and Wüstholz is then the following:

THEOREM 4.12 [Faltings and Wüstholz 1994, Theorem 8.1]. *With the notation above, assume that the data $s_{w,\alpha}$ and $c_{w,\alpha}$ define a jointly semistable fibration on V_L. Assume furthermore that*

$$\sum_{w \in S} \mu_w(V_L) > [L : K].$$

Then the set of all points $x \in \mathbb{P}^n(K)$ satisfying

$$\|s_{w,\alpha}(x)\|_w \leq H_K(x)^{-c_{w,\alpha}} \qquad \text{for all } w \in S, \ \alpha \in I_w$$

is finite.

We remark that this result proves finiteness, not just that the set of points x lies in a finite union of proper linear subspaces. M. McQuillan and R. Ferretti (unpublished) have translated it into a statement for holomorphic curves.

Approximation to other divisors on $\mathbb{P}^n$. Approximation to divisors of higher degree on $\mathbb{P}^n$ is a trickier question. At present, no results approaching the bounds of Conjecture 3.1 are known, but weaker bounds can be obtained either from the methods of Faltings and Wüstholz [1994] (over number fields), or by using a d-uple embedding as noted in [Shiffman 1979] (for holomorphic curves, but the translation to number fields is immediate).

5. The Complement of Curves in $\mathbb{P}^2$

Conjecture 3.1 implies that, if D is a normal crossings divisor on $\mathbb{P}^2$ of degree at least 4, then any holomorphic curve in $\mathbb{P}^2 \setminus D$ must lie in a fixed divisor E depending only on D. Moreover, it is conjectured that if $\deg D \geq 5$, then we may take $E = 0$ for a suitably generic choice of D. More specifically, let $d_1, \ldots, d_k$ be positive integers with $d_1 + \cdots + d_k \geq 5$. Then it is conjectured that there exists a dense Zariski-open subset of the space of all divisors with irreducible components of degrees $d_1, \ldots, d_k$, respectively, such that if D is a divisor corresponding to a point in that open subset, then we may take $E = 0$. Partial results for the latter conjecture are as follows:

If D consists of five or more lines in general position, then the conjecture follows from Corollary 4.9. If D consists of any five components such that no three intersect, the conjecture was proved by Babets [1984] and by Eremenko and Sodin [1991]; more generally, their proof applies to any divisor D on $\mathbb{P}^n$ with at least $2n + 1$ irreducible components, such that any $n + 1$ of them have empty intersection.

The case of a quadric and four lines was proved by M. Green [1975]. If D is composed of at least three irreducible components, none of which is a line, then the conjecture was proved by Grauert [1989] and by Dethloff, Schumacher and Wong [Dethloff et al. 1995a; 1995b].

The case in which D is irreducible and smooth is much harder; in this case the only known general results are due to Zaĭdenberg and to Siu and Yeung:

THEOREM 5.1. *For positive integers d, let Σ_d denote the set of complex curves in $\mathbb{P}^2$ of degree d.*

(a) [Zaĭdenberg 1988] *If $d \geq 5$ then the set of points in Σ_d corresponding to smooth curves whose complement is (Kobayashi) hyperbolic and hyperbolically embedded, is nonempty and open in the classical topology.*

(b) [Siu and Yeung 1996b] *If $d \geq 5 \times 10^{13}$ then there exists a Zariski-dense open subset of Σ_d such that, if C is a curve corresponding to a point in that subset, then $\mathbb{P}^2 \setminus C$ is Brody hyperbolic.*

Recall that Kobayashi hyperbolicity implies Brody hyperbolicity, but that the converse holds only on compact manifolds. Thus the above theorem provides partial answers to a question posed by S. Kobayashi [1970, p. 132]: Is the complement in $\mathbb{P}^n$ of a generic hypersurface of high degree hyperbolic?

See also [Masuda and Noguchi 1996], for hypersurfaces defined by polynomials with few terms.

Of these results, only those relying on the Borel lemma translate to the number field case. See [Ru and Wong 1991] for the result concerning five lines in general position, and [Ru 1993] for the result of Babets and Eremenko and Sodin.

In addition, Ru [1995] has shown that the complement of a collection of hyperplanes in $\mathbb{P}^n$ has no nontrivial holomorphic curves if and only if it has only finitely many integral points.

6. Refinements of the Error Term

Motivated by Khinchin's theorem on approximation to arbitrary real numbers by rational numbers, Lang [1971] conjectured that the error term in Roth's theorem could be strengthened considerably. See also [Lang and Cherry 1990, page 10, including the footnote], for references to earlier, weaker conjectures.

No progress has been made on this, but corresponding questions in Nevanlinna theory have been solved; these questions were motivated by the conjecture in number theory and the dictionary with Nevanlinna theory. For example, the lemma on the logarithmic derivative has been strengthened by Miles as follows:

THEOREM 6.1 [Miles 1992]. *Let f be a meromorphic function on $\mathbb{C}$, and let $\phi : [1, \infty) \to [1, \infty)$ be a continuous function such that $\phi(x)/x$ is nondecreasing*

and

$$\int_1^\infty \frac{dx}{\phi(x)} < \infty.$$

Then

$$m(r, f'/f) <_{\text{exc}} \log^+ \left(\frac{\phi(T_f(r))}{r} \right) + O(1),$$

where in this case the notation $<_{\text{exc}}$ means that the inequality holds for all r outside a set E with

$$\int_E \frac{dt}{t} < \infty.$$

Corresponding versions of the Second Main Theorem for meromorphic functions and for hyperplanes in $\mathbb{P}^n$ were proved by Hinkkanen [1992] and by Ye [1995], respectively (although the details of the error terms vary somewhat).

7. Slowly Moving Targets

Early on, Nevanlinna conjectured that the Second Main Theorem should remain valid if the constants a_i being approximated were replaced by meromorphic functions $a_i(z)$, provided that these functions move *slowly;* that is, $T_{a_i}(r) = o(T_f(r))$, where f is the meromorphic function that is doing the approximating.

THEOREM 7.1. *Let f be a meromorphic function and let $a_1, \ldots, a_q$ be meromorphic functions with $T_{a_j}(r) = o(T_f(r))$ for all j. Then, for all $\varepsilon > 0$,*

$$\sum_{j=1}^q m_f(a_j, r) \leq_{\text{exc}} (2 + \varepsilon) T_f(r).$$

Nevanlinna proved this when $q \leq 3$. The general case was proved by Osgood [1981] (motivated, surprisingly, by the proof of Roth's theorem, and Osgood's own analogy with Nevanlinna theory). Soon after that, Steinmetz [1986] found a simple, elegant proof. It was generalized to the case of moving hyperplanes in $\mathbb{P}^n$ by Ru and Stoll [1991a]. In that case, extra care is necessary: since the hyperplanes are moving, the diagonal hyperplanes are also moving, and one needs to make sure that the holomorphic curve does not stay within such a diagonal (or other linear subspace, as in Theorem 4.6), because then the inequality would no longer hold. Therefore a stronger condition than linear nondegeneracy is needed.

Describing this stronger condition requires some additional notation. Let $n > 0$ and let $H_1, \ldots, H_q : \mathbb{C} \to (\mathbb{P}^n)^*$ be moving hyperplanes. For each j choose holomorphic functions $a_{j0}, \ldots, a_{jn}$ such that H_j is the hyperplane determined by the vanishing of the linear form $a_{j0}x_0 + \cdots + a_{jn}x_n$. For such a collection $\mathscr{H} := \{H_1, \ldots, H_q\}$, let $\mathscr{R}_{\mathscr{H}}$ denote the field of meromorphic functions generated over $\mathbb{C}$ by all ratios a_{jk}/a_{jl} such that $a_{jl} \neq 0$, where $j = 1, \ldots, q$, $k = 0, \ldots, n$, $l = 0, \ldots, n$.

DEFINITION 7.2. Let $f : \mathbb{C} \to \mathbb{P}^n$ be a holomorphic curve, written in homogeneous coordinates as $f = [f_0 : \cdots : f_n]$, where $f_0, \ldots, f_n$ are holomorphic functions with no common zero. Then we say that f is **linearly nondegenerate** over $\mathscr{R}_{\mathscr{H}}$ if the functions $f_0, \ldots, f_n$ are linearly independent over the field $\mathscr{R}_{\mathscr{H}}$.

THEOREM 7.3 [Ru and Stoll 1991a]. *Let* n, $H_1, \ldots, H_q$ *and* f *be as above. Assume that*:

(i) *for at least one value of z (and hence for almost all z), $H_1(z), \ldots, H_q(z)$ are in general position*;

(ii) $T_{H_j}(r) = o(T_f(r))$ *for all j (where $T_{H_j}(r)$ is defined via the isomorphism $(\mathbb{P}^n)^* \cong \mathbb{P}^n$); and*

(iii) f *is linearly nondegenerate over $\mathscr{R}_{\mathscr{H}}$.*

Then, for all $\varepsilon > 0$,

$$\sum_{j=1}^{q} m_f(H_j, r) \leq_{\mathrm{exc}} (n + 1 + \varepsilon) T_f(r).$$

In [Ru and Stoll 1991b] this theorem was generalized to the case of Cartan's conjecture (Theorem 4.7).

These results were carried over to the number field case by Bombieri and van der Poorten [1988] and by the author [Vojta 1996b] for Roth's theorem; by Ru and Vojta [1997] for Schmidt's theorem and Cartan's conjecture; and by Tucker [1997] for approximation to moving divisors on an elliptic curve.

A representative sample of such a statement is that of Schmidt's theorem with moving targets. We begin with some definitions.

DEFINITION 7.4. Let I be an infinite index set.

(i) A **moving hyperplane** indexed by I is a function $H : I \to (\mathbb{P}^n)^*(k)$, denoted $i \mapsto H(i)$.

(ii) Let $H_1, \ldots, H_q$ be moving hyperplanes. For each $j = 1, \ldots, q$ and each $i \in I$ choose $a_{j,0}(i), \ldots, a_{j,n}(i) \in k$ such that $H_j(i)$ is cut out by the linear form $a_{j,0}(i)X_0 + \cdots + a_{j,n}(i)X_n$. Then a subset $J \subseteq I$ is **coherent** with respect to $H_1, \ldots, H_q$ if, for every polynomial

$$P \in k[X_{1,0}, \ldots, X_{1,n}, \ldots, X_{q,0}, \ldots, X_{q,n}]$$

that is homogeneous in $X_{j,0}, \ldots, X_{j,n}$ for each $j = 1, \ldots, q$, either

$$P(a_{1,0}(i), \ldots, a_{1,n}(i), \ldots, a_{q,0}(i), \ldots, a_{q,n}(i))$$

vanishes for all $i \in J$, or it vanishes for only finitely many $i \in J$.

(iii) We define $\mathscr{R}_I^0$ to be the set of equivalence classes of pairs (J, a), where $J \subseteq I$ is a subset with finite complement; $a : J \to k$ is a map; and the equivalence relation is defined by $(J, a) \sim (J', a')$ if there exists $J'' \subseteq J \cap J'$ such that J'' has finite complement in I and $a\big|_{J''} = a'\big|_{J''}$. This is a ring containing k as a subring.

(iv) Let $H_1, \ldots, H_q$ be moving hyperplanes, denoted collectively by $\mathscr{H}$. If J is coherent with respect to $\mathscr{H}$, and if $a_{j,\alpha}(i) \neq 0$ for some $i \in J$, then $a_{j,\beta}/a_{j,\alpha}$ defines an element of $\mathscr{R}_J^0$. Moreover, by coherence, the subring of $\mathscr{R}_J^0$ generated by all such elements is *entire*. We define $\mathscr{R}_{J,\mathscr{H}}$ to be the field of fractions of that entire ring.

Thus, a little additional work is needed in order to define something having a property that comes automatically with meromorphic functions: that is, a meromorphic function either vanishes identically, or it is nonzero almost everywhere.

Given a hyperplane H defined by the linear form $a_0 X_0 + \cdots + a_n X_n$ and a point $\boldsymbol{x} \notin H$ with homogeneous coordinates $[x_0 : \cdots : x_n]$, we define a more precise Weil function at a place $v \in M_k$ by

$$\lambda_{H,v}(\boldsymbol{x}) = -\log \frac{\|a_0 x_0 + \cdots + a_n x_n\|_v}{\max_{0 \leq \alpha \leq n} \|a_\alpha\|_v \cdot \max_{0 \leq \alpha \leq n} \|x_\alpha\|_v}. \tag{7.5}$$

The extra term $\max_{0 \leq \alpha \leq n} \|a_\alpha\|_v$ ensures that $\lambda_{H,v}(\boldsymbol{x})$ depends only on H and $\boldsymbol{x}$, and not on $a_0, \ldots, a_n$ or on the choice of homogeneous coordinates $[x_0 : \cdots : x_n]$.

Schmidt's subspace theorem with moving targets can now be stated as follows.

THEOREM 7.6. *Let k be a number field, let S be a finite set of places of k, let $n > 0$, let I be an index set, and let $H_1, \ldots, H_q$ be moving hyperplanes in $\mathbb{P}^n_k$, denoted collectively by $\mathscr{H}$. Also let $\boldsymbol{x} : I \to \mathbb{P}^n(k)$ be a sequence of points, and let $[x_0 : \cdots : x_n]$ be homogeneous coordinates for $\boldsymbol{x}$. Suppose that*

(i) *for all $i \in I$, the hyperplanes $H_1(i), \ldots, H_q(i)$ are in general position;*
(ii) *for each infinite coherent subset $J \subseteq I$, $x_0\big|_J, \ldots, x_n\big|_J$ are linearly independent over $\mathscr{R}_{J,\mathscr{H}}$; and*
(iii) *$h_k(H_j(i)) = o(h_k(\boldsymbol{x}(i)))$ for all $j = 1, \ldots, q$ (that is, for all $\delta > 0$,*

$$h_k(H_j(i)) \leq \delta h_k(\boldsymbol{x}(i))$$

for all but finitely many $i \in I$).

Then for all $\varepsilon > 0$ and all $C \in \mathbb{R}$,

$$\frac{1}{[k : \mathbb{Q}]} \sum_{v \in S} \sum_{j=1}^q \lambda_{H_j(i),v}(\boldsymbol{x}) \leq (n + 1 + \varepsilon) h(\boldsymbol{x}(i)) + C$$

for all but finitely many $i \in I$.

Theorem 7.6 is proved using an extension of Steinmetz's method; in the end it reduces to reducing the problem to Schmidt's subspace theorem with fixed targets, but in a space of much higher dimension. As M. Ru points out [1997], it is more convenient to use the variant Theorems 4.3 and 4.4 instead of the formulation of Theorem 4.1. By the same token, it would be better to phrase Theorem 7.6 in these terms as well; in fact, the proof of [Ru and Vojta 1997] actually gives the following stronger result.

THEOREM 7.7. *Let k, S, I, $\mathscr{H}$, $H_1, \ldots, H_q$, $\boldsymbol{x}$, and $[x_0 : \cdots : x_n]$ be as in the first two sentences of Theorem 7.6. Also let $\mathscr{L}$ be a collection of subsets of $\{1, \ldots, q\}$. Suppose that*

(i) *for all $i \in I$ and all $L \in \mathscr{L}$, the hyperplanes $H_j(i)$, $j \in L$ are in general position;*

(ii) *for each infinite coherent subset $J \subseteq I$, $x_0\big|_J, \ldots, x_n\big|_J$ are linearly independent over $\mathscr{R}_{J,\mathscr{H}}$; and*

(iii) *$h_k(H_j(i)) = o(h_k(\boldsymbol{x}(i)))$ for all $j = 1, \ldots, q$.*

Then for all $\varepsilon > 0$ and all $C \in \mathbb{R}$,

$$\frac{1}{[k : \mathbb{Q}]} \sum_{v \in S} \max_{L \in \mathscr{L}} \sum_{j \in L} \lambda_{H_j(i),v}(\boldsymbol{x}) \leq (n + 1 + \varepsilon)h(\boldsymbol{x}(i)) + C$$

for all but finitely many $i \in I$.

Similarly, we can define a more precise Weil function in the context of holomorphic curves as

$$\lambda_H(\boldsymbol{x}) = -\log \frac{|a_0 x_0 + \cdots + a_n x_n|}{\max_{0 \leq \alpha \leq n} |a_\alpha| \cdot \max_{0 \leq \alpha \leq n} |x_\alpha|}$$

(using the notation of (7.5)). Then the methods of [Ru and Stoll 1991a] immediately give:

THEOREM 7.8. *Let $n > 0$ be an integer, let $H_1, \ldots, H_q$ be moving hyperplanes in $\mathbb{P}^n$, and let f be a holomorphic curve in $\mathbb{P}^n$. Assume that:*

(i) *$T_{H_j}(r) = o(T_f(r))$ for all j (where $T_{H_j}(r)$ is defined via the isomorphism $(\mathbb{P}^n)^* \cong \mathbb{P}^n$); and*

(ii) *f is linearly nondegenerate over $\mathscr{R}_{\mathscr{H}}$.*

Then, for all $\varepsilon > 0$,

$$\int_0^{2\pi} \max_L \sum_{j \in L} \lambda_{H_j}(f(re^{i\theta})) \frac{d\theta}{2\pi} \leq_{\mathrm{exc}} (n + 1 + \varepsilon)T_f(r),$$

where L varies over all subsets of $\{1, \ldots, q\}$ for which $(H_j)_{j \in L}$ lie in general position (for at least one value of z).

8. Discriminant Terms

Instead of working with rational points in inequalities such as (1.3) and Conjecture 3.2, one may conjecture more generally that the inequalities hold for *algebraic* points, provided that the inequalities are modified appropriately. This modification involves the discriminant of the number field generated by the algebraic point in question. To justify this suggestion, we begin with Nevanlinna theory.

Inequality (1.2) may be rewritten, via the definition $T_D(r) = m(D, r) + N(D, r)$, as

$$N(D, r) \geq_{\text{exc}} T_{K+D}(r) - \varepsilon T_A(r) - O(1).$$

This may be sharpened to

$$N^{(1)}(D, r) \geq_{\text{exc}} T_{K+D}(r) - \varepsilon T_A(r) - O(1), \tag{8.1}$$

where the counting function is replaced by the **truncated counting function**, which is defined for effective divisors D by

$$N^{(1)}(D, r) := \sum_{w \in \mathbb{D}_r^\times} \min\{1, \operatorname{ord}_w f^* s\} \cdot \log \frac{r}{|w|}.$$

Of course, one can make the same definition in the context of number fields:

$$N^{(1)}(D, x) = \frac{1}{[k : \mathbb{Q}]} \sum_{v \notin S} \min\{1, \operatorname{ord}_v s(x)\} \cdot \log(R : \mathfrak{p}) \tag{8.2}$$

where as usual $\mathfrak{p}$ is the prime ideal in R corresponding to the non-archimedean place v. One may then conjecture that (1.3) can be replaced by the stronger inequality

$$N^{(1)}(D, x) \geq h_{K+D}(x) - \varepsilon h_A(x) - C. \tag{8.3}$$

In Nevanlinna theory there is a stronger statement than (8.1):

$$m(D, r) + T_K(r) + N_{\text{Ram}}(r) \leq_{\text{exc}} \varepsilon T_A(r) + O(1), \tag{8.4}$$

Here N_{Ram} is the **ramification term**: it counts the ramification of the holomorphic curve f; that is, it counts the zeroes of f' (in local coordinates) in the same way that $N_f(\infty, r)$ counts the poles of a meromorphic function f. It is well known that (8.4) implies (8.1).

We claim that, in the notation of (1.3), if $x \in X(\overline{\mathbb{Q}})$, then the analogue of $N_{\text{Ram}}(r)$ should be $-d(x)$, where $d(x)$ is the **discriminant term**

$$d(x) := \frac{1}{[k : \mathbb{Q}]} \log |D_{K(x)}|.$$

Thus the arithmetic equivalent of (8.4) is the following:

CONJECTURE 8.5. *Let X be a smooth projective curve over a number field k, let D be an effective divisor on X with no multiple points, let K be a canonical divisor on X, let A be an ample divisor on X, let r be a positive integer, let $\varepsilon > 0$, and let C be a constant. Then the inequality*

$$m(D, x) + h_K(x) \leq d(x) + \varepsilon h_A(x) + C \tag{8.5.1}$$

holds for all but finitely many $x \in X(\overline{\mathbb{Q}})$ with $[K(x) : k] \leq r$.

Because of the sign change, it may seem unusual to suggest that $-d(x)$ is an analogue of $N_{\text{Ram}}(r)$. To support this assertion, however, we point out that (8.5.1) implies (8.3), corresponding to the fact that (8.4) implies (8.1):

PROPOSITION 8.6. *If Conjecture 8.5 holds, then (8.3) holds as well.*

This has not been proved elsewhere, so a proof appears in Appendix A.

In higher dimensions, there is likewise a modification of Conjecture 3.2:

CONJECTURE 8.7. *Let X, D, K, A, and ε be as in Conjecture 3.2, and let r be a positive integer. Then there exists a proper Zariski-closed subset $Z \subsetneq X$, depending only on X, D, A, and ε, such that for all algebraic points $x \in X(\bar{k})$ with $x \notin Z$ and $[K(x):k] \leq r$,*

$$m(D,x) + T_K(x) \leq d(x) + \varepsilon\, h_A(x) + O(1).$$

To conclude this section, we mention the "abc conjecture" of Masser and Oesterlé. As was first observed by J. Noguchi [1996, (9.5)], it is the number-theoretic counterpart to Nevanlinna's Second Main Theorem with truncated counting functions, applied to the divisor $[0] + [1] + [\infty]$ on $\mathbb{P}^1$. In its simplest form it reads as follows.

CONJECTURE 8.8. *Let $\varepsilon > 0$. Then there exists a constant C, depending only on ε, such that for all relatively prime integers $a, b, c \in \mathbb{Z}$ with $a + b + c = 0$,*

$$\max\{|a|, |b|, |c|\} \leq C \prod_{p \mid abc} p^{1+\varepsilon}.$$

This conjecture, if proved, would have far-reaching consequences; for example, it would imply a weak form of Fermat's Last Theorem (now proved by Wiles).

In [Vojta 1987, pp. 71–72] it is shown that Conjecture 8.5 implies the abc conjecture.

It is also possible, via the variety $X \subseteq \mathbb{P}^2 \times \mathbb{P}^2$ defined by $ux^4 + vy^4 + wz^4 = 0$, to obtain from Conjecture 3.2 a weak form of the abc conjecture, for $\varepsilon > 26$. Here X is a rational three-fold. Thus, versions of the Second Main Theorem, applied even to rational varieties, would give highly nontrivial consequences.

Appendix A: Proof of Proposition 8.6

This appendix gives a proof of Proposition 8.6, because a proof has not appeared elsewhere. It will necessarily be more technical than the rest of the paper.

Recall that we are proving that the inequality

$$m(D,x) + h_K(x) \leq d(x) + \varepsilon\, h_A(x) + O(1) \tag{A.1}$$

for algebraic points of bounded degree on a curve implies the inequality

$$N^{(1)}(D,x) \geq h_{K+D}(x) - \varepsilon\, h_A(x) - O(1). \tag{A.2}$$

for rational points on a curve.

This implication is proved by taking a cover X' of X, highly ramified over D but unramified elsewhere, and applying (A.1) to the pull-back of everything to X'. The counting functions end up being truncated because of the fact that the ramification of a number field is limited over any given place. The details of this construction are as follows.

First, we define a slightly different truncated counting function.

DEFINITION A.3. If D is a divisor on a curve, then the **modified truncated counting function** is the function $N^\flat(D, x)$, defined for prime divisors D by (8.2) and for arbitrary D by linearity. (We will not define $N^\flat(D, x)$ on varieties of higher dimension, because the support of D may be singular in that case.)

Next, we give an improved lemma of Chevalley-Weil type; compare [Vojta 1987, Thm. 5.1.6].

LEMMA A.4. *Let $\pi : X' \to X$ be a finite morphism of smooth projective curves over a global field k of characteristic zero, let R be the ramification divisor of π, and let r be a positive integer. Then*

$$d(y) - d(\pi(y)) \le N^\flat(R, y) + O(1) \qquad (A.4.1)$$

for all $y \in X'(\bar{k})$ with $[K(y) : k] \le r$; here the constant in $O(1)$ depends on π, r, and the model used in defining $N^\flat$, but not on y.

Moreover, if for all $y \in X'$ the ramification index of π at y depends only on $\pi(y)$, so that $R = \pi^* B$ for some $\mathbb{Q}$-divisor B on X, then

$$d(y) - d(\pi(y)) \le N^\flat(B, \pi(y)) + O(1) \qquad (A.4.2)$$

for all y as before.

PROOF. To simplify the notation, we will assume that k is a number field. Let A be its ring of integers.

Let $\mathscr{X}'$ and $\mathscr{X}$ be regular models for X' and X, over $\operatorname{Spec} A$, such that π extends to a morphism $\mathscr{X}' \to \mathscr{X}$, also denoted π. Let R also denote the ramification divisor of $\mathscr{X}'$ over $\mathscr{X}$. Let S be the set of places of k containing:

(i) all archimedean places;
(ii) all places of bad reduction of $\mathscr{X}'$ and $\mathscr{X}$;
(iii) all places where $\pi(\operatorname{Supp} R)$ is not étale over $\operatorname{Spec} A$;
(iv) all places where $\pi^{-1}(\pi(\operatorname{Supp} R))$ is not étale over $\operatorname{Spec} A$; and
(v) all places where π fails to be a finite morphism.

This is a finite set. For places $v \in S$ the contribution to $d(y)$ is bounded; hence it suffices to show that the contribution to each side of (A.4.1) from places not in S obeys the inequality. This will be done place by place, without any $O(1)$ term.

By making a base change, we may assume that $\pi(y)$ is rational over k.

Let v be a place of k not in S, and let w be a place of $K(y)$ lying over v. Let v also denote the point of $\operatorname{Spec} A$ corresponding to v; similarly let C be the ring of integers of $K(y)$ and let w also denote the point of $\operatorname{Spec} C$ corresponding to w. Let σ be the section of the map $\mathscr{X} \to \operatorname{Spec} A$ corresponding to $\pi(y)$, and let $\tau : \operatorname{Spec} C \to \mathscr{X}'$ be the map corresponding to y. If $\tau(w)$ does not meet R, then the contribution at w to the right-hand side of (A.4.1) is zero, but the contribution to the left-hand side is also zero, by [Vojta 1987, Lemma 5.1.8]. Thus we may assume that $\tau(w) \in \operatorname{Supp} R$.

Write $\xi = \sigma(v)$ and $\eta = \tau(w)$, so that $\xi = \pi(\eta)$. Let n be the degree of π. After base change of $\pi : \mathscr{X}' \to \mathscr{X}$ to $\operatorname{Spec} \widehat{\mathscr{O}}_{\xi, \mathscr{X}}$, we have a finite morphism $\pi' : \operatorname{Spec} C' \to \operatorname{Spec} \widehat{\mathscr{O}}_{\xi, \mathscr{X}}$, where C' is a semilocal ring. Let $\mathfrak{m}$ be the maximal ideal of $\widehat{\mathscr{O}}_{\xi, \mathscr{X}}$ and $\mathfrak{m}_1, \dots, \mathfrak{m}_r$ the maximal ideals of C'. For sufficiently large e, we have

$$(\mathfrak{m}_1 \cdots \mathfrak{m}_r)^e \subseteq \mathfrak{m} \subseteq \mathfrak{m}_1 \cdots \mathfrak{m}_r;$$

therefore C' is $\mathfrak{m}_1 \cdots \mathfrak{m}_r$-adically complete, and by [Matsumura 1986, Theorem 8.15] it follows that C' is the product of the completions of its local rings at the maximal ideals. Thus

$$C' = \prod_{\alpha \in \pi^{-1}(\xi)} \widehat{\mathscr{O}}_{\alpha, \mathscr{X}'}.$$

Let e be the degree of $\operatorname{Spec} \widehat{\mathscr{O}}_{\eta, \mathscr{X}'}$ over $\operatorname{Spec} \widehat{\mathscr{O}}_{\xi, \mathscr{X}}$. There is a unique branch of $\pi(\operatorname{Supp} R)$ passing through ξ and a unique branch of $\pi^{-1}(\pi(\operatorname{Supp} R))$ passing through η. Therefore the multiplicity of that latter branch in $\pi^{-1}(\pi(\operatorname{Supp} R))$ (pulling back $\pi(\operatorname{Supp} R)$ as a divisor), must also equal e. If $e > 1$ then R has a component with multiplicity $e - 1$ passing through η; otherwise, $e - 1 = 0$ and R does not pass through η. Then the contribution at w to the right-hand side of (A.4.1) is $(e - 1)(\log q_w)/[K(y) : \mathbb{Q}]$, where q_w is the number of elements of the residue field at w. But $\operatorname{Spec} \widehat{\mathscr{O}}_{\eta, \mathscr{X}'}$ has degree e over $\operatorname{Spec} \widehat{\mathscr{O}}_{\xi, \mathscr{X}}$, so the local field $K(\eta)_w$ has degree at most e over k_v. Thus the contribution at η to the left-hand side is also at most $(e - 1)(\log q_w)/[K(y) : \mathbb{Q}]$ (since wild ramification cannot occur). This is sufficient to imply (A.4.1).

Next we show (A.4.2). Again, we prove the inequality place by place, for places $v \notin S$, without the $O(1)$ term. We again assume that $\pi(y)$ is rational over k. Pick $v \in M_k \setminus S$. As before, we may assume that $\sigma(v)$ meets $\operatorname{Supp} B$. Then the contribution at v to the right-hand side of (A.4.2) is

$$\frac{1}{[k : \mathbb{Q}]} \cdot \frac{e - 1}{e} \log q_v.$$

For a place w of $K(y)$ over v, let $e_{w/v}$ and $f_{w/v}$ denote the ramification index and residue field degree, respectively. Then the contribution at v to the left-hand side of (A.4.2) is

$$\frac{1}{[K(y) : \mathbb{Q}]} \sum_{w|v} (e_{w/v} - 1) \log q_w.$$

Thus it will suffice to show that

$$\frac{1}{[K(y) : k]} \sum_{w|v} (e_{w/v} - 1) \log q_w \leq \frac{e - 1}{e} \log q_v.$$

But this follows from the easy facts

$$\sum_{w|v} e_{w/v} f_{w/v} = [K(y):k] \qquad \text{and} \qquad \log q_w = f_{v/w} \log q_v,$$

and from the inequality $e_{v/w} \leq e$ proved earlier. $\qquad\square$

We may now continue with the proof of Proposition 8.6.

If X has genus 0 and if $\deg D < 2$, the right-hand side of (A.2) is negative, and the result is trivial. Hence we may assume that $2g(X) - 2 + \deg D \geq 0$. Let e be an integer, chosen large enough so that

$$h^0(X, n([e\varepsilon]A - D)) > 0 \qquad\qquad (A.5)$$

for some $n > 0$, where $[e\varepsilon]$ denotes the greatest integer function. By [Vojta 1989a, Lemma 3.1], there is a cover $\pi : X' \to X$, where X' is also nonsingular, which is unramified outside $\pi^{-1}(D)$ and ramified exactly to order e at all points of X' lying over D.

The $\mathbb{Q}$-divisor

$$D' := \frac{1}{e}\pi^*D$$

is an integral divisor on X' with no multiple points. The ramification divisor R of π is given by

$$R = (e-1)D' = \frac{e-1}{e}\pi^*D . \qquad\qquad (A.6)$$

Thus the canonical divisor on X' satisfies the linear equivalence

$$K_{X'} \sim \pi^*K_X + \frac{e-1}{e}\pi^*D ,$$

so that

$$K_{X'} + D' \sim \pi^*(K_X + D) . \qquad\qquad (A.7)$$

LEMMA A.8. *In this situation, points $y \in X'(\bar{k})$ of bounded degree over k satisfy*

$$N(D', y) + d(y) \leq N^{(1)}(D, \pi(y)) + d(\pi(y)) + \varepsilon h_{\pi^*A}(y) + O(1) , \qquad (A.8.1)$$

where the constant in $O(1)$ depends on X, X', π, the models used to define the counting functions, and the bound on the degree, but not on y.

PROOF. By (A.6), we may apply (A.4.2) to X' with

$$B = \frac{e-1}{e}D$$

to obtain the inequality

$$d(y) \leq d(\pi(y)) + N^\flat(B, \pi(y)) + O(1) . \qquad\qquad (A.8.2)$$

By (A.5) we have $h^0(X', n([e\varepsilon]\pi^*A - \pi^*D)) > 0$ for some $n > 0$, so

$$\varepsilon\, h_{\pi^*A}(y) \geq h_{D'}(y) + O(1)$$

and therefore

$$N(D', y) \leq h_{D'}(y) + O(1)$$
$$\leq \varepsilon\, h_{\pi^* A}(y) + O(1)\ .$$

By definition of $N^\flat$ we then have

$$N^\flat(B, \pi(y)) = \frac{e-1}{e} N^\flat(D, \pi(y))$$
$$= \frac{e-1}{e} N^{(1)}(D, \pi(y)) + O(1)$$
$$\leq N^{(1)}(D, \pi(y)) + O(1)$$
$$\leq N^{(1)}(D, \pi(y)) - N(D', y) + \varepsilon h_{\pi^* A}(y) + O(1)\ .$$

Combining this with (A.8.2) then gives (A.8.1). $\square$

Since π is a finite map and A is ample, $\pi^* A$ is ample on X'. Thus, (A.1) applies to points y on X' lying over rational points on X, relative to the divisor D', giving

$$m(D', y) + h_{K_{X'}}(y) \leq d(y) + \varepsilon h_{\pi^* A}(y) + O(1)\ .$$

By the First Main Theorem, this is equivalent to

$$N(D', y) + d(y) \geq h_{K_{X'} + D'}(y) - \varepsilon\, h_{\pi^* A}(y) - O(1)\ . \tag{A.9}$$

By (A.8.1), (A.9), and (A.7), we then have

$$N^{(1)}(D, \pi(y)) + d(\pi(y)) \geq N(D', y) + d(y) - \varepsilon\, h_{\pi^* A}(y) - O(1)$$
$$\geq h_{K_{X'} + D'}(y) - 2\varepsilon\, h_{\pi^* A}(y) - O(1)$$
$$\geq h_{K_X + D}(\pi(y)) - 2\varepsilon\, h_A(\pi(y)) - O(1)\ .$$

Since $\pi(y)$ is rational, $d(\pi(y))$ is bounded; hence (A.2) follows after adjusting ε. Thus, Proposition 8.6 is proved.

References

[Babets 1984] V. A. Babets, "Theorems of Picard type for holomorphic mappings", *Sibirsk. Mat. Zh.* **25**:2 (1984), 35–41. In Russian; translated in *Siberian Math. J.* **25** (1984), 195–200.

[Baily and Borel 1966] J. Baily, W. L. and A. Borel, "Compactification of arithmetic quotients of bounded symmetric domains", *Ann. of Math.* (2) **84** (1966), 442–528.

[Bloch 1926] A. Bloch, "Sur les systèmes de fonctions uniformes satisfaisant à l'équation d'une variété algébrique dont l'irrégularité dépasse la dimension", *J. de Math.* **5** (1926), 19–66.

[Bombieri and van der Poorten 1988] E. Bombieri and A. J. van der Poorten, "Some quantitative results related to Roth's theorem", *J. Austral. Math. Soc. Ser. A* **45**:2 (1988), 233–248.

[Cartan 1933] H. Cartan, "Sur les zéros des combinaisons linéaires de p fonctions holomorphes données", *Mathematica (Cluj)* **7** (1933), 5–29.

[Dethloff et al. 1995a] G. Dethloff, G. Schumacher, and P.-M. Wong, "Hyperbolicity of the complements of plane algebraic curves", *Amer. J. Math.* **117**:3 (1995), 573–599.

[Dethloff et al. 1995b] G. Dethloff, G. Schumacher, and P.-M. Wong, "On the hyperbolicity of the complements of curves in algebraic surfaces: the three-component case", *Duke Math. J.* **78**:1 (1995), 193–212.

[Dufresnoy 1944] J. Dufresnoy, "Théorie nouvelle des familles complexes normales. Applications à l'étude des fonctions algébroïdes", *Ann. Sci. École Norm. Sup.* (3) **61** (1944), 1–44.

[Erëmenko and Sodin 1991] A. È. Erëmenko and M. L. Sodin, "The value distribution of meromorphic functions and meromorphic curves from the point of view of potential theory", *Algebra i Analiz* **3**:1 (1991), 131–164. In Russian; translated in *St. Petersburg Math. J.* **3** (1992), 109–136.

[Faltings 1991] G. Faltings, "Diophantine approximation on abelian varieties", *Ann. of Math.* (2) **133**:3 (1991), 549–576.

[Faltings 1994] G. Faltings, "The general case of S. Lang's conjecture", pp. 175–182 in *Barsotti Symposium in Algebraic Geometry* (Abano Terme, 1991), edited by V. Christante and W. Messing, Perspectives in Math. **15**, Academic Press, San Diego, 1994.

[Faltings and Wüstholz 1994] G. Faltings and G. Wüstholz, "Diophantine approximations on projective spaces", *Invent. Math.* **116**:1-3 (1994), 109–138.

[Grauert 1989] H. Grauert, "Jetmetriken und hyperbolische Geometrie", *Math. Z.* **200**:2 (1989), 149–168.

[Green 1975] M. L. Green, "Some Picard theorems for holomorphic maps to algebraic varieties", *Amer. J. Math.* **97** (1975), 43–75.

[Green and Griffiths 1980] M. Green and P. Griffiths, "Two applications of algebraic geometry to entire holomorphic mappings", pp. 41–74 in *The Chern Symposium* (Berkeley, 1979), edited by W.-Y. Hsiang et al., Springer, New York, 1980.

[Griffiths and King 1973] P. Griffiths and J. King, "Nevanlinna theory and holomorphic mappings between algebraic varieties", *Acta Math.* **130** (1973), 145–220.

[Hinkkanen 1992] A. Hinkkanen, "A sharp form of Nevanlinna's second fundamental theorem", *Invent. Math.* **108**:3 (1992), 549–574.

[Kobayashi 1970] S. Kobayashi, *Hyperbolic manifolds and holomorphic mappings*, Pure and Applied Mathematics **2**, Marcel Dekker, New York, 1970.

[Lang 1970] S. Lang, *Algebraic number theory*, Addison-Wesley, Reading, MA, 1970. Reprinted by Springer, 1986.

[Lang 1971] S. Lang, "Transcendental numbers and diophantine approximations", *Bull. Amer. Math. Soc.* **77** (1971), 635–677.

[Lang 1983] S. Lang, *Fundamentals of Diophantine geometry*, Springer, New York, 1983.

[Lang 1991] S. Lang, *Number theory, III: Diophantine geometry*, Encyclopaedia of Mathematical Sciences **60**, Springer, Berlin, 1991.

[Lang and Cherry 1990] S. Lang and W. Cherry, *Topics in Nevanlinna theory*, Lecture Notes in Math. **1433**, Springer, Berlin, 1990.

[Masuda and Noguchi 1996] K. Masuda and J. Noguchi, "A construction of hyperbolic hypersurface of $\mathbb{P}^n(\mathbb{C})$", *Math. Ann.* **304**:2 (1996), 339–362.

[Matsumura 1986] H. Matsumura, *Commutative ring theory*, Cambridge Stud. Adv. Math. **8**, Cambridge University Press, Cambridge, 1986.

[Miles 1992] J. Miles, "A sharp form of the lemma on the logarithmic derivative", *J. London Math. Soc.* (2) **45**:2 (1992), 243–254.

[Nochka 1982] E. I. Nochka, "Defect relations for meromorphic curves", *Izv. Akad. Nauk Moldav. SSR Ser. Fiz.-Tekhn. Mat. Nauk* **1982**:1 (1982), 41–47, 79.

[Nochka 1983] E. I. Nochka, "On the theory of meromorphic curves", *Dokl. Akad. Nauk SSSR* **269**:3 (1983), 547–552. In Russian; translated in *Soviet Math. Dokl.* **27** (1983), 377–381.

[Noguchi 1981] J. Noguchi, "Lemma on logarithmic derivatives and holomorphic curves in algebraic varieties", *Nagoya Math. J.* **83** (1981), 213–233.

[Noguchi 1996] J. Noguchi, "On Nevanlinna's second main theorem", pp. 489–503 in *Geometric complex analysis* (Hayama, 1995), edited by J. Noguchi et al., World Sci. Publishing, Singapore, 1996.

[Noguchi 1998] J. Noguchi, "On holomorphic curves in semi-abelian varieties", *Math. Z.* **228**:4 (1998), 713–721.

[Osgood 1981] C. F. Osgood, "A number theoretic–differential equations approach to generalizing Nevanlinna theory", *Indian J. Math.* **23**:1-3 (1981), 1–15.

[Ru 1993] M. Ru, "Integral points and the hyperbolicity of the complement of hypersurfaces", *J. Reine Angew. Math.* **442** (1993), 163–176.

[Ru 1995] M. Ru, "Geometric and arithmetic aspects of $\mathbb{P}^n$ minus hyperplanes", *Amer. J. Math.* **117**:2 (1995), 307–321.

[Ru 1997] M. Ru, "On a general form of the second main theorem", *Trans. Amer. Math. Soc.* **349**:12 (1997), 5093–5105.

[Ru and Stoll 1991a] M. Ru and W. Stoll, "The second main theorem for moving targets", *J. Geom. Anal.* **1**:2 (1991), 99–138.

[Ru and Stoll 1991b] M. Ru and W. Stoll, "The Cartan conjecture for moving targets", pp. 477–508 in *Several complex variables and complex geometry* (Santa Cruz, CA, 1989), vol. 2, edited by E. Bedford et al., Proc. Sympos. Pure Math. **52**, Amer. Math. Soc., Providence, RI, 1991.

[Ru and Vojta 1997] M. Ru and P. Vojta, "Schmidt's subspace theorem with moving targets", *Invent. Math.* **127**:1 (1997), 51–65.

[Ru and Wong 1991] M. Ru and P.-M. Wong, "Integral points of $\mathbb{P}^n - \{2n + 1$ hyperplanes in general position$\}$", *Invent. Math.* **106**:1 (1991), 195–216.

[Schmidt 1980] W. M. Schmidt, *Diophantine approximation*, Lecture Notes in Math. **785**, Springer, Berlin, 1980.

[Shiffman 1979] B. Shiffman, "On holomorphic curves and meromorphic maps in projective space", *Indiana Univ. Math. J.* **28**:4 (1979), 627–641.

[Shiffman 1996] B. Shiffman, "The second main theorem for log canonical $\mathbb{R}$-divisors", pp. 551–561 in *Geometric complex analysis* (Hayama, 1995), edited by J. Noguchi et al., World Sci. Publishing, Singapore, 1996.

[Siu 1995] Y.-T. Siu, "Hyperbolicity problems in function theory", pp. 409–513 in *Five decades as a mathematician and educator: on the 80th birthday of Professor Yung-Chow Wong*, edited by K.-Y. Chan and M.-C. Liu, World Sci. Publishing, Singapore, 1995.

[Siu and Yeung 1996a] Y.-T. Siu and S.-K. Yeung, "A generalized Bloch's theorem and the hyperbolicity of the complement of an ample divisor in an abelian variety", *Math. Ann.* **306**:4 (1996), 743–758.

[Siu and Yeung 1996b] Y.-T. Siu and S.-K. Yeung, "Hyperbolicity of the complement of a generic smooth curve of high degree in the complex projective plane", *Invent. Math.* **124**:1-3 (1996), 573–618.

[Soulé 1992] C. Soulé, *Lectures on Arakelov geometry*, Cambridge Stud. Adv. Math. **33**, Cambridge University Press, Cambridge, 1992. With the collaboration of D. Abramovich, J.-F. Burnol and J. Kramer.

[Steinmetz 1986] N. Steinmetz, "Eine Verallgemeinerung des zweiten Nevanlinnaschen Hauptsatzes", *J. Reine Angew. Math.* **368** (1986), 134–141.

[Tucker 1997] T. J. Tucker, "Moving targets on elliptic curves", preprint, University of Georgia, 1997.

[Vojta 1987] P. Vojta, *Diophantine approximations and value distribution theory*, Lecture Notes in Math. **1239**, Springer, Berlin, 1987.

[Vojta 1989a] P. Vojta, "Dyson's lemma for products of two curves of arbitrary genus", *Invent. Math.* **98**:1 (1989), 107–113.

[Vojta 1989b] P. Vojta, "A refinement of Schmidt's subspace theorem", *Amer. J. Math.* **111**:3 (1989), 489–518.

[Vojta 1992] P. Vojta, "A generalization of theorems of Faltings and Thue–Siegel–Roth–Wirsing", *J. Amer. Math. Soc.* **5**:4 (1992), 763–804.

[Vojta 1996a] P. Vojta, "Integral points on subvarieties of semiabelian varieties, I", *Invent. Math.* **126**:1 (1996), 133–181.

[Vojta 1996b] P. Vojta, "Roth's theorem with moving targets", *Internat. Math. Res. Notices* **1996**:3 (1996), 109–114.

[Vojta 1997] P. Vojta, "On Cartan's theorem and Cartan's conjecture", *Amer. J. Math.* **119**:1 (1997), 1–17.

[Vojta 1999] P. Vojta, "Integral points on subvarieties of semiabelian varieties, II", *Amer. J. Math.* **121**:2 (1999), 283–313.

[Ye 1995] Z. Ye, "On Nevanlinna's second main theorem in projective space", *Invent. Math.* **122**:3 (1995), 475–507.

[Zaĭdenberg 1988] M. G. Zaĭdenberg, "Stability of hyperbolic embeddedness and the construction of examples", *Mat. Sb.* (*N.S.*) **135(177)**:3 (1988), 361–372, 415. In Russian; translated in *Math. USSR Sbornik* **63** (1989), 351–361.

PAUL VOJTA
DEPARTMENT OF MATHEMATICS
UNIVERSITY OF CALIFORNIA
970 EVANS HALL #3840
BERKELEY, CA, 94720-3840
UNITED STATES
  vojta@math.berkeley.edu